Introduction to Mineralogy

Introduction to
MINERALOGY

SECOND EDITION

WILLIAM D. NESSE

University of Northern Colorado

New York   Oxford
OXFORD UNIVERSITY PRESS

For Carl and Erik

Oxford University Press, Inc., publishes works that further Oxford University's objective of excellence in research, scholarship, and education.

Oxford   New York
Auckland   Cape Town   Dar es Salaam   Hong Kong   Karachi
Kuala Lumpur   Madrid   Melbourne   Mexico City   Nairobi
New Delhi   Shanghai   Taipei   Toronto

With offices in
Argentina   Austria   Brazil   Chile   Czech Republic   France   Greece
Guatemala   Hungary   Italy   Japan   Poland   Portugal   Singapore
South Korea   Switzerland   Thailand   Turkey   Ukraine   Vietnam

For titles covered by Section 112 of the US Higher Education Opportunity Act, please visit www.oup.com/us/he for the latest information about pricing and alternate formats.

Published by Oxford University Press, Inc.
198 Madison Avenue, New York, New York 10016
www.oup.com

Oxford is a registered trademark of Oxford University Press

Library of Congress Cataloging-in-Publication Data
Nesse, William D.
Introduction to mineralogy / William Nesse.—2nd ed.
    p. cm.
Summary: "Undergraduate textbook for mineralogy students in the department of geology"—Provided by publisher.
ISBN 978-0-19-982738-1
1. Mineralogy. I. Title.
QE363.2.N48 2011
549—dc23   2011019290

Printing number: 9 8 7 6 5 4 3 2 1

Printed in the United States of America
on acid-free paper

BRIEF CONTENTS

CONTENTS

PREFACE TO THE SECOND EDITION

This book is written to provide a text for teaching mineralogy to undergraduate students in the geosciences. The challenge is to provide a survey of mineralogy that is sufficiently comprehensive to meet the needs of students in a wide range of different curricula and to present the information in a concise, well-organized, and clear manner.

In preparing the second edition of *Introduction to Mineralogy*, the main objectives have been to bring the book up to date and to improve the clarity and ease of use of the text. Students and faculty from around the world have provided thoughtful suggestions for improvement, and I have attempted to address their concerns. Few pages survived without some revisions, and many figures were refined to improve clarity.

The organization used in the first edition has been retained; there are three sections—Crystallography and Crystal Chemistry (Part I), Mineral Properties, Study, and Identification (Part II), and Mineral Descriptions (Part III). Placing crystallography and crystal chemistry first allows that foundational material to be used in the discussions of Part II. However, a summary of mineral physical properties has been included in Chapter 1 for instructors who wish to get students actively involved with mineral identification early in the syllabus.

An exciting development over the past decade has been growth of studies of minerals and human health. The section in Chapter 1 on minerals and human health has therefore been expanded, and additional information has been included with the mineral descriptions in Part III. In particular, I want to acknowledge Ulli Limpitlaw for her detailed work in documenting the ways in which minerals have been and continue to be used as medicinal materials.

A second important development in mineralogy is the rapid growth of geobiology and geomicrobiology. While the study of biology has always been integral to geology, there has been an impressive growth of our understanding of the role that organisms, particularly microbes, play in mineralization processes in the near-surface environment. The literature on the subject has expanded rapidly in the past decade; the Geobiology and Geomicrobiology Section has been established by the Geological Society of America; and study groups have been established at numerous universities and research institutes. In reviewing the literature for the new section on biomineralization in Chapter 5 and the mineral descriptions in Part III, it was discovered that around 100 different minerals are formed by biologic activity. I trust that this new understanding will finally put to rest the archaic notion that minerals are inorganic.

A third important development is the recognition that the Earth's mineralogy has evolved over geologic time. A new section has been included in Chapter 5 summarizing some of the important findings.

Additional significant revisions include the following:

- Expansion of the section on mineral stability and phase diagrams in Chapter 5
- Addition of a section on electrical conductivity in Chapter 6
- Addition of a section on the scanning electron microscope in Chapter 9
- Addition of sections on evaporite deposits and carbonate rocks in Chapter 17
- Addition of sections on paleoclimate and sedimentary iron formations in Chapter 18
- New two-color design to improve hierarchy and presentation
- Improved photo reproduction throughout
- Creation of a new interference color chart with improved color accuracy

Graham Baird at the University of Northern Colorado thoughtfully reviewed the new sections and also provided a detailed critique of the first edition that was very useful. I also must thank the numerous students of mine who, through their suggestions and comments, have provided very useful direction to me as I prepared the second edition. In addition, I would like to thank the following people for their careful reviews of the first edition and thoughtful suggestions for improvement. Their contributions were invaluable to the revision.

Olivier Bachman, University of Washington
Jeffrey Byrnes, Oklahoma State University
Mark Colberg, Southern Utah University
Jamey Jones, University of Arkansas Little Rock
Meagen Pollock, College of Wooster
John C. White, Eastern Kentucky University
Terri L. Woods, East Carolina University

Above all, I remain deeply indebted to my wife, Marianne Workman-Nesse, who edited and critiqued the entire manuscript. If, despite the efforts of these individuals, errors, omissions, and inconsistencies remain, they are solely my responsibility and I request that they be brought to my attention.

Greeley, Colorado
WDN

PART ONE

Crystallography and Crystal Chemistry

CHAPTER 1

Introduction

Almost every human endeavor is influenced by minerals. Many natural resources used in the manufacture of goods on which civilization and our lives depend begin as minerals in the ground. The crops that we eat are grown in soil composed of minerals. The safety and stability of structures such as buildings, roads, and bridges depend on the mechanical properties of the minerals that comprise the rocks and soil on which they are built. In addition, the chemical composition, structure, and texture of the minerals that comprise the rocks beneath our feet provide a myriad of clues that guide geoscientists as they attempt to decipher the history of the earth.

MINERALS

The term **mineral** is used in a variety of ways. In the economic sense it often means any valuable material extracted from the earth, including coal, oil, sand and gravel, iron ore or other mined commodity, even groundwater. Nutritionists use the term "mineral" to mean any of a variety of chemical compounds or elements that are important for health. In common usage, anything that is neither animal nor vegetable might be considered mineral. As used in the geosciences, however, a different and more restrictive definition is applied.

A mineral is a naturally occurring crystalline solid.

A material that is **naturally occurring** is formed without the benefit of human action or intervention. It must be possible to find samples of it formed in the natural environment. Many crystalline solids with the same chemical and physical properties as their natural mineral counterparts may be synthesized in the laboratory. These materials are **synthetic minerals**.

Minerals must be **crystalline solids**. The atoms and/or ions that comprise crystalline materials are arranged and chemically bonded in a regular and repeating long-range pattern. The beautiful, symmetrically arranged crystal faces that adorn many mineral samples are a consequence of this internally ordered atomic structure. Solids such as glass, lacking long-range atomic order, are considered **amorphous**. They are not minerals. To be considered crystalline, a material must be a solid, although crystalline

materials may deform in a ductile manner under appropriate temperature–pressure conditions.

As a necessary consequence of being crystalline solids, minerals have a definite, but not necessarily fixed, chemical composition. A chemical formula may be written for any mineral. An example is the common mineral quartz (SiO_2), which is composed of silicon and oxygen in a ratio of 1:2. The composition of many mineral species may vary within certain limits. An example is the mineral olivine, which may be iron rich (Fe_2SiO_4) or magnesium rich (Mg_2SiO_4), or may have an intermediate composition. However, the proportions always work out so that the ratio (Fe + Mg):Si:O remains 2:1:4. Hence, different samples of a mineral species may have different compositions, but the variability is limited. Because minerals are crystalline and have a definite chemical composition, they also have definite physical properties. These physical properties may also vary within limits because they are controlled by the variation in chemical properties.

Some definitions of what constitutes a mineral require that it be formed by inorganic processes. This historical encrustation, which probably has its roots in Aristotle's division of matter into animal, vegetable, and mineral, should have been jettisoned long ago. Minerals constitute an integral part of biologic structures and processes. Obvious examples are the minerals calcite and aragonite (both $CaCO_3$), which are secreted to form the shells of many marine invertebrates and whose remains are a major component of limestone layers found in the stratigraphic record. The mineral apatite [$Ca_5(PO_4)_3(OH, F, Cl)$] makes up a substantial portion of the teeth and bones of vertebrates. If desired, these biologically produced minerals may be called **biominerals**.

Bacteria are an integral part of many geochemical processes at or near the surface of the Earth and directly influence the growth of many minerals. For example, the pyrite (FeS_2) found in many shale and coal beds is produced by the action of sulfate-reducing bacteria. The interaction of microbes and geologic processes is the focus of the emerging and exciting field of study called **geomicrobiology**. Numerous different minerals and mineraloids are now recognized to be produced by biologically induced and controlled mineralization processes (Weiner and Dove, 2003), a number that undoubtedly will rise as our understanding of these processes improves. Further, biologic processes have profoundly affected the Earth's near-surface chemical environment and therefore also the types and distribution of the minerals found there.

MINERALOIDS

Mineraloids are mineral-like materials that lack a long-range crystalline structure. They include amorphous solids and glasses.

Amorphous solids lack long-range atomic order but may possess short-range (~10–100 Å) order. Opal is probably the best-known example and consists of silica gel, often arranged in small spherical masses. The crystalline structure of U- and Th-bearing minerals such as zircon may be extensively disrupted by radioactive bombardment. The term **metamict** is applied to these disrupted structures, and once a mineral's structure becomes metamict, it is properly considered a mineraloid.

Natural glasses also may be considered mineraloids. Volcanic glass is the most common example. Frictional melts called **pseudotachylite** may be produced in fault zones in response to intense shearing. Meteorite impacts may, if large enough, release enough energy to melt the rocks that they strike, producing an **impact melt**. Small masses of glass called **tektites** are usually interpreted to be samples of now-solidified impact melt ejected from an impact crater. A lightning strike may heat soil or rock sufficiently to melt some of it and produce a **fulgurite**. Burning coal beds may generate enough heat to fuse the surrounding rock, forming scoriaceous or slaglike glasses referred to as ash glass or **clinker**.

MINERALOGY

Mineralogy is the study of minerals. The beginning of this particular branch of science extends well back to prehistoric times, for our ancestors surely knew about and used many minerals. Evidence of mining and smelting minerals to extract useful metals such as copper, lead, and zinc is found in many ancient civilizations.

The modern study of mineralogy can be traced back to Theophrastus (ca. 387–272 BC) who wrote the earliest-preserved book dealing with minerals and rocks, titled *On Stones*. Some 400 years later, Pliny the Elder, who met his death at Pompeii, provided us with an encyclopedic review of mineralogy as it applied to the metallic ores, gemstones, and pigments in use in the Roman Empire circa AD 77. Some 1500 years later (1556) German physician and mining engineer Georg Bauer, known to us by his Latinized name Georgius Agricola, provided detailed descriptions and defined physical properties such as hardness and cleavage that continue to provide the basis for hand-sample identification of minerals.

Through the seventeenth, eighteenth, and nineteenth centuries a number of notable scholars provided significant advances to the science:

Niels Stensen (Nicholas Steno), Denmark. Demonstrated the law of constancy of interfacial angles (1669).

A. G. Werner, 1750–1817, Germany. Standardized nomenclature for mineral descriptions.

René-Juste Haüy, 1743–1822, France. "Father of mathematical crystallography." Showed that crystals were constructed by stacking together identical building blocks that we now call unit cells and developed the idea that crystal faces have rational orientations relative to these building blocks.

J. J. Berzelius 1779–1848, Sweden. Recognized that minerals are chemical compounds and provided the foundation for the chemical classification of minerals.

William Nicol, 1768–1851. Invented the Nicol prism (1828), which allowed the anisotropic behavior of light passing through minerals to be studied and provided the foundation for optical mineralogy.

James D. Dana, 1813–1895, Yale University. Published the first edition of *A System of Mineralogy* in 1837. The fourth edition (1854) introduced the chemical classification of minerals that is still in use.

Henry Clifton Sorby, 1826–1908. Developed the use (along with Cloizeaux in France) of the petrographic microscope for studying rocks and minerals.

Perhaps the most dramatic progress in understanding minerals came with the discovery of X-rays. In 1912 Max von Laue (1879–1960) demonstrated that crystals would diffract X-rays, thus proving that minerals possess a regular and repeating internal arrangement of atoms. By 1914, W. H. Bragg (1862–1942) and his son W. L. Bragg (1890–1971) in Cambridge, England had used X-rays to determine the crystal structure of minerals.

During the twentieth century a wide variety of instrumentation was developed to improve our ability to determine the chemical composition of minerals and to refine our understanding of their crystal structures. In addition, petrologists and chemists immensely expanded our knowledge of the chemical and petrologic behavior of minerals in a wide range of geologic environments.

MINERAL NOMENCLATURE

A **mineral species** is distinguished from other minerals by a unique combination of composition (Chapter 3) and crystal structure (Chapter 4). Approximately 4400 mineral species have been identified, described, and named, although less than a hundred mineral species are at all common. The **Commission on New Minerals, Nomenclature and Classification** of the **International Mineralogical Association** provides criteria by which new minerals are recognized. A list of approved mineral names is compiled by Ernest H. Nickel and Monte C. Nichols and it is available as a free Internet download from Materials Data Inc. New minerals are regularly discovered and a summary of new minerals is included in each issue of the *American Mineralogist*, published by the Mineralogical Society of America.

The criteria to be satisfied before a new mineral species is approved include the following.

- It must be a mineral as defined above.
- It must not previously have been described and named.
- The crystallography (Chapter 2), composition (Chapter 3), and crystal structure (Chapter 4) must be determined.
- Physical (Chapter 6) and optical (Chapter 7) properties must be described.
- The geologic and geographic setting in which the mineral was found must be described.
- A **type sample** of the mineral must be preserved in an appropriate repository, such as a museum or the collection of a research institute.

A mineral may be named after an individual, a place where it is found, or in reference to its chemical composition or a significant physical property. Some minerals have names whose origins are lost in the mists of antiquity. A detailed listing of the sources of mineral names can be found in Mitchell (1979). Gains and others (1997) also describe the origin of mineral names. A glossary of mineral synonyms (~35,000 of them) has been prepared by de Fourestier (1999) and a compilation of obsolete mineral names has been published by Bayliss (2000).

A mineral species may have more than one **mineral variety**. Different varieties of a mineral are distinguished by differences in color, habit (shape), or other properties. For example, the mineral corundum (Al_2O_3, Chapter 18) is typically gray and fairly mundane looking. However, ruby and sapphire are varieties that form beautiful gems. Ruby is red because it contains small amounts of Cr, and sapphire is blue because it contains small amounts of Ti + Fe. The terminology for mineral varieties is fairly haphazard; mostly it follows historical and colloquial practice.

A **mineral series** is two or more minerals among which there is a range of chemical compositions. An example is in the common plagioclase mineral series (Chapter 12). The two end members of the plagioclase series are the minerals albite ($NaAlSi_3O_8$) and anorthite ($CaAl_2Si_2O_8$). Natural plagioclase typically has a composition intermediate between albite and anorthite.

A **mineral group** is a set of minerals with the same basic structure but different compositions. The group is generally named for one of the constituent minerals. An example is the calcite group (Chapter 17) whose general chemical formula is XCO_3, where X is a metal cation (Chapter 3).

Calcite	$CaCO_3$
Magnesite	$MgCO_3$
Rhodochrosite	$MnCO_3$
Siderite	$FeCO_3$
Smithsonite	$ZnCO_3$

Members of a mineral group may also form a mineral series. For example, magnesite and siderite form a mineral series because intermediate compositions are common.

A **crystal**, in a restricted sense, is a piece of a mineral bounded, at least in part, by regular crystal faces produced as the crystal grew. The presence of crystal faces is a direct consequence of the fact that minerals are crystalline solids—the atoms are arranged in a regular and repeating manner. In a broader sense, however, mineralogists also may use the term *crystal* to mean a piece of a mineral regardless of whether it has nicely formed crystal faces. The bounding surfaces may be crystal faces, fractures, cleavages, or contacts with adjacent mineral grains. It might be better to use the term "crystal" only when crystal faces are present, and use the term **mineral grain** when the nature of the bounding surfaces is not specified. In any event, the reader should be aware of the ambiguous way "crystal" is used in the literature. In this book, the term *crystal* is used in the restricted sense.

GENERAL REFERENCES ON MINERALOGY

This book is intended as an introduction to the study of minerals and, of necessity, cannot include the depth of coverage provided in more specialized resources. Reference will be provided in each chapter to sources for additional information. However, there are a number of important sources with which all readers should be aware.

Among the most important is the series of volumes by Deer and others titled *Rock-Forming Minerals*, which provides extensive information on the structure, chemistry, properties, and occurrence of common minerals. The same authors also have compiled a one-volume summary (Deer and others, 1992) of *Rock-Forming Minerals* that is highly recommended.

The seventh edition of *Dana's Manual of Mineralogy*, in three volumes (1942–1962), provides coverage of the physical and related properties of nonsilicate minerals, and quartz and its polymorphs. Unfortunately, much of the chemical and crystal-structure information is now out of date and volumes dealing with the silicates are not published. The eighth edition (Gains and others, 1997) provides an exhaustive compilation of minerals in one volume.

The *Handbook of Mineralogy*, in five volumes, by Anthony and others (1990—2003), provides another exhaustive compilation of mineralogical data. Separate files for each of the 3734 minerals included in the handbook may be downloaded from the Mineralogical Society of America. Struntz and Nickel (2001) provide additional data. Mineral data are widely available on the Internet. **Mindat.org** has one of the most comprehensive compilations, and other sources can be readily found with normal search tools.

The Mineralogical Society of America publishes the *Reviews in Mineralogy & Geochemistry* series, which now totals over 70 volumes. New volumes are published every year, often based on a short course offered at the society's annual meeting. Each volume typically deals in depth with a specific group of minerals or some mineral-related topic. They are highly recommended. The Mineralogical Society of America also publishes *Elements* quarterly. Each issue presents a group of articles addressing a topical issue related to mineralogy.

The literature of mineralogy is published in many journals in a variety of languages. The most widely circulated English-language journals are *American Mineralogist*, published by the Mineralogical Society of America; *Canadian Mineralogist*, published by the Mineralogical Society of Canada; *Mineralogical Magazine*, published by the Mineralogical Society of Great Britain; and *European Journal of Mineralogy*, published by Schweizerbart'sche Verlagsbuchhandlung (Nägele u. Obermiller).

MINERALS AND SOCIETY

Natural Resources

Civilization depends very directly on the materials obtained from minerals. A few metals, and the minerals from which they are extracted, are listed in Table 1.1. Because the distribution of economic mineral deposits—geologic concentrations of minerals that can be economically extracted—is uneven, it is essential for nations to establish a global trade in these resources. The United States, for example, has large economic resources of molybdenum and gold but little nickel and tungsten; and for many other resources, production is insufficient to meet domestic requirements. American manufacturers must, therefore, purchase raw materials from worldwide sources. The same applies to manufacturers in England, China, Germany, Japan, and all the other countries in the world. This international trade allows the generation of wealth that supports our society.

Because minerals are valuable, it is not surprising that wars are fought over access to mineral resources. Further, trade in minerals also can support brutal and repressive regimes. A good recent example is provided by the conflicts in Angola, Democratic Republic of the Congo, Liberia, Côte d'Ivoire, and adjacent areas in Africa that have ties to the global trade in diamonds. Most conflicts in human history have involved competition for natural resources, either directly or as a strategic consideration.

World War II is instructive. Iron is absolutely essential to support modern industry and to support a strong military. At the beginning of the war Germany had only modest iron ore production, whereas France, Sweden, the Soviet Union, and the United Kingdom each had significantly more production (Table 1.2). Even tiny Luxembourg had nearly as much iron ore production as Germany. If Germany was to succeed with its military ambitions, it was

Table 1.1 Selected Metals and Some Minerals from Which They Are Extracted

Metal	Mineral	Chemical Formula	Major World Sources
Iron	Magnetite Hematite	Fe_2O_3 Fe_3O_4	Australia, Brazil, China
Copper	Chalcopyrite	Cu_5FeS_4	Chile, Peru, United States
Molybdenum	Molybdenite	MoS_2	Chile, United States, China
Lead	Galena	PbS	China, Australia, United States
Tungsten	Scheelite Wolframite	$CaWO_4$ $(Fe,Mn)WO_4$	China, Russia, Canada
Nickel	Pentlandite	$(Fe,Ni)_9S_8$	Russia, Canada, Indonesia
Aluminum	Bauxite	Mixed oxides of Al	Australia, China, Brazil
Titanium	Rutile Ilmenite	TiO_2 $FeTiO_3$	Australia, Norway, China
Zinc	Sphalerite	ZnS	China, Peru, Australia

Source: U.S. Geological Survey Minerals Yearbook 2007, 2008.

Table 1.2 Iron Ore Production in 1937 from Selected Countries

Country	Amount (million metric tons)
Germany	10
France	38
Soviet Union	26
Sweden	15
United Kingdom	14
Luxembourg	8
United States	73

Source: U.S. Geologic Survey Minerals Yearbook 1939.

essential to acquire at least some of the iron ore production from other sources in Europe. The German invasion of France and Luxembourg served to lock up the supplies of iron from those countries. Also of great importance was the Swedish supply of iron ore in Kiruna, which was all transshipped through Narvik, on the west coast of Norway. The German invasion of Norway, and particularly Narvik, in 1940, had as one strategic goal the acquisition of those supplies from neutral Sweden. Fortunately for Great

Britain and the other allies, the United States had abundant supplies of iron ore that were available for the war effort both before and after the United States entered the conflict on December 8, 1941.

Minerals and Health

Over 2000 years ago the Greek physician Hippocrates established a correlation between disease and location, demonstrating that geological environment influences the incidence of disease. We now recognize that some of these spatial variations in disease are tied to the mineralogical composition of the soils and rocks on which we live. Selenium provides a good example because it is an essential nutrient and is taken up by plants from the soil. Thus humans and animals eating plants grown in soil that is deficient in selenium may develop health problems from a selenium deficiency. Too much selenium, however, may also lead to significant health problems. Many other metals ultimately derived from minerals in the soil and bedrock are essential nutrients in small amounts but toxic in large amounts; examples include arsenic, chromium, cobalt, copper, iron, manganese, molybdenum, nickel, tin, tungsten, vanadium, and zinc.

Other chemical elements, such as hexavalent chromium, lead, and mercury, do not have known nutritional benefits and can be quite toxic. While poisoning from natural sources of these elements has occurred, most cases of toxicity come from anthropogenic sources. The metals are mined and used in products or industrial processes that make them available to people. Lead in paint, gasoline, and solder is a good example.

Our bodies are generally well equipped to deal with the routine and unavoidable inhalation of environmental mineral dust. However, chronic inhalation of significant quantities of any mineral dust can be hazardous to one's health. Several minerals appear to pose particular problems.

One of the most widely recognized problems is with the group of minerals referred to as asbestos, that form thin flexible fibers that are useful for a number of products and industrial processes. The commonly used asbestos minerals are chrysotile (Chapter 13) and amphibole (Chapter 14). While many uses of asbestos have now been banned, in the past asbestos minerals were used in insulation, flooring, automotive brakes, and many other products. Inhalation of the fibrous mineral particles has been documented to cause cancer and a variety of other lung-related pathologies. In response to the real and imagined threat posed by asbestos, this material has been removed at considerable expense from homes, schools, and other buildings.

Quartz (Chapter 12) dust also has been documented to cause silicosis, a pathology most commonly found in miners and mill workers exposed to high levels of dust in the work environment. The accumulation of high levels of quartz particles in the lungs leads to scarring and decreased lung function. Other pulmonary diseases, including cancer, also have been linked to silicosis. However, it seems unlikely that routine casual environmental exposure presents significant health risks because life has evolved for billions of years in close contact with quartz.

Chronic inhalation of other mineral dusts (e.g., kaolinite, pyrophyllite, montmorillonite, zircon) has also been documented to cause medical problems. Chronic inhalation of almost any particulate material, mineral or otherwise, should be avoided because of the potential to cause problems.

Medical Uses of Minerals

Minerals have probably been used for medical purposes from before the start of recorded history. Before the development of modern pharmaceuticals, physicians and healers had to rely on a relatively limited collection of natural materials. This included minerals in addition to materials from plants and animals. Pliny, who died at Pompeii in AD 79, mentions the medical uses of minerals (Bostock and Riley, 1857), and the earliest textbook on mineralogy (Agricola, 1546) includes a section in which the medical uses of minerals are described. Limpitlaw (2006) has documented that about 100 different minerals have been used for medicinal purposes and that some continue to be an integral part of the modern pharmacy.

Many minerals are used in folk remedies, and in homeopathic and other preparations for which scientific documentation of their efficacy is notably lacking. In many cases, these uses probably do no harm; and are only successful to the degree that they trigger the placebo effect. The placebo effect is based on the observation in medical trials that placebos (pills/preparations with no active ingredient) can produce benefits to patients simply because the patient believes that they will. This example of the nearly infinite human capacity for self-delusion probably helps explain the common historical use of gemstones, beautiful crystals, and other precious or rare minerals for medical purposes. These uses still flourish, as an Internet search will readily confirm.

The most common means of application include ingestion of powders, preparing a "tea" or extract by soaking the mineral in water, beer, wine, or other liquid, and then drinking the liquid, preparing a poultice of the powdered mineral, and carrying the mineral in an amulet. Homeopathic preparations are commony diluted to the point that only traces (or none) of the original material is present. The modern practice of wearing gemstones derives, in part, from ancient beliefs that the gems provided protection from, or a cure for various ailments.

Some of the historical medical uses of minerals were definitely hazardous to the patient's health. Among the minerals that did considerable harm was calomel, a mercury chloride (Hg_2Cl_2), used to treat gastrointestinal problems, infections, and other maladies from at least 1500 up to the 1860s. It also produced mercury poisoning. Preparations made with gold were claimed to energize and assure longevity. Diane de Poitiers, mistress to King Henry II of France in the sixteenth century, routinely drank an elixir made from gold and probably died from gold poisoning (Charlier and others, 2009). Lead oxides such as minium (Pb_3O_4) continue to be used as a folk remedy to treat diarrhea in certain Central American societies and can lead to serious lead poisoning.

The reader should not infer that all medical uses of minerals are bunk. A significant number of minerals are routinely and productively used for modern medical purposes These include the clay minerals (Chapter 13), zeolites (Chapter 12), and calcite (Chapter 17).

Medical uses of some minerals are mentioned as part of the descriptions in Chapters 12 through 20. The reader is cautioned that the author does not endorse any of these uses and that qualified physicians and pharmacists should be consulted about the appropriate treatment for any medical condition. The reader is also cautioned that mineral uses that now seem improbable may yet prove to be beneficial.

GETTING STARTED

Students are understandably eager to get their hands on minerals in the lab as they start a mineralogy course. However, the detailed discussion of the physical properties that provides the basis for identification of unknown samples is not presented until Chapter 6. The reason for this organization is that an understanding of mineral physical properties (habit, color, luster, streak, hardness, specific gravity, cleavage, fracture, etc.) requires an understanding of crystallography and crystal chemistry, the topics of Chapters 2 through 5. The objective of studying minerals is not simply to collect, identify, and catalog them. Much of the Earth's geologic history is recorded in the compositions, textures, associations, and spatial and temporal distributions of the minerals that comprise the rocks, sediments, and soils that make up the solid Earth. Without a working knowledge of crystallography and crystal chemistry, it is not possible to document and interpret the information that minerals have to provide.

It is assumed here that students using this text will have completed at least one course in introductory geology to provide them with basic information about minerals, rocks, and geologic processes and environments. A review of the relevant chapters in any introductory geology textbook is strongly recommended for all readers and will make the study of the following chapters easier and more informative. A brief review of the physical properties that are routinely used to identify hand samples of minerals is provided in Box 1.1.

Box 1.1 Getting Started: Mineral Physical Properties

The properties used to identify minerals can be grouped into those that involve shape, mass (density and specific gravity), mechanical properties (hardness, cleavage, fracture), interaction with light (luster, color, streak), and other properties (magnetism, taste, smell, reaction with acid).

Mineral Shape
Minerals commonly form beautiful crystals; each mineral forms crystals with a specific set of shapes and crystal faces. For example, quartz crystals form prisms with six sides, and pyrite commonly forms cubes with striated faces. These shapes can be described with common terminology—elongate, stubby, fibrous, cubic, prismatic, and so forth. However, the detailed study of crystals has revealed symmetrical relationships among crystal faces and has developed into the subject of crystallography. Chapter 2, on crystallography, explores symmetry in minerals and establishes a coordinate system of crystal axes and an extensive nomenclature for crystals that allows us to systematically describe and discuss the properties of minerals.

Mass-Related Properties
The mass-related properties of **density** and **specific gravity** are closely related. Density (ρ) is defined as mass (m) per volume (v):

$$\rho = m/v$$

The usual units are grams per cubic centimeter (g/cm^3). Specific gravity (G) is the ratio of the density of a material (ρ) divided by the density of water at 4°C (ρ_{H_2O}).

$$G = \frac{\rho}{\rho_{H_2O}}$$

G is unitless because it is a ratio of densities. Because the density of water at 4°C is essentially 1 g/cm^3, the numerical value for specific gravity is the same as the specific gravity expressed in terms of grams per cubic centimeter. Details of how specific gravity is measured are provided in Chapter 6, but a rough estimate can be made by simply hefting a sample of a mineral in one's hand and comparing it to a sample whose specific gravity is known, such as quartz ($G = 2.65$). The controls of density are chemical composition (Chapter 3) and the manner in which the atoms that comprise the mineral are packed and chemically bonded together (Chapter 4).

Mechanical Properties
Hardness (H) is a measure of the resistance of a mineral to being scratched. The Mohs scale of hardness ranks minerals from 1 for very soft, easily scratched minerals, to 10 for the hardest mineral, diamond (Table 1.3). To provide a standard of comparison, specific minerals are assigned to each hardness number. A mineral with hardness 5 will scratch a mineral with hardness 4 and be scratched by a mineral with hardness 6, for

Continued

Box 1.1 *Continued*

Table 1.3 Mohs Scale of Hardness

1	Talc
2	Gypsum
3	Calcite
4	Fluorite
5	Apatite
6	Orthoclase
7	Quartz
8	Topaz
9	Corundum
10	Diamond

example. To make comparisons in the lab and field easier, mineral hardness is routinely tested with common materials of known hardness—a fingernail (2+), a steel nail or knife blade (~5), window glass (~6), and a piece of quartz (7). The identification tables at the end of this book (Appendix B) use groupings based on these values.

Hardness is fundamentally controlled by the nature of the chemical bonds among the atoms that comprise a mineral. The detailed discussion of hardness in Chapter 6 depends on an understanding of chemical bonding (Chapter 3) and crystal structure (Chapter 4). Both of these are intimately related to a mineral's symmetry (Chapter 2).

Cleavage and **fracture** both refer to the manner in which a mineral breaks. All minerals fracture when hit with a hammer or otherwise are forced to break. When a mineral fractures, it breaks on an irregular surface. If the surface is a smooth curved surface, like broken glass, the fracture is **conchoidal**. **Irregular**, **hackly**, and **splintery** fracture are terms that indicate increasing degree of relief ("pointyness") on the fracture surface. Only some minerals display cleavage, where the breaks are on smooth planar surfaces that are planes of weakness in a mineral. The cleavage may be described as perfect, good, fair, or poor depending on the ease with which the cleavage can be produced, as well as how smooth and continuous the surfaces are. Minerals with cleavage may have one, two, three, or more different directions along which they break. Table B.1 in Appendix B groups minerals based on the number of cleavages. Because cleavages are planes of weakness in a mineral's structure (Chapter 4) they have crystallographic control and are identified with the nomenclature of crystallography (Chapter 2). A detailed discussion is provided in Chapter 5.

Color Properties

A mineral's **luster** and **color** are perhaps the first things that an observer notices. Luster is generally categorized as **metallic** or **nonmetallic**. Whether we perceive a mineral to be metallic or nonmetallic is not a function of color—rather, it is a function of how much of the incident light is reflected, which is discussed in Chapters 6 and 7. Separate identification tables are provided in Appendix B for metallic and nonmetallic minerals. Color for minerals with a metallic luster is generally quite consistent for different samples of the mineral and is therefore a useful diagnostic property. For nonmetallic minerals,

however, color can be quite variable. Quartz, for example, can be almost any color, and K-feldspar, another very common mineral, can be pink, white, gray, green, or blue. An understanding of the controls of a mineral's color requires knowledge of chemical bonding and the interaction of light with matter. These topics are covered in Chapters 3, 6, and 7. Fluorescence and related properties are covered in Chapter 6.

For most minerals, it turns out that the streak is quite consistent. Streak color is the color of the powdered mineral. In most mineral laboratories, the streak color is obtained by rubbing a mineral sample on an unglazed white porcelain tile, called a streak plate. The surface of the tile is rough, and some of the mineral rubs off where its streak color can easily be seen. If a streak plate is not available, a knife blade, rock hammer, or other similar implement can be used to produce a little powder of an unknown mineral. While the streak color of nonmetallic minerals is usually quite consistent regardless of variations in mineral color, it is not a particularly useful diagnostic property because the streak color of most nonmetallic minerals usually is white, gray, or some other pale color. Only a few nonmetallic minerals have a distinctly colored streak (Appendix B: Table B.2).

Other Properties

A few common minerals are **magnetic**, meaning that they experience obvious attraction to a magnet (Appendix B: Table B.6). However, magnetism is much more complicated than simple attraction to a magnet and includes the properties of **diamagnetism**, **paramagnetism**, **ferromagnetism**, **ferrimagnetism**, and **antiferromagnetism**. All minerals have one or another of these magnetic properties, which are described in Chapter 6. Understanding these properties requires knowledge of chemical bonding and the nature of the chemical elements (Chapter 3).

Other properties such as **taste** and **smell** are unique to certain minerals. Halite (NaCl) and sylvite (KCl) are readily soluble in water and have a salty taste; most clay minerals yield an earthy odor if moistened a little. Some carbonate minerals such as calcite ($CaCO_3$) and dolomite [$CaMg(CO_3)_2$] will effervesce in dilute hydrochloric acid, though dolomite must first be powdered. These properties are discussed at more length in Chapter 6. Table B.7 in Appendix B lists minerals that effervesce in dilute HCl.

Other Tools

The study of minerals is not restricted to the physical properties outlined in this section. Many mineral

Continued

Box 1.1 *Continued*

samples cannot be identified based on physical properties, either because the sample is too small or because the physical properties that can be measured do not lead to a unique answer. Physical properties cannot be used to determine the details of a mineral's chemistry. Other analytical tools are required and are routinely used. For purposes of identification, the petrographic microscope (Chapter 7) and X-ray diffractometer (Chapter 8) are workhorses in most mineral studies, and each has strengths and weaknesses. Some common techniques of mineral chemical analysis are described in Chapter 9.

REFERENCES CITED AND SUGGESTIONS FOR ADDITIONAL READING

Agricola, G. 1546. *De natura fossilum.* Translated by M. C. Bandy, and J. A. Bandy, 1955. Geological Society of America Special Paper 63, 240 p.

Agricola, G. 1556. *De re metallica.* Translated by H. Hover, 1950. Dover Publications, New York, 638 p.

Anthony, J. W., Bideaux, R. A., Bladh, K. W., and Nichols, J. C. 1990–2003. *Handbook of Mineralogy* (5 volumes). Mineral Data Publishing, Tucson, AZ.

Back, M., and Mandarino, J. A. 2008. *Fleischer's Glossary of Mineral Species 2008.* The Mineralogical Record, Tucson, AZ, 346 p.

Bayliss, P. 2000. *Glossary of Obsolete Mineral Names.* The Mineralogical Record, Tucson, AZ, 235 p.

Blackburn, W. H., and Dennen, W. H. 1997. *Encyclopedia of Mineral Names.* Canadian Mineralogist Special Publication 1, 368 p.

Bostock, J., and Riley, H. T. 1857. *The Natural History of Pliny.* Henry G. Bohn, London, 529 p.

Carr, D. D. (ed.). 1994, *Industrial Minerals and Rocks,* 6th ed. Society for Mining, Metallurgy, and Exploration, Littleton, CO, 1196 p.

Chang, L. L. Y., Howie, R. A., and Zussman, J. 1996. *Rock-Forming Minerals,* Volume 5B: *Nonsilicates: Sulfates, Carbonates, Phosphates, Halides,* 2nd ed. The Geological Society, London, 383 p.

Charlier, P., Pupon, J., Huynh-Charlier, I., Saliege, J.-F., Favier, D., Keyser, C., and Ludes, B. 2009. A gold elixir of youth in the 16th century French court. British Medical Journal 339, 5311.

de Fourestier, J. 1999. *Glossary of Mineral Synonyms.* Canadian Mineralogist Special Publication 2, 448 p.

Deer, W. A., Howie, R. A., and Zussman, J. 1992. *An Introduction to Rock-Forming Minerals.* Longman Group Limited, London, 696 p.

Deer, W. A., Howie, R. A., and Zussman, J. 1997. *Rock-Forming Minerals,* Volume 2B: *Double-Chain Silicates,* 2nd ed. The Geological Society, London, 784 p.

Deer, W. A., Howie, R. A., and Zussman, J. 2001. *Rock-Forming Minerals,* Volume 1A: *Orthosilicates,* 2nd ed. The Geological Society, London, 919 p.

Deer, W. A., Howie, R. A., and Zussman, J. 2001, *Rock-Forming Minerals,* Volume 1B: *Disilicates & Ring Silicates,* 2nd ed. The Geological Society, London, 630 p.

Deer, W. A., Howie, R. A., and Zussman, J. 2001. *Rock-Forming Minerals,* Volume 2A: *Single-Chain Silicates,* 2nd ed. The Geological Society, London, 668 p.

Deer, W. A., Howie, R. A., and Zussman, J. 2001. *Rock-Forming Minerals,* Volume 4A: *Feldspars,* 2nd ed. The Geological Society, London, 992 p.

Deer, W. A., Howie, R. A., and Zussman, J. 2009. *Rock-Forming Minerals,* Volume 3B: *Layered Silicates: Excluding Micas and Clay Minerals,* 2nd ed. The Geological Society, London, 320 p.

Fleet, M. E., and Howie, R. A. 2004. *Rock-Forming Minerals,* Volume 3A: *Micas,* 2nd ed. The Geological Society, London, 780 p.

Frondel, C. 1962. *The System of Mineralogy*: Volume III, 7th ed. John Wiley & Sons, New York, 334 p.

Gains, R. V., Skinner, H. C. W., Foord, E. E., Mason, B., and Rosenzweig, A. 1997. *Dana's New Mineralogy,* 8th ed. John Wiley & Sons, New York, 1819 p.

Hazen, R. M. 1984. Mineralogy: A historical review. Journal of Geological Education 32, 288–298.

Howie, R. A., Deer, W. A., Wise, W. S., and Zussman, J. 2004. *Rock-Forming Minerals,* Volume 4B: *Silica Minerals,* 2nd ed. The Geological Society, London, 982 p.

Kraus, E. H. 1941. Mineralogy. *Geological Society of America 50th Anniversary volume, Geology 1888–1938,* 307–332.

Limpitlaw, U. G. 2006. *Palliative and Curative Earth Materials.* Unpublished MA thesis, University of Northern Colorado, Greeley, 384 p.

Mitchell, R. S. 1979. *Mineral Names: What Do They Mean?* Van Nostrand Reinhold, New York, 229 p.

Palache, C., Berman, H., and Frondel, C. 1944. *The System of Mineralogy*: Volume I, 7th ed. John Wiley & Sons, New York, 834 p.

Palache, C., Berman, H., and Frondel, C. 1951. *The System of Mineralogy*: Volume II, 7th ed. John Wiley & Sons, New York, 1124 p.

Reviews in Mineralogy & Geochemistry, 1974–2010, Volumes 1–71. Mineralogical Society of America, Washington, DC.

Sahai, N., and Schoonen, M. A. A. (eds.). 2006. *Medical mineralogy and geochemistry.* Reviews in Mineralogy & Geochemistry 64, 332 p.

Strunz, H., and Nickel, E. H. J. 2001. *Strunz Mineralogical Tables,* 9th ed. E. Schweizerbart'sche Verlagsbuchhandlung (Nägele u. Obermiller), Stuttgart, 621 p.

Weiner, S., and Dove, P. M. 2003. An overview of biomineralization processes and the problem of the vital effect. Reviews in Mineralogy & Geochemistry 54, 1–30.

CHAPTER 2

Crystallography

INTRODUCTION

The descriptive nomenclature of crystallography has evolved over a period of several centuries to describe the shape, symmetry, and crystal structure of minerals. The discussion that follows contains only an introduction to crystallography and omits development of the theoretical and mathematical foundation on which the descriptive material depends. For more extensive coverage of crystallography refer to Bloss (1971), Smith (1982), or O'Keefe and Hyde (1996). The mathematical foundations of crystallography are described by Boisen and Gibbs (1985).

Some reluctance to dive right into a discussion of crystallography in the second chapter of a mineralogy textbook is understandable because crystallography is often considered to be among the most challenging parts of mineralogy. It is perhaps more exciting to immediately begin reading about specific minerals and learning to use analytical tools such as the X-ray diffractometer, petrographic microscope, and scanning electron microscope. However, it is important to establish a framework within which to describe minerals and to develop the specialized nomenclature that enables us to communicate clearly and succinctly about the structure and the physical and chemical properties of minerals.

The broad questions to be addressed in this chapter include the following.

What is the symmetry of crystals and crystal structures?

What is the descriptive nomenclature of crystals and crystal structures?

In studying this chapter, it will be useful to keep the organization in mind. The subjects of symmetry and the descriptive nomenclature are addressed in the following order.

Symmetry

- Start with **translational symmetry**, first in two dimensions and then three to introduce **unit cells**, **crystal axes**, **crystal systems,** and **Bravais lattices**.

- Present the elements of **point symmetry** (reflection, rotations, inversion) that can be identified in hand samples of crystals and define the 32 **point groups/ crystal classes** that are members of the six crystal systems.

- Combine translational symmetry with point symmetry operations to define **space groups.**

Descriptive Nomenclature of Crystals

- Establish that **crystal faces** have rational orientations relative to the crystal lattices.

- Develop a nomenclature (**Miller indices**) that allows individual crystal faces or crystallographic planes (e.g., cleavage) to be identified and described.

- Develop a nomenclature for identifying **crystallographic directions** and **zones** in minerals.

- Develop the nomenclature that allows the collection of faces that make up a crystal to be described. These **crystal forms** include shapes such as cubes and octahedrons as well as a variety of other shapes such as prisms, pinacoids, and pyramids.

- Introduce some general terminology describing **crystal habit** and the way that mineral grains are intergrown in rocks and mineral deposits.

A conventional view is that an object with symmetry consists of two halves that are mirror images of each other. However, symmetry involves far more than just mirror images; it involves all of the ways that something can be systematically repeated in three dimensions. Because minerals have a regular and repeating atomic structure, it follows that they must display symmetry. This symmetry is manifested in the symmetrical arrangements of crystal faces and the internal structure of crystals that control cleavage and the diffraction of X-rays. It is not surprising, then, that the concepts of crystallography are built on an understanding symmetry.

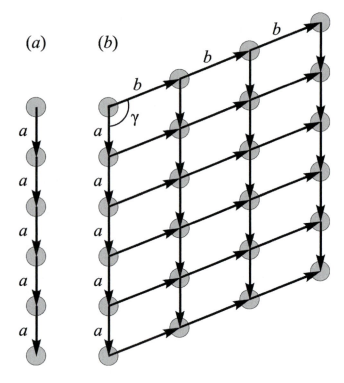

(a)   *(b)*

Figure 2.1 Simple translation to form a plane lattice. (*a*) A spot, representing a collection of atoms, is repeated by translation distance *a* toward the bottom of the page. (*b*) Repeated translation of these spots, distance *b* at angle γ, produces a two-dimensional plane lattice. The spots represent the lattice nodes.

TRANSLATIONAL SYMMETRY: TWO DIMENSIONS

Consider a spot, which could represent a collection of atoms (Figure 2.1a). That spot can be repeated by translating parallel to vector *a* toward the bottom of the page again and again. The result is a row of dots extending to infinity. The dots also can be repeated by translating parallel to vector *b* at angle γ from the *a* direction (Figure 2.1b). Repeated translations parallel to *a* and *b* produce a continuous repeating pattern of dots known as a **plane lattice** that extends to infinity in the *ab* plane. The center of each spot at the intersection of the lattice lines is called a **lattice node.**

nt plane lattices that can be produced in two dimensions (Figure 2.2, left) gle, diamond or centered rectangle, plane lattices. These five plane lat- lamentally different shapes, called right): square, rectangle (for both plane lattices), rhombus (or hex- llelogram. All but the diamond (p) unit meshes because lat- corners. The diamond plane

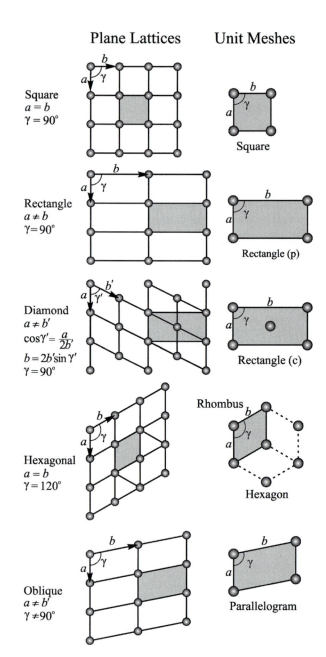

Plane Lattices   Unit Meshes

Square
$a = b$
$γ = 90°$
    Square

Rectangle
$a \neq b$
$γ = 90°$
    Rectangle (p)

Diamond
$a \neq b'$
$\cos γ' = \dfrac{a}{2b'},$
$b = 2b' \sin γ'$
$γ = 90°$
    Rectangle (c)

Hexagonal
$a = b$
$γ = 120°$
    Rhombus   Hexagon

Oblique
$a \neq b'$
$γ \neq 90°$
    Parallelogram

Figure 2.2 Plane lattices and unit meshes. Only five different plane lattices can be produced by combinations of translation in two directions. The four different unit meshes are shaded and are identified on the right. Note that both the rectangle and diamond lattice can be repeated on the page by using a rectangular unit mesh. Unit mesh axes *a* and *b* at angle γ are parallel to the edges of the unit mesh.

lattice has a **centered** (c) unit mesh because a lattice node is at its center.

Unit mesh axes parallel to the edges of the unit mesh are labeled *a* and *b* even if they are the same length. The angle between them is γ. So that the unit mesh axes in the rectangle and diamond lattices can be the same ($a \neq b$, $γ = 90°$), the second translation vector in the diamond lattice is b' at angle γ', and $b = 2b' \sin γ'$.

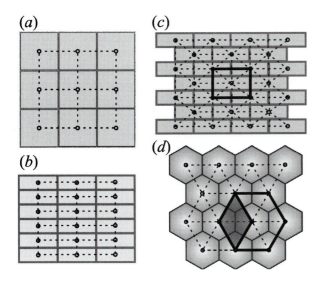

Figure 2.3 Plane lattices and unit meshes in everyday life. (*a*) Square ceramic tile laid in a square lattice. (*b*) Rectangular bricks laid in a rectangular lattice. (*c*) Rectangular bricks laid in a centered rectangular (diamond) lattice. The unit mesh includes pieces sufficient to make two bricks. (*d*) Hexagonal ceramic tile laid in a hexagonal lattice. The rhombic unit mesh includes pieces sufficient to make one tile.

The type of translational symmetry described here is routinely used by craftsmen in building brick walls and laying ceramic tile on floors and walls, and it must be present in the repeating pattern of wallpaper used to decorate homes. Bricks typically have a rectangular unit mesh and are most commonly laid using a centered rectangular lattice. Square ceramic tiles are laid on floors and walls using a square lattice; hexagonal ceramic tiles must be laid using a hexagonal lattice. Wallpaper typically uses either square, rectangular, or centered rectangular plane lattices so that the patterns on adjacent pieces of wallpaper will match. Oblique lattices are rarely used because surfaces such

as walls and floors typically have square or rectangular shapes (Figure 2.3).

TRANSLATIONAL SYMMETRY: THREE DIMENSIONS

Space Lattices and Unit Cell

If the two-dimensional plane lattices are systematically repeated one above the other, to allow for a translation vector in the third dimension, the result is the formation of a **space lattice** (Figure 2.4) in which lattice nodes are repeated in all three dimensions. The volume outlined by lattice nodes is known as the **unit cell**, analogous to the unit mesh in two dimensions.

The edges of the unit cells are parallel to **crystal axes**, identified as *a*, *b*, and *c* (Figure 2.4). The axes, which intersect at a point called the **origin**, have positive and negative ends: positive *a* is to the front, positive *b* is to the right, and positive *c* is up. The dimensions of the unit cell along the *a*, *b*, and *c* axes are *a*, *b*, and *c*; the angles between the axes are α, β, and γ. Note that α is the angle between *b* and *c*, β is the angle between *a* and *c*, and γ is the angle between *a* and *b*.

Bravais Lattices and Crystal Systems

The five plane lattices can be repeated in three dimensions to produce 14 different space lattices known as the **Bravais lattices** (Figures 2.5–2.9), named for Auguste Bravais (1811–1863). These 14 Bravais lattices are divided into six groups based on the shape of the **unit cell** (Figure 2.10). Many unit cell shapes are familiar to anyone who has played with wooden blocks. They represent the shapes that can be regularly stacked together without leaving voids between the blocks. The six unit cell shapes are identified

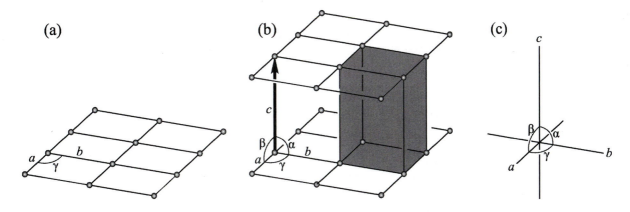

Figure 2.4 Space lattice. (*a*) A place lattice with axes *a* and *b* at angle γ. (*b*) Translation of the lattice at angles α and β to form a repeating pattern of lattice nodes in three dimensions. A unit cell (shaded) is the volume outlined by lattice nodes. (*c*) Crystal axes *a*, *b*, and *c* are parallel to the edges of the unit cell at angles α, β, and γ.

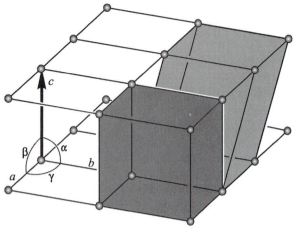

Figure 2.5 Triclinic Bravais lattice. An oblique plane lattice is repeated by translation distance c so that the α and β angles are not equal to 90°. Light and dark shading shows two alternative unit cell choices.

with the six *crystal systems*: **triclinic, monoclinic, orthorhombic, hexagonal, tetragonal**, and **isometric (cubic)**. **Primitive** (P) unit cells contain lattice nodes only at the corners. **Body-centered** (I) unit cells contain an additional lattice node at the center. **Face-centered** (C) unit cells contain lattice nodes on the corners and on two opposite sides. **Face-centered** (F) unit cells contain lattice nodes at the corners and at the center of each face.

Except for triclinic, each crystal system has more than one Bravais lattice. In hand sample, it is not possible to distinguish among the Bravais lattices within a crystal system—X-ray diffraction techniques are required. Hence, when studying the Bravais lattices, pay particular attention to the differences among the crystal systems and the geometries of the six different unit cells.

The single triclinic Bravais lattice (Figure 2.5) is produced by translating an oblique plane lattice distance c in a direction not at right angles to either a or b. The unit cell dimensions (Figure 2.10a) are of different lengths and

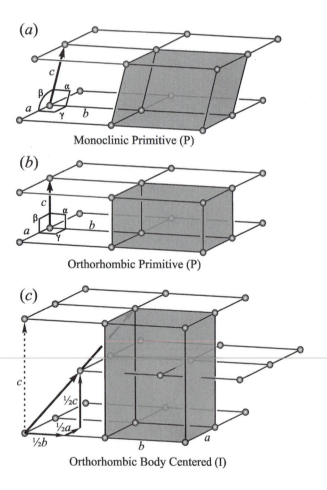

Monoclinic Primitive (P)

Orthorhombic Primitive (P)

Orthorhombic Body Centered (I)

Figure 2.6 Bravais lattices produced by translation of rectangular plane lattice where a and b are different lengths and angle $\gamma = 90°$. (*a*) Primitive (P) monoclinic lattice produced by translation distance c so angles $\alpha = 90°$ and $\beta > 90°$. (*b*) Primitive (P) orthorhombic lattice produced by translation distance c so that angles α and β are 90°. (*c*) Translation ½ b to the right, ½ a back, and ½ c up produces a body-centered orthorhombic (I) lattice.

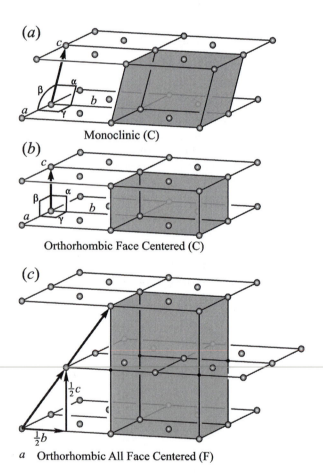

Monoclinic (C)

Orthorhombic Face Centered (C)

Orthorhombic All Face Centered (F)

Figure 2.7 Bravais lattices produced by translation of a centered rectangular plane lattice where a and b are different lengths and $\gamma = 90°$. (*a*) Translation distance c so that $\alpha = 90°$ and $\beta > 90°$ yields a face-centered (C) monoclinic lattice. (*b*) Translation distance c so that $\alpha = \beta = 90°$ produces a face-centered (C) orthorhombic lattice. (*c*) Translation ½ b to the right and ½ c up produces an all face-centered (F) orthorhombic lattice.

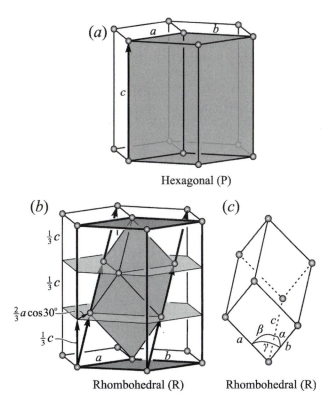

(a) a b

c

Hexagonal (P)

(b) $\frac{1}{3}c$ $\frac{1}{3}c$ $\frac{2}{3}a\cos30°$ $\frac{1}{3}c$ a b

(c) β c α a γ b

Rhombohedral (R)   Rhombohedral (R)

Figure 2.8 Bravais lattices produced by translation of a hexagonal plane lattice. (*a*) Vertical translation by distance *c* produces a primitive hexagonal (P) lattice. The simplest unit cell is shaded. (*b*) Translation 1/3 *c* up and 2/3 *a* cos 30° back at right angles to the *b* axis produces a rhombohedral (R) lattice. The translation vector is parallel to two edges of the rhombohedron defined by the lattice nodes as shown. A unit cell that is the same as the primitive hexagonal lattice is shown with heavy lines and shading on the top and bottom. (*c*) A unit cell and crystal axes also can be defined by the edges of the rhombohedron ($a = b = c$, $\alpha = \beta = \gamma \neq 90°$), but this unit cell is almost never used.

none of the angles between the axes is 90° ($a \neq b \neq c$, $\alpha \neq \beta \neq \gamma \neq 90°$). No universally-adopted convention exists to guide the assignment of crystal axes to the edges of the unit cell. Two different possible unit cells are shown with light and dark shading in Figure 2.5. Generally the *c* axis is placed parallel to some prominent elongation (a **zone** that will be described later) defined by crystal faces seen in a hand sample. The *b* axis is placed down and to the right, and the *a* axis is placed down and to the front.

The two monoclinic Bravais lattices (Figure 2.10b) are derived from either a rectangular (P) or rectangular (C) plane lattice. All unit cell dimensions are of different length ($a \neq b \neq c$) and any axis may be the longest. The *b* and *c* axes are at right angles, and the *a* axis is at right angles to *b* but not to *c* ($\alpha = \gamma = 90°$, $\beta > 90°$). Note that the axes are arranged so that angle between the positive ends of the *a* and *c* axes is greater than 90°.

The monoclinic primitive (P) lattice is produced by translating a primitive rectangular plane lattice a distance equal to c so that angle $\beta > 90°$ and angle $\alpha = 90°$ (Figure 2.6a).

The monoclinic centered (C) lattice is produced by translating a centered rectangular plane lattice a distance equal to c so that angle $\beta > 90°$ and angle $\alpha = 90°$ (Figure 2.7a).

The four orthorhombic Bravais lattices (Figure 2.10c) are derived from the rectangular (P) plane lattice and rectangular (C) plane lattice whose unit mesh dimensions are *a* and *b*. All three unit cell dimensions are of different length. A common convention is to assign the axes so that the *c* unit cell dimension is the smallest, *b* is the longest, and *a* is intermediate ($c < a < b$). Other conventions have been used and in some minerals either *c* or *a* may be the longest unit cell dimension. The axes are all at right angles so $\alpha = \beta = \gamma = 90°$.

The orthorhombic primitive (P) lattice is produced by vertical translation of a primitive rectangular lattice a distance equal to c (Figure 2.6b)

The orthorhombic body-centered (I) lattice is produced by translating a primitive rectangular lattice a distance equal to ½ b right, ½ a back, and ½ c up. Every second lattice plane is aligned vertically at distance c, and the intermediate lattice plane places a node in the center of the unit cell (Figure 2.6b).

The orthorhombic face-centered (C) lattice has nodes in the center of the top and bottom faces, and is produced by vertical translation of a centered rectangular plane lattice a distance equal to c at right angles to the a–b plane. Note that if the faces with the centered node were on the front and back or on right and left faces, this also could be called an (A) or (B) face-centered lattice. The differences among A, B, and C face-centered lattices are only in how the crystal axes are labeled, which, in the orthorhombic crystal system, is not done consistently.

The orthorhombic face-centered (F) lattice has nodes in the center of all faces. It is produced by translating a centered rectangular plane lattice a distance equal to ½ b right, ½ a back, and ½ c up. Every second lattice plane is aligned vertically at a distance c (Figure 2.7c).

The two hexagonal Bravais lattices (Figure 2.10e) are derived from the hexagonal plane lattice whose lattice dimensions are $a = b$. The unit cell dimension perpendicular to the plane lattice is $c \neq a$. Because *a* and *b* are interchangeable, some authors prefer to designate them a_1 and a_2. See the discussion on indices and crystal axes in the hexagonal crystal system later in this chapter for an alternative crystal axis geometry used in hexagonal minerals.

The primitive lattice (P) is produced by vertical translation of a hexagonal plane lattice a distance equal to c (Figure 2.8a). The unit cell (shaded) has a rhomb-shaped cross section with internal angles of 60° and 120°.

The rhombohedral (R) lattice is produced by translating a hexagonal plane lattice on a diagonal a distance equal to $\frac{1}{3}$ c up and $\frac{2}{3}$ a cos 30° back at right angles to

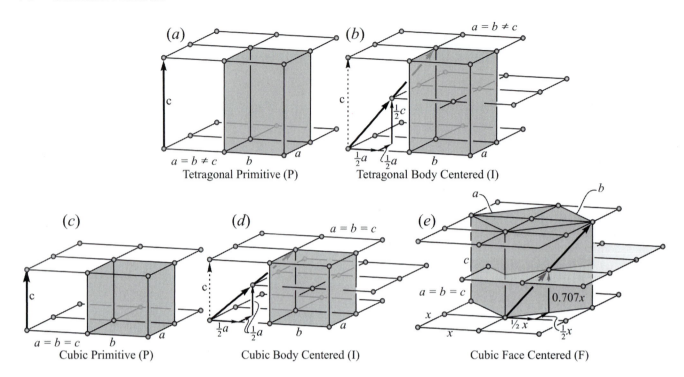

Figure 2.9 Bravais lattices produced by translation of a square plane lattice whose dimensions are $a = b$. (*a*) Vertical translation distance $c \neq a$ produces a primitive (P) tetragonal lattice. (*b*) Translation ½ *a* right, ½ *a* back, and ½ *c* up where $c \neq a$ produces a body-centered (I) tetragonal lattice. (*c*) Vertical translation distance a yields a primitive (P) cubic (isometric) lattice. (*d*) Translation ½ *a* right, ½ *a* back, and ½ *a* up yields a body-centered (I) cubic lattice. (*e*) Given a square lattice whose dimensions are x, translation ½ *x* to the right, ½ *x* back, and 0.707 *x* up produces a face-centered (F) cubic lattice whose unit cell outline is at 45° to the lattice edges of the original plane lattice.

the *b* axis (Figure 2.8b). A primitive rhombohedral unit cell, which looks like a cube stretched or shortened along a diagonal through the center, can be constructed by connecting lattice nodes as shown. It is possible to define crystal axes along the edges of the rhombohedron (Figure 2.8c) in a manner similar to how the axes are defined in the isometric system except that the angles are not 90° ($a = b = c$, $\alpha = \beta = \gamma \neq 90°$). It is far more common, however, to define the unit cell by using the same convention as for the primitive hexagonal lattice with two axes (*a* and *b*) at 120° with *c* at right angles, as shown in Figure 2.8b. This convention is used consistently in this book and in most other references because it allows the same unit cell convention to be used for all hexagonal minerals.

The tetragonal crystal system has two different Bravais lattices derived from the square plane lattice whose lattice dimensions are $a = b$. The unit cell dimension perpendicular to the original plane lattice is $c \neq a = b$ (Figure 2.10d). All three axes are at right angles ($\alpha = \beta = \gamma = 90°$), and the *c* axis may be either longer or shorter than the *a* and *b* axes. Note that the *a* and *b* axes are interchangeable, and some authors prefer to designate them a_1 and a_2.

The primitive (P) lattice is produced by vertically translating a square plane lattice distance *c* (Figure 2.9a).

The body-centered (I) lattice is produced by translating a square plane lattice a distance equal to ½ *a* right, ½ *a* back, and ½ *c* up. Alternate plane lattices are aligned vertically at distance *c* (Figure 2.9b).

The isometric crystal system has three different Bravais lattices derived from the square plane lattice. Unit cell dimensions along all three axes are the same, and the axes are all at right angles (Figure 2.10f); $a = b = c$, $\alpha = \beta = \gamma = 90°$. Note that the three axes are interchangeable. For this reason, some authors prefer to use a_1, a_2, and a_3 for the three axes.

The primitive (P) lattice is produced by translating the square plane lattice distance *a* perpendicular to the plane lattice (Figure 2.9c).

The body-centered (I) lattice is produced by translating the square plane lattice along a diagonal a distance equal to $\frac{1}{2}a$ to the right, $\frac{1}{2}a$ to the rear, and $\frac{1}{2}a$ up. Alternate plane lattices are aligned vertically at distance *a* (Figure 2.9d).

The face-centered (F) lattice is produced by translating the square plane lattice (whose unit mesh dimension is *x*) a distance equal to $\frac{1}{2}x$ to the right, $\frac{1}{2}x$ to the rear, and $x/\sqrt{2}$ vertically (Figure 2.9e). The cubic unit cell thus produced has a lattice node in the center of each face and is at 45° to the original plane lattice. The length of each side of the unit cell is $a = b = c = \sqrt{2}x$.

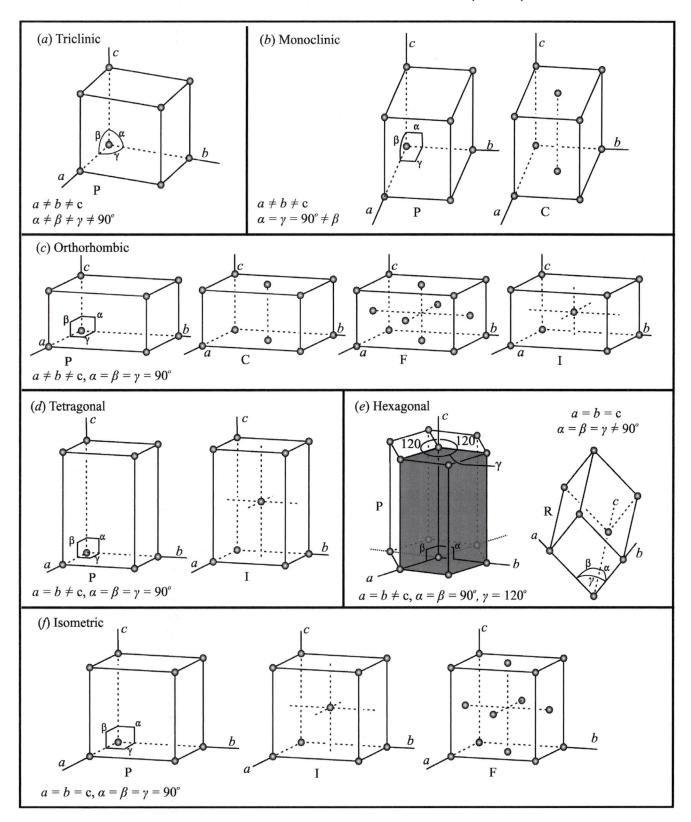

Figure 2.10 The 14 Bravais lattices define six different three-dimensional volumes (*a–f*) that correspond to the unit cells of the six crystal systems. The lengths of the three unit cell axes are *a*, *b*, and *c*, and the angles between them are α, β, and γ. In the notations, the ≠ sign indicates that equality of the axis lengths or angles is not required, although occasionally equality may occur by chance. The hexagonal (R) lattice shown in (*e*) is based on the rhombohedral axes shown in Figure 2.8c. A unit cell with the same geometry as the hexagonal (P) lattice shown in Figure 2.8b is far more commonly used.

POINT SYMMETRY

Point symmetry deals with how a motif can be repeated about a point. In minerals, the motifs that are repeated may be crystal faces or a particular arrangement of atoms that makes up the structure of a mineral. The possible point symmetry operations are reflection, rotation, and inversion. For the discussion that follows, the point about which symmetry is recognized is the center of the crystal or the origin of the unit cell.

Reflection

A reflection is produced by a mirror plane, indicated by the letter "m" that passes through a crystal structure so that the pattern on one side is a mirror image of the pattern on the other. Figure 2.11 shows a monoclinic mineral with a mirror plane that contains the a and c crystal axes and is perpendicular to the b axis.

In a given mineral, only planes in specific orientations can be mirror planes. In the monoclinic crystal system, only one mirror plane is possible, and it is usually taken as the ac plane. This is similar to the human body, in which the right side is a mirror image (more or less) of the left but no other plane of symmetry is present. Triclinic minerals have no mirrors, and isometric minerals may have as many as nine different mirror planes.

Rotation

Rotational symmetry involves repeating a motif by a set of uniform rotations about an axis. The notation for a rotation is a capital letter "A." The repeated motif is a wedge-shaped segment of crystal including crystal faces, lattice, and atoms. Consider the tetragonal crystal shown in Figure 2.12. The entire crystal can be produced by repeating a 90° wedge four times by rotations of 90° about the c crystal axis. This is a 4-fold rotation (A_4). The other possible rotational symmetry operations are 6-fold, 3-fold, 2-fold, and 1-fold rotations (A_6, A_3, A_2, A_1) (Figure 2.12), which involve rotations of 60°, 120°, 180°, and 360°, respectively. Note that 1-fold rotational symmetry is essentially no symmetry because it is simply rotation of an object through 360°. Because rotation of 360° brings everything back to its original position, every object has an infinite number of 1-fold symmetry axes.

Inversion

If a crystal has **inversion** or **center** symmetry, any line drawn through the origin will find identical features equidistant from the origin on opposite sides of the crystal. Inversion symmetry is indicated with the letter "i." Consider the simple crystal shown in Figure 2.13a. A line from point h through the center of the crystal finds an equivalent point h' on the other side. Similarly lines from i, j, and k all find equivalent points i', j', and k' on opposite sides of the crystal. The crystal shown in Figure 2.13b

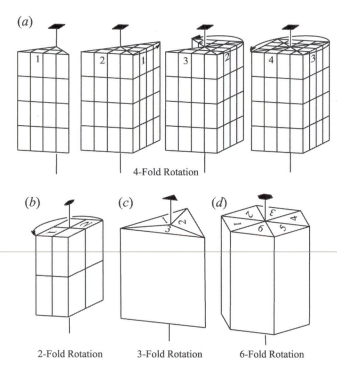

Figure 2.12 Rotational symmetry. (*a*) Four-fold rotation. The full crystal can be produced by repeating four wedge-shaped segments by rotations of 90° about the rotation axis. (*b*) Two-fold rotation. The lattice and crystal faces are repeated every 180°. (*c*) Three-fold rotation. The lattice and crystal faces are repeated every 120°. (*d*) Six-fold rotation. The lattice and crystal faces are repeated every 60°. Note the different symbols used for rotation axes.

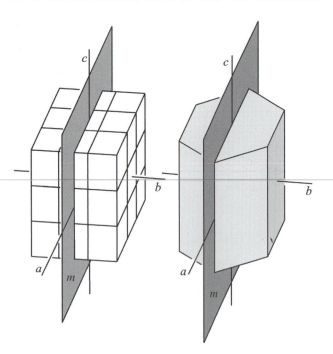

Figure 2.11 Symmetry by a mirror (*m*) plane or reflection. The mirror plane operates on the lattice, structure, and crystal faces of the mineral.

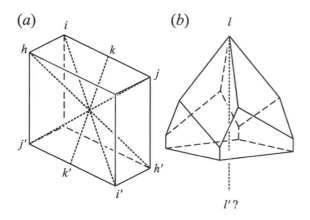

Figure 2.13 Inversion. (*a*) Crystal with inversion. Points *h,i, j*, and *k* all have equivalent points *h'*, *i'*, *j'*, and *k'* on the opposite side of the crystal. (*b*) Crystal without inversion. Points such as the *l* are not duplicated on the opposite side of the crystal.

lacks center symmetry because lines through the crystal's center from points such as *l* do not find equivalent points on the opposite side of the crystal.

Compound Symmetry Operations

Rotation axes may be combined with inversion to produce symmetry elements called 1-, 2-, 3-, 4-, and 6-fold **rotoinversion axes**.

The simplest is a 1-fold rotoinversion $\bar{A}_1$ ("A bar one"). Consider point *a* on the crystal shown in Figure 2.14a. It is repeated by rotation of 360° followed by inversion through the center to produce point *a'*. All other points on the top of the crystal are similarly reproduced on the bottom by the same process. This combination is trivial because it is the same as a simple inversion ($\bar{A}_1 = i$).

A 2-fold rotoinversion ($\bar{A}_2$) is produced by a rotation of 180° followed by inversion through the center (Figure 2.14b). Consider point *b*. Rotation of 180° followed

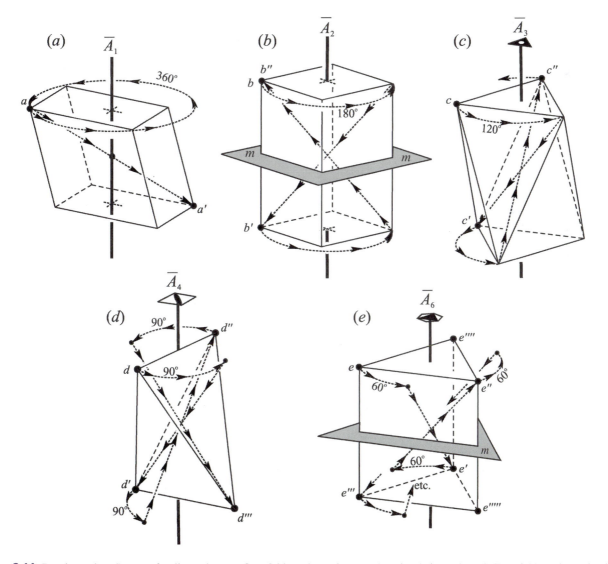

Figure 2.14 Rotoinversion. See text for discussion. (*a*) One-fold rotoinversion equals a simple inversion. (*b*) Two-fold rotoinversion is the same as a mirror at right angles to the rotation axis. (*c*) Three-fold rotoinversion is equivalent to a simple 3-fold rotation and center symmetry. (*d*) Four-fold rotoinversion is not replicated by other simpler symmetry operations. (*e*) Six-fold rotinversion is equivalent to 3-fold rotation with a mirror at right angles.

by inversion through the center produces point b' directly below b. The second 180° rotation followed by inversion produces point b'' coincident with the original position of b. This result turns out to be the same as a mirror plane at right angles to the rotoinversion axis ($\bar{A}_2 = m$).

Figure 2.14c shows a 3-fold rotoinversion ($\bar{A}_3$). Point c is rotated 120° and inverted through the center to c'. A second rotation/inversion produces c'' and so forth. The net result is to produce a crystal with a conventional 3-fold rotation axis that also possesses center symmetry ($\bar{A}_3 = A_3 + i$).

The symmetry produced by 4-fold rotoinversion ($\bar{A}_4$) cannot be duplicated with combinations of other simpler symmetry operations. It involves 90° of rotation followed by inversion through the center (Figure 2.14d). Point d gets rotated 90° and inverted to d', an additional 90° rotation and inversion leads to d'', and so forth. The result is to produce four faces (or sets of faces) at 90° to each other that alternate between being upright and upside down.

Figure 2.14e shows that a 6-fold rotoinversion ($\bar{A}_6$) produces the same result as a 3-fold rotation axis at right angles to a mirror ($\bar{A}_6 = A_3 + m$). Consider point e. Rotation of 60° and inversion through the center produces e'. A second rotation and inversion produce e'' and so forth. When completed, each point on the top of the crystal has a mirror image on the bottom.

Symmetry Notation

The notation used for symmetry is shown in Table 2.1. In the foregoing discussion of symmetry, **symmetry symbols** have been used to refer to symmetry operations. For example, the orthorhombic mineral shown in Figure 2.15 contains the following symmetry.

$$i, 3A_2, 3m$$

This means that the crystal possesses center symmetry (i), three mirror planes ($3m$), and three 2-fold rotation axes ($3A_2$). The symmetry elements may be listed in any order, although for the sake of consistency it is convenient to list center first (if present), followed by rotation axes and then mirrors.

Hermann–Mauguin symbols, named for Carl Hermann (1898–1961) and Charles-Victor Mauguin (1878–1958), recognize that the symmetry elements in a mineral must be related to each other and that some symmetry operations are implied by the presence of others, and therefore, do not need to be explicitly listed. The Hermann–Mauguin notation for the symmetry of the orthorhombic mineral shown in Figure 2.15 is $2/m\ 2/m\ 2/m$. This means that there are three 2-fold axes, each perpendicular to a mirror. The presence of the three mirrors at right angles requires that the mineral have center symmetry, so it is not necessary to explicitly state that center symmetry is present. The order in which Hermann–Mauguin symbols are listed is specific to each crystal system and is described in a later section.

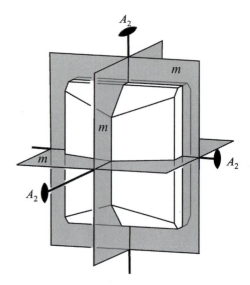

Figure 2.15 This crystal has center symmetry (i), three 2-fold axes ($3A_2$), and three mirrors ($3m$).

Table 2.1 Nomenclature for Symmetry Elements[a]

Symmetry Operation	Symmetry Symbol	Hermann-Mauguin Symbol
Mirror	m	m
Rotation axis	A_1, A_2, A_3, A_4, A_6	1, 2, 3, 4, 6
Rotoinversion axis	$\bar{A}_1 = i, \bar{A}_2, \bar{A}_3, \bar{A}_4, \bar{A}_6$	$\bar{1}, m, \bar{3}, \bar{4}, \bar{6}$

[a] Center (i) is equivalent to a one-fold rotoinversion axis ($\bar{A}_1$).

In the sections that follow, symmetry symbols are used to describe the specific symmetry operations and axes that a mineral may have. Hermann–Mauguin symbols are used to specifically identify each of the 32 different possible combinations of symmetry known as point groups, which are described in the next section and are used in the mineral descriptions in Chapters 12 through 20.

THE 32 POINT GROUPS

As the preceding discussion implies, the symmetry elements of inversion, rotations, and mirrors may be combined in a variety of ways, although combinations are limited because the symmetry elements must be compatible with each other.

In two dimensions the possible symmetry elements are mirrors (m) and rotations (A_1, A_2, A_3, A_4, and A_6) (inversion is possible only in three dimensions). The symmetry elements may each be taken individually or may be combined to produce a total of 10 different two-dimension point groups (Figure 2.16).

With the addition of the third dimension, inversion becomes possible and the number of different combinations of symmetry increases to 32. These symmetry combinations are known as the **32 point groups**. The 32 point groups also are known as the **32 crystal classes**. These point groups may be grouped into the six different crystal systems based on common symmetry elements. The six crystal systems were defined earlier based on unit cell geometry generated from the Bravais lattices. Table 2.2 shows the symmetry elements possessed by each of the different crystal classes and the crystal system to which each belongs. By convention, the point groups are identified with Hermann–Mauguin (H–M) representations of the symmetry.

It should be no surprise that the positions of the crystal axes for each crystal system must be in rational orientations relative to the symmetry elements. The conventions adopted for orienting crystal axes relative to the symmetry axes and planes are summarized in Table 2.2. More thorough descriptions of the conventions used to orient crystal axes are included in the descriptions of the crystal classes that follow in a later section of this chapter.

The relation between the 32 point groups/crystal classes and the 14 Bravais lattices is not immediately obvious. Clearly, minerals displaying isometric point groups must utilize an isometric Bravais lattice, tetragonal point groups must utilize a tetragonal Bravais lattice, and so forth. While it is beyond the scope of this discussion to demonstrate it, the point groups from each crystal system can use any Bravais lattice from that crystal system. The single exception occurs in the hexagonal system, where only members of the trigonal division may use the rhombohedral (R) Bravais lattice. While point group symmetry can be identified from well-developed crystals, Bravais lattices cannot be identified without X-ray diffraction studies.

Steno's Law

Knowledge of an unknown mineral's point group and/or crystal system often can be of great help when attempting to identify it. The point group may be determined by carefully examining well-formed crystals to identify symmetry elements. Allowance often has to be made for the fact that

crystals are rarely perfect and that equivalent faces may have different sizes.

Fortunately, the angles between crystal faces are remarkably consistent. This observation was originally made by Nicolas Steno (Niels Stensen) in 1669. He demonstrated that the angles between the same two faces on quartz crystals are the same for different specimens of the mineral, regardless of sample size, the size of the crystal faces, or where the crystals may have been collected. This observation has been formalized into **Steno's Law** or the **Law of Constancy of Interfacial Angles** because it applies to all minerals.

> The angles between equivalent faces on crystals of the same mineral are the same.

Measurement of Crystal Angles

A simple contact goniometer (Figure 2.17) allows crystal angles to be measured with reasonable accuracy. This instrument is basically a protractor with a movable arm. The bottom edge of the goniometer is placed on one crystal face and the movable arm is aligned with an adjacent face. The plane of the goniometer must be perpendicular to the crystal edge formed by the two faces. The goniometer scale has two sets of numbers, one for the internal angle between the faces and one for the complement to that angle. The internal angle is the angle between lines oriented at right angles to the two faces, and it is the angle normally reported. Note that the sum of internal angles going all the way around a crystal must equal 360°.

Determining Crystal System and Crystal Class

Assuming that well-formed crystals of a mineral are available, the procedure to ascertain the class is as follows.

- Determine whether the mineral possesses center symmetry. To do this, select several points, such as

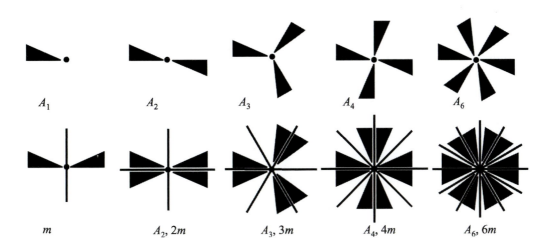

Figure 2.16 The ten two-dimension point groups.

Table 2.2 The 32 Point Groups

Crystal System	Point Group/ Crystal Class	Symmetry Content	Common Symmetry Elements	Crystal Axes
Triclinic	1	None ($= A_1$)	One-fold rotation with or without inversion.	No symmetry restrictions; prominent zone $= c$.
	$\bar{1}$	$i = \bar{A}_1$		
Monoclinic	2	$1A_2$	Two-fold rotation and/or a single mirror.	Two-fold axis or line perpendicular to mirror $= b$, prominent zone $= c$. $a \wedge c > 90°$.
	m	m		
	$2/m$	$i, 1A_2, m$		
Orthorhombic	222	$3A_2$	Three 2-fold rotation axes and/or three mirrors.	The 2-fold axes or lines perpendicular to mirrors coincide with the three mutually perpendicular crystal axes; the first H–M symbol refers to the a crystal axis, the second to the b crystal axis, and the third to the c crystal axis.
	$mm2$	$1A_2, 2m$		
	$2/m\ 2/m\ 2/m$	$i, 3A_2, 3m$		
Tetragonal	4	$1A_4$	Single 4-fold rotation or rotoinversion axis.	The 4-fold axis $= c$; the a and b crystal axes are perpendicular to c and coincide with A_2 axes or are perpendicular to mirrors, if present: the first H–M symbols refers to the c axis; if present, the second refers to the a and b crystal axes, and the third to symmetry axes that bisect the angles between a and b [110].
	$\bar{4}$	$1\bar{A}_4$		
	$4/m$	$i, 1A_4, m$		
	422	$1A_4, 4A_2$		
	$4mm$	$1A_4, 4m$		
	$\bar{4}\ 2m$	$1\bar{A}_4, 2A_2, 2m$		
	$4/m\ 2/m\ 2/m$	$i, 1A_4, 4A_2, 5m$		
Hexagonal (trigonal division)	3	$1A_3$	One 3-fold axis.	The 3-fold or 6-fold axis $= c$, the a and b crystal axes are parallel to A_2 axes or perpendicular to mirrors, if present: the first H–M symbol refers to the c axis; if present, the second refers to the a and b axes, and the third to symmetry axes that bisect the angles between a and b [210].
	$\bar{3}$	$1\bar{A}_3 (= i + 1A_3)$		
	32	$1A_3, 3A_2$		
	$3m$	$1A_3, 3m$		
	$\bar{3}\ 2/m$	$1\bar{A}_3 (= i + 1A_3), 3A_2, 3m$		
Hexagonal (hexagonal division)	6	$1A_6$	One 6-fold axis.	
	$\bar{6}$	$1\bar{A}_6 (= 1A_3 + m)$		
	$6/m$	$i, 1A_6, m$		
	622	$1A_6, 6A_2$		
	$6mm$	$1A_6, 6m$		
	$\bar{6}\ m2$	$1\bar{A}_6 (= 1A_3 + m), 3A_2, 3m$		
	$6/m\ 2/m\ 2/m$	$i, 1A_6, 6A_2, 7m$		
Isometric	23	$3A_2, 4A_3$	Four 3-fold axes; these axes are oriented to go diagonally through the center of the cubic unit cell from corner to corner.	The three 2-fold or 4-fold axes coincide with the interchangeable a, b, and c crystal axes; symmetry requires that these axes be at 90° to each other, the first H–M symbol refers to the symmetry on the crystal axes, the second to the long diagonals through the unit cell [111]. the third to edge-to-edge diagonals through the unit cell [110].
	$2/m\ \bar{3}$	$3A_2, 3m, 4\bar{A}_3 (= i + 4A_3)$		
	432	$3A_4, 4A_3, 6A_2$		
	$\bar{4}3m$	$3\bar{A}_4, 4A_3, 6m$		
	$4/m\ \bar{3}\ 2/m$	$3A_4, 4\bar{A}_3 (= i + 4A_3), 6A_2, 9m$		

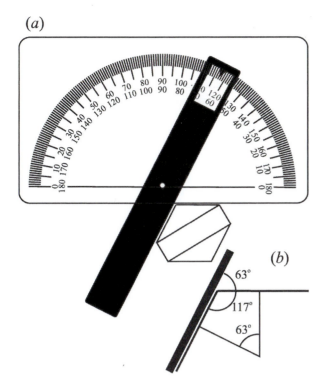

(a)

Figure 2.17 Contact goniometer. (*a*) The base of the goniometer is placed on one crystal face, and the cursor is aligned with the adjacent face. The edge between the faces must be perpendicular to the plane of the goniometer. The internal angle between the faces is read from the scale. (*b*) The internal angle (63°) is the angle between lines at right angles to the crystal faces.

corners, on one side of the crystal (Figure 2.13a). The crystal has center symmetry if an imaginary line from each point finds an equivalent point on the opposite side of the crystal when projected through the center.

- Identify mirror planes. Mirrors divide crystals into two mirror-image halves (Figure 2.18a). Angles on opposite sides of the mirror must be identical.

- Identify rotation axes. A rotation axis can be considered an axle about which the mineral rotates to repeatedly bring equivalent collections of faces into view (Figure 2.18b). Two-fold axes require 180° of rotation to bring duplicate sets of faces into view. Other rotations are 3-fold axes = 120°, 4-fold axes = 90°, and 6-fold axes = 60°. Refer to Figure 2.14d to see a 4-fold rotoinversion.

- Compile the combination of symmetry. The crystal in Figure 2.18 has i, $1A_2$, m and is a member of the $2/m$ crystal class in the monoclinic system (Table 2.2).

Minerals are not evenly distributed among the crystal systems and crystal classes (Table 2.3). Nearly half of all minerals are in just two crystal classes—monoclinic $2/m$ and orthorhombic $2/m$ $2/m$ $2/m$.

SPACE GROUPS

The symmetry elements of mirror, rotations, and inversion that in combination produce the 32 point groups can completely describe the symmetry observed in the distribution of crystal faces and related crystallographic features arrayed around the center of a crystal. The symmetry of simple translation in three dimensions yields the 14 Bravais lattices. The 230 **space groups** represent all the combinations of the point symmetry with translational symmetry.

Groupings of atoms that possess the symmetry of one of the 32 point groups are placed at the lattice nodes of a compatible Bravais lattice. These groupings are then duplicated over and over at each lattice node in the Bravais lattice to create the structure of the mineral. The point group symmetry must be consistent with the symmetry of the lattice. Isometric point groups can be used only with isometric lattices, tetragonal point groups with tetragonal lattices, and so forth. These combinations of point symmetry and simple translation yield 73 of the space groups.

This is not yet the full story because two additional symmetry operations called **glides** and **screws** are possible in

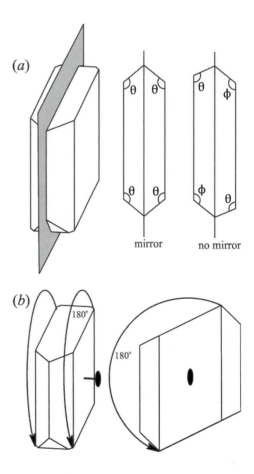

(a)

mirror   no mirror

(b)

Figure 2.18 Determining crystal symmetry. (*a*) Mirrors split the crystal into two mirror images. Angles on opposite sides of the mirror are the same. (*b*) Rotations repeat the same collection of faces every 60°, 90°, 120°, or 180° of rotation (A_6, A_4, A_3, A_2). A 2-fold rotation is shown.

Table 2.3 Percentage of Minerals in Each Crystal System and Crystal Class

Crystal System	Crystal Class	Percent	Crystal System	Crystal Class	Percent
Triclinic 9.1%	1	1.4%	Hexagonal (trigonal division) 13.2%	3	1.2%
	$\bar{1}$	7.7%		$\bar{3}$	2.4%
Monoclinic 36.6%	2	1.5%		32	0.8%
	m	2.0%		3m	4.0%
	2/m	33.1%		$\bar{3}2/m$	4.8%
Orthorhombic 18.6%	222	1.7%	Hexagonal (hexagonal division) 6.2%	6	0.7%
	mm2	3.1%		$\bar{6}$	0.1%
	2/m 2/m 2/m	13.8%		6/m	1.4%
Tetragonal 6.9%	4	0.1%		622	0.2%
	$\bar{4}$	0.1%		6mm	0.9%
	4/m	1.3%		$\bar{6}m2$	0.5%
	422	0.4%		6/m 2/m 2/m	2.4%
	4mm	0.1%	Isometric 9.4%	23	0.3%
	$\bar{4}2m$	1.0%		$2/m\,\bar{3}$	1.7%
	4/m 2/m 2/m	3.9%		432	0.2%
				$\bar{4}3m$	1.7%
				$4/m\,\bar{3}\,2/m$	5.5%

Source: Based on data compiled by Mindat.org.

three dimensions. Each is a compound symmetry operation consisting of a translation and either reflection or rotation.

A glide is produced by a combination of translation a specific distance and direction followed by reflection across a mirror plane called the **glide plane** (Figure 2.19a). The pattern made by footprints is the same as that produced by a glide. Footprints follow each other at a regular distance, but they alternate between the right and left feet, which are mirror images of each other.

A typical way in which glides operate in minerals is seen in the chains of silicon tetrahedra found in pyroxenes (Figure 2.19b). A silicon tetrahedron consists of a silicon ion surrounded by four oxygen ions in a tetrahedral arrangement. Silicon tetrahedra are strung together in long chains by sharing oxygen ions between adjacent tetrahedra (see Chapter 11). Successive tetrahedra are related to each other by translation *t* followed by reflection *m*. Glide planes, if present in a mineral, must be parallel to mirrors represented by the point group symmetry.

A screw is produced by a combination of translation a specific distance and direction followed by a rotation (Figure 2.20). The axis about which rotation occurs is the **screw axis**. The possible rotations are 2-fold (180°), 3-fold (120°), 4-fold (90°), and 6-fold (60°). A 4-fold screw axis, for example, repeats a 90° wedge of the crystal structure every 90° in a spiraling stepwise fashion, somewhat like the treads on a spiral staircase. The other symmetry operations present in the mineral repeat the wedges of crystal to fill in the full structure. Screw axes, if present, must be parallel to and consistent with rotation axes present in the point group symmetry.

The addition of screw axes and glide planes to the mix of symmetry yields an additional 157 space groups for a total of 230 possible ways of repeating a motif in three dimensions. A complete tabulation of all 230 space groups and discussion of the nomenclature used to enumerate them can be found in Bloss (1971) or O'Keeffe and Hyde (1996). Henry and Lonsdale (1952) provide an authoritative

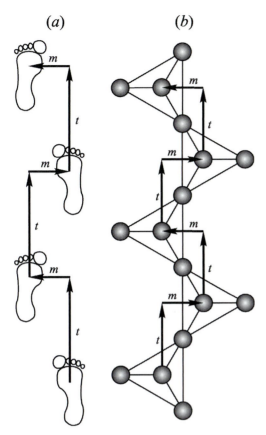

(a)   *(b)*

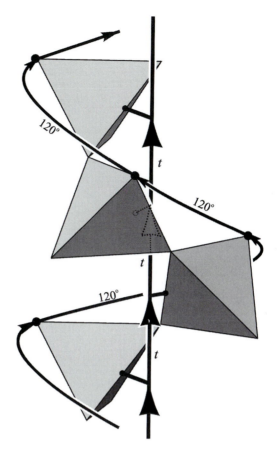

Figure 2.19 Glide symmetry. (*a*) Footprints repeated by a translation *t* and reflection *m*. (*b*) Chain of silicon tetrahedra (single silicon, hidden, surrounded by four oxygen ions) produced by translation *t* followed by reflection *m*.

Figure 2.20 Screw axis. A 3-fold screw axis is shown. In quartz, a silicon tetrahedron is translated distance *t* and rotated 120° in spiral-staircase fashion to form a helix of tetrahedra extending along the *c* axis. Screw axes may also be 2-fold, 4-fold, or 6-fold.

compilation and description of space groups. All minerals belong to one or another of the space groups. Although a thorough understanding of space groups is essential for anyone attempting to decipher the structural complexity of a mineral, it is beyond the scope of this text. Interested readers are encouraged to study the references cited above for additional enlightenment.

CRYSTAL FACES

Crystal faces always grow in rational orientations relative to the crystal lattice. The most common faces are often parallel to the surfaces of the unit cell, that is, parallel to the principal planes through the crystal lattice. The result is that minerals with cubic unit cells often form cubic crystals, minerals with hexagonal unit cells often have hexagonal cross sections, and so forth. Crystal faces also often follow simple diagonals through the lattice.

Laws of Haüy and Bravais

The tendency of crystal faces to follow simple, rational orientations through the crystal lattice has resulted in recognition of two related laws of crystallography. The first is called the **Law of Haüy**.

> Crystal faces make simple rational intercepts on crystal axes.

The second is called the **Law of Bravais**:

> Common crystal faces are parallel to lattice planes that have high lattice-node density.

Figure 2.21 shows a primitive monoclinic lattice with several potential crystal faces. Planes A, B, and C (Figure 2.21a) are principal planes in the lattice and have high lattice-node density (many lattice nodes per unit of area). Based on the laws of Bravais and Haüy, crystal faces should be parallel to all three lattice planes. Plane T (Figure 2.21b) cuts through the lattice on a simple diagonal ($a:c = 1:1$). It makes rational intercepts on the crystal axes and has a fairly high lattice-node density, so crystal faces parallel to T should be fairly common. In contrast, plane Q (Figure 2.21c) cuts at an angle through the lattice so that it goes up one cell along *c* for every two unit cells along *a* ($a:c = 2:1$). Although it has rational intercepts on the crystal axes, the lattice-node density is low. Crystal faces parallel to Q are less likely to develop or will be minor. A simple crystal with these faces is shown in Figure 2.21d.

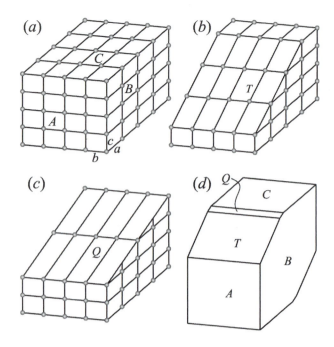

Figure 2.21 Primitive monoclinic lattice. Prominent crystal faces are parallel to rational planes in the lattice that have high density of lattice nodes. (*a*) planes *A, B,* and *C* are parallel to the faces of the unit cell, have high lattice-node density, and are likely crystal faces. (*b*) Plane *T* has fairly high lattice-node density and is a likely crystal face. (*c*) Plane *Q* has low lattice-node density and is not a likely crystal face. (*d*) A simple crystal with these faces.

Miller Indices

Because crystal faces have rational orientations relative to the crystal lattice, it is possible to develop an elegant short-hand system to describe the orientation of crystal faces and crystallographic planes. A **Miller index**, named for William H. Miller (1801–1880), consists of a series of coprime integers that are inversely proportional to the intercepts of the crystal face or crystallographic plane with the edges of the unit cell. A Miller index has the general form (*hkl*), where *h*, *k*, and *l* are integers related to the *a*, *b*, and *c* crystal axes, respectively. Parentheses ( ) enclose the Miller index for a specific crystal face or crystallographic plane. The parentheses are omitted from some illustrations to avoid clutter. Examples of Miller indices are (001), (110), or (112). These are read "*zero zero one, one one zero, one one two.*" Negative values are indicated with a bar over the top of the number. A negative one would be shown as $\bar{1}$ and read "*bar one*" or "*negative one.*" A Miller index of ($1\bar{1}0$) would be read "*one bar-one zero*" or "*one negative-one zero.*" Because the values in the Miller index are coprime integers, values like (224) are not used because all the integers are divisible by 2. As will be shown shortly, this index would be reported as (112).

Consider a monoclinic mineral that forms a crystal with face *t* that, if extended, intersects all three crystal axes (Figure 2.22a). This face cuts through the lattice as shown in Figure 2.22b so that the numbers of unit cells out from

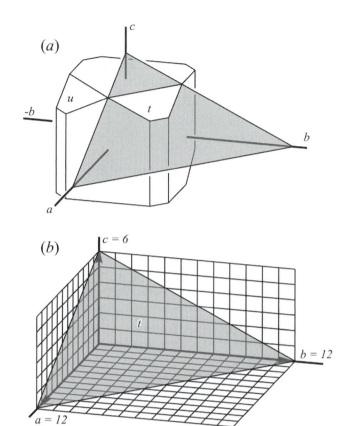

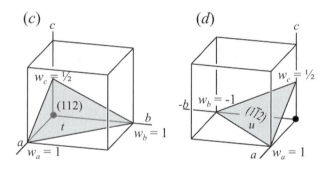

Figure 2.22 Miller index. (*a*) Face *t*, when extended, intersects the positive ends of all three crystal axes. (*b*) Face *t* cuts through the lattice so that it intersects the *a* and *b* crystal axes at the same number of unit cell from the origin (12 in this case) and intersects the *c* crystal axis at half that number (six). (*c*) When referenced to a single unit cell, face *t* intersects the crystal axes so that $w_a = 1$, $w_b = 1$, and $w_c = \frac{1}{2}$. (*d*) Face *u*, which intersects the negative end of *b*, intersects the crystal axes of a single unit cell at $w_a = 1$, $w_b = -1$, and $w_c = \frac{1}{2}$. See text for discussion.

the crystal center on *a* and *b* are both twice the number of unit cells on *c*. For purposes of describing Miller indices, it is easiest to reference the crystal face or crystallographic plane to a single unit cell as shown in Figure 2.22c. The cell intercepts w_a, w_b, and w_c for face *t* through the unit cell are

$$w_a = 1, \quad w_b = 1, \quad \text{and} \quad w_c = \frac{1}{2}.$$

This means that the face cuts the cell at one unit cell dimension along a, one unit cell dimension along b, and one-half a unit cell dimension along c. The Miller index for face t is simply the inverse of these values or

$$(hkl) = \frac{1}{w_a}\frac{1}{w_b}\frac{1}{w_c} = \left(\frac{1}{1}\frac{1}{1}\frac{1}{\frac{1}{2}}\right) = (112). \qquad (2.1)$$

The Miller index is read "one one two."

Some faces intersect the negative end of one or more crystal axes. Consider face u on the crystal shown in Figure 2.22. It intersects the positive ends of the a and c axes and the negative end of the b axis with cell intercepts of $w_a = 1$, $w_b = -1$, and $w_c = \frac{1}{2}$ (Figure 2.22d). Its Miller index is ($1\bar{1}2$), which is read as "one negative-one two" or "one bar-one two."

Many common crystal faces are parallel to one or two of the crystal axes. Consider face v shown in Figure 2.23.

This face intersects the a and b axes with a cell intercept of $w_a = 1$ and $w_b = 1$, but is parallel to the c axis. The value for w_c must therefore be infinity ($w_c = \infty$), because no matter how far the face goes, it never intersects the c axis. Note that anything divided by infinity is zero, so the index for this face calculated from Equation 2.1 is

$$(hkl) = \frac{1}{w_a}\frac{1}{w_b}\frac{1}{w_c} = \left(\frac{1}{1}\frac{1}{1}\frac{1}{\infty}\right) = (110)$$

$$= (\text{one one zero}).$$

Face w, which is parallel to a and c, cuts the positive end of the b axis ($w_a = \infty$, $w_b = 1$, $w_c = \infty$). Its Miller index must therefore be (010) (zero one zero). Faces x, y, and z get Miller indices of (100), ($1\bar{1}0$), and (001) by similar calculation. Note that if $h = 0$, the face is parallel to a, if $k = 0$, the face is parallel to b, and if $l = 0$, the face is parallel to c.

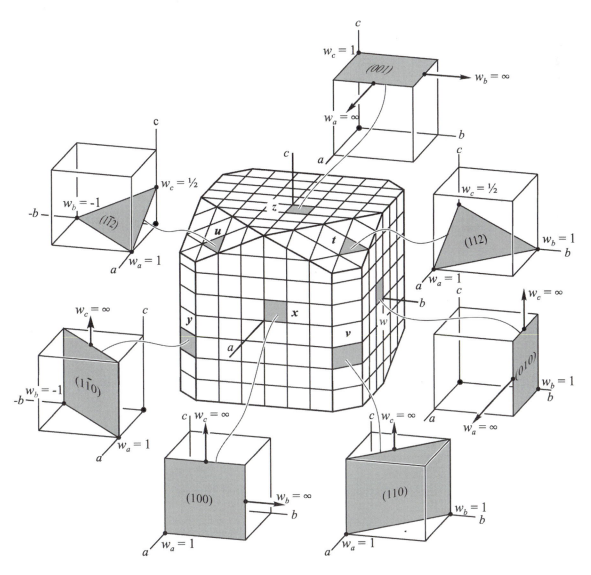

Figure 2.23 Miller indices for all visible faces on the crystal shown in Figure 2.22. The traces of unit cells are sketched on the crystal. See text for discussion.

Indices and Crystal Axes in the Hexagonal Crystal System

Crystal axes and indices in the hexagonal system require some additional explanation because two common conventions are in use. One, which was described earlier (Figures 2.8 and 2.10, and Table 2.2) utilizes three axes (a, b, and c), and the other utilizes four axes (a_1, a_2, a_3, and c).

Consider the hexagonal crystal shown in Figure 2.24a. The a, b, and c crystal axes are located according to the three-axis convention. Figure 2.24b shows the unit cell intercepts of the lighter-shaded face – $w_a = 1$, $w_b = \infty$, and $w_c = \infty$. By Equation 2.1, the Miller index for this face is (100). The relation of the ($1\bar{1}0$) face (dark shading) to a unit cell is shown in Figure 2.24c; note that this face cuts the negative end of the b axis. The Miller indices for the other vertical faces shown in cross section (Figure 2.24d)

can be obtained by similar considerations. The top face of the crystal is (001) and the bottom face is ($00\bar{1}$).

An alternative convention uses four crystal axes in the hexagonal crystal system (Figure 2.25). Hexagonal crystals have three directions at 120° from each other that are typically parallel to 2-fold rotation axes and/or perpendicular to mirrors. Three crystal axes identified as a_1, a_2 and a_3 are placed parallel to these directions. The a_1 and a_2 axes are parallel to the a and b crystal axes as defined in the preceding discussion of the hexagonal unit cell; the a_3 is at 120° to them. The c axis is the same in both axis conventions. The unit cell based on the four-axis convention therefore has a hexagonal cross section. The four-axis convention is widely used in many older reference works.

A modification of Miller indices was developed by Auguste Bravais to accommodate the four-axis convention, so these indices are referred to as **Miller–Bravais** indices. Because there are four crystal axes, the Miller–Bravais

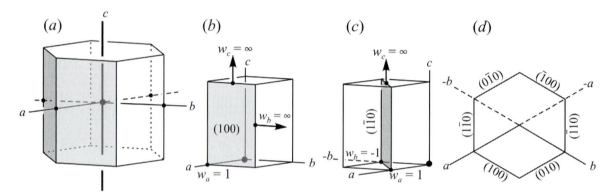

Figure 2.24 Miller indices in the hexagonal crystal system. (*a*) Hexagonal crystal with *a, b,* and *c* crystal axes. (*b*) Relation of light-shaded face to the unit cell. Based on unit cell intercepts, the Miller index is (100). (*c*) Relation of the dark-shaded face to a unit cell. Based on unit cell intercepts, the Miller index is ($1\bar{1}0$). (*d*) View down the *c* axis with the Miller indices for all vertical faces shown.

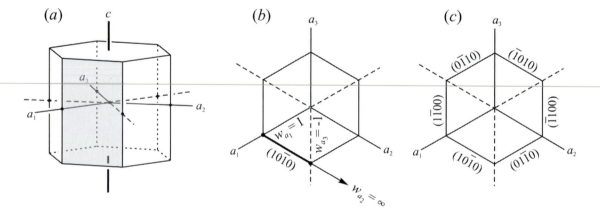

Figure 2.25 Miller–Bravais indices in the hexagonal crystal system. (*a*) Hexagonal crystal with a_1, a_2, a_3, and *c* crystal axes. The *a* axes are parallel to 2-fold rotation axes and are 120° from each other. The shaded face cuts the a_1 and $-a_3$ axes and is parallel to a_2 and *c*. (*b*) Top view of the unit cell showing the unit cell intercepts that yield a Miller–Bravais index of ($10\bar{1}0$). (*c*) Top view of the crystal showing the Miller–Bravais indices for all the vertical faces. The top and bottom faces are (0001) and ($000\bar{1}$), respectively. See text for discussion.

index contains four integers (*hkil*), which refer to the a_1, a_2, a_3, and c axes, respectively. Consider the shaded face on the hexagonal crystal shown in Figure 2.25a. This face is parallel to c and a_2 so $w_c = \infty$ and $w_{a_2} = \infty$. It cuts the positive a_1 axis at one unit cell dimension $w_{a_1} = 1$, and the negative a_3 axis at one unit cell ($w_{a_3} = -1$). By Equation 2.1, the Miller index must be

$$(hkil) = \left(\frac{1}{1}\ \frac{1}{\infty}\ \frac{1}{\bar{1}}\ \frac{1}{\infty}\right) = (10\bar{1}0).$$

Because the three a axes are at 120°, it always works out that

$$h + k + i = 0 \tag{2.2}$$

for all Miller–Bravais indices in the hexagonal system. The mathematical proof of this assertion can be found in Bloss (1971, pp. 53–54). Either all three of the a axes have 0 for an index, or at least one will have a negative index and one a positive index.

Miller–Bravais indices can be converted to Miller indices by deleting the third (*i*) digit. Miller–Bravais indices may be derived from Miller indices with reference to Equation 2.2.

Determining Miller Index

Determining the Miller index for a crystal face on a real crystal requires two pieces of data: the **axial intercepts** for the crystal face and either the unit cell dimension or the **axial ratio**. Consider the crystal face (*x*) shown on the augite crystal in Figure 2.26.

The axial intercepts (i_a, i_b, i_c) for the face are the distances from the origin where the crystal face, when extended, crosses the crystal axes. In this case $i_a = 3.82$ cm, $i_b = 3.5$ cm, and $i_c = 2.07$ cm (ignoring the difficulty of actually making these measurements).

The axial ratio is calculated by dividing each unit cell dimension by the b unit cell dimension.

$$\text{axial ratio} = a:b:c = \frac{a}{b}:\frac{b}{b}:\frac{c}{b} \tag{2.3}$$

For augite, the unit cell dimensions are $a = 9.73$ Å, $b = 8.91$ Å, and $c = 5.25$ Å so the axial ratio is 1.09:1:0.59. The Miller index is obtained by dividing the axial ratio values (or the unit cell dimensions) by the intercept values and clearing fractions.

$$(hkl) \alpha \left(\frac{a}{i_a}\ \frac{b}{i_b}\ \frac{c}{i_c}\right) = \left(\frac{1.09}{3.82}\ \frac{1}{3.50}\ \frac{0.59}{2.07}\right) \tag{2.4}$$

$$= (0.29\ 0.29\ 0.29).$$

Dividing by 0.29 produces the smallest available integers to yield the Miller index

$$(hkl) = (111).$$

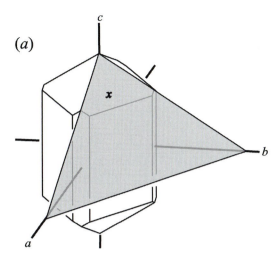

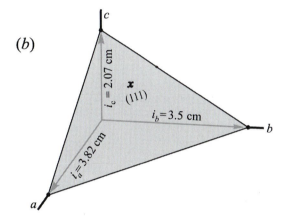

Figure 2.26 Calculating the Miller index for face *x* on an augite crystal. (*a*) Face *x* is extended out until it intersects all three crystal axes, as shown by the shaded plane. (*b*) The dimensions of these intersections measured from the center of the crystal are $i_a = 3.82$ cm, $i_b = 3.5$ cm, and $i_c = 2.07$ cm. The Miller index for this face is (111). See text for discussion.

If the Miller index and axial intercepts of a crystal face are known, the axial ratio can be calculated by rearranging Equation 2.4.

$$a:b:c = i_a h : i_b k : i_c l \tag{2.5}$$

Crystallographic Planes

Miller indices for crystallographic planes within a crystal, such as cleavage, are assigned with reference to the unit cell (Figure 2.23) in the same way that crystal faces are assigned indices. Consider the crystal shown in Figure 2.27a. A cleavage surface parallel to the (001) crystal face is given a Miller index of (001) because it has the same relation to the unit cell as the face. The three principal crystallographic planes in the crystal are shown in Figure 2.27b. They are parallel to the (100), (010), and (001) crystal faces and are assigned those Miller indices. The convention is to use Miller indices with positive

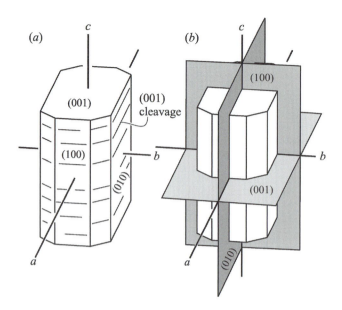

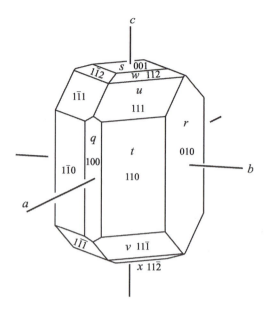

Figure 2.27 Crystallographic planes are assigned the same Miller index as the crystal face to which they are parallel. (*a*) A cleavage (shown with thin lines) parallel to the (001) crystal face is identified as a (001) cleavage. (*b*) The three principal crystallographic planes are parallel to the surfaces of the unit cell and the (100), (010), and (001) crystal faces. They therefore have Miller indices of (100), (010), and (001).

Figure 2.28 Assigning Miller indices by inspection. See text for discussion.

integers where possible. The *bc* crystallographic plane is therefore (100) and not ($\bar{1}$00).

Assigning Miller Indices by Inspection

Fortunately, the Miller indices for the major crystal faces on many minerals can be assigned without any calculation. Because of the relationships observed in the Laws of Bravais and Haüy, the integers found in Miller indices are small. Integers larger than 2 are unlikely for prominent crystal faces. Integers larger than 4 are rare even for minor faces. This makes it relatively easy to assign Miller indices to major crystal faces by inspection, at least for crystals without too many different faces.

Prominent faces that cut just one of the crystal axes have indices (100), (010), (001), ($\bar{1}$00), (0$\bar{1}$0), or (00$\bar{1}$). The most prominent faces that cut two axes but not the third are usually (110), (011), (101), (0$\bar{1}$1), ($\bar{1}\bar{1}$0), and so forth; less prominent faces are likely to have indices such as (120) or (210). The most prominent faces that cut all three axes are usually (111), (1$\bar{1}$1), (1$\bar{1}\bar{1}$), and so forth. The (111) face is known as the **unit face**. Note that because Miller indices are inverses of the intercepts, a larger index number for a given axis means that the intercept (distance from the origin) will be smaller (Figure 2.22).

Consider the orthorhombic crystal shown in Figure 2.28. Face *q* cuts the positive *a* axis and is parallel to both *b* and *c* so its index must be (100). Similarly, face *r* cuts only the positive end of the *b* axis so its index must be (010), and face *s* cuts only the positive end of *c* axis so its index must be (001). Prominent face *t* cuts the positive *a* and *b* axes and

is parallel to *c*, so its index is (110). Face *u*, if extended out, would cut the positive ends of all three axes and is prominent, so its index is (111). Face *v*, the counterpart of *u* on the bottom of the crystal, cuts the negative *c* axis, so its index is (11$\bar{1}$). Face *w* cuts the *c* axis closer to the origin than face *u*, so the index for *w* works out to (112). Face *x*, the counterpart to *w* on the bottom of the crystal, has index (11$\bar{2}$).

Assigning Miller indices by inspection is, of course, not always foolproof. Some faces with larger integers in their Miller index may, because of accidents of growth or crystal chemical requirements, be more prominent than the laws of Bravais and Haüy suggest. Further, the unit cell and crystal axes may be assigned in more than one way for some minerals, allowing published sources to assign different Miller indices to the same features on some minerals.

CRYSTALLOGRAPHIC DIRECTIONS

Crystallographic directions or linear features are identified by using integers within brackets [*uvw*] similar to Miller indices. For purposes of developing the nomenclature, consider several lines starting at the origin of the crystal lattice as shown in Figure 2.29. The index for each direction is determined by the coordinates of the lattice nodes through which it passes. Coordinates for lattice nodes are given in terms of unit cell dimensions (*a*, *b*, and *c*).

Consider line *p* that passes through lattice coordinates 1, 1, 1; 2, 2, 2; etc. Its crystallographic direction is obtained by dividing the coordinates by their common factor. Coordinate 2, 2, 2 can be divided by 2 to yield crystallographic direction [111] for line *p*. Line *q* passes through lattice coordinates 1, 0, 1; 2, 0, 2; etc.; its index is [101]. The crystal axes are *a* = [100], *b* = [010], and *c* = [001].

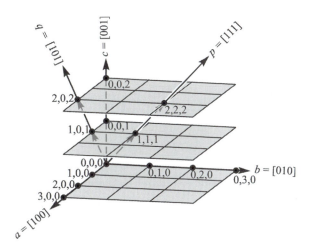

Figure 2.29 Crystallographic directions. Crystallographic direction [111] (line p) goes through lattice nodes 1,1,1 and 2,2,2, and so forth. Direction [101] (q) goes through lattice nodes 1,0,1 and 2,0,2. The three crystal axes a, b, and c are the [100], [010], and [001] crystallographic directions.

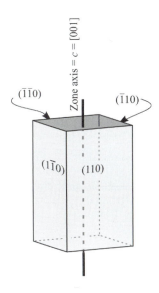

Figure 2.30 Crystal zone. The (110), ($1\bar{1}0$), ($\bar{1}10$), and ($\bar{1}\bar{1}0$), faces constitute a [001] zone because they intersect in edges parallel to [001] (the c axis), which is the zone axis.

ZONES

A **zone** is a collection of crystal faces all of which are parallel to a common line called a **zone axis**. A zone is identified with the index of the zone axis [uvw]. For example, a tetragonal prism (Figure 2.30) constitutes a zone because the faces are all parallel to a common line (the c axis) through the middle of the prism. The zone axis is therefore [001]. The zone axis also is parallel to the edges defined by the intersection of the faces.

If the Miller indices of two intersecting faces (hkl) and (qrs) in the zone are known, a simple mathematical technique allows the index of the zone axis to be calculated. Consider the ($1\bar{1}0$) and (110) faces on the prism in Figure 2.30. Write out the index for the ($1\bar{1}0$) face twice, and immediately beneath, write out the index for the (110) face twice.

$$h\,k\,l\,h\,k\,l \quad \text{or} \quad 1\,\bar{1}\,0\,1\,\bar{1}\,0 \tag{2.6}$$
$$q\,r\,s\,q\,r\,s \qquad\qquad 1\,1\,0\,1\,1\,0$$

Discard (cross off) the first and last number in each row.

$$\cancel{h}\,k\,l\,h\,k\,\cancel{l} \quad \text{or} \quad \cancel{1}\,\bar{1}\,0\,1\,\bar{1}\,\cancel{0} \tag{2.7}$$
$$\cancel{q}\,r\,s\,q\,r\,\cancel{s} \qquad\qquad \cancel{1}\,1\,0\,1\,1\,\cancel{0}$$

The zone index [uvw] is given by cross multiplying as follows.

$$
\begin{array}{cccc}
k & l & h & k \\
\times & \times & \times & \\
r & s & q & r \\
\hline
u & v & w &
\end{array}
\quad \text{or} \quad
\begin{array}{cccc}
\bar{1} & 0 & 1 & \bar{1} \\
\times & \times & \times & \\
1 & 0 & 1 & 1 \\
\hline
0 & 0 & 2 &
\end{array}
$$

$$u = k*s - l*r \quad \text{or} \quad -1*0 - 0*1 = 0$$
$$v = l*q - h*s \quad \text{or} \quad 0*1 - 1*0 = 0$$
$$w = h*r - k*q \quad \text{or} \quad 1*1 - 1*-1 = 2$$

Divide by 2 to reduce to the smallest possible integers:

$$[uvw] = [001]$$

CRYSTAL FORMS

A **crystal form** is a collection of equivalent crystal faces that are related to each other by the symmetry of the mineral. Figure 2.31a shows a simple orthorhombic crystal of the crystal class $2/m\ 2/m\ 2/m$ that is composed of two sets of equivalent faces. One set consists of the four faces like the (011) face: (011) and ($0\bar{1}1$) on the top and ($01\bar{1}$) and ($0\bar{1}\bar{1}$) on the bottom. The (100) and ($\bar{1}00$) faces are the second set. Each of these sets of equivalent faces is a crystal form. For convenience, each crystal form is identified by braces { } around the hkl Miller index of one of the faces that comprise the form. For example, {011} identifies the four faces equivalent to (011), and {100} identifies the two faces equivalent to (100).

Given one face of a crystal form, the point symmetry operations of the crystal will replicate the face to produce all the faces of the form. Consider the (011) face on Figure 2.31a, which is shown isolated in Figure 2.31b. A vertical mirror m parallel to (010) produces a mirror image of this face with index ($0\bar{1}1$) (Figure 2.31c). The horizontal mirror parallel to (001) duplicates these two faces to create the ($01\bar{1}$) and ($0\bar{1}\bar{1}$) faces (Figure 2.31d). The form thus produced is identified with the original face and is called a {011} rhombic prism. The (100) and ($\bar{1}00$) faces are related to each other by the third mirror in the orthorhombic crystal, which is not shown, and create a separate form called a {100} pinacoid. Prisms, pinacoids, and other forms are discussed in the following sections.

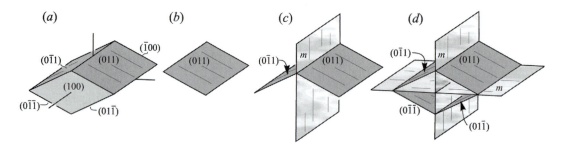

Figure 2.31 Crystal forms are produced by a repetition of crystal faces by symmetry operations. (*a*) A simple orthorhombic crystal. (*b*) The (011) face. (*c*) The (0$\bar{1}$1) face is produced by reflection on the vertical mirror (*m*). (*d*) A horizontal mirror produces the (01$\bar{1}$) and (0$\bar{1}\bar{1}$) faces to complete the four faces of the {011} rhombic prism.

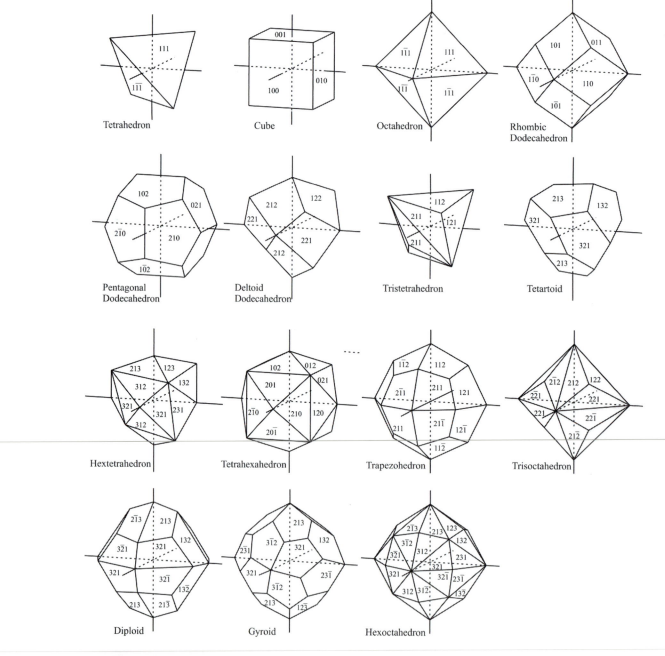

Figure 2.32 Crystal forms in the isometric system.

A form may be either an **open form** or a **closed form**. A closed form such as a cube entirely encloses a volume. An open form such as a prism does not entirely enclose a volume.

Isometric Forms

A total of 15 different forms, all of them closed, is possible in the isometric system (Figure 2.32). Of these, the most common are the cube {001}, octahedron {111}, tetrahedron {111}, and rhombic dodecahedron {110}. Note that both the octahedron and the tetrahedron have the same form symbol {111}. Although it might seem that this would produce confusion, it turns out that tetrahedrons and octahedrons are found in crystals with different point group symmetry (see later).

Nonisometric Forms

The crystal forms possible in the remaining crystal systems (tetragonal, hexagonal, orthorhombic, monoclinic, and triclinic) consist of **pedions, pinacoids, dihedrons, prisms, pyramids, dipyramids, scalenohedrons, trapezohedrons, rhombohedrons**, and **tetrahedrons** (Figures 2.33, 2.34, and 2.35). The definition of each of these different types of forms follows. Open forms are indicated with (O) and closed forms are indicated with (C).

> **PEDION (O):** A single face with no geometrically equivalent face elsewhere on the crystal. Also known as a **monohedron**. No symmetry element repeats the face (Figure 2.33a).

> **PINACOID (O):** Two parallel faces on opposite sides of the crystal related by inversion or a mirror (Figure 2.33b). Also known as a **parallelohedron**.

> **DIHEDRON (O):** Two nonparallel faces. Also known as a **dome** if the two faces are related only by a mirror plane or a **sphenoid** if the two faces are related by a 2-fold rotation axis (Figure 2.33c).

> **PRISM (O):** A collection of 3, 4, 6, 8, or 12 faces that intersects in a set of mutually parallel edges forming a tube. Faces that may close off the ends of the prism are not part of the form. Prisms are named (rhombic prism, tetragonal prism, etc.) based on the shape of the cross section (Figure 2.34).

> **PYRAMID (O):** A collection of 3, 4, 6, 8, or 12 nonparallel faces that can intersect at a point. The different types of pyramids are named based on the shape of the cross section with the same nomenclature used for prisms (Figure 2.34). A pyramid may be either top or bottom depending on whether it occurs on the top of the crystal or the bottom.

> **DIPYRAMID (C):** Two pyramids, one on each end of the crystal, with a total of 6, 8, 12, 16, or 24 faces. The two pyramids are related to each other by reflection across a mirror plane between them (Figure 2.34). The different types of dipyramids are named based on the shape of the cross section with the same nomenclature used for prisms.

> **TRAPEZOHEDRONS (C):** Forms consisting of 6, 8, or 12 faces, each of which is a trapezoid (Figure 2.35). The faces on one end of the crystal are offset relative to the faces on the other end. Tetragonal trapezohedrons have eight faces, four each on the top and bottom. Trigonal trapezohedrons have six faces, three each on the top and bottom. Both positive and negative forms are possible (see later). Hexagonal trapezohedrons have 12 faces, six each on the top and bottom.

> **SCALENOHEDRONS (C):** Forms consisting of 8 or 12 faces, each of which is a scalene triangle (Figure 2.35). The faces appear to be arranged in pairs.

> **RHOMBOHEDRONS (C):** A form consisting of six faces, each of which is rhomb shaped (Figure 2.35). A

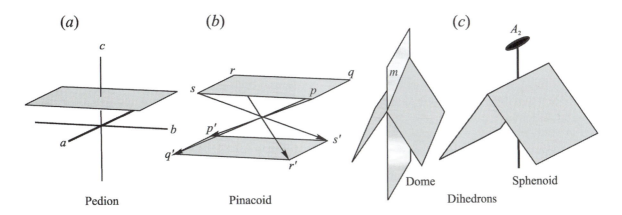

Figure 2.33 Crystal forms in the nonisometric crystal systems. (*a*) Pedion. (*b*) Pinacoid. (*c*) Dihedrons (dome and sphenoid).

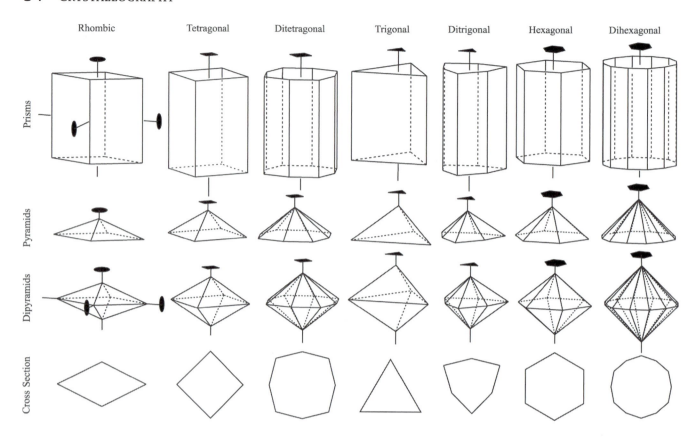

Figure 2.34 Crystal forms in the nonisometric crystal systems: prisms, pyramids, and dipyramids. Each is named based on the shape of the cross section (hexagonal prism, tetragonal pyramid, ditrigonal dipyramid, etc.).

rhombohedron looks like a cube that has been either stretched or shortened along an axis going from corner to corner through the center.

TETRAHEDRONS (C): A form consisting of four triangular faces. In isometric crystals (Figure 2.32), each face is an equilateral triangle. In the tetragonal system (tetragonal tetrahedron) the four faces are identical isosceles triangles. In the orthorhombic system (rhombic tetrahedron) the faces form two pairs of different isosceles triangles. The nonisometric forms are also known as **disphenoids** (Figure 2.35).

Combining Crystal Forms

Simple crystals may consist of only a single closed form such as a cube or a dipyramid; more complex crystals may include several different open and/or closed forms. If a crystal consists only of open forms, at least two and sometimes three or more different open forms must be present to enclose the volume represented by the crystal.

All the forms on any given crystal must be compatible with one another. A mineral that crystallizes in the isometric crystal system may include only isometric forms, tetragonal minerals may include only tetragonal forms, and so forth. **A mineral can have only the forms that**
are consistent with the symmetry of the crystal class to which it belongs.** These are described shortly.

Avoid the mistake of assuming that a multifaced form is composed of several simpler forms. For example, a pinacoid is not composed of two pedions. A pinacoid consists of two and only two parallel faces related to each other by the symmetry of the mineral. Similarly, a cube consists of six equivalent faces; it is not composed of three pinacoids.

Enantiomorphous Forms and Crystals

Enantiomorphous forms lack center symmetry and mirrors. This makes it possible for each form to have left- and right-handed versions, like left- and right-handed gloves. The trigonal trapezohedron shown in Figure 2.36a is a good example. The crystal on the right displays the right-hand form $\{211\}$ and the one on the left displays the left-hand form $\{3\bar{1}1\}$. Note that the crystals are mirror images of each other, and the individual trapezoidal faces of the right- and left-hand versions are also mirror images of each other. On a single crystal, only the right- or the left-hand form may be present, never both together. Different samples of the same mineral may have either form.

The presence of an enantiomorphous form indicates that the mineral has a screw axis that is a repeating structure analogous to a spiral staircase that may spiral either to the right or left. Common quartz, for example, has a

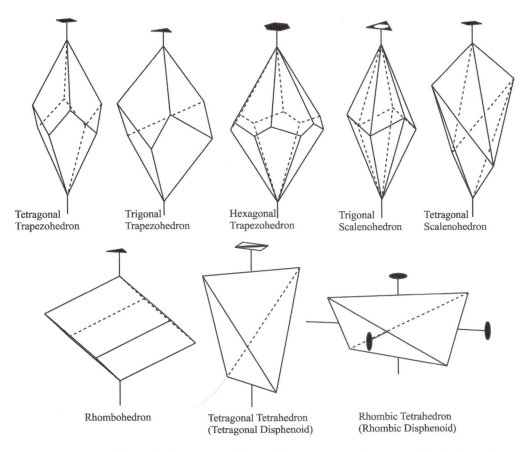

Figure 2.35 Crystal forms in the nonisometric crystal systems: trapezohedrons, scalenohedrons, rhombohedrons, and tetrahedrons (rhombic and tetragonal). Tetragonal trapezohedrons have eight faces, four each on the top and bottom. Trigonal trapezohedrons have six faces, three each on the top and bottom. Both positive and negative forms are possible, as shown later (Figure 2.37). Hexagonal trapezohedrons have 12 faces, 6 each on the top and bottom.

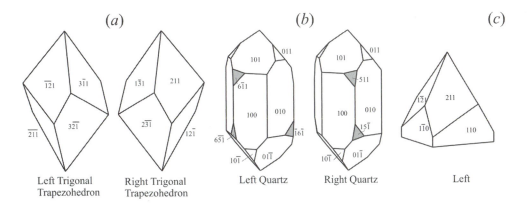

Figure 2.36 Enantiomorphous forms and crystals. (*a*) The right-handed {211} and left-handed {3$\overline{1}$1} trigonal trapezohedron lacks center or mirrors. (*b*) The right {511} and left {6$\overline{1}$1}trigonal trapezohedron (small shaded faces) on right- and left-handed quartz. (*c*) Right- and left-handed crystals in the 4 crystal class. The {211} and {2$\overline{1}$1} pyramids are not themselves enantiomorphous, but when combined with the {110} prism and {00$\overline{1}$} pedion, they form right- and left-handed crystals.

structure with either a right- or left-hand spiral. Right-hand quartz can display the {511} right-hand trigonal trapezohedron; left-hand quartz can display the {6$\overline{1}$1} left-hand trapezohedron (Figure 2.36b). These faces are not commonly found, however.

Forms that may be enantiomorphous and the classes in which they are found are listed in Table 2.4. Enantiomorphous forms are the characteristic forms in the crystal classes in which they are found, all of which lack center, rotoinversion, and mirrors. The 1, 2, 3, 4, and

Table 2.4 Enantiomorphous Forms (Figures 2.32, 2.35, and 2.36) and the Point Groups in Which They Occur

Form	Crystal Class/ Point Group
Rhombic tetrahedron (disphenoid)	222
Trigonal trapezohedron	32
Hexagonal trapezohedron	622
Tetragonal trapezohedron	422
Gyroid	23
Tetartoid	432

6 crystal classes also lack center, rotoinversion, and mirrors and may have enantiomorphous crystals, although they will be composed of forms that are not individually enantiomorphous. Figure 2.36c shows enantiomorphous crystals in the 4 crystal class.

Positive and Negative Forms

Positive and negative varieties of a form, such as a rhombohedron, differ only in that the negative form is rotated relative to the positive form. Figure 2.37a shows a positive {111} and a negative {1$\bar{1}$1} tetrahedron. Note that the negative tetrahedron can be made coincident with the positive version by rotation of 90° about the c axis. Rhombohedrons (Figure 2.37b) have a single 3-fold rotation axis coincident with the c axis. Each face is repeated

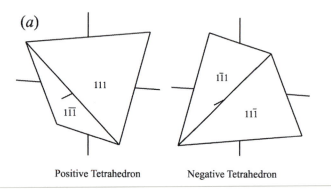

Positive Tetrahedron   Negative Tetrahedron

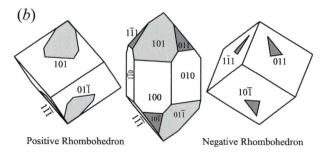

Positive Rhombohedron   Negative Rhombohedron

Figure 2.37 Positive and negative forms. (a) Tetrahedrons. (b) Quartz crystal with both positive {101} and negative {011} unit rhombohedrons.

once every 120°. The positive rhombohedron is {101}. The equivalent negative rhombohedron {011} is produced by rotating the faces of the positive rhombohedron 60° about c. Note that the negative version of a form is produced by rotation of 60° on a 3-fold rotation axis, or 90° on a 2-fold axis or 4-fold rotoinversion axis. Forms that may have positive and negative varieties are indicated later (see Tables 2.6, 2.7, and 2.8).

The same crystal may have both a positive and negative version of the same form. Common quartz, for example (Figure 2.37b), displays both the positive {101} and negative {011} unit rhombohedrons. One rhombohedron typically has larger faces than the other. Crystals are usually oriented so the larger set of faces is assigned to the positive rhombohedron.

Relating positive and negative crystal forms can often be done by reversing the sign of the last number of the Miller index: {101} versus {10$\bar{1}$} for rhombohedrons or {111} versus {11$\bar{1}$} for tetrahedrons. However, the negative unit rhombohedron and tetrahedron are usually identified as {011} and {1$\bar{1}$1} respectively, because those faces cut the positive c axis. In some crystal classes, positive and negative forms are related by reversing two numbers in the Miller index—{102} versus {012} for pyritohedrons for example.

FORMS IN THE SIX CRYSTAL SYSTEMS

The forms that may occur in each of the 32 point groups (32 crystal classes) within the six crystal systems are limited by the symmetry of the point group. For example, since tetragonal prisms have a single 4-fold rotation axis they can be found only in tetragonal point groups that have a single 4-fold rotation to match. Similarly, pedions, which consist of only one face, never occur in minerals with center symmetry because the inversion operation requires that an equivalent face be on the opposite side of the crystal, producing a pinacoid.

In the illustrations that follow, crystals are systematically oriented so that the c axis is vertical, b is to the right, and a is to the front.

Triclinic Crystal System

There are no symmetry constraints on the orientation of the crystal axes. Generally the c axis is chosen to be parallel to the dominant zone axis of the mineral, with the b and a axes parallel to crystal edges or in other rational orientations. Axes are generally oriented so that both the α and β angles are greater than 90°; this places the b axis down to the right and the a axis down to the front as these minerals are conventionally illustrated.

It is common in triclinic minerals to find that several different crystal axis orientations and unit cell geometries are in use. In some cases, crystal axis orientation and unit cell shape are established by the external geometry of the crystal faces. In other cases, crystal axis orientation is determined by detailed study of the crystal structure. The different axis conventions for a given mineral can be related to each other by appropriate mathematical or graphic manipulation. Confusion is common, however, when structural analysis is tied to one set of axes and unit cell, and descriptions of crystal faces, cleavage, or optical properties are tied to a different set of axes and unit cell. Beware—even standard reference works are not free of this particular problem.

The forms that are found in the two triclinic classes are listed in Table 2.5. Figure 2.38a shows a 1 point group crystal with **positive front** {100}, **side** {010}, and **basal** {001} pedions; negative front {$\bar{1}$00}, side {0$\bar{1}$0}, and basal {00$\bar{1}$} pedions; and the {011} pedion. Figure 2.38b shows a $\bar{1}$ point group crystal with **front** {100}, **side** {010}, and

basal {001} pinacoids; and a {011} pinacoid. Note that the two faces of a pinacoid have the same Miller indices but with opposite signs. A variety of common minerals, among them plagioclase and microcline, crystallize with $\bar{1}$ symmetry.

Monoclinic Crystal System

The crystal axes in the monoclinic system are most commonly oriented so that the *b* axis coincides with the 2-fold rotation axis and/or is perpendicular to the mirror plane. The *c* axis is placed parallel to a prominent zone axis and *a* is placed down and to the front so that the β angle is greater than 90°. Cleavage also may assist in establishing the orientation of the *a* and *c* axes. If a single prominent cleavage is parallel to the *b* axis, it generally is assumed to be parallel to the basal pinacoid {001}, and hence also parallel to the *a* axis. If two equivalent cleavages are symmetrically related by the mirror plane, they are usually taken to be parallel to the *c* axis. This convention for orienting the crystal axes is known as the **second setting**. The **first setting** convention places the *c* axis parallel to the 2-fold axis and *b* is parallel to a prominent zone. The second setting for monoclinic minerals will be used throughout this book and is used in all modern mineralogic literature. The forms found in monoclinic classes are listed in Table 2.5. Only minerals with 2/*m* symmetry are common (Table 2.3), and they include amphiboles, pyroxenes, micas, and many other minerals. Figure 2.39 shows crystals with monoclinic symmetry.

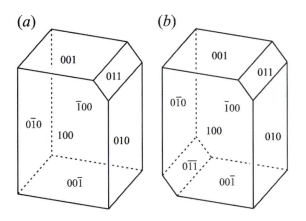

Orthorhombic Crystal System

The symmetry elements in the orthorhombic crystal system combine to produce three unique directions at right angles to each other that are parallel to the three crystal axes. The

Figure 2.38 Forms in the triclinic crystal system. (*a*) Pedial 1 class crystal consisting of {100}, {$\bar{1}$00}, {010}, {0$\bar{1}$0}, {001}, {00$\bar{1}$}, and {011} pedions. Note the lack of center symmetry. (*b*) Pinacoidal $\bar{1}$ class crystal consisting of {100}, {010}, {001}, and {011} pinacoids.

Table 2.5 Forms Found in Triclinic, Monoclinic, and Orthorhombic Point Groups[a]

		Triclinic		Monoclinic			Orthorhombic		
Number of Faces	Name of Form	1 Pedial	$\bar{1}$ Pinacoidal	*m* Domatic	2 Sphenoidal	2/*m* Prismatic	*mm2* Rhombic Pyramidal	222 Rhombic Dispenoid	2/*m* 2/*m* 2/*m* Rhombic Dipyramidal
1	Pedion	X		X	X		X		
2	Pinacoid		X	X	X	X	X	X	X
2	Dihedron (dome)			X			X		
2	Dihedron (sphenoid)				X				
4	Prism					X	X	X	X
4	Rhombic tetrahedron (disphenoid)							X	
4	Rhombic pyramid						X		
8	Rhombic dipyramid								X

[a]The name of each crystal class is listed beneath the point group.

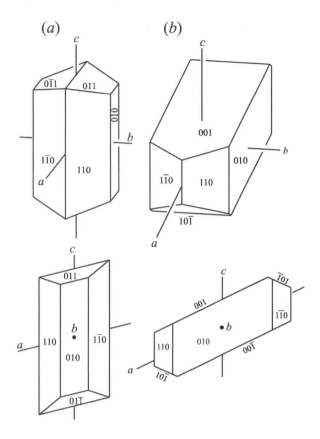

Figure 2.39 Monoclinic crystals. The views down the *b* axis show that the *a* and *c* axes are not at right angles. (*a*) 2/*m* Point group amphibole crystal with a side {010} pinacoid and {110} and {011} prisms. (*b*) 2/*m* Point group sanidine crystal with side {010}, basal {001}, and {10$\bar{1}$} pinacoids; and {110} prism.

crystal axes are therefore either parallel to a 2-fold rotation axis and/or perpendicular to a mirror. Hermann–Mauguin symbols for each crystal class are written so that the first element refers to the *a* axis, the second to *b*, and the third to *c*. In the *mm*2 point group, for example, the *a* axis [100] is at right angles to a mirror, the *b* axis [010] is at right angles to a mirror, and the *c* axis [001] is a 2-fold rotation axis.

Except for the *mm*2 class, where the 2-fold rotation axis is taken as the *c* axis, no universally accepted convention determines which crystal axis coincides with which symmetry axis. A common procedure is to orient the crystal axes so that the unit cell dimensions are *c* < *a* < *b*. However, if a crystal has an elongate habit, the long axis may be assigned to *c* regardless of unit cell dimensions. Similarly, the *c* axis may be placed perpendicular to a single prominent cleavage or to a prominent pinacoid that produces a distinctly tabular habit. Minerals with two equivalent cleavage directions (not generally at 90°) are oriented so that these prismatic cleavages are parallel to *c* and have the general form {*hk*0}. The reader is cautioned to confirm crystal axis conventions when reading the literature on orthorhombic minerals. The forms possible in the orthorhombic classes are listed in Table 2.5. Figure 2.40 shows crystals that illustrate the different symmetry of the orthorhombic point groups. The common orthorhombic minerals all have 2/*m* 2/*m* 2/*m* symmetry (Table 2.3).

Tetragonal Crystal System

All minerals in the tetragonal crystal system have a single 4-fold axis of rotation or rotoinversion that coincides with

Table 2.6 Forms Found in the Point Groups of the Tetragonal Crystal System[a]

Number of Faces	Name of Form	Form Symbols	4 Tetragonal Pyramidal
1	Pedion	{001}, {00$\bar{1}$}	X
2	Pinacoid	{001}	
4	Tetragonal prism	{110}, {010}	X
4	Tetragonal pyramid	{*hhl*}, {0*kl*}	X
4	Tetragonal tetrahedron (disphenoid)	+{*hhl*}; −{*hh̄l*}	
8	Ditetragonal prism	{*hk*0}	
8	Tetragonal dipyramid	{*hhl*}, {0*kl*}	
8	Tetragonal trapezohedron	{*hkl*}	
8	Tetragonal scalenohedron	{*hkl*}	
8	Ditetragonal pyramid	{*hkl*}	
16	Ditetragonal dipyramid	{*hkl*}	

[a]The name of each crystal class is listed beneath the point groups.

the *c* crystal axis. The *a* and *b* axes are at right angles to each other and coincide with 2-fold rotation axes or are perpendicular to mirror planes. The first Hermann–Mauguin symbol for the point group refers to symmetry on the *c* axis. The second symbols refer to symmetry tied to the *a* and *b* axes ([100] and [010]) and the third to symmetry tied to [110] and [1$\bar{1}$0] axes at 45° to the *a* and *b* axes. In the $\bar{4}2m$ point group, the $\bar{4}$ indicates that the *c* axis [001] is a 4-fold rotoinversion axis, the 2 indicates that the *a* and *b* crystal axes ([100] and [010]) are 2-fold rotation axes, and the *m* indicates that [110] and [1$\bar{1}$0] directions in the crystal lattice that bisect the angles between the *a* and *b* axes are perpendicular to mirrors. The forms found in the tetragonal crystal classes are listed in Table 2.6. Most common

tetragonal minerals crystallize with 4/*m* 2/*m* 2/*m* symmetry (Table 2.3). Figure 2.41 shows a selection of tetragonal crystals illustrating the symmetry found in each point group.

Hexagonal Crystal System

The hexagonal crystal system is divided into hexagonal and trigonal (rhombohedral) divisions based on the presence of 6-fold or 3-fold symmetry, respectively. The *c* axis always is parallel to the 6- or 3-fold rotation axis. The *a* and *b* axes are arranged so that their positive ends are 120° from each other and either parallel to 2-fold rotation axes or perpendicular to mirror planes if these symmetry elements are

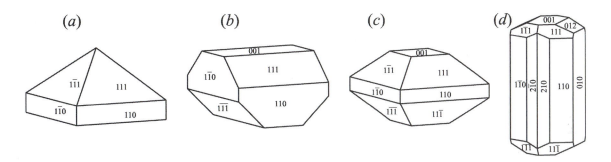

Figure 2.40 Orthorhombic crystals. Note how the (111), (110), and (00$\bar{1}$) faces are duplicated to produce different forms based on the symmetry content. (*a*) *mm*2 Point group crystal with a {110} rhombic prism, {00$\bar{1}$} bottom pedion, and {111} rhombic pyramid. (*b*) 222 Point group crystal with {111} rhombic tetrahedron (disphenoid), {110} rhombic prism, and {001} pinacoid. (*c*) 2/*m* 2/*m* 2/*m* Point group crystal with a {001} pinacoid, {110} rhombic prism, and {111} rhombic dipyramid. (*d*) 2/*m* 2/*m* 2/*m* topaz crystal with {010} and {001} pinacoids; {110}, {210}, and {012} rhombic prisms, and {111} rhombic dipyramid.

			Point Group		
$\bar{4}$ Tetragonal Disphenoid	4/*m* Tetragonal Dipyramidal	422 Tetragonal Trapezohedal	4*mm* Ditetragonal	$\bar{4}2m$ Tetragonal Scalenohedral	4/*m* 2/*m* 2/*m* Ditetraganol Dipyramidal
			X		
X	X	X		X	X
X	X	X	X	X	X
			X		
X				X	
		X	X	X	X
	X	X		X	X
		X			
				X	
			X		
					X

present. The first number in the Hermann–Mauguin symmetry notation refers to symmetry tied to the *c* axis, the second to symmetry elements associated with the *a* and *b* axes ([100] and [010]), and the third to directions perpendicular to *a* and *b* ([210] and [120]).

Forms possible in each of the different hexagonal crystal classes are listed in Table 2.7. Many hexagonal

minerals display the hexagonal prism because it is possible in all but the $\bar{6}$ and 3 point groups. Hexagonal minerals not dominated by a hexagonal prism are usually dominated by a rhombohedron or a trigonal prism. This makes distinguishing minerals in the hexagonal crystal system from other crystal systems relatively straightforward. Distinguishing among the different hexagonal

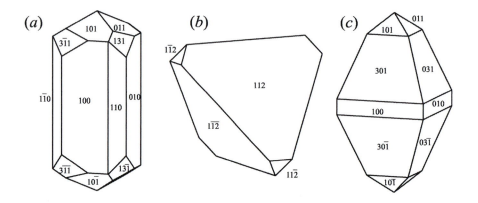

Figure 2.41 Tetragonal crystals. (*a*) 4/*m* Scapolite crystal with {100} and {110} prisms, and {011} and {131} dipyramids. (*b*) $\bar{4}2m$ Chalcopyrite crystal consisting of a positive {112} tetragonal tetrahedron (disphenoid) and a negative {1$\bar{1}$2} tetragonal tetrahedron (smaller faces). (*c*) 4/*m* 2/*m* 2/*m* Zircon crystal with {100} prism and {011} and {031} dipyramids.

Table 2.7 Forms Found in Point Groups of the Hexagonal Crystal System

			Trigonal (Rhombohedral) Division				
Faces	**Name of Form**	**Form {Generic}: {Example}**	**3 Trigonal Pyramidal**	**$\bar{3}$ Rhombohedral**	**32 Trigonal Trapezohedral**	**3m Ditrigonal Pyramidal**	**$\bar{3}$ 2/m Hexagon Scalenohedral**
1	Pedion	{001}:{00$\bar{1}$}	X			X	
2	Pinacoid	{001}		X	X		X
3	Trigonal prism	(*hk*0) : {100}	X		X	X	
3	Trigonal pyramid	{*hkl*} : {101}	X			X	
6	Ditrigonal prism	{*hk*0}: {120}			X	X	
6	Hexagonal prism	(*hk*0) : {100}		X	X	X	X
6	Trigonal dipyramid	(*hkl*) : {101}			X		
6	Rhombohedron	(*hkl*) : {101}		X	X		X
6	Trigonal trapezohedron	(*hkl*) : {211}			X		
6	Ditrigonal pyramid	{*hkl*} : {101}				X	
6	Hexagonal pyramid	{*hkl*} : {101}				X	
12	Hexagonal dipyramid	{*hkl*} : {101}					X
12	Hexagonal scalenohedron	{*hkl*}: {211}					X
12	Dihexagonal prism	{*hk*0} : {210}					X
12	Ditrigonal dipyramid	{*hkl*} : {211}					
12	Hexagonal trapezohedron	{*hkl*} : {211}					
12	Dihexagonal pyramid	{*hkl*} : {211}					
24	Dihexagonal dipyramid	{*hkl*} : {211}					

point groups is often difficult, however, because faces that display reduced symmetry compared with the 6/m 2/m 2/m symmetry may be small or lacking. Visual inspection of real crystals often will allow differentiating only between the hexagonal and trigonal divisions based on the presence of 6-fold versus 3-fold symmetry parallel to c. Figure 2.42 shows several common minerals that crystallize in the trigonal division. Figure 2.43 shows examples of minerals that crystallize in the hexagonal division.

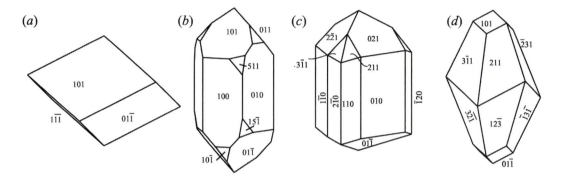

Figure 2.42 Crystals in the hexagonal crystal system, trigonal division. (a) Dolomite {101} rhombohedron in the $\bar{3}$ crystal class. The lack of vertical mirrors is not belied by these crystal faces. (b) 32 Point group. Right-handed quartz with {100} prism, positive {101} and negative {011} rhombohedrons, and {511} trigonal scalenohedron. (c) 3m Point group. Tourmaline with {021} trigonal pyramid and {211} ditrigonal pyramid on top, {110} and {010} prisms in the middle, and {01$\bar{1}$} trigonal pyramid on the bottom. (d) $\bar{3}2m$ Point group. Calcite crystal with {211} scalenohedron and {101} positive rhombohedron.

Hexagonal Division						
6 Hexagonal Pyramidal	$\bar{6}$ Trigonal Dipyramidal	6/m Hexagonal Dipyramidal	622 Hexagonal Trapezohedral	6mm Dihexagonal Pyramidal	$\bar{6}$m2 Ditrigonal Dipyramidal	6/m 2/m 2/m Dihexagonal Dipyramidal
X				X		
	X	X	X		X	X
	X				X	
					X	
X		X	X	X	X	X
	X				X	
X				X		
	X		X		X	X
			X	X		X
					X	
			X			
				X		
						X

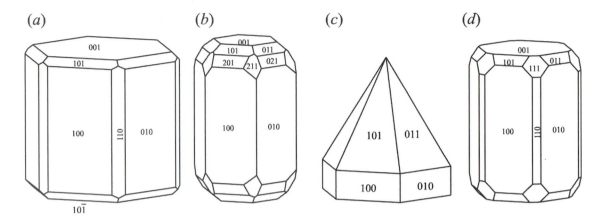

Figure 2.43 Crystals in the hexagonal crystal system, hexagonal division. (*a*) Nepheline (6 point group) crystals are rare, composed of {100} and {110} prisms, upper {101} and lower {10$\bar{1}$} pyramids, and top {001} and bottom {00$\bar{1}$} pedions. The crystal habit appears to have higher symmetry than the nepheline actually possesses. (*b*) 6/*m* Apatite crystal composed of {100} prism; {101}, {201}, {211} dipyramids; and {001} pinacoid. Only the {211} dipyramid indicates that the symmetry is 6/*m*. (*c*) 6*mm* Zincite crystal consists of {100} prism, {101} pyramid, and 00$\bar{1}$ pedion. (*d*) 6/*m* 2/*m* 2/*m* Beryl crystal with {100} and {110} prisms, {101} and {111} dipyramids, and {001} pinacoid.

Table 2.8 Forms Found in the Point Groups of the Isometric Crystal System

Number of Faces	Name of Form	Form Symbols	Point Group				
			23 Tetratoidal	432 Gyroidal	2/*m* $\bar{3}$ Diploidal	$\bar{4}$3*m* Hextetrahedral	4/*m* $\bar{3}$2/*m* Hexoctahedral
4	Tetrahedron	+{111}, −{1$\bar{1}\bar{1}$}	X			X	
6	Cube (hexahedron)	{001}	X	X	X	X	X
8	Octahedron	{111}		X	X		X
12	Rhombic dodecahedron	{101}	X	X	X	X	X
12	Pentagonal dodecahedron (pyritohedron)	+{102}, −{201}	X		X		
12	Tristetrahedron	+{112}, −{1$\bar{1}$2}	X			X	
12	Deltoid dodecahedron	+{122}, −{1$\bar{2}$2}	X			X	
12	Tetartoid	{213}	X				
24	Tetrahexahedron	{201}		X		X	X
24	Trapezohedron	{211}		X	X		X
24	Trisoctahedron	{122}		X	X		X
24	Hextetrahedron	+{321}, −{3$\bar{2}$1}				X	
24	Diploid	+{321}, −{231}			X		
24	Gyroid	{321}		X			
48	Hexoctahedron	{321}					X

Isometric Crystal System

Crystals in the isometric crystal system have a cubic unit cell with three mutually perpendicular crystal axes a, b, and c. Note that these axes are sometimes referred to as a_1, a_2, and a_3. Three equivalent symmetry axes at right angles to each other coincide with the crystallographic axes. In three crystal classes these axes are 4-fold or 4-fold rotoinversion axes; in the remaining two classes they are 2-fold axes. Hermann–Mauguin symbols for the crystal classes consist of up to three parts.

The first symbol refers to the symmetry operations tied to the three crystallographic axes. The second refers to the three diagonal axes [111] through the cubic unit cell. The third refers to the six directions that run from the center of one edge of the unit cell to the center of the opposite edge [110]. Crystal forms found in each of the isometric crystal classes are listed in Table 2.8 and are shown individually in Figure 2.32. Symmetry elements and selected crystal forms found in the $2/m\overline{3}$, $\overline{4}3m$, and $4/m\overline{3}2/m$ point groups are shown in Figures 2.44 through 2.47.

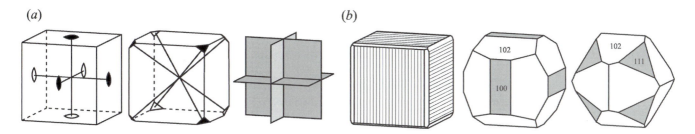

Figure 2.44 $2/m\overline{3}$ Point group. (*a*) Three 2-fold axes are parallel to the crystal axes, four 3-fold axes are parallel to the long diagonals through the unit cell, and three mirror planes are parallel to the cube faces. (*b*) Striations on pyrite cube (left) are produced by interaction between the shaded cube faces {001} and pyritohedron {102} (center). Shaded octahedron {111} faces modify pyritohedron (right).

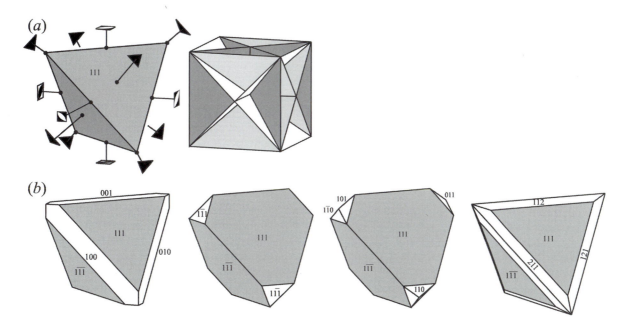

Figure 2.45 $\overline{4}3m$ Point group. (*a*) Crystal axes are parallel to the three 4-fold rotoinversion axes. Four 3-fold axes are parallel to the long diagonals through the cube and perpendicular to the tetrahedron crystal faces. Six mirrors are parallel to the body diagonals through the cubic unit cell. (*b*) Tetrahedrons {111} (dark) combined with (left to right) cube {001}, negative tetrahedron {1$\overline{1}$1}, dodecahedron {011}, and tristetrahedron {121}.

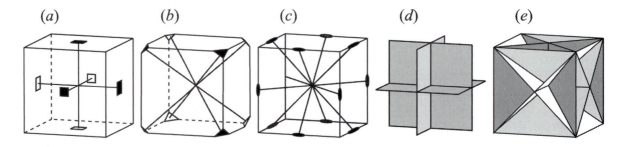

Figure 2.46 Symmetry elements in the $4/m\bar{3}2/m$ point group. (*a*) Three 4-fold rotation axes are parallel to the *a* crystal axes. (*b*) Four 3-fold axes are parallel to the long diagonals through the cube. (*c*) Six 2-fold axes are parallel to edge-to-edge diagonals. (*d*) Three mirrors are parallel to cube faces. (*e*) Six mirrors parallel to diagonals through the cube.

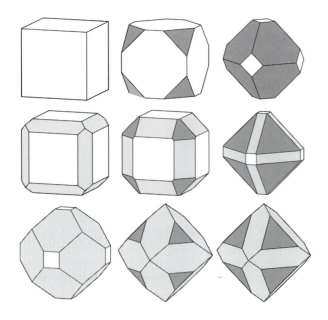

Figure 2.47 Combination of {001} cube (white), {011} dodecahedron (light), and {111} octahedron (dark) in the $4/m\bar{3}2/m$ point group. See Table 2.8 and Figure 2.32 for additional forms that may appear.

CRYSTAL HABIT

In addition to the formal crystallographic nomenclature, a variety of descriptive terminology has developed to help describe minerals and mineral aggregates. Some commonly used terms are defined here. A wide variety of additional terminology is used to describe textural features of various rock types. For a coherent description of much of this terminology, peruse Williams and others (1982).

Minerals may or may not display crystal faces depending on conditions of growth. Samples that display well-formed crystal faces are **euhedral**, and those without crystal faces are **anhedral**. If crystal faces are present, but are not well formed, the term **subhedral** is used (Figure 2.48).

Figure 2.49 shows the range of terminology used to describe the relative dimensions of individual crystals

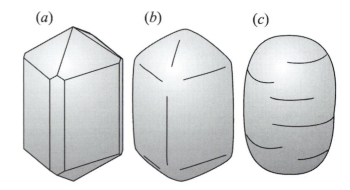

Figure 2.48 Degree to which crystal faces are well developed: (*a*) euhedral, (*b*) subhedral, (*c*) anhedral.

or mineral grains. **Equant** crystals have length, width, and height that are about the same. **Elongate columnar, acicular**, and **fibrous, filiform**, or **capillary** crystals are progressively more elongate. **Stubby columnar, platy**, and **scaly** or **micaceous** grains are progressively flatter or thinner. **Tabular** crystals are roughly book shaped, and **bladed** crystals are relatively thin and elongate like a knife blade.

Some of the systematic patterns by which mineral grains are intergrown are shown in Figure 2.50. Equant minerals typically intergrow in a **granular** texture and elongate minerals can form **parallel** or **radiating** aggregates. Platy or micaceous minerals that grow in parallel alignment are **foliated**; those that form a radiating aggregate are **plumose**.

When minerals precipitate and grow on a preexisting rock surface, in a fault or fracture zone for example, the pattern of growth may take on distinctive forms (Figure 2.51).

Drusy   An encrustation of euhedral crystals
Colloform   Rounded forms usually consisting of radiating aggregates of elongate mineral grains or crystals

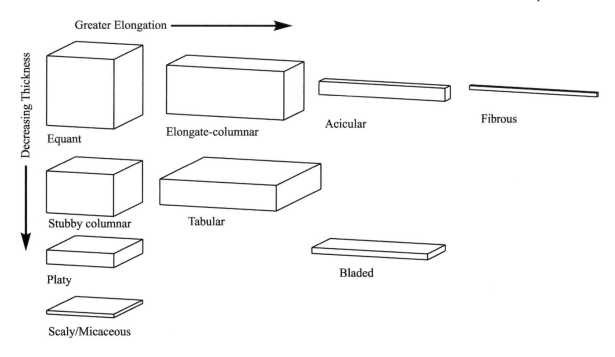

Figure 2.49 Nomenclature for describing the habit of individual mineral grains as a function of relative dimensions.

Globular   A colloform mass that is roughly spherical

Botryoidal   Overlapping colloform masses that look like a bunch of grapes

Mammillary   Overlapping colloform masses that have the smooth rounded forms of mammary glands (e.g., a cow's udder)

Reniform   A colloform mass that is kidney shaped

Botryoidal, mammillary, and reniform are listed in order of decreasing "lumpiness"—the dividing lines between them are poorly defined. Other terms that describe crystal groups or mineral aggregates include the following.

Banded   Bands or layers of mineral with different color or texture.

Dendritic   Mineral growth in a branching form somewhat like a fern (Figure 18.10).

Lamellar   Composed of layers, like the pages of a book.

Massive   An aggregate of mineral grains without distinctive form or crystal faces. If individual grains are visible: granular massive. If very fine grained: compact massive.

Oolitic/Pisolitic   Roughly spherical mineral grains formed of concentric layers. Pisolites are roughly pea sized, oolites are like BB shot.

Reticulated   Crystals intergrown to form an open framework.

The descriptive nomenclature is by no means restricted to these terms—almost any descriptive term may be used.

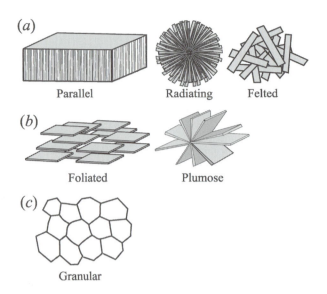

Figure 2.50 Patterns of mineral growth. (*a*) Elongate minerals can form parallel, radiating, or felted masses. (*b*) Platy or micaceous minerals can form foliated or plumose arrangements. (*c*) Equant minerals can form fine-to coarse-grained granular masses.

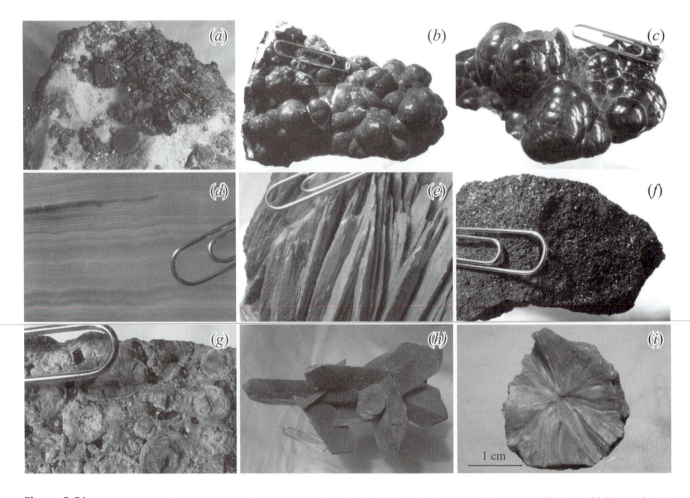

Figure 2.51 Habits of crystalline masses. (*a*) Drusy encrustation of sphalerite and galena on white quartz. (*b*) Botryoidal hematite. (*c*) Mammillary hematite. (*d*) Banded chalcedony. (*e*) Lamellar intergrowth of tabular barite crystals. (*f*) Granular massive chromite. (*g*) Pisolitic aluminum oxides in bauxite. (*h*) Reticulated crystals of gypsum. (*i*) Radiating concretion of barite.

References Cited and Suggestions for Additional Reading

Bloss, F. D. 1971. *Crystallography and Crystal Chemistry.* Holt, Rinehart & Winston, New York, 545 p.

Boisen, M. B., Jr., and Gibbs, G. V. 1985. *Mathematical Crystallography. Reviews in Mineralogy*, Volume 15. Mineralogical Society of America, Washington, DC, 406 p.

Burke, J. G. 1966. *Origins of the Science of Crystals.* University of California Press, Berkeley and Los Angeles, 198 p.

Hahn, T. (ed.). 1983. *International Tables for X-ray Crystallography*, Volume A, *Space Group Symmetry*, D. Reidel, Boston, 854 p.

Henry, N. F. M., and Lonsdale, K. (eds.). 1952. *International Tables for X-Ray Crystallography*, Volume 1: *Symmetry Groups.* Kynoch Press, Birmingham, England, 558 p.

O'Keeffe, M., and Hyde, B. G. 1996. *Crystal Structures I. Patterns and Symmetry.* Mineralogical Society of America, Washington, DC, 453 p.

Smith, J. V. 1982. *Geometrical and Structural Crystallography.* John Wiley & Sons, New York, 450 p.

Williams, H., Turner, F.J., and Gilbert, C.M. 1982. *Petrography: An Introduction to the Study of Rocks in Thin Section*, 2nd ed. W.H. Freeman, San Francisco, 626 p.

CHAPTER 3

Crystal Chemistry

INTRODUCTION

Because minerals are crystalline solids, they must have definite, but not necessarily fixed, chemical compositions. In this chapter some introductory chemistry is explored to provide a basis for understanding mineral compositions. The following questions will be addressed.

What is the nature of chemical elements?

What is the abundance of chemical elements available to make minerals?

What mechanisms cause chemical elements to bond and form solids?

What is the size of atoms and ions?

THE NATURE OF CHEMICAL ELEMENTS

All minerals are composed of chemical elements in various combinations. The atoms that comprise each of the elements are composed of the fundamental physical building blocks of matter: **protons, neutrons**, and **electrons** (Table 3.1). Other particles such as neutrinos and quarks are of great interest to atomic physicists but are not important in the present context. A periodic table of the elements is on the inside of the back cover.

Nucleus

Protons and neutrons are held together in the nucleus by poorly understood strong attractive forces that overcome the otherwise normal tendency of the positively charged protons to repel each other. Each chemical element has a specific number of protons in the nucleus. The number of protons is the **atomic number** (Z) of that particular element. For example, the nucleus of hydrogen has a single proton, oxygen has 8 protons, and aluminum has 13 protons.

The number of neutrons in the nucleus of each atom is about the same as the number of protons for elements with a small atomic number, and larger than the number of protons for higher atomic numbers. The number of neutrons in the nucleus, however, varies for each element. Each isotope of an element has a different number of neutrons. The sum of the number of neutrons and protons is the **mass number** of an atom. To distinguish among the different isotopes of an element, the mass number of the isotope is written as a superscript to the left of the atomic symbol. For example, potassium (K) is atomic number 19 ($Z = 19$), so the nucleus of every potassium atom contains 19 protons. It has three main isotopes with different numbers of neutrons in the nucleus: ^{39}K has mass number 39 with 20 neutrons, ^{40}K has mass number 40 and 21 neutrons, and ^{41}K has mass number 41 and 22 neutrons.

The **atomic mass** of an atom is the mass of the atom divided by one-twelfth the mass of a ^{12}C atom. Except for ^{12}C, the atomic mass is just a trifle different from the mass number for the atom. Because each isotope of an element has a different number of neutrons, each isotope must have a different atomic mass. The **atomic weight** of an element is the weighted average of the atomic masses of the isotopes. The weighted average takes into account the different abundances of the isotopes. For potassium, the three main isotopes (^{39}K, ^{40}K, and ^{41}K) have atomic masses of 38.964, 39.964, and 40.963 and abundances of 93.26%, 0.01%, and 6.73%, respectively. The weighted average of the atomic masses gives the atomic weight of 39.098. These atomic weights are reported on the periodic table of the elements (inside back cover).

The atomic weight of an element depends on the isotopic composition of the sample being considered. The Commission on Isotopic Abundances and Atomic Weights (Wieser and Berglund, 2009) periodically reviews these values and recommends revisions based both on refinements of the atomic masses of the isotopes and isotopic abundances. The isotopic abundances used by the

Table 3.1 The Principal Atomic Particles

Particle	Electric Charge	Atomic Mass Units[a]
Proton	+1	1.00728
Neutron	0	1.00867
Electron	−1	0.00055

[a] 1 AMU = 1/12 the weight of a ^{12}C atom.

commission are based on material likely to be available in the laboratory or used for industry and explicitly are **not** based on the composition of the Earth or its crust or any other terrestrial material. The actual isotopic composition and atomic weight of any element found in rocks and minerals may differ somewhat from the "standard" values reported by the commission. For many purposes, the variations are not significant; in some cases, however, variations in the isotopic compositions of some elements in minerals are important and can provide very useful geologic information. An obvious example is radiometric dating, in which radioactive decay of isotopes of uranium to isotopes of lead, for example, changes the isotopic composition of those elements in a mineral and allows the time of crystallization of the mineral to be determined.

Only 83 chemical elements are available to make minerals. A total of 112 different chemical elements have been recognized by the International Union of Pure and Applied Chemistry. Additional chemical elements with atomic numbers greater than 112 have been reported in the literature but are not yet formally recognized. Of the 112 elements, only numbers 94 and below occur naturally; numbers 95 and above are made only in nuclear laboratories. However, 11 of the 94 naturally occurring elements are **geologically ephemeral elements**. These elements are technetium (Tc, $Z = 43$), promethium (Pm, $Z = 61$), polonium (Po, $Z = 84$), astatine (At, $Z = 85$), radon (Rn, $Z = 86$), francium (Fr, $Z = 87$), radium (Ra, $Z = 88$), actinium (Ac, $Z = 89$), protactinium (Pa, $Z = 91$), neptunium (Np, $Z = 93$), and plutonium (Pu, $Z = 94$). They occur only in very small amounts as short-lived radioactive isotopes produced by neutron capture or radioactive decay involving other elements; subsequently, they decay rapidly to yet other elements down the decay series. Because the radioactive decay and nuclear processes that make and subsequently destroy these elements is continuous, some of these elements are always present, but only in minute quantities.

The rarest of these geologically ephemeral elements is astatine, whose isotopes are produced in the decay series of both uranium and thorium. Decay of uranium in a mineral such as uraninite (UO_2), yields atoms of astatine, which then promptly radioactively decay to form yet other chemical elements down the decay chain. The longest-lived isotope of astatine has a half-life of 56 seconds. It has been estimated that the total amount of natural astatine present in the Earth's crust at any one time is about 30 grams, or about a teaspoonful.

Ephemeral elements with long half-lives include radium ($Z = 88$) and protactinium ($Z = 91$). The isotope ^{226}Ra, produced by ^{238}U decay, has a half-life of 1602 years, so small amounts are always present in uranium-bearing minerals. Another isotope, ^{231}Pa, produced by decay of ^{235}U, has a half-life of 32,760 years, so it could potentially be a primary chemical element in a mineral. No mineral with Pa as an essential chemical constituent has been identified, so Pa is grouped here with the geologically ephemeral elements.

It is worth noting that lead ($Z = 82$) is the element with the highest atomic number to have stable isotopes. Of the elements with atomic numbers greater than 82, only bismuth ($Z = 83$), thorium ($Z = 90$), and uranium ($Z = 92$) have long-lived isotopes that become normal constituents of minerals. Long considered to be a stable isotope, ^{209}Bi has recently been found to have a half-life of 1.9×10^{19} years.

Electrons

Electrons orbit about the nucleus. In an uncharged atom, the number of electrons is equal to the number of protons. The electrons do not orbit the nucleus randomly; rather, they are systematically organized into discrete energy levels identified with four different quantum numbers: n, l, m_l, and m_s (Table 3.2). The **Pauli exclusion principle** informs us that no two electrons in an atom can have the same four quantum numbers. The four quantum numbers are something like an address, equivalent to city, street, building, and room. No two electrons can, in effect, occupy the same room.

The **principal quantum number** n can have any positive integer value. The energy of an electron principally depends on n; higher n means higher energy. Because higher energy is associated with greater distance from the nucleus, the principal quantum numbers are correlated with **shells** that traditionally are identified with letters (K, L, M, N, etc.)

$$n \quad 1 \quad 2 \quad 3 \quad 4 \ldots$$
$$\text{Shell} \quad K \quad L \quad M \quad N \ldots$$

Table 3.2 Relationship among Quantum Numbers[a]

Principal Quantum Number		Angular Momentum Quantum Number		Magnetic Quantum Numbers	Spin Quantum Number	
n	Shell	l	Subshell	m_l	m_s	Number of Electrons
1	K	0	s	0	$+\frac{1}{2}$	2
					$-\frac{1}{2}$	
2	L	0	s	0	$+\frac{1}{2}$	2
					$-\frac{1}{2}$	
		1	p	−1	$+\frac{1}{2}$	6
					$-\frac{1}{2}$	
				0	$+\frac{1}{2}$	
					$-\frac{1}{2}$	
				+1	$+\frac{1}{2}$	
					$-\frac{1}{2}$	

[a]Only the K and L shells ($n = 1$ and 2) are shown.

The **angular momentum quantum number** l has values between 0 and $n - 1$ and distinguishes **subshells** with different shapes (Figure 3.1). Although an electron can, in principal, be in any location around the nucleus of an atom, on a statistical basis it will spend most of its time within the volume represented by its subshell. The subshells are conventionally identified with letters (s, p, d, and f).

$$l \quad\quad 0 \quad 1 \quad 2 \quad 3$$

$$\text{Subshell} \quad s \quad p \quad d \quad f$$

The maximum subshell number within a shell is limited to $n - 1$. The K shell ($n = 1$) has only the s subshell ($l = 0$), the L shell ($n = 2$) can have only s and p subshells ($l = 1$), and so forth. The subshells are commonly identified with the principal quantum number of the shell and the letter associated with the subshell. For example, the p subshell in the L shell ($n = 2$) is the $2p$ subshell.

The **magnetic quantum number** m_l may have integer values between $-l$ and $+l$. It distinguishes among different **orbitals** with different orientations within a subshell. The number of orbitals within a subshell is $2l + 1$. For the s subshell $l = 0$, so $m_l = 0$ and the number of orbitals is 1. Because only one orientation is possible, the orbital is spherical (Figure 3.1a). In the p subshell $l = 1$, so the magnetic quantum numbers m_l are -1, 0, and $+1$. These are associated with three different bilobate (dumbell-shaped) orbitals (Figure 3.1b) that lie along three mutually perpendicular axes. In the d subshell $l = 2$, so there are five orbitals (Figure 3.1c) distinguished with magnetic quantum numbers -2, -1, 0, $+1$, and $+2$. Four of the five orbitals within

the d subshell are quadralobate; the fifth is bilobate with a torus in a plane at right angles. The f orbitals have more complex geometries, which are not readily illustrated.

Each orbital can contain two electrons that are distinguished by their **spin quantum number** m_s. The possible values of m_s are $+\frac{1}{2}$ and $-\frac{1}{2}$. Electrons behave as though they were spinning on an axis, so the spin quantum numbers are generally taken to indicate a spin to the right and left, respectively. Because magnetic fields are generated by the movement of electric charge, each spinning negatively charged electron behaves like a simple magnet with north and south poles. If electron spins are balanced so that the number of right and left spins is the same, no net magnetic moment is formed. However, if electron spins are not balanced, a net magnetic moment may result, as is seen in metallic iron and the mineral magnetite.

The relative energy of the different orbitals is shown diagrammatically in Figure 3.2. Note that the energy level is not shown to scale. The energy difference between the inner orbitals ($1s$ vs. $2s$, for example) is large compared to the energy difference between the outer orbitals. Note also that the range of energy levels found in the subshells of the different shells overlaps. The $4s$ orbital has lower energy than the $3d$ orbital, for example. The general progression of increasing energy level is

$$1s\,2s\,2p\,3s\,3p\,4s\,3d\,4p\,5s\,4d\,5p\,6s\,4f\,5d\,6p\,7s\,6d\,5f.$$

The electron configuration of uncharged atoms of the first 92 chemical elements is shown in Table 3.3. Note that with increasing atomic number, electrons systematically

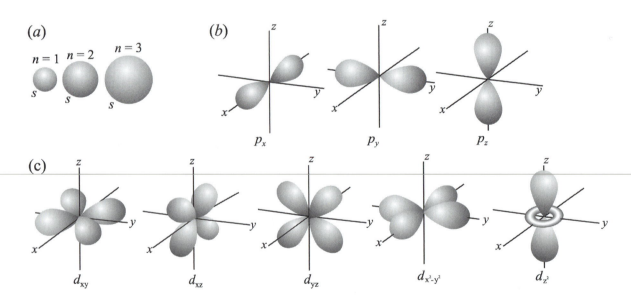

Figure 3.1 Geometry of orbitals in the s, p, and d subshells. Orbitals represent the volume of space around a nucleus in which an electron is most probably located. (*a*) Orbitals in the s subshells are spherical in all shells. (*b*) The p subshell contains three different bilobate orbitals oriented along orthogonal x, y, and z axes. (*c*) The d subshell contains five orbitals. The d_{xy}, d_{xz}, and d_{yz} orbitals are quadralobate and lie in the xy, xz, and yz planes, respectively, so that they bisect the angles between the orthogonal axes. The $d_{x^2-y^2}$ orbital forms a quadralobate shape with lobes aligned along the x and y axes. The d_{z^2} orbital forms a torus with a bilobate shape aligned along the z axis.

fill orbitals in order of energy level. The 1s orbital must be filled before electrons may occupy the 2s orbital, which must be filled before electrons may occupy the 2p orbitals, and so forth. Similarly, the 4s subshell has lower energy than the 3d subshell, so the 4s is filled first.

As every chemistry student learns, the elements are conveniently arranged in the periodic table, in a pattern reflecting the order in which electrons fill the orbitals in the ground state of uncharged atoms of the elements. Elements in column 1 have one electron in the outer s subshell, those in column 2 have both electron positions in the outer s subshell filled, and so forth. The elements at the far right side of the table (column 18) have all of their subshells filled (Table 3.3) and are the chemically unreactive **noble gasses**.

Table 3.3 shows that the electron configuration of elements exclusive of the noble gasses consists of an inner shell configuration, called a core, in which all the orbital positions in each shell (K, L, M, etc.) are entirely filled. Additional electrons called **valence electrons** occupy subshells within shells that are not entirely filled. If the core is identical to the electron configuration of a noble gas, it is called a **noble-gas core**. For example, Na has a Ne noble-gas core plus a single 3s valence electron, and Ca has an Ar noble-gas core plus two 4s valence electrons. A **pseudo-noble-gas core** consists of a noble gas core plus an entirely filled d and/or f subshell. Arsenic (As), for example, has a pseudo–noble gas core consisting of entirely filled 1s, 2s, 2p, 3s, 3p, 4s, and 3d subshells, plus three 4p valence electrons. Not surprisingly, the noble gasses have noble-gas cores and no valence electrons.

Formation of Ions

Ions are atoms that have an excess or deficiency of electrons compared to the number of protons in the nucleus. **Anions** have a net negative charge because they have more electrons than protons; **cations** have a net positive charge because they have fewer electrons than protons. The charge of an ion is known as its **valence** or **oxidation state**.

Whether an element will form an anion or cation can usually be inferred from the configuration of the valence electrons. Elements will tend to gain or lose electrons to acquire the stable electron configuration displayed by the noble gasses.

Neutral atoms of the metals have noble gas or pseudo–noble gas cores and one or more valence electrons. Little energy is required to strip away the valence electrons to acquire the electron configuration of a noble gas, so the metals typically form cations.

Neutral atoms of the nonmetals, found on the right side of the periodic chart, have valence subshells that need only a few electrons to be filled. Hence the nonmetals have a strong affinity for electrons to fill the outer subshells and typically form anions.

Figure 3.3 shows the common oxidation states of the first 38 elements. For elements 1–20 and 31–38, the common oxidation states have a systematic pattern and reflect the pattern of filling the orbitals in the s and p subshells. For example, the increase in oxidation state from Li^+ to N^{5+} is caused by the loss of electrons from the 2s and 2p subshells required to achieve the filled-shell configuration of He. In contrast, O^{2-} and F^{1-} gain electrons to fill the remaining sites in the 2p subshell and thereby acquire the stable electron configuration of Ne.

The systematic pattern breaks down for the transition metals (elements 21–30). This is because the valence electrons are in the 4s and 3d subshells. With the exception of Cr, the 4s subshell fills first, then the 3d. When electrons are removed, however, they come first from the subshell with the highest quantum number, which is the 4s subshell, even though this subshell has lower energy.

For Sc, Ti, and V, it is possible to remove both 4s electrons and one, two, and three electrons, respectively, from the 3d subshell to produce Sc^{3+}, Ti^{4+}, and V^{5+}, all with the electron configuration of Ar.

For atomic numbers 24–30 (Cr–Zn), both 4s and 3d electrons are available to be removed. When these elements ionize they typically lose their 4s electrons first. One or more of the 3d electrons also may be lost. As a consequence, all of these elements may have different valence states depending on how many 3d electrons are removed. Iron is the most notable example. It occurs both as divalent Fe^{2+} known as **ferrous iron**, and as trivalent Fe^{3+}, known as **ferric iron**. Although Mn and Cr are most commonly

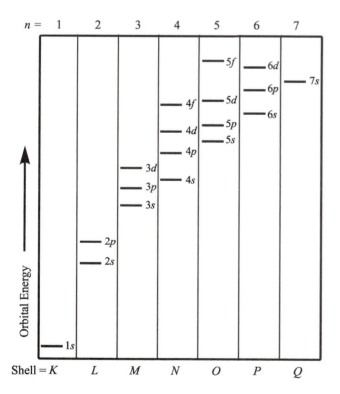

Figure 3.2 Relative energy level of electron subshells. Electrons fill subshells in order of increasing energy level (compare Table 3.3).

Table 3.3 Electron Configuration of Uncharged Atoms[a]

Atomic Number	Chemical Element	Shell/Subshell																			
		K	L		M			N				O					P			Q
		1s	2s	2p	3s	3p	3d	4s	4p	4d	4f	5s	5p	5d	5f	5g	6s	6p	6d	7s
1	H	1																		
2	He	2																		
3	Li	2	1																	
4	Be	2	2																	
5	B	2	2	1																
6	C	2	2	2																
7	N	2	2	3																
8	O	2	2	4																
9	F	2	2	5																
10	Ne	2	2	6																
11	Na	2	2	6	1															
12	Mg	2	2	6	2															
13	Al	2	2	6	2	1														
14	Si	2	2	6	2	2														
15	P	2	2	6	2	3														
16	S	2	2	6	2	4														
17	Cl	2	2	6	2	5														
18	Ar	2	2	6	2	6														
19	K	2	2	6	2	6		1												
20	Ca	2	2	6	2	6		2												
21	Sc	2	2	6	2	6	1	2												
22	Ti	2	2	6	2	6	2	2												
23	V	2	2	6	2	6	3	2												
24	Cr	2	2	6	2	6	5	1												
25	Mn	2	2	6	2	6	5	2												
26	Fe	2	2	6	2	6	6	2												
27	Co	2	2	6	2	6	7	2												
28	Ni	2	2	6	2	6	8	2												
29	Cu	2	2	6	2	6	10	1												
30	Zn	2	2	6	2	6	10	2												
31	Ga	2	2	6	2	6	10	2	1											
32	Ge	2	2	6	2	6	10	2	2											
33	As	2	2	6	2	6	10	2	3											
34	Se	2	2	6	2	6	10	2	4											

(Continued)

Table 3.3 Continued

Atomic Number	Chemical Element	Shell/Subshell																	
		K	L		M			N				O					P			Q
		1s	2s	2p	3s	3p	3d	4s	4p	4d	4f	5s	5p	5d	5f	5g	6s	6p	6d	7s
35	Br	2	2	6	2	6	10	2	5											
36	Kr	2	2	6	2	6	10	2	6											
37	Rb	2	2	6	2	6	10	2	6			1								
38	Sr	2	2	6	2	6	10	2	6			2								
39	Y	2	2	6	2	6	10	2	6	1		2								
40	Zr	2	2	6	2	6	10	2	6	2		2								
41	Nb	2	2	6	2	6	10	2	6	4		1								
42	Mo	2	2	6	2	6	10	2	6	5		1								
43	Tc	2	2	6	2	6	10	2	6	5		2								
44	Ru	2	2	6	2	6	10	2	6	7		1								
45	Rh	2	2	6	2	6	10	2	6	8		1								
46	Pd	2	2	6	2	6	10	2	6	10										
47	Ag	2	2	6	2	6	10	2	6	10		1								
48	Cd	2	2	6	2	6	10	2	6	10		2								
49	In	2	2	6	2	6	10	2	6	10		2	1							
50	Sn	2	2	6	2	6	10	2	6	10		2	2							
51	Sb	2	2	6	2	6	10	2	6	10		2	3							
52	Te	2	2	6	2	6	10	2	6	10		2	4							
53	I	2	2	6	2	6	10	2	6	10		2	5							
54	Xe	2	2	6	2	6	10	2	6	10		2	6							
55	Cs	2	2	6	2	6	10	2	6	10		2	6				1			
56	Ba	2	2	6	2	6	10	2	6	10		2	6				2			
57	La	2	2	6	2	6	10	2	6	10		2	6	1			2			
58	Ce	2	2	6	2	6	10	2	6	10	1	2	6	1			2			
59	Pr	2	2	6	2	6	10	2	6	10	2	2	6	1			2			
60	Nd	2	2	6	2	6	10	2	6	10	3	2	6	1			2			
61	Pm	2	2	6	2	6	10	2	6	10	4	2	6	1			2			
62	Sm	2	2	6	2	6	10	2	6	10	5	2	6	1			2			
63	Eu	2	2	6	2	6	10	2	6	10	6	2	6	1			2			
64	Gd	2	2	6	2	6	10	2	6	10	7	2	6	1			2			
65	Tb	2	2	6	2	6	10	2	6	10	8	2	6	1			2			
66	Dy	2	2	6	2	6	10	2	6	10	9	2	6	1			2			
67	Ho	2	2	6	2	6	10	2	6	10	10	2	6	1			2			
68	Er	2	2	6	2	6	10	2	6	10	11	2	6	1			2			

(Continued)

Table 3.3 Continued

Atomic Number	Chemical Element	K	L		M			N				O					P			Q
		1s	2s	2p	3s	3p	3d	4s	4p	4d	4f	5s	5p	5d	5f	5g	6s	6p	6d	7s
69	Tm	2	2	6	2	6	10	2	6	10	12	2	6	1			2			
70	Yb	2	2	6	2	6	10	2	6	10	13	2	6	1			2			
71	Lu	2	2	6	2	6	10	2	6	10	14	2	6	1			2			
72	Hf	2	2	6	2	6	10	2	6	10	14	2	6	2			2			
73	Ta	2	2	6	2	6	10	2	6	10	14	2	6	3			2			
74	W	2	2	6	2	6	10	2	6	10	14	2	6	4			2			
75	Re	2	2	6	2	6	10	2	6	10	14	2	6	5			2			
76	Os	2	2	6	2	6	10	2	6	10	14	2	6	6			2			
77	Ir	2	2	6	2	6	10	2	6	10	14	2	6	7			2			
78	Pt	2	2	6	2	6	10	2	6	10	14	2	6	9			1			
79	Au	2	2	6	2	6	10	2	6	10	14	2	6	10			1			
80	Hg	2	2	6	2	6	10	2	6	10	14	2	6	10			2			
81	Tl	2	2	6	2	6	10	2	6	10	14	2	6	10			2	1		
82	Pb	2	2	6	2	6	10	2	6	10	14	2	6	10			2	2		
83	Bi	2	2	6	2	6	10	2	6	10	14	2	6	10			2	3		
84	Po	2	2	6	2	6	10	2	6	10	14	2	6	10			2	4		
85	At	2	2	6	2	6	10	2	6	10	14	2	6	10			2	5		
86	*Rn*	*2*	*2*	*6*	*2*	*6*	*10*	*2*	*6*	*10*	*14*	*2*	*6*	*10*			*2*	*6*		
87	Fr	2	2	6	2	6	10	2	6	10	14	2	6	10			2	6		1
88	Ra	2	2	6	2	6	10	2	6	10	14	2	6	10			2	6		2
89	Ac	2	2	6	2	6	10	2	6	10	14	2	6	10			2	6	1	2
90	Th	2	2	6	2	6	10	2	6	10	14	2	6	10			2	6	2	2
91	Pa	2	2	6	2	6	10	2	6	10	14	2	6	10	2		2	6	1	2
92	U	2	2	6	2	6	10	2	6	10	14	2	6	10	3		2	6	1	2

[a]The noble gasses are indicated in bold italics. The noble-gas core of elements is shown with light shading. The pseudo-noble-gas core of elements is shown with darker shading.

divalent and trivalent, respectively, both may occur in higher oxidation states.

A measure of the propensity of an element to gain or lose electrons is conveniently provided by a value called **electronegativity**. It was defined in 1932 by Linus Pauling, who used an arbitrary scale so that lithium (Li) had an electronegativity of 1, carbon (C) an electronegativity of 2.5, and fluorine (F) an electronegativity of 4.0. Elements with low electronegativity lose their outer valence electrons readily to form cations, whereas those with high electronegativity have a strong affinity for extra electrons and tend to form anions. Table 3.4 shows electronegativity values from Pauling (1960) for the first 38 elements.

At least four other electronegativity scales have been developed, but they all yield similar results. In a later section, electronegativity values are used to estimate the nature of chemical bonds. Pauling's values are more than adequate for this purpose.

ABUNDANCE OF THE ELEMENTS

Various attempts have been made to estimate the relative abundance of the chemical elements in the crust. The most influential is a compilation by Clarke and Washington

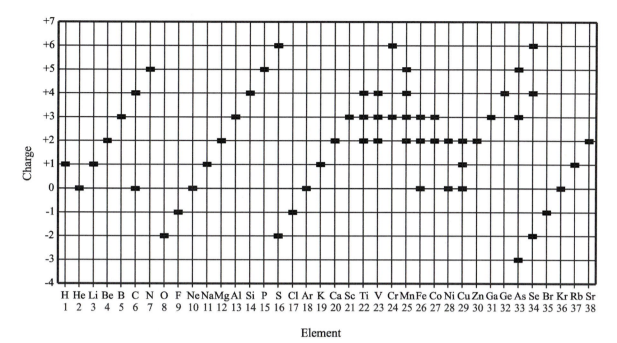

Figure 3.3 Common oxidation states of elements 1 through 38.

Table 3.4 Electronegativity of the First 38 Elements in Their Most Common Oxidation State(s)

Z	Element	Electronegativity	Z	Element	Electronegativity	Z	Element	Electronegativity
1	H	2.1	14	Si	1.8	27	Co	1.8
2	He	N/A	15	P	2.1	28	Ni	1.8
3	Li	1.0	16	S	2.5	29	Cu^+/Cu^{2+}	1.9/2.0
4	Be	1.5	17	Cl	3.0	30	Zn	1.6
5	B	2.0	18	Ar	N/A	31	Ga	1.6
6	C	2.5	19	K	0.8	32	Ge	1.8
7	N	3.0	20	Ca	1.0	33	As	2.0
8	O	3.5	21	Sc	1.3	34	Se	2.4
9	F	4.0	22	Ti	1.5	35	Br	2.8
10	Ne	N/A	23	V	1.6	36	Kr	N/A
11	Na	0.9	24	Cr	1.6	37	Rb	0.8
12	Mg	1.2	25	Mn	1.5	38	Sr	1.0
13	Al	1.5	26	Fe^{2+}/Fe^{3+}	1.8/1.9			

Source: Adapted from Pauling (1960). Bloch and Schatteman (1981) have somewhat different values.

(1924) who calculated an average composition based on 5159 high-quality analyses of igneous rocks. The analyzed rocks came from worldwide locations but were dominated by European and North American sources and included few samples representative of the ocean basins. With some revision and the addition of minor elements (e.g., Taylor, 1969; Mason and Moore, 1982), their values have provided the basis for the abundances shown in Table 3.5. Of the 83 available elements, only eight are present in substantial amounts (Table 3.6). These eight elements, O, Si, Al, Fe, Ca, Na, K, and Mg, collectively comprise the large majority of the Earth's crust, and are the elements from which most common minerals are composed.

The oceanic crust is somewhat richer in iron and magnesium than the average crust, and these values also are shown in Table 3.6. Additional elements present in the oceanic crust in significant abundance include Mn (1.0%) and Ti (0.9%).

Determining the composition of the entire Earth is even more difficult because the mantle and core cannot be sampled directly. Estimates are obtained by

- considering the Earth's mass and density distribution as determined by geophysical means,

Table 3.5 Abundance of the Elements in the Earth's Crust[a]

Element	Abundance	Element	Abundance	Element	Abundance	Element	Abundance	Element	Abundance
1 H	**0.14**	20 Ca	**3.6**	39 Y	33	58 Ce	60	77 Ir	0.001
2 He	??	21 Sc	22	40 Zr	165	59 Pr	8.2	78 Pt	0.01
3 Li	20	22 Ti	**0.4**	41 Nb	20	60 Nd	28	79 Au	0.004
4 Be	2.8	23 V	135	42 Mo	1.5	61 Pm	??	80 Hg	0.08
5 B	10	24 Cr	100	43 Tc	??	62 Sm	6	81 Tl	0.5
6 C	200	25 Mn	**0.1**	44 Ru	0.01	63 Eu	1.2	82 Pb	13
7 N	20	26 Fe	**5.0**	45 Rh	0.005	64 Gd	5.4	83 Bi	0.2
8 O	**46.6**	27 Co	25	46 Pd	0.01	65 Tb	0.9	84 Po	??
9 F	625	28 Ni	75	47 Ag	0.07	66 Dy	3	85 At	??
10 Ne	??	29 Cu	55	48 Cd	0.2	67 Ho	1.2	86 Rn	??
11 Na	**2.8**	30 Zn	70	49 In	0.1	68 Er	2.8	87 Fr	??
12 Mg	**2.1**	31 Ga	15	50 Sn	2	69 Tm	0.5	88 Ra	??
13 Al	**8.1**	32 Ge	1.5	51 Sb	0.2	70 Yb	3.4	89 Ac	??
14 Si	**27.7**	33 As	1.8	52 Te	0.01	71 Lu	0.5	90 Th	7.2
15 P	**0.1**	34 Se	0.1	53 I	0.5	72 Hf	3	91 Pa	??
16 S	260	35 Br	2.5	54 Xe	??	73 Ta	2	92 U	1.8
17 Cl	130	36 Kr	??	55 Cs	3	74 W	1.5	93 Np	??
18 Ar	??	37 Rb	90	56 Ba	**0.04**	75 Re	0.001	94 Pu	??
19 K	**2.6**	38 Sr	375	57 La	30	76 Os	0.005		

Source: Adapted from Mason and Moore (1982). [a]Bold numbers are weight percent; others are parts per million.

Table 3.6 The Eight Most Abundant Elements in the Earth's Crust

Element	Common Oxidation State	Proportion of Earth's Crust			Oceanic Crust (wt %)	Total Earth (wt %)
		(wt %)	(at %)	(vol %)		
O	−2	46.6	62.5	91.7	40.9	29.5
Si	+4	27.7	21.2	0.2	23.1	15.2
Al	+3	8.1	6.5	0.5	8.5	1.1
Fe	+2/+3	5.0	1.9	0.5	8.2	34.6
Ca	+2	3.6	1.9	1.5	8.1	1.1
Na	+1	2.8	2.6	2.2	2.1	0.6
K	+1	2.6	1.4	3.1	1.3	0.1
Mg	+2	2.1	1.8	0.4	4.6	12.7
Total		98.5	99.8	~100	96.8	94.9

Source: Adapted from Mason and Moore (1982).

- studying the composition of the usually basaltic magmas derived from the mantle and the samples of the mantle that sometimes arrive with the magma,
- analyzing the composition of meteorites that are presumed to, in some way, represent the material from which the Earth accreted, and
- applying appropriate cosmological, geochemical, and petrophysical models to evaluate these data.

The values shown in Table 3.6 are approximations at best, but they clearly illustrate that the entire Earth is much richer in iron and magnesium than the crust, and lower in silicon, aluminum, potassium, sodium, and calcium. The Earth as a whole also is presumed to have substantially more nickel and sulfur than the crust.

CHEMICAL BONDING

The chemical bonds that hold the atoms that form minerals together can be grouped into two categories: bonds that involve valence electrons and bonds that do not. The bonds that involve valence electrons include ionic, covalent, and metallic bonds. Bonds that do not involve valence electrons include van der Waals and hydrogen bonding.

Valence-Related Bonding

Of the 83 available elements, all but the noble gasses (He, Ne, Ar, Kr, Xe, and Rn) routinely bond chemically with themselves or other elements. The feature that the noble gasses have in common is that all subshells are entirely filled with electrons—they lack valence electrons (Table 3.3). The lack of chemical reactivity suggests that the electron configuration displayed by the noble gasses is a low-energy or stable configuration. If the other elements were to acquire an electron configuration the same as one of the noble gasses by gaining, losing, or sharing valence electrons, they too would be in a lower-energy, more stable configuration. All the valence-related bond types involve mechanisms that allow elements to acquire the electron configuration of the noble gasses, or failing that, a pseudo-noble-gas configuration.

Ionic Bonding

An **ionic bond** is a chemical bond formed by the electrostatic attraction between positive and negative ions. Consider the mineral halite (NaCl). The electron configuration of a Na atom shown in Table 3.3 has the noble-gas core of Ne with a single electron in the $3s$ subshell. If it loses that $3s$ electron it acquires the stable electron configuration of Ne and a net charge of +1. But the Na cannot simply discard the extra electron—charge neutrality must be maintained That electron must go somewhere. That "somewhere" is a Cl atom, which has the $1s$, $2s$, $2p$, and $3s$ orbitals completely filled but only five electrons in the $3p$ subshell. The electron lost by the Na is readily accepted by the Cl so that it can have a full complement of six $3p$ electrons. The addition of an electron to the $3p$ orbital gives Cl the electron configuration of argon, and it also acquires a net −1 charge. The exchange of an electron from sodium to chlorine produces a stable electron configuration for both elements, and the sodium and chlorine wind up with opposite charges and are attracted to each other.

The nature of the attractive force F_a between the oppositely charged Na$^+$ and Cl$^-$ is given by the **Coulomb law** and is inversely proportional to the distance d between the center of the ions

$$F_A \propto (q^+)(q^-)/d^2 \qquad (3.1)$$

where q^+ and q^- are the magnitudes of the charges on cation and ion, respectively. Figure 3.4a shows the magnitude of the attraction between Na$^+$ and Cl$^-$. This attraction is balanced by a **Born repulsion** produced when the electron shells of anion and cation begin to overlap. The magnitude of the Born repulsive force F_R is a negative inverse function of the distance between anion and cation

$$F_R \propto -n/d^{1+n} \qquad (3.2)$$

where n is an integer dependent on the number of filled electron shells. The repulsion exists because the electrons in the stable shells of both elements have the same charge and cannot occupy the same space. Distortion of the electron subshells involves an energy increase and therefore resistance. The magnitude of F_R for the Na$^+$–Cl$^-$ bond is shown by the lower curve in Figure 3.4a.

The net force F is the sum of the attractive and repulsive forces

$$F = F_A + F_R \qquad (3.3)$$

(a)

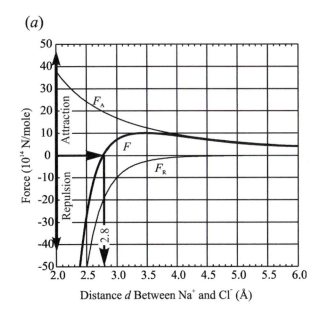

(b)

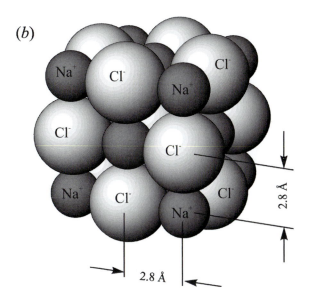

Figure 3.4 Ionic bonding between Na$^+$ and Cl$^-$. (*a*) The attractive F_A and repulsive F_R forces between Na$^+$ and Cl$^-$ ions. The net force F is zero when the centers of the ions are at a distance of about 2.8 Å. At greater distances there is a net attraction and at a smaller distance a net repulsion. (*b*) Na$^+$ and Cl$^-$ ions pack together in a face-centered cubic lattice with a spacing of about 2.8 Å.

The equilibrium distance between Na^+ and Cl^- is given where $F = 0$, which occurs at about 2.8 Å as shown on Figure 3.4a. At greater distances, the attractive force is larger, and at smaller distances, the repulsive force is larger.

Ions bond together in ratios that ensure that the net positive charge and the net negative charge are equal. Charges must balance. For halite, this means one Na^+ for every Cl^- and the formula is written NaCl. In the mineral fluorite there must be two F^- for every Ca^{2+}, and the formula is written CaF_2.

When bonding, ions act like charged spheres and pack together in a systematic and symmetrical manner. Positive and negative charges alternate to form an electrically neutral crystalline solid. This represents a low-energy configuration, so ionic bonds are fairly strong. Na and Cl pack together in a face-centered cubic lattice (Figure 3.4b) in which each Na is surrounded by six Cl and each Cl is surrounded by six Na.

Ionic-bonded crystals tend to be brittle. Attempts to deform the crystal structure and slide one part of the crystal past another will juxtapose cations against cations and anions against anions. Because like charges repel each other, ionic crystals have a strong resistance to sliding different parts of the crystal past each other. Forcing the issue may result in rupture rather than ductile deformation. Because the structures are often quite orderly, failure commonly occurs along cleavage planes.

Covalent Bonding

Covalent bonds are chemical bonds formed by the sharing of electrons between atoms. Covalent bonding occurs when the orbitals of two atoms overlap. Provided the overlapping orbitals have no more than two electrons combined, the electrons in the orbitals begin to move about both atoms. Because the electrons are attracted to the nuclei of both atoms, a bond is established. The strength of covalent bonds is a function of the degree to which the orbitals of adjacent atoms overlap—more overlap yields stronger bonds. With diamond, the high degree of overlap produces very strong bonds, which is reflected in its great hardness.

Diamond is composed of carbon that has a base-state electron configuration of $1s^2 2s^2 2p^2$ (the superscript indicates the number of electrons in the subshell). A stable noble-gas electron configuration can be obtained by either gaining four electrons to form a Ne configuration or losing four electrons to form a He configuration (Table 3.3). However, ionic bonding is not possible because all the carbon atoms have identical electron configurations and identical affinity for valence electrons. One carbon cannot steal an electron from another to form negative and positive ions. Adjacent carbon atoms can, however, share electrons so that each acquires a stable noble-gas electron configuration.

To share four electrons, four unpaired orbitals must be available. This poses a problem because in the ground-state configuration of carbon (Figure 3.5a), only the three $2p$ orbitals are available to share electrons; two have one electron each, and the third has none. The $1s$ and $2s$ orbitals cannot share electrons because each already contains two. To create four unpaired orbitals, **hybrid orbitals** are formed as follows.

- One of the electrons in the $2s$ orbital is promoted to the vacant $2p$ orbital; this provides four

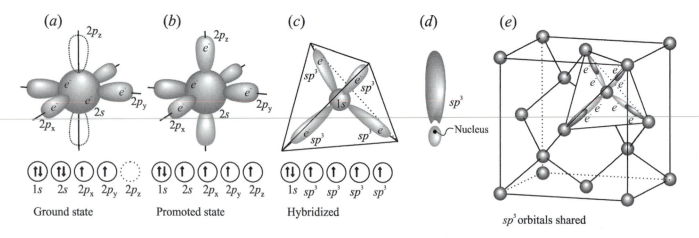

Figure 3.5 Formation of covalent sigma (σ) bonds with carbon in diamond. (*a*) Ground-state electron configuration of carbon. The $2s$ orbital contains two electrons, the $2p_x$ and $2p_y$ orbitals each contain one electron, and the $2p_z$ orbital is vacant. (*b*) One electron from the $2s$ orbital is promoted to the $2p_z$ orbital so that four orbitals are partially filled. (*c*) Hybridization of the half-filled orbitals into four identical sp^3 orbitals arrayed to point to the corners of a tetrahedron. (*d*) Each sp^3 orbital consists of a large lobe and a small lobe on the opposite side of the nucleus. The small lobes are not shown in (*c*). (*e*) Bonding of carbon atoms in diamond by sharing electrons between hybridized $2sp^3$ orbitals on adjacent atoms. Four pairs of overlapping sp^3 orbitals reaching a carbon atom are shown; similar bonds occur along each of the solid lines. The small lobes are omitted for clarity.

unpaired orbitals that can therefore be shared (Figure 3.5b).

- Because all four bonds to adjacent carbon atoms must be identical, the orbitals in the $2s$ and $2p$ subshells are hybridized into four identical sp^3 orbitals (Figure 3.5c). Each of these orbitals consists of a large lobe pointing in one direction and a small lobe in the opposite direction (Figure 3.5d). They are positioned around the nucleus so that the large lobes define the corners on a tetrahedron.

Each of the four sp^3 hybrid orbitals is shared with an identical hybrid orbital on each of four adjacent carbon atoms to form the continuous crystal structure of diamond (Figure 3.5e). Each of these covalent bonds, termed **sigma bonds** (σ), has a high degree of symmetry about an axis parallel to the length of the orbitals that overlap end to end.

Graphite displays a different pattern of hybrid orbitals. It too is made entirely of carbon, but only $2p_x$ and $2p_y$ orbitals hybidize with the $2s$ orbital to form sp^2 hybrids. The $2p_z$ orbital is not hybridized. The carbon atoms bond to three neighbors by using the three sp^2 hybrid orbitals forming continuous sheets of carbon (Figure 3.6). These three bonds are sigma bonds. The $2p_z$ orbitals, oriented at right angles to the plane of the carbon sheets, share electrons differently and form what is known as **pi bonds** (π). With pi bonds, the orbitals overlap side to side rather than end to end, so the electrons are shared laterally.

Note that each carbon atom can form pi bonds with each of its three neighbors, forming what amounts to continuous bands of electrons from the $2p_z$ orbitals above and below the sheet of carbon atoms. The electrons involved in these pi bonds are able to move from atom to atom with relative ease, effectively forming metallic bonds, which are described in the next subsection. This accounts for the fact that graphite is an electrical conductor. An applied voltage can cause these electrons to move, producing a flow of electricity along the sheets. The pi and sigma bonding within the sheets of carbon is very strong, stronger in fact than the bonds in diamond. That this is so is attested by the observation that C-C bonds in graphite are shorter than in diamond. The low hardness of graphite is a consequence of the weak van der Waals bonds that hold adjacent sheets together. Van der Waals bonds are discussed in a following section.

Some additional common hybrid orbital geometries are shown in Figure 3.7. All the conventional (Figure 3.1) and hybrid orbitals have specific spatial geometries. As a consequence, covalent bonding requires that atoms be placed in specific geometries so that the orbitals will overlap. The bonds are highly directional and the structures produced are often not readily described in terms of simple packing of spheres.

Metallic Bonding

Metallic bonds can be considered to be a type of covalent bonding in which the valence electrons are delocalized and are free to move from atom to atom throughout the crystal structure. In the ideal case the metal atoms consist of positively charged spherical cores packed together to form the structure of the solid metal. Valence electrons form a mobile glue that holds it all together.

The formation of metallic bonds depends on several factors. The first is that the valence electrons are held weakly. Metals, which have low electronegativity (Table 3.4), relinquish their valence electrons easily. Second, metallic bonding is favored when the number of electrons that must be shared to form a noble-gas configuration is large. Sodium in its metallic form, for example, needs to share an additional seven electrons to acquire the argon configuration. The nonmetals, in contrast, need to gain only one or two electrons to acquire a noble-gas configuration and therefore typically form covalent or ionic bonds. A third consideration is the availability of vacant energy

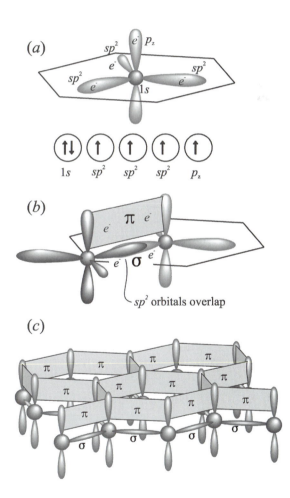

Figure 3.6 Formation of covalent σ and π bonds with carbon in graphite. The ground and promoted states of carbon are shown in Figure 3.5a and b. (a) Hybridization of orbitals forms three sp^2 orbitals with the $2p_z$ orbital at right angles. (b) σ Bonds form by sharing electrons between sp^2 orbitals from adjacent C atoms. π Bonds are formed by sharing electrons laterally between the $2p_z$ orbitals. (c) Carbon bonds to form continuous sheets. The π bonds form on both the top and bottom of the sheet but are shown only on the top.

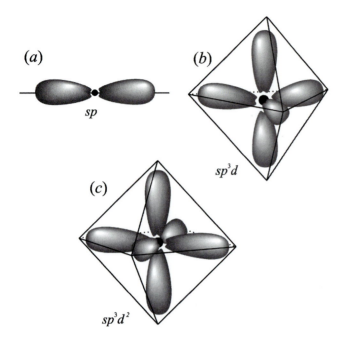

Figure 3.7 Hybrid orbitals. Each consists of a large and small lobe (Figure 3.5d); only the large lobe is shown. (*a*) Hybrid *sp* orbital formed from an *s* orbital and a single *p* orbital. (*b*) Hybrid *sp³d* orbitals formed from an *s* three *p* and a *d* orbital. (*c*) Hybrid *sp³d²* orbitals formed from an *s*, three *p,* and two *d* orbitals.

levels into which valence electrons can readily move. This is controlled, in part, by the spacing between atoms.

Assume that the spacing between identical atoms that make up a crystal is infinite. The energy levels of the electrons within each subshell and orbital would be the same for all atoms. If the atoms are brought into proximity, however, the Pauli exclusion principle conspires to ensure that no two electrons can occupy exactly the same energy state—each electron must have its own energy level or combination of quantum numbers. The result is that the energy level represented by a given subshell (1*s*, 2*s*, 2*p*, etc.) must spread out into a band (Figure 3.8). Given the large number of atoms found even in small crystals (~10²²), the energy levels within each band are essentially continuous.

The width of the energy band produced by each subshell increases as the proximity of the atoms increases. Electrons become free to migrate through the structure if the energy level of an unfilled band overlaps that of a filled band or if the outer valence subshell is only partially filled.

Magnesium illustrates the overlap of an unfilled band with a filled band (Figure 3.8a). When Mg atoms are placed in close proximity, as they are in metallic magnesium, the widths of the energy bands increase so that the 3*s* and 3*p* bands overlap. Because the 3*s* orbitals are filled on all the Mg atoms, every available 3*s* energy level is occupied. The

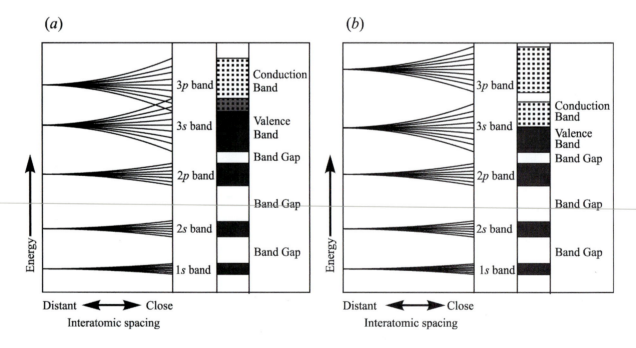

Figure 3.8 Energy bands form when atoms are brought into close proximity. (*a*) Magnesium. Energy levels of the 3*s* and 3*p* subshells spread into bands that overlap. Electrons from the 3*s* subshell (the valence band) can move into unoccupied energy levels of the 3*p* subshell and therefore move freely throughout the structure. (*b*) Sodium. Energy levels of electrons in the 3*s* subshell spread into a band when sodium atoms are packed together to form sodium metal but do not overlap with other bands. However, orbitals in the 3*s* subshell are only half-filled, so electrons can move into vacant energy levels and therefore move freely throughout the structure. The 3*s* subshell serves as both the valence and conduction bands.

$3p$ bands, however, are vacant, so electrons from the $3s$ band are free to move into the $3p$ band, where the abundance of available energy levels allows the electrons to migrate easily through the structure. The $3p$ band is known as the **conduction band** because more energy levels are available in the band than there are electrons (none in isolated Mg atoms), and it can therefore conduct electrons through the crystal if a voltage is applied. The $3s$ band is known as the **valence band** because it provides the electrons that move into the conduction band. The inner subshells do not contribute to electrical conduction because an energy gap, known as the **band gap**, is present between bands.

The situation with Na is slightly different, because its $3s$ subshell is only half occupied. The $3s$ band that forms when Na atoms are brought together to form metallic sodium can therefore serve both as the valence band and the conduction band. Only half of the energy levels in the $3s$ band can be occupied at any one time so the remaining energy levels are available for conduction.

The transition metals, such as iron, utilize both mechanisms to produce metallic bonds. The d subshell typically contains unfilled orbitals, so conduction can occur within the d bands. In addition, the energy level of the d band overlaps the higher s and p bands.

Because metallic bonds do not involve matching specific orbitals on adjacent atoms, the bonds are nondirectional. Metals atoms pack together in a highly symmetrical manner and are held together by weak bonds provided by the valence electrons migrating throughout the structure in the conduction band. Because the bonds are weak, metals tend to be relatively soft and malleable. Further, because the metal atoms have their valence requirements satisfied by the sea of valence electrons, it is very easy for one metal atom to substitute for another in the crystal structure. That is why many different metal alloys, which are mixtures of different metals, can be made and why it is possible to weld together pieces of metal by melting them with a torch.

The width of the energy bands increases as the atoms are brought closer together. This suggests that at sufficiently high pressure, most solids should display metallic conduction as the energy levels of the highest-filled band and next unfilled band spread sufficiently to overlap. The electrical conduction observed in the Earth's deep crust and mantle may, in part, be produced by this mechanism operating in silicate minerals, which at surface pressures would be electrical insulators.

Relation among the Valence-Dependent Bonds

The three valence-dependent bond types—ionic, covalent, and metallic—all depend on acquisition of a stable electron configuration for the atoms involved. It turns out that many chemical bonds have characteristics that are intermediate between the ionic, covalent, and metallic end members. This variation can be schematically illustrated by placing

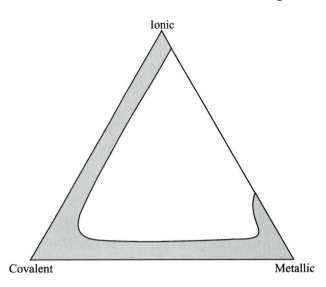

Figure 3.9 Range of bond character (shaded) found in valance-dependent bonding. Continuous variation exists between ionic and covalent bonds and between covalent and metallic bonds. Limited ionic character is found in metallic bonds between different metals.

the three bond types at the corners of a triangle (Figure 3.9). As an approximation, values of electronegativity can be used to estimate the nature of a chemical bond.

Ionic bonds require the formation of cations and anions with opposite charges. Hence, ionic bonding should be expected if the electronegativity of the elements bonding together is quite different. Covalent and metallic bonds, however, involve sharing of electrons among atoms with similar affinity for electrons and therefore a small or zero difference in electronegativity between atoms. Using this logic, Pauling (1960) devised a largely empirical relationship between the difference in electronegativity between elements and the degree of ionic character produced between those elements, shown in Figure 3.10. The mathematical relationship is

$$\text{Percent ionic character} = 1 - e^{-0.25(x_a - x_c)^2} \times 100. \quad (3.4)$$

Of the eight most abundant elements in the crust (Table 3.6), only oxygen forms anions, and it has an electronegativity of 3.5. The other seven elements (Si, Al, Fe, Ca, Na, K, Mg) all form cations when bonding with oxygen and have electronegativities of between 1.8 (Si) and 0.8 (K) (Table 3.4). The electronegativity difference between these cations and oxygen therefore ranges from 1.7 (Si) to 2.7 (K) and, based on Figure 3.10, bonding ranges from about 50% ionic (Si–O) to 80% ionic (K–O). The ionic-bonding model therefore can serve as a reasonable first approximation for most common minerals in which oxygen is the anion.

The native elements, including diamond and graphite (C), sulfur (S), gold (Au), and silver (Ag), form bonds in which the electronegativity difference is zero. These bonds cannot be ionic but may be intermediate in character

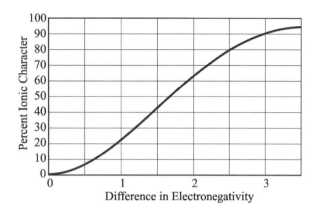

Figure 3.10 Curve showing the relationship between difference in electronegativity and degree of ionic character. See Table 3.4 for values of electronegativity.

between covalent and metallic. A simple number is not available to provide a measure of the relative degree of covalent versus metallic bonding. However, those native elements that have a high degree of metallic bonding (copper, gold, silver, etc.) typically have relatively low electronegativities (1.8–2.4) in addition to overlapping orbital band energies and the other factors discussed earlier. The native nonmetals (S, C) have higher electronegativities (~2.5) and favor covalent bonding.

The difference between an ionic and a covalent bond is fundamentally a statistical question of where the electrons are located. With ideal covalent bonding, there is an equal probability that the valence electrons will be found on either of a sharing pair of atoms. As the degree of ionic character increases, the probability of finding the valence electron(s) in cation orbitals decreases as the anions exercise greater and greater claim on the electrons. With ideal ionic bonding, the valence electrons occupy orbitals on the anions, they do not retain an identity associated with the cation. It should be evident from Figure 3.10 and Equation 3.4 that ideal ionic bonding is not possible because the difference in electronegativity cannot be infinite. Even the strongest ionic bonds retain some covalent character.

As the degree of metallic character increases, the effective number of atoms over which valence electrons are shared increases from two atoms for covalent bonding, to the number of atoms in the crystal for metallic bonding. A limited range of bond properties extends from metallic toward ionic, usually in alloys of different metallic elements. If the two metals that form an alloy have significantly different electronegativities, the element with the higher affinity for electrons will have a higher claim on some valence electrons.

Crystals with ionic and covalent bonding have low electrical conductivity because the valence electrons are held tightly in specific orbitals. The electrical conductivity typically increases if these crystals are heated because the

thermal energy increases the energy of some electrons so that they can be bumped into potential conduction bands.

Crystals with metallic bonds have high electrical conductivity because electrons are free to migrate throughout the structure in the conduction bands. The electrical conductivity of metallic bonds decreases as temperature increases because the greater thermal vibrational energy of the lattice tends to impede the migration of electrons.

Bonds Not Involving Valence Electrons

The bonds that do not involve valence electrons depend on relatively weak electrostatic forces that can develop because of asymmetric charge distribution. These bonds are sometimes referred to as **molecular** or **intermolecular bonds**. Two mechanisms by which asymmetric charge distribution is developed are illustrated by **hydrogen** and **van der Waals bonding**.

Hydrogen Bonding

Ice is the most familiar mineral in which hydrogen bonding is prominent. It is composed of molecules of H_2O bonded together to form snow, hail, glaciers, and the ice cubes that cool soft drinks. Bonding occurs because the H_2O molecule is polar. Two hydrogen atoms covalently bond to each oxygen atom by sharing electrons with two of the oxygen's $2p$ orbitals (Figure 3.11a). Because oxygen is highly electronegative, it places a greater claim on the shared electrons than do the hydrogen atoms. This produces an electrical polarity to the water molecule (Figure 3.11b)—positive near the two H nuclei and negative at two nodes on the opposite side of the O. The positive and negative charge concentrations are located at the corners of a tetrahedron. If temperatures are sufficiently low (<0°C), weak electrostatic attractions between the polar molecules can bond them together. In ice, this produces a hexagonal framework with each water molecule bonding to four neighbors through hydrogen bonds.

Van der Waals Bonds

Van der Waals bonds also depend on asymmetrical charge distribution, but the asymmetry is produced in a different way. Consider graphite, whose structure consists of sheets of carbon atoms that are covalently bonded using both sigma and pi bonds. When the electron positions within one of these sheets of carbon atoms are averaged over a period of time, the charge on both sides of the sheet are the same. At any given instant, however, more electrons will be on one side of the sheet than the other, producing a polarization (Figure 3.12)—positive on one side and negative on the other. The net positive charge on one side of the sheet attracts electrons from the neighboring sheet so that the neighboring sheet also becomes polarized in the same sense. The result is a weak electrostatic attraction between the sheets produced by the short-term polarization of the sheets.

(a)

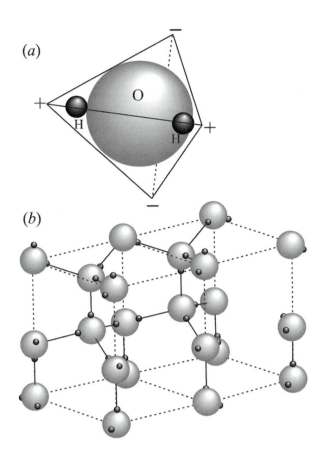

(b)

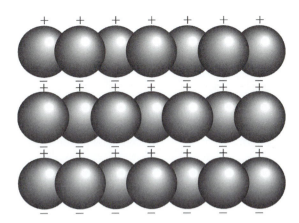

Figure 3.12 Van der Waals bonding. Sheets of carbon are bonded internally by π and σ covalent bonds (Figure 3.6). At any instant, more negatively charged electrons are on one side of a sheet than the other, producing a polarization. The polarization of one sheet produces a similar polarization in its neighbors. Van der Waals bonds are produced by the weak electrostatic attraction between the opposite charges on the surfaces of adjacent sheets.

Figure 3.11 Hydrogen bonding in ice. *(a)* A polar H_2O molecule is formed by covalent/ionic bonds between H and O using the $2p$ orbitals in O. The O has greater claim on the electrons, so the H consists of little more than its positively charged nucleus, a single proton. Charge is tetrahedrally distributed—positive at the H atoms and negative at nodes on the opposite side of the O. *(b)* Each negative node on the molecule attracts a positive node (an H atom) on an adjacent H_2O molecule to form hydrogen bonds that hold the water molecules together. Two unit cells are outlined with dashed lines.

Van der Waals bonds are quite weak, and minerals that have them, such as graphite and talc, are typically quite soft and have a greasy feel. These properties may be used to advantage. Because graphite is soft and black, it is used to make pencil lead, and the ease with which the sheets of carbon slide past each other makes graphite a useful lubricant. Talc [$(Mg_2Si_4O_{10}(OH)_2)$] is widely used as a body powder (talcum powder) because it is very soft and helps prevent chafing of tender skin. The weakness of these bonds found in certain clay minerals also leads to a major engineering problem, known as swelling soil, which will be described in Chapter 13.

It should be evident from the preceding discussion that a mineral may have several different bond types. Graphite, for example, has covalent sigma bonds, pi bonds with intermediate covalent/metallic character, and van der Waals bonds.

SIZE OF ATOMS AND IONS

In the section discussing chemical bonds, it was tacitly assumed that the ions behave as small hard spheres that pack together like so many differently sized marbles or ball bearings. This conceptual model for atoms and ions, although useful, has significant flaws. The volume occupied by the protons, neutrons, and electrons that comprise an isolated atom or ion is actually defined by the probability of finding electrons at specific locations around the nucleus. Those probabilities are not spherical, except for electrons in *s* shells, and do not have crisply defined boundaries. Some probability, however slight, always exists of finding an electron at a great distance from the nucleus. Further, the chemical bonds found in most minerals typically involve some degree of covalent bonding in which orbitals from adjacent atoms/ions are inferred to overlap, to allow for sharing of electrons.

Regardless of these difficulties, it is clear that atoms and ions do pack together in crystalline solids with specific geometries and at regular distances. Within limits, they behave as though they were little spheres. It is therefore convenient to define the sizes of atoms and ions in terms of their **effective radius** based on the distance between the centers of adjacent atoms or ions in the crystal structure. The effective radius of an atom or ion is, in effect, the size it would have if it behaved like a small hard sphere. Appendix A contains a tabulation of effective radii for the elements under various conditions. The effective radius of an uncharged atom is the **atomic radius**, and the effective radius of an anion or cation is the **ionic radius.**

Consider a covalent- or metallic-bonded crystalline solid made of a single element. The **bond length** L between two atoms is equal to twice the effective radius R of the atoms (Figure 3.13a).

$$L = 2R \qquad (3.5)$$

The techniques of X-ray diffraction (Chapter 8) make it possible to measure the bond length and therefore the sum

of the effective radii of the constituent atoms. Figure 3.14 shows the effective atomic radius for the elements exclusive of the noble gasses. There are two sets of data depending on the nature of the chemical bonds involved. For the metals and semimetals, the radii are based on metallic bonding in structures in which the atoms are closely packed so that each atom is in contact with 12 other atoms. For the nonmetals, the radii are based on single covalent bonds.

In an ionic-bonded crystalline solid, the bond length between the negatively charged anion and the positively charged cation is equal to the sum of the effective ionic radii of the cation R_c and of the anion R_a (Figure 3.13b).

$$L = R_a + R_c \qquad (3.6)$$

X-ray techniques allow measurement of the length of bonds between anions and cations but cannot tell us how much of the bond length to allot to the cation and how much to the anion. However, by comparing many different structures and carefully evaluating cation—anion and anion—anion distances, it is possible to make reliable estimates of the effective ionic radii of the anions and cations.

The effective ionic radii of the elements in their ionic configurations are shown in Figures 3.15 and 3.16 and are listed in Appendix A. These values are based on the assumption that O^{2-} has an effective ionic radius of 1.26 Å when in contact with six cations. Effective ionic radii based on 1.40 Å for O^{2-} have been widely used, but the values based on an O^{2-} radius of 1.26 Å are probably a

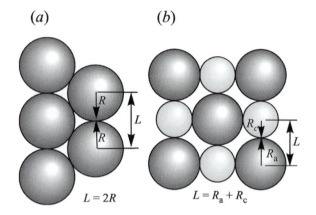

Figure 3.13 Bond length and effective radius. (*a*) Close-packed metal atoms of uniform size. The bond length (L) is twice the effective atomic radius R. (*b*) Anions and cations of different size. The bond length (L) is the sum of the effective ionic radii of the cation (R_c) and anion (R_a).

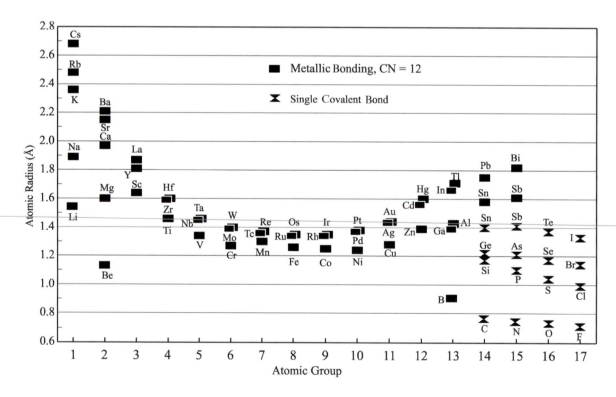

Figure 3.14 Effective atomic radii for the elements. The lanthanide and actinide series and noble gasses are omitted. Elements with metallic bonding are in a close-packed arrangement. Elements that do not form metallic bonds are shown for a single covalent bond. Data from Zhdanov (1965).

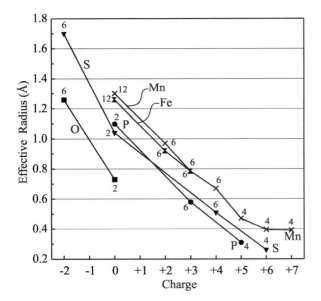

Figure 3.15 Effective radii as a function of oxidation state (charge) for oxygen (O), sulfur (S), phosphorus (P), manganese (Mn), and iron (Fe), elements that occur in different oxidation states. The coordination number is indicated by each symbol. Data from Appendix A and Figure 3.14.

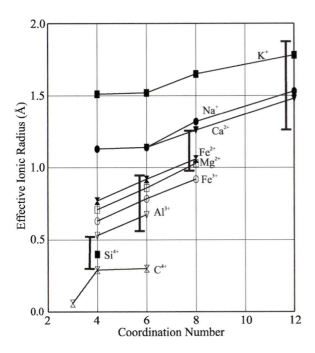

Figure 3.16 Effective ionic radii of the common cations as a function of coordination number. The normal ranges of cation sizes that will fit into 4-, 6-, 8-, and 12-fold coordination with oxygen are shown by bars (compare Table 4.1). Data from Appendix A.

better representation of the actual sizes of the ions and do not lead to the improbable result that some cations have negative sizes. To obtain effective ionic radii based on 1.40 Å for O^{2-}, subtract 0.14 Å from all values (except for O^{2-}) in Figure 3.15 and Appendix A.

The principal variables that influence effective ionic radius include the oxidation state (valence) of the ion and the number of anions in contact with the cation. The number of anions packed around the cation is called the **coordination number** (CN) and is discussed at more length in Chapter 4.

Oxidation State

Figure 3.15 clearly shows that effective radius is strongly controlled by charge. For a given element that may occur in more than one oxidation state, size decreases as the positive charge increases. All cations are smaller than the equivalent uncharged atoms, and anions are larger. An increase in positive charge produces smaller ions because the remaining electrons are held more tightly and more closely to the nucleus. Anions, with an excess of electrons, are larger than the equivalent uncharged atom because the nucleus is less able to hold the additional electrons as tightly.

Coordination

Interatomic distances, and effective ionic radii, also are influenced by the number of anions with which a cation is in contact. Anions are larger than cations, so, in many minerals, the structure can be considered to consist of a framework of regularly packed anions with smaller cations tucked into the interstices, or holes, between the anions. The size of the interstices is determined by the number of anions that surround it, which is the coordination number. An extensive discussion of coordination can be found in the next chapter; but for current purposes, note that a large hole requires more anions to define its boundaries and a smaller hole, fewer. A coordination number of 4, for example, involves four anions arranged at the corners of a tetrahedron with a small space in the middle in which a small cation may reside. A coordination number of 12 involves 12 anions surrounding and in contact with a cation that is about the same size as the anions. Figure 3.16 shows that as the coordination number increases, so does the effective ionic radius of the cations occupying the hole defined by the anions. In a sense, cations expand or shrink to fill the space available. The effective ionic radii of anions also are influenced by the number of cations with which they are in contact (coordinate), but not as much as anions.

An additional complexity that influences ionic radii is the **spin state**, which is a factor in transition metals with d orbitals containing four, five, six, or seven electrons. **High spin** ions have minimum pairing of electrons in the d orbitals; **low spin** ions have maximum pairing and are smaller (Figure 3.17). In general, the high spin state is preferred because having unpaired electrons represents a lower energy configuration than having paired electrons within the d orbital. However, some bonding geometries require hybridization that favors the low-spin configuration. Iron is the most important element that exhibits different spin

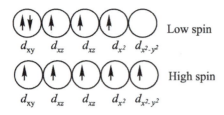

Figure 3.17 Spin configuration in Fe^{3+}. (*a*) Low spin. Electrons are paired in the d_{xy} orbital even though the $3d_{x^2-y^2}$ orbital is vacant. (*b*) High spin. No electrons are paired in the orbitals.

states. Under crustal conditions, the high-spin configuration dominates in most minerals.

REFERENCES CITED AND SUGGESTIONS FOR ADDITIONAL READING

Ahrens, L. A. (ed.). 1967. *Origin and Distribution of the Elements*. Pergamon Press, Oxford, 1178 p.

Bloch, A., and Schatteman, G. 1981. Quantum-defect orbital radii and the structural chemistry of simple solids, in M. O'Keefe, and A. Navrotsky, (eds.), *Structure and Bonding in Crystals*, Volume 1, p. 49–72. Academic Press, New York.

Clarke, F. W., and Washington, H. S. 1924. *The Composition of the Earth's Crust*. U.S. Geological Survey Professional Paper 127, 117 p.

Evans, R. C. 1966. *An Introduction to Crystal Chemistry*, 2nd ed. Cambridge University Press, Cambridge, 410 p.

Mason, B., and Moore, C. B. 1982. *Principles of Geochemistry* 4th ed. John Wiley & Sons, New York, 344 p.

Murrell, J. N., Kettle, S. F., and Tedder, J. M. 1985. *The Chemical Bond*. John Wiley & Sons, Chichester, 333 p.

Pauling, L. 1960. *The Nature of the Chemical Bond*, 3rd ed. Cornell University Press, Ithaca, NY, 644 p.

Shannon, R. D. 1976. Revised effective ionic radii and systematic studies of interatomic distances in halides and chalcogenides. Acta Crystallographica 32, 751–767.

Taylor, S. R. 1969. Abundance of chemical elements in the continental crust: A new table. Geochimica et Cosmochimica Acta 28, 1273–1285.

Wieser, M. E., and Berglund, M. 2009. Atomic weights of the elements 2007 (IUPAC Technical Report). Pure and Applied Chemistry 81, No. 11, 2131–2156.

Zhdanov, G.S. 1965. *Crystal Physics*. Academic Press, New York, 500 p.

CHAPTER 4

Crystal Structure

INTRODUCTION

In Chapter 3 we addressed the structure of matter and the various bonding mechanisms that hold atoms and ions together to form minerals. In this chapter we will address the following topics:

What determines how elements combine and pack together to form minerals?

How are mineral structures illustrated?

What are the structural variations in minerals?

How are minerals classified?

What controls the compositional variation of minerals?

How are chemical formulas written for minerals?

How can the chemical composition of minerals be graphically represented?

CONTROLS OF CRYSTAL STRUCTURE

The first-order control on the structure of a mineral is the nature of the chemical bonding that holds the elements together. As we saw in Chapter 3, any given mineral may display several different chemical bonding mechanisms and some of the mechanisms represent ideal end members of what in reality is a continuum. Regardless, it is convenient to consider the controls on crystal structures for the three major types of bonding: metallic, covalent, and ionic. Brief consideration will also be given to molecular crystals in which van der Waals or hydrogen bonding is found.

Structure Controls with Metallic Bonding

The bonding between metals, which have low and nearly equal electronegativity, typically is metallic. Because the valence electrons are in conduction bands and are free to migrate throughout the structure, the metal atoms tend to pack together in highly ordered arrangements that minimize void space. The configurations most commonly found

are based on **hexagonal closest packing** or **cubic closest packing** of spheres.

Hexagonal and cubic closest packing are the most compact ways to pack spheres of equal size together. In both cases, it is convenient to consider that the spheres lie in close-packed layers (Figure 4.1a). The layers are stacked so that the spheres in one layer nestle in the hollows between spheres in adjacent layers (Figure 4.1b). The difference between hexagonal and cubic closest packing is in how the layers are arranged. In both cases, each sphere is in contact with 12 neighboring spheres, six from its own layer and three each from the layers above and below.

In hexagonal closest packing (Figure 4.1c), the layers alternate between two positions. If the initial layer is considered to be in the A position, the next is in the B position, followed by A, then B, and so forth, forming what can be called an AB structure. This produces hexagonal symmetry with the planes of atoms parallel to the {001} crystallographic plane in the hexagonal lattice.

In cubic closest packing (Figure 4.1d) every third layer is aligned. The initial layer is A, followed by B and C, and finally A, each shifted in the same direction. The positions of the spheres in the A layers are directly above each other. This forms an ABC structure in which the planes of atoms are parallel to the {111} crystallographic plane of a face-centered cubic lattice; the spheres are located at lattice nodes.

Many of the metals that occur as native elements, including gold, silver, and copper, adopt the cubic closest packed arrangement. Metals with similar size and chemistry can be readily interchanged in this structure, forming **alloys**. Gold and silver, for example, have about the same size (Appendix A) and are in the same group in the periodic table of the elements (inside back cover), so it is possible to form gold–silver alloys with any composition intermediate between pure gold and pure silver.

A somewhat less-dense arrangement, known as a **body-centered cubic packing**, is used by iron (Figure 4.2). In this arrangement, the atoms are placed at the nodes of a body-centered cubic lattice. Each sphere is in contact with 8 rather than 12, neighbors.

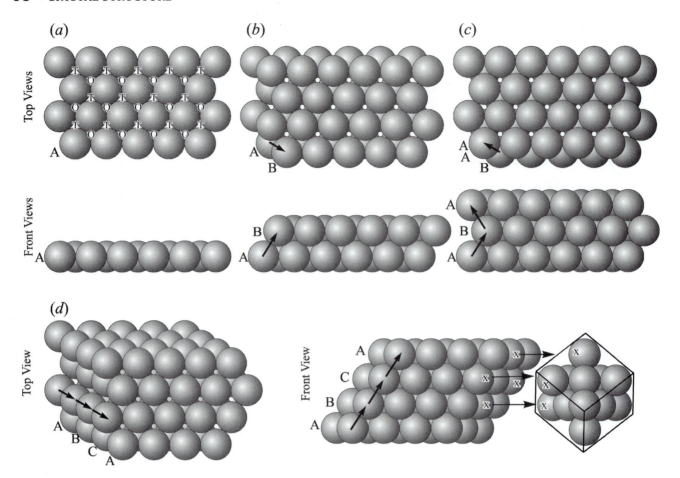

Figure 4.1 Closest packing of spheres. (*a*) Close-packed (A) layer of spheres. Each sphere is in contact with six neighbors. The T and O sites that will be formed when the B layer is placed on top of A are indicated. (*b*) Second (B) layer of spheres stacked on top of the first so that the spheres nestle into the T sites. (*c*) Hexagonal close packing is produced when a third layer is stacked directly above the first. The layers of spheres therefore alternate ABAB . . . and are parallel to the {001} plane in the hexagonal lattice. (*d*) Cubic close packing of spheres. Instead of shifting back and forth between two positions, each successive layer is shifted in the same direction. The first two layers of spheres (A and B) are the same as with hexagonal close packing. The C layer is shifted in the same direction as B, so the C spheres are directly above the O sites in the A layer. An additional shift in the same direction places the spheres in the final layer (A) directly above those in the initial A layer. The layers of spheres therefore alternate ABCABC . . . and are parallel to the {111} plane in a face-centered cubic lattice. The spheres marked *x* define the corners on one face of the cube.

Because of the closeness of packing and relatively high atomic weights, most metal crystals have high density compared to ionic-bonded crystals. Minerals with metallic bonding also conduct electricity easily and are relatively malleable.

Structure Controls with Covalent Bonding

Covalent bonds are typically quite strong, so covalent crystals are hard and have high melting points. The strong directional nature of covalent bonds requires that the atoms be placed in specific orientations so that orbitals can overlap. This often prevents the atoms from acquiring a close-packed arrangement. Diamond, described in Chapter 3 (Figure 3.5), is a good example.

Structure Controls with Molecular Crystals

Molecular crystals consist of discrete molecules packed together in a systematic way. Each molecule in the structure is typically bonded internally with covalent bonds and lacks a residual charge or valence with which to bond with other atoms. The forces that hold the molecules to each other are limited to relatively weak van der Waals and hydrogen bonds. Molecular crystals are therefore quite soft. The geometry of the structure is controlled by the shape and charge distribution of the individual molecules and the pattern by which they can be systematically packed together.

Two examples have previously been described. In ice (H_2O) (Figure 3.11) hydrogen bonding holds water

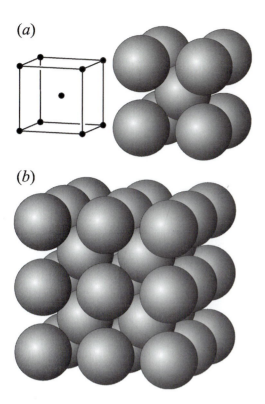

Figure 4.2 Body-centered cubic packing. (*a*) Body-centered lattice (left) and with atoms at each lattice node (right). (*b*) Eight unit cells. Each sphere is in contact with eight neighbors.

molecules in a hexagonal structure that is reflected in the hexagonal shape of snow flakes. Graphite (C) (Figure 3.12) is constructed of sheets of covalent/metallic bonded carbon that can be considered extremely large molecules. The sheets of carbon are bonded to each other by van der Waals bonds.

Structure Controls with Ionic Bonding

One of the most successful approaches to understanding the structure of many common minerals is based on the assumption that the anions and cations pack together as different-sized spheres. We have seen in Chapter 3 that bonding in many common minerals in which oxygen is the anion is dominantly ionic. The abundant silicate minerals are excellent examples. The bond between oxygen and silicon has about half ionic character, and the bonds between oxygen and the other common cations (Al, Fe, Mg, Ca, Na, K) are even more ionic. Therefore the assumption of ionic character is reasonable.

Because ionic bonds are not directional, it also is reasonable, as a first approximation, to look at the manner in which cations and anions bond together in purely geometric terms, unencumbered by the complexities of aligning orbitals in the specific orientations required for covalent bonds. Based on these assumptions, a set of five rules, now

collectively known as **Pauling's Rules**, was developed in the late 1920s:

> **RULE 1:** The Coordination Principle: A coordination polyhedron of anions forms around each cation. The cation-anion distance is determined by the sum of the cation and anion radii, and the number of anions coordinating with the cation is determined by the relative size of the cation and anion.

> **RULE 2:** Electrostatic Valency Principle: In a stable ionic structure, the total strength of the valency bonds that reach an anion from all neighboring cations is equal to the charge of the anion.

> **RULE 3:** Sharing of Polyhedral Elements I: The existence of edges, and particularly of faces, common to coordination polyhedra decreases the stability of ionic structures.

> **RULE 4:** Sharing of Polyhedral Elements II: In a crystal containing different cations, those with large valence and small coordination number tend not to share polyhedral elements.

> **RULE 5:** Principle of Parsimony: The number of essentially different kinds of constituents in a crystal tends to be small.

Rule 1. The Coordination Principle

The coordination principle states that how anions and cations pack together in a crystal structure depends on their relative sizes. As mentioned in Chapter 3, a cation's **coordination number** (**CN**) is the number of anions with which it is in contact. When the bonding is dominantly ionic, each cation will attract or coordinate with as many anions as it can pack around itself. The **coordination polyhedron** is the shape defined by anions coordinating with a cation. Twelve-fold coordination (Figure 4.3a) is the largest common coordination number, but the geometric shape defined by the 12 anions is not a regular polyhedron. The common regular coordination polyhedra are the cube, octahedron, tetrahedron, triangle, and line, which have coordination numbers of 8, 6, 4, 3, and 2, respectively.

The number of anions with which a cation coordinates is determined by the relative sizes of the anions and cations. For a given anion size, large cations will be able to coordinate with more anions than will a small cation. A convenient way of expressing the relative size of cation and anion is the **radius ratio** (*RR*):

$$RR = R_c / R_a \tag{4.1}$$

where R_c is the cation radius and R_a is the anion radius. The minimum number of anions that will coordinate with a cation is not strictly limited; but as a general rule, cations will bond with as many anions as they can. The maximum

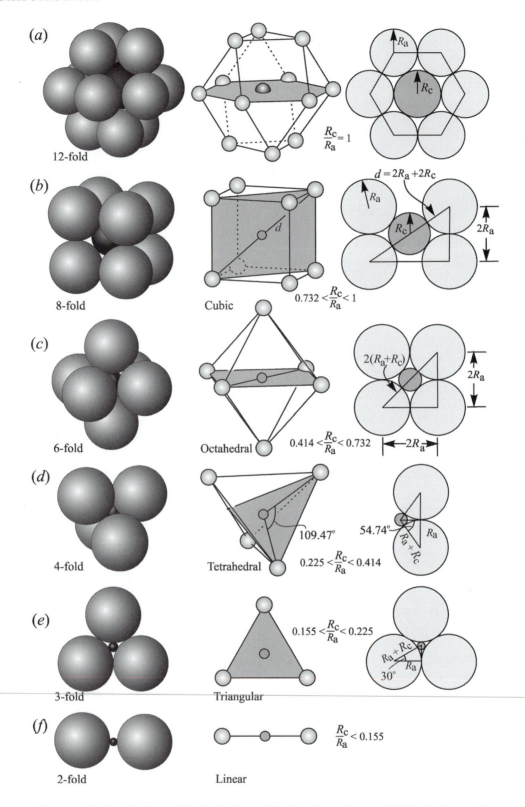

Figure 4.3 Coordination polyhedra. Anions with radius R_a are shown with light shading, cations with radius R_c with dark. Left view shows cation and anions drawn to scale. Center view shows the coordination polyhedra. Right view shows the plane through the polyhedron (shaded in center view) from which the radius ratio is calculated. See text for additional discussion. (*a*) 12-fold coordination (based on cubic closest packing). The coordination polyhedron is not a regular shape. (*b*) 8-fold, or cubic, coordination. (*c*) 6-fold, or octahedral, coordination. (*d*) 4-fold, or tetrahedral, coordination. (*e*) 3-fold, or triangular, coordination. (*f*) 2-fold, or linear, coordination.

Table 4.1 Radius Ratio for Common Regular Coordination Geometries

Coordination Number	Polyhedron	Radius Ratio
12	Not regular, based on cubic or hexagonal close packing	~1
8	Cube	0.732–1
6	Octahedron	0.414–0.732
4	Tetrahedron	0.225–0.414
3	Triangle	0.155–0.225
2	Line	< 0.155

number of anions that will coordinate with a cation, however, is limited by the need to maintain contact between the cation and surrounding anions. This keeps all the bonds of about equal strength. The range of radius ratios for the regular coordination geometries (CN = 12, 8, 6, 4, 3, and 2) is shown in Table 4.1.

12-Fold Coordination If the anion and cation have about the same size, then $RR \sim 1$. Each cation can be surrounded by no more than 12 anions (CN = 12), so each anion touches the cation. The two different ways in which 12-fold coordination can be achieved are based on cubic closest packing and hexagonal closest packing. Cubic closest packing is shown in Figure 4.3a.

8-Fold Coordination If the radius ratio is less than 1, the cation is too small to maintain contact with 12 anions. With regular packing, the next smaller number of anions that can surround and touch the cation is eight (Figure 4.3b). The eight anions arrange themselves to form a cube, so 8-fold coordination is known as **cubic coordination**. The lower limit of 8-fold coordination is achieved when the cation is sufficiently small that the anions come in contact with each other. The radius ratio at which this happens can be calculated with the help of basic geometry. If the eight anions that define the cube are in contact, the length of each edge of the cube is $2R_a$. The Pythagorean theorem can be used to demonstrate that the length of the long diagonal d through the center of the cube is equal to

$$d = \sqrt{3} \times 2R_a. \tag{4.2}$$

The length of d also is equal to $2R_a + 2R_c$ (Figure 4.3b) so

$$\sqrt{3} \times 2R_a = 2R_c + 2R_a. \tag{4.3}$$

Rearranging and solving for the radius ratio yields

$$\frac{R_c}{R_a} = \sqrt{3} - 1 = 0.732.$$

6-Fold Coordination If the radius ratio is less than 0.732, the cation is too small to maintain contact with eight anions, but it can maintain contact with six, down to a limiting

radius ratio of 0.414. The six anions arrange themselves to form the corners of an octahedron (Figure 4.3c) so 6-fold coordination is known as **octahedral coordination** (not to be confused with 8-fold coordination). Note that the cation is in the center of a square defined by four anions with a single anion above and below. The length of the diagonal through the square is equal to $2R_a + 2R_c$, and is also (by Pythagorus) equal to the square root of 2 times the length of the sides ($2R_a$).

$$2R_a + 2R_c = \sqrt{2} \times 2R_a \tag{4.4}$$

Rearranging to solve for the radius ratio yields

$$\frac{R_c}{R_a} = \sqrt{2} - 1 = 0.414.$$

4-Fold Coordination If the radius ratio is smaller than 0.414, down to a limit of 0.225, cations are too small to maintain contact with six anions; rather they coordinate with four anions (Figure 4.3d). Because the four anions define the corners of a tetrahedron, 4-fold coordination is known as **tetrahedral coordination**. The lower limit of 4-fold coordination is achieved when the four anions come in contact. Because the coordination polyhedron is a tetrahedron, a triangle drawn through the centers of two anions and the cation will have an angle of 109.47° between its two equal legs. The sine of half that angle is equal to the radius of the anion divided by the sum of the anion and cation radii.

$$\sin\frac{109.47°}{2} = \frac{R_a}{R_a + R_c}. \tag{4.5}$$

Rearranging to solve for the radius ratio yields

$$\frac{R_c}{R_a} = \frac{1}{\sin 54.74°} - 1 = 0.225.$$

3-Fold Coordination If the radius ratio is less than 0.225, down to a limit of 0.155, only three anions can maintain contact with the cation. Three-fold coordination is known as **triangular coordination**. The lower limit of 3-fold coordination is achieved when the cation is so small that the three anions come in contact (Figure 4.3e). The triangle produced with 3-fold coordination is an equilateral triangle so

$$\frac{R_a}{R_a + R_c} = \cos 30°. \tag{4.6}$$

Rearranging to solve for the radius ratio yields

$$\frac{R_c}{R_a} = \frac{1}{\cos 30°} - 1 = 0.155.$$

2-Fold Coordination Only two anions can maintain contact in **linear coordination** with a cation that is so small that the radius ratio is less than 0.155 (Figure 4.3f).

Other Coordination Geometries While the emphasis here is on regular coordination polyhedra, in many minerals, cations coordinate with 5, 7, 9, 10, or 11 oxygen anions. In addition, it also is common to find that the 3-, 4-, 6-, 8-, and 12-fold sites are somewhat distorted compared to the ideal geometry described above. Although the reasons that minerals may have structures with nonregular coordination polyhedra are numerous, they often are related either to the fact that some chemical bonds have substantial covalent character and are therefore somewhat directional, or to the requirements of accommodating the geometry of anionic groups such as CO_3^{2-}, SO_4^{2-}, and SiO_4^{4-}.

Coordination of Common Cations Given that O^{2-} is by far the most abundant anion and that it has an effective ionic radius of about 1.26 Å, it is possible to predict the coordination in which the common cations will be found. Table 4.2 shows the range of cation sizes that would normally go into the regular coordination polyhedra and the common cations that have the appropriate sizes when coordinating with oxygen. The same ranges also are shown in Figure 3.16. Note that some of the cations fit comfortably in more than one coordination: Al^{3+} fits both in 4-fold and 6-fold coordination, Fe^2 and Mg^{2+} have sizes appropriate for either 6-fold or 8-fold coordination, and Na^+ and Ca^{2+} are appropriate for either 8-fold or 12-fold coordination.

Many minerals have structures that are based, at least approximately, on close-packed anions. The interstices between the anions form tetrahedral and octahedral sites, some of which become occupied by cations. Figure 4.4 shows how these sites are formed.

Rule 2. Electrostatic Valency Principle

For an ionic bond, the bonding capacity of an ion is proportional to its oxidation state (charge). The strength of each **electrostatic valence bond** (evb) reaching a cation is therefore equal to the charge of the cation divided by the number of anions with which it coordinates, that is, the coordination number (CN).

$$evb = cation\ charge/CN \qquad (4.7)$$

Although this relationship seems straightforward, it leads to some interesting consequences.

Uniform Bond Strength First, consider the case in which all ionic bonds have the same strength. The term for this is **isodesmic**. This will be the case for all ionic minerals with a single anion and single cation, as well as some minerals with multiple cations and/or anions. If all the cation–anion bonds are the same strength, it is impossible for any bond to occupy more than half of the anion's charge because the smallest possible coordination number is 2. Because the anion–cation bonds are all uniform, the anions tend to pack together in a highly symmetrical arrangement, and the cations occupy polyhedra formed by the interstices among the anions. Some of the common structures are based on cubic and hexagonal closest packing of the anions. It should be no surprise that most of these minerals are highly symmetrical and crystallize in the isometric, tetragonal, or hexagonal crystal systems.

Minerals with uniform bond strengths are conveniently classified as oxides, fluorides, chlorides, and so forth based on the identity of the anion.

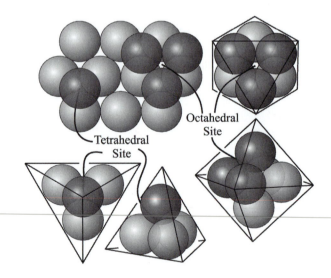

Figure 4.4 Tetrahedral and octahedral sites in close-packed structures. A layer of anions (light) is shown with just four anions from an adjacent layer (dark). Tetrahedral, or 4-fold, coordination sites are formed where an anion from one layer nestles into the hollow formed by three anions from the next layer. The tetrahedral site is the little opening hidden in the middle of the four anions. Octahedral, or 6-fold, coordination sites are formed where three anions from one layer nestle into three from the next. Compare Figure 4.1a, where all the tetrahedral (T) and octahedral (O) sites between two layers of anions are identified.

Table 4.2 Cation Sizes Appropriate for Regular Coordination Polyhedra When Coordinating with O^{2-} (~1.26 Å)

Coordination	Radius Ratio	Minimum Radius (Å)	~Maximum Radius (Å)	Common Cations
2	< 0.155	N/A	0.20	None
3	0.155–0.225	0.20	0.28	C^{4+} [a]
4	0.225–0.414	0.28	0.52	Si^{4+}, Al^{3+}, S^{6+}, P^{5+}
6	0.414–0.732	0.52	0.92	Al^{3+}, Fe^{2+}, Fe^{3+}, Mg^{2+}
8	0.732–1.00	0.92	1.26	Fe^{2+}, Ca^{2+}, Na^+, Mg^{2+}
12	~1.00	1.26	N/A	K^+, Ca^{2+}, Na^+

[a] According to Appendix A, C^{4+} has an ionic radius of 0.06 Å when in 3-fold coordination, substantially smaller than theory would suggest is appropriate. The discrepancy is probably a consequence of the fact that the C–O bond has substantial covalent character, and the simple model used here of ionically bonding spheres of different sizes is not entirely appropriate.

Nonuniform Bond Strength In some structures certain anion–cation bonds are stronger than others. The cations forming stronger bonds are small and have a high charge; C^{4+}, S^{6+}, P^{5+}, and Si^{4+} are good examples (Table 4.2 and Appendix A). Because these cations strongly bond to the anions (usually O^{2-}) the result is the formation of CO_3^{2-}, SO_4^{4-}, PO_4^{3-}, and SiO_4^{4-} anionic groups that serve as structural elements in their mineral's structures.

Minerals with nonuniform bond strengths typically have lower symmetry than those in which bond strengths are uniform because of the geometric complexity involved with packing the anionic groups together and coordinating with other cations. Coordination polyhedra formed between these anionic groups often are irregular. The two types of nonuniform bonding are **anisodesmic** and **mesodesmic**. With anisodesmic bonding, some anion–cation bonds take more than half of the anion charge; with mesodesmic bonding, some anion–cation bonds take exactly half of the anion charge.

Anisodesmic bonding is found in minerals containing small, highly charged cations. Consider the mineral calcite ($CaCO_3$) described in Chapter 17. Carbon is a small cation (Table 4.2) with a +4 charge that coordinates with three O^{2-} (Figure 4.5). The electrostatic valence bond (evb) strength of the C–O bonds is therefore 4/3 = 1.33. The Ca^{2+} is in 6-fold coordination with these same O^{2-} anions with an evb of 2/6 = 1/3. Each O^{2-} bonds with two Ca^{2-} cations, so the total evb reaching each oxygen is 2. The oxygen anions are more strongly bonded to the carbon than to the calcium. As a consequence, distinct CO_3^{2-} structural anionic groups are formed. S^{6+} and P^{5+} form anisodesmic SO_4^{2-} and PO_4^{3-} anionic groups.

Mesodesmic bonding represents a special case in which the anion–cation bond takes up exactly half of the available anion charge. The silicates provide the best example. Si^{4+} almost always coordinates with four O^{2-} in what is known as a **silicon tetrahedron** (Figure 4.6a). The evb strength is therefore 4/4 = 1, which is exactly half of

the –2 charge on oxygen. This leaves a –1 charge on each oxygen available to bond with a second Si^{4+} by sharing an oxygen between tetrahedra (Figure 4.6b). Each silicon tetrahedron can potentially share oxygen ions with one, two, three, or four adjacent tetrahedra depending on how much silicon is available. The degree to which silicon tetrahedra share oxygen ions with adjacent tetrahedra and polymerize to form pairs, chains, rings, sheets, and three-dimensional frameworks provides the basis for classification of the silicate minerals (Chapter 11). For example, in the micas (Chapter 13) silicon tetrahedra polymerize by sharing three oxygen anions with each of three neighbors to form continuous sheets. The single perfect cleavage and tabular crystal habit of the micas reflects this sheet structure.

The oxygen in a silica tetrahedron also may bond with other cations such as Al^{3+}, $Fe^{2+/3+}$, Mg^{2+}, and Na^+, depending on the relative amounts of silicon and the other cations that are available. In forsterite (Mg_2SiO_4), no oxygen from one silicon tetrahedron is shared with another, so each oxygen anion satisfies its remaining –1 charge by bonding with Mg^{2+}. The Mg^{2+} coordinates with six oxygens, so its evb will be 2/6, or 1/3. Because each Mg–O bond takes up just 1/3 of a charge, each oxygen must bond with

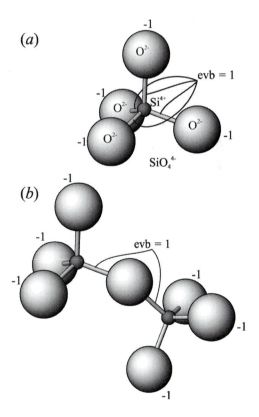

Figure 4.6 Mesodesmic silicon tetrahedra. (*a*) Si–O bonds occupy 1 (evb = 1) of the available –2 charge on oxygen, which is left with a net –1 charge unsatisfied. (*b*) Each oxygen can therefore bond to another Si^{4+}, so that the oxygen anion is equally shared between silicon tetrahedra. A silicon tetrahedron may share one, two, three, or all four of its oxygen anions with adjacent tetrahedra.

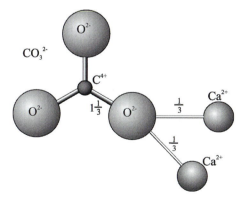

Figure 4.5 Anisodesmic carbonate (CO_3^{2-}) group. Each O–C bond occupies 1.33 of the available –2 charge on oxygen; only two-thirds of a charge on the oxygen is available to bond with other cations such as Ca^{2+}.

three Mg^{2+}. Of the two cations, the Si^{4+}, with evb = 1, has the stronger claim on the oxygen anions, so the silicon tetrahedron is considered the primary structural element in the mineral.

Minerals with nonuniform bond strength are classified based on the identity of the anionic group formed around the small highly charged cation. Anisodesmic examples include the carbonates, sulfates, and phosphates. The silicates are the only common mesodesmic group. The structure of these minerals is controlled by the geometry of the anionic group and the requirement for additional cations to coordinate with the anions in the anionic groups to satisfy any remaining negative charge.

Rule 3. Sharing of Polyhedral Elements I

It is a direct consequence of the electrostatic nature of ionic bonds that adjacent polyhedra prefer to share only single anions, or less commonly, pairs of anions. If adjacent cation polyhedra share only a single anion, that is, have one corner in common, this ensures that the positively charged cations, which repel each other, are kept at the greatest possible distance. Figure 4.7 shows these relations for tetrahedra. If only a corner (single anion) is shared, the cations are 2.45 R_a apart, where R_a = anion radius. If an edge (two anions) is shared, the cations are brought to within 1.414 R_a of each other. This reduces the stability of the structure because energy is required to bring

the cations into closer proximity. The lowest stability configuration involves sharing a face (three anions) between tetrahedra. This brings the cations to within 1.15R_a of each other. Similar relations are obtained for other coordination polyhedra.

Rule 4. Sharing of Polyhedral Elements II

In structures with differently charged cations, the high-charged cations minimize the number of anions that they share between polyhedra so that the high-charged cations can be kept a maximum distance apart. The corollary is that low-charged cations are more likely to share two or more anions on edges or faces of their coordination polyhedra. Perovskite ($TiCaO_3$) illustrates this nicely (Figure 4.7d). The Ti^{4+} occupies octahedral (6-coordination) sites defined by six O^{2-} anions. These octahedra share only a single O^{2-} at a corner, thereby keeping the Ti^{4+} cations well away from each other. The lower-charged Ca^{2+} is much larger and occupies 12-fold coordination sites that share four O^{2-} anions on a face. An additional consideration in some minerals follows from the second rule. Small high-charged cations C^{4+}, S^{6+}, and P^{5+} form low-coordination-number anisodesmic anionic groups (CO_3^{2-}, SO_4^{4-}, PO_4^{3-}) in which more than half the anion charge is occupied. This means that the anions in these anionic groups can bond only with lower-charged cations that occupy sites with higher coordination numbers.

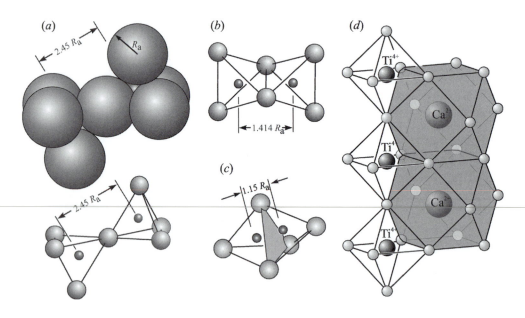

Figure 4.7 Sharing anions between polyhedra. (*a*) Sharing a single anion (a corner) between tetrahedra keeps the positively charged cations at a maximum distance. The upper diagram shows the anions (large) and cations (small) to scale. The lower diagram reduces the ion size to allow the tetrahedral coordination polyhedra to be more easily seen. The cations are 2.45R_a apart where R_a is the radius of the anion. (*b*) Sharing two anions (an edge) places the cations 1.414R_a apart. (*c*) Sharing three anions (a face, shaded) places the cations 1.15 R_a apart. (*d*) In perovskite ($CaTiO_3$), the Ti^{4+} occupies octahedra that share only a single oxygen (small spheres), and Ca^{2+} occupies 12-fold sites that share four oxygen anions on common faces.

Rule 5. Principle of Parsimony

This is a statement of the general observation (or perhaps fervent hope) that nature tends toward simplicity. The number of fundamentally different structural sites in most minerals, even those with many different chemical constituents, tends to be small. Usually the cations systematically array themselves into no more than four (and usually fewer) different types of coordination polyhedra and are allocated among the different polyhedra based on size and charge. Although detailed studies of some crystal structures may reveal subtle differences among different groups of tetrahedral or other sites, the observation that the anions and cations in the chemical formulas of minerals occur in small integer ratios is strong support for Rule 5.

Application of Pauling's Rules

Pauling's second rule indicates that ionic minerals may be grouped into three broad categories: those with uniform bonding between anions and cations (isodesmic), those with anionic groups in which more than half of the anion charge is occupied (anisodesmic), and those with anionic groups in which half of the anion charge is occupied (mesodesmic).

Uniform Isodesmic Bonding

Because the anion–cation bonds are all uniform, the anions tend to pack together in symmetrical arrangements and the cations occupy polyhedra formed by the interstices between the anions. Some of the common structures are based on cubic closest packing, hexagonal closest packing, and body-centered cubic packing of the anions. These structures and variants on them are found in many minerals with simple anions such as the halides, oxides, and many sulfides. In addition, the structures of other minerals are often conveniently described with reference to these simple structures.

The mineral halite (NaCl) provides a convenient example of a structure based on cubic close packing of Cl^- anions. The Cl^- anions are arranged in layers parallel to {111} in a cubic close-packed arrangement. In the NaCl structure (Figure 4.8a) the coordination of Na^+ is dictated by the first rule. The ionic radii of Na^+ and Cl^- are 1.16 and 1.81 Å, respectively, yielding a radius ratio of 0.64. Based on Table 4.1, Na^+ should coordinate with six Cl^-. To maintain charge balance, every one of the octahedral (6-fold) sites between the {111} layers of Cl^- must be occupied. This allows every Na^+ to bond with six Cl^- and vice versa, thus satisfying the electrostatic valency principle (second rule). Because Na^+ occupies all the available octahedral sites between the layers of Cl^-, it is necessary for the Na^+ octahedra to share edges. Because Na^+ has only a single charge, this makes edge sharing more acceptable in the third rule. The fourth rule does not apply because only a single cation is involved; and the structure is simple with only a single cation site, so the fifth rule is satisfied.

Nonuniform Bonding

Anisodesmic   The structure of anhydrite ($CaSO_4$), which is a common constituent of many evaporite deposits, has an anisodesmic structure (Figure 4.9). S^{6+} tetrahedrally coordinates with O^{2-}. The second rule informs us that the strength of the electrostatic valence bonds reaching the S^{6+} is $6/4 = 1.5$, meaning that of the -2 charge available on each oxygen, 1.5 is occupied by the sulfur and only $\frac{1}{2}$ is available to bond with the Ca^{2+}. We know from the previous

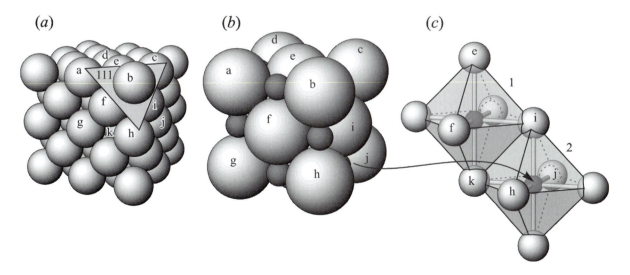

Figure 4.8 NaCl structure. (*a*) Layers of Cl^- anions are arranged parallel to the (111) crystallographic plane. (*b*) One unit cell showing Na^+ (dark spheres) in octahedral interstices between the Cl^-. (*c*) Octahedron 1 is formed by the Cl^- on the center of each face of the unit cell (the ions are shown at a reduced size). Octahedron 2 surrounds the Na^+ on the bottom right edge of the unit cell and includes Cl^- from adjacent unit cells. The octahedra share an edge.

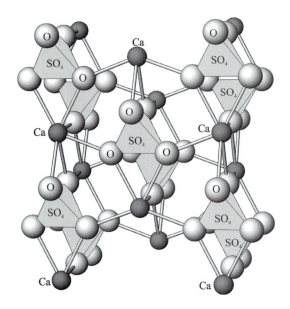

Figure 4.9 Anhydrite structure. Isolated SO_4^{2-} groups are bonded laterally through Ca^{2+} (dark spheres).

discussion on coordination that Ca^{2+} is most commonly in 8-fold coordination with oxygen. The strength of each electrostatic valence bond reaching Ca^{2+} is $2/8 = 0.25$. This means that each oxygen must bond with two Ca^{2+} to satisfy its remaining $\frac{1}{2}$ charge. The requirement in the anhydrite structure is to somehow arrange the SO_4^{2-} tetrahedra in such a way that Ca^{2+} can bond with eight O^{2-}, each of which must bond with two Ca^{2+}. It turns out that ideal cubic 8-fold sites are not possible, so Ca^{2+} must settle for a somewhat distorted 8-fold site.

Mesodesmic The silicates provide an excellent example of mesodesmic bonding. As described earlier, it is possible to share O^{2-} between silicon tetrahedra. This allows silicon tetrahedra to string together to form rings, chains, sheets, and other polymerized structures. Other cations (Ca, Mg, Fe, etc.) coordinate with oxygen ions of the silicon tetrahedra to bond the polymerized tetrahedra together. Silicate structures are described at length in Chapters 11 through 16.

ILLUSTRATING MINERAL STRUCTURES

Mineral structures are conventionally illustrated in a variety of different ways, including spheres to scale, stick-and-ball models, polyhedra, and by using coordinates.

Perhaps the most realistic, but often least informative, way of illustrating mineral structures is to represent atoms/ions as spheres drawn to scale (Figure 4.10a). This method has the distinct disadvantage that the atoms/ions occupy most of the volume, so the first layer obscures whatever is behind. The stick-and-ball method (Figure 4.10b) deals

with this problem by shrinking the size of the atoms/ions and showing the bonds between atoms with lines or "sticks."

With the polyhedral method (Figure 4.10c), the anions and cations are not shown at all. Rather, the structure is represented by the coordination polyhedra in which the cations reside. Anions are inferred to be located at each polyhedral corner with the cations in the centers of the polyhedra. A common tactic is to use a hybrid of the polyhedral and the stick-and-ball methods (Figure 4.10d). Some cation sites are shown as balls and others are represented by coordination polyhedra. Anions may be omitted and are inferred to occupy polyhedral corners or kinks in the sticks.

The most accurate means of illustrating a mineral structure is to map it by using unit cell coordinates. In Figure 4.10e, the olivine structure is viewed down the *a* axis with the positions of the ions projected onto the back surface of the unit cell; this effectively gives them their *b* and *c* coordinates, just like putting a spot on a map defines the longitude and latitude. The position of each atom above the back of the unit cell (the *z* coordinate) is indicated by a number representing a percentage of the unit cell. Zero means that the atom is at the back surface of the unit cell, 49 indicates that the atom is 49% of the way up from the back of the unit cell, and so forth. Full characterization of the structure requires that the position of each atom/ion within the unit cell be specified.

Software such as Atoms, from Shape Software, allows illustrations of crystal structures to be readily created by using data from the online database of mineral structures maintained by the Mineralogical Society of America.

ISOSTRUCTURAL MINERALS

In the preceding discussion it was implied that different minerals may have essentially identical structures. Two or more minerals whose atoms are arranged in the same type of crystal structure are said to be **isostructural**. Halite (NaCl) and galena (PbS), for example, are isostructural because the arrangement of Pb and S in galena is identical to the arrangement of Na and Cl in halite. The minerals are dramatically different in many physical and chemical properties, but the fact that they are isostructural is reflected in an identical symmetry, cleavage (three perfect cleavages at right angles) and crystal habits (cubes). Isostructuralism is sometimes referred to as **isomorphism**, or less commonly as **isotypism**, but these terms are not recommended.

An **isostructural group** of minerals is a group of minerals that is isostructural and also chemically related by having a common anion or anionic group. The minerals within an isostructural group may display extensive ionic or atomic substitution. An example of an isostructural group (Table 4.3) is the calcite group of carbonates that is described in Chapter 17.

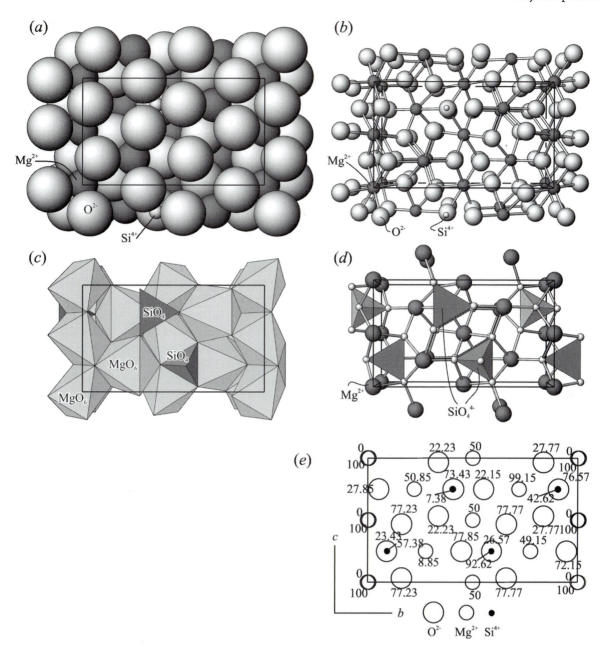

Figure 4.10 Illustration of olivine (Mg_2SiO_4) crystal structure viewed down the *a* axis. Each unit cell (outlined) contains four formula units. (*a*) Spheres to scale: O^{2-}, Mg^{2+}, and Si^{4+} are large, medium, and small spheres, respectively. (*b*) Stick and ball. Sphere size is reduced and the sticks show bonds. (*c*) Polyhedral. Tetrahedra (dark) contain Si^{4+} and somewhat distorted octahedra contain Mg^{2+}. The O^{2-} are not shown. (*d*) Hybrid polyhedra and stick and ball: O^{2-} are shown as small spheres at corners of the tetrahedra. (*e*) "Mapped" structure. The height of the ions above the back surface of the unit cell are given as a percentage of the *a* unit cell dimension. The outlines of the Mg^{2+} at the corners of the unit cell are offset slightly so that all will show.

POLYMORPHISM

The ability of a chemical compound to crystallize with more than one structure is known as **polymorphism**. The different structures of a given chemical compound are known as **polymorphic forms** or **polymorphs**, and the set of different minerals with the same chemical composition is known as a **polymorphic group**. Examples of polymorphic groups are listed in Table 4.4.

Minerals form polymorphs because most mineral structures represent a collection of compromises that balance conflicting requirements related to attraction and repulsion of cations and anions, the fit of cations in their coordination sites, geometric requirements related to partial covalent

Table 4.3 Calcite Isostructural Group[a]

Mineral	Formula
Calcite	$CaCO_3$
Magnesite	$MgCO_3$
Siderite	$FeCO_3$
Rhodochrosite	$MnCO_3$
Smithsonite	$ZnCO_3$

[a] All crystallize in the $\overline{3}2/m$ point group and differ only in the cation in the structure.

Table 4.4 Common Polymorphic Mineral Groups

Chemical Composition	Mineral Name
SiO_2	α-Quartz
	β-Quartz
	α-Tridymite
	β-Tridymite
	Cristobalite
	Coesite
	Stishovite
FeS_2	Pyrite
	Marcasite
C	Graphite
	Diamond
$AlAlOSiO_4$	Andalusite
	Sillimanite
	Kyanite
$KAlSi_3O_8$	Sanidine
	Orthoclase
	Microcline

bonding, and so forth. At a given temperature and pressure, one structure may represent the lowest energy configuration; at other temperatures and pressures, a different structure may be more stable. In general, high pressures favor structures in which the atoms and ions are more tightly packed, reflected in higher mineral density. High temperature tends to favor somewhat more open, lower density structures and structures that allow greater diversity in the occupancy of specific structural sites. Because different polymorphs of the same substance are stable under different sets of conditions, the presence of a given polymorph in a rock may indicate something about the conditions under which the rock was formed or to which it has been subjected.

The four different structural relationships between polymorphic forms are **reconstructive polymorphism, displacive polymorphism, order–disorder polymorphism**, and **polytypism**.

Reconstructive Polymorphism

With reconstructive polymorphism, conversion from one polymorph to another involves a major reorganization of the crystal structure. The chemical bonds that hold one structure together must be broken so that the atoms/ions can be rearranged and bonded into the new structure. Reconstructive polymorphs need not have symmetry or structural elements in common, although the fact that the polymorphs have identical chemical compositions may lead to some structural similarities. The relationship between the two carbon polymorphs is typical.

The structures of both diamond and graphite are described in Chapter 3 (Figures 3.5 and 3.12). Diamond has a structure in which each carbon atom is bonded with covalent σ bonds to four neighboring carbon atoms forming extremely hard cubic or octahedral crystals. In their most prized gem form, diamonds are beautifully clear and nearly colorless. In graphite, each carbon atom is bonded to three neighbors using a combination of σ and π bonds to form continuous sheets that are bonded to each other with van der Waals bonds. Crystals are typically very soft, black, opaque hexagonal plates. Graphite is used to make the "lead" in pencils. The structure of diamond cannot be derived from graphite except by breaking chemical bonds and then reassembling the carbon atoms in a completely different structure.

The temperature–pressure range over which each polymorph is stable is shown in Figure 4.11. Note that diamond is stable only at pressures found in the mantle and that graphite is the polymorph stable at the Earth's surface. The igneous rock **kimberlite** in which diamond is normally found must therefore have its origins within the mantle. The diamond found at or near the Earth's surface that is mined and used in jewelry and industrial cutting tools is **metastable**. That means that it persists outside of its normal stability field and does not spontaneously convert to graphite, which is actually the more stable, lower energy configuration for carbon at the Earth's surface. Diamond does not automatically convert to graphite on cooling because it takes a significant amount of energy to break the chemical bonds holding the carbon atoms together in the diamond structure. Cooling, particularly fast cooling, removes energy from the system that would be needed to break the bonds. The fact that diamond does not automatically convert to graphite reflects one of the important features of reconstructive polymorphs. The reactions that convert one polymorph to another are **quenchable**; reconstructive polymorphs do not automatically convert from one structure to another with changing conditions.

Displacive Polymorphism

Displacive polymorphic inversions do not involve breaking chemical bonds; the difference between polymorphs is simply a distortion or bending of the crystal structure. The relation between α-quartz and β-quartz is a good example (Figure 4.12). At one atmosphere pressure, β-quartz (**high quartz**) is stable above 573°C and has a structure consisting of a framework of silicon tetrahedra that includes 6-fold spirals (actually 6-fold screw axes). On cooling below 573°C, the structure undergoes an immediate, unquenchable conversion to the α-quartz (**low-quartz**) structure in which the spirals become distorted to have 3-fold symmetry. Every time the sample is heated or cooled through the inversion temperature, the structure distorts from one structure to the other; the inversion is not quenchable.

High-temperature forms typically have higher symmetry than low-temperature polymorphs. The crystal shape of

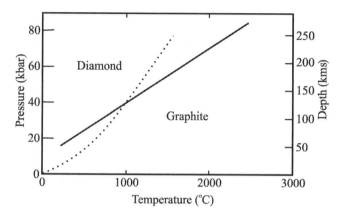

Figure 4.11 Diamond and graphite stability fields. The dotted line shows an approximate geothermal gradient beneath continental shields. For this geothermal gradient, diamond is stable only at pressures greater than ~40 kilobars, which corresponds to a depth of about 125 kilometers in the Earth's mantle.

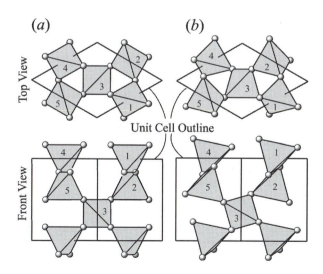

Figure 4.12 Structures of α-quartz and β-quartz. The top view looks down the c axis, the bottom view is from the side. The unit cell is outlined. (a) β-Quartz is the high-temperature polymorph. (b) α-Quartz is the low-temperature polymorph. The inversion from β-quartz to α-quartz is accomplished by a distortion that reduces the symmetry.

the high-temperature polymorph will be retained on inversion to the low-temperature polymorph, though internal strains in the crystal lattice may lead to the formation of transformation twins (segments of the crystal with different crystallographic orientation). Twinning is described in Chapter 5.

Order–Disorder Polymorphism

With **order–disorder polymorphism**, the mineral structure remains more or less the same. What changes is the cation distribution within structural sites. If two cations X and Y can occupy two equivalent structural sites A_1 and A_2, the structure is considered disordered if there is an equal probability of finding X in either A_1 or A_2. If all X cations are located in site A_1 and all Y cations are in site A_2, the structure is considered fully ordered.

Consider the mineral K-feldspar ($KAlSi_3O_8$). The three Si and one Al in the formula unit are found in two T_1 and two T_2 tetrahedral sites shown schematically in Figure 4.13a. In the completely disordered structure, represented by the **high-sanidine** polymorph, Al is equally likely to be in any of the four sites (Figure 4.13b); 25% of each of the sites are filled with Al and 75% with Si.

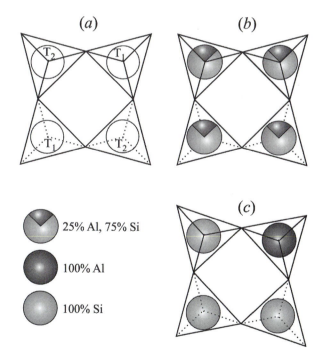

Figure 4.13 Order–disorder in K-feldspar ($KAlSi_3O_8$) polymorphs. (a) Schematic view of two up-pointing and two down-pointing tetrahedra that contain the three Si and one Al per formula unit. Two tetrahedra are T_1 sites, two are T_2 sites. (b) High sanidine. Al is equally likely to be in any of the four sites, so on average each contains 25% Al. (c) Maximum microcline: Al is preferentially placed in one T_1 site; Si occupies the other three sites, producing a distortion to the structure.

In the fully ordered structure, represented by the **low-** or **maximum-microcline** polymorph, all of the Al is placed in just one of the T_1 sites; the remaining T_1 and both T_2 sites contain only Si (Figure 4.13c). This causes a distortion to the structure and reduces the symmetry from monoclinic to triclinic. Additional ordering arrangements are possible and are discussed in the section on K-feldspar in Chapter 12.

The degree of order in most minerals is strongly influenced by crystallization temperature and cooling history. In general, high temperatures favor crystallization with a high degree of disorder, low temperature favors order. Slow cooling allows time for ions to migrate and diffuse through the crystal structure so that the different chemical elements parse themselves into the sites favored in the ordered structure; rapid cooling prevents this ordering. It should be no surprise, then, that sanidine is found in volcanic rocks, which combine high temperatures and rapid cooling. K-feldspar in plutonic igneous rocks initially crystallizes as sanidine, but slow cooling allows ordering to occur. Deep-seated intrusives usually have sufficiently slow cooling to allow extensive ordering, so they contain microcline.

Polytypism

Polytypism is a variety of polymorphism in which the polymorphs differ only in the stacking sequence of identical sheets. Figure 4.14 illustrates several different polytypes that can be created using a single layer. The difference between hexagonal and cubic closest packing of spheres (previously described) is another example. The sheet silicates, which include the micas and clay minerals, are an important group of minerals in which polytypism is common.

MINERAL CLASSIFICATION

Minerals other than the native elements are conventionally classified based on the identity of the major anion or anionic group (Table 4.5). This approach is consistent with normal practice in chemistry concerning the classification of inorganic chemical compounds. It also accords well with the observation that minerals with the same anion or anionic group often have clear family relations in terms of structure, and physical and chemical properties. Further, the cation content of many minerals may be quite variable, whereas the anion content is typically quite restricted.

This classification scheme also follows from Pauling's rules for ionic-bonded solids. The first, third, and fourth rules inform us that we can expect the anions to define the basic structure of minerals. The second rule provides the basis for separating minerals based on whether they contain anionic groups.

Table 4.5 Chemical Classification of Minerals

Mineral Group	Anion or Anionic Group	Mineral Group	Anion or Anionic Group
Native elements	N/A	Carbonates	CO_3
Oxides	O	Nitrates	NO_3
Hydroxides	OH	Borates	BO_3, BO_4
Halides	Cl, Br, F	Chromates	CrO_4
Sulfides	S	Tungstates	WoO_4
Arsenides	As	Molybdates	MoO_4
Antimonides	Sb	Phosphates	PO_4
Selenides	Se	Arsenates	AsO_4
Tellurides	Te	Vanadates	VO_4
Sulfates	SO_4	Silicates	SiO_4

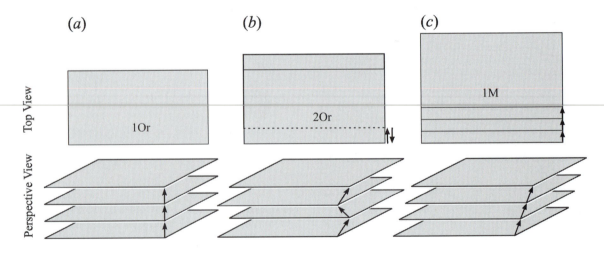

Figure 4.14 Polytypism involves different patterns of stacking identical sheets. The symbols used here are adapted from those used in the sheet silicates. The number in the symbol indicates the number of layers per unit cell, and the letters indicate the crystal system that results (Or = orthorhombic, M = monoclinic). (*a*) 1Or polytype with a single stacking vector at right angles to the sheets. (*b*) 2Or polytype with two stacking vectors with opposite senses. (*c*) 1M polytype with a single stacking vector at an angle to the sheets.

COMPOSITIONAL VARIATION IN MINERALS

When chemical analyses of different samples of a mineral are performed, it is routinely found that they do not all have the same chemical composition. The variations may be substantial, but they always fall within limits. Recall from Chapter 1 that minerals have a definite but not fixed chemical composition. The term applied to this compositional variation is **solid solution**.

The term *solid solution* is somewhat misleading because it is not really analogous to liquid solution with which we are familiar from chemistry and everyday experience. In a liquid solution one or more materials, the **solutes**, are mixed in variable amounts with a liquid **solvent** to form a homogeneous mixture. The solutes can be solids that dissolve, other liquids, or gasses. Beer is a good example of a liquid solution. The solvent is water and the solutes are liquid alcohol, gaseous CO_2, and various organic and other compounds derived from grain, hops, and yeast that provide the flavor and color. Different beers have different amounts of the various solutes.

With minerals, however, compositional variation is not a matter of mixing various solutes into a solid solvent. Rather, the compositional variation is a consequence of the ability of different elements, mostly cations, to substitute for each other within the crystal structure. The structure of an ionic-bonded mineral may be considered to be constructed of a framework of anions that form a number of distinct structural sites into which the cations fit. Additional sites may be vacant. Compositional variation is possible because different cations can interchangeably occupy the various sites among the anions. Additionally, in some cases different anions may be interchanged for each other. The three principal types of solid solution are **substitution solid solution, omission solid solution, and interstitial solid solution**.

The range of compositions produced by solid solution in a given mineral is known as a **substitution series** or **solid solution series**. The compositional extremes of a substitution series are known as the **end members**. End-member compositions are typically written with integer subscripts for all elements in the formula (see later). A **continuous** or **complete substitution series** is one in which all intermediate compositions are possible. An **incomplete or discontinuous substitution series** is one in which only a restricted range of compositions between the end members is found.

Substitution Solid Solution

Substitution involves interchanging one cation for another in a structural site. Two requirements control whether substitution can occur.

- Ion sizes must be similar.
- Charge neutrality must be maintained.

Size

The size requirement follows logically from Pauling's first rule and the discussion of coordination. Table 4.2 provides a useful basis for determining which of the common elements can substitute for each other based on size. For example, Si^{4+} and Al^{3+} may substitute in tetrahedral sites, Mg^{2+}, Fe^{2+}, Fe^{3+} and Al^{3+} may substitute in octahedral sites, and Na^+ and Ca^{2+} may substitute in 12-fold sites. Two or more elements, such as Mg^{2+} and Fe^{2+}, that can occupy the same structural site in a crystal structure are **diadochic**. They can proxy or substitute for each other. As a general rule, if the difference in ion size is less than 15%, extensive substitution is usually possible; if the difference in ion size is greater, substitution becomes limited.

Temperature has a substantial influence on the degree to which elements of different sizes may substitute for each other. For example, when in 8-fold coordination with oxygen, Na^+ and K^+ have effective ionic radii of 1.32 and 1.65 Å, a difference of about 25%. Only limited substitution of K^+ for Na^+ is likely in minerals crystallized at temperatures normally found at or near the Earth's surface. At high temperatures, however, substantially greater tolerance of size difference is allowed. Minerals such as K-feldspar crystallized from lavas at high temperatures may display extensive Na^+–K^+ substitution.

Because the extent of ionic substitution of dissimilar-sized cations is strongly influenced by temperature, the degree of that substitution can therefore be used to imply something about the temperature of crystallization or recrystallization of a mineral. Further, because the extent of substitution of dissimilar-sized cations decreases with lowered temperature, samples that grew with intermediate compositions at high temperature can, on cooling, be expected to unmix into two phases—one rich in the larger cation and the other rich in the smaller cation. This type of unmixing is known as **exsolution** and is described in more detail in Chapter 5.

Charge

Charge neutrality must be maintained in any substitution mechanism. The ions that substitute for each other in a crystal structure may have the same or different charges. However, the charge difference of substituting cations is rarely greater than 1, if only because the size difference would become too great. Simple substitution is possible if the cations have the same charge. Coupled, omission, or interstitial substitution is invoked if cation charges are different.

Simple Substitution

Simple substitution involves substituting cations with identical charges. If the cations are also about the same size, substitution is generally extensive. The mineral olivine (Figure 4.15a) forms a continuous solid solution series

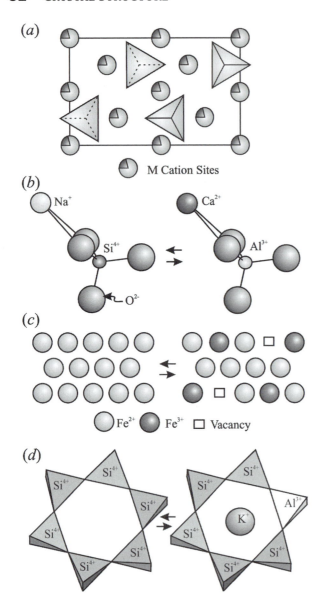

Figure 4.15 Solid solution. (*a*) Simple substitution in olivine, whose structure is viewed down the *a* axis. The octahedral M sites between tetrahedra may be occupied by either Mg^{2+} or Fe^{2+}. The shaded wedge shown on each M cation graphically indicates the percentage of the M sites that are occupied by Fe^{2+} (22% in this case). (*b*) Coupled substitution in plagioclase shown schematically. To maintain charge balance, substitution of Ca^{2+} for Na^+ is balanced by substitution of Al^{3+} for Si^{4+}. (*c*) Omission substitution of $3Fe^{2+} \Leftrightarrow 2Fe^{3+} + \square$ in pyrrhotite shown schematically. (*d*) Interstitial ionic substitution in beryl. Insertion of large monovalent cations (K^+, Na^+, Rb^+, etc.) in the open channelways defined by the rings of silicon tetrahedra is balanced by substitution of Al^{3+} for Si^{4+} in the tetrahedra; H_2O, CO_2, and other species also may occupy the channelways.

whose end members are forsterite (Mg_2SiO_4) and fayalite (Fe_2SiO_4). Note that the subscripts for all elements in the end members are integers. The formula for olivine is conventionally written $(Mg,Fe)_2SiO_4$. The parentheses indicate that Mg^{2+} and Fe^{2+} are interchangeable in

octahedral sites known as the *M* sites. Most olivine has compositions intermediate between forsterite and fayalite. Fe^{2+} and Mg^{2+} are diadochic: they readily substitute for each other because their radii in octahedral coordination with O^{2-} are 0.75 and 0.86 Å, respectively, and the charges are the same.

Coupled Substitution

Coupled substitution maintains charge balance by coupling one substitution that increases the charge with another that reduces the charge. The mineral plagioclase is a good example (Figure 4.15b). The end members are albite ($NaAlSi_3O_8$) and anorthite ($CaAl_2Si_2O_8$). The Ca^{2+} and Na^+ both occupy a distorted 8-fold coordination site and Al^{3+} and Si^{4+} both occupy tetrahedral coordination sites. Note that size considerations allow these substitutions (Table 4.1). To substitute a Ca^{2+} for a Na^+ it is necessary to also substitute an Al^{3+} for a Si^{4+}:

$$Ca^{2+} + Al^{3+} \Leftrightarrow Na^+ + Si^{4+}$$

Note that charge balance is maintained—both $Ca^{2+} + Al^{3+}$ and $Na^+ + Si^{4+}$ have a net charge of +5.

In some cases coupled substitution involves only a single structural site. In the mineral corundum, $Fe^{2+} + Ti^{4+}$ may substitute for $2Al^{3+}$ in octahedral sites. Coupled substitution also may involve both cations and anions. For example, in both biotite and hornblende, Fe^{2+} may be replaced by Fe^{3+}, balanced by replacement of OH^- by O^{2-}:

$$Fe^{2+} + OH^- \Leftrightarrow Fe^{3+} + O^{2-}$$

Omission Substitution

Omission substitution maintains charge balance when ions of different charge substitute for each other by leaving structural sites vacant or unfilled. The general relation for cations is

$$(n+1)M^{n+} \Leftrightarrow nM^{(n+1)+} + \square$$

where M^{n+} and $M^{(n+1)+}$ are the two different cations substituting for each other, n is the charge of lower charged cation, and $\square$ is a vacant site that normally would be occupied by M^{n+}. Omission substitution is found in **pyrrhotite**. Its structure consists of hexagonal close-packed sulfur, with iron in octahedral coordination sites between the layers of sulfur. If all of the octahedral sites are occupied by Fe^{2+} the formula is FeS. However, in most samples some of the Fe^{2+} is replaced by Fe^{3+}. To maintain charge balance, three Fe^{2+} can be replaced by only two Fe^{3+} leaving one site vacant (Figure 4.15c).

$$3Fe^{2+} \Leftrightarrow 2Fe^{3+} + \square$$

The amount of Fe^{3+} that can substitute for Fe^{2+} in pyrrhotite is limited to less than about 20%, so the formula for pyrrhotite is $(Fe^{2+}_{1-3x}Fe^{3+}_{2x})\square_x S$ where x is the number of vacant octahedral sites per formula unit.

Interstitial Substitution

Interstitial solid solution is a variation of coupled substitution in which charge balance is maintained by placing ions in sites that normally are vacant. Minerals whose structures provide large openings are particularly susceptible to interstitial substitution. The mineral beryl ($Al_2Be_3Si_6O_{18}$), for example, is constructed of rings of silicon tetrahedra that are stacked atop each other to form channel-like cavities in the center of the rings (Figure 4.15d). Various monovalent cations, including K^+, Rb^+, and Cs^+, can be inserted into these cavities. Charge balance is maintained by substitution of Al^{3+} and Be^{2+} for Si^{4+} in tetrahedral sites:

$$\square + Si^{4+} \Leftrightarrow Al^{3+} + (K^+, Rb^+, Cs^+) \quad \text{or}$$
$$\square + Si^{4+} \Leftrightarrow Be^{2+} + 2(K^+, Rb^+, Cs^+)$$

Other species, including H_2O and CO_2, also may be found in the cavities.

MINERAL FORMULAS

Mineral formulas are typically written to provide structural information. The basic rules for ionic solids include the following:

- Cations are written first, followed by the anion(s) or anionic group.
- Charges must balance. The total charge of cations must equal the total charge of anions.
- Cations in the same structural site are grouped together.
- Cations in different structural sites are listed in order of decreasing coordination number.

For historical or other reasons, these rules are not always followed, but they are nonetheless quite useful. The idealized formula for a common pyroxene (described in Chapter 14) illustrates how these rules are applied.

$$CaMgSi_2O_6$$

The cations are Ca^{2+}, Mg^{2+}, and Si^{4+}, and they coordinate with O^{2-}, which is the only anion. That the charges balance can be determined by tallying the charges of cations and anions as shown in Table 4.6.

Based on knowledge of cation size we can anticipate that Si will be in 4-fold coordination, Mg in 6-fold coordination, and Ca in 8-fold coordination, consistent with the order in which they are written. If desired, the coordination of each cation in the formula may be explicitly indicated by placing the coordination number in Roman numerals as a superscript immediately preceding the cation in the formula as follows.

$$^{VIII}Ca\,^{VI}Mg\,^{IV}Si_2O_6$$

Table 4.6 Charge Balance Calculation for Diopside ($CaMgSi_2O_6$)

Cation	Charge	Stoichiometric Coefficient	Total Charge[a]
Ca	2+	1	+2
Mg	2+	1	+2
Si	4+	2	+8
O	2–	6	–12
Total			0

[a] The total charge provided by each ion is equal to the charge of the ion times the stoichiometric coefficient for the ion in the formula. The formula is balanced if the sum of the positive and negative charges equals zero.

If a structural site may interchangeably be occupied by different cations as part of a substitution series, the interchangeable cations are grouped within parentheses. The pyroxene formula can be modified to indicate that the octahedral sites can hold either Mg or Fe as follows.

$$^{VIII}Ca\,^{VI}(Mg,Fe)\,^{IV}Si_2O_6$$

An alternate way of indicating composition variation related to a substitution series is to write the formula to explicitly state the degree of substitution possible. The formula of olivine can be written

$$^{VI}(Mg_{2-x}Fe_x)\,^{IV}SiO_4 \quad (0 \le x \le 2)$$

where the range of possible values for the coefficient x is specified. In this case complete ionic substitution is indicated. If, in a particular olivine, 22% of the octahedral sites are occupied by Fe^{2+} and 78% are occupied by Mg^{2+}, $x = 0.44$ and the formula is

$$Mg_{1.56}Fe_{0.44}SiO_4.$$

It often is convenient to identify a mineral's composition in terms of the relative amounts of the end members. The names of the end member compositions are often given two-letter abbreviations. For olivine, the end members forsterite (Mg_2SiO_4) and fayalite (Fe_2SiO_4) are abbreviated Fo and Fa, respectively. The olivine sample that contains 78% Mg and 22% Fe can be presented as $Fo_{78}Fa_{22}$. In minerals with just two end members this is sometimes shortened to report just one end member (e.g., Fo_{78}) because the sum of both end members (forsterite plus fayalite in olivine) must equal 100%.

GRAPHIC REPRESENTATION

The techniques of chemical analysis of minerals and the procedures by which chemical analyses are converted to mineral formulas are described in Chapter 9. The procedures will be familiar to anyone who has taken a laboratory-based chemistry course. It is convenient in many cases to graphically represent mineral compositions derived from those analyses. Although the possibilities are limited only

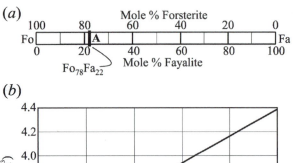

(a)

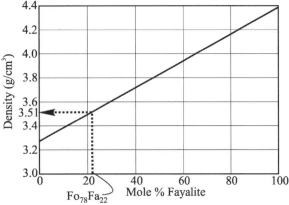

(b)

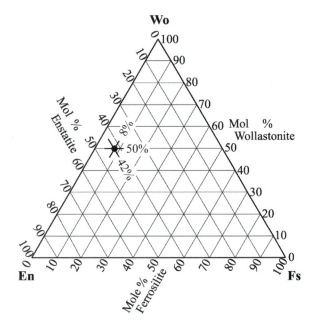

Figure 4.16 Binary diagram of olivine. (a) The end members, forsterite (Mg_2SiO_4) and fayalite (Fe_2SiO_4), are plotted at the ends of a line. The scale is mole percent of the end members. Point **A** indicates a composition with 22% fayalite and 78% forsterite ($Fo_{78}Fa_{22}$). (b) Use of a binary composition diagram to illustrate the variation of density as a function of composition in olivine. Olivine A has a density of 3.51 g/cm^3.

Figure 4.17 Ternary diagram of pyroxene. The end members enstatite (En, $Mg_2Si_2O_6$), ferrosilite (Fs, $Fe_2Si_2O_6$), and wollastonite (Wo, $Ca_2Si_2O_6$) are plotted at the corners of a triangle. Three sets of lines on the diagram allow compositions to be plotted. Horizontal lines indicate the percentage of Wo, lines parallel to the left side of the triangle indicate percentage of Fs, and lines parallel to the right side of the triangle indicate percentage of En. The composition $En_{42}Fs_8Wo_{50}$ is plotted where the 50% Wo line intersects the 8% Fs and 42% En lines.

by the imagination (and the reader's ability to understand), binary and ternary diagrams are widely used.

Binary Diagrams

A binary diagram is used for minerals whose composition can be described in terms of two end members; compositions are represented on a straight line. The ends of the line are commonly selected to be the end-member compositions of a substitution series or other mineral or chemical species of interest. The scale along the line is mole percent of the end members. Figure 4.16a shows a binary diagram for olivine. The left end of the line is taken as forsterite (Mg_2SiO_4) and the right end is fayalite (Fe_2SiO_4). The line is scaled so that the left end represents 100% forsterite and 0% fayalite, the right end is 0% forsterite and 100% fayalite, and the middle is 50% of each. The composition of olivine just discussed (78% forsterite) is shown on the diagram. Figure 4.16b shows use of a binary olivine diagram to illustrate the variation of density as a function of composition.

Ternary Diagrams

A ternary diagram is used when there are three end-member compositions. The diagram is plotted as an equilateral triangle with an end member placed at each corner (Figure 4.17). Each side of the triangle is a binary diagram, and the three sets of lines that cross the diagram indicate the percentage of each end member. For example, the pyroxenes can be

represented by three end members: enstatite ($Mg_2Si_2O_6$), ferrosilite ($Fe_2Si_2O_6$), and wollastonite ($Ca_2Si_2O_6$), which are represented by the symbols En, Fs, and Wo, respectively. The horizontal lines indicate the percentage of Wo: 100% at the top, 50% across the middle of the triangle, and 0% along the bottom edge of the triangle. The lines going up to the right indicate the percentage of Fs, and the lines going up to the left indicate the percentage of En. A composition that is 42% enstatite, 8% ferrosilite, and 50% wollastonite ($En_{42}Fs_8Wo_{50}$) is shown as the point where the 42% En line intersects the 8% Fo and 50% Wo lines.

REFERENCES CITED AND SUGGESTIONS FOR ADDITIONAL READING

Bloss, F. D. 1971. *Crystallography and Crystal Chemistry, an Introduction.* Holt, Rinehart & Winston, New York, 545 p.

Evans, R. C. 1966. *An Introduction to Crystal Chemistry.* Cambridge University Press, Cambridge, 410 p.

Pauling, L. 1960. *The Nature of the Chemical Bond*, 3rd ed. Cornell University Press, Ithaca, NY, 644 p.

Putnis, A. 1992. *Introduction to Mineral Science.* Cambridge University Press, Cambridge, 457 p.

Smyth, J. R., and Bish, D. L. 1988. *Crystal Structures and Cation Sites of the Rock-Forming Minerals.* Allen & Unwin, Boston, 332 p.

CHAPTER 5

Mineral Growth

INTRODUCTION

Many, if not most, geologic processes are fundamentally involved with or influenced by changes in the mineralogic makeup of Earth materials. The rock cycle (Figure 5.1), which is introduced to all introductory geology students, is about mineralogical changes in response to changing geological environments. A magma, when it cools and crystallizes, forms an igneous rock that may consist of feldspars, quartz, and other minerals. When exposed by erosion to the environment at the Earth's surface, these minerals are converted via weathering to yield clay minerals and detrital quartz that become, after transport, deposition, and lithification, a layer of shale. The shale may in turn be buried and subjected to metamorphic conditions of higher temperatures and pressure, which convert the quartz and clay to micas, feldspars, and other minerals. When subjected to sufficiently high temperatures, those minerals may melt to form magmas, completing the cycle. Igneous rocks may similarly be converted to metamorphic rocks, and metamorphic rocks may be converted into sediments and sedimentary rocks.

It follows that the mineralogy of Earth materials must be directly related to the nature of the geologic environment in which those materials were formed. As a generalization, temperature, pressure, chemical composition, rates of heating/cooling, and so forth exert primary controls on mineralogy. Because the Earth has changed over the course of its history, we can anticipate that minerals will document those environmental changes. Similarly, because biologic processes have profoundly influenced Earth's environments, we can anticipate that biologic processes will influence the Earth's mineralogy.

The following questions are therefore addressed in this chapter.

What determines which minerals are formed?

How are phase diagrams used to interpret mineral stability and growth?

How do minerals nucleate and grow?

What types of defects are produced in mineral structures?

What changes occur in minerals after crystallization?

How has the Earth's mineralogy changed through time?

What is the role of biologic processes in mineral growth?

MINERAL STABILITY

To form a mineral, the first and obvious requirement is that the constituent elements be available. The formation of quartz (SiO_2) requires the availability of Si and O, and pyrite (FeS_2) requires Fe and S. Hence, the mineralogy of any rock is fundamentally limited by the bulk composition of the rock.

Given a particular bulk composition, with specific amounts of O, Si, Al, Mg, Fe, and so forth, the list of minerals that might conceivably be formed is large; hundreds can be constructed from the five elements named.

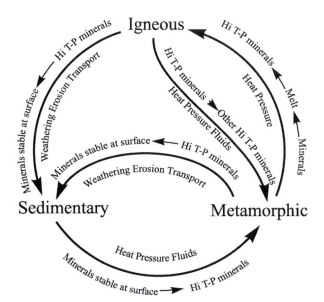

Figure 5.1 The rock cycle, a mineralogist's view. The rock cycle fundamentally involves mineralogic changes in response to different temperature–pressure (T–P) conditions.

To understand which combination of minerals among the many possibilities is actually present, it is necessary to introduce two closely related concepts: **stability** and **energy**.

Stability

If two states are compared, one will be more stable if it represents a lower energy configuration. Note that the concept of stability is always based on a comparison among different possibilities. Consider two alternative positions for this book (Figure 5.2): the first on the floor and the other 2 meters above the floor. The location on the floor is more stable because it represents a lower energy position. The relevant measure of energy for this example is gravitational potential energy. The position 2 meters above the floor has higher gravitational potential energy than on the floor. You can determine that this is true by observing that the higher book will spontaneously drop to the floor, but a book on the floor will not spontaneously leap into the air. Spontaneous processes always involve changing from an energy configuration that is higher to one that is lower.

A higher energy configuration does not always spontaneously change to a lower energy configuration. For example, a book placed on a bookshelf has higher

Figure 5.2 Stability. The book two meters above the floor is unstable relative to a position on the floor. If not restrained, the book will spontaneously fall to the floor. The book on the shelf is metastable relative to a position on the floor. While the floor represents a lower energy position, getting there requires a nudge to push the book off the shelf.

gravitational potential energy than one on the floor but will not spontaneously leap off the shelf to acquire the lower energy position—it is **metastable** with respect to the floor. This means that the floor represents lower energy, but the book cannot drop to the floor unless energy is added in the form of a nudge to dislodge it. The energy required to nudge the book from the shelf is called the **activation energy**.

Gibbs Free Energy

Measuring the energy involved in mineralogical/chemical changes is more complex than measuring the gravitational potential energy of a book. A complete derivation of the relevant relationships is well beyond the scope of this book; only a general outline of a few important relations will be provided here. For a thorough treatment, peruse Nordstrom and Munoz (1994) or Fletcher (1993).

The measure of energy with which to judge the stability of minerals is **Gibbs free energy**, whose symbol is G. Units of Gibbs free energy are energy (**calories** or **joules**) per mole. One calorie is the amount of heat energy required to change the temperature of 1 g of water from 15 to 16°C. A joule (J) is 0.2390 calorie.

To compare the Gibbs free energy of two possible mineral arrangements of the same chemical elements to decide which is more stable, it is necessary to have a common basis for comparison. The widely adopted convention in chemistry is to compare Gibbs free energy levels based on the **free energy of formation from the elements**, ΔG_f. The free energy of formation from the elements is the energy difference between the elements in their standard states (e.g., 298 K and 1 atm pressure) and the elements when they are chemically bonded to form a mineral at the temperature and pressure of interest. Given two minerals with the same composition, such as quartz and tridymite (SiO_2), the stable mineral under specified temperature and pressure conditions is the mineral with the lowest ΔG_f. Because ΔG_f varies as a function of both temperature and pressure, under some conditions quartz has lower Gibbs free energy and is more stable, and under others tridymite has lower Gibbs free energy and is more stable.

Mineral Reactions

Mineral transformations involve a conversion of one mineral or set of minerals to form another mineral or set of minerals. They are chemical reactions and can be dealt with by using the normal conventions of chemistry. The polymorphic transition of tridymite to form quartz can be written

$$\text{tridymite} = \text{quartz}$$
$$SiO_2 = SiO_2.$$

A more complex reaction is the formation of K-feldspar and sillimanite at the expense of muscovite and quartz at fairly high temperatures in metamorphic rocks such as mica schist.

muscovite + quartz = K-feldspar
+ sillimanite + water

$$KAl_2(AlSi_3O_{10})(OH)_2 + SiO_2 = KAlSi_3O_8 \\ + Al_2OSiO_4 + H_2O$$

Both reactions are chemically balanced; the number of atoms of each element is the same on reactant and product sides of the reaction.

The stability of a chemical reaction can be expressed in terms of Gibbs free energy:

$$\Delta G_{reaction} = \Delta G_{f(products)} - \Delta G_{f(reactants)} \qquad (5.1)$$

That is, the free energy of a reaction is the difference in the free energy of formation of the product minerals and the reactant minerals. The product minerals are taken to be those to the right of the equal sign in a chemical reaction. If, at a specific temperature–pressure condition, $\Delta G_{reaction} <$ 0, the products are more stable and the reaction can spontaneously proceed. If $\Delta G_{reaction} > 0$, the reactants are more stable; minerals to the right of the equal sign would break down to form the reactant assemblage. The equilibrium condition is obtained if $\Delta G_{reaction} = 0$; the products and reactants have the same free energy and no net reaction will proceed.

Rocks are in most cases made of a number of different minerals. For a given bulk chemical composition of rock and specified conditions of temperature and pressure, the stable collection of minerals in a rock is the collection whose aggregate free energy of formation is the lowest of all the possible mineral combinations that might be made from the chemical elements in the rock. In practice it is difficult to determine the lowest energy mineral configuration for a particular bulk composition of rock at a specified temperature and pressure. Detailed knowledge of the free energy of formation of all relevant minerals as a function of all their possible compositional variations generally is not available or not of sufficient quality to allow accurate calculations.

Further, there can be no assurance that the mineral assemblage in any particular rock actually represents the lowest energy assemblage. Minerals may persist metastably for billions of years unless activation energy is provided by heating, deformation or by the introduction of chemically reactive fluids. Recall that the activation energy is equivalent to the energy required to push a book off a shelf so that it can fall to the lower energy position on the floor. The activation energy triggers the conversion of metastable minerals to lower-energy minerals. Further, rocks are subjected to changes in temperature and pressure as a consequence of tectonic activity, erosion, and so forth, and the mineralogy typically does not continually change to reflect all those activities, particularly at lower temperatures. Most igneous and metamorphic rocks, for example, consist of minerals that are metastable in the weathering environment at the Earth's surface.

Regardless, all mineralogical changes are in the direction of lower energy, more stable mineral assemblages. Therefore, experimental studies of mineral stability provide important information that allows geologists to infer the conditions to which rocks have been subjected and the processes that have helped shape the mineralogy and textures that we can now examine. One useful way of presenting this information is to use phase diagrams.

PHASE DIAGRAMS

To illustrate the stability of minerals and mineral reactions, it is convenient to use **phase diagrams**, which show, for a given set of conditions, which **phase** (or phases) represents the lowest energy configuration of a given composition. A phase can be a mineral, melt, or gas. Ice, for example is a single phase, whereas ice in water represents two phases. We will look only at single-component and binary systems. In single-component systems, all the phases (minerals, melts, etc.) can be described with a single chemical composition. In a binary system, all the phases can be described by combinations of two chemical compositions. While rock and mineral systems with three or more components have been extensively studied, single component and binary systems illustrate many of the important processes involved in mineral growth and stability.

In the examples that follow, it is assumed that equilibrium is maintained. That means that the real-world problems of crystal nucleation and growth are ignored. To actually get crystals to grow requires that the equilibrium conditions be significantly overstepped to provide the activation energy needed to get the chemical reactions to proceed. Before crystals can begin to grow, for example, melts must be cooled significantly below the equilibrium temperature (free energy of formation of crystals and melt are the same). Crystal nucleation and growth are discussed in later sections.

Single-Component Systems

With single-component systems, a common diagram showing the stability of different phases is a temperature–pressure (T–P) diagram. A good example is shown in Figure 5.3 for the aluminum silicates. The aluminum silicates andalusite, sillimanite, and kyanite are polymorphs with the chemical formula $AlAlOSiO_4$. They are common minerals in mica schists produced by metamorphism of shale. The free energy of formation of each of these minerals from aluminum, silicon, and oxygen varies as a function of both temperature and pressure and can be shown by surfaces in temperature–pressure–free energy space (Figure 5.3a). These surfaces are shown as planes for ease of illustration, but they actually are curved. Under some conditions andalusite has the lowest free energy, and under others, sillimanite or kyanite is the lowest-energy form of $AlAlOSiO_4$.

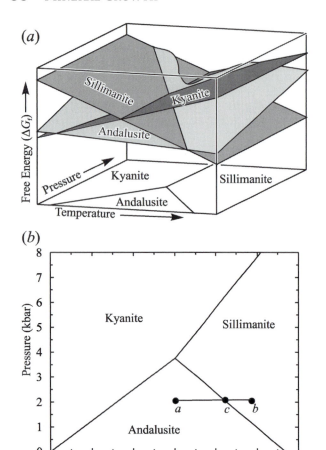

(a)

(b)

Figure 5.3 Aluminum silicate stability relations. (*a*) Schematic illustration of Gibbs free energy of formation of andalusite (light), sillimanite (intermediate), and kyanite (dark) as a function of pressure and temperature. The mineral with the lowest free energy at a specific temperature–pressure condition is the stable phase at those conditions. (*b*) Temperature–pressure stability fields of the aluminum silicates. See text for discussion. Adapted from Hemingway and others (1991).

The T–P field in which each mineral is the most stable is indicated in Figure 5.3b. The stability fields are separated by lines that indicate the T–P conditions along which both minerals have the same ΔG_f; their free energy surfaces intersect. Along each line, the minerals on adjacent sides are in equilibrium. The one T–P condition at which all three minerals have the same ΔG_f and are in equilibrium lies at the point where the three lines join.

If a metamorphic rock contains andalusite, the conditions of metamorphism can be inferred to have been somewhere within the andalusite stability field, indicating low pressures and moderate temperatures such as point *a* (500°C and 2 kbar) in Figure 5.3b. If this same andalusite-bearing rock were progressively heated, say to a temperature indicated by point *b* in the sillimanite field, sillimanite should grow at the expense of the andalusite because sillimanite is the lowest energy form of AlAlOSiO$_4$ under the

new conditions. Andalusite and sillimanite have the same ΔG_f along the line separating their stability fields (point *c*), and neither would break down to form the other at this temperature.

Binary Systems

Binary systems have two components. The components may be the chemical composition of minerals, or they may be another convenient chemical species. In the examples that follow, the components are identified with mineral species. Three examples of crystallization from a melt will be examined: one with no solid solution among the minerals, and two in which solid solution is present.

Figure 5.4 shows the binary system diopside–anorthite at one atmosphere pressure. The diagram shows the melting and crystallizing stability relations for mixtures of diopside, which is a pyroxene (Chapter 14), and anorthite, which is the calcium-rich end of the plagioclase solid solution series (Chapter 12). Diopside and anorthite do not have significant solid solution with each other. Composition is plotted on the horizontal axis, and temperature is plotted on the vertical axis. The pressure is fixed at one atmosphere.

Pure diopside melts at 1392°C and pure anorthite melts at 1553°C. A mixture of diopside and anorthite, however, melts at a lower temperature. The two curved line segments labeled **liquidus** show the composition of liquid in equilibrium with crystals at a particular temperature. Above the liquidus curves, the system is entirely melt. The heavy lines on each side of the diagram are the **solidus** lines, and they show the composition crystals in equilibrium with liquid at a particular temperature. The solidus lines are vertical here because the compositions of diopside and anorthite are presumed to not vary; that is, there's no solid solution. If there is solid solution, these solidus lines will be curved, as will be seen in a later section.

The roughly triangular area labeled *anorthite + melt* shows the temperature–composition combinations in which both melt and crystals of anorthite can coexist. The area labeled *diopside + melt* shows the temperature–composition combinations in which both melt and crystals of diopside coexist. **The ends of a horizontal tie-line drawn from solidus to liquidus within one of the crystals + melt areas at a particular temperature define the compositions of the melt and crystals that are in equilibrium with each other at that temperature.**

The area on the diagram below 1273°C shows where all mixtures of diopside and anorthite are entirely solid. While real rocks are more complicated than this simple system, Figure 5.4 does help illustrate some of the features found in nature with crystallization of minerals from a melt, and the melting of rocks to produce magma.

Consider crystallization of a melt whose composition is 75% anorthite (An) and 25% diopside (Di) beginning at 1600°C shown at point *a* on Figure 5.4 (the percentage of the anorthite component in the melt will be identified as

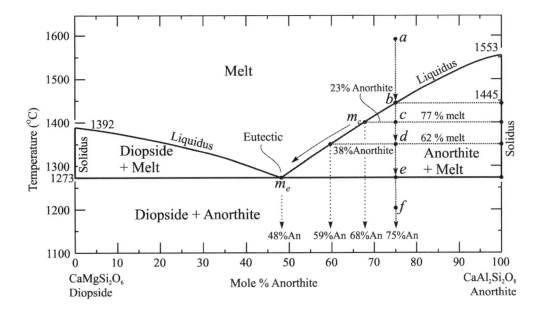

Figure 5.4 Crystallization in the system diopside (Di)–anorthite (An). After Osborn (1942). See text for discussion

X% An). To determine the crystallization history of melt *a*, begin by drawing a vertical dotted line through point *a* so that we can keep track of the bulk composition of the system. At different temperatures, the system may contain various combinations of crystals and melt, but the overall bulk chemical composition of the system will remain constant. We will look at the changes that occur as the system cools under conditions of **equilibrium crystallization**. This means that, as cooling progresses, the crystals and melt remain in equilibrium with each other and represent the lowest energy configuration available to the system under the specified conditions; kinetic issues of mineral nucleation and growth (discussed in a later section) are ignored.

• Point *a*. At point *a* only one phase (the melt) is present. This melt will be allowed to cool until it reaches point *b*.

• Point *b* is right on the liquidus line at 1445°C, which represents the equilibrium temperature at which crystals have the same free energy of formation as the melt. At temperatures below this line, the anorthite + melt field is entered. That means that anorthite crystals will begin to crystallize as heat is removed if the system is to be in its lowest energy configuration. A temperature line drawn through *b* from the solidus to the liquidus shows the compositions of melt and anorthite in equilibrium with each other. Anorthite is presumed to have no solid solution with diopside, so its composition remains fixed for all temperatures.

• Point *c*. Continued cooling down to 1400°C will be accompanied by progressive crystallization and growth of more anorthite, and the residual melt must become progressively enriched in the diopside component. The composition of the melt (68% An) is given by drawing a horizontal

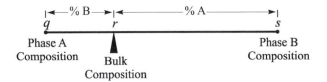

Figure 5.5 Lever rule. Given the bulk composition and a tie-line *qrs* between the compositions of phase A and phase B, the relative amounts of A and B are given by the relative lengths of the two line segments: %A = *rs/qs* and %B = *qr/qs*.

line through *c* to the point at which it intersects the liquidus at m_c and then projecting a line down to read the composition off the horizontal axis. By the **lever rule** (Figure 5.5), we find 77% melt and 23% anorthite crystals. Continued cooling to *d* causes more anorthite to precipitate, so that at 1350°C the system consists of 38% anorthite crystals and 62% melt having a composition of 59% An.

• Point *e* at 1273°C marks the point at which diopside crystals begin to grow from the melt. Three phases (melt, anorthite crystals, and diopside crystals) are present, and these three phases can be together only at the **eutectic temperature**, which, at one atmosphere pressure, is 1273°C. The composition of the melt (m_e = 48% An) is called the **eutectic composition** because it is defined by the **eutectic**, where the two liquidus curves meet. With continued removal of heat from the system, both anorthite and diopside crystals grow as the melt is used up. However, the temperature does not drop below 1273°C until the last of the liquid has been converted into crystals. The proportion of anorthite and diopside crystals being crystallized is the same as the composition of the eutectic melt—48% anorthite crystals and 52% diopside crystals.

• Point *f* at 1200°C is in the anorthite + diopside field, meaning that the magma is now completely crystallized and consists of 75% anorthite crystals and 25% diopside crystals.

If crystallization begins with a diopside-rich composition to the left of the eutectic, crystallization begins with diopside, and the melt composition becomes progressively enriched in the anorthite component until it reaches the eutectic.

Equilibrium melting is the reverse of the crystallization sequence. Starting with a solid rock made up of 75% anorthite and 25% diopside crystals at 1200°C and adding heat will melt the rock as follows.

• Point *f*. The rock is solid anorthite and diopside crystals. Heating brings it to point *e*.
• Point *e*. This is at the eutectic temperature (1273°C). Melting of both diopside and anorthite crystals begins to produce the eutectic melt whose composition is 48% An. Melting will continue until one or the other of the minerals is used up. In this case, diopside crystals are used up first because the bulk composition is to the right of the eutectic. The temperature remains at 1273°C as long as diopside, anorthite, and melt are present. Once all the diopside has been used up, the temperature can continue to increase with addition of heat.
• Point *d*. Anorthite continues to melt and progressively enriches the bulk melt composition in the anorthite

component. At point *d* the melt is 59% An, and the amounts of anorthite crystals and melt are 38% and 62%, respectively. Continued heating to point *c* reduces the amount of anorthite crystals to 23%, and the melt composition is now 68% An.
• Point *b*. At this point the last crystals of anorthite have been melted and the bulk composition of the melt is now 75% An, the same as the initial bulk composition. Continued heating up to point *a* just heats up an entirely molten magma.

Figure 5.6 shows a case with continuous solid solution between two end members. In this case the mineral is olivine, which has complete solid solution between forsterite (Fo, Mg_2SiO_4) and fayalite (Fa, Fe_2SiO_4). The two curves in this binary diagram are the liquidus and the solidus. At a given temperature, the ends of a tie-line from the solidus to the liquidus indicate the composition of coexisting melt and olivine crystals. Above the liquidus, only melt is present; and below the solidus, only olivine crystals are present. Between the solidus and liquidus is the olivine + melt area.

Consider an initial melt consisting of 70% forsterite composition and 30% fayalite composition at 1900°C (point *a*). Begin by drawing a vertical dotted line through point *a* so that the bulk composition can be tracked. Equilibrium crystallization proceeds as follows.

• Point *a* represents the composition 70% Fo and 30% Fa at a temperature of 1900°C. Because it is above the

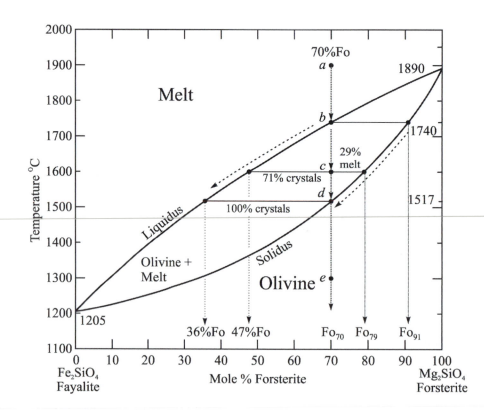

Figure 5.6 Olivine crystallization at 1 atmosphere pressure. After Bowen and Schairer (1935). See text for discussion.

liquidus, it is entirely melt. Progressive cooling will bring the system to point *b*.

• Point *b* is on the liquidus curve, and it marks the temperature (1740°C) at which crystallization of olivine begins if equilibrium is maintained. The small, first-formed crystals of olivine have a composition 91% Mg and 9% Fe, or Fo_{91}, given where the temperature tie-line intersects the solidus. Because these crystals are enriched in Mg relative to the melt, growth of these crystals requires that the residual melt be depleted of Mg and become richer in Fe.

• Progressive cooling to point *c* at 1600°C causes additional crystallization of olivine from the melt. The temperature line going from liquidus to solidus through *c* shows that the melt now has only 47% of the forsterite component, and the crystals are 79% forsterite (Fo_{79}). If equilibrium crystallization is maintained, the first-formed crystals, and all subsequently crystallized olivine, must exchange Fe and Mg with the melt via diffusion to also acquire the composition Fo_{79}. The relative amounts of olivine crystals and melt are given by the lever rule: 71% crystals and 29% melt.

• Still more cooling down to 1517°C at point *d* brings the composition of all the olivine to Fo_{70}, the same as the original composition of the melt—70% forsterite. At this point, crystallization is complete. The composition of the last drop of melt is given where the temperature tie-line intersects the liquidus—36% Fo in this case.

• Point *e* is in the olivine field, so the system is now entirely made of solid crystals of olivine whose composition is Fo_{70}.

Equilibrium melting would trace the melt and crystal compositions in the reverse direction. First-formed melt would have a composition of 36% Fo and melting would progress until all the olivine had been consumed at 1740°C to produce a melt whose composition is 70% Fo.

Figure 5.7 shows a case of two minerals that display solid solution with the added complexity of a **solvus**, a curve that defines the temperature–pressure field where two separate minerals represent the lowest–energy configuration rather than a single homogeneous solid solution. In this case, the minerals are K-feldspar (Ks, $KAlSi_3O_8$) and albite (Ab, $NaAlSi_3O_8$); both are feldspars (Chapter 12), and together they are referred to as alkali feldspars. The presence of water results in a substantial reduction in silicate melting temperatures compared with dry conditions (compare Figures 5.4 and 5.6), but no water enters into either mineral.

From the discussion in Chapter 4, we know that K^+ and Na^+ have sizes that allow them to substitute for each other at high temperatures, but that at low temperatures, substitution is restricted, as is shown by the presence of the solvus. The position of the solvus relative to the solidus and liquidus varies with pressure.

Figure 5.7a shows the system at 1.96 kbar water pressure. The solidus and liquidus curves for both albite and K-feldspar define a K-feldspar + melt area and an albite + melt area. Each of these can be treated like a simple solid

solution system shown in Figure 5.6. The point at which the two segments meet in the middle at 770°C is called a **binary minimum** or **eutectic minimum**. The alkali feldspar field below the solidus shows that complete solid solution is possible between albite and K-feldspar at 1.96 kbar at temperatures below the binary minimum and the top of the solvus. The solvus curve outlines the albite + K-feldspar field. A temperature tie-line from one side of the solvus to the other defines the equilibrium compositions of coexisting albite and K-feldspar at that temperature.

In Figure 5.7a consider a melt (point *a*) at 950°C whose composition is 60% K-feldspar component and 40% albite component. As described in Figure 5.6, crystallization will proceed as follows.

• Crystallization of Ks_{92} crystals begins at point *b*.

• At 850°C (point *c*), Ks_{87} crystals are in equilibrium with 50% Ks melt, and the system consists of 28% crystals and 72% melt.

• Crystallization is complete at 786°C (point *d*), and the crystals have the composition Ks_{60}, the same as the original melt. The last drops of melt have a composition of 33% Ks. Continued cooling of the solid, homogeneous K-feldspar crystals brings the system to the solvus at point *e*.

• Point *e* is on the solvus, which bounds the albite + K-feldspar field, where a single homogeneous feldspar is no longer stable and will unmix into separate albite and K-feldspar phases. The first-formed albite to separate from the K-feldspar has a composition of 14% K and 86% Na—$Ks_{14}Ab_{86}$. Continued cooling to point *f* allows continued growth of progressively more Na-rich albite from K-feldspar, which becomes enriched in K. This unmixing, or exsolution, is described at more length in a later section.

Figure 5.7*b* shows the same system but at 5 kbars water pressure; at this pressure, the solvus intersects the liquidus/solidus curves. Complete solid solution from albite to K-feldspar is no longer possible, and separate fields are present for both albite and K-feldspar. Consider an initial melt (*a*) with 50% Ks and 50% Ab at 850°C. Equilibrium crystallization would proceed as follows.

• Crystallization of K-feldspar (Ks_{91}) begins at point *b*.

• At point *c* at 750°C, Ks_{85} crystals and 37% Ks melt are in equilibrium, and the system consists of 28% crystals and 72% melt. Further cooling brings the system to the eutectic at 703°C.

• Point *d* is at the eutectic temperature (compare Figure 5.4), so albite crystals start to grow in addition to K-feldspar. The last melt has composition 28.5% Ks, and the albite and K-feldspar have compositions of Ks_{19} and Ks_{53}, respectively. When crystallization is complete, the system will consist of 91% K-feldspar crystals and 9% albite crystals.

• Cooling to point *e* allows the already formed crystals of albite and K-feldspar to unmix. The albite becomes progressively Na rich and the K-feldspar becomes progressively

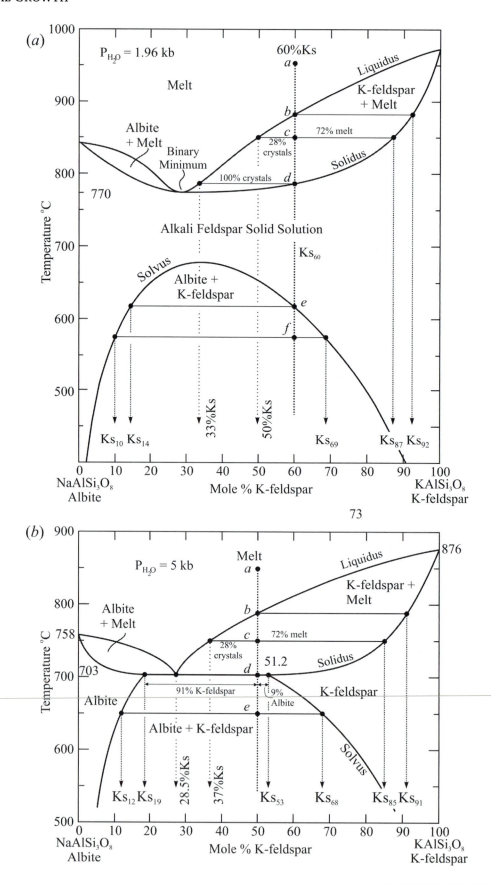

Figure 5.7 Alkali feldspar crystallization. (*a*) The system at 1.96 kbar water pressure (after Bowen and Tuttle, 1950; the solvus is from Thompson and Waldbaum, 1969). A small field of stability for leucite at extreme potassium-rich compositions is ignored. (*b*) The system at 5 kbar water pressure (Morse, 1970). See text for discussion.

K rich as Na and K are exchanged between the already crystallized minerals. This process is described in a later section on exsolution.

MINERAL NUCLEATION

For a new mineral grain to form, the first few atoms/ions from which the mineral is to be constructed must chemically bond to form the nucleus from which the mineral will grow. The formation of mineral nuclei may take place by processes of either homogeneous nucleation or heterogeneous nucleation.

Homogeneous Nucleation

With homogeneous nucleation, the growth of a mineral grain requires that the appropriate atoms and ions find each other and then chemically bond to form what will become the nucleus of a crystal. The nucleus must then grow by progressively adding additional atoms/ions to its surface.

In any chemical system, such as a melt or an aqueous solution, the atoms are constantly moving about, bumping into neighbors, and forming a variety of chemical combinations. Some of these combinations, called embryos, will, by chance, have the composition and structure of a mineral that could crystallize from the melt. Most of the embryos will be small, consisting of only a few atoms, but some will be larger. In general, the size distribution of embryos decreases exponentially with size (Figure 5.8a). Whether these embryos will ultimately grow to form mineral grains

depends on both the stability of the mineral and the size of the embryos.

The embryos cannot grow if the new mineral has a higher Gibbs free energy of formation (is less stable) than the same elements as part of a melt, remaining in solution, or tied up in other minerals. The embryos will therefore break apart, and the constituent atoms/ions will return to the lower energy configuration represented by the melt, solution, or other minerals.

At some point, however, the temperature, pressure, or fluid composition (or all three) may change so that the crystalline structure of a new mineral represents a lower energy configuration for the chemical elements. This does not, however, mean that all the embryos will immediately begin to grow into mineral grains. The complicating factor is the surface energy of the embryos. The disrupted chemical bonds at the surface of each embryo represent a higher energy configuration. The magnitude of the surface energy is proportional to the surface area.

In the discussion that follows, only crystal growth from melts is considered, but the principles derived here apply also to growth from an aqueous solution or from previously existing minerals. Based on the preceding paragraphs, it can be surmised that the energy change involved in forming a crystal nucleus from a melt includes two terms, one related to the free energy of formation of the crystal nucleus and the other to the surface energy of the crystal nucleus.

The first term is the free energy of formation of the crystal nucleus (ΔG_v) whose volume is v

$$\Delta G_v = (\Delta G_{f(xl)} - \Delta G_{f(melt)})v \qquad (5.2)$$

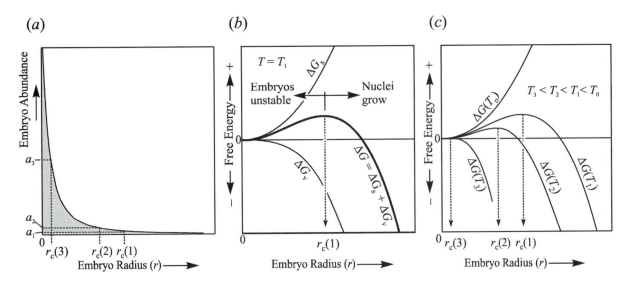

Figure 5.8 Free energy of formation of crystal nuclei from a melt as a function of size. (*a*) Abundance of different-sized crystal embryos in melt. Small embryos are much more abundant than large embryos. (*b*) The Gibbs free energy of formation ΔG of crystal nuclei is the sum of the surface energy ΔG_s and volume energy ΔG_v. Only nuclei larger than the critical growth radius r_c are stable with respect to the melt; additional growth reduces the energy level. Crystal nuclei smaller than r_c will be resorbed spontaneously by the melt because they represent a higher energy configuration than melt and additional growth involves an increase in energy. (*c*) The critical growth radius for different degrees of undercooling. See text for discussion.

where $\Delta G_{f(xl)}$ and $\Delta G_{f(melt)}$ are the free energy of formation of crystal and melt expressed in terms of calories or joules per unit volume of crystal. If $\Delta G_v < 0$, the chemically bonded crystalline solid represents a lower energy configuration than the same atoms in the melt. If $\Delta G_v = 0$, the melt and the crystalline configurations have the same energy levels; they are in equilibrium. If $\Delta G_v > 0$, the melt is the lower energy configuration.

The second term is the surface energy of the crystal nucleus (ΔG_s)

$$\Delta G_s = \gamma a \qquad (5.3)$$

where γ is the surface energy per unit area of crystal and a is the surface area of the crystal nucleus. If we assume a cubic crystal whose edges are length c, the area is $6c^2$ and volume is c^3. The free energy of formation of a crystal is therefore

$$\Delta G = \Delta G_v + \Delta G_s$$
$$= (\Delta G_{f(xl)} - \Delta G_{f(melt)})c^3 + \gamma 6c^2. \qquad (5.4)$$

Figure 5.8b shows the balance between the surface energy (ΔG_s) term, which is positive, and the volume term (ΔG_v), which in this case is negative because the temperature (T_1) is slightly below the equilibrium temperature. If the temperature selected is above the equilibrium temperature, ΔG_v will be positive and the net free energy must be positive for all sizes of embryos.

Figure 5.8c schematically shows the free energy of formation of crystal nuclei as a function of the embryo size for different degrees of undercooling below the equilibrium temperature.

- T_0 = no undercooling. This is the ideal equilibrium condition; the melt and crystalline states have exactly the same energy: $\Delta G_v = 0$. While embryos may form by chance, the surface energy ensures that ΔG of crystal growth is positive for all embryo sizes. All crystal embryos, regardless of size, are therefore unstable with respect to the melt and will be destroyed.
- T_1 = small degree of undercooling. The crystalline state has very slightly lower energy than the melt, but the surface energy is very significant (Figure 5.8b). Embryos smaller than the **critical growth radius**, r_c, are resorbed by the melt because the melt represents lower energy and additional crystal growth requires an even higher energy configuration. Embryos larger than r_c can continue to grow as crystal nuclei to form crystals because additional growth yields a lower energy configuration. For crystals to grow, therefore, the chance collision and combination of atoms in the melt must yield embryos larger than the critical growth radius $r_c(1)$ as shown in Figure 5.8b,c. The abundance of stable nuclei (a_1) is low (Figure 5.8a).
- T_2 = moderate undercooling. ΔG_v is more negative, so the surface energy term becomes less important. The critical growth radius $r_c(2)$ is smaller, and more nuclei (a_2) can be stable.

- T_3 = strong undercooling. ΔG_v is quite negative, so the critical growth radius $r_c(3)$ is small and many nuclei (a_3) can be stable.

The fact that crystals do not begin forming as soon as the temperature drops slightly below the equilibrium temperature indicates that melt is remaining metastable with respect to crystals. To get crystals growing, it is necessary to supercool the melt to provide the activation energy (equivalent to nudging the book off the shelf) to overcome the difficulties related to surface energy. The magnitude of the required supercooling, and therefore the activation energy, is time dependent. For conditions in which cooling is slow and lots of time is available, the activation energy is small. Where cooling is rapid, the activation energy is larger.

The significance of these observations for igneous rocks should be clear. In a magma that cools very slowly at deep crustal levels, so that the degree of undercooling is small, chance collisions of atoms/ions in the melt will form few nuclei that are large enough to become stable. Those few nuclei, however, will have the opportunity to grow over the extended duration of cooling to form the large mineral grains typically found in plutonic igneous rocks. Volcanic rocks and small, shallow intrusives, in contrast, cool rapidly, so the degree of undercooling is large. The much more abundant small nuclei formed by chance collisions of the atoms/ions can therefore become stable and continue to grow. Because many crystals are growing, the size of each is limited and the resulting grain size is smaller.

Growth of a mineral from an aqueous fluid or in the solid state during metamorphism follows similar kinetic constraints. Chance combinations of elements either in solution or along the surfaces between mineral grains must form stable nuclei once the stability field of a new mineral has been reached and then grow by the addition of successive layers of atoms to the surface. To get stable nuclei requires that the equilibrium condition be overstepped to some degree.

With metamorphic rocks, grain size is not simply a matter of cooling rate. Grain size of minerals in metamorphic rocks is principally controlled by three processes. The first is the growth of new minerals as temperature/pressure conditions change. The rate of change of temperature and pressure in most metamorphic environments is quite slow. That means that the degree of overstepping of equilibrium conditions is small, so relatively few nuclei of new mineral(s) can become stable. Newly crystallized minerals in metamorphic rocks often form relatively few large crystals called porphyroblasts, rather than many smaller grains. However, much of the crystallization that produces new minerals occurs as the temperature increases, not decreases. The second is recrystallization. Higher temperatures allow the fine grains of the parent sediment or low-grade metamorphic rock to recrystallize and form fewer, but larger grains. This happens as temperature increases during metamorphism. The third process is deformation, which may distort the crystal lattice of minerals and trigger recrystallization of larger grains into smaller grains to relieve that strain. With strong deformation, grains may be mechanically broken.

Although they might not have realized it, grade school children and other individuals who have grown salt or sugar crystals, or used crystal-growing kits available from nature stores, often have experience with the nucleation issues described here. Rapid evaporation of the water from the solution produces a relatively high degree of supersaturation and allows many embryos to become stable. The result is the formation of many small crystals. If, however, evaporation is restricted so that the degree of supersaturation is quite low, large crystals are more likely to be formed because only a very few embryos are large enough to become stable and grow.

Heterogeneous Nucleation

With heterogeneous nucleation, new minerals nucleate by taking advantage of the structure of an existing mineral. Some of the energetic problems associated with homogeneous nucleation are avoided. If an existing mineral has a surface or structure that is similar to a new mineral, the existing mineral can serve as the nucleus for growth. The need to form embryos is largely eliminated.

Epitaxial nucleation is one variety of heterogeneous nucleation in which the new mineral selectively nucleates on certain crystallographic surfaces of an existing mineral with structural similarities. A common example is the overgrowth of hematite (Fe_2O_3) on magnetite (Fe_3O_4) (Figure 5.9). The magnetite structure is based on cubic close-packed oxygen anions with Fe in octahedral and tetrahedral sites between close-packed layers of oxygen that extend parallel to (111). The hematite structure is based on hexagonal close-packed oxygen anions with Fe cations similarly situated between close-packed layers of oxygen parallel to (001). Recall from Figure 4.1 that hexagonal and cubic close packing differ only in the order of stacking close-packed layers of anions. Hence, hematite can readily nucleate and grow on the (111) faces of magnetite because the layer of oxygen exposed on that face is essentially identical to the layers of oxygen that are parallel to the (001) face of hematite.

Heterogeneous nucleation also may take place on structural defects such as grain boundaries and various imperfections in crystal structures (described later). In each case, formation of a nucleus on these features destroys part of the defect and yields a reduction of energy level that can reduce the size of the activation energy required to get a nucleus growing.

CRYSTAL GROWTH

Once nuclei have become stable, growth must proceed by adding atoms/ions to the surface of the crystal. This entails kinetic problems similar to those of nucleation. Consider the process of adding atoms/ions or some larger building block to the face of a crystal. Adding a new unit to a crystal with a planar surface (Figure 5.10a) is energetically difficult because the new unit stands proud of the surface and bonds only at the surface of contact. The chemical bonds that normally would stabilize the unit in the middle of the crystalline solid are missing, so there is a significant likelihood that the unit will simply be redissolved back into the liquid. In contrast, adding a new unit along the edge of a half-completed layer (Figure 5.10b) on the surface of a crystal is easier because bonds can be made on several sides. This suggests that the limiting factor in crystal growth is overcoming the kinetic difficulties involved in starting a new growth layer on a crystal face. Theoretical considerations dealing with adding layers one at a time suggest that a substantial degree of undercooling or supersaturation should be needed to allow the first growth units of each new layer to be added so that growth can progress.

It turns out that real crystals grow much more easily and rapidly than this theoretical analysis suggests. The solution to this conflict appears to be that real crystals do not grow one layer at a time. Rather, growth often takes advantage of **screw dislocations** (Figure 5.10c). A screw dislocation is a structural defect (described in more detail in a following section) in which one part of the crystal is displaced relative to the other such that the displacement terminates in a line normal to the crystal face. The offset produces a geometry equivalent to a helix or a spiral staircase. As growth progresses, the continuously exposed edge of the screw dislocation spirals upward to always provide a location on which to add the next segment of crystal.

Rate of Growth

Although it seems counterintuitive, the **slowest growing crystal faces are the faces most likely to be prominant on a crystal**. The reason is that fast-growing faces grow themselves out of existence.

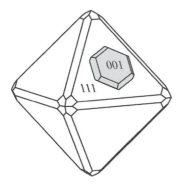

Figure 5.9 Epitaxial growth. Hematite crystal (shaded) may nucleate and grow on the (111) face of magnetite. See text for discussion.

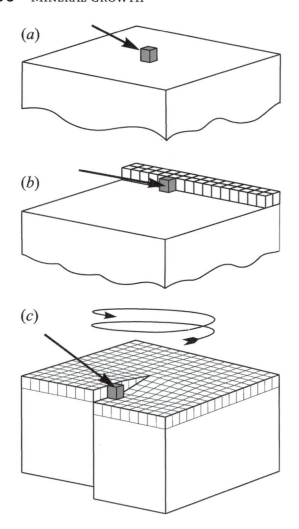

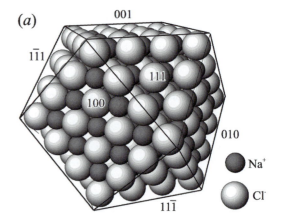

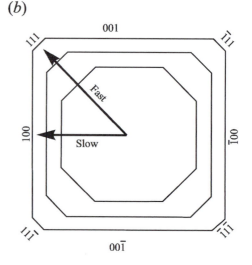

Figure 5.10 Growth on a crystal face. (*a*) Adding a new growth unit to a crystal face is energetically difficult because the new unit stands proud of the surface and has many unsatisfied chemical bonds. (*b*) Adding a new growth unit along the edge of a partially completed layer is easier because more chemical bonds are satisfied. (*c*) Growth on a screw dislocation ensures that an edge always remains available for growth. The crystal grows by adding a continuous spiraling layer.

Figure 5.11 Slow-growing faces become larger. (*a*) Potential crystal faces on halite such as (111) are covered either with all Na^+ or all Cl^- and have a net charge and high surface energy. A very strong attraction for successive layers of Na^+ and Cl^- on the {111} faces produces high growth rates. Faces such as (100) have equal numbers of Na^+ and Cl^-, are uncharged, and have low surface energy. Growth rates are slow. (*b*) Section through the halite crystal showing {001} and {111} faces. Each successive layer on the {111} faces is thicker than on the slow-growing {100} faces. The {111} faces rapidly grow out of existence.

Consider the halite shown in Figure 5.11a, which shows potential octahedral {111} and cubic {001} faces. The octahedral {111} faces are parallel to layers of Na^+ and Cl^- in the structure, so growth consists of adding a layer of Na^+, a layer of Cl^-, another layer of Na^+, and so forth. Growth should be very fast. If the face is covered with Cl^- as shown, the strong negative charge ensures that Na^+ are strongly attracted and easily bonded. If the face is covered with Na^+, Cl^- will similarly be strongly attracted and bonded. In contrast, the {001} faces have equal numbers of Na^+ and Cl^- exposed; they are electrically neutral and lack a net electrical attraction for cations or anions. Adding a new Na^+ to the face, for example, depends on random movement of a Na^+ in solution to a position close to the face, where the Na^+ can be attracted and bonded by the

local negative charge of a Cl^-. The growth rate therefore is relatively slow.

Figure 5.11b shows a cross section through a halite crystal with {001} and {111} faces. If the {111} faces grow faster, each successive layer will be thicker than the growth layers on the {001} faces. The result is that the {111} faces become progressively smaller with each additional layer and ultimately cease to be present.

The fastest-growing faces also have the highest surface energy. The surface of a crystal contains unsatisfied or distorted chemical bonds and therefore represents a higher energy configuration of atoms/ions than the same elements chemically bonded within the center of the crystal. Adding additional atoms/ions to the surface satisfies the chemical bonds and allows distorted bonds to realign themselves, thereby reducing the energy level of the atoms/ions on the old surface. The new surface then bonds the next layer of atoms/ions to reduce its energy level, and so forth, as the crystal grows. The surfaces with the most unsatisfied or distorted chemical bonds, that is, with the highest surface energy, should grow fastest because the greatest energy reduction is achieved there. Consider the halite crystal described in Figure 5.11. Potential {111} faces have a net positive or negative charge and represent a much higher surface energy than the {001} faces, which are uncharged.

All of this is entirely consistent with the principles of thermodynamics. Natural systems tend to change in directions that minimize their energy level. Rapid growth of high-energy faces reduces their size. The faces with low surface energy grow more slowly and are therefore larger. Minimizing the area of high-surface-energy faces and maximizing the area of low-surface-energy faces tend to ensure that the total surface energy is minimized.

The **Law of Bravais** (Chapter 2) also does a good job of predicting the dominant crystal faces for many minerals. Recall that the Law of Bravais holds that the most prominent faces are those that cut the greatest density of lattice nodes—that is, the lattice nodes are most closely spaced. This ensures that the Miller indices for prominent faces usually consist of ones and zeros. Figure 5.12a shows a primitive rectangular lattice looking down the b axis, where the c unit cell dimension (d_{001}) is the smallest and the a unit cell dimension (d_{100}) is larger. A face parallel to (100) has lattice nodes on that face spaced d_{001} apart, which is the closest spacing possible. Lattice nodes on faces parallel to (001) are spaced d_{100} apart, which is more than for (100). All other faces have more widely spaced lattice nodes on the face and the spacing between lattice planes is smaller. For example, lattice planes parallel to (102) are d_{102} apart, the smallest of the three faces, and lattice nodes on the face are spaced $(4d_{100}^2 + d_{001}^2)^{1/2}$ apart, the largest of the three faces. Of these three faces, (102) should be fastest growing and smallest, (001) should be slower growing and larger, and (100) should be slowest growing and largest. Based on these observations, the Law of Bravais can be restated.

> The growth rate of a crystal face is, in general, inversely proportional to the interplanar spacing of that face.

Note that this leads to another seemingly counterintuitive relationship: **the longest dimension of a crystal typically is associated with the shortest unit cell dimension**. Consider the unit cell of sillimanite, shown in Figure 5.12b, whose c unit cell dimension is significantly smaller than the a and b dimensions. Because the interplanar spacing along c is less than along a and b, growth is fastest along the c axis and crystals become elongate parallel to c. This relationship does not always hold, but a perusal of the unit cell dimensions of elongate minerals such as zircon,

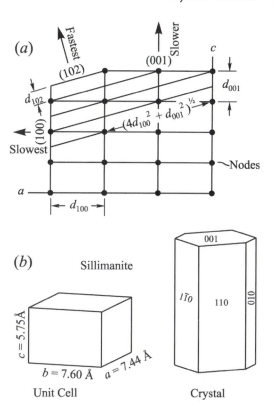

Figure 5.12 Growth rates of crystal faces are inversely proportional to interplanar (d) spacing. (*a*) Growth is fastest normal to the (102) face, slower normal to the (001) face, and slowest normal to the (100) face because $d_{102} < d_{001} < d_{100}$. Crystals should therefore be elongate parallel to the c axis, and the {102} faces should be small or nonexistent. (*b*) Sillimanite. The c unit cell dimension is smallest, so crystals are elongate parallel to the c axis.

tourmaline, apatite, amphibole, and andalusite, in later chapters, shows that the long dimension of most minerals correlates with the smallest unit cell dimension.

Zoned Crystals

Chapters 1 and 3 emphasize that minerals have definite, but not necessarily fixed chemical compositions. For example, the relative amounts of Fe and Mg in olivine may vary in different samples. This is generally indicated in a mineral formula by placing the elements that may substitute for each other in parentheses, as in olivine: $(Mg,Fe)_2SiO_4$. The composition within a single crystal of a mineral that displays solid solution also may vary in a similar manner, and it is very common to find that the center of a crystal has a different composition than the rim.

Because chemical composition may control mineral properties such as color, index of refraction, or other optical properties, chemical zoning may be recognized in thin sections of rock and mineral samples, and sometimes in hand samples as well. A thin section is a very thin (0.03 mm) slice of rock or mineral mounted on a

Figure 5.13 Photomicrograph of a thin section (see Chapter 7) showing zoned crystals of pyroxene (P) (crossed polarizers). The chemical zoning is indicated by a change in the color from the center of crystals to the rim. The slender crystals are plagioclase. See also Figure 12.12 for zoned plagioclase.

glass microscope slide. Thin sections are examined with a petrographic microscope using techniques described in Chapter 7. Figure 5.13 shows chemically zoned crystals of pyroxene. The presence of chemical zoning in a crystal generally indicates that chemical and/or physical conditions changed as the crystal grew.

Plagioclase, which is the most abundant mineral in the Earth's crust, forms a continuous ionic substitution series whose end members are albite ($NaAlSi_3O_8$) and anorthite ($CaAl_2Si_2O_8$) (Chapter 4). Coupled substitution allows Na^+ + Si^{4+} in albite to substitute for Ca^{2+} + Al^{3+} in anorthite.

A phase diagram at 5 kbar pressure for plagioclase in the presence of water is shown in Figure 5.14. Compare with Figure 5.6, which shows a similar solid solution series for olivine. Consider a sample of melted plagioclase whose composition is 55% An at a temperature of 1200°C, as shown in Figure 5.14a. Under equilibrium conditions, crystallization begins at 1100°C with growth of An_{77} plagioclase. At 1050°C, An_{68} crystals are growing from a melt whose composition is 43% An. Crystallization is completed at 980°C, and all the plagioclase has the composition An_{55}, the same as the initial bulk composition.

With **fractional crystallization**, crystals are isolated from the melt as soon as they grow, so that no chemical interaction between melt and old crystals is allowed. For the system shown in Figure 5.14b, crystallization again begins at 1100°C (or slightly below, to allow for nucleation) and the first-formed crystals are An_{77}. Successively more Na-rich layers of plagioclase are then grown on these cores as cooling progresses. At 1050°C, layers of An_{68} are being applied to the exterior of the crystals whose composition becomes more Ca rich inward to the An_{77} core. Because crystals do not interact with the melt, each increment of cooling is equivalent to starting anew with a slightly more Na-rich bulk composition and no crystals. It is therefore possible for the melt composition to progress all the way to

pure albite, from which the last thin layer of albite grows on the crystals. Crystals therefore become zoned from a relatively Ca-rich core to albite on the rim. The bulk composition of the crystals, however, will average out to be the same as the original bulk composition of the melt, which in this case is An_{55}.

In almost all cases, real crystallization of solid solutions is intermediate between ideal equilibrium and ideal fractional crystallization. An approach to equilibrium crystallization is more likely in deep-seated, slowly cooled intrusives, and fractional crystallization is more likely in rapidly cooled, shallow intrusives and volcanics. Whether zoned crystals are formed depends upon the diffusion rate of the relevant elements through the crystals and the amount of time available for diffusion. Diffusion through plagioclase is slow, so zoned plagioclase crystals are common, even for relatively deep-seated intrusives for which cooling is slow. The diffusion of Fe and Mg through the crystal lattice of olivine is relatively fast, so zoned olivine crystals usually are found only in volcanic rocks and shallow intrusives.

The preceding discussion of zoned crystals emphasizes crystallization from magmas. Zoned minerals also are common both in metamorphic rocks and in minerals grown from aqueous fluids, such as in hydrothermal mineral deposits. In all these cases, chemical zoning documents progressive growth of the mineral during periods of changing temperature, pressure, and/or chemical conditions. Zoning can therefore provide important information concerning the changing conditions under which the mineral grew. Deciphering the chemical story may be difficult but often is very productive.

STRUCTURAL DEFECTS

The ideal model of crystalline structures assumes that the regular and repeating pattern of atoms/ions continues without flaw throughout an entire crystal. Real crystals, however, always contain a variety of irregularities or defects that can influence their properties. These defects are conveniently grouped into point defects, line defects, and planar defects.

Point Defects

All real crystals have point defects consisting of vacant sites, atoms out of their correct position, extraneous atoms or ions, or substituted atoms or ions. In general, the density of these defects is higher for crystals grown at high temperature. Defects represent disorder, and high temperatures favor disordered arrangements.

Schottky Defects

Schottky defects are vacancy defects (Figure 5.15a). In ionic crystals, it is necessary for vacant cation sites to be

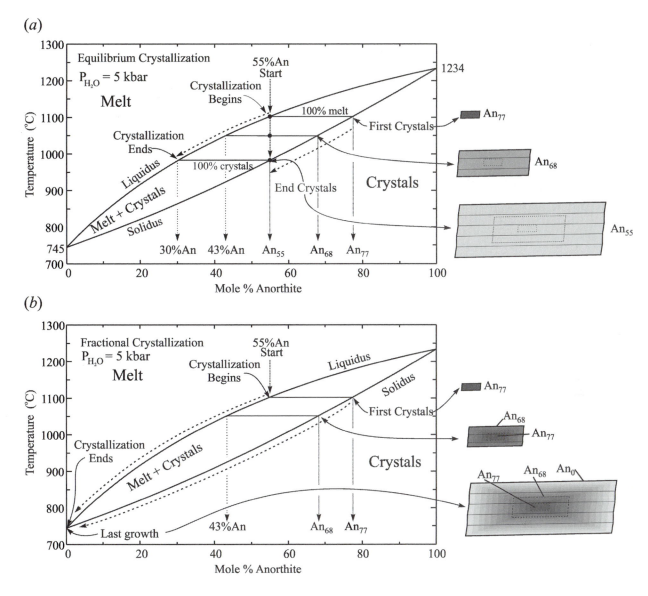

Figure 5.14 Plagioclase phase diagram at 5 kbar water pressure. Adapted from Yoder and others (1957). (*a*) Equilibrium crystallization. (*b*) Fractional crystallization. See text for discussion.

balanced by vacant anion sites for charge balance to be maintained. In a simple structure, such as NaCl, every vacant Na^+ site is balanced by a vacant Cl^- site. In crystals with more complex compositions, charge balance still must be maintained, but the balance between anion and cation vacancies depends on the charges of the ions. The presence of Schottky defects does not influence the stoichiometry of the mineral; the ratio of anions and cations remains the same.

Frenkel Defects

Frenkel defects are mislocation defects (Figure 5.15b). A cation (or anion) is moved from its normal position to an alternative site that normally is not occupied. Cations are more commonly involved in Frenkel

defects because they are smaller than anions. Frenkel defects do not affect charge balance or stoichiometry of minerals.

Impurity Defects

No mineral is pure in the sense of consisting only of the elements present in the ideal chemical formula. Other elements always are present, even if only in trace amounts. **Interstitial defects** are foreign atoms or ions that occupy sites not normally used in the crystal (Figure 5.15c). Their charge is balanced elsewhere, for example, by substituting a lower charge cation for a higher charge cation. A **substitution defect** (Figure 5.15d) is an ion that substitutes for one of the normally present ions in the ideal or pure crystal structure.

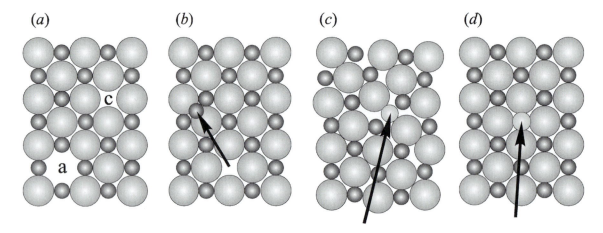

Figure 5.15 Point defects. (*a*) Schottky defects. A missing cation is balanced by a missing anion. (*b*) Frenkel defects. A cation, missing from its normal structural site, is located elsewhere in the structure. (*c*) Interstitial defect. A foreign cation is located in the structure, presumably balanced elsewhere. (*d*) Substitution defect. A foreign cation substitutes for a cation normally in the structure.

It should be evident that the identity of an impurity defect depends on first defining what constitutes the "pure" chemical composition and ideal structure for the mineral. Substitution defects are chemically and structurally the same as simple ionic substitution and interstitial substitution described earlier. The choice of whether to describe the compositional variation of a mineral in terms of impurity defects or of normal ionic substitution depends mostly on the expectation of the mineralogist regarding the normal chemical variability of the mineral. High temperatures allow more impurities to be present because thermal vibrations of the atoms/ions are greater, allowing atoms and ions to occupy positions that would be denied them at lower temperature.

Schottky and Frenkel defects were originally proposed to explain diffusion of atoms and ions through crystal structures. For an ion to move, it must either move into a vacant site, such as a Schottky defect, or at least temporarily occupy an interstitial position, forming a Frenkel defect. The abundance of these defects is a function of temperature; more defects are present at high temperatures. This helps explain the observation that **diffusion** of atoms or ions through a crystalline solid is more rapid at high temperatures.

Line Defects

In tectonic environments that involve high temperatures and pressures and/or slow deformation, it is regularly observed that rocks deform in a ductile manner to form folds that range in scale from microscopic to the dimensions of mountain ranges. To allow deformation of rocks to occur, the constituent mineral grains must be deformed.

Based on extensive studies of the ductile deformation of crystalline metals, such as copper wire or the steel used in automobile bodies, it was observed that crystals deform by slip along favored crystallographic planes and in specific

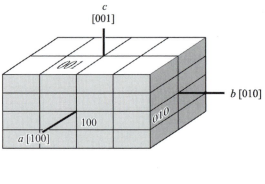

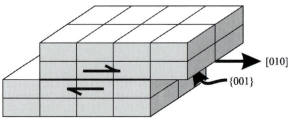

Figure 5.16 Slip system in a crystal lattice. Slip occurs on a crystal plane parallel to (001) and in a direction parallel to [010] (the *b* axis), so the slip system is {001}[010].

directions. The combination of the crystallographic plane and the slip direction together define a **slip system**. In the simple orthorhombic lattice shown in Figure 5.16, the slip system can be described as {001}[010], meaning that slip occurs on crystallographic planes parallel to the (001) face and that the slip direction is parallel to [010]: that is, parallel to the *b* crystal axis.

Producing slip in a crystal by having all the chemical bonds on a slip plane break simultaneously requires orders of magnitude more energy than what is actually required experimentally. This suggests that the method of slip does not involve simultaneously breaking all chemical bonds along the slip plane. Rather, slip propagates through a

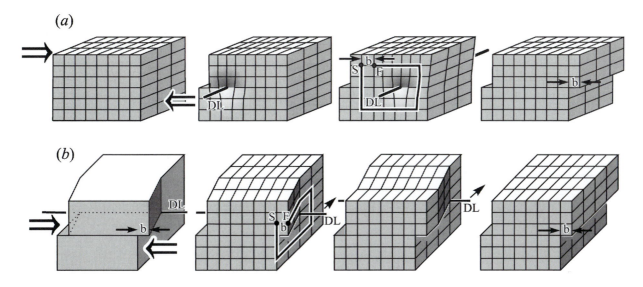

Figure 5.17 Dislocations. The Buergers vector **b** is defined by completing a Buergers circuit. Start at lattice position S and make a box around the dislocation, counting the same number of lattice units in opposite directions, to reach finish position F. Vector **b** between S and F is the Buergers vector. (*a*) An edge dislocation is produced by deforming the crystal as shown. The dislocation line (DL) is perpendicular to the Buergers vector and moves to the right (parallel to the Buergers vector) as deformation progresses. When the dislocation line exits the crystal on the right, the top segment of the crystal is offset relative to the bottom by a distance equal to the Buergers vector. (*b*) Screw dislocation. The dislocation line is parallel to the Buergers vector and moves to the back (perpendicular to the Buergers vector) as deformation progresses. When the dislocation line exits the crystal to the back, the top of the crystal is offset relative to the bottom by a distance equal to the Buergers vector.

crystal so that bonds are broken only along a line defining the leading edge of the slip surface. The edges of the propagating slip surface, where chemical bonds are being broken, are known as **dislocation lines**. They define the boundary between where slip has already occurred and where it has not. There are two varieties of dislocation lines: **edge dislocations** and **screw dislocations**.

Edge Dislocation

With edge dislocations, the dislocation direction known as a **Buergers vector** is at right angles to the dislocation line (Figure 5.17a). The Buergers vector can be recognized by starting at a point S on a lattice node and tracing a circuit around the dislocation, taking an equal number of lattice steps in opposite directions. If a dislocation is enclosed within the Buergers circuit, the circuit will not close and will finish at point F. The vector **b** between F and S is the Buergers vector, and as shown, it is at right angles to the dislocation line. With progressive deformation, the dislocation line moves through the crystal parallel to the Buergers vector, as shown in the sequence.

Screw Dislocation

With screw dislocations, the Buergers vector is parallel to the dislocation line (Figure 5.17b). The applied stress produces a shear parallel to the dislocation plane and parallel to the Buergers vector. However, as the deformation progresses, as shown in the sequence, the dislocation line

moves at right angles to the Buergers vector. Note that if the Buergers circuit were continued, it would define a spiral, like the threads on a screw or a spiral staircase.

Both ends of a dislocation line must either terminate at the surface of a crystal (Figure 5.17), or form a continuous loop within a crystal (Figure 5.18). The nature of the dislocation line changes from edge to screw along its length, but the Buergers vector remains constant and is parallel to the sense of shear that produces the dislocation. Students of structural geology will note that a dislocation is geometrically identical to a fault.

Mineral samples that are strongly deformed typically contain many screw and edge dislocations. One of the consequences of a high density of dislocations is that materials like metals **work harden** as dislocations in different orientations interfere with each other and cease to propagate. Heating a material allows the crystal structure to rearrange itself to remove the dislocations in a process called **annealing**.

In this discussion, edge and screw dislocations have been described in terms of formation as a result of deformation. However, screw dislocations play an important part in facilitating mineral growth as was described in an earlier section on crystal growth.

Planar Defects

Planar defects involve a mismatch of the crystal structure across a surface. The surfaces may be planar, or they may

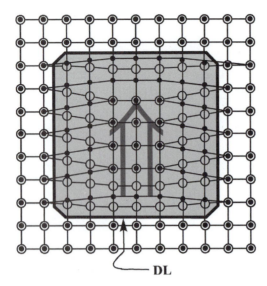

DL

Figure 5.18 Unless terminated at the edge of a crystal, a dislocation line (DL) forms a continuous loop outlining a surface, equivalent to a fault, with movement parallel to the Buergers vector. The "fault surface" (shaded) between the lower layer of circles and upper layer of solid dots is bounded by screw dislocations on the sides and edge dislocations at the top and bottom.

be curved in complex ways. The planar defects described here include grain boundaries, stacking faults, and antiphase domain boundaries. Twinning is described in the following section.

Grain Boundaries

In the mathematical model of a crystalline structure, the regular pattern of atoms and ions continues to infinity in all directions. To an atom sitting in the middle of a crystal that is a few millimeters across, the surface of the crystal is many millions of atoms away, so it behaves as though the structure goes on forever. The boundaries are there, however. The mineral's surface may be a crystal face, a fracture or cleavage surface, or the boundary at which two mineral grains are juxtaposed.

Stacking Faults

Stacking faults are produced when a structure that can be described in terms of layers contains a layer that is out of sequence or position. In the hexagonal closest packed structure shown in Figure 4.1 the layers of spheres alternate between the A and B positions to form an ABABABAB repeating pattern, whereas in cubic closest packing the layers use a third C position to produce a repeating sequence of ABCABCABC, etc. A stacking fault would be present if a layer in the C position were inserted in a hexagonal closest packed structure to produce a pattern ABABCABAB. Minerals that display polytypism, such as the micas and clay minerals, also commonly have stacking faults when the repeating pattern of layers is disrupted.

Antiphase Boundaries

Antiphase boundaries separate segments of a crystal, called **antiphase domains**, that are related to each other by a simple translation. Adjacent antiphase domains have the same crystallographic orientation, but there is a translational mismatch between the domains. Most antiphase boundaries are produced when crystals undergo a displacive polymorphic transition. Figure 5.19a schematically shows the structure of pigeonite, which is a pyroxene (Chapter 14) commonly found in mafic volcanics. The high-temperature structure contains chains of silicon tetrahedra that all have the same geometry. At lower temperatures, pigeonite goes through a displacive polymorphic transition in which the chains distort and the symmetry is reduced. The distortion, however, is accomplished by alternate layers of chains kinking in alternate directions. Ideally, alternate layers of chains kink in opposite directions throughout the crystal. In practice, however, the kinking in one domain follows one pattern and in an adjacent domain it follows the opposite pattern (Figure 5.19b). The result is the formation of antiphase domains bounded by antiphase boundaries. The pattern in one domain can be made to match its neighbor by a simple translation.

Because antiphase domains have identical crystallographic orientations, they cannot be seen in hand samples or by using conventional microscopic techniques, even if the domains are sufficiently large. Imaging of natural and synthetic materials by means of transmission electron microscope techniques shows that the domains are usually highly irregular.

Twinning

Twinning represents a symmetrical intergrowth of two or more crystal segments of the same mineral. Because the segments are joined along a surface, twinning can be considered a variety of planar structural defect. Twinning can take a variety of forms, from the beautiful intergrown cross pattern found in twinned staurolite crystals to the finely laminated twinning seen in plagioclase. Twinning is not the same as the random intergrowth of mineral grains and crystals that invariably takes place in rock and mineral deposits.

In every case the two or more twin segments are related to each other by a symmetry operation called a **twin operation**. There are three possible twin operations: reflection, rotation, and inversion. The **twin law** that describes the twin operation includes specification of the twin operation and identification of the crystallographic plane or axis associated with the twinning.

Twinning by reflection produces two segments of crystals that are related to each other by a mirror parallel to a crystallographic plane common to both segments (Figure 5.20a). The mirror plane is identified by its Miller index. It must not be a plane that is normally a mirror in the untwinned crystal. The twin law is usually expressed

(a)

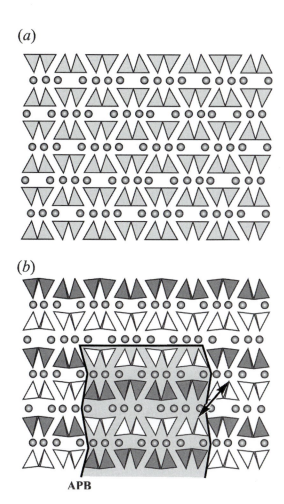

(b)

APB

Figure 5.19 Antiphase boundary in pigeonite $(Mg,Fe,Ca)_2$ Si_2O_6. (a) The high-temperature structure consists of chains of silicon tetrahedra extending in and out of the page. The chains are bonded to each other through the Mg, Fe, and Ca (circles). (b) Upon cooling, the chains distort, with alternate layers of chains kinking inward (darker) and outward (lighter). The pattern of kinking in one part of the crystal may differ from that in adjacent parts, resulting in the formation of an antiphase domain (shaded) bounded by an antiphase boundary (APB). The antiphase domain is related to the surrounding crystal by a translation (arrow).

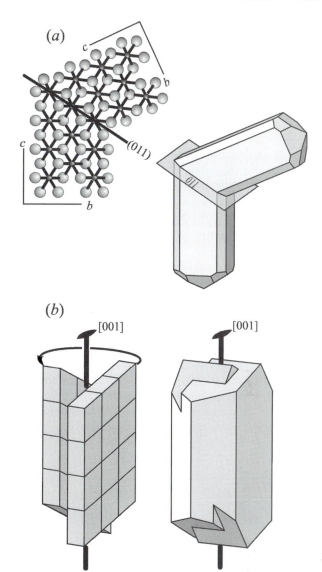

Figure 5.20 Symmetry operations in twinning. (a) Twinning by reflection on {011} in rutile. (b) Twinning by rotation on [001] in K-feldspar to produce a Carlsbad twin.

as "reflection on {hkl}" or "twins on {hkl}," where {hkl} is the form symbol of the mirror plane.

Rotation produces two or more crystal segments related to each other by rotation on a crystallographic axis common to all segments (Figure 5.20b). Almost always the rotation is 2-fold (180°), although other rational rotations are possible. The rotation responsible for the twinning must not duplicate a rotation axis present in the untwinned crystal, although a twin axis may be parallel to a symmetry axis if the rotation required for twinning is not the same as the rotation required by the symmetry axis. For example, the [111] body diagonal through a cube is typically a 3-fold rotation axis. Twinning by 2-fold rotation on [111] is possible because it does not duplicate the 3-fold rotation. The

twin law for a typical 180° rotation is usually expressed as "2-fold rotation on [uvw]," where [uvw] is the crystallographic axis on which rotation occurs. This may be shortened to "twinning on [uvw]" if 2-fold rotation is presumed. If the twin axis is defined as being perpendicular to a specific crystal plane (hkl), the twin law would be expressed as "2-fold rotation ⊥ (hkl)."

With inversion twinning, the two crystal segments are related to each other by inversion through the center. In most cases, however, the processes of rotation or reflection can be used to express the same twin.

The surface along which two twin segments are joined is called the **composition surface**. In some cases the composition surface is irregular and in others it follows a rational crystallographic plane and can be indexed. If the composition surface is a smooth rational crystallographic plane, it is called a **composition plane** and is identified with its Miller index (*hkl*). If the twinning is produced by reflection, the twin plane will be the composition plane. Rotation twins may produce either a composition plane or an irregular composition surface.

Twins may be characterized by whether the twin segments appear to be intergrown or not. **Contact twins** (Figure 5.21) are joined on a rational composition plane, and the twin segments do not appear to be intergrown. Contact twins may be produced by either reflection or rotation. Twin segments that appear to be intergrown (Figure 5.22) are known as **penetration twins** and typically have irregular composition surfaces. Penetration twins are usually produced by rotation.

Simple twins, such as those shown in Figures 5.20–5.22 are composed of just two twin segments. **Multiple twins**, however, are composed of three or more segments repeated by the same twin law. If the twin segments are joined on successive parallel composition planes, the result is the formation of **polysynthetic twins** (Figure 5.23a). Plagioclase is probably the most widely recognized mineral with polysynthetic twins. Plagioclase is triclinic, and twinning by reflection on {010} is repeated over and over, producing thin lamellae in one orientation alternating with lamellae in the reflected orientation. The striations on plagioclase cleavage surfaces that introductory geology students are taught to recognize are the outward manifestation

of this twinning. If the successive composition surfaces are not parallel, a **cyclic twin** may result (Figure 5.23b).

Recognition of twinning, as opposed to random intergrowths of mineral grains and crystals, is not always easy. A pattern of reentrants on crystal faces may indicate the presence of twinning, as does a systematic intergrowth pattern seen on multiple samples of the same mineral. Striations on cleavage surface also may indicate the presence of twinning. Different twin segments on the same crystal or mineral grain often can be recognized when a thin section of a mineral sample is examined with a petrographic microscope (see Chapter 7) because the different segments have different optical orientations.

Common types of twins in the different crystal systems are listed next. Certain twins in some minerals have been given names: Carlsbad twins (Figure 5.20b) in the feldspars and spinel twins (Figure 5.21a) in the spinel group of

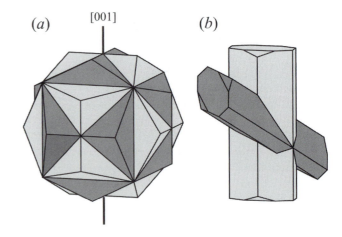

Figure 5.22 Penetration twins. (*a*) Pyrite "Iron Cross" twin by 90° rotation on [001]. (*b*) Staurolite twin by reflection on {231}.

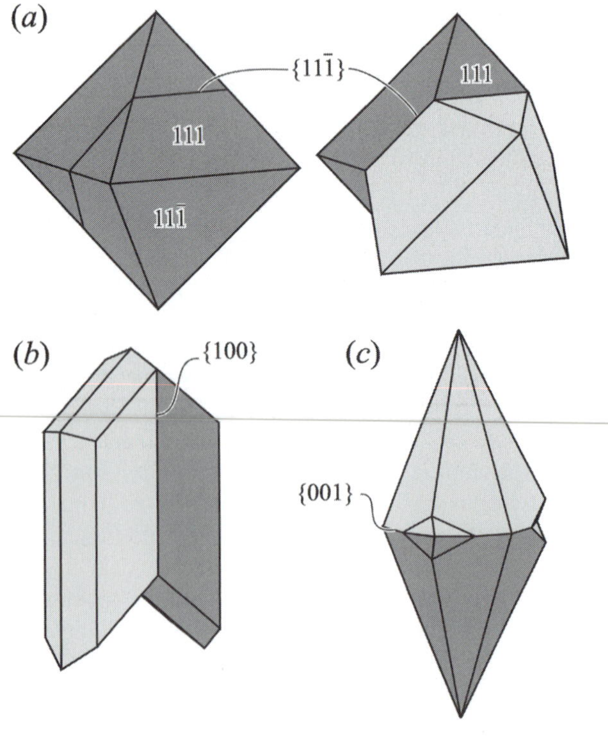

Figure 5.21 Contact twins. (*a*) Octahedron of spinel twinned by reflection on {11$\bar{1}$} (spinel law). (*b*) Gypsum twinned by reflection on {100}. (*c*) Calcite twin with {001} composition plane.

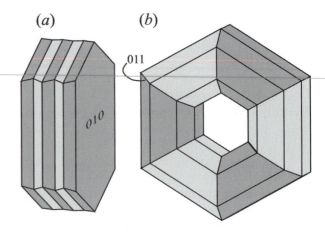

Figure 5.23 Multiple twins. (*a*) Polysynthetic twinning in plagioclase by repeated reflection on {010}. These twins are known as albite twins. (*b*) Cyclic twinning in rutile by repeated reflection on {011}.

minerals, for example. These names usually apply only to the specific minerals from which they were derived.

TRICLINIC SYSTEM: There are few symmetry restrictions regarding twinning in triclinic minerals. Albite twins in plagioclase (Figure 5.23a) are an example.

MONOCLINIC SYSTEM: Twinning by reflection on the {001} and {100} planes is common, although just about any plane except {010} is possible. Because [010] is usually a 2-fold axis, twins on this axis are not found, but rotation on the c axis as in Carlsbad twins (Figure 5.20b) is possible.

ORTHORHOMBIC SYSTEM: Twins by reflection are common parallel to a prism face such as {110}.

TETRAGONAL SYSTEM: Typical twins involve reflection parallel to a pyramid face such as {011} producing elbow-shaped twinned crystals (Figure 5.20a).

HEXAGONAL SYSTEM: Twins by reflection parallel to one of the rhombohedral crystal face {101} or {111} are common. Twins by reflection on {001} or 2-fold rotation on {001} may be found in classes with appropriate symmetry (Figure 5.21c).

ISOMETRIC SYSTEM: Twins by reflection on one of the octahedral faces are known as spinel twins because they are common in spinel (Figure 5.21a). Two-fold rotation on the [111] or [001] axes also may be found (Figure 5.22a).

The mechanisms by which twinned crystals may be produced include growth, transformation, and deformation.

Growth Twinning

There are many examples of penetration twins and contact twins that could have occurred only by growing that way. It is clear in many cases that the twinning was established at, or soon after, the initiation of mineral growth because the twin segments are continuous all the way to the center of crystals. The mechanisms that favor the growth of twinned crystals as opposed to untwinned crystals are poorly understood, but the systematic development of twins in some minerals suggests that the twinned geometry may represent a lower energy or kinetically favored configuration.

Transformation Twinning

Transformation twinning is produced as a consequence of a displacive polymorphic transition. The mineral leucite ($KAlSi_2O_6$), which is common in certain K-rich mafic lavas, displays transformation twinning. Leucite typically forms trapezohedral crystals and crystallizes with isometric symmetry with a cubic unit cell (Figure 5.24a) under the conditions of growth in the lava. When the temperature

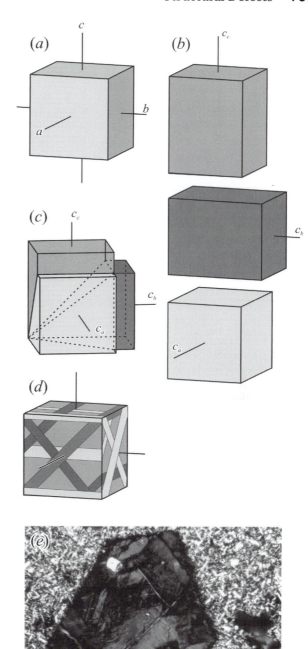

Figure 5.24 Transformation twinning in leucite. (*a*) The unit cell is isometric at temperatures above 665°C. (*b*) Below 665°C, the transition from isometric to tetragonal allows three equally probable orientations (c_a, c_b, c_c) for the tetragonal c axis that are parallel to the original isometric a, b, and c axes. Dimensions are exaggerated. (*c*) The three orientations are related by reflection on the {101} dodecahedral faces. (*d*) Twinned crystals contain lamellae in each of the three different orientations. (*e*) Photomicrograph of a leucite phenocryst showing twin lamellae related to each other by reflection on {101}.

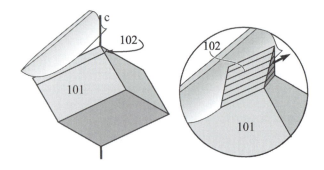

Figure 5.25 Deformation twinning in calcite can be produced by glide on {102} crystallographic planes.

drops below about 665°C, the structure undergoes a displacive polymorphic transition to tetragonal symmetry with a change in unit cell dimensions.

Because the three isometric *a*, *b*, and *c* axes are identical, there is an equal probability that any one of them will become the *c* axis of the tetragonal unit cell (Figure 5.24b). If the entire crystal transforms to a single tetragonal orientation, the crystal would change dimensions consistent with the new tetragonal unit cell. However, the crystal is constrained by the solid rock around it. To maintain the same external shape, different parts of the crystal convert to each of the three possible tetragonal *c* axis orientations (Figure 5.24b,c). The result is to form twin lamellae with each of the three different crystallographic orientations so that the dimensional changes average out, and the macroscopic shape of the crystal does not change (Figure 5.24d).

Deformation Twinning

Deformation twins can be produced in many minerals by the appropriate application of shear stress. In response to the stress, the crystal lattice is distorted to a new orientation by coordinated displacement along successive planes (Figure 5.25). The process is sometimes referred to as a **glide**, and the crystallographic plane along which displacement occurs is a **glide plane**. Deformation twins are always mirror twins, and the glide plane is the mirror. Calcite is particularly easy to twin by deformation; samples may readily be twinned by the pressure of a knife blade on the edges of cleavage rhombs. Calcite in deformed rocks typically displays twin lamellae produced by glide on {102}.

POSTCRYSTALLIZATION PROCESSES

Once a mineral has crystallized, it may undergo a variety of changes in response to changing conditions and/or the passage of time. These changes include ordering, twinning, exsolution, recrystallization, and damage from radioactive decay.

Ordering

The concept of order–disorder in a crystal structure was discussed at some length in relation to order-disorder polymorphism in Chapter 4. Recall that in many minerals, two or more different cations may occupy the same structural sites in a crystal's structure. For example, both Al^{3+} and Si^{4+} are found in the tetrahedral sites in many silicates. In K-feldspar ($KAlSi_3O_8$), Al^{3+} and Si^{4+} are randomly distributed among tetrahedral sites at high temperature (Figure 4.13). At lower temperature, however, the Al^{3+} becomes preferentially segregated into one or two of the four available tetrahedral sites (see Chapter 4 for more detailed discussion). Ordering of Al^{3+} and Si^{4+} among the tetrahedral sites is accompanied by changes in unit cell dimension and, in some cases, symmetry. Many other minerals display order–disorder changes accompanied by unit cell changes, among them the pyroxenes and amphiboles in which Fe^{2+} and Mg^{2+} display varying degrees of ordering into octahedral sites.

Twinning

Twinning was described earlier. Transformation twinning may occur when a mineral goes through a polymorphic phase transition. Leucite (Figure 5.24), for example, becomes polysynthetically twinned on {101} dodecahedral planes when it converts from the high-temperature isometric symmetry to tetragonal symmetry at lower temperatures. The microcline polymorph of K-feldspar ($KAlSi_3O_8$) develops polysynthetic twins in two directions when the symmetry changes from monoclinic to triclinic in response to Al^{3+}–Si^{4+} ordering that occurs with slow cooling.

Recrystallization

Recrystallization means changes to the size and shape of already formed mineral grains and to the collection of processes that remove defects from the crystal structure.

As was previously discussed in relation to crystallization and mineral growth, the surface of a mineral represents a high-energy arrangement because the crystal lattice is terminated and the continuous pattern of chemical bonds is disrupted. We therefore should anticipate that minerals will change or recrystallize in ways that minimize the total amount of surface area. In general, this is accomplished by forming smooth bounding surfaces and by increasing grain size.

If the bounding surfaces between mineral grains in a rock are irregular, there is more surface area than if the boundary were smooth (Figure 5.26a). Hence, it is energetically advantageous to rearrange the atoms/ions on the surfaces along which mineral grains come in contact to form a smooth boundary. Provided temperatures are high enough to allow easy migration, the atoms/ions will diffuse to smooth out the surface, removing high points and filling in low areas. The required temperature conditions are commonly found during metamorphism, and also in magmatic environments. Quartz and feldspars that are allowed to recrystallize at high temperatures typically develop a granular texture in which grains are bounded by smooth surfaces.

One of the most dramatic consequences of contact and regional metamorphism is the systematic increase of grain size with temperature. Figure 5.26b shows the distribution of grain size of calcite grains in marble that have been recrystallized in a contact metamorphic aureole. Marble adjacent to the contact was subjected to higher temperature than at greater distances. Increasing the grain size reduces the surface area per unit volume of rock/mineral. For example, total surface area of a thousand grains whose radius is 0.1 mm is 126 mm^2. If combined into a single grain whose radius is 1 mm the surface area decreases to only 12.6 mm^2. In metamorphic rocks, recrystallization progresses as some grains grow at the expense of their neighbors. The number of grains decreases, their size increases, and surface area and surface energy decrease.

Other structural defects such as the screw and edge dislocations described previously also represent a higher energy configuration than if the lattice were defect-free. The distortion of crystal lattices produced by deformation also represents higher energy. Provided that temperatures are sufficiently high, atoms/ions will tend to rearrange themselves within the crystal lattice to remove these defects.

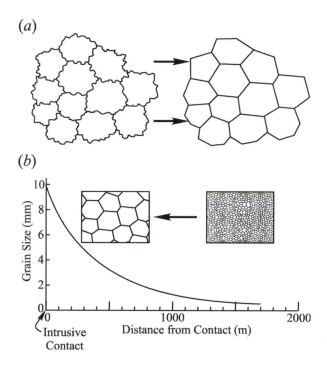

(a)

(b)

Figure 5.26 Recrystallization. (*a*) Recrystallization allows irregular grain boundaries to become smooth, reducing surface energy. (*b*) Average grain size of calcite in marble in a contact metamorphic aureole increases systematically toward the intrusive. Adapted from Grigor'ev (1965). Increase in grain size reduces mineral surface area and surface energy.

Exsolution

In Chapter 4, the conditions for solid solution or substitution of one cation for another in a crystal structure are described. These include the requirement that the substitution must maintain electrical neutrality and that the sizes of the substituting cations must be similar. Recall that crystal structures at high temperatures are somewhat more accommodating of different-sized cations than at low temperatures. At lower temperatures, the range of possible substitution of different-sized cations is much reduced.

The alkali feldspars, whose crystallization was introduced in the section on phase diagrams earlier in this chapter (Figure 5.7), provide a good example of exsolution. The end-member compositions are albite (NaAlSi$_3$O$_8$) and K-feldspar (KAlSi$_3$O$_8$). At high temperatures, Na$^+$ and K$^+$ are readily substituted for each other, and solid solution may extend across the complete compositional range from albite to K-feldspar in the field below the solidus, as shown in Figure 5.7a. At lower temperatures a solvus is present because only limited amounts of K$^+$ may substitute for Na$^+$ in the albite structure, and vice versa in the K-feldspar structure. Figure 5.27a shows just the solvus at 1 kbar of water pressure. Crystallization from a melt takes place at higher temperature at this pressure. As was described earlier, the solvus shows the temperature–composition field in which two feldspars, albite and K-feldspar, represent the lowest energy configuration. In the field outside the solvus, homogeneous alkali feldspar is the low energy configuration.

Begin with an alkali feldspar that crystallized from a granitic magma and now is at 700°C. Its composition, Ab$_{40}$Ks$_{60}$ (40% Na, 60% K), is stable because it lies outside the solvus. If this alkali feldspar is allowed to cool, the solvus will be reached at 590°C. Below that temperature, the alkali feldspar structure cannot accommodate all the Na$^+$. Provided time is available to allow for nucleation, domains of albite will crystallize and grow within the feldspar to take the excess Na$^+$. A tie-line at 590°C from one side of the solvus to the other shows the composition of albite (Ab$_{85}$Ks$_{15}$) in equilibrium with the K-feldspar. Because the structure of K-feldspar is essentially the same as that of albite, formation of albite domains within the K-feldspar requires only diffusion of Na$^+$ into those domains and migration of K$^+$ out. The albite typically forms irregular lamellae within the K-feldspar (Figure 5.27b).

At lower temperatures, the amount of Na$^+$ allowed in the K-feldspar host progressively decreases, as does the amount of K$^+$ allowed in albite. Hence, the albite-rich lamellae become larger and progressively more Na$^+$ rich, and the K-feldspar becomes more K rich as cooling progresses. At 500°C, the K-feldspar composition is Ab$_{28}$Ks$_{72}$ and the albite is Ab$_{93}$Ks$_7$. The relative amounts of K-feldspar host (81%) and albite lamellae (19%) are given by the lever rule (Figure 5.5).

(a)   *(b)*

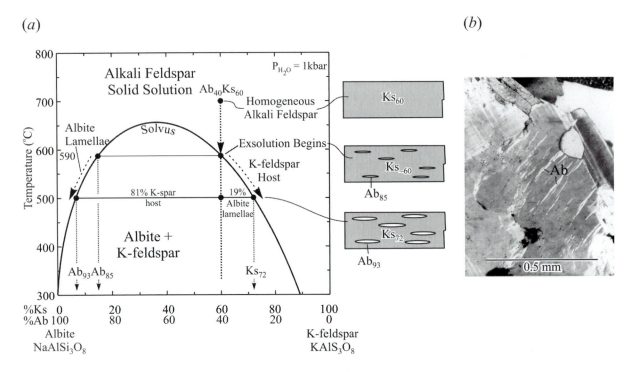

Figure 5.27 Exsolution in alkali feldspar. (*a*) The solvus for alkali feldspar. See text for discussion. Adapted from Smith and Parsons (1974). (*b*) Photomicrograph of a thin section of perthitic K-feldspar with exsolution lamellae of albite (Ab) that extend NE–SW.

Exsolution requires that ions migrate through the crystal structure; therefore the process is temperature and time dependent. If cooling is rapid, as in a volcanic rock or shallow intrusive, time may not be available to allow exsolution to progress to a significant degree. With slow cooling, as in deep-seated intrusives or regional metamorphism, albite exsolution lamellae within K-feldspar may be well developed.

K-feldspar with microscopic or larger albite exsolution lamellae is **perthitic**, or sometimes referred to as **perthite**. A photomicrograph is shown in Figure 5.27b. In some K-feldspar, particularly from granitic pegmatites, the exsolution lamellae may be visible in a hand sample. The terms **antiperthitic** or **antiperthite** are applied to albite from which lamellae of K-feldspar have exsolved as it cools.

Exsolution textures are common in other minerals, notably the pyroxenes (Chapter 14) and Fe–Ti oxides (Chapter 18).

Pseudomorphism

In addition to the other postcrystallization processes described here, one mineral may simply be replaced by another while preserving the external form of the original. Goethite [$FeO(OH)$], for example, may pseudomorphically replace pyrite (FeS_2) crystals and retain the external shape of the original pyrite. The secondary material in a **pseudomorph** may be either a single mineral or a collection of several minerals. **Paramorphs** result from polymorphic phase tranformations in which the replacing mineral is a polymorph (same chemical composition) as the mineral being replaced.

Radioactivity and Minerals

Radioactive isotopes of the elements are routinely incorporated into the crystal structure of many minerals, just like nonradioactive elements. Some of the common or important radioisotopes are listed in Table 5.1. Of these, ^{40}K is the most abundant. It is found in all minerals containing potassium, such as K-feldspar, the micas, and some amphiboles. Radioactive isotopes of uranium and thorium also are widely distributed, though only infrequently is either abundant enough to form minerals with U or Th as major elements. In most cases, U and Th are found as minor constituents in minerals such as zircon ($ZrSiO_4$). ^{14}C is a common isotope found in biological material, including the calcite, aragonite, and apatite that may form the dentition, shells, and skeletal material of mollusks, vertebrates, and other critters. Because of its short half-life, ^{14}C is not preserved in significant amounts in geological materials much older than late Pleistocene.

The nuclei of radioactive isotopes will, given enough time, decay to new daughter elements through processes that include alpha, beta, and electron-capture decay, and fission. The rate of decay is described in terms of the **half-life**, that is, the time it takes one-half of the parent atoms to decay to daughter atoms.

Table 5.1 Selected Radioactive Isotopes Found in Common Minerals

Isotope	Half-Life (years)	Ultimate Daughter Element(s)
^{14}C	5730	^{14}N
^{40}K	1.248×10^9	^{40}Ar, ^{40}Ca
^{87}Rb	48.8×10^9	^{87}Sr
^{232}Th	1.405×10^{10}	^{208}Pb, ^{4}He
^{235}U	7.038×10^8	^{207}Pb, ^{4}He
^{238}U	4.468×10^9	^{206}Pb, ^{4}He

Decay Processes

In **alpha decay**, an alpha particle (α) is ejected from the nucleus of the parent atom, along with the emission of gamma (γ) rays and the release of heat. An alpha particle consists of two protons and two neutrons, identical to the nucleus of a helium atom. Decay of the parent, therefore, produces a daughter element whose atomic number (Z) is two less than the parent, because two protons are lost. The atomic weight of the daughter is reduced by four because two neutrons also are lost. A typical decay is as follows.

$$^{238}_{92}\text{U} \rightarrow \, ^{234}_{90}\text{Th} + \alpha + \gamma + \text{heat}$$

Ultimately the thorium may decay through a series of additional steps to form $^{206}_{82}$Pb. This decay series is utilized in the U–Pb radiometric dating method

Beta decay (also called **negatron decay**) can be considered the breakdown of a neutron into a proton and an electron. The electron, or beta particle (β^-), is ejected from the nucleus along with a particle called an antineutrino ($\overline{v}$), whose mass is very small or zero. The net result is for the daughter to have an atomic number that is one larger than the parent's and to have the same atomic weight because the neutron and proton have the same weight. A beta decay of geological importance is as follows.

$$^{87}_{37}\text{Rb} \rightarrow \, ^{87}_{38}\text{Sr} + \beta^- + \overline{v} + \text{heat} \tag{5.7}$$

This decay provides the basis for the Rb–Sr radiometric dating method.

Positron decay is closely related to beta decay and involves breakdown of a proton into a neutron and a positron (β^+) plus a neutrino. A positron is a particle like an electron but with a positive charge. The positron ultimately combines with a normal electron and is annihilated with the release of gamma radiation. The net result is that the atomic number decreases by one and the atomic weight remains the same. A small amount (~0.001%) of $^{40}_{19}$K decays by positron decay to form $^{40}_{18}$Ar as follows.

$$^{40}_{19}\text{K} \rightarrow \, ^{40}_{18}\text{Ar} + \beta^+ + \overline{v} + \text{heat} \tag{5.8}$$

Electron capture (ec) involves capture of an electron by the nucleus so that a proton is converted to a neutron. The net change is a decrease in the atomic number (Z) by one, with no net change in atomic weight, the same as with positron decay. About 11% of $^{40}_{19}$K decays to $^{40}_{18}$Ar to provide the basis for the K–Ar radiometric dating technique.

$$^{40}_{19}\text{K} + \text{ec} \rightarrow \, ^{40}_{18}\text{Ar} + \text{heat} \tag{5.9}$$

The remaining ~89% of $^{40}_{19}$K decays by beta (negatron) decay to form $^{40}_{20}$Ca.

An additional decay mechanism, called **fission**, is found in ^{238}U and, to a lesser extent, ^{235}U and ^{232}Th. The heavy nuclei of these heavy elements may spontaneously split or fission into two nuclei of approximately equal mass. As a consequence of the fission, a large amount of energy is released and the new nuclei fly apart at high velocity, tearing through and disrupting the crystal lattice of the host mineral.

If the temperature is sufficiently high, the damaged portion of the crystal lattice will be repaired as atoms migrate back to their original positions. At temperatures below the **closing temperature**, however, the crystal lattice does not get repaired and the disrupted trails in the crystal lattice are preserved. These disrupted trails, called **fission tracks**, may be observed under the microscope after the polished mineral surface his been etched with a caustic solution. The abundance of the fission tracks depends on both the fissionable isotopes and how long the mineral has been below its closing temperature. The **fission track dating method** is based on careful analysis of fission tracks found in minerals such as titanite, epidote (Chapter 15), and apatite (Chapter 17). Because different minerals have different closing temperatures, the method is particularly useful in documenting the thermal history of uplifts, basins, and orogenic belts.

Radioactive Decay and Mineral Structure

One way in which radioactive decay influences minerals is that new elements are formed. In many cases, the daughter element is not the same size as the parent, nor does it have the same oxidation state. The decay of ^{40}K, for example, produces both ^{40}Ar and ^{40}Ca. In neither case is the daughter element of the same size, nor does it have the same oxidation state as the original ^{40}K; but both elements wind up occupying structural sites that originally contained the parent ^{40}K. The ^{40}Ar is an inert gas that does not chemically bond with other elements, and its atomic radius is about 1.96 Å. The ^{40}Ca is divalent and has an ionic radius that is significantly smaller than that of K^+ in equivalent coordination (Figure 3.16). The ^{40}Ar and ^{40}Ca in the K sites are **substitution defects** (Figure 5.15d). Only if the temperature is high enough and sufficient time is available can either element diffuse out of the crystal.

If temperatures are below a temperature known as the closing temperature (which varies for different minerals), diffusion of ^{40}Ar is extremely slow and the ^{40}Ar will be effectively trapped in the crystal lattice. Chemical analysis to measure the amounts of ^{40}K and ^{40}Ar can be used to determine when the mineral was last held at temperatures above the closing temperature.

Alpha decay has an additional structural consequence. The alpha particle, which has an atomic mass of four, is propelled at high velocity away from the decaying nucleus for distances up to about 10,000 nm (0.1 mm). Much of its energy is expended in ionizing the matter through which it passes, but as it nears the end of its trajectory, it typically dislodges several hundred atoms to form **Frenkel defects**. Recall that a Frenkel defect consists of an ion that is dislodged from its normal structural position. The decayed atom also recoils for distances up to 10 nm and may dislodge its neighboring atoms, producing a small volume of disrupted crystal structure within a few atoms of the original site.

A mineral such as zircon that contains significant amounts of uranium and thorium, both of which display alpha decay, can experience over 10^{15} alpha decay events per cubic centimeter over a period of several hundred million years. The result may be that the entire crystal structure of the zircon is completely disrupted. Minerals whose structures have been thoroughly disrupted by alpha decay are considered **metamict**. They do not possess long-range crystallographic order and, for all practical purposes, are glasses. Heating to sufficiently high temperatures may allow a metamict mineral to anneal and reacquire its original crystal structure.

Metamict and partially metamict minerals may be recognized because, by comparison to unaltered minerals, their density is lower, their color may be darker, and they may be less hard. Index of refraction and other optical properties also may be influenced (see Chapter 7). Most importantly, metamict minerals do not yield characteristic X-ray diffraction peaks (see Chapter 8) because they lack long-range order. The accumulation of structural defects in the steel components of nuclear reactors from alpha (and other) radiation is also known to cause the metal to become more brittle.

Because alpha particles may travel for up to about 0.1 mm within crystalline solids, they can easily travel into whatever mineral is enclosing the host of the radioactive element. The result is often the formation of a disrupted halo around a small radioactive-element-bearing mineral such as a zircon. These radioactively disrupted zones often are visible in a thin section (a 0.03 mm thick slice of rock/mineral) examined with a petrographic microscope (see Chapter 7) and are known as **pleochroic halos** (Figure 5.28).

MINERAL EVOLUTION

The suite of minerals that make the solid Earth, and particularly the minerals at or near the Earth's surface, has changed through time because the Earth itself has changed

Figure 5.28 Photomicrograph of a thin section of mica schist showing dark pleochroic halos around radioactive zircon (Z) inclusions in biotite (B).

through time. Uniformitarianism and the Earth's various cycles (rock cycle, hydrologic cycle, etc.) which are routinely taught as part of introductory geology courses, can leave the impression that the Earth as we see it now is basically the same as it has always been. However, no important aspect of the Earth's history gets repeated. At the most basic level, the Earth formed only once, from the accumulation of matter orbiting the sun, and its segregation into core, mantle, and crust happened only once. Life developed in the Archean and has evolved ever since. The tectonic evolution in the Hadean, Archean, and Proterozoic eons, which remains poorly understood, has not been repeated over and over. Rather, that early tectonic history set the stage for the development of plate tectonics that now dominates. At some point in the future, the plate tectonic process will cease to operate and another tectonic paradigm will take over. Regardless of the debate over when plate tectonics started, it is important to recognize that time has an arrow, and each day brings changes that irreversibly alter the Earth's environment.

The Earth's mineralogical evolution has been the result of progressive separation and concentration of elements into different geologic environments from their relative homogeneity in the presolar nebula. This increasingly diverse range of bulk compositions also was subjected to a growing range of temperatures and pressures, as well as different activities of water, oxygen, pH, and so forth. The development of life also profoundly changed the chemical environment at the Earth's surface. As geochemical conditions changed, environmental niches in which new minerals could form were opened up.

Earth's mineralogical history can be divided into three major eras: planetary accretion (>4.55 Ga), crust and mantle reworking (4.55–2.5 Ga), and biologically mediated mineralogy (<2.5 Ga) (Ga = gigaannum = billion years) (Hazen and Ferry, 2010; Hazen and others, 1998).

During planetary accretion, the mineralogy of the primordial material that accumulated to form the Earth is probably represented by chondritic meteorites whose major mineralogy includes olivine (Chapter 16) and pyroxene (Chapter 14), as well as a variety of less abundant minerals and carbonaceous material. Some of these minor minerals are known only from meteorites because the chemical and physical conditions needed for their formation are not found on Earth. The accumulation of chondritic material into asteroids and planetesimals necessarily changed the chemical and physical environment that the chondritic material experienced. New and different minerals became the stable phases and, where possible, chemical reactions needed to produce the more stable minerals proceeded. For the Earth, the accumulation of material was sufficient to initiate melting and segregation of the Earth into its core and mantle. Meteorites that document this melting and segregation in planetesimals consist of olivine crystals in an iron–nickel matrix and are called **pallasites**.

The next stage of crust and mantle reworking is still going on and involves melting, magmatic differentiation and fractional crystallization, volcanism, metamorphism, erosion, deposition, and so forth, processes that lead to the generation of the continental crust. Violent impact events, including one that apparently separated the moon from the Earth around 4.55 Ga, served to melt large portions of the Earth. It is possible that the Earth was covered, at least in part, by vast oceans of basaltic magma. Crystal settling and floating in these magma bodies following the principles originally outlined by N. L. Bowen (and now summarized in the Bowen Reaction Series, which every introductory geology student learns) would serve to segregate different compositions of rocks. The generation of granitic rocks, probably by the melting of wet basaltic crustal rocks, represented a substantial segregation of silicon-rich material into the Earth's crust and allowed for the production of quartz, K-feldspar, plagioclase, micas, amphiboles, and other minerals in large volumes. The presence of water allowed the formation of hydroxide, hydrate, carbonate, and evaporite minerals. The oceans probably were developed by 4.3 Ga, and their presence has been critical to many of the processes of crust and mantle reworking.

The development of life on Earth has profoundly affected the Earth's mineralogy. Hazen and Ferry (2010) argue that two-thirds of all known mineral species owe their existence to the environmental changes caused by life on Earth. When life first developed sometime between 3.5 and 4 billion years ago, the atmosphere was anoxic (no free oxygen) and may have consisted of gasses similar to those emitted by modern volcanos (H_2O, CO_2, SO_2, CO, S_2, Cl_2, N_2, H_2) plus ammonia (NH_3) and methane (CH_4). The earliest life appears to have included bacteria that utilized photosynthesis and released O_2 into the atmosphere as a waste product.

It took more than a billion years for the oxygen to accumulate in the atmosphere in substantial amounts. Initially, it appears that bacterially generated oxygen was consumed by weathering processes and reaction with other gasses in the atmosphere. However, around 2.5 Ga, which roughly coincides with the beginning of the Proterozoic, the oxygen content of the atmosphere began to increase substantially in what has been called the Great Oxidation Event. It is perhaps ironic that we are clear beneficiaries of what, from the cyanobacteria's point of view, was pollution and environmental catastrophe. What triggered the change is unclear; but the oxygen derived from cyanobacteria accumulated in the atmosphere and changed the near-surface environment, making possible the development of a host of new mineral species. Of the roughly 4400 mineral species, more than half are oxidized and hydrated weathering products of other minerals. Few of these would be stable in the environment that preceded the Great Oxidation Event.

The origin of life itself seems to be intimately tied to minerals in ways that remain poorly understood (see, e.g., Hazen, 2005). The critical observation is that minerals have a regular and repeating structure, and proteins and other biologic building blocks also must have a regular and repeating structure. The ability of mineral surfaces to adsorb and organize organic molecules may have provided one of the critical early steps in the emergence of life.

It has been argued (e.g., by Hazen and Ferry, 2010) that the number of mineral species has systematically increased over the course of geologic time from as few as 60 in the chondritic meteoritic material from which the Earth formed to the current number of roughly 4400. While this is probably true, it is not rigorously testable because no one was around to compile a list of minerals in the Hadean, and the rock record documenting the Earth's early history is absent. The oldest rock reliably dated so far is just slightly older than 4 billion years (Bowring and Williams, 1999). It is likely that the list of early minerals would be much larger if a better rock record of those early environments existed.

Further, the possibility of **"extinct" mineral species** can be considered. An extinct mineral would be one that was present in some early geochemical environment but has since been lost because the conditions needed for its formation no longer exist and no samples have survived normal geologic processes. This speculation notwithstanding, it is clear that geologic environments have changed through time. The suite of mineral species that could be stable at any one time and their relative abundances also must have changed through time.

BIOMINERALIZATION

The literature documenting the intimate relationship between many minerals and biologic processes is now quite extensive. In the preceding section it was argued that perhaps two-thirds of all mineral species owe their existence, either directly or indirectly, to biologic processes, mostly because of changes to the environment begun with the Great Oxidation Event near the end of the Archean.

However, for present purposes, *biomineralization* refers to the processes by which organisms are directly involved with mineral growth. Even with this restricted definition, over 55 phyla in all five kingdoms are involved in mineralization, and about 100 biominerals and mineraloids (Table 5.2) have been identified. It is likely that the number of recognized biominerals will grow as more research is done.

Biomineralization can be separated into two broad categories––**biologically induced mineralization** and **biologically controlled mineralization**. Biologically induced mineralization involves mineral precipitation as a consequence of an organism's interaction with the surrounding aqueous environment or as a by-product of metabolic activity. Biologically controlled mineralization involves the growth of minerals that serve some physiological purpose.

Table 5.2 Biominerals

Mineral	Formula
Carbonates	
Calcite	$CaCO_3$
Aragonite	$CaCO_3$
Vaterite	$CaCO_3$
Monohydrocalcite	$CaCO_3 \cdot H_2O$
Dolomite	$CaMg(CO_3)_2$
Hydrocerussite	$Pb_3(CO_3)_2(OH)_2$
Rhodochrosite	$MnCO_3$
Siderite	$FeCO_3$
Strontianite	$SrCO_3$
Lansfordite	$MgCO_3 \cdot 5H_2O$
Nesquehonite	$MgCO_3 \cdot 3H_2O$
Magnesite	$MgCO_3$
Phosphates	
Brushite	$CaHPO_4.2H_2O$
Francolite	$Ca_{10}(PO_4)_6F_2$
Carbonate apatite	$Ca_5(PO_4,CO_3)_3(OH)$
Whitlockite	$Ca_{18}H_2(Mg,Fe^{2+})_2(PO_4)_{14}$
Struvite	$Mg(NH_4)(PO_4) \cdot 6H_2O$
Bakhchisaraitsevite	$Na_2Mg_5(PO_4)_4.7H_2O$
Hannayite	$Mg_3(NH_4)_2H_4(PO_4)_4 \cdot 8H_2O$
Monetite	$Ca[PO_3(OH)]$
Newberyite	$Mg[PO_3(OH)] \cdot 3H_2O$
Octocalcium phosphate	$Ca_8H_2(PO_4)_6$
Hazenite	$KNaMg_2(PO_4)_2 \cdot 14H_2O$
Vivianite	$Fe^{2+}_3(PO_4)_2 \cdot 8H_2O$
Sulfates	
Gypsum	$CaSO_4 \cdot 2H_2O$
Barite	$BaSO_4$
Celestine	$SrSO_4$
Jarosite	$KFe^{3+}_3(SO_4)_2(OH)_6$
Melanterite	$FeSO_4 \cdot 7H_2O$
Aphthitalite	$K_3Na(SO_4)_2$
Ardealite	$Ca_2[PO_3(OH)](SO_4) \cdot 4H_2O$
Hexahydrite	$MgSO_4 \cdot 6H_2O$
Schwertmannite	$Fe_8O_8SO_4(OH)_6$
Epsomite	$MgSO_4 \cdot 7H_2O$
Sulfides and Sulfosalts	
Pyrite	FeS_2
Hydrotroilite	$FeS \cdot nH_2O$
Marcasite	FeS_2

Mineral	Formula
Chalcopyrite	$CuFeS_2$
Cubic FeS	FeS
Sphalerite	ZnS
Wurtzite	ZnS
Galena	PbS
Greigite	Fe_3S_4
Mackinawite	$(Fe,Ni)_9S_8$
Acanthite	Ag_2S
Pyrrhotite	$Fe_{1-x}S$
Millerite	NiS
Proustite	Ag_3AsS_3
Pearceite	$(Ag,Cu)_{16}As_2S_{11}$
Arsenates	
Orpiment	As_2S_3
Realgar	AsS
Chlorides	
Atacamite	$Cu_2Cl(OH)_3$
Fluorides	
Fluorite	CaF_2
Hieratite	K_2SiF_6
Oxides	
Magnetite	αFe_2O_3
Anatase	TiO_2
Ilmenite	$FeTiO_3$
Maghemite	γFe_2O_3
Uraninite	UO_2
Hydroxides and Hydrous Oxides	
Goethite	$\alpha FeO(OH)$
Lepidocrocite	$\gamma FeO(OH)$
Ferrihydrite	$5Fe_2O_3 \cdot 9H_2O$
Todorokite	$(Mn,Ca,Na,K)(Mn,Mg)_6O_{12} \cdot 3H_2O$
Birnessite	$Na_{0.5}Mn_2O_4 \cdot 1.5H_2O$
Other Minerals	
Hoganite	$Cu(CH_3COO)_2 \cdot H_2O$
Idrialite	$C_{22}H_{14}$
Earlandite	$Ca_3(C_6H_5O_7)_2 \cdot 4H_2O$
Glushinskite	$MgC_2O_4 \cdot 4H_2O$
Manganese oxalate	$Mn_2C_2O_4 \cdot 2H_2O$
Abelsonite	$Ni^{2+}C_{31}H_{32}N_4$

(Continued)

Table 5.2 *Continued*

Mineral	Formula
Carpathite	$C_{24}H_{12}$
Hartite	$C_{20}H_{34}$
Sodium urate	$C_5H_3N_4NaO_3$
Ca tartrate	$C_4H_4CaO_6$
Ca malate	$C_4H_4CaO_5$
Kratochvilite	$C_{13}H_{10}$
Lindbergite	$MnC_2O_4 \cdot 2H_2O$
Moolooite	$CuC_2O_4 \cdot nH_2O$
Paceite	$CaCu(CH_3COO)_2 \cdot 6H_2O$
Sodium urate	$C_5H_3N_4NaO_3$
Urea	$CO(NH_2)_2$
Uricite	$C_5H_4N_4O_3$
Weddellite	$CaC_2O_4 \cdot 2H_2O$
Whewellite	$CaC_2O_4 \cdot H_2O$
Guanine	$C_5H_3(NH_2)N_4O$
Native Elements	
Gold	Au
Sulfur	S
Silicates	
Clay	Smectite, nontronite, chlorite
Mineraloids	
Amorphous calcium carbonate (at least 5 types)	$CaCO_3 \cdot H_2O$ or $CaCO_3$
Amorphous calcium phosphate (at least 6 types)	variable
Amorphous calcium pyrophosphate	$Ca_2P_2O_7 \cdot 2H_2O$
Amorphous pyrrhotite	$Fe_{1-x}S$ $(x = 0 - 0.17)$
Amorphous silica/Opal	$SiO_2 \cdot nH_2O$
Amorphous ilmenite	$FeTiO_3$
Amorphous iron oxide	Fe_2O_3
Amorphous manganese oxide	Mn_3O_4

Source: Adapted from Lowenstam and Weiner (1989), Zierenberg and Schiffman (1990), Weiner and Dove (2003), Konhauser (2007), Hazen and others (2008), and sources cited in Chapters 12–20 of this book.

Undoubtedly overlap exists between the two categories, but they provide a useful frame of reference.

Biologically Induced Mineralization

Biologically induced mineralization has the same requirements as any other mineral growth—the requisite chemical elements must be present, the mineral must be thermodynamically more stable than its precursors, and the activation energy to allow nucleation and growth must be available. Most biologically induced mineralization is associated with microorganisms and operates in two ways—either the cells passively provide nucleation sites for mineral growth, or the cells metabolically influence the local chemical environment to promote mineral nucleation.

When a cell provides a nucleation site for a mineral, it provides a site for heterogeneous nucleation and minimizes the activation energy needed to induce crystal growth. Cell surfaces have an abundance of positively and negatively charged sites (ligands) that are capable of attracting and holding ions. The details of the chemical structure of the organic material dictate the spacing of positive and negative charges on the cell surface and therefore which anions/cations are preferentially sorbed. The process is similar to epitaxic nucleation (Figure 5.9). Good examples are provided by iron hydroxide minerals.

In the simplest case, Fe^{2+}-bearing water reacts with oxygen dissolved from the atmosphere and precipitates ferrihydrite $(5Fe_2O_3) \cdot 9H_2O$. Bacteria in the water provide nucleation sites and trigger precipitation of crusts of this ochre-colored material, which is one of the constituents of limonite (Chapter 18). It appears that many bacterial species are capable of providing nucleation sites for mineral growth.

A microorganism's metabolism may directly promote mineralization. One example is provided by *Chlorobium ferrooxidans*. This bacterium utilizes photosynthesis and oxidizes Fe^{2+} to Fe^{3+} as part of its metabolism. Once oxidized, the Fe^{3+} combines with oxygen and water to precipitate as ferric hydroxide. Bog iron is a product of this type of bacterial precipitation in springs. Bog iron served as an important source of iron in medieval Europe and was important in colonial America because other local sources of iron were lacking. Another example involves the reduction of iron to form magnetite (Chapter 18). The bacteria *Geobacter metallireducens* and *Shewanella putrefaciens* both oxidize organic material in sediment and reduce the Fe^{3+} in ferric hydroxides to Fe^{2+} as part of their metabolism. The reduced iron is then available to crystallize as magnetite (Fe_3O_4), which forms very small grains outside the cells and can contribute to the magnetism that gets preserved in sedimentary rocks.

The $CaCO_3$ polymorphs calcite and aragonite (Chapter 17) are minerals whose growth in sedimentary environments is both biologically-induced and controlled. Cyanobacteria utilize photosynthesis to biologically induce precipitation of calcite or aragonite by the general reaction

$$Ca^{2+} + 2HCO_3^- = CaCO_3 + CH_2O + O_2.$$

CH_2O represents the organic material from which the cells are made. In environments where other divalent cations are available, the carbonate could be dolomite $[CaMg(CO_3)_2]$, strontianite $(SrCO_3)$, or magnesite $(MgCO_3)$. Photosynthetic fixation of inorganic carbon increases the pH of the water (makes it less acidic) which decreases the solubility of calcium carbonate so that it becomes oversaturated. The surfaces of the cyanobacteria have ligands that attract and hold Ca^{2+} (or other cations) and promote heterogeneous nucleation of the carbonate minerals. Cyanobacteria and other types of algae can produce large quantities of fine-grained calcite or aragonite mud that can accumulate as limestone layers. **Stromatolites** are produced by microbial mats of cyanobacteria and other organisms that precipitate calcium carbonate.

Other minerals and mineraloids whose growth is biologically induced include dolomite, sulfates (Chapter 17), pyrite and other sulfides (Chapter 19), manganese oxides (Chapter 18), and clay minerals (Chapter 13). It is likely that many mineral alteration processes in the supergene zone of hydrothermal sulfide deposits are influenced by bacterial processes.

Biologically Controlled Mineralization

Organisms most commonly utilize biologically controlled mineralization to create skeletal structures. The cation of choice is usually calcium, and the most common minerals produced are carbonates and phosphates; others include oxides and sulfides. While not a mineral, amorphous silica is abundantly produced by diatoms, radiolarians, and silicoflagellates and may subsequently crystallize as quartz, usually as a microcrystalline form, such as chert. In many cases of biologically controlled mineralization, the organism expends cellular energy to concentrate the constituents (Ca, P, Fe, Mn, Si, F, etc.) needed to precipitate a mineral from an environment in which those chemical elements may be present in low abundance.

Coccolithophores and **foraminifera** utilize biologically controlled mineralization to produce large quantities of $CaCO_3$. Along with mineralization by cyanobacteria and algae, these organisms are responsible for a large portion of carbonate precipitation from marine environments. Coccolithophores are unicellular algae that have been abundant in the oceans since the Jurassic and, in the Cretaceous, experienced a population explosion that was associated with worldwide accumulation of limestone. Coccolithophores grow microscopic **coccoliths** (Figure 5.29), which form micritic mud in limestone. Most species are planktonic, living in the photic zone in mid- to high-latitude coastal waters. Foraminifera are both planktonic and benthic (bottom dwelling); they secrete beautiful shells that range in size from ~30 μm to >1 mm. They first appeared in the Cambrian and occupy a wide range of environments.

Carbonate apatite (Chapter 17) is the biomineral of most importance to vertebrates (Pasteris and others, 2008) because it forms the mineral content of bones and teeth. It also forms the shells of inarticulate brachiopods. Phosphorus and calcium from dietary sources are combined in bone- and teeth-forming cells to precipitate as small platelets of apatite. The structure of bone and tooth is a composite of mineral and cellular material. Bone consists of 45 to 70 wt % apatite and 10 wt % water, with the remainder being collagen and other proteins. Tooth enamel, in contrast, consists of almost pure apatite. The mineral content of bones is in a constant state of dissolution and precipitation: it is entirely replaced every 5 to 10 years. Dissolution of the apatite is promoted if calcium and phosphorus are needed elsewhere in the body and dietary sources lack these elements.

Several magnetic minerals are produced by biologically controlled mineralization. One case is the presence

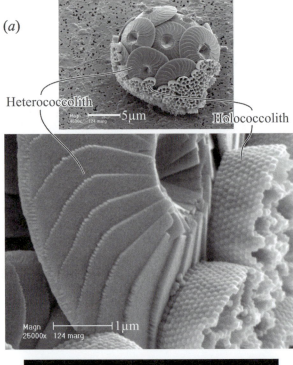

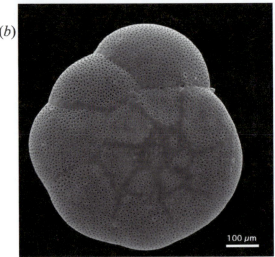

Figure 5.29 Coccoliths and foraminifera. (*a*) A combination coccosphere of the coccolithophore *Calcidiscus leptoporus* sp. *quadriperforatus*. Heterococcoliths and holococcoliths are formed during different phases of coccolithophore growth. Each of the individual plates in a heterococcolith consists of a single calcite crystal. Holococcoliths are made of numerous granules, each of which is a single calcite crystal. It is rare to find coccospheres with both heterococcoliths and holococcoliths. Images © Jeremy Young and Markus Geisen, The Natural History Museum London, and used by permission. (*b*) The foraminifera *Ammonia beccarii* (Thomas and others, 2000).

of fine grains (nanoparticles) of magnetite on the radular teeth of the chiton (a marine mollusk), which the animal uses to grind algae and other food. The scaly-footed snail, found near hydrothermal vents in the Indian Ocean, also precipitates greigite (Fe_3S_4) on its foot, evidently as a sort of protective armor. In fact, magnetite (Fe_3O_4) and maghemite (Fe_2O_3) nanoparticles are found in a wide range of organisms including bacteria, algae, mollusks, insects, and vertebrates. In some cases, such as bacteria and insects, the magnetic particles appear to assist with orientation and navigation, but in others the function is unknown. Even the human brain contains very small particles of magnetite and maghemite, although the function, if any, remains unknown (Pósfai and Dunin-Borkowski, 2009).

One of the exciting developments coming from research on biomineralization is a realization that microbes can be employed to help clean up waters contaminated with heavy metals. Under appropriate geochemical conditions, microbes can induce crystallization of oxides and hydroxides of heavy metals and remove the metals from surface waters. Microbes also can be employed to improve the extraction of metals from ores. The September 2005 issue of *Elements* magazine, published by the Mineralogical Society of America, has a number of excellent review articles.

REFERENCES CITED AND SUGGESTIONS FOR ADDITIONAL READING

Banfield. J. F., and Nealson, K. H. (eds.). 1997. Geomicrobiology: Interactions between microbes and minerals. Reviews in Mineralogy 35, 448 p.

Bard, J. P. 1986. *Microstructures of Igneous and Metamorphic Rocks.* Part I: *Nucleation and Crystal Growth.* D. Reidel, Dordrecht, 263 p.

Barker, A.J. 1998. *Introduction to Metamorphic Textures and Microstructures,* 2nd ed. Stanley Thornes Publishers, Cheltenham, United Kingdom, 264 p.

Bowen, N. L., and Schairer, J. F. 1935. The system MgO–FeO–SiO_2. American Journal of Science, Series 5, 29, 170, 151–217.

Bowen, N. L., and Tuttle, O. F. 1950. The system NaAlSi_3O_8–KAlSi_3O_8–H_2O. Journal of Geology 58, 489–517.

Bowring, S. A., and Williams I. S. 1999. Priscoan (4.00-4.03 Ga) orthogneisses from northwestern Canada. Contributions to Mineralogy and Petrology 134, 3–16.

Cashman, K. V., and Ferry, J. M. 1988. Crystal size distribution (CSD) in rocks and the kinetics and dynamics of crystallizations: III. Metamorphic crystallization. Contributions to Mineralogy and Petrology 99, 401–415.

Dickin, A. P. 2005. *Radiogenic Isotopic Geology,* 2nd ed. Cambridge University Press, Cambridge, 492 p.

Dove, P. M. 2010. The rise of skeletal biominerals. Elements 6, 37–42.

Dove, P. M., De Yoreo, J. J., and Weiner, S. (eds.). 2003. Biomineralization. Reviews in Mineralogy 54, 381 p.

Faure, G., and Mensing, T. M. 2005. *Isotopes, Principles and Applications,* 3rd ed. John Wiley & Sons, Hoboken, NJ, 896 p.

Fletcher, P. 1993. *Geochemical Thermodynamics for Earth Scientists.* John Wiley & Sons, New York, 464 p.

Grigor'ev, D. P. 1965. *Ontogeny of Minerals.* Israel Program for Scientific Translations, Jerusalem, 250 p.

Hazen, R. M. 2005. Genesis: Rocks, minerals, and the geochemical origin of life. Elements 1, 135–137.

Hazen, R. M., and Ferry, J. M. 2010. Mineral evolution: Mineralogy in the fourth dimension. Elements 6, 9–17.

Hazen, R. M., Papineau, D., Bleeker, W., Downs, R. T., Ferry, J. M., McCoy, T. J., Sverjensky, D. A., and Hexiong, Y. 2008. Mineral evolution. American Mineralogist 93, 1693–1720.

Hemingway, B. S., Robie, R. A., Evans, H. T., Jr., and Kerrick, D. M. 1991. Heat capacities and entropies of sillimanite, fibrolite, andalusite, kyanite, and quartz and the Al_2SiO_5 phase diagram. American Mineralogist 76, 1597–1613.

Hibbard, M. J. 1995. *Petrography to Petrogenesis.* Prentice Hall, Englewood Cliffs, NJ, 587 p.

Konhauser, K. 2007. *Introduction to Geomicrobiology.* Blackwell Publishing, Malden, MA, 425 p.

Lowenstam, H. A., and Weiner, S. 1989. *On Biomineralization.* Oxford University Press, New York, 324 p.

Mann, S. 2001. *Biomineralization: Principles and Concepts in Bioinorganic Minerals Chemistry.* Oxford University Press, New York, 216 p.

McCoy, T. J. 2010. Mineralogical evolution of meteorites. Elements 6, 19–23.

Morse, S. A. 1970. Alkali feldspars with water at 5 kb pressure. Journal of Petrology 11, 221–251.

Nordstrom, D. K., and Munoz, J. L. 1994. *Geochemical Thermodynamics.* Blackwell Scientific, New York, 493 p.

Osborn, E. F. 1942. The system CaSiO_3–diopside–anorthite. American Journal of Science 240, 751–788.

Papineau, D. 2010. Mineral environments in the earliest Earth. Elements 6, 25–30.

Pasteris, J. D., Wopenka, B., and Valsami-Jones, E. 2008. Bone and tooth mineralization: Why apatite? Elements 4, 97–104.

Pósfai, M., and Dunin-Borkowski, R. E. 2009. Magnetic nanocrystals in organisms. Elements 5, 235–240.

Price, G. D., and Ross, N. L. (eds.). 1992. *The Stability of Minerals.* Chapman & Hall, London, 368 p.

Smith, J. V. and Parsons, I. 1974. The alkali–feldspar solvus at 1 kilobar water vapour pressure. Mineralogical Magazine 39, 747–767.

Sverjensky, D. A., and Lee, H. 2010. The Great Oxidation Event and mineral diversification. Elements 6, 31–36.

Thomas, E., Gapoichenko, T., Varekamp, J.C., Mecray, E. L., and Buchholtz ten Brink, J.R. 2000. Maps of benthic foraminiferal distribution and environmental changes in Long Island Sound between 1940s and 1990s. U.S. Geological Survey Open File Report OF00-304.

Thompson, J. B., Jr., and Waldbaum, D. R. 1969. Mixing properties of sanidine crystalline solutions: III. Calculations based on two-phase data. American Mineralogist 54, 811–838.

Wagner, G., and Van den Haute, P. 1992. *Fission-Track Dating.* Kluwer Academic Publishers, Boston, 285 p.

Weiner, S., and Dove, P. M. 2003. An overview of biomineralization processes and the problem of the vital effect. Reviews in Mineralogy and Geochemistry 54,1–29.

Yoder, H. S., Stewart, D. B., and Smith J. R. 1957. Ternary feldspars. Carnegie Institute Geophysical Laboratory Yearbook 56, 206–214.

Zierenberg, R. A., and Schiffman, P. 1990. Microbial control of silver mineralization at a sea-floor hydrothermal site in the northern Gorda ridge. Nature 348, 155–157.

PART TWO

Mineral Properties, Study, and Identification

CHAPTER 6

Physical Properties of Minerals

INTRODUCTION

The physical properties of minerals are fundamentally controlled by structure and chemical composition. This means that within the limits of the structural and chemical variation, different samples of the same mineral should display similar properties. It follows also that physical properties can be used to identify minerals.

The physical properties of many minerals make them valuable. For example, graphite and diamond, both polymorphs of carbon, have properties that make them useful for industrial purposes. Diamond, because of its great hardness, is widely used as an abrasive and in cutting tools, and it also can be a prized gemstone because of its clarity and brilliance. Graphite, in contrast, is very soft and is used as a lubricant and in the "lead" in pencils.

It is convenient to combine physical properties into groups that depend on the mass (density and specific gravity), mechanical cohesion (hardness, tenacity, cleavage, fracture, and parting), interaction with light (luster, color, streak, and luminescence), and other properties (magnetism, electrical behavior, taste, odor, and reaction with acid).

MASS-DEPENDENT PROPERTIES

Density

Density (ρ) is defined as mass (m) per unit volume (v).

$$\rho = m/v \qquad (6.1)$$

Density is usually expressed in units of grams per cubic centimeter (g/cm³).

Specific Gravity

Specific gravity (G) is defined as the density of a material divided by the density of water at 4°C.

$$G = \frac{\rho}{\rho_{H_2O}} \qquad (6.2)$$

G is unitless because it is a ratio of densities. Because the density of water at 4°C is essentially 1 g/cm³ (actually

0.999973 g/cm³), the numerical values of specific gravity and density are essentially identical if the latter is expressed in units of grams per cubic centimeter.

Controls on Density and Specific Gravity

Specific gravity of a mineral depends on the chemical composition and on how tightly the atoms/ions are packed together.

The **packing index** (PI) is a convenient means of measuring how tightly the ions in a mineral are packed together. It is defined as

$$PI = \frac{V_I}{V_C} \times 100 \qquad (6.3)$$

where V_C = the unit cell volume and V_I = the total volume of the ions in the unit cell based on their ionic radii (Appendix A). For most minerals, the packing index is between about 35 and 74, meaning that between 35 and 74% of the crystal structure consists of ions. The remaining space is equivalent to the porosity. Close-packed structures of uniform spheres have packing indexes of 74. The fact that most ionic minerals have packing indices less than this is an indication that the anion framework is more open to accommodate cations.

As a general rule, the packing index for minerals formed at high pressure is higher than that for minerals formed at lower pressure. **Kyanite** is the high-pressure polymorph of AlAlOSiO$_4$; it has a packing index of 60.1 and a specific gravity of 3.6. **Andalusite**, which is formed at lower pressure, has a packing index of 52.3 and a specific gravity of about 3.1.

Composition also controls density and specific gravity. Minerals containing high atomic weight elements typically also have high ρ and G. If a mineral displays extensive ionic substitution, specific gravity typically varies systematically between the end-member values. In olivine [(Mg,Fe)$_2$SiO$_4$], for example, the specific gravities of the Mg and Fe end members are 3.26 and 4.39, respectively. Iron-rich varieties typically have higher specific gravity than the magnesium-rich varieties because the atomic weights of iron and magnesium are 55.85 and 24.31, respectively. The relationship is not necessarily linear, however. Intermediate compositions have specific gravities slightly higher than a straight-line relation would suggest (Figure 6.1).

Measurement of Specific Gravity

Most routine measurements of specific gravity involve measurement of the weight of a sample in the air and while suspended in a container of water. A commonly used device that speeds these measurements for samples weighing a few grams is called a **Jolly balance** (Figure 6.2a). A sample of a mineral is first weighed in air (m_a) and then while suspended in water (m_w). The difference in these measurements ($m_a - m_w$) is, by Archimedes' principle, the weight of the water displaced by the sample. The specific gravity of the sample is

$$G = \frac{m_a}{m_a - m_w}. \qquad (6.4)$$

A **pychnometer** (Figure 6.2b) may be used to measure specific gravity for samples too small for the Jolly balance. It is a small flask fitted with a ground-glass stopper that contains a fine hole. It is used as follows:

- Weigh the mineral grain(s) to obtain its weight (m_a).
- Weigh the pychnometer completely filled with water (m_p).
- Place samples in the pychnometer and replace the stopper, forcing water equal to the volume of the samples out of the hole.
- Weigh the pychnometer with sample(s) (m_{p+s}).

The weight of the water displaced from the pychnometer by the sample is equal to $m_a + m_p - m_{p+s}$, so the specific gravity is

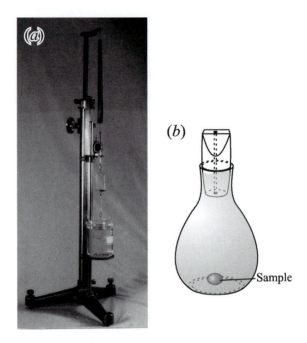

Figure 6.2 Apparatus to measure specific gravity. (*a*) Jolly balance: a sample of a mineral is weighed first in the upper pan and then in the lower pan, which is suspended in water. (*b*) Pychnometer.

$$G = \frac{m_a}{m_a + m_p - m_{p+s}}. \qquad (6.5)$$

Another routine means of estimating specific gravity is by comparison to liquids whose specific gravity is known. Commonly used **heavy liquids** include **diiodomethane** ($G = 3.31$), **bromoform** ($G = 2.90$), and **lithium metatungstate** ($G = 3.0$). A sample of an unknown mineral is placed in one of these heavy liquids. If it sinks, its specific gravity is greater than that of the liquid, and if it floats, its specific gravity is less. If by chance the sample neither sinks nor floats, but remains suspended, liquid and sample have the same specific gravity.

The density of the heavy liquids may be reduced by adding an appropriate solvent to obtain a liquid whose density matches the sample. Acetone is used for diiodomethane and bromoform, and water for lithium metatungstate. If a known volume of this liquid is weighed, its density and specific gravity, and hence the specific gravity of the sample, may be calculated directly. Calibrated samples of known specific gravity also may be used to determined the specific gravity of diluted heavy liquids. Please refer to Material Safety Data Sheets for information concerning safe handling of these materials.

With samples that are sufficiently large, a crude but effective way of evaluating the specific gravity is by simply "hefting" a sample in the hand and comparing the weight to the sample's size. An average specific gravity is taken as quartz ($G = 2.65$); samples that seem to be significantly

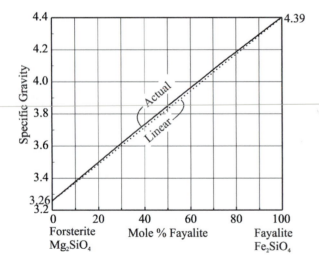

Figure 6.1 Specific gravity of olivine. The dotted line assumes a linear variation in specific gravity between 3.26 for forsterite (Mg_2SiO_4) and 4.39 for fayalite (Fe_2SiO_4). The solid line is the actual specific gravity for intermediate compositions.

heavy in comparison to their size have a higher specific gravity, and those that seem to be light have lower. With a little practice and after some comparisons to known samples, reasonable estimates of specific gravity can be made.

PROPERTIES RELATED TO MECHANICAL COHESION

All mechanical cohesion properties are related to the strength of chemical bonds that hold the mineral together. They include hardness, tenacity, cleavage, fracture, and parting.

Hardness

The ease with which a mineral may be scratched is a measure of its **hardness** (H). The most widely used scratch hardness scale in mineralogy is known as the **Mohs scale of hardness**. It is a unitless 10-point scale, with each hardness number represented by a common mineral (Table 6.1). Each mineral on the list will scratch a mineral whose hardness is less, and be scratched by a harder mineral. No attempt is made to be more precise than half a hardness unit. A mineral whose hardness is intermediate between calcite (3) and fluorite (4), for example, would have a hardness of $3\frac{1}{2}$.

Hardness test kits consisting of scribes containing samples of minerals 1 through 9 are available from some suppliers. For rapid testing in the field or laboratory, common objects such as a fingernail (2+), penny (3), knife blade (~5), window glass (~$5\frac{1}{2}$), and a small fragment of quartz (7) can be quite serviceable. The determinative tables in Appendix B are set up based on these latter values.

In practice, a mineral's hardness is determined by scratching it with materials of known hardness and scratching materials of known hardness with the mineral. Materials of equal hardness will, with difficulty, scratch each other. Beware that a softer mineral may rub off on the surface of a harder material to leave a mark that looks like a scratch, but can be rubbed off. Hardness also may be affected by weathering and other alteration, so it is important to test a fresh sample.

Scratch hardness may vary substantially with direction and crystallographic plane in some minerals (Figure 6.3). Kyanite ($AlAlOSiO_4$) is a notable example. It typically forms bladed crystals that are elongate parallel to chains of edge-sharing Al octahedra. Hardness when scratched parallel to the length of crystals is 5 and at right angles is 7, reflecting weaker chemical bonding between adjacent chains than along the length of chains. Based on symmetry considerations, the hardness of all nonisometric minerals should vary somewhat with direction, although the variation may not be easily noticed. Even isometric minerals may display some variation. Halite, for example, is slightly softer when scratched parallel to the trace of cleavage than when scratched at an angle. In all cases, the symmetry implied by hardness variation is consistent with the point group symmetry of the mineral.

An alternative and more quantitative hardness measure is **indentation hardness**. It is measured by placing a plunger with a carefully machined tip of diamond or other hard material in contact with a polished mineral surface and applying a constant load. The indentation is observed with a microscope after the plunger is removed. The depth to which the plunger penetrates the mineral is a function of the indentation hardness. Several different indentation hardness scales have been developed, based on the shape of the indenter's tip. Of these, **Vickers hardness**, which uses a pyramid-shaped indenter with a square cross section, is the most widely reported. It is commonly measured in conjunction with petrographic analysis of opaque minerals, using a reflected-light microscope. The Vickers hardness number (VHN) is the load applied divided by the

Table 6.1 Mohs Scale of Hardness

Hardness	Mineral
1	Talc
2	Gypsum
3	Calcite
4	Fluorite
5	Apatite
6	Orthoclase
7	Quartz
8	Topaz
9	Corundum
10	Diamond

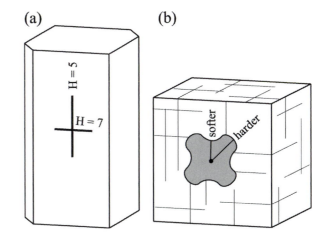

Figure 6.3 Variation of mineral hardness with direction. (*a*) Kyanite. Scratch hardness parallel to the length of crystals is 5 and at right angles is 7. (*b*) Halite. Scratch hardness parallel to the trace of cleavage is slightly less than hardness at 45°. The length of a line drawn from the central dot to the edge of the shaded figure schematically shows the hardness parallel to that line.

surface area of the indentation; the units are kilograms per millimeter squared. VHN is somewhat variable on a given mineral depending on crystallographic direction and the size of the load.

The relation between VHN and Mohs scratch hardness is shown in Figure 6.4. Note that for a given Mohs hardness, the VHN has a significant range, indicated by the width of the shaded band. Despite this variability, VHN increases systematically for Mohs hardness numbers 1 through 9 but jumps dramatically for hardness 10, reflecting the great hardness of covalently bonded diamond.

Tenacity

Tenacity refers to the behavior of minerals when deformed or broken. The terms commonly applied to minerals include the following:

BRITTLE: The mineral breaks or powders easily. Most ionic-bonded minerals are brittle.

MALLEABLE: The mineral may be pounded out into thin sheets. Metallic-bonded minerals may be malleable.

DUCTILE: The mineral may be drawn into a wire. Obviously not easy to test. Malleable materials also may be ductile.

SECTILE: May be cut smoothly with a knife. Relatively few minerals are sectile.

ELASTIC: If bent, will spring back to its original position when the stress is released.

FLEXIBLE: If bent, will not spring back to its original position when the stress is released. It stays bent.

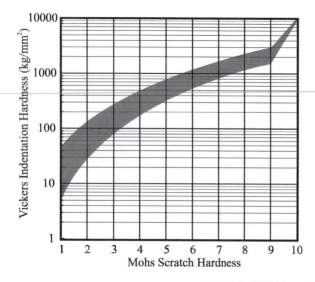

Figure 6.4 Approximate range of Vickers indentation hardness (shaded) as a function of Mohs scratch hardness.

Cleavage

Many minerals have certain crystallographic planes along which chemical bonding is weaker than others. These planes of weakness along which a mineral may break are called **cleavage planes**, or **cleavages**. Because cleavages are controlled by structure and symmetry, they are always crystallographic planes and may be identified by **Miller indices**. For example, a cleavage plane parallel to a (100) face on a mineral is identified as a (100) cleavage surface.

Because cleavages are planes of weakness in the crystal structure, it should be possible, at least in principle, to cleave minerals into layers as thin as a unit cell that are only a few angstroms thick. Unfortunately, our fingers are not sufficiently dextrous, nor would our eyes be able to see these sheets of matter a few atoms thick.

Cleavages are typically described in terms of the crystal form (Chapter 2) to which they are parallel. Because a **crystal form** is a collection of essentially identical crystal faces related by the point group symmetry of the mineral, a **cleavage form** is a collection of cleavage planes related by the point group symmetry of the mineral. For example, in isometric minerals {001} cleavage signifies three cleavage directions parallel to the faces of a {001} cube [i.e., (100), (010), (001)]. In monoclinic minerals, {010} cleavage represents a single cleavage parallel to the {010} side pinacoid. The symmetry of the cleavage must, of course, be consistent with the mineral's symmetry; a monoclinic mineral will not have cubic cleavage. Figure 6.5 shows typical cleavages for several crystal systems. **Cleavage surfaces need not be parallel to the crystal faces that a mineral may acquire when it grows.** For example, fluorite grows as {001} cubes but has {111} octahedral cleavage.

In the mineral descriptions found in the latter part of this book, cleavage is described with reference to the cleavage form as described above. If necessary, the number of cleavages and the angles between them also may be given.

In addition to citing the Miller index symbol for the cleavage form, it is customary to describe cleavages in terms of the quality. *Quality* refers both to the ease with which the mineral cleaves and the degree of perfection of the cleavage surface. Cleavage is **perfect** if the mineral breaks easily to form continuous flat cleavage surfaces that easily reflect light. **Good** cleavage is relatively easy to produce, but cleavage surfaces are interrupted by other fractures and the surfaces are less continuous. **Distinct, indistinct,** and **poor** cleavage represent progressively less well-developed cleavage surfaces. The term **fair** cleavage is used to designate cleavage whose quality is between good and poor.

Fracture

When broken or crushed, a mineral may break along **fracture surfaces** without obvious crystallographic control. No clear line of demarcation exists between indistinct cleavage

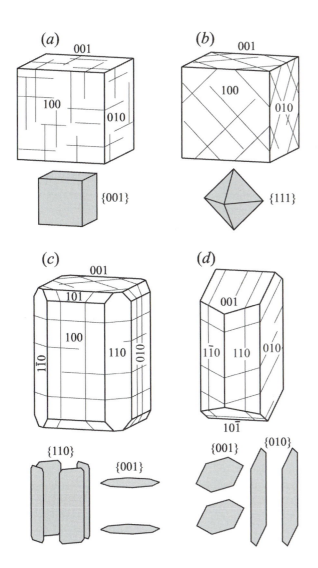

Figure 6.5 Cleavage. (*a*) Isometric {001} cleavage (cubic). (*b*) Isometric {111} cleavage (octahedral). (*c*) Tetrahedral {110} (prism) and {001} (pinacoid) cleavage. (*d*) Monoclinic {001} (pinacoid} and {010} (pinacoid) cleavage.

and fracture; it is part of a continuum of properties. Even though fracture does not display obvious planar surfaces, the nature of the fracture surfaces is often characteristic for some minerals. The common terms used to describe fracture surfaces, listed in order of increasing relief on the surface, include the following

CONCHOIDAL: Smoothly curved as on broken plate glass.

IRREGULAR OR UNEVEN: A rough surface or one with random irregularities.

HACKLY: A surface with sharp-edged irregularities.

SPLINTERY: Resembles the splintery end of a broken piece of wood.

Parting

A parting resembles a cleavage in that a mineral breaks along a smooth, often crystallographically controlled, surface. However, with parting, the mineral breaks because of discrete discontinuities that may be up to a millimeter or so apart. The mineral will not break at closer spacings between the parting surfaces. Further, not all samples of the mineral necessarily display a parting. The most common discontinuities that can produce a parting are twinning composition planes and planar exsolution lamellae (Chapter 5).

It is fairly common for minerals with polysynthetic twinning to display a parting on the composition planes, which are crystallographic surfaces. The fact that the twin segment on one side of the composition plane has a different crystallographic orientation than the segment on the other may cause the composition plane to be a plane of weakness, so that it breaks like a cleavage. However, the thickness of the parting flakes is controlled by the thickness of twin lamellae. If the twin lamellae are 0.5 mm thick, the parting flake can be no thinner than that 0.5 mm.

As described in Chapter 5, certain minerals may display extensive solid solution at high temperature but restricted solid solution at lower temperature. Under appropriate conditions of slow cooling, it is often possible for originally homogeneous crystals to unmix and form planar exsolution lamellae. The boundary between lamellae and host mineral may be a plane of weakness, particularly if incipient alteration is present. Exsolution lamellae may have strict crystallographic control or may be in irrational crystallographic orientations.

Partings are described in much the same manner as cleavages. A Miller index specifies the parting form and terms such as perfect and good may be applied to indicate the ease with which the mineral parts.

If a parting lacks strict crystallographic control, it is identified with the form to which it is most closely parallel (e.g., ~{100}).

COLOR AND LUSTER

When light strikes an object, some portion of the light is absorbed, some is transmitted, and some is reflected. The balance among absorbed, transmitted, and reflected light determines our perception of color and luster. To understand how we perceive these properties, it is first necessary to acquire some background on the nature of light and how our eyes function. The nature of light will be explored in more depth in Chapter 7.

Light

Visible light is just a small portion of the spectrum of **electromagnetic radiation**, which includes radio waves,

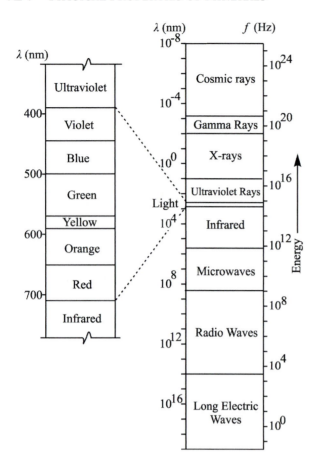

Figure 6.6 The electromagnetic spectrum. Visible light, with wavelengths between about 400 and 700 nm, is a small portion of this spectrum (1 nm = 10^{-9}m = 10Å).

cosmic rays, X-rays, and so forth (Figure 6.6). Although light has properties of both particles (**photons**) and waves, for present purposes it can be described by using the nomenclature and mathematics of waves. It is convenient to identify different parts of the spectrum with the wavelength of the radiation in a vacuum. Visible light has wavelengths between about 400 nm (violet light) and 700 nm (red light). The energy of light increases with decreasing wavelength.

The behavior of light as it passes through minerals will be explored at length in Chapter 7. For present purposes it is necessary to note only that the velocity of light in a vacuum is its maximum, about 3×10^{17} nm/sec. If light enters any other material, it slows down. As a generalization, light velocity decreases as mineral density increases. The velocity of light in a mineral is usually expressed in terms of the mineral's **index of refraction** (n), which is the ratio of the velocity of light in a vacuum (V_v) to the velocity in the mineral (V_m).

$$n = \frac{V_v}{V_m} \qquad (6.6)$$

The velocity of light in most minerals is in the range of 2 to 1.5×10^{17} nm/sec, so most minerals have indices of refraction in the range of 1.5 to 2.

Perception of Color

The human eye is constructed to discriminate the different wavelengths of visible light. Light whose wavelength in a vacuum is about 670 nm is perceived as red, 530 nm light is perceived as green, and so forth. The eye accomplishes this feat by having three different color receptors that are most sensitive to wavelengths of about 660, 500, and 420 nm, corresponding to red, green, and blue-violet.

Our eyes interpret **monochromatic light**, which consists of a single wavelength, as one of the spectral colors. For example, if yellow monochromatic light, whose wavelength is 570 nm, reaches the eye, it stimulates the red and green receptors about equally and the blue-violet receptors very little. The brain interprets those signals to indicate that a wavelength about halfway between red and green is present and we perceive the color as yellow.

Polychromatic light consists of more than one wavelength. If polychromatic light stimulates the color receptors of the eye, the combination of wavelengths is still perceived as a single color, even though the wavelength associated with that color may not actually be present in the light. For example, the perception of yellow can be produced by stimulating the red, green, and blue-violet color receptors with appropriate combinations of wavelengths that mimic the stimulation produced by yellow light. Television sets, computer monitors, and color film take advantage of the way the eye works to allow us to perceive colors by using only combinations of red, green, and blue-violet light or dye. To confirm this, examine a white area on a color computer monitor with a hand lens to see the individual spots of red, green, and blue-violet. Color printing typically uses different combinations of cyan, yellow, magenta, and black ink to produce the same results.

The eye also is capable of perceiving combinations of wavelengths as colors that are not in the normal spectrum. If the entire visible spectrum is present, such as in light from the sun, the eye perceives it as white. The eye also will perceive as white various combinations of two colors called complementary light colors, because those two colors stimulate the red, green, and blue-violet color receptors about equally. Many other color sensations, such as purple and brown, have no counterpart in the visible spectrum. The sensation of purple is produced by combinations of red and violet light, whereas the sensation of brown is produced by combinations of red, blue, and yellow.

About 4% of the population, mostly male, have forms of color blindness that affect their perception of color because one or more of the color receptors is absent or does not function normally. For most activities, color blindness poses no significant difficulty. However, the perception of color is important in certain areas of mineralogy, particularly optical mineralogy discussed in Chapter 7, so the inability to perceive color conventionally may pose a hardship. Typically, problems with color blindness can be dealt with once they are recognized by paying closer attention to other mineral properties or by recognizing color subtleties that are not part of the normal color lexicon.

Mineral Luster

Minerals can conveniently be grouped into those that are opaque and those that are transparent. Opaque materials do not transmit light, even if quite thin, whereas transparent minerals do. In general, minerals dominated by ionic and covalent bonding are transparent, whereas minerals with metallic bonding are opaque. In hand samples of minerals, we perceive this distinction in terms of **luster**, which is divided into two broad categories: **nonmetallic** and **metallic**.

Minerals with a nonmetallic luster may be quite shiny, brilliant even, but they lack the appearance of metals because they are transparent, at least in thin pieces, and as little as 5% of the incident light may be reflected; most of the light passes into the mineral. The different qualities of nonmetallic luster depend on the index of refraction and texture. High index of refraction (low velocity) is associated with a more brilliant luster (Table 6.2).

Minerals with a metallic luster reflect light like metals and typically are opaque. If between roughly 20 and 50% of the incident light is reflected, we perceive a normal metallic luster, like a piece of polished gold or steel. If more than about 50% of the light is reflected, the metallic luster may take on a splendent appearance. If less than about 20% of the light is reflected, the luster is **submetallic**, transitional to an adamantine nonmetallic luster. The details of which

wavelengths are more strongly or less strongly reflected determine the color.

Mineral Color

The color of a mineral (or any other object) is our perception of the wavelengths of light that are reflected from or pass through the material to reach our eye. Color depends on the wavelengths that the mineral does *not* absorb. A red mineral looks red because it reflects or transmits red light and absorbs a substantial part of the blue end of the spectrum. A white mineral reflects essentially all of the visible spectrum to our eye. A black mineral absorbs all wavelengths.

Recall from Chapter 3 that every electron about each atom/ion in a crystal structure is in a specific energy level. The electrons in outer shells are typically in higher energy levels than those in the inner shells. Electromagnetic radiation (like light) passing through a mineral can be absorbed if the energy level of the radiation, which is a function of frequency, corresponds to the difference in the energy level (Δ) of an electron and some open higher energy orbital. When the radiation is absorbed, its energy bumps the electron into the vacant higher energy level, leaving a lower energy vacancy (Figure 6.7a). When the promoted electrons fall back to their base energy levels, they reemit electromagnetic radiation, some of which may be in the visible spectrum (Figure 6.7b,c). This process is described in a later section of photoluminescence. Processes in minerals that yield energy levels to which electrons can be promoted

Table 6.2 Varieties of Nonmetallic Luster

Luster	Description
Subvitreous	Somewhat less brilliant luster than glass; $n < 1.5$
Vitreous	The luster of glass; $1.5 < n < 2.0$
Adamantine	Exceptionally brilliant luster like that of diamond; $n > 2.0$
Resinous	Has the luster of resin; transitional to adamantine
Earthy	Dull surface that lacks any shine; generally seen with aggregates of very fine mineral grains (like soil)
Greasy	Luster as if covered with oil or grease; generally caused by microscopically rough surface texture
Pearly	Has an iridescent pearl-like luster; often produced by incipient development of cleavage surfaces parallel to mineral surface
Silky	Has a subtly textured shiny appearance similar to silk; generally produced by reflection of light from the surface of aggregates composed of parallel mineral fibers

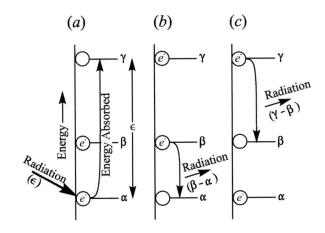

Figure 6.7 Absorption of electromagnetic radiation. (*a*) Electromagnetic radiation whose energy level is ϵ excites an electron in its base energy level α. If the electron can be bumped to an open energy level γ where the difference between γ and α is equal to ϵ, the electromagnetic radiation energy will be absorbed. (*b*) An electron whose base energy level is β may cascade into the now-vacant α site, releasing electromagnetic radiation whose energy level is $\beta - \alpha$. (*c*). The high-energy γ electron now can cascade into the lower energy β site, releasing electromagnetic radiation whose energy is equal to $\gamma - \beta$.

when visible light energy is absorbed are described by **crystal field theory, band theory, charge transfer transitions**, and **color centers**.

Crystal Field Theory

Color produced by crystal field transitions is common in ionically bonded minerals that contain cations with partially filled $3d$ orbitals. These cations include Ti, V, Cr, Mn, Fe, Co, Ni, and Cu (Table 3.3) whose outer orbitals contain unpaired electrons. Of these, Fe is the most abundant and therefore is responsible for color in many minerals. Because these elements are effective in absorbing visible light energy and producing mineral colors, they are known as **chromophore elements**.

In a single isolated cation of one of these transition elements, all the electrons in the five orbitals in the $3d$ subshell have the same energy level (but different magnetic and spin quantum numbers) (Figure 6.8). When this metal cation is placed in a coordination polyhedron with anions (usually oxygen), the negative charge of the anions produces an electric field known as a **crystal field**, which interacts with the negative charge of the electrons in the orbitals around the cation. The result is that some of the orbitals in the $3d$ subshell wind up with higher energy and some with lower energy. This process is called **crystal field splitting**.

Orbitals that are directed toward the anions acquire a higher energy because the negative charge of the electron(s) within these orbitals is repelled by the anions. Orbitals directed between the anions wind up with lower energy. When these coordination polyhedra are combined to construct a mineral, each of the energy levels produced by crystal field splitting is spread into an energy band because the Pauli exclusion principle ensures that no two electrons in the structure can have exactly the same energy level. The difference in energy between the low and high energy levels produced by crystal field splitting of $3d$ orbitals matches the energy level of portions of the visible light spectrum. Precisely which energy levels are available depends on the element, the details of the shape of the coordination polyhedra, and the oxidation state of the element.

The color of ruby is caused by crystal field splitting. Pure corundum is Al_2O_3; but in ruby, small amounts of Cr^{3+} substitute for Al^{3+}. Crystal field splitting provides excited energy levels B, C, and D, shown schematically in Figure 6.9a, that are about 1.9 eV, 2.2 eV, and 3.0 eV above the ground state for electrons. Incident yellow-green and violet light (λ = 550 and 400 nm, respectively) has these energy levels and can be absorbed by bumping electrons from the ground state to the excited levels C and D. Level B is not available because of quantum mechanics considerations. The remaining wavelengths pass to our eyes and are dominated by red with some blue light. We perceive these wavelengths as a rich red color. Fluorescence produced when the excited electrons return to their ground state is described below.

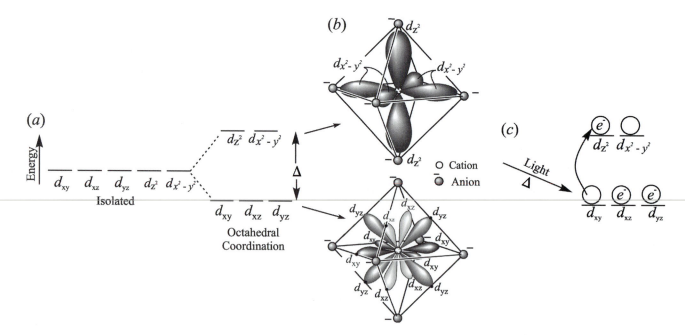

Figure 6.8 Crystal field splitting. (a) All five d orbitals (d_{xy}, d_{xz}, d_{yz}, d_{z^2}, and $d_{(x^2-y^2)}$) are identical in energy in an isolated transition metal cation (left). Compare Figure 3.1. When placed in regular octahedral coordination with six anions, the d orbitals are split into two energy levels with an energy difference = Δ. (b) The d_{z^2} and $d_{(x^2-y^2)}$ orbitals are directed toward the position of the anions (top) and are forced to have higher energy because the negative charges of the anions repel the electrons. The d_{xy}, d_{xz}, and d_{yz} orbitals are directed between the anions (bottom), and have lower energy. (c) If incident electromagnetic energy has energy equal to Δ, it can be absorbed by bumping a lower energy electron into a vacant higher energy position.

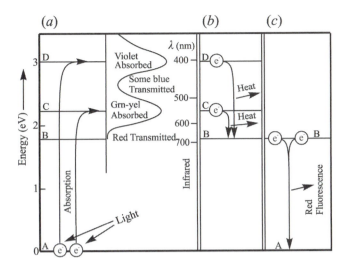

Figure 6.9 Light absorption in ruby. (*a*) Light with energies of about 2.2 eV and 3.0 eV can be absorbed by bumping electrons from their ground state (A) to energy levels C and D produced by crystal field splitting. When the electrons return to their ground states, they do so by stages. (*b*) Electrons promoted to C and D fall back to the vacant B level with emission of heat. (*c*) Electrons from the B level subsequently return to A with emission of red light. Adapted from Nassau (1978).

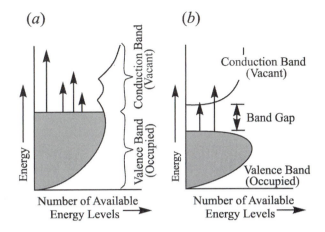

Figure 6.10 Band theory. (*a*) Valence bands and conduction bands do not have an energy gap between them with metallic bonding. All light is absorbed because a continuum of energy levels are available to which electrons in the valence band can be promoted. The light that is emitted as electrons return to their base energy state determines the mineral color. The emitted light is perceived as being reflected. (*b*) In semiconductors, an energy gap exists between the valence band and the conduction band. Light whose energy is greater than the band gap can be absorbed by bumping valence electrons into the conduction band.

Band Theory

Minerals with a significant degree of **metallic bonding** typically absorb a wide range of wavelengths and are therefore opaque. Recall from Chapter 3 that metallic bonding involves spreading energy levels available to electrons into bands because of the Pauli exclusion principle. Energy levels of the **valence band** that contains electrons are continuous with those of the **conduction band**. If excited, electrons can easily move from the valence band to the conduction band (Figure 6.10a). Because the conduction band contains what amounts to a continuum of available vacant energy levels, the entire visible spectrum can be absorbed, and no light passes through these minerals. The color is determined by the light that is radiated from the mineral as electrons fall back into the lower energy positions. The efficiency of this process depends on the density of energy levels within the conduction band and on complex quantum mechanics considerations.

Minerals with a substantial degree of covalent bonding may behave as semiconductors. With semiconductors, (Figure 6.10b) a gap in available electron energy levels known as a **band gap** (E_g) exists between the conduction and valence bonds. Light whose energy is greater than that of the band gap will be absorbed because it can promote electrons from the top of the valence band to vacant levels in the conduction band. If the band gap is smaller than the energy range of visible light, all the light is absorbed, and the mineral will be black or gray, like galena (PbS). If the band gap is intermediate, high-energy, short-wavelength light from the violet end of the spectrum will be absorbed,

but light from the lower energy red end of the spectrum will not. Minerals such as cinnabar (HgS) owe their red color to this process. If the band gap is large, as in diamond ($E_g \sim 5.5$ eV), none of the visible spectrum is absorbed and the mineral is white or colorless.

Impurities in minerals with large band gaps may influence color in some cases. The blue color of some diamonds is due to the presence of trace amounts of boron. Boron provides an energy level a short distance above the top of the valence bond to which electrons can be bumped. This allows some of the light at the red end of the spectrum to be absorbed.

Charge Transfer Transitions

Charge transfer transitions, also known as **molecular orbital transitions**, absorb electromagnetic radiation when valence electrons are bumped to higher energy levels on adjacent ions. With charge transfer transitions, differently charged cations occupy adjacent sites A and B. Electromagnetic radiation is absorbed by an electron on the lower charged cation in site A, and that electron is transferred to a higher energy orbital on the higher charged cation in site B. Two common examples are $Fe^{2+} \rightarrow Fe^{3+}$ and $Fe^{2+} \rightarrow Ti^{4+}$. In both cases the Fe^{2+} can be considered site A. When the Fe^{2+} absorbs electromagnetic radiation energy and ejects an electron, it becomes Fe^{3+}. The Fe^{3+} or Ti^{4+} in site B receives the electron, which reduces its charge by one. When the electrons fall back to their lower energy positions, the cations revert to their original charges.

In both these examples, the energy difference often matches the energy of light at the red end of the spectrum, so minerals that utilize this mechanism tend to be blue. A notable example is sapphire, which is a blue gem variety of corundum (Al_2O_3) in which small amounts of Fe and Ti substitute for Al.

Color Centers

Whereas crystal field transitions and charge transfer transitions depend on having electrons on specific elements present in a mineral to absorb parts of the visible spectrum, **color centers** depend on having electrons mislocated in the structure. Some minerals whose color depends on color centers are listed in Table 6.3.

Electron color centers (Figure 6.11a) depend on structural defects such as **Frenkel defects** to absorb light. Recall from Chapter 5 that Frenkel defects are flaws in a crystal structure in which an ion is missing from its normal location. If the missing ion is an anion, its position may be occupied by an electron whose charge is required to maintain electrical neutrality. The electron is held in place by the crystal field supplied by the adjacent anions and cations. These electrons, like all other electrons in the structure, have a variety of specific energy levels to which they may be excited. If the steps between these energy levels match the energy of light, some of the light can be absorbed.

Hole color centers are produced if an electron is missing from its normal location (Figure 6.11b). The purple color of amethyst, which is a variety of quartz (SiO_2), is produced when Si^{4+} is replaced by Fe^{3+}. Charge balance is maintained by including a monovalent cation such as H^+ elsewhere in the structure. The color center is produced when high-energy radiation ejects an electron from an oxygen anion bonded to the Fe^{3+}. The ejected electron gets trapped elsewhere in the structure, leaving the oxygen with an unpaired electron. The hole has energy levels to which other electrons in the oxygen can be bumped when light at the red end of the spectrum is absorbed.

Because color centers are defects in the crystal structure, they may be produced both when minerals grow and by the disruption produced by radiation. The defects can be removed by heating, which allows ions to migrate through the structure. Minerals such as fluorite (CaF_2), whose blue-violet color is produced by electron color centers, may be rendered nearly colorless by heating. In some cases, exposure to ultraviolet or visible light is sufficient to trigger removal of hole color centers. Because color centers may be produced by radiation, gems such as topaz and diamond may be irradiated to enhance their color and value.

Color from Mechanical Causes

In addition to being colored because of their chemical and/or structural makeup, minerals may be colored because they contain finely dispersed inclusions of other materials. Hematite (Fe_2O_3) is a widely distributed mineral that, if included in other minerals, may cause them to be various shades of red or brown. **Jasper**, which is a red-brown variety of fine-grained quartz (SiO_2) widely used in lapidary work, is a good example. Similarly, graphite (C) may cause normally white or colorless calcite ($CaCO_3$) to be black. The milky-white color of milky quartz is caused by the presence of abundant microscopic to submicroscopic inclusions of fluid, mostly water, from which the quartz originally grew.

A special mechanical phenomena is **chatoyancy**, which is seen as a silky appearance. It is produced in minerals such as satin spar gypsum ($CaSO_4 \cdot 2H_2O$) that consist of parallel fine fibers, or minerals containing abundant parallel fibrous inclusions. Closely related **asterism** is sometimes seen in corundum and quartz. Both minerals are hexagonal and may contain inclusions of fine fibrous rutile (TiO_2) that are aligned parallel to the 2-fold symmetry axes at right angles to c. Light scattering from these fine inclusions may be perceived to form a six-pointed star pattern (Figure 6.12).

Some minerals display flashes of varied internal color that change as the sample is moved about in the light. This "**play of color**" is caused by diffraction of light on

Table 6.3 Minerals Whose Color Is due to Color Centers

Mineral	Color
Amythyst (var. quartz)	Purple
Fluorite	Purple
Smoky quartz	Brown, black
Diamond	Green, yellow, brown, blue, pink
Topaz	Blue
Halite	Blue, yellow

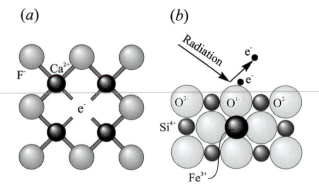

(a)   *(b)*

Figure 6.11 Color centers. (*a*) Electron color center. A F^- anion in fluorite is missing and replaced with an electron (e^-) that has energy levels to which it may be promoted with absorption of light energy. (*b*) Hole color center. Radiation displaces an electron on an oxygen anion bonded to Fe^{3+} that has substituted for Si^{4+} in quartz (shown schematically). The vacant electron site on the oxygen provides energy levels to which other electrons may be promoted with absorption of light energy.

Figure 6.12 Photograph of star sapphire whose long dimension is 8 mm. The star pattern is due to numerous fibrous rutile (TiO$_2$) inclusions parallel to the hexagonally arranged 2-fold rotation axes in the host corundum.

closely spaced features within the structure of the mineral. In precious opal (fire opal), the diffraction is produced by the structure, which consists of roughly uniform spheres of silica gel that are regularly packed together. The dimension of the spheres corresponds with the wavelength of visible light, so they act as a diffraction grating, resolving the incident polychromatic light into its spectral components. Some plagioclase, called moonstone, also displays a play of color produced by diffraction on a closely spaced lamellar structure produced by exsolution (see Chapter 12).

Iridescence is seen as a play of color on the mineral's surface. It is common in minerals such as pyrite, chalcopyrite, and bornite, whose surface has been oxidized to produce a thin layer of material that interacts with the incoming light in the same way that a film of oil on water produces interference colors. Cracks inside transparent minerals such as quartz also may produce iridescent colors.

Consistency of Mineral Color

Minerals that contain chromophore elements as an integral part of their chemical composition or are opaque typically have consistent colors. These **idiochromatic** (self-colored) minerals may display a range of color consistent with the compositional variation displayed by the mineral.

Minerals whose color may be quite variable are termed **allochromatic** (foreign-colored) minerals. These minerals typically do not have chromophore elements as an integral part of their chemical composition and typically are fairly light colored, at least if pure. The color comes from color centers, trace amounts of chromophore elements, inclusions, and so forth. **Quartz** (SiO$_2$), for example, is clear and colorless when pure but may be purple (amethyst) if it contains trace amounts of Fe^{3+}, pink (rose quartz) if it contains trace amounts of Ti^{4+}, or almost any color if it contains appropriately colored inclusions.

Streak

Streak is the color of a mineral when it is finely powdered. Minerals with dominantly ionic and covalent bonding tend to have fairly light-colored streaks, even if the mineral color is dark in large pieces. This is because these minerals are all relatively transparent, so only a small amount of the incident light is absorbed on passing through the fine particles of the powder. Nearly the complete spectrum is reflected back to our eye, to be perceived as white or a pale color. Even though the mineral color of these minerals may vary significantly, the streak color is usually quite consistent.

The streak of minerals with metallic bonding may be richly colored or black, depending on the mineral. These minerals are typically opaque, so the incident light is strongly absorbed, even in the fine particles of a powder. Streak color may be darker than mineral color because the rough surface represented by the powder does a better job of absorbing the incident light.

Streak color is conveniently observed by rubbing a mineral sample on a **streak plate**, which is an unglazed white porcelain tile. Streak plates have a hardness of about 7, so they cannot be used for harder minerals. For harder minerals, or if a streak plate is not available, the streak color may be observed by pulverizing a sample with a knife blade, rock hammer, or other convenient implement.

Luminescence

Luminescence is a phenomenon in which a material absorbs one form of energy and then reemits the energy as visible light. The energy source may be mechanical, thermal, or electromagnetic.

Triboluminescence

Some materials may become faintly luminous if struck, crushed, scratched, or rubbed. This **triboluminescence** is typically quite faint and requires near total darkness to be seen.

Thermoluminescence

If a material emits visible light energy as a consequence of being heated, it is **thermoluminescent** or **incandescent**. Thermoluminescence generally requires that the sample be exposed to light or other radiation at some point, so that electrons can be excited to higher energy positions. These electrons may be trapped in their higher energy levels (equivalent to throwing an object into the air and having it land on a bookshelf or in a tree rather than on the floor or ground). Heating can provide sufficient activation energy (shaking the tree or bookshelf) to allow the electrons to fall back to their ground states. Thermoluminescence is generally strongest between 50 and 100°C and ceases at temperatures above 475°C. Thermoluminescence also requires near total darkness to be seen. At temperatures of about

550°C, materials become incandescent and begin to glow dull red. Incandescence is caused when thermal excitation causes electrons to be bumped to energy levels sufficiently high that the energy released when they drop back to their ground state yields light in the visible spectrum. With increases in temperature, the incandescent color changes to yellow and eventually becomes white, as light is emitted in nearly the entire visible spectrum.

Photoluminescence

Materials that emit light in response to being exposed to visible or ultraviolet light are **photoluminescent**. The light energy that is absorbed by crystal field transitions, charge transfer transitions, color centers, and so forth does not just go away. The process of absorbing the energy and bumping an electron to a higher energy position leaves vacant the lower energy orbital from which the electron came. Because natural systems tend toward conditions of lowest energy, a higher energy electron will drop into the vacant orbital, and another will fill its position, and so forth, until electrons occupy the lowest set of orbitals available.

When a higher energy electron drops into a vacant lower energy level, energy in the form of electromagnetic radiation is released (Figure 6.7). If this radiation is in the visible spectrum, the mineral is said to be **fluorescent** or **phosphorescent**. Fluorescence and phosphorescence differ only in how long it takes the electrons to return to their ground states. With fluorescence, the vacant lower energy electron positions are filled within 10^{-8} second; phosphoresence takes longer. When the incident radiation is turned off, fluorescent materials cease emitting light. Phosphorescent materials continue emitting light energy significantly after the exciting radiation is turned off, sometimes for hours or more.

The fluorescent/phosphorescent light has lower energy (longer wavelength) than the original exciting radiation. Hence, wavelengths in the ultraviolet range (Figure 6.6) tend to be effective in producing fluorescence in the lower energy visible spectrum. Fluorescence is routinely observed by illuminating a mineral sample in a darkened room with either a short-wave ($\lambda \approx 254$ nm) or long-wave ($\lambda \approx 366$ nm) ultraviolet lamp. Because our eyes are insensitive to ultraviolet light, the color of the fluorescent light is readily observed.

Visible light also may trigger fluorescence. The absorption of light by ruby was described earlier (Figure 6.9a). Small amounts of Cr^{3+} absorb green-yellow and violet light via crystal field transitions. When the excited electrons revert to their ground state, they cannot do so directly for quantum mechanics reasons. Rather, they first drop to energy level B at about 1.9 eV above the ground state, with emission of infrared radiation (heat) whose energy is about 1.1 and 0.3 eV (Figure 6.9b). Light emission occurs when the electron drops from B back to the ground state at A (Figure 6.9c). This energy drop is about 1.9 eV and

is the energy of red light. Hence, in addition to transmitting red light, some of the absorbed energy is reradiated as additional red light, which tends to give the ruby extra color depth and the appearance of glowing. This warmth or glow adds considerably to the appeal of ruby as a gemstone.

MAGNETISM

Magnetic fields are produced by the movement of electrons. At the atomic level, the most important movement is **electron spin**, denoted by the **spin quantum number**. Recall from Chapter 3 that each orbital may contain two electrons, but because no two electrons in an atom may have precisely the same set of quantum numbers, those two electrons must have opposite spins. Each spinning electron produces a magnetic field, whose **Bohr magnetic moment** (μ_B) can be canceled by an electron in the same orbital spinning in the opposite direction.

The magnetic behavior of a mineral, therefore, depends on whether atoms/ions have orbitals with unpaired electrons. If no element in the mineral contains unpaired electrons, the mineral is **diamagnetic**. If unpaired electrons in one or more orbitals are present, the possibilities include **paramagnetism**, **ferromagnetism**, **ferrimagnetism**, and **antiferromagnetism** (Figure 6.13), depending on how the magnetic moments of the atoms/ions are oriented within the crystal structure. The most important elements that have unpaired electrons include the transition metals whose $3d$ orbitals are only partially filled (Table 3.3). Of these Fe, Mn, Ti, and Cr are the most abundant. The strength of the magnetic moment of these elements depends on how many electrons are unpaired. Fe^{3+} and Mn^{2+} have the largest magnetic moments with five unpaired $3d$ electrons each. Minerals that are ferromagnetic and ferrimagnetic display an obvious attraction to a magnet, and a list is provided in Appendix B, Table B.6.

Diamagnetism

In **diamagnetic** materials, all orbitals of the atoms/ions contain paired electrons. Quartz (SiO_2) is diamagnetic. Silicon, by losing four electrons to become Si^{4+}, acquires the electron configuration of Ne, in which the $1s$, $2s$, and $2p$ orbitals are entirely filled (Figure 6.13a). Each oxygen gains two electrons to become O^{2-}, which also has the electron configuration of Ne. The number of "up" spin electrons is exactly balanced by the "down" spin electrons, so each ion has zero net magnetic moment and the mineral cannot have a net magnetic moment. Minerals whose atoms/ions have acquired electron configurations of the inert gasses, or whose orbitals otherwise do not contain unpaired electrons are diamagnetic. In general, this excludes the transition metals, which have partially filled d orbitals. When placed in a magnetic field, diamagnetic minerals will actually be slightly repelled. The external magnetic field causes

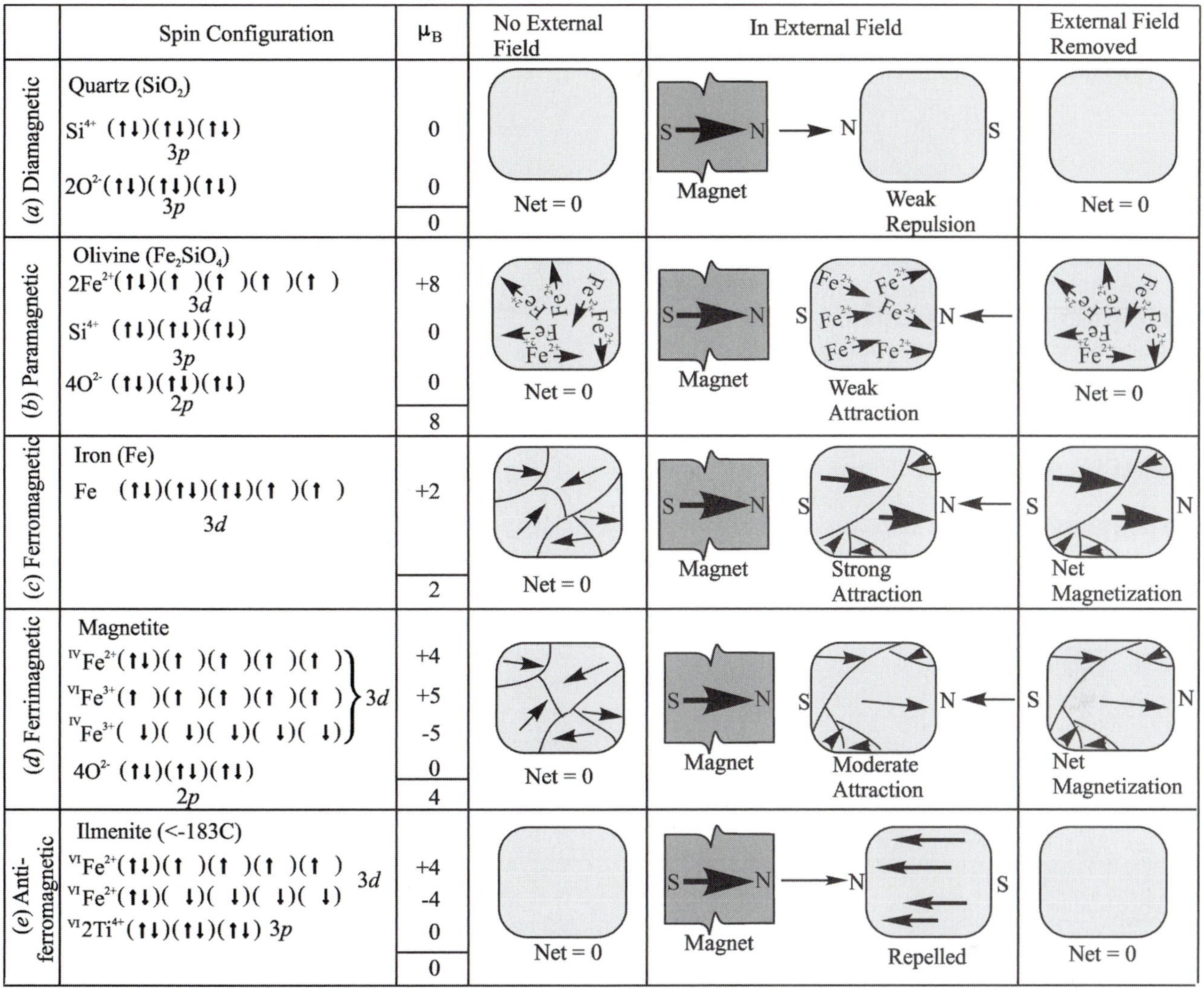

	Spin Configuration	μ_B	No External Field	In External Field	External Field Removed
(a) Diamagnetic	Quartz (SiO$_2$) Si^{4+} (↑↓)(↑↓)(↑↓) 3p 2O^{2-}(↑↓)(↑↓)(↑↓) 3p	0 0 — 0	Net = 0	Magnet S→N → N ⋯ S Weak Repulsion	Net = 0
(b) Paramagnetic	Olivine (Fe$_2$SiO$_4$) 2Fe^{2+}(↑↓)(↑)(↑)(↑)(↑) 3d Si^{4+} (↑↓)(↑↓)(↑↓) 3p 4O^{2-} (↑↓)(↑↓)(↑↓) 2p	+8 0 0 — 8	Net = 0	Magnet S→N Weak Attraction	Net = 0
(c) Ferromagnetic	Iron (Fe) Fe (↑↓)(↑↓)(↑↓)(↑)(↑) 3d	+2 2	Net = 0	Magnet S→N Strong Attraction	Net Magnetization
(d) Ferrimagnetic	Magnetite ^{IV}Fe^{2+}(↑↓)(↑)(↑)(↑)(↑) ^{VI}Fe^{3+}(↑)(↑)(↑)(↑)(↑)}3d ^{IV}Fe^{3+}(↓)(↓)(↓)(↓)(↓) 4O^{2-} (↑↓)(↑↓)(↑↓) 2p	+4 +5 -5 0 — 4	Net = 0	Magnet S→N Moderate Attraction	Net Magnetization
(e) Anti-ferromagnetic	Ilmenite (<-183C) ^{VI}Fe^{2+}(↑↓)(↑)(↑)(↑)(↑) 3d ^{VI}Fe^{2+}(↑↓)(↓)(↓)(↓)(↓) ^{VI}2Ti^{4+}(↑↓)(↑↓)(↑↓) 3p	+4 -4 0 — 0	Net = 0	Magnet S→N → N ⋯ S Repelled	Net = 0

Figure 6.13 Magnetism. See text for discussion. (*a*) Diamagnetism in quartz. Neither Si^{4+} nor O^{2-} has unpaired electrons, so neither has a Bohr magnetic moment μ_B. (*b*) Paramagnetism in olivine. The Fe^{2+} has a net Bohr magnetic moment; but in the absence of an external magnetic field, the orientations are random. (*c*) Ferromagnetism in native iron. Each iron atom has two unpaired electrons and a Bohr magnetic moment of 2. Domains within the mineral have uniform magnetic orientation. (*d*) Ferrimagnetism in magnetite. Both ^{IV}Fe^{3+} and ^{VI}Fe^{3+} have five unpaired electrons but with opposite spins, so their net magnetic moments cancel. The ^{IV}Fe^{2+} has a net magnetic moment and causes the magnetite to have ferromagnetic properties. (*e*) Antiferromagnetism in ilmenite below −183°C. Fe^{2+} in adjacent sites have antiparallel spins. Net magnetic moments cancel. Antiferromagnetic materials are repelled by a magnet.

electrons in the mineral to move, which in turn produces a magnetic field opposing the external field.

Paramagnetism

Minerals in which the magnetic moments of the constituent atoms/atoms are not mutually aligned are **paramagnetic**. Olivine [(Mg,Fe)$_2$SiO$_4$], for example, contains Fe^{2+} in variable amounts. Each Fe^{2+} has a magnetic moment because each contains four unpaired electrons in the 3d orbitals (Figure 6.13b). These four electrons all have spins in the same direction because of **Hund's Rule**, which states that each orbital within a subshell gets one electron with a $+\frac{1}{2}$

quantum number (up spin) before any orbital gets a second electron with a $-\frac{1}{2}$ quantum number (down spin).

Although each Fe^{2+} has a magnetic moment, the moments do not align with each other. The net magnetic moment of all the randomly oriented Fe^{2+} is zero. If olivine is placed in a magnetic field, the orientations of the magnetic moments of the Fe^{2+} will tend to align parallel to the external field so that the sample is attracted to the magnet. Normal thermal motion of the ions in the structure conspires to randomize this alignment so the net magnetic moment and the magnetic attraction are weak. Paramagnetic minerals are not noticeably attracted to a handheld magnet.

The strength of the magnetic moment produced in a paramagnetic mineral is a function of the **magnetic susceptibility**, which depends on chemical composition and mineral structure. When the external magnetic field is removed, the magnetic moment of the mineral rapidly dissipates, as the magnetic moments of the constituent Fe^{2+} ions revert to random orientations. Most iron-bearing silicate minerals are paramagnetic.

Because of differences in magnetic susceptibility, minerals can be separated from each other by placing them in a strong magnetic field. A **Franz isodynamic separator** is an instrument that uses a strong magnetic field to separate diamagnetic from paramagnetic minerals, and to separate paramagnetic minerals with different magnetic susceptibilities from each other. Its use is briefly described in Chapter 10.

Ferromagnetism

A material that is **ferromagnetic**, such as a piece of iron, is capable of retaining a magnetic polarity because the magnetic moments of the constituent atoms with unpaired electrons become systematically aligned (Figure 6.13c). An effect known as **exchange coupling** locks the magnetic moment of adjacent ions into parallel alignment, for reasons tied to quantum mechanics. Within a single grain of a ferromagnetic material, parallel alignment is generally restricted to microscopic volumes known as **domains**. Adjacent domains need not have parallel alignment. If the magnetic moments in different domains are randomly oriented, the net magnetic moment will be zero. Attraction to a magnet, however, is strong because the external field will tend to rearrange the alignment of the magnetic domains.

Exposure to an external magnetic field of sufficient strength will cause domains whose magnetization is parallel to the external field to expand, as other domains in unfavored orientations shrink. The crystal therefore becomes permanently magnetized. The attraction between a ferromagnetic material and an applied magnetic field is many orders of magnitude greater than is the case for paramagnetic materials. When the external field is removed, the parallel alignment of the magnetic domains remains and the material retains its magnetic properties. Magnetizing a steel nail by consistently rubbing it with a strong magnet is a good example.

Ferrimagnetism

Ferrimagnetism (Figure 6.13d) differs from ferromagnetism in that some atoms/ions in adjacent structural sites have antiparallel magnetic moments because of exchange coupling. The magnetic moments of these atoms/ions therefore cancel each other. Additional atoms/ions may have unpaired electrons that can produce a magnetic moment. As long as some atoms/ions do not have an antiparallel partner, the mineral will display ferromagnetic properties. **Magnetite**, which is by far the most common magnetic mineral, is ferrimagnetic. Its formula can be written $^{IV}Fe^{3+}\uparrow\,^{VI}(Fe^{3+}\downarrow Fe^{2+}\uparrow) O_4$. The arrows in the formula indicate the magnetic spins. It consists of cubic close-packed oxygen anions with the first Fe^{3+} in tetrahedral coordination, and the second Fe^{3+} and the Fe^{2+} in octahedral coordination. Exchange coupling forces the tetrahedral and octahedral Fe^{3+} to have opposite spins, so their magnetic moments cancel. The magnetic moments of the octahedral Fe^{2+} align in adjacent sites, so magnetite can have a net magnetic moment.

The term **antiferromagnetism** is applied to materials in which antiparallel spins completely cancel to yield zero net magnetic moment. It can be considered a special case of ferrimagnetism. Among the Fe–Ti oxides, pure ilmenite ($Fe^{2+}Ti^{4+}O_3$) is antiferromagnetic below $-183°C$ (Figure 6.13e). The Ti^{4+} lacks a magnetic moment, and the moments of Fe^{2+} in adjacent lattice planes are oriented antiparallel. Above $-183°C$ (the **Néel temperature**) the antiparallel ordering is destroyed and ilmenite becomes paramagnetic. The magnetic behavior of ilmenite at room temperatures is due to nearly ubiquitous exsolution blebs and inclusions of magnetite. Hematite (Fe_2O_3) also is antiferromagnetic below $-10°C$, but above that temperature the opposing spin orientations of Fe^{3+} in adjacent lattice planes are not quite antiparallel. It therefore has ferrimagnetic properties and can retain a net magnetic moment.

If a magnetized ferromagnetic or ferrimagnetic mineral is heated, the strength of the magnetization gradually decreases, as the randomness imposed by thermal vibration decreases the effectiveness of exchange coupling. When a temperature known as the **Curie temperature** (T_c) is reached, all the exchange coupling is destroyed and ferromagnetic/ferrimagnetic properties are lost. Above the Curie temperature, these materials behave paramagnetically. When cooled back through the Curie temperature, minerals such as **magnetite** reacquire their ferromagnetic properties.

If a magnetic field, such as the Earth's, is present when ferromagnetic or ferrimagnetic minerals cool through their Curie temperature, the substances may acquire a magnetization parallel to that field. These minerals also may acquire a net magnetic moment if they crystallize below the Curie temperature in the presence of a magnetic field.

These seemingly inconsequential observations have had a profound influence on our understanding of the Earth's history. Magnetism (called **remnant magnetism**) acquired by magnetite grains when an igneous rock cools or by hematite during diagenesis of sediments may be preserved for millions or billions of years. Detailed study of remnant magnetism has helped document that the continents have moved relative to each other and has established that the Earth's magnetic field periodically reverses polarity. These two observations provide key parts of the foundation on which our understanding of plate tectonics has been built.

ELECTRICAL PROPERTIES

Electrical Conductivity

The ability of minerals to conduct electricity depends directly on the nature of the chemical bonding. Minerals with a significant degree of metallic bonding, in which electrons can migrate freely through the crystal structure, typically conduct electricity with ease, as was described in the section on metallic bonding in Chapter 3. The valence electrons are not tightly bonded to specific ions; rather, they are free to migrate through the structure. The application of an external voltage, or a changing magnetic field, provides a driving force to cause the electrons to migrate in one direction, thereby producing an electrical current. The conductivity of electrical conductors typically decreases somewhat with increasing temperature.

Minerals with dominantly covalent or ionic bonds are electrical insulators because the electrons are tightly constrained to particular atoms/ions in the structure and, therefore, are not free to move in response to an applied electrical voltage. While the conductivity of these minerals is very low, it is not zero. Current flow in electrical insulators generally does not rely on the direct movement of electrons, as is the case with metallic bonding. Rather, current flow is accomplished by the migration of point defects in the crystal structure. Recall from Figure 5.15, that point defects include Schottky, Frenkel, and impurity defects. The movement of a cation to fill a Schottky defect, for example, involves both the transport of the charge of the cation (i.e., an electrical flow) and also the generation of a new Schottky defect at the site from which the cation came.

Electrical conductivity related to migration of point defects increases with increasing temperature because higher temperature makes it much easier for ions to migrate through the crystal structure. Figure 6.14 shows the electrical conductivity of olivine and K-feldspar. Conductivity increases exponentially with increasing temperature but remains very low by comparison to conductors. Copper, for example, has a conductivity of $\sim 60 \times 10^6$ ohm^{-1}-meter^{-1}, which is more than 8 orders of magnitude greater than the highest conductivity measured for either K-feldspar or olivine.

Piezoelectricity

When deformed, certain minerals will generate a voltage, so that they are positively charged on one side and negatively charged on the other. These minerals are **piezoelectric**. Common minerals that are piezoelectric include **quartz** (SiO_2), **topaz** [$Al_2SiO_4(F,OH)$], and **tourmaline** [$NaFe_3Al_6(BO_3)_3Si_6O_{18}(OH)_4$]. Although the piezoelectric property of these minerals does not appear to have geological significance, it has great practical use. Synthetic quartz, in particular, is widely used in electronic components and transducers.

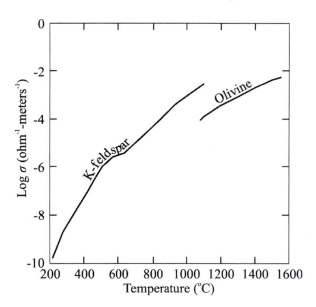

Figure 6.14 Electrical conductivity σ in units of reciprocal ohm-meters (ohm^{-1}-meter^{-1}) of olivine and K-feldspar. Olivine data at 70 kbar pressure from Yousheng and others (1998); K-feldspar data from Guseinov and Gargatsev (2002).

If quartz is shortened along a 2-fold axis at right angles to c, it generates a voltage, positive at one end of the 2-fold axis, and negative on the other; stretching reverses the polarity. Figure 6.15 shows schematically how this works on a single silicon tetrahedron. If the tetrahedron is deformed by pushing downward on the O^{2-} labeled p, the Si^{4+} in the middle of the tetrahedron is forced downward distance d and the three O^{2-} on the base are spread outward so that the lengths of the four Si–O bonds can remain about the same. The result is a net movement of charge. One O^{2-} and the Si^{4+}, whose net charge is +2, move down distance d relative to the three O^{2-} on the base of the tetrahedron. Relative to the original configuration, the tetrahedron becomes positively charged on the bottom and negatively charged above. Because quartz lacks center symmetry, the voltage produced by this silicon tetrahedron is not counterbalanced by a reverse voltage on other tetrahedra. The entire crystal becomes positive on one side and negative on the other.

A simple transducer can be made by cutting a thin slice of quartz and mounting it so that vibrations, such as those from an old-style phonograph needle, cause the quartz to distort. Wires attached to the quartz lead to an amplifier that detects and amplifies the voltage signal.

Just as deformation of a piezoelectric crystal produces a voltage, application of a voltage can produce a deformation. This phenomenon, called the **electrostriction** or **converse piezoelectric effect**, provides the basis for precise timing in quartz watches and tuning of modern radios. If an alternating voltage is applied to an appropriately cut slice of quartz, the quartz will alternately expand and

(a)   *(b)*

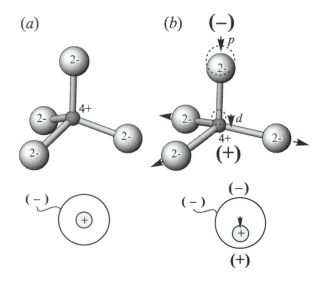

Figure 6.15 Piezoelectricity. (*a*) A silicon tetrahedron consists of a Si^{4+} centered within four O^{2-}. The positive charge is centered within the negative charge, as shown schematically with the two circles below. (*b*) If deformed by pressing downward at *p*, an O^{2-} and the Si^{4+} are moved downward distance *d* and the O^{2-} on the base are spread outward. Because only one O^{2-} moves down, the center of the positive charge on the Si^{4+} is moved downward relative to the center of mass of the negative charge on the four O^{2-}. This charge displacement produces a voltage, positive on the bottom, negative on top.

Table 6.4 The 20 Piezoelectric Point Groups and 10 Pyroelectric Point Groups[a]

Crystal System	Point Group	Piezoelectric	Pyroelectric
Triclinic	1	x	x
Monoclinic	2	x	x
	m	x	x
Orthorhombic	222	x	
	*mm*2	x	x
Tetragonal	4	x	x
	$\bar{4}$	x	
	422	x	
	4*mm*	x	x
	$\bar{4}2m$	x	
Hexagonal	3	x	x
	32	x	
	3*m*	x	x
	6	x	x
	$\bar{6}$	x	
	622	x	
	6*mm*	x	x
	$\bar{6}m2$	x	
Isometric	23	x	
	$\bar{4}3m$	x	

[a] All the groups lack a center of symmetry.

contract. The rate at which the quartz vibrates in response to the alternating current depends on how it is mounted and its thickness. The thinner the quartz slice, the greater the frequency of vibration. By connecting the quartz with other electronic components, it can be used to precisely control the frequency of an alternating current. Quartz watches keep track of the time by counting the oscillations of the alternating current whose frequency is fixed by the oscillating quartz. Radio tuners allow only signals whose frequency matches the oscillation of the quartz to be amplified.

Piezoelectric effects occur only in crystals that lack a center of symmetry (Table 6.4). The one exception is the 432 crystal class, which lacks center, but whose other symmetry precludes piezoelectricity. For a voltage to be produced, the charge distribution resulting from deformation must be **acentric**, otherwise a charge movement produced in one direction would be canceled by a symmetrically equivalent charge movement in the opposite direction. Lack of a center of symmetry allows the 20 piezoelectric crystal classes to possess **polar directions**. A polar direction is one whose end [*uvw*] is not symmetrically related to the opposite end [$\bar{u}\bar{v}\bar{w}$]. Recall that a center of symmetry requires that the ends of any line through a crystal must be symmetrically related; minerals with center symmetry may not have polar directions. Because voltage is a polar phenomena (i.e., it has a direction and a polarity defined by

where the positive and negative charges are), a voltage can be generated only along polar directions.

Pyroelectricity

A crystal is **pyroelectric** if a change in temperature causes displacement of positive and negative charges and the development of a voltage. Pyroelectricity works because heat causes distortion of the crystal lattice that displaces positive and negative charges relative to each other, much like deformation does in piezoelectricity.

Pyroelectricity is possible only in minerals with a single polar direction. These minerals lack a center of symmetry and have either no symmetry, a single mirror, or a single rotation axis (+/− mirrors parallel to that axis) (Table 6.4). This phenemenon is present only in crystals with a unique polar direction, because heat, unlike pressure, is nondirectional. Pressure can be directed along just one of several symmetrically equivalent directions, like the *a* axes in quartz, to produce the piezoelectric effect. Applying heat to quartz is equivalent to applying pressure equally in all directions, which, because of the higher symmetry of quartz, will result in the piezoelectric voltages canceling each other out. If there is a unique polar direction, heating or uniform pressure in all directions will trigger charge displacement associated with that direction and produce a voltage.

MISCELLANEOUS PROPERTIES

A variety of other properties may be useful in identifying minerals. They include taste, odor, feel, and reaction with acid.

Minerals that are readily soluble in water may have a perceptible taste. Taste is perceived if some of the mineral dissolves in saliva and activates taste receptors on the tongue. Common minerals with a distinct taste include **halite** (NaCl), which is salty, and **sylvite** (KCl), which also is salty but somewhat more bitter than halite. Mineral samples routinely handled in a student laboratory may acquire a salty taste from salt in the perspiration on hands.

Even at room temperature, thermal vibration may cause elements or molecules to break away from the surface of weakly bonded minerals and be carried in the air to odor sensors in the nose. Most ionic-, covalent-, and metallic-bonded minerals have bonding that is too strong to allow perceptible odor without significant heating. Minerals with van der Waals bonds may have an odor. A common example is the **clay** group of minerals, many of which have van der Waals bonding. Clay is perceived to have an earthy or argillaceous odor.

Feel includes a host of perceptions, most of which involve various properties of grain size and surface texture, and therefore are only loosely related to crystal structure and/or composition. One property of feel that is associated with crystal structure is the greasy feel associated with van der Waals bonds. Recall from Chapter 3 that van der Waals bonds are very weak. A finger rubbed on the surface of a mineral such as graphite breaks these bonds and allows the mineral to smear out, producing the greasy feel.

A characteristic of many carbonate minerals is that they may visibly react with dilute HCl. The acid is usually prepared by mixing one part concentrated HCl with nine parts water. The reaction with calcite is typical.

$$CaCO_3 + 2H^+ = Ca^{2+} + H_2O + CO_2$$

The CO_2 released by the reaction causes the dilute acid to fizz or bubble. Some sulfide minerals, such as sphalerite, will release H_2S when attacked by dilute acid. H_2S smells like rotten eggs.

Many field geologists carry a small bottle of hydrochloric acid with which to test sedimentary rocks for the presence of calcite ($CaCO_3$) and dolomite [$CaMg(CO_3)_2$].

Calcite reacts vigorously with the dilute acid; dolomite only if it is powdered to increase the surface area for reaction. The dilute acid solution is weak enough that it does not pose a serious hazard to flesh, but clothing does not survive well. It is not uncommon to find that a shirt pocket in which an acid bottle is kept has developed holes after a few washings. When using any caustic material, always use caution, protect your eyes, and follow Material Safety Data Sheet precautions.

In the latter part of the nineteenth century an array of qualitative chemical tests was developed that aided geologists and mineralogists with identification of minerals in the field and laboratory. These have collectively been called **blowpipe tests** because many were based on the behavior of minerals in a flame produced by blowing through a small metal pipe to direct a stream of air into a candle or other small flame. With the ready availability of X-ray diffraction and other modern analytical tools, and the ease with which samples can be identified with optical techniques, blowpipe tests are largely obsolete.

REFERENCES CITED AND SUGGESTIONS FOR ADDITIONAL READING

Bannerjee, S. K. 1991. Magnetic properties of Fe-Ti oxides. *Reviews in Mineralogy*, 25, 107–128.

Burns, R. G. 1993. *Mineralogical Applications of Crystal Field Theory*, 2nd ed. Cambridge University Press, Cambridge 551 p.

Guseinov, A. A., and Gargatsev, I. O. 2002. Electrical conductivity of alkaline feldspars at high temperature. Izvestiya, Physics of the Solid Earth 38, 520–523.

Harrison, R. J., and Feinberg, J. M. 2009. Mineral magnetism: Providing new insights into geoscience processes. Elements 5, 29–215.

Nassau, K. 1978. The origins of color in minerals. American Mineralogist 63, 219–229.

Nye, J. F. 1985. *Physical Properties of Crystals*. Oxford University Press, Oxford, 329 p.

O'Reilly, W. 1984. *Rock and Mineral Magnetism*. Chapman & Hall, New York, 220 p.

West, A. R. 1984. *Solid State Chemistry and Its Applications*. John Wiley & Sons, Chichester, England, 734 p.

Yousheng, X., Poe, B. T., Shankland, T. J., and Rubie, D. C. 1998. Electrical conductivity of olivine, wadsleyite, and ringwoodite under upper-mantle conditions. Science 5368, 1415–1418.

Zussman, J. 1967. *Physical Methods in Determinative Mineralogy*. Academic Press, New York, 514 p.

CHAPTER **7**

Optical Mineralogy

INTRODUCTION

The petrographic microscope provides one of the primary means of studying minerals and the rocks they compose. It is a specialized instrument that utilizes polarized light to allow measurement of a variety of optical properties. These properties provide a means to rapidly identify unknown minerals. The samples examined with the petrographic microscope are grain mounts and thin sections.

Grain mounts are prepared by crushing a mineral sample. A few dozen grains that pass a 140-mesh sieve and do not pass a 200-mesh sieve, producing a size range from 0.105 to 0.075 mm, are placed on a microscope slide and covered with a coverslip (Figure 7.1a). A dropper is used to introduce a liquid called an **immersion oil** between the slide and coverslip to surround and cover the grains. This grain mount is then placed on the microscope stage for examination. If a permanent grain mount is desired, an epoxy or other cement can be used instead of immersion oil. See Chapter 10 for more information on preparing grain mounts.

Thin sections are thin slices of rock or mineral mounted on a microscope slide (Figure 7.1b). They are prepared by

cementing a piece of rock to a microscope slide and then grinding it to its final thickness, usually 0.03 mm. The thin section is completed by cementing a coverslip in place to protect the sample. The coverslip is a thin (0.17 mm) piece of glass. In some cases, the coverslip is omitted and the sample is completed by carefully polishing the surface to a mirror like shine. Chapter 10 contains more information on preparing thin sections.

LIGHT

Before discussing the details of the petrographic microscope and the optical properties measured with it, it is essential to provide background on the nature and properties of light.

Light is conveniently described either as a wave phenomenon or a particle phenomenon. When atoms are sufficiently heated, or otherwise excited, the outer electrons are forced into higher-than-normal energy levels. When these electrons revert to their normal energy level, a small amount of energy, which we perceive as light, can be released. Depending on how this energy is forced to interact with matter, it may behave as either a particle or a wave. In the particle paradigm, this energy is described in terms of subatomic particles called photons that possess energy but not mass. In the wave paradigm, the energy is described in terms of electromagnetic radiation (Figure 6.6) with properties of a wave. Modern quantum mechanical theories of matter and energy have reconciled the seemingly contradictory particle and wave theories. Because the wave theory very effectively describes the phenomena of polarization, reflection, refraction, and interference, dealt with in this chapter, we shall treat light as a wave phenomenon.

Light Waves

In the wave theory, radiant energy like light has both electrical and magnetic properties and is therefore called **electromagnetic radiation**. Visible light is just a small part of the electromagnetic spectrum (Figure 6.6). It consists of electric and magnetic vectors that vibrate at right angles to the direction in which the energy is propagating

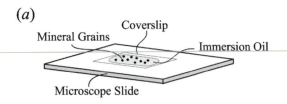

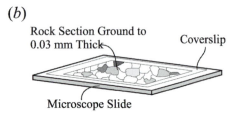

Figure 7.1 Sample preparations for the petrographic microscope. (*a*) Grain mount. (*b*) Thin section.

136

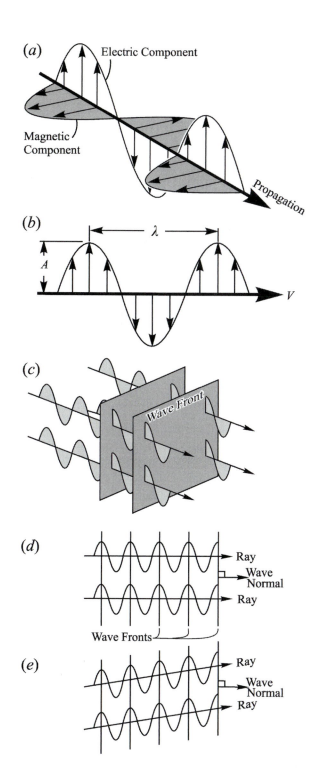

Figure 7.2 Electromagnetic radiation. (*a*) Electric and magnetic components of electromagnetic radiation vibrate transverse to the direction of propagation. (*b*) Wave nomenclature. The wave travels to the right with velocity *V*. The wavelength (*λ*) is the distance between successive wave crests. Frequency (*f*) is the number of wave crests that pass some point per second, expressed in units of cycles per second, or hertz (Hz). Brightness of the light is proportional to the square of the amplitude (*A*). (*c*) Wave fronts connect equivalent points on adjacent waves. (*d*) In isotropic materials, the wave normal and ray directions are coincident. (*e*) In anisotropic materials wave normal and ray directions are not parallel except for certain directions.

(Figure 7.2a). For present purposes it is necessary to consider only the vibration of the electric vector that interacts with the electric character of atoms and chemical bonds in minerals. Forces arising from the magnetic vector of light are generally very small and can be ignored here.

The oscillating electric field can be described using the mathematics of any wave; it has **velocity** (*V*, nm/sec) and **wavelenth** (*λ*, nm) (Figure 7.2b) that are related to the **frequency** [*f*, cycles/sec, or hertz (Hz)].

$$f = \frac{V}{\lambda} \qquad (7.1)$$

A **wave front** is a surface that connects the same points on adjacent waves (Figure 7.2c). A line drawn at right angles to the wave front is a **wave normal** and represents the direction that, could we see it, the wave would move. A **light ray** is the propagation direction of the light energy, equivalent to the path followed by a photon, were light being described by using the particle theory. In general, the wave normal and the path of the light ray coincide only in materials in which light velocity is uniform in all directions (Figure 7.2d). If velocity varies with direction, as it does in all transparent minerals except those that crystallize in the isometric system, the path followed by light rays generally does not coincide with the wave normal (Figure 7.2e) as we shall see in later sections.

Two light waves that vibrate at an angle to each other can be resolved into a resultant vibration direction by means of vector addition. The resultant vibration direction *R* in Figure 7.3a is obtained by constructing a parallelogram whose sides are parallel to the vibration directions of waves *A* and *B*. Similarly, a component of a single wave *X* (Figure 7.3b) may be resolved into a new vibration direction *V* at angle *θ*. The magnitude of *V* is given by the equation

$$V = X \cos (\theta). \qquad (7.2)$$

Polarized Light

Ordinary light coming directly from the sun or an incandescent lightbulb vibrates in all directions at right angles to the direction of propagation (Figure 7.4a). It is unpolarized. If the vibration of the light is constrained to lie in a single plane, the light is **plane polarized**, and the vibration can be represented by a simple sine wave (Figure 7.4b).

Modern petrographic microscopes provide polarized light by placing a piece of **polarizing film** in the optical path. Polarizing film consists of a sheet of plastic that is optically anisotropic. When unpolarized light enters this plastic, it is split into two plane-polarized rays that vibrate at right angles to each other (much more about this is given in a later section); about half of the light energy is resolved into each ray. One of the rays is strongly absorbed by the plastic and is eliminated. The other is only weakly absorbed and passes through the plastic to emerge as plane-polarized light.

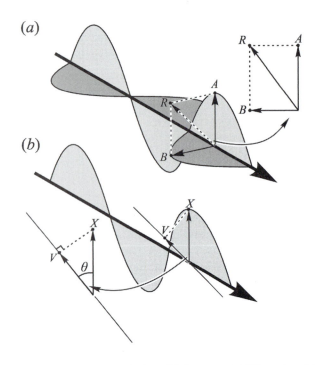

(a)

(b)

Figure 7.3 Vector resolution of light waves. (*a*) Waves *A* and *B* form resultant *R*. (*b*) A component *V* of wave *X* can be resolved in a new direction at angle θ from *X*.

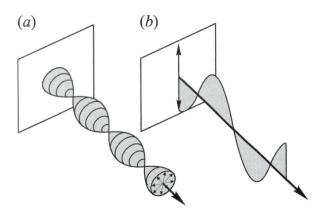

(a)   *(b)*

Figure 7.4 Polarized light. (*a*) Unpolarized light vibrates in all directions at right angles to the direction of propagation. (*b*) Plane-polarized light. The electric vector vibrates in a single plane.

INTERACTION OF LIGHT AND MATTER

The velocity of light depends on the nature of the material through which it travels and the wavelength of the light. The maximum possible velocity of 3.0×10^{17} nm/sec is achieved only in a vacuum. In all other materials, light moves more slowly. The detailed explanation of why light's velocity is slower is beyond the scope of this book, but it involves the interaction of the electric vector of the light and the electric environment around each atom/ion in the structure.

When light passes from one material to another, the frequency remains constant. When velocity decreases on passing from air into a mineral, the wavelength must also decrease (Figure 7.5) to keep the frequency constant (Equation 7.1). A physical analog is a group of cars traveling down the freeway. At high speed, the cars are far apart, but when slowed down, they bunch up. Whether fast or slow, however, the number of cars passing an observer per unit of time (the frequency) remains the same.

Optically Isotropic versus Anisotropic Materials

An **optically isotropic** material is one in which the velocity of light is the same in all directions. The isotropic rock-forming materials include volcanic glass and minerals belonging to the isometric crystal system. In these materials, electron density is the same in all directions, at least on average, so the strength of the electric field with which the electric vector of light interacts also is the same regardless of direction. Light velocity is therefore the same in all directions.

An **optically anisotropic** material is one in which the velocity of light is different in different directions. The anisotropic rock-forming materials include minerals in the tetragonal, hexagonal, orthorhombic, monoclinic, and triclinic crystal systems. Minerals in these crystal systems have lower symmetry than those in the isometric system, and the electron density varies with direction. The electrons of the atoms/ions in these minerals are not able to interact with light in the same way in all directions, so the velocity and absorption characteristics (color) of light vary with direction. If normally isotropic solids are unevenly strained, by bending for example, some chemical bonds are stretched and others are shortened. This changes the electron density, and these materials become optically anisotropic.

Reflection and Refraction

Light reaching the boundary between two transparent materials will be both **reflected** from the surface and **refracted** into the new material. For the reflected light,

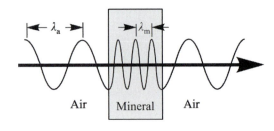

Figure 7.5 Light entering a mineral slows down and the wavelength (λ) decreases.

the angle of incidence and angle of reflection are identical (Figure 7.6a). The path taken by the light entering a transparent mineral is bent, or refracted, when the incident light does not strike the boundary at 90° (Figure 7.6b).

The **index of refraction** (n) provides a measure of how much the light will be bent and is a function of velocity

$$n = \frac{V_v}{V_m} \qquad (7.3)$$

where V_v is the velocity in a vacuum and V_m is the velocity in the material. The angle that the light is bent upon passing from material 1 to material 2 is given by **Snell's law**

$$\frac{\sin \theta_1}{\sin \theta_2} = \frac{n_2}{n_1} \qquad (7.4)$$

where n_1 and n_2 are the indices of refraction of light in materials 1 and 2 and θ_1 and θ_2 are the angles shown in Figure 7.6b. Note that $\theta_1 = 0°$ for normal incidence to the surface and θ_1 is nearly 90° where the light glances off the surface at a very low angle.

Snell's law applies for both isotropic and anisotropic materials, provided the angles θ_1 and θ_2 are measured between the normal to the interface between the materials and the wave normals, not the rays. Recall that the wave normal and ray directions coincide for isotropic materials but diverge for anisotropic materials (Figure 7.2d,e).

The relative amounts of light that are reflected from and refracted into a mineral depend both on the index of refraction of the material and the angle of incidence of the light. The amount that is reflected increases with higher index of refraction. Recall from Chapter 6 that mineral luster depends, in part, on the amount of reflection, and a more brilliant luster is associated with a higher index of refraction. The angle of incidence also controls how much light is reflected from a surface. If the light strikes normal to the surface (angle of incidence $\theta_1 = 0°$), reflection is a minimum, and only when the light comes in at a very shallow angle (θ_1 is close to 90°), does most of the light get reflected. The relative amounts of light that are reflected

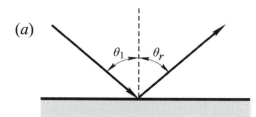

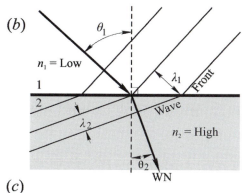

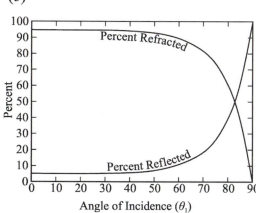

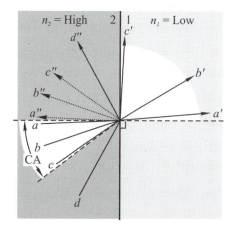

Figure 7.6 Reflection and refraction. (a) Reflection from a smooth surface. The angles of incidence (θ_1) and reflection (θ_r) are identical. (b) Refraction. Wave fronts one wavelength apart and the wave normal (WN) are shown for light passing from material 1 (n_1 = low index of refraction, higher velocity) to material 2 (n_2 = high index of refraction, low velocity). The wave fronts and wave normal are bent at the interface because the wavelength λ_2 in the low-velocity material is shorter than the wavelength in the high-velocity material. (c) Relative amount of light coming from air (n = 1) that is reflected and refracted from the surface of a material with n = 1.55 as a function of the angle of incidence.

Figure 7.7 Critical angle and total internal reflection. Some of the light following wave normals a, b, and c within the unshaded arc in high-index material 2 can be refracted to a', b' and c' in low-index material 1 because the angle of incidence is less than the critical angle (CA). Wave normal d, which has an angle of incidence greater than the critical angle, is totally reflected to d" at the boundary. A portion of the light following wave normals a, b, and c is reflected to a", b", and c" as is shown with dotted lines. The amount of reflection increases from a to c and is 100% for angles of incidence greater than the CA.

and refracted can be calculated from the **Fresnel equation** (Dyer and others, 2008, p. 428). Figure 7.6c shows the amount of light that is reflected and refracted at the surface of an isotropic colorless material with an index of refraction of 1.55, such as a piece of window glass. For this material, whose luster is vitreous, light striking normal to the surface ($\theta_1 = 0°$) is only about 5% reflected, so that 95% of the light enters the material. To get more than 20% of the light to be reflected with normal incidence, the minimum to be perceived as a metallic luster, the index of refraction must be greater than 2.6.

Light can always be refracted from a low-index material into a high-index material because the angle of refraction is smaller than the angle of incidence. Light traveling from a high-index material into a low-index material cannot pass through the boundary if the angle of incidence is greater than the **critical angle** (CA). The critical angle is the angle of incidence that yields an angle of refraction of 90°. Consider Figure 7.7. Some light following wave normals *a, b,* and *c* is refracted into the low-index material with angles of refraction that are larger than the angles of incidence according to Equation 7.4. Wave normal *c* has an angle of incidence that yields an angle of refraction slightly less than 90°. For any angle of incidence larger than CA, the angle of refraction would have to be greater than 90°. Because this is not possible, wave normals such as *d*, whose angles of incidence exceed CA, experience **total internal reflection**. If the indices of refraction are known, the critical angle can be calculated as follows

$$\frac{n_1(\text{low})}{n_2(\text{high})} = \sin \text{CA} \qquad (7.5)$$

which can be derived from Snell's law by setting $\theta_1 = \text{CA}$ and $\theta_2 = 90°$.

Dispersion

In most transparent minerals violet light is more strongly refracted than red light (Figure 7.8). This relationship, in which the index of refraction is higher for short wavelengths and lower for long wavelengths of light, is known as **normal dispersion of the refractive indices**. To describe the dispersion of a material, it is necessary to report the index of refraction at several wavelengths. By convention, indices of refraction n_F, n_D, and n_C are reported for light whose wavelengths are 486, 589, and 656 nm, respectively. These wavelengths are chosen because they correspond to absorption lines in the sun's spectrum known as **Fraunhofer lines** that are due to the presence of various gasses and because they can easily be produced in the laboratory. When a single index of refraction is reported for a material, it is n_D that is meant. The value of n_D is selected because 589 nm light is in the middle of the visible spectrum and is easily produced in the laboratory with a sodium vapor lamp. This index of refraction may be identified either n_D or n.

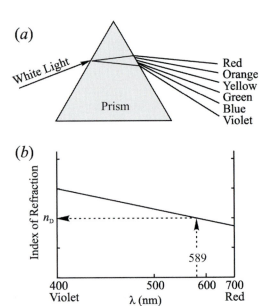

Figure 7.8 Dispersion of the refractive indices. (*a*) White light is spread into its spectral colors when passed through a prism because the index of refraction for violet light is larger than for red light. (*b*) Normal dispersion of the refractive indices. Index of refraction decreases with increasing wavelength. Strongly colored minerals may display abnormal dispersion of the refractive indices (index of refraction increases with increasing wavelength) for wavelengths that are strongly absorbed by the mineral.

PETROGRAPHIC MICROSCOPE

Numerous high-quality petrographic microscope designs are available from manufacturers such as Nikon, Zeiss, and Leitz. Although they differ in detail, all have fundamentally the same design and construction (Figure 7.9). From the bottom up they consist of an illuminator, substage assembly, stage, objective lenses, upper polarizer, Bertrand lens, and ocular lens(es). Focus is achieved with a focusing mechanism mounted in the microscope frame, and a slot is provided above the objective lenses for insertion of optical accessory plates.

Illuminator

The illuminator, mounted in the microscope's base, consists of an incandescent lightbulb that may have a rheostat control for brightness. The light is directed upward with a combination of mirrors and/or lenses and is filtered so that the color balance more closely approximates natural sunlight.

Substage Assembly

The substage assembly typically consists of **lower polarizer, condensing lens, auxiliary condensing lens,** and **aperture diaphragm**. A gear mechanism may allow the substage assembly to be raised or lowered.

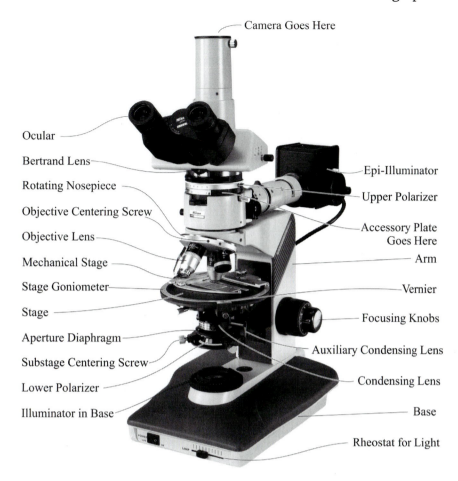

Camera Goes Here

Ocular

Bertrand Lens

Rotating Nosepiece

Objective Centering Screw

Objective Lens

Mechanical Stage

Stage Goniometer

Stage

Aperture Diaphragm

Substage Centering Screw

Lower Polarizer

Illuminator in Base

Epi-Illuminator

Upper Polarizer

Accessory Plate Goes Here

Arm

Vernier

Focusing Knobs

Auxiliary Condensing Lens

Condensing Lens

Base

Rheostat for Light

Figure 7.9 The petrographic microscope. Photo courtesy of Nikon Instruments Inc., Melville, New York.

The lower polarizer typically consists of optical-quality polarizing film. In many microscopes the lower polarizer may be rotated to adjust the vibration direction of the light coming up through the instrument. Polarizers may be referred to as **Nicols**, because Nicol prisms, constructed of calcite, were used on early microscopes to provide polarized lights. The lower polarizer is typically aligned so that it passes light that vibrates east–west as viewed down the microscope tube. In some older instruments the lower polarizer may be oriented to pass light vibrating north–south.

The condensing lens serves to concentrate and focus the light from the illuminator onto an area of the sample immediately beneath the objective lens. The illumination provided by the condensing lens is termed **orthoscopic illumination** because the light reaching the sample is nearly parallel.

The auxiliary condensing lens is mounted on a pivot so that it may be swung into or out of the optical path. Its function is to provide **conoscopic illumination**, which consists of strongly converging light. Conoscopic illumination is used to observe optical phenomena called interference figures, which are examined with the aid of the Bertrand lens and the high-power objective.

The aperture diaphragm is an iris diaphragm that is adjusted to control the size of the cone of light passing up through the microscope. Closing the aperture diaphragm decreases the size of the cone of light and increases the contrast in the image seen through the microscope. The aperture diaphragm is not intended to be used to adjust the intensity of illumination. The rheostat control on the illuminator should be used for that purpose.

Some microscopes are equipped with slots or holders in the substage assembly intended to hold color filters. Appropriate filters may be used to adjust the color balance of the light or to provide light of a specific wavelength.

Microscope Stage

The circular stage of the petrographic microscope is mounted on bearings so that it can be rotated smoothly. Angles of stage rotation may be measured with the **stage goniometer** and **vernier** on the outside edge of the stage. A thumbscrew or lever can be engaged on some instruments to lock the stage in one position.

Objective Lenses

The objective lenses provide the primary magnification of the optical system. Most student-model microscopes are equipped with three objective lenses with magnifications of around 2.5, 10, and 40×, mounted in a rotating nosepiece. Markings typically found on objective lenses include the magnification, numerical aperture, length of the microscope tube for which it is designed, and, for high-power lenses, the coverslip thickness that provides optimum optical performance (Figure 7.10a). Lenses designed for use with an optical oil between the lens and sample will be marked "oil" or "oel."

The **numerical aperture** (NA) of a lens is a measure of the size of the cone of light that it can accommodate. It is given by the equation

$$NA = n \sin \frac{AA}{2} \qquad (7.6)$$

where n is the index of refraction of the medium between the objective lens and the sample and AA is the **angular aperture** (Figure 7.10b). A large numerical aperture

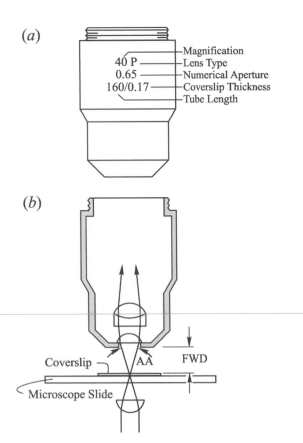

Figure 7.10 Objective lens. (a) Typical objective lens markings. P indicates that the lens is intended for use with polarized light. (b) The free working distance (FWD) is the distance between the end of the lens and top of the sample. Angular aperture (AA) is the size of the cone of light that the lens can accept.

provides greater resolving power and is possible only on lenses with a small **free working distance**, which is the distance between the end of the lens and the top of the sample. Knowledge of the numerical aperture of the high-power (e.g., 40×) lens is useful in the interpretation of optical phenomena known as interference figures. Most student-model microscopes have high-power lenses with NA = 0.65.

Upper Polarizer

The **upper polarizer**, also called the **analyzer**, is located above the objective lenses and is mounted on a slide or pivot so that it can be easily inserted into or removed from the optical path. The vibration directions of the lower and upper polarizers are conventionally set so that they are at right angles to each other. When the upper polarizer is inserted, the polarizers are said to be **crossed**; with nothing on the microscope stage, the field of view is dark because the plane-polarized light that passes the lower polarizer is absorbed by the upper polarizer. If the upper polarizer is removed, the view through the microscope is with **plane light** because light from the lower polarizer is plane polarized.

Bertrand Lens

The Bertrand lens (also called the **Amici–Bertrand** lens) is a small optical element mounted just below the ocular on a pivot or slide. It is introduced into the optical path to allow the observer to view optical phenomena called interference figures that are seen near the top surface of the objective lens. Interference figures are described in later sections. The Bertrand lens may be equipped with an iris diaphragm or a pinhole to restrict the field of view, and some manufacturers provide a focusing mechanism.

Oculars

Oculars (eyepieces) are lenses that slide into the upper end of the microscope tube. They magnify the image provided by the objective lens and focus the light so that it can be accepted by the human eye. The magnification is usually in the range of 5 to 12×. The total magnification of the microscope is the magnification of the objective times the magnification of the ocular. With a 40× objective and a 10× ocular, the total magnification is 400×. Many microscopes are equipped with binocular heads and therefore use two ocular lenses; one with fixed focus and the other adjustable to accommodate small differences between the user's eyes. Adjustment also is provided to match the spacing between binocular oculars to the interpupillary distance of the user.

Reticle markings that may include crosshairs and a micrometer scale are mounted in the ocular. The upper part of this lens may be screwed in or out to bring the markings into crisp focus. On binocular microscopes, the ocular with fixed focus is usually equipped with the

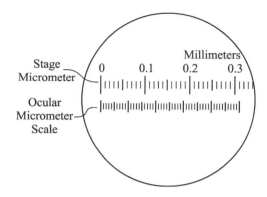

Figure 7.11 Calibration of an ocular micrometer scale. The 50 divisions of the ocular micrometer scale (bottom scale) subtend 0.31 mm on the stage micrometer. Each ocular micrometer division must therefore equal 0.0062 mm. A separate calibration must be made for each objective lens.

reticle markings. Detents are provided in the top of the microscope tube that match a small peg on the side of the ocular. They allow the crosshairs to be accurately oriented N–S and E–W or in the 45° positions. A micrometer scale may be provided to allow measurement of the size of grains in the sample. The micrometer scale must be calibrated for each different objective lens by comparison with a **stage micrometer**, which is a microscope slide on which a millimeter scale has been etched (Figure 7.11).

Conventional oculars are designed for use without eyeglasses, and the microscope focus is adjusted to accommodate near- or farsightedness. If eyeglasses are worn, the field of view will be restricted because the eyes are kept too far from the ocular(s). Many newer microscopes are equipped with **high-eye-point oculars** that are designed for use with eyeglasses. They typically have rubber eyecups that can be flipped up for use without eyeglasses.

Focusing Mechanism

Focusing is accomplished in most cases by raising or lowering the stage by means of a screw or gear mechanism that provides both coarse and fine adjustment. The mechanism may use two knobs or a single knob that incorporates both functions. If the microscope is focused for one objective lens, it should be very nearly focused for the other lenses. Particularly when using the high-power lens, where the free working distance is very small, care must be exercised to avoid smashing the sample into the end of the objective lens while attempting to focus.

Accurately focus all elements of the microscope to avoid eyestrain. Begin by focusing the crosshair with nothing on the stage of the microscope in plane light and then focus on a sample. On binocular microscopes, focus first while viewing the sample thorough the fixed focus ocular; then manipulate the adjustable ocular to bring it into focus.

Accessories

A slot is provided immediately above the objective lens for insertion of **accessory** or **compensator plates**. They consist of a metal or plastic frame in which an optical element is mounted. The common accessory plates are the **gypsum (full-wave)** and **mica (quarter-wave) plates**. They are used to assist with measuring a variety of optical properties described in later sections.

Direction Conventions

Compass directions are routinely used to describe orientations in the microscope image. North is to the top, south to the bottom, and east and west to the right and left, respectively. The 45° positions are NE–SW and NW–SE.

ISOTROPIC MATERIALS

Materials of geological interest that are optically isotropic include volcanic glass and minerals that crystallize in the isometric crystal system. Appendix B lists the common isotropic minerals.

Isotropic materials can be distinguished from anisotropic materials by inserting the upper polarizer so that the sample is between crossed polarizers. **All isotropic minerals are dark (black) between crossed polarizers** because isotropic minerals do not affect the polarization direction of the light coming from the lower polarizer. Recall that the lower and upper polarizers are arranged so that their vibration directions are at 90° to each other. From Equation 7.2 it can be seen that the vector component of incident light resolved into a direction at 90° is zero. Because the light passing through an isotropic mineral retains its original polarization, the vector component parallel to the upper polarizer is zero, so no light passes the upper polarizer. In contrast to isotropic minerals, **anisotropic minerals generally display a variety of colors between crossed polarizers**, except in certain orientations, because the anisotropic minerals affect the polarization of the incident light, as will be described in subsequent sections.

Measuring the index of refraction is a convenient optical means of identifying an isometric mineral. The section on refractometry later in this chapter describes the techniques used to measure index of refraction using grain mounts. Appendix B lists the indices of refraction of common isometric minerals.

For many minerals, the index of refraction varies depending on chemical composition, and different minerals may have the same index of refraction. Other information may therefore be needed to confirm an identification. Color in grain mount may be useful but should be used with a certain amount of care because it may be variable, as described in Chapter 6. Preparation of grain mounts involves crushing the sample, so cleavage, if a mineral has

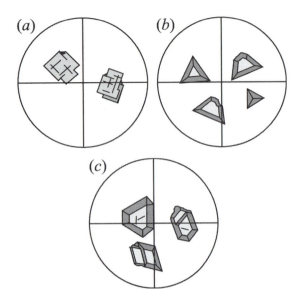

Figure 7.12 Cleavage in isometric minerals as seen in grain mount. (*a*) Cubic {001} cleavage. (*b*) Octahedral {111} cleavage. (*c*) Dodecahedral {011} cleavage.

it, may control fragment shape. The common cleavages found in isometric minerals are cubic {001}, octahedral {111}, and dodecahedral {011} (Figure 7.12). If the sample is large enough to observe physical properties such as color, luster, cleavage, and hardness, that information can aid in the identification. If hand-sample and optical techniques do not provide an identification, X-ray diffraction techniques may be employed (Chapter 8).

In thin sections, accurate measurement of the index of refraction is not practical. Color, however, should be visible, and if crystals are euhedral or subhedral, crystal shape may be evident. The shape seen in thin section is the shape of a random slice through the crystal. Sections through cubic crystals are typically four-sided or sometimes three-sided shapes. Sections through octahedrons are usually four- or six-sided shapes, and sections through dodecahedra usually are six- or eight-sided shapes. Cleavage also may be visible as straight lines or fine cracks in the grains.

ANISOTROPIC MINERALS

Optically anisotropic minerals are those in the tetragonal, hexagonal, orthorhombic, monoclinic, and triclinic crystal systems. Because the electronic environment varies with direction, light velocity also must vary with direction. As a consequence, anisotropic minerals show **double refraction**. What this means is that light entering an anisotropic mineral in most directions is split into two rays with different velocities. The two rays vibrate at right angles to each other. This can be demonstrated by placing a cleavage rhomb of clear calcite on some marks on a piece of paper (Figure 7.13). Two images corresponding to the two

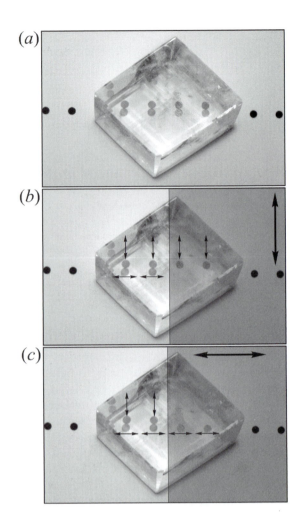

Figure 7.13 Double refraction of calcite. (*a*) Cleavage rhomb of clear calcite on a row of dots. Two images are produced because the light is split into two rays that vibrate at right angles to each other. (*b*) A polarizing film whose vibration direction is parallel to the short diagonal of the rhomb passes one set of dots and absorbs the other. (*c*) The polarizing film is rotated 90°. The first set of dots is absorbed and the other passes. In intermediate orientations, both sets of dots are visible with subdued brightness.

rays are produced. If the rhomb is viewed through a piece of polarizing film whose vibration direction is parallel to the short axis of the rhomb, only one image is seen. If the polarizing film is rotated 90°, only the other image is seen. The two rays must therefore be plane polarized and vibrating at right angles to each other.

If the light velocity of the two rays is determined by measuring indices of refraction using techniques described later, it will be found that one ray is faster than the other. The ray with the lower index of refraction and greater velocity is called the **fast ray**, and the ray with the higher index of refraction and lower velocity is called the **slow ray**.

Every anisotropic mineral has either one or two directions, called **optic axes**, along which the light is not split into two rays. Minerals in the hexagonal and tetragonal crystal systems have one optic axis and are optically **uniaxial**. Minerals in the orthorhombic, monoclinic, and triclinic crystal systems have two optic axes and are optically **biaxial**. Uniaxial and biaxial are the two categories of **optical character**.

Interference Phenomena

When an anisotropic mineral is placed on the microscope stage between crossed polarizers, in most cases light passes the upper polarizer and the mineral displays a color. The colors seen between crossed polarizers, called **interference colors**, are produced as a consequence of light being split into two rays on passing through the mineral. It is convenient to first consider interference with monochromatic light and then extend the discussion to polychromatic light.

Monochromatic Illumination

Consider a ray of plane-polarized light from the lower polarizer that enters a plate of an anisotropic mineral (Figure 7.14). When light enters the mineral, it is split into two rays that vibrate at right angles to each other and have different indices of refraction (different velocities). Because of the difference in index of refraction, the slow ray lags behind the fast ray. In the time it takes for the slow ray to pass through the mineral, the fast ray will have passed through the mineral plus an additional distance called **retardation** (Δ). The retardation remains constant after the slow and fast rays have exited into the air above the mineral grain because both have the same velocity there. The magnitude of the retardation depends on the thickness of the mineral (d) and on the difference in index of refraction of the slow ray (n_s) and fast ray (n_f) in the mineral

$$\Delta = d\,(n_s - n_f) = d\,(\delta) \qquad (7.7)$$

where (δ) is the **birefringence** equal to the value ($n_s - n_f$). The numerical value of birefringence depends on the direction followed by the light through the mineral. Directions parallel to an optic axis show zero birefringence, other directions show a maximum birefringence, and most show an intermediate value. **The maximum birefringence is a useful diagnostic property of minerals**.

Interference colors are produced when the slow and fast rays reach the upper polarizer and are resolved into its vibration direction. What happens depends on the magnitude of the retardation and whether the slow and fast rays are in or out of phase.

In Figure 7.15a the slow ray is retarded an integer number of wavelengths relative to the fast ray or

$$\Delta = i\,\lambda \qquad (7.8)$$

where i is an integer and λ is the wavelength of the monochromatic light. The components of the two rays resolved into the vibration direction of the upper polarizer are of equal magnitude but in opposite directions, so they cancel

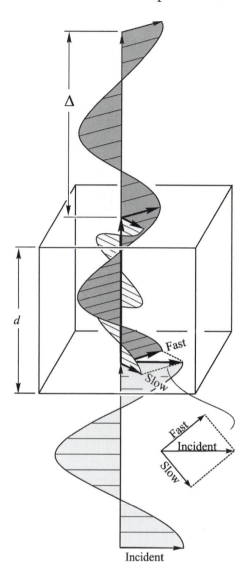

Figure 7.14 Development of retardation. Light entering the mineral with thickness d is split into slow and fast rays. In the time it takes the slow ray to pass through the mineral, the fast ray has passed through the mineral and traveled an additional distance Δ, which is the retardation.

each other. No light passes the upper polarizer, and the mineral grain appears dark.

In Figure 7.15b, the retardation is equal to one-half wavelength.

$$\Delta = \left(i + \frac{1}{2}\right)\lambda \qquad (7.9)$$

As before, the waves are resolved into the vibration direction of the upper polarizer. In this case, however, the resolved components are both in the same direction, so the light constructively interferes, and light passes the upper polarizer.

These relationships, in which in-phase light cancels and out-of-phase light constructively interferes, appear to be at odds with the conventional presentation of interference taught in physics and mathematics classes (in-phase adds,

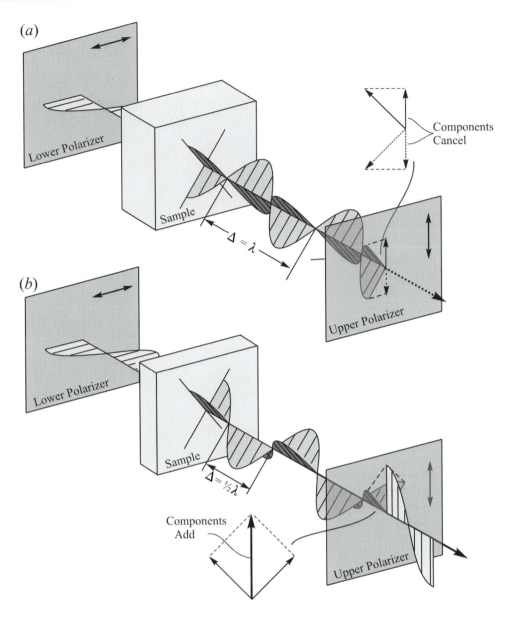

Figure 7.15 Interference at the upper polarizer with monochromatic light (a) The retardation (Δ) is one wavelength. Vector components of the two rays resolved into the vibration direction of the upper polarizer are in opposite directions, so they cancel. No light passes and the mineral appears dark. (b) The retardation is one-half wavelength. Vector components of the two waves resolved into the vibration direction of the upper polarizer are in the same direction, so they constructively interfere. Light passes the upper polarizer, and the mineral appears bright.

out-of-phase cancels). The relationships presented here are, in fact, correct because the two rays vibrate at right angles after passing through the mineral, not in the same plane. When resolved into the vibration direction of the upper polarizer they destructively or constructively interfere as described.

If the sample is wedge shaped instead of flat (Figure 7.16a), the thickness and retardation vary from zero at the thin end to a maximum at the thick end. When placed between crossed polarizers, areas along the wedge where retardation is equal to 0, 1, 2, 3, etc. wavelengths ($i\lambda$)

are dark, and areas where the waves have other retardations are light. The brightest illumination is where the retardation is $(i + \frac{1}{2})\lambda$. For a colorless mineral and no losses from reflection or absorption, the percentage of light that passes the upper polarizer if the stage is rotated so that the mineral vibration directions are in the 45° positions (NE–SW, NW–SE), is given by

$$T = \left(\sin^2 \frac{180° \Delta}{\lambda} \right) \times 100 \qquad (7.10)$$

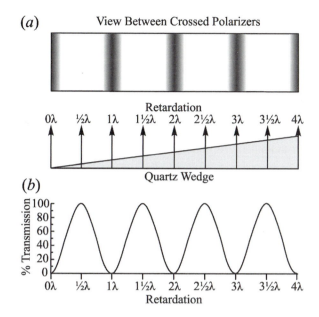

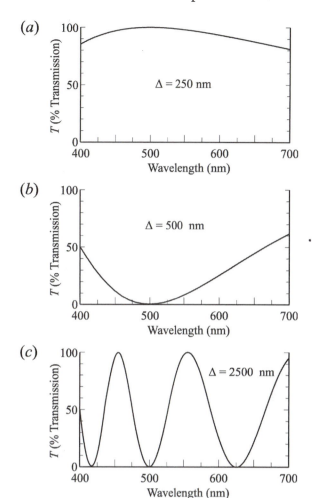

(a) View Between Crossed Polarizers

Retardation
0λ ½λ 1λ 1½λ 2λ 2½λ 3λ 3½λ 4λ

Quartz Wedge

(b)
% Transmission
0λ ½λ 1λ 1½λ 2λ 2½λ 3λ 3½λ 4λ
Retardation

Figure 7.16 Interference pattern formed with a quartz wedge in monochromatic light. (*a*) Where retardation is an integer number of wavelengths, slow and fast rays destructively interfere at the upper polarizer and a dark band is seen. Where retardation is $i + \frac{1}{2}$ wavelength, the two rays constructively interfere and light passes with maximum intensity. (*b*) Percentage transmission of light by the upper polarizer, assuming ideal optical conditions. Computed from Equation 7.10.

Figure 7.17 Formation of interference colors with polychromatic light. (*a*) Retardation for all wavelengths is 250 nm. Slow and fast rays are largely out of phase for all wavelengths, so over 80% of all wavelengths pass the upper polarizer to form a first-order white interference color. (*b*) Retardation for all wavelengths is 500 nm. Portions of the violet and red ends of the spectrum are transmitted and 500 nm light is entirely canceled, producing a color perceived as first-order red. (*c*) Retardation for all wavelengths is 2500 nm. Wavelengths around 417, 500, and 625 nm are canceled, and wavelengths near 455, 555, and 714 nm are passed with maximum intensity. This combination of transmitted light is perceived as an upper-order creamy white.

where T is percentage transmission, Δ is the retardation (Equation 7.7), and λ is the wavelength of light being used. Figure 7.16b graphically illustrates this relationship for the monochromatic light passing through the quartz wedge shown in Figure 7.16a.

Polychromatic Illumination

With the use of white light instead of monochromatic light, all the different wavelengths are present, and each is split into slow and fast rays. For a given thickness of mineral, approximately the same amount of retardation is produced for all wavelengths. The two rays for some wavelengths reach the upper polar in phase and are canceled, and the two rays for other wavelengths reach the upper polar out of phase and are transmitted. The combination of wavelengths that passes the upper polarizer produces what we perceive as interference colors.

If an optical accessory called a **quartz wedge**, which consists of a wedge-shaped piece of quartz mounted in a holder, is placed between crossed polarizers, the sequence of colors seen in the **interference color chart**, also known as a **Michel-Lévy Chart** (Plate 1), is produced. At the thin edge of the wedge, the thickness is zero so retardation also is zero; all wavelengths cancel at the upper polarizer ($\Delta = 0\lambda$) and the color is black. As the thickness increases, the color changes to gray, white, yellow, red and then a repeating sequence of blue, green, yellow, red, with each

repetition becoming progressively paler. The color produced at any point along the wedge depends on which wavelengths pass the upper polarizer and which are canceled. The percentage transmission for a given thickness of mineral can be calculated for each wavelength of light from Equation 7.10. These values are shown in Figure 7.17 for three different thickness of quartz whose birefringence is 0.009.

At the point on the quartz wedge where the thickness is 0.0278 mm (1 mm = 10^6 nm), retardation for all wavelengths is about 250 nm (Equation 7.7) and the slow and fast rays for all wavelengths are substantially out of phase.

Over 80% of every wavelength passes the upper polar (Figure 7.17a), and the interference color appears white, with a yellowish tinge because small amounts of the red and violet ends of the spectrum are canceled at the upper polarizer.

Where the quartz is 0.056 mm thick, retardation for all wavelengths is about 500 nm. Light whose wavelength is 500 nm is canceled, as is most of the blue and green light in the middle part of the spectrum (Figure 7.17b). The remaining light that passes the upper polarizer is perceived as red with a purplish cast.

For a piece of quartz 0.278 mm thick, which is thicker than that found on most quartz wedges, retardation is about 2500 nm for all wavelengths. Light whose wavelengths are 417, 500, and 625 nm is entirely canceled, and other wavelengths pass with varying intensity (Figure 7.17c). This interference color is perceived as a creamy white because a fair amount of each part of the spectrum is allowed to pass.

The interference color chart (Plate 1) shows the interference colors produced for retardations (path differences) between 0 and 1800 nm. These values are plotted along the lower edge of the diagram. This color sequence is conveniently divided into **orders**, with the breaks between orders occurring every 550 nm of retardation. First- and second-order colors are most vivid; higher-order colors are progressively more and more washed out so that above the fourth order, the colors degenerate into a creamy white.

Some minerals display interference colors not shown on the interference color chart. These **anomalous interference colors** are produced because the dispersion of refractive indices is substantially different for the slow and fast rays. This results in substantially different retardation for short-wavelength light than for long-wavelength light. A different complement of wavelengths reaches the eye than if retardation were uniform for all wavelengths, and the eye perceives the color differently. Common minerals that display anomalous interference colors are listed in Appendix B. Interference color also may be influenced by mineral color. Minerals with a distinct green color, for example, will lend a green cast to interference colors that they produce.

Use of the Interference Color Chart

The primary function of the interference color chart (Plate 1) is to allow rapid determination of retardation between the slow and fast rays. Retardation is determined by observing the interference color produced by a mineral grain in either grain mount or thin section and finding the same color on the color chart. The numerical value of retardation is read from the bottom of the chart beneath the color. If, for example, the interference color displayed by a mineral is second-order green, the retardation between slow and fast rays must be about 700 nm.

Distinguishing among similar colors in the different orders may pose some difficulty. The most direct way of determining whether a yellow interference color, for example, is first, second, or third order is to carefully observe the thin edge of the grain displaying the color. Because grains come to zero thickness at the edge, retardation also must be zero (Equation 7.7). It is therefore often possible to count interference colors inward from the edge of the grain to determine which order interference color is displayed in the middle. The red and blue colors at the transition from one order to the next form a dark band that is easily identified. Accessory plates described in a following section also may be used to determine which order of color is displayed.

Knowledge of retardation can be used to determine the thickness of a thin section if a mineral with known birefringence is present, or to determine birefringence of an unknown mineral if thickness is known.

Determining Thickness

If birefringence and retardation are known, Equation 7.7 provides the basis for determining thickness of a thin section. Quartz is an easily identified mineral that is present in many different rocks. Its maximum birefringence is a consistent 0.009. The retardation that any particular quartz grain in a thin section will display depends on orientation. A grain whose c axis is parallel to the microscope stage (horizontal) will display the maximum retardation and thus the 0.009 birefringence. If the c axis is vertical, retardation and birefringence are zero. In all other orientations, retardation and briefringence are intermediate. Hence, to find the thickness of a quartz-bearing thin section proceed as follows.

- Scan the slide to find grains of quartz that display the highest interference color, meaning the color furthest to the right on the interference color chart. Only these grains are oriented to display the known 0.009 birefringence.

- Identify the highest interference color displayed by the quartz and find it on the interference color chart. Read the retardation corresponding to that color from the bottom edge of the chart. If, for example, the color is first-order white with a tinge of yellow, the retardation is 270 nm.

- Calculate thickness from Equation 7.7, using 0.009 for the birefringence and the retardation determined here.

$$d = \frac{270\,\text{nm}}{0.009} = 30{,}000\,\text{nm} = 0.03\,\text{mn}$$

The calculation can be done graphically, directly on the interference color chart, by using the three sets of lines. Vertical lines are retardation lines, diagonal lines are birefringence lines, and horizontal lines are thickness lines (Figure 7.18a). Follow the vertical retardation line (i.e., the color) up to where it intersects the diagonal birefringence line (0.009). The horizontal line (0.03 mm) that passes through this intersection indicates the thickness.

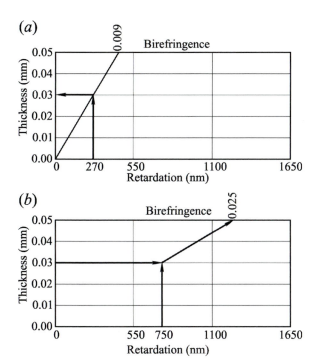

Figure 7.18 Use of the interference color chart (Plate 1). Horizontal lines indicate thickness, diagonal lines are for birefringence, and vertical lines indicate retardation (interference color). (*a*) Thickness is indicated by the horizontal line that goes through the point where birefringence and interference color intersect. (*b*) Birefringence is indicated by the diagonal line that goes through the intersection of interference color (retardation) and thickness.

Other minerals whose maximum birefringence is known may be used to determine thickness if quartz is not present.

Thin sections used for routine mineralogical and petrographic examination are typically 0.03 mm thick. In these thin sections, the maximum interference color displayed by quartz is first-order white with a hint of yellow. It is not uncommon for thin sections to have nonuniform thickness, thin at the edges or thick on one side, for example. Some variation can be tolerated for routine work.

Determining Birefringence

The maximum birefringence of a mineral is a very useful diagnostic property. It is easy to measure in thin sections whose thickness is known. Recall that the retardation displayed by a grain of a given mineral depends on exactly how it happens to be oriented on the stage of the microscope. A few grains will display zero retardation, a few will display the maximum retardation if they are cut to have maximum birefringence, and most will display intermediate values. To measure the maximum birefringence, it is necessary to search out grains that display maximum retardation as follows.

• Scan the thin section to find a sample of the unknown mineral whose interference color is highest, meaning farthest to the right on the interference color chart. Only these grains are oriented to display the maximum birefringence.

• Identify that interference color, find it on the interference color chart (Plate 1), and read the retardation corresponding to that interference color from the bottom of the chart. If, for example, the interference color is second-order yellowish green, the retardation is about 750 nm.

• Calculate the birefringence (δ) from Equation 7.7 by using the known thickness of the thin section (e.g., 0.03 mm = 3×10^4 nm), and the retardation determined here.

$$\delta = \frac{\Delta}{d} = \frac{750\,nm}{3 \times 10^4\,nm} = 0.025$$

The calculation can be carried out graphically by using the lines on the interference color chart (Figure 7.18b). Follow the vertical retardation line corresponding to the interference color up to where it intersects the thickness line. The diagonal birefringence line that goes through this intersection indicates the birefringence.

The reader should recognize that there is room for error with this method related to the imprecision with which thin section thickness is known, the inability to accurately recognize interference colors, and inaccurate color rendition of the interference color chart. Further, none of the grains in the sample may be fortuitously cut so that the birefringence is the maximum value. Regardless of these limitations, the method provides sufficient accuracy for routine mineral identification work with the petrographic microscope. If a more accurate numerical value for birefringence is needed, the relevant indices of refraction may be measured using grain mounts and the techniques described in a later section on refractometry of uniaxial and biaxial minerals.

Extinction

Unless an optic axis is vertical (recall that an optic axis is a direction in an anisotropic mineral along which light is not doubly refracted), all anisotropic minerals go dark or **extinct** between crossed polarizers once in every 90° of stage rotation. Extinction occurs when the vibration directions of light passing through the mineral are parallel to the vibration directions of the lower and upper polarizer (Figure 7.19a). No component of the incident plane-polarized light from the lower polarizer can be resolved into the mineral vibration direction that is parallel to the upper polarizer (Equation 7.2). All the light passing through the mineral therefore retains its original vibration direction and is entirely absorbed at the upper polarizer.

If the stage is rotated to place the vibration directions of the mineral in the 45° positions (Figure 7.19b), the plane-polarized light from the lower polarizer can be resolved into both slow and fast rays in the mineral with

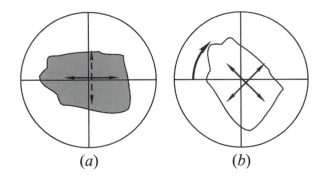

Figure 7.19 Extinction. (*a*) When the vibration directions of a mineral grain are parallel to the vibration directions of the lower and upper polarizer, the mineral is dark or extinct between crossed polarizers. (*b*) If the stage is rotated so that the mineral's vibration directions are not parallel to the polarizers, vector components of both rays may be resolved into the N–S vibration of the upper polarizer so that they may interfere to form interference colors.

equal amplitude. When these rays reach the upper polarizer they interfere with each other and produce an interference color as described above. The interference color does not change with rotation other than to get brighter and dimmer, because the retardation between the two rays remains constant.

The practical significance of extinction is that it allows the observer to determine the vibration directions of the two rays passing through the mineral. When extinct, mineral vibration directions are oriented N–S and E–W. An accessory plate can be used to determine which ray is slow and which is fast.

Function of Accessory Plates

The primary function of accessory plates is to determine which of the rays coming through a mineral is the fast ray and which is the slow ray. This information is used to

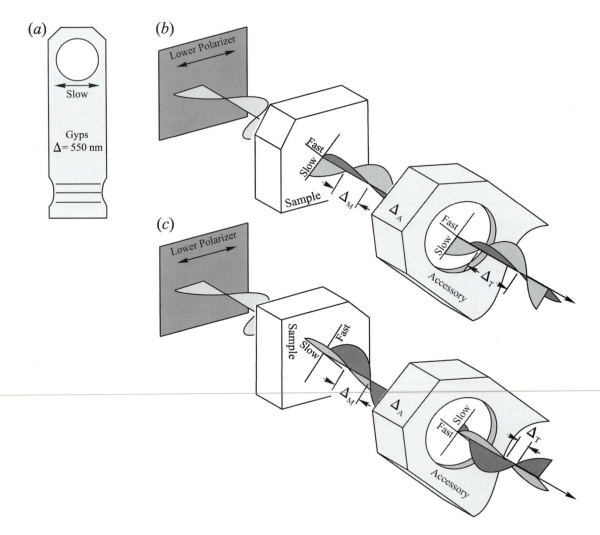

Figure 7.20 Compensation with an accessory plate. (*a*) Typical gypsum plate. The optical element is a piece of gypsum or quartz whose thickness produces a retardation $\Delta_A = 550$ nm and whose slow vibration direction is as indicated. (*b*) The mineral sample, which produces a retardation $= \Delta_M$, is oriented so that its slow and fast ray vibration directions coincide with those of the accessory plate. The total retardation produced by sample and accessory is Δ_T, which equals $\Delta_M + \Delta_A$. (*c*) The sample is rotated so that its slow ray vibration is parallel to the fast ray vibration direction in the accessory plate. The total retardation is Δ_T, which equals $|\Delta_M - \Delta_A|$.

determine the **sign of elongation** (described shortly) and the **optic sign** described in later sections on uniaxial and biaxial minerals. In addition, the accessory plates may help distinguish between different orders of interference colors.

The common accessory plates are the gypsum and mica plates (Figure 7.20a). They consist of pieces of quartz, muscovite, or gypsum mounted in a plastic or metal holder. Because these minerals are anisotropic, light passing through them is split into slow and fast rays. The optical element is carefully ground so that the accessory plate produces a known amount of retardation and is mounted so that the slow ray vibration direction is across the width of the holder and the fast ray vibration direction is parallel to the length. In most microscopes, the accessory plates slide into the optical path in a slot aligned NW–SE so that the accessory slow ray vibrates NE–SW and the fast ray vibrates NW–SE.

The **gypsum plate** also is known as a **full-wave, one-wavelength, quartz-sensitive tint**, or **first-order red plate**, and may be marked Gips, Gyps, Rot I, 1λ, Δ = 550 nm, or Δ = 537 nm. It produces 537 or 550 nm of retardation (depending on manufacturer) and yields a distinct magenta interference color seen at the transition from first to second order (Plate 1).

The **mica plate** also is called the **quarter-wave** and **quarter-wavelength plate** and may be marked Mica, Glimmer, 1/4 λ, or Δ = 147 nm. It produces around 150 nm of retardation (depending on the manufacturer) and yields a first-order white interference color.

Consider the mineral grain shown in Figure 7.20b. The grain is oriented on the stage so that its vibration directions are in the 45° positions, slow NE–SW, fast NW–SE. Light passing through the mineral is split into two rays and, after passing through the mineral, the slow ray is retarded Δ_M relative to the fast ray. If the accessory plate is inserted, its slow ray vibration direction will be parallel to the slow ray vibration direction in the sample, and fast will be parallel to fast. When the slow and fast rays from the mineral enter the accessory plate, the slow ray will be retarded an additional distance equal to the retardation of the accessory plate (Δ_A), so the total retardation (Δ_T) of mineral and accessory is the sum of the two

$$\Delta_T = \Delta_M + \Delta_A \qquad (7.11)$$

and the interference color increases. If the mineral produces 250 nm (first-order white) of retardation and the gypsum plate is used (Δ_A = 550 nm), the total retardation is 800 nm and the interference color observed will increase to second-order yellow. Hence,

Retardations add = slow on slow.

In Figure 7.20c, the mineral grain is rotated so that its fast ray vibration direction is parallel to the slow ray vibration direction of the accessory plate and vice versa. The ray that was slow in the mineral becomes the fast ray

in the accessory plate. The total retardation produced by both mineral and accessory plate is the absolute value of the difference between the retardations of mineral and accessory.

$$\Delta_T = |\Delta_M - \Delta_A| \qquad (7.12)$$

If the mineral produces 250 nm of retardation and the gypsum plate is used (Δ_A = 550 nm), the total retardation will be 300 nm, producing an interference color of first-order white with a yellow cast. Hence,

Retardations subtract = slow on fast.

Fast and Slow Vibration Directions

In a number of different optical measurements it is necessary to determine which of the two vibration directions for light coming through a mineral is the fast ray and which is the slow ray. To distinguish the vibration directions, proceed as follows with polarizers crossed.

• Rotate the stage to place the selected grain in an extinction position (Figure 7.21a). This places the mineral's two vibration directions (p and q) N–S and E–W. In your mind's eye, draw a line on the grain parallel to the N–S crosshair, which also is parallel to the p vibration direction in the grain. This vibration direction could be for either the fast or the slow ray.
• Rotate the stage 45° clockwise (Figure 7.21b). This rotates the p vibration direction into the NE–SW position. Note the interference color displayed by the grain and, from the interference color chart, record the retardation Δ_M associated with that color. In this example, let Δ_M = 400 nm, which is first-order yellow-orange.
• Insert an accessory plate (slow NE–SW) whose retardation is Δ_A. Note the new interference color. If the retardations add, p is the slow vibration direction (Equation 7.10), and if they subtract, p is the fast vibration direction (Equation 7.11). If the accessory is the gypsum plate (Δ_A = 550 nm), addition will yield second-order yellow-orange (Δ_T = 950 nm) and subtraction will yield first-order gray (Δ_T = 150 nm). If retardations added with p NE–SW, they will subtract if the stage is rotated 90° to place q NE–SW.

Determining Order of Interference Color

In the section on formation of interference colors, it was mentioned that accessory plates can be used in some cases to determine the order of an interference color. The gypsum plate can be used to quickly distinguish between second- and higher-order colors and to distinguish between first-order white and a high-order white. Recall that the gypsum plate either adds or subtracts retardation equal to one order of interference colors (550 nm). As the microscope stage is rotated with the gypsum plate inserted, the retardation and interference color alternates between addition and subtraction as the slow ray of the mineral is first parallel to

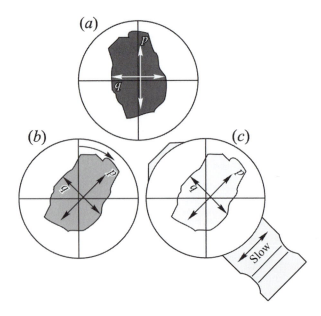

Figure 7.21 Determining slow and fast ray vibration directions between crossed polarizers. (*a*) Mineral grain at extinction between crossed polarizers. Vibration directions *p* and *q* in the mineral are N–S and E–W. (*b*) Mineral grain rotated 45° clockwise so that vibration direction *p* is NE–SW. (*c*) Accessory plate inserted. If retardation and interference colors add, *p* is the slow ray vibration direction; if retardations subtract, *p* is the fast ray

the slow ray of the accessory plate and then parallel to the fast ray 90° later. When the mineral vibration directions are N–S and E–W, the mineral displays the magenta interference color of the gypsum plate because only one ray comes through the mineral in these orientations. A mineral displaying first-order colors between crossed polarizers will, with insertion of the gypsum plate, alternately display second- and first-order colors as retardations add and subtract as the stage is rotated. A mineral with second-order colors will, with insertion of the gypsum plate, alternately display first- and third-order colors with rotation. A high-order white interference color will not noticeably change because the different orders of high-order colors are all washed-out white.

OPTICAL INDICATRIX

A geometric figure that shows the index of refraction and vibration direction for light passing in any direction through a material is called an **optical indicatrix**. An indicatrix is constructed so that **indices of refraction are plotted as radii that are parallel to the vibration direction of the light**. Consider ray *p* with index of refraction $n_p = 2.0$ traveling along the *Y* axis in Figure 7.22a and vibrating parallel to the *Z* axis. The numerical value

of this index of refraction is plotted along the positive and negative ends of the *Z* axis as shown. Ray *q* traveling along the *X* axis and vibrating parallel to *Y* has an index of refraction $n_q = 1.5$. This value is plotted along the positive and negative ends of the *Y* axis. If the indices of refraction for all possible light paths through the material are plotted in a similar manner, the surface of the optical indicatrix is defined. The shape of the indicatrix (Figure 7.22b) depends on mineral symmetry, as will be described shortly.

The primary use of the indicatrix is to determine the indices of refraction and vibration directions of the slow and fast rays given the wave normal direction followed by the light through the mineral. The basic steps are as follows.

• Construct a section through the indicatrix at right angles to the wave normal (WN) (Figure 7.22c). This section is parallel to the wave front. In the general case, the section through the indicatrix is an ellipse.
• The axes of the elliptical section are parallel to the vibration directions of the slow and fast rays, and the lengths of radii parallel to those axes are equal to the indices of refraction.
• To find the ray paths, which are the paths followed by an image through the mineral such as those seen with the calcite experiment (Figure 7.13), tangents to the indicatrix are constructed parallel to the vibration directions of the slow and fast rays (Figure 7.22d). In the general case in which the indicatrix is a triaxial ellipsoid, both rays diverge from their associated wave normals.

Isotropic Indicatrix

Optically isotropic minerals all crystallize in the isometric crystal system. One unit cell dimension (*a*) is required to describe the unit cell and one index of refraction (*n*) is required to describe the optical properties because light velocity is uniform in all directions for a particular wavelength of light. The indicatrix is therefore a sphere. All sections through the indicatrix are circles, and light is not split into two rays. Birefringence may be considered zero.

Uniaxial Indicatrix

Minerals that crystallize in the tetragonal and hexagonal crystal systems have two different unit cell dimensions (*a* and *c*) and a high degree of symmetry about the *c* axis. Two indices of refraction are required to define the dimensions of the indicatrix, which is an ellipsoid of revolution whose axis is the *c* crystal axis (Figure 7.23). The radius of the indicatrix measured parallel to the *c* axis is called n_ϵ, and the radius at right angles is called n_ω. The maximum birefringence of uniaxial minerals is always $|n_\epsilon - n_\omega|$.

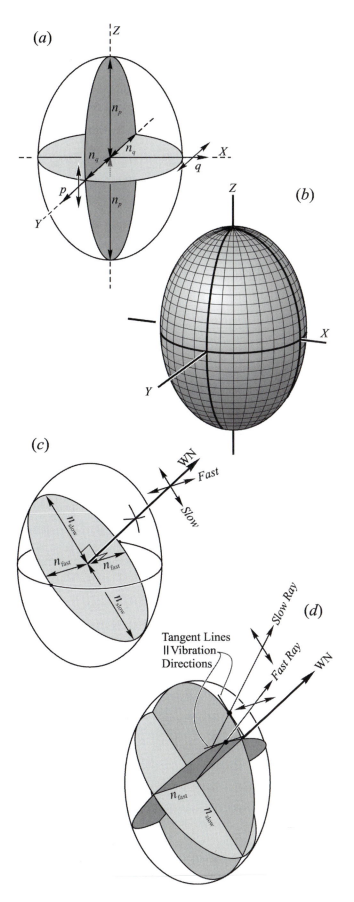

All vertical sections through the indicatrix that include the c axis are identical ellipses called **principal sections** whose axes are n_ω and n_ϵ. Random sections are ellipses whose dimensions are n_ω and n_ϵ' where n_ϵ' is between n_ω and n_ϵ. The section at right angles to the c axis is the **circular section** whose radius is n_ω. Because this section is a circle, light propagating along the c axis is not doubly refracted; it is following an **optic axis**. Because hexagonal and tetragonal minerals have a single optic axis, they are called **optically uniaxial**.

Ordinary and Extraordinary Rays

As described earlier (Figure 7.13), if a cleavage rhomb of calcite is placed on a dot or other image on a piece of paper, two images appear, each composed of plane-polarized light vibrating at right angles to the other. The light passing up through the calcite can be considered to be incident at right angles. Based on Snell's law (Equation 7.4), the wave normal for this light is not bent; it remains perpendicular to the bottom surface of the rhomb.

One image (Figure 7.24a) is composed of light rays that refract as though they were in an isotropic material, and the ray path and wave normal coincide. These rays are called **ordinary rays** or **ω rays**. All ordinary rays have the same velocity and index of refraction, which is n_ω, regardless of propagation direction. In calcite, $n_\omega = 1.658$. The ordinary ray vibration vector is always parallel to the (001) plane in uniaxial minerals, which is the only plane in which electron density is uniform. Regardless of propagation direction, one of the two rays produced as a consequence of double refraction in uniaxial minerals always is an ordinary ray. If the calcite rhomb is rotated about a vertical axis, the position of the ordinary ray image remains fixed.

Figure 7.22 Optical indicatrix. (a) The indicatrix is constructed by plotting indices of refraction as radii parallel to the vibration direction of the light. Ray p, propagating along Y, vibrates parallel to the Z axis, so its index of refraction (n_p) is plotted as radii along Z. Ray q, propagating along X, vibrates parallel to Y so its index of refraction (n_q) is plotted as radii along Y. (b) The indicatrix is an ellipsoid whose shape depends on the mineral symmetry. (c) An indicatrix showing a wave normal direction (WN) along which the light propagates. An elliptical section through the indicatrix perpendicular to the wave normal is parallel to the wave front. The long axis of this elliptical section is parallel to the slow ray vibration direction, and the radius parallel to this direction is equal to the slow ray index of refraction (n_{slow}). The short axis of the elliptical section is parallel to the fast ray vibration direction, and the radius parallel to this direction is equal to the fast ray index of refraction (n_{fast}) (d) Ray paths are determined by constructing tangents to the indicatrix as shown.

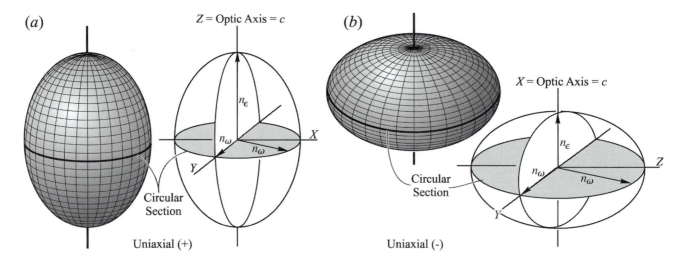

(a) Z = Optic Axis = c   *(b)*   X = Optic Axis = c

Circular Section

Uniaxial (+)   Uniaxial (-)

Figure 7.23 Uniaxial indicatrix. The circular section and a principal section are shown for each indicatrix. X, Y, and Z axes are included to be consistent with the convention used in biaxial minerals. (*a*) Uniaxial positive. The indicatrix is a prolate spheroid whose axes are n_ϵ and n_ω where $n_\epsilon > n_\omega$. (*b*) Uniaxial negative. The indicatrix is an oblate spheroid whose axes are n_ϵ and n_ω where $n_\epsilon < n_\omega$.

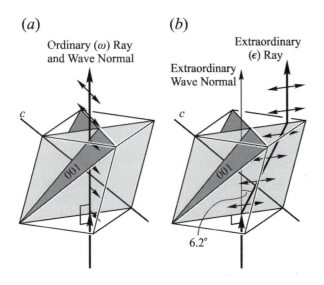

(a)

Ordinary (ω) Ray and Wave Normal

(b)

Extraordinary (ε) Ray

Extraordinary Wave Normal

6.2°

Figure 7.24 Ordinary and extraordinary rays. Light incident at right angles to the bottom surface of a calcite rhomb is split into two rays (Figure 7.13). (*a*) The vibration vector of ordinary ω rays is always parallel to (001) planes, and the wave normal and ray direction coincide. If the calcite rhomb is rotated about a vertical axis, the image composed of ordinary rays (ω) remains stationary. (*b*) The extraordinary (ε′) ray direction diverges from its wave normal, and its vibration direction lies in a plane (light shading) containing the c axis and the two rays. If the calcite rhomb is rotated, the position of the extraordinary ray image rotates about the ordinary ray image.

The image deflected to the side is composed of light rays that behave very differently from light in isotropic materials. They are called **extraordinary rays** or ε rays because the ray path and wave normal do not coincide.

The geometry shown in Figure 7.13 and 7.24b has normal incidence, so Snell's law (Equation 7.4) requires that the wave normal for the extraordinary component is not bent when incident light enters the bottom surface of the calcite rhomb; it is parallel to the ordinary ray and wave normal. However, the extraordinary ray path diverges from the wave normal. The index of refraction of extraordinary rays varies with direction between n_ω and n_ϵ, where n_ϵ may be either higher or lower than n_ω. In calcite $n_\epsilon = 1.486$. For any propagation direction except perpendicular to the c axis, the index of refraction of the extraordinary ray is designated $n_{\epsilon'}$ and is between n_ω and n_ϵ. For light whose wave normal is perpendicular to a cleavage surface of calcite (Figure 7.24b), the path followed by the extraordinary ray diverges at an angle of 6.2° from the wave normal direction and has an index of refraction $n_{\epsilon'} = 1.566$. Extraordinary rays are labeled ε′ if their index of refraction is $n_{\epsilon'}$ and ε if their index of refraction is n_ϵ.

Optic Sign

The dimension of the indicatrix along the c axis may be either greater or less than the dimension at right angles. This provides the basis for defining **optic sign** in uniaxial minerals.

In optically positive minerals, n_ϵ is greater than n_ω; extraordinary rays are slow rays

In optically negative minerals, n_ϵ is less than n_ω; extraordinary rays are fast rays.

Hence, the indicatrix for optically positive uniaxial minerals is a prolate spheroid (shaped like a rugby ball) (Figure 7.23a) and the indicatrix for optically negative uniaxial minerals is an oblate spheroid (shaped like a discus in its extreme form) (Figure 7.23b).

Use of the Indicatrix

We can now examine the behavior of light passing through grains of a uniaxial mineral in different orientations either in a thin section or a grain mount as shown in Figure 7.25. Orthoscopic illumination is used (auxiliary condenser lens removed) so the light strikes the bottom surface of the sample more or less normal to the surface. This means that the wave normal of the light entering the mineral is not bent (Equation 7.4), and the wave front is parallel to the bottom surface of the mineral.

In Figure 7.25a the mineral sample is oriented so that its optic axis is horizontal. The mineral's indicatrix, which for convenience of illustration is assumed to be uniaxial positive, is constructed so that it is bisected by the bottom surface of the mineral grain. A wave normal is constructed through the center of the indicatrix. Because the light is incident normal to the bottom surface of the mineral, the section through the indicatrix along the bottom surface of the mineral is the section that defines the indices of refraction and vibration directions for the ordinary and extraordinary rays. Because the optic axis is horizontal, this section is a principal section, which is an ellipse whose axes are n_ω and n_ϵ as shown. The ordinary ray therefore has index of refraction n_ω. The extraordinary ray has index of refraction n_ϵ, which is its maximum value because the mineral is optically positive. The extraordinary ray vibrates parallel to the trace of the optic axis (c axis) and the ordinary ray vibrates at right angles. Birefringence, and hence interference color, display maximum values. Note that for this specific orientation, the extraordinary ray does not behave in

an extraordinary manner. The wave normal and ray directions coincide.

In Figure 7.25b, the mineral sample is oriented so that the optic axis is vertical. The section through the indicatrix perpendicular to the wave normal is the circular section whose radius is n_ω. Light coming from below is not doubly refracted. Birefringence is zero, and the light preserves whatever vibration direction it initially had. Between crossed polarizers, this mineral grain should behave like an isotropic mineral and remain dark as the stage is rotated. In practice, however, the light from the substage condenser is moderately converging, so some light passes at a small angle to the optic axis and experiences double refraction. If the mineral has moderate or high birefringence, the mineral may display interference colors, but they will be the lowest order color found in the sample for that mineral.

In Figure 7.25c, the mineral sample is in a random orientation so that the light path is at angle θ to the optic axis. The section through the indicatrix parallel to the bottom surface of the mineral is an ellipse whose axes are n_ω and n'_ϵ. The extraordinary ray vibrates parallel to the trace of the optic axis as seen from above; the ordinary ray vibrates at right angles. Birefringence and interference color is intermediate because n'_ϵ is intermediate between n_ω and n_ϵ. The numerical value of n'_ϵ is given by

$$n'_\epsilon = \frac{n_\omega}{\sqrt{1+\left(\frac{n_\omega^2}{n_\epsilon^2}-1\right)\sin^2\theta}} \qquad (7.13)$$

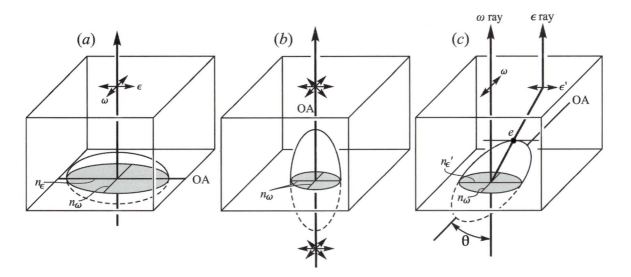

Figure 7.25 Behavior of uniaxial minerals in orthoscopic illumination. The optic axis is labeled OA. (*a*) Mineral grain oriented with the optic axis horizontal. The section through the indicatrix is a principal section with axes n_ω and n_ϵ. The mineral displays maximum birefringence and interference color. (*b*) Mineral grain oriented with the optic axis vertical. The section through the indicatrix is the circular section; all light passes as ordinary ray and preserves its original vibration direction. This grain displays zero birefringence and behaves like an isotropic mineral. (*c*) Mineral grain with a random orientation. The section through the indicatrix is an ellipse whose axes are n_ω and n'_ϵ. Birefringence is intermediate and the ω and ϵ' rays diverge.

The path followed by the extraordinary ray can be found by constructing a tangent to the indicatrix parallel to the extraordinary ray vibration direction (Figure 7.26). The extraordinary ray, which carries the image through the calcite rhomb shown in Figures 7.13 and 7.24, goes through point E at angle ψ to the optic axis given by the equation

$$\tan\psi = \frac{n_\omega^2}{n_\epsilon^2}\cot(90-\theta) \qquad (7.14)$$

where θ is the angle between the optic axis and the wave normal. This construction helps explain the unusual behavior of the image carried by the extraordinary rays coming through a mineral such as calcite. Its application to the petrographic microscope is very limited because reflection, refraction, and interference of the light passing through minerals are best described with reference to the wave normal.

Biaxial Indicatrix

Minerals that crystallize in the orthorhombic, monoclinic, and triclinic crystal systems require three dimensions (a, b, and c) to describe their unit cells and three indices of refraction to define the shape of their indicatrix (Figure 7.27). The three **principal indices of refraction** are n_α, n_β, and n_γ, where $n_\alpha < n_\beta < n_\gamma$. Other conventions that may be encountered to identify the principal indices of refraction include

α, β, γ; n_x, n_y, n_z; n_a, n_b, n_c; n_1, n_2, n_3; n_p, n_m, n_g; N_x, N_y, N_z; nX, nY, nZ; and X, Y, Z. **The maximum birefringence (δ) of a biaxial mineral is always $n_\gamma - n_\alpha$.**

The biaxial indicatrix is constructed and used much like the uniaxial indicatrix except that three indices of refraction are plotted, rather than two. The convention is to plot the three principal indices of refraction in an X Y Z coordinate axis system (Figure 7.27). The value of n_α is plotted along the X axis, n_β is plotted along the Y axis, and n_γ is plotted along the Z axis. In every case n_α is the smallest index of refraction and n_γ is the largest. The indicatrix is therefore a triaxial ellipsoid elongate along the Z axis and shortened along the X axis.

While three indices of refraction are required to describe biaxial optics, light that enters biaxial minerals is still split into two rays. As we shall see later, both of these rays behave as extraordinary rays for most propagation paths through the mineral. The wave normal and the ray diverge like the extraordinary ray in uniaxial minerals, and their indices of refraction vary with direction. We, therefore, revert to the terminology used earlier and call

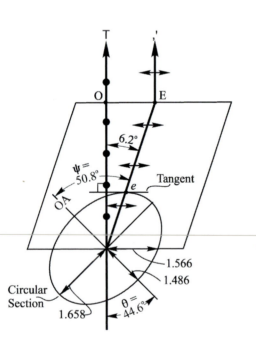

Figure 7.26 Calcite indicatrix construction. The ϵ' ray path is found by constructing a tangent to the indicatrix parallel to the ϵ' ray vibration direction at point e. The extraordinary image, which follows the ray, emerges from the top of the rhomb at E. A similar tangent for the ordinary ray (not shown in and out of the page) yields an ordinary ray path that coincides with the wave normal so the ordinary ray emerges at O. The dimensions of the indicatrix are not drawn to scale.

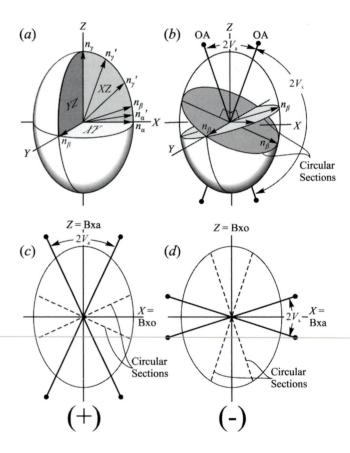

Figure 7.27 Biaxial indicatrix. (a) The indices n_α, n_β, and n_γ are plotted along the X, Y, and Z axes, respectively. Principal sections are the XY, XZ, and YZ planes. Within the XZ plane, radii of the ellipse vary from n_α to n_γ with certain radii equal to n_β. (b) Circular sections have radius n_β. The optic axes (OA) are perpendicular to the circular sections and lie in the XZ, or optic, plane. (c) Optic plane of a biaxial positive indicatrix. (d) Optic plane of a biaxial negative indicatrix.

them the fast ray and the slow ray. The index of refraction of the fast ray is identified as n'_α where $n_\alpha < n'_\alpha < n_\beta$, and the index of refraction of the slow ray is n'_γ where $n_\beta < n'_\gamma < n_\gamma$. Rays whose indices of refraction are n_α, n_β, and n_γ are labeled α β, and γ. Rays whose index of refraction is n'_α are labeled *fast* or α', and rays whose index of refraction is n'_γ are labeled *slow* or γ'.

The biaxial indicatrix contains three principal sections: the *XY, XZ,* and *YZ* planes. The *XY* section is an ellipse with axes n_α and n_β, the *XZ* section is an ellipse with axes n_α and n_γ, and the *YZ* section is an ellipse with axes n_β and n_γ. Random sections through the indicatrix are ellipses whose axes are n'_α and n'_γ.

The indicatrix has two **circular sections** with radius n_β that intersect in the *Y* axis. Consider the *XZ* plane (Figure 7.27a). It is an ellipse whose radii vary between n_α and n_γ, so radii equal to n_β must be present. Radii shorter than n_β are n'_α and those longer are n'_γ. The radius of the indicatrix along the *Y* axis also is n_β. The *Y* axis and the n_β radii in the *XZ* plane define the two circular sections (Figure 7.27b). In uniaxial minerals we saw that the circular section was perpendicular to the optic axis. In biaxial minerals, the same applies. The two **optic axes** are perpendicular to the two circular sections. The term "biaxial" is derived from the fact that two optic axes are present.

Because both optic axes lie in the *XZ* plane of the indicatrix, that plane is called the **optic plane**. The angle between the optic axes bisected by the *X* axis is called the **$2V_x$** angle, and the angle between the optic axes bisected by the *Z* axis is called the **$2V_z$** angle where $2V_x + 2V_z = 180°$. The *Y* axis, which is perpendicular to the optic plane, is called the **optic normal** or **ON**.

Optic Sign

The acute angle between the optic axes is called the **optic angle**, or **2V angle**. The axis (either *X* or *Z*) that bisects the optic angle is the **acute bisectrix** or **Bxa** (Figure 7.27c, d). The axis (either *Z* or *X*) that bisects the obtuse angle between the optic axes is the **obtuse bisectrix** or **Bxo**. The optic sign of biaxial minerals is defined to be consistent with the convention used with uniaxial minerals and depends on whether the *X* or *Z* indicatrix axis bisects the acute angle between the optic axes.

If the acute bisectrix is the *X* axis, the mineral is ***optically negative***; $2V_x$ is less than 90°.

If the acute bisectrix is the *Z* axis, the mineral is ***optically positive;*** $2V_z$ is less than 90°.

If 2V is exactly 90° so neither *X* nor *Z* is the acute bisectrix, the mineral is ***optically neutral.***

The numerical values of n_α, n_β, n_γ and $2V_z$ are mathematically related by the following equation:

$$\cos^2 V_z = \frac{n_\alpha^2(n_\gamma^2 - n_\beta^2)}{n_\beta^2(n_\gamma^2 - n_\alpha^2)} \quad (7.15)$$

where V_z is half the angle $2V_z$.

Uniaxial indicatrixes may be considered special cases of the biaxial indicatrix. Note that if the value of n_β approaches n_α, the two circular sections converge into a single circular section in the *XY* plane and the two optic axes converge on the *Z* indicatrix axis, forming a uniaxial positive indicatrix (Figure 7.23a) where $n_\alpha = n_\omega$ and $n_\gamma = n_\epsilon$. If the value of n_β is increased so that it approaches n_γ, the two circular sections diverge away from the *XY* plane and close on each other in the *YZ* plane, and the two optic axes converge on the *X* axis, forming a uniaxial negative indicatrix (Figure 7.23b), where $n_\alpha = n_\epsilon$ and $n_\gamma = n_\omega$.

Crystallographic Orientation of Indicatrix Axes

The term **optic orientation** refers to the relationship between indicatrix axes and crystal axes. Because the optical properties of minerals are directly controlled by the symmetry of the crystal structure, optic orientation must be consistent with mineral symmetry.

Orthorhombic minerals have three mutually perpendicular crystallographic axes of unequal length. These crystal axes must coincide with the three indicatrix axes and the symmetry planes in the mineral must coincide with principal sections in the indicatrix (Figure 7.28a). Any crystal axis may coincide with any indicatrix axis, however. The optic orientation is defined by indicating which indicatrix axis is parallel to which mineral axis. For example, in aragonite, *X = c, Y = a,* and *Z = b,* and in anthophyllite *X = a, Y = b,* and *Z = c.*

In monoclinic minerals, the *b* crystallographic axis coincides with the single 2-fold rotation axis and/or is perpendicular to the single mirror plane. The *a* and *c* axes are perpendicular to *b* and intersect in an obtuse angle (β). One indicatrix axis, which could be either *X, Y,* or *Z,* is always parallel to the *b* crystallographic axis, and the other two lie in the {010} plane and are not parallel to either *a* or *c* except by chance (Figure 7.28b). The optic orientation is defined by specifying which indicatrix axis coincides with the *b* axis, as well as the angles between the other indicatrix axes and the *a* and *c* crystal axes. By convention, the angle between an indicatrix axis and either the *a* or *c* axis is a positive angle if the indicatrix axis lies in the obtuse angle between *a* and *c*, and a negative angle if the indicatrix axis lies in the acute angle. By this convention, the β angle between *a* and *c* is

$$\beta = 90° + a \wedge (X, Y, \text{or } Z) \\ + c \wedge (X, Y, \text{or } Z). \quad (7.16)$$

The optic orientation of the sample in Figure 7.28b is $X \wedge a = -5°$, $Y = b$, $Z \wedge c = +15°$. The β angle must therefore be $90° - 5° + 15° = 100°$. The reader is cautioned that other sources may define the sign of the optic orientation angles differently.

Triclinic minerals have three crystallographic axes of different lengths, none of which is at right angles.

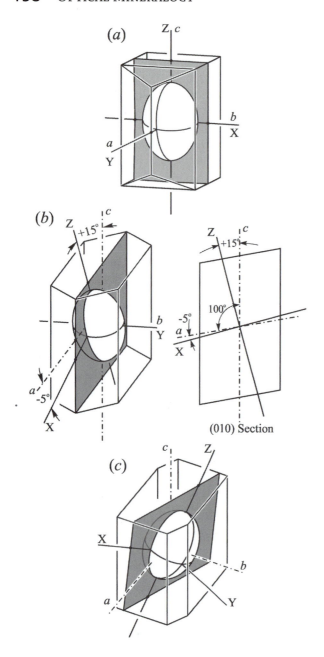

Figure 7.28 Relationship between crystal axes and indicatrix axes. The optic planes are shaded. (*a*) Orthorhombic. Crystal axes and indicatrix axes coincide. In this case the optic orientation is X = b, Y = a, Z = c. (*b*) Monoclinic. The *b* crystal axis coincides with one of the indicatrix axes. The other axes do not coincide except by chance. In this case the $X \wedge a = -5°$, Y = b, Z $\wedge c = +15°$. (*b*) Triclinic. None of the crystal axes coincide with indicatrix axes except by chance.

Because the only possible symmetry is center, the indicatrix axes are not constrained to be parallel to any crystal axis (Figure 7.28c). Although precise mathematical ways are available to specify the angles between indicatrix and crystal axes, in most cases the optic orientation is specified by indicating the approximate angle between indicatrix and crystal axes. Unfortunately, triclinic minerals may be described by using a variety of different crystal axis settings. The optic orientation described in different sources may therefore vary depending on which crystal axis setting is selected as a reference.

Use of the Indicatrix

The biaxial indicatrix is used in the same way as the uniaxial indicatrix. It provides information about the indices of refraction and vibration direction given the wave normal direction that light is following through a mineral. Figure 7.29 shows a biaxial mineral cut in three different orientations. Birefringence depends on how the sample is cut. Birefringence is a maximum if the optic normal is vertical, minimum if an optic axis is vertical, and intermediate for random orientations.

MINERAL COLOR AND PLEOCHROISM

Pleochroic minerals change color as the stage is rotated when the sample is observed in plane light (Figure 7.30). The color changes because the slow and fast rays are absorbed differently as they pass through the mineral and therefore have different colors. When the fast ray vibration direction is parallel to the lower polarizer, all light passes as fast ray, so the mineral displays that color. When the slow ray vibration direction is parallel to the lower polarizer, the mineral displays the color of the slow ray. If the stage is rotated to allow both slow and fast rays to come through, the perceived color is typically intermediate.

Isotropic Minerals

Isotropic minerals are not pleochroic because they do not experience double refraction. In plane light, isotropic minerals display a uniform color as the stage is rotated.

Uniaxial Minerals

Colored uniaxial minerals are usually pleochroic. To describe the pleochroism, it is sufficient to identify the color of both the ω and ϵ rays. For example, the pleochroic formula of common **tourmaline (schorl)** is described as ω = dark green and ϵ = pale green. An alternative convention is to identify which ray is most strongly absorbed and therefore darker colored: $\omega > \epsilon$. Pleochroism may be described as strong or weak, depending on the extent of the color change and intensity of the colors. A grain oriented to display minimum retardation between crossed polarizers will, in plane light, yield the color of ω. A second grain oriented to display maximum retardation between crossed polarizers will, when rotated in plane light, alternately display the colors of ϵ and ω. Because the color of ω is already known from the first grain, the color of ϵ can be determined by default.

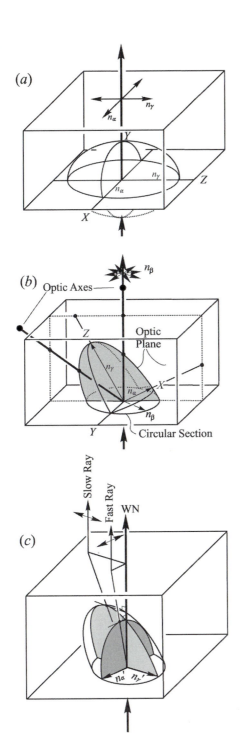

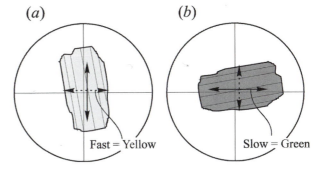

Figure 7.30 Pleochroism is seen in plane light. (*a*) The fast ray is yellow. When its vibration direction is parallel to the lower polarizer (E–W), the mineral is yellow. (*b*) The slow ray is green. When its vibration direction is parallel to the lower polarizer, the mineral appears green. In intermediate positions, both rays come through the mineral and color is intermediate.

Biaxial Minerals

To describe the pleochroism of biaxial minerals completely, it is necessary to specify three colors. One color is for light vibrating parallel to the X indicatrix axis, the second for light vibrating parallel to Y, and the third for light vibrating parallel to Z. For example, the pleochroism of hornblende can be described as X = yellow, Y = pale green, and Z = dark green. An alternate, and less informative, convention is to specify which is most strongly absorbed and therefore darkest, such as $Z > Y > X$.

A grain oriented to show minimum retardation between crossed polarizers, will, in plane light, display the color associated with Y. A grain oriented to display maximum retardation between crossed polarizers will show the colors of X and Z as the stage is rotated in plane light. Use the accessory plate to identify the slow and fast ray vibration directions. With the slow ray vibration direction parallel to the lower polar, the mineral will, in plane light, display the color associated with Z. Rotation by 90° yields the color associated with X.

EXTINCTION ANGLE AND SIGN OF ELONGATION

Extinction Angle

The angle between the length or a prominent cleavage in a mineral and a vibration direction is a diagnostic property called the **extinction angle**. It is measured in either grain mount or thin section with crossed polarizers as follows.

- Rotate the stage of the microscope until the length or cleavage of the mineral is aligned with the north–south crosshair (Figure 7.31a). Record the reading from the stage goniometer (g_1).
- Rotate the stage of the microscope (either clockwise or counterclockwise) until the mineral

Figure 7.29 Behavior of biaxial minerals in orthoscopic illumination. (*a*) Optic normal (*Y*) vertical. Axes of the indicatrix section are X and Z, indices of refraction are n_α and n_γ, and birefringence is maximum. The slow ray vibrates parallel to Z and the fast ray vibrates parallel to X. (*b*) Optic axis vertical. The indicatrix section is a circle with radius n_β, birefringence is zero, and the mineral would be dark between crossed polarizers. (*c*) Random orientation. The indicatrix section yields indices of refraction n'_α and n'_γ and vibration directions as shown. Because n'_α is greater than n_α and n'_γ is less than and n'_γ birefringence is intermediate. Ray paths for both the slow and fast rays diverge from the wave normal, so both slow and fast rays are extraordinary.

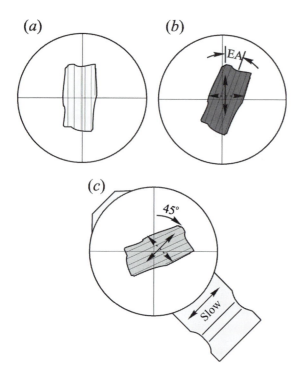

Figure 7.31 Measurement of extinction angle and sign of elongation. Polarizers are crossed. (*a*) Grain oriented so that cleavage or length is parallel to the N–S crosshair. (*b*) Stage rotated so the grain is extinct and one vibration direction is parallel to the N–S crosshair. The extinction angle (EA) is the angle through which the stage was rotated to go from (*a*) to (*b*). (*c*) Stage rotated 45° clockwise to place the vibration direction nearest the mineral length parallel to NE–SW. Insert the accessory plate. If retardations add, the mineral is length slow; if retardations cancel, the mineral is length fast.

goes extinct (Figure 7.31b). This places either the slow or fast ray vibration direction parallel to the north–south crosshair. If the mineral is already extinct, no rotation is required. Record the new reading from the stage goniometer (g_2). The extinction angle is the difference between g_1 and g_2. If the extinction angle measured with clockwise stage rotation is EA, the extinction angle measured with counterclockwise rotation is 90° − EA. Generally it is the smaller of these two angles that is reported, although in some cases it is necessary to specifically measure the extinction angle to the slow or the fast ray vibration direction. The technique to determine which ray is which has been discussed earlier.

The extinction angle measured on a specific mineral in thin section or grain mount depends on exactly how the grain happens to be oriented in the sample. Hence, to be diagnostic, it is necessary to specify the orientation of the mineral grain on which the measurement is made. In many cases the diagnostic extinction angle is measured on grains that display maximum retardation (optic axes horizontal).

In others, extinction angles are specified for certain orientations, such as resting on a cleavage surface or a cut through the mineral parallel to a specific crystallographic plane.

Sign of Elongation

The terms **length fast** and **length slow** are repeatedly encountered in the mineral descriptions in the latter part of this text. Length fast means that the fast ray vibrates more or less parallel to the length of an elongate mineral; length slow means that the slow ray vibrates more or less parallel to the length. Length fast is also called **negative elongation**; length slow is called **positive elongation**. Sign of elongation is not the same as optic sign. Optic sign is discussed in the sections on uniaxial and biaxial indicatrixes. Determine sign of elongation as follows.

• Start with the mineral extinct and with the mineral elongation or trace of cleavage so that it is less than 45° from the N–S crosshair. This generally is the position obtained after rotating to measure the extinction angle (Figure 7.31b).

• Rotate the stage 45° clockwise (Figure 7.31c). This places the vibration direction closest to the length or prominant cleavage NE–SW. Insert an accessory plate. If retardations add, the ray whose vibration direction is closest to the length or cleavage is the slow ray and the mineral is length slow. If retardations subtract, the ray whose vibration direction is closest to the length or cleavage is the fast ray, and the mineral is length fast.

Categories of Extinction

The four categories of extinction are parallel extinction, inclined extinction, symmetrical extinction, and no extinction angle.

With **parallel extinction**, the mineral is extinct when the cleavage or length is aligned parallel to one of the crosshairs (Figure 7.32a). The extinction angle is 0°. Either the slow ray or fast ray vibration direction is parallel to the trace of cleavage or length of the mineral.

With **inclined extinction**, the mineral is extinct when the cleavage or length is at some angle to the crosshairs (Figure 7.32b). The extinction angle is greater than 0° and is measured as previously described. Neither vibration direction is aligned parallel to the trace of cleavage or the length of the mineral. If the slow ray vibration direction is closest to the length or trace of cleavage, the mineral is length slow, if the fast ray is closest, the mineral is length fast.

Symmetrical extinction may be observed in minerals that display either two cleavages or two distinct crystal faces (Figure 7.32c). If the extinction angles EA_1 and EA_2, measured from the two cleavages or crystal faces to the same vibration direction, are the same, extinction is symmetrical. To measure the extinction angles, it is most convenient to rotate the mineral to an extinction position and

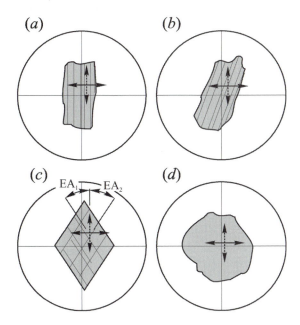

Figure 7.32 Categories of extinction. All grains are in extinction orientations. (*a*) Parallel extinction. The grain is extinct when the trace of cleavage or length is parallel to a crosshair. (*b*) Inclined extinction. The grain is extinct when the cleavage or length is at an angle to a crosshair. (*c*) Symmetrical extinction. Extinction angles EA_1 and EA_2 measured to the two cleavages are the same. (*d*) No extinction angle. The grain lacks cleavage or elongation from which to measure an extinction angle.

record the stage goniometer reading. Then rotate clockwise to measure the extinction angle to one cleavage/crystal face. Return to the extinction position and then rotate counterclockwise to measure the second extinction angle.

Many minerals lack distinct cleavages or do not display an elongation or crystal faces. Although they go extinct once in every 90° of stage rotation, there is no cleavage, elongation, or crystal face from which to measure an extinction angle. Hence, these minerals have **no extinction angle** (Figure 7.32d).

In addition to these four categories of extinction angles, some mineral grains may not go uniformly extinct at the same point of stage rotation because either of **deformation** or **chemical zoning**. In many deformed rocks, mineral grains may be bent or otherwise strained so that different parts of a grain go extinct at different points of stage rotation. If the extinction follows an irregular or wavy pattern it is called **undulatory extinction**. Many minerals, including plagioclase, grow so that they are chemically zoned. Because extinction angle may be controlled by chemical composition in monoclinic and triclinic minerals, extinction angle may vary systematically with composition, so that the center of a grain may display one extinction angle and the rim another. A special term is not applied to this type of extinction, but minerals that display it are said to be zoned.

Extinction in Uniaxial Minerals

Tetragonal and many hexagonal minerals are prismatic and either elongate or stubby parallel to the *c* axis. The common forms are prisms parallel to *c*, pinacoids perpendicular to *c*, pyramids, and related forms. Hexagonal minerals that crystallize in the trigonal group are commonly rhombohedral. Cleavages may be parallel to any of these forms (Chapter 2). A simple tetragonal mineral shown in Figures 7.33 is cut in a variety of different orientations such as might be found in a thin section. In grain mount, individual grains are most likely to rest on one of the cleavage surfaces, although all orientations are possible. A sample with the highest birefringence will have its *c* axis parallel to the microscope stage; it will display parallel extinction to prismatic cleavage and inclined or symmetrical extinction to a rhombohedral or pyramidal cleavage. Extinction is parallel to {001} cleavage in all grain orientations.

Extinction in Biaxial Minerals

Orthorhombic minerals display parallel or symmetrical extinction in sections cut parallel to (100), (010), and (001), and inclined extinction in random orientations. Grains cut to yield maximum retardation always display parallel or symmetrical extinction. Monoclinic minerals display parallel or symmetrical extinction if {010} happens to be vertical, and inclined extinction in most other orientations (Figure 7.34). A common indicatrix orientation is for the *Y* indicatrix axis to be parallel to *b*. In minerals with this orientation, such as the monoclinic pyroxenes and amphiboles, grains oriented to display maximum retardation also have extinction angles that indicate the relation between the *X* and *Z* indicatrix axes and *a* and *c* crystal axes as shown. Triclinic minerals display inclined extinction in most sections because indicatrix and crystal axes need not be parallel.

INTERFERENCE FIGURES

An **interference figure** is obtained to rapidly determine **optical character**: that is, whether a mineral is uniaxial or biaxial, and to determine the **optic sign**. If the mineral is biaxial, the 2*V* angle also may be measured. An interference figure also may be used to confirm certain mineral orientations. To view an interference figure:

- Focus on a single mineral grain with the high-power objective lens.
- Flip in the auxiliary condensing lens. Refocus if needed, and open the aperture diaphragm. Insert the upper polarizer.
- Insert the Bertrand lens. The interference figure also may be observed without the Bertrand lens by removing the ocular and looking directly down the microscope tube.

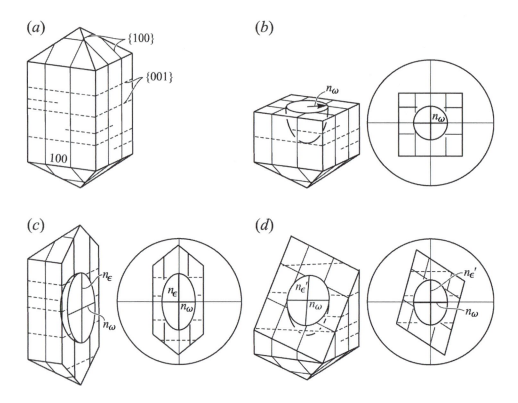

Figure 7.33 Extinction in a typical tetragonal mineral. (a) Mineral showing traces of pinacoid {001} (dashed) and prismatic {100} (solid) cleavages. (b) Mineral resting on the {001} cleavage with the optic axis vertical. The indicatrix section is circular, so all light passes as ordinary rays. The mineral behaves like an isotropic mineral and remains extinct with stage rotation. (c) Mineral oriented with optic axis horizontal. The indicatrix section is a principal section, so indices of refraction are n_ω and n_ϵ, and retardation is maximum. Extinction is parallel to the traces of both {001} and {100} cleavage. (d) A random cut yields a parallelogram-shaped section. Birefringence is intermediate, and extinction is parallel to the trace of {001} and inclined to the traces of the {100} cleavages.

The interference figure is formed near the top surface of the objective lens and consists of a pattern of interference colors called **isochromes** on which dark bands called **isogyres** are superimposed. The nature of the interference figure and its behavior as the stage is rotated depends on the orientation of the mineral grain and whether the mineral is uniaxial or biaxial.

Uniaxial Interference Figures

Optic Axis Interference Figure

A **uniaxial optic axis interference figure** (Figure 7.35) is produced if a uniaxial mineral's optic axis is perpendicular to the microscope stage. A grain with a vertical optic axis should display the lowest interference color (retardation) of all grains in the sample. The interference figure consists of **isogyres** forming a black cross superimposed on a circular pattern of **isochromes**. The point in the center where the isogyres cross is called the **melatope**, and it marks the point of emergence of the optic axis. The interference colors increase in order outward from the melatope; those nearest the melatope

are low first order. If the optic axis is exactly vertical, the interference figure does not appear to change as the stage is rotated. **The presence of a single melatope indicates that the mineral is uniaxial**. Biaxial minerals produce interference figures with two melatopes, as will be described below.

The formation of isochromes is shown in Figure 7.36. The auxiliary condensing lens provides strongly convergent light that passes through the mineral and is collected by the objective lens. Light following path 1, parallel to the optic axis, is not split into two rays and exits the mineral with zero retardation to form the melatope. Light following path 2 experiences moderate retardation because the value of n_ϵ is close to n_ω. Light following path 3, at a greater angle to the optic axis, encounters higher birefringence and must traverse a greater distance through the mineral, so retardation is proportionately greater. Because optical properties are symmetric about the optic axis, rings of equal retardation and interference color are formed around the melatope. Minerals that are thick or have high birefringence show more isochromes than thin or low-birefringence minerals.

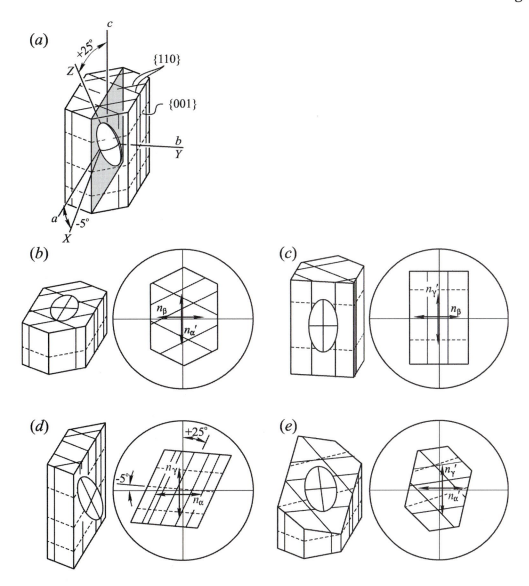

Figure 7.34 Extinction in monoclinic minerals. (*a*) Crystal habit showing trace of prismatic {110} (solid lines) and pinacoidal {001} (dashed lines) cleavages. The indicatrix orientation is $Y = b$, $Z \wedge c$ = +25°, $X \wedge a$ = –5°. (*b*) Fragment resting on the {001} cleavage. Extinction is symmetrical. (*c*) Section parallel to {100} shows parallel extinction to the trace of both cleavages. (*d*) Section parallel to {010} displays maximum retardation. Extinction is 25° (the $Z \wedge c$ angle) measured to the trace of the {110} cleavage and 5° (the $X \wedge a$ angle) measured to the trace of the {001} cleavage. (*e*) Random section: inclined extinction to all cleavage traces.

Isogyres are formed where vibration directions in the interference figure are N–S and E–W. They are areas of extinction. Figure 7.37a schematically shows the vibration directions for strongly convergent light that pierces the uniaxial indicatrix. The vibration directions are determined as described in Figure 7.22. Ordinary (ω) rays vibrate parallel to lines of latitude and extraordinary (ϵ') rays vibrate parallel to lines of longitude on the indicatrix. These vibration directions are carried up into the interference figure (Figure 7.37b). Extraordinary (ϵ') rays vibrate along radial lines symmetric about the melatope and ordinary (ω) rays vibrate tangent to the circular isochromes. The isogyres are formed where the vibration directions in the interference

figure are N–S and E–W. Isogyres become wider and have more diffuse edges further from the melatope.

Off-Center Figure

If the optic axis is inclined from the vertical, the interference figure will no longer be centered in the field of view. As long as the optic axis is within about 30° of being vertical (Figure 7.38a), the melatope will be visible in the field of view and the interference figure is called an **off-center optic axis figure**. The isogyres still form a NS–EW cross centered on the melatope that swings in an arc around the field of view as the stage is rotated. As long as the melatope

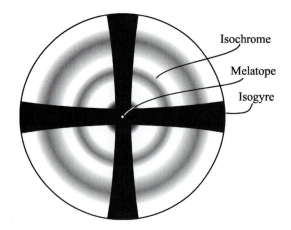

Figure 7.35 Uniaxial optic axis interference figure.

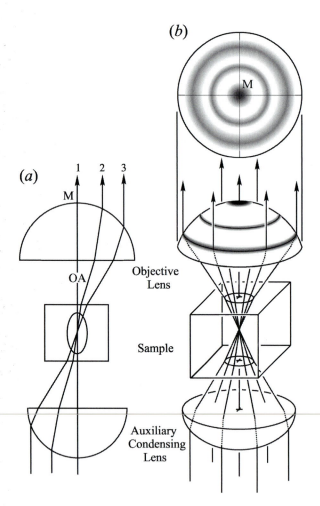

Figure 7.36 Formation of uniaxial isochromes. (*a*) Light following path 1, which emerges at the melatope (M), experiences zero retardation because it follows the optic axis (OA). Paths 2 and 3 produce progressively higher retardation because both birefringence and path length through the sample increase as the inclination of the path to the optic axis increases. (*b*) Optical properties are symmetric about the optic axis, so rings of equal retardation are produced about the melatope.

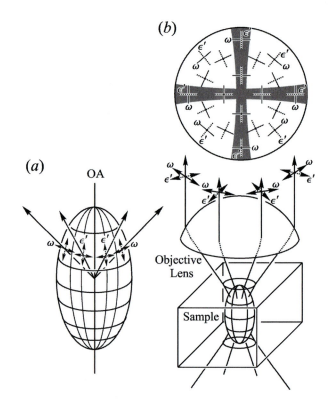

Figure 7.37 Formation of isogyres. (*a*) Vibration directions for light emerging from the center of a uniaxial indicatrix. Ordinary (ω) rays vibrate parallel to lines of latitude and extraordinary (ϵ) rays vibrate parallel to lines of longitude. (*b*) Strongly convergent light that passes through a mineral whose optic axis is vertical exits with a vibration pattern that is symmetric about the melatope. Extraordinary rays (ϵ) vibrate parallel to radial lines (like lines of longitude as seen from above), and ordinary rays (ω) vibrate tangent to the circular isochromes (like lines of latitude as seen from above). Isogyres form between crossed polarizers where vibration directions in the figure are parallel to the vibration directions of the lower and upper polarizers. They are areas of extinction.

is present, the figure may be used to determine optical character and optic sign as described shortly.

If the optic axis is inclined more than about 30° from vertical (Figure 7.38b), the melatope will not be visible in the field of view and the figure is called an **off-center figure**. The four isogyre arms sweep across the field of view sequentially as the stage is rotated (Figure 7.38c). This is the interference figure most commonly encountered if grains are randomly selected. Because uniaxial and biaxial minerals may produce similar off-center interference figures, they generally cannot be used to determine either optical character or optic sign.

Optic Normal (Flash) Figure

If a mineral grain is oriented with the optic axis parallel to the microscope stage, an **optic normal interference figure** is produced (Figure 7.39). Mineral grains oriented to yield optic normal figures display maximum retardation. Optic

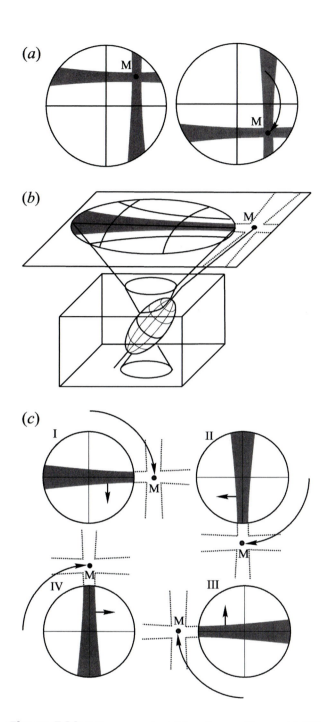

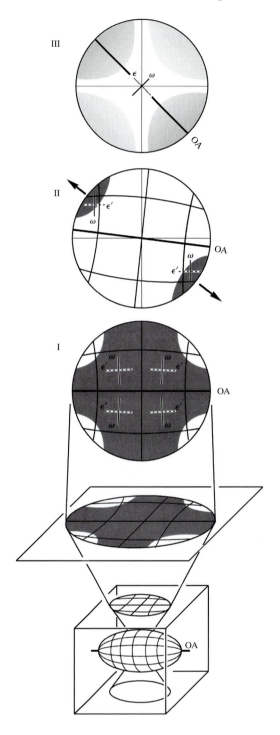

Figure 7.38 Off-center uniaxial interference figures. (*a*) Off-center optic axis figure. As the stage is rotated, the melatope sweeps around the field of view. Isogyres and isochromes remain centered on the melatope and the isogyres retain their essential N–S and E–W orientations. (*b*) Mineral sample with optic axis more than 30° from vertical. The melatope is out of the field of view. (*c*) Stage rotation causes isogyre arms to sweep across the field of view in sequence. The thin end of an isogyre arm points to the melatope.

normal figures are characterized by broad fuzzy isogyres that occupy nearly the entire field of view when the trace of the optic axis is either E–W or N–S. With only a few degrees of stage rotation, the isogyre splits into two segments that

Figure 7.39 Uniaxial optic normal (flash) figure. The optic axis is parallel to the microscope stage. Vibration directions in the figure are derived as shown in Figure 7.37. In view I, the optic axis is exactly E–W. The isogyre is a broad, fuzzy cross because vibration directions of ω and ϵ' rays in all but the outer parts of the four quadrants are parallel to the lower and upper polarizers. In view II, the stage has been rotated a few degrees clockwise. The isogyres quickly split and exit the field of view from the quadrants into which the optic axis (OA) is being rotated (isochromes are not shown). View III shows the figure with the optic axis NW–SE. Diffuse isochromes are concave outward. The retardation decreases outward toward the NW and SE and increases outward toward the NE and SW.

exit the field of view from the quadrants into which the optic axis is being rotated. Because the isogyres appear and disappear rapidly with stage rotation, these figures also are known as **flash figures**. An optic normal figure confirms that the optic axis is approximately horizontal. However, it is not routinely used to determine either optical character or optic sign because an interference figure with nearly identical appearance is produced by biaxial minerals.

Determining Optic Sign

The optic axis interference figure is used to determine **optic sign** because the vibration directions of ordinary and extraordinary rays are known at each point in the figure (Figure 7.37). Note the interference colors in the four quadrants of the figure (Figure 7.40a) and determine the retardation (Δ_1) from the interference color chart (Plate 1). When an accessory plate with retardation Δ_A (slow NE–SW) is inserted, the retardations will add in two quadrants and subtract in two quadrants. Where retardations add, the new interference color will have retardation $\Delta_1 + \Delta_A$. Where retardations subtract, the new interference color will have a retardation that is the absolute value of $\Delta_1 - \Delta_A$. Consider the NW and SE quadrants where the ordinary rays vibrate NE–SW, parallel to the accessory plate slow ray. If the retardations subtract in these quadrants with insertion of the accessory plate (Figure 7.40b), the ordinary ray is the fast ray and the mineral is optically positive. If the retardations in the NW and SE quadrants add (Figure 7.40c), the ordinary ray is the slow ray and the mineral is optically negative. The interference colors in the NE and SW quadrants must change in the opposite sense because the ordinary rays vibrate NW–SE and the extraordinary rays vibrate NE–SW there.

Biaxial Interference Figures

Biaxial minerals can be distinguished from uniaxial minerals by examining an interference figure on an appropriately oriented grain. The optic sign and the $2V$ angle also may be determined.

Acute Bisectrix Figure

An **acute bisectrix interference figure** (Figure 7.41) is obtained if the acute bisectrix (X or Z depending on optic sign) is oriented perpendicular to the microscope stage. Grains in this orientation display intermediate retardation. Provided that $2V$ is less than 45° to 55°, the melatopes marking the points of emergence of the optic axes will be in the field of view. The interference figure consists of isogyres that change as the stage is rotated superimposed on a field of isochromes.

The isochromes form an oval or figure-eight pattern about the melatopes (Figure 7.42). Only light that follows the optic axes experiences zero retardation. Light experiences increasing birefringence for paths inclined to the optic axes, so retardation must increase away from

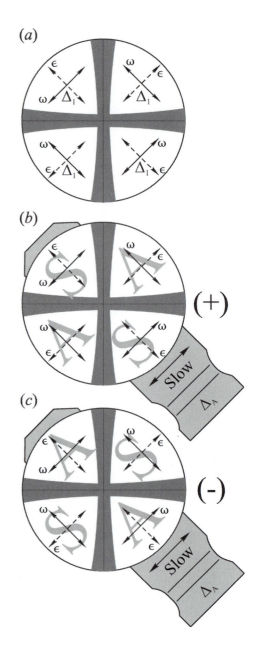

Figure 7.40 Determining optic sign in a uniaxial optic axis interference figure. (*a*) Retardation Δ_1 is displayed in the four quadrants and the vibration directions of ordinary (ω) and extraordinary (ϵ) rays are shown. When an accessory plate with retardation Δ_A (slow NE–SW) is inserted, retardations in two quadrants will add (A) to become $\Delta_1 + \Delta_A$, and retardations in the other two quadrants will subtract (S) to become $|\Delta_1 - \Delta_A|$. (*b*) Optically positive. (*c*) Optically negative.

the melatopes to form the oval pattern. Minerals that are thick or have high birefringence display more isochromes than thin or low birefringence minerals; this is because retardation and interference color are dependent on both thickness and birefringence (Equation 7.7). The pattern of isochromes remains fixed relative to the melatopes as the stage is rotated.

The isogyres form a pattern that changes as the stage is rotated. Vibration directions in the figure can be

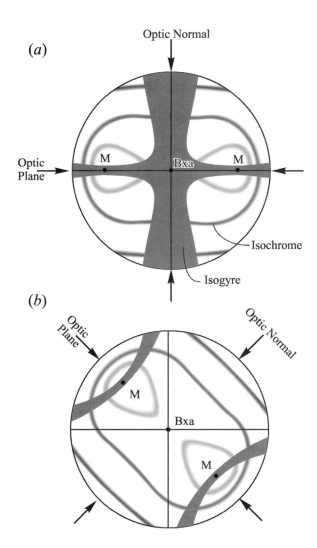

Figure 7.41 Biaxial acute bisectrix interference figure. The melatopes (M) mark the points of emergence of the optic axes, and the acute bisectrix (Bxa) is in the center of the field of view. Isochromes form an oval or figure-eight pattern centered on the melatopes. The isogyre pattern depends on stage rotation. (*a*) The isogyres form a cross-shaped pattern when the trace of the optic plane is oriented E–W. (*b*) Optic plane rotated to a 45° position. With rotation from (*a*) to (*b*), the isogyres split and move across the field of view to form the two arc-shaped segments centered on the melatopes.

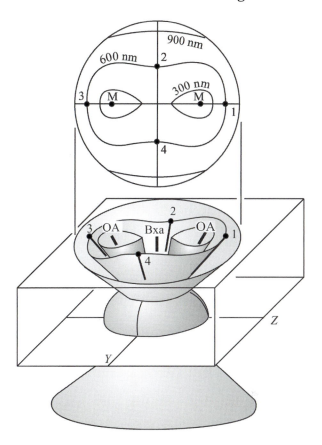

Figure 7.42 Formation of biaxial isochromes. Light following the optic axes emerges at the melatopes (M) with zero retardation. Light following paths within the flattened cone outlined by rays 1, 2, 3, and 4 develops the same retardation by the time each ray has exited the top of the mineral. Light inclined further from the optic axes develops greater retardation, and light following paths inclined less to the optic axes develops less. Isochromes are formed in the interference figure along bands of equal retardation.

derived in the same manner employed for uniaxial figures. Figure 7.43a shows the indicatrix for a biaxial negative mineral. Note that if the optic angle (2*V*) approaches zero, the indicatrix and the vibration directions approach those of a uniaxial negative mineral. Vibration directions for a number of light paths are projected onto the top surface of the mineral and into the interference figure (Figure 7.43b).

If the optic plane is oriented E–W (Figure 7.43c), the isogyres, defined by areas in the figure where vibration directions are N–S and E–W, form a cross. The arm parallel to the trace of the optic normal is somewhat wider than the isogyre parallel to the trace of the optic plane. The

positions of the melatopes are marked by a narrowing of the isogyres. This narrowing may be obscured in minerals with low birefringence because the low first-order gray interference color in the vicinity of the melatope may be indistinguishable from the extinction area of the isogyre.

If the optic plane is rotated away from E–W (or N–S), the cross-shaped isogyre splits into two separate isogyre segments that appear to pivot about the positions of the melatopes (Figure 7.43d). When the trace of the optic plane is placed in a 45° position, the isogyres form hyperbolic arcs whose vertices are the melatopes (Figure 7.43e). The isogyres are narrowest at the melatopes and fan out somewhat toward the edge of the field of view.

Optic Axis Figure

An optic axis interference figure is produced when one of the optic axes is vertical. Grains in this orientation display zero or minimum retardation. The melatope corresponding to the optic axis is centered in the field of view. The

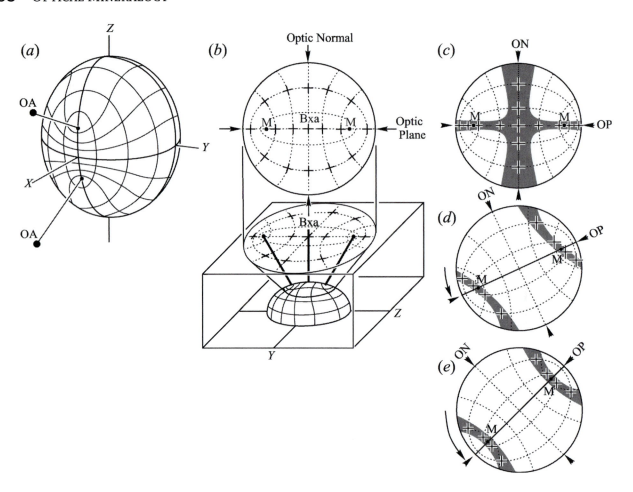

Figure 7.43 Formation of biaxial isogyres. (*a*) Vibration directions for all paths emerging from the biaxial indicatrix are determined as described in Figure 7.22 and are projected onto the indicatrix. (*b*) Vibration directions for a number of light paths have been projected onto the upper surface of the mineral grain and are shown in the interference figure. Isogyres are areas of extinction where vibration directions in the mineral are N–S and E–W. (*c*) Optic plane oriented E–W. The isogyres form a cross that is wider along the trace of the optic normal (ON) and is thin at the melatopes (M). (*d*) With a few degrees of stage rotation, the isogyres split into two segments that pivot about the position of the melatopes. (*e*) Optic plane NE–SW. Isogyres form hyperbolas centered on the melatopes.

other melatope may be in the field of view if $2V$ is less than roughly 25°, otherwise it will be outside the field of view. If $2V$ is small, the interference figure looks like an off-center acute bisectrix figure (Figure 7.44a).

If $2V$ is greater than about 50°, the pattern shown in Figure 7.44b is seen. When the optic plane is oriented N–S or E–W, only a single arm of the cross pattern is visible in the field of view. It narrows somewhat at the position of the melatope and is straight, parallel to the trace of the optic plane. If the stage is rotated clockwise, the isogyre pivots about the melatope counterclockwise and vice versa. When the trace of the optic plane is in a 45° position, the isogyre shows a maximum curvature. The position of the acute bisectrix lies on the convex side of the isogyre. For $2V$ between 25° and 50°, the Bxa will be in the field of view so some portion of the broad isogyre along the trace of the optic normal will be seen when the optic plane is N–S or E–W.

Obtuse Bisectrix Figure

Obtuse bisectrix interference figures (Figure 7.45) are produced if the obtuse bisectrix is oriented perpendicular to the microscope stage. Grains with this orientation display intermediate retardation. Because the angle between the Bxo and the optic axes must be greater than 45°, the melatopes will be outside the field of view. The pattern of isochromes and geometry of vibration directions are essentially the same as for the acute bisectrix figure, except that the melatopes are well out of the field of view. The isogyres form a cross pattern if the optic plane is E–W or N–S. Typically 5° to 15° of stage rotation is required to cause the isogyres to leave the field of view. Note that for $2V$ of nearly 90°, the obtuse and acute bisectrix figures are very similar. If $2V$ is small, the obtuse bisectrix figure looks much like an optic normal figure.

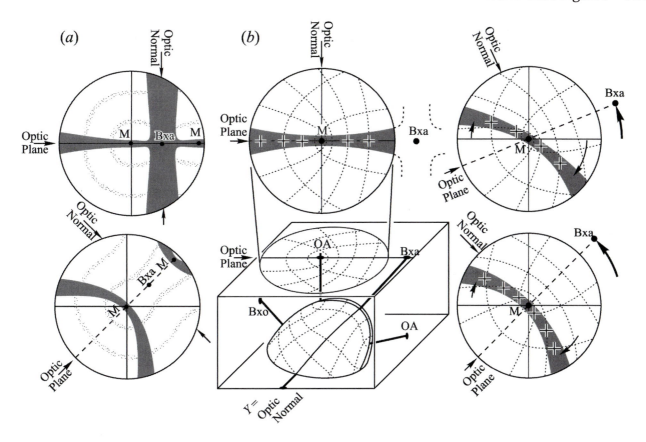

Figure 7.44 Biaxial optic axis interference figure. (*a*) 2*V* less than about 30°. The second melatope is in the field of view, and the figure looks like an off-center acute bisectrix figure. Isochromes are shown. (*b*) Larger 2*V*. The second melatope is out of the field of view. The dashed lines show vibration directions; isochromes are not shown. The view on the left shows the indicatrix orientation and interference figure with the optic plane oriented E–W. The views on the right show the isogyre motion as the stage is rotated counterclockwise. Note that isogyres pivot in a sense opposite to the stage rotation direction.

Optic Normal (Flash) Figure

Optic normal or flash figures are produced when the optic normal is vertical. Grains oriented to produce this figure display maximum retardation because the *X* and *Z* indicatrix axes are horizontal. The pattern of vibration directions in the figure is nearly rectilinear (Figure 7.46a) and is very similar to that seen with uniaxial flash figures (Figure 7.39). When the *X* and *Z* indicatrix axes are N–S and E–W, the field of view is occupied by a broad, fuzzy isogyre because vibration directions in all but the outer edges of the quadrants are N–S and E–W. If the stage is rotated a few degrees (Figure 7.46b), the isogyre splits into two curved segments that exit the field of view from the quadrants into which the acute bisectrix is being rotated. For minerals with 2*V* of nearly 90°, the diffuse cross-shaped isogyre simply dissolves as the stage is rotated. The amount of rotation required to cause the isogyres to completely leave the field of view is typically less than 5°.

Off-Center Figure

Grains in random orientations display off-center interference figures whose pattern of isogyres and isochromes depends on the details of orientation. A typical figure is shown in Figure 7.47. As the stage is rotated, the isochrome pattern pivots about the center and isogyres sweep sequentially across the field of view. The wide end of the isogyre sweeps across the field more rapidly than the thin end, and the sense of isogyre rotation is counter to the direction of stage rotation. Unless a melatope or the acute bisectrix is, by chance, in the field of view, off-center figures are of no particular value in interpreting 2*V* or determining the optic sign. However, the fact that the isogyre does not remain parallel to the N–S or E–W as it sweeps across the field of view does indicate that the mineral is biaxial, not uniaxial (compare Figure 7.38c).

Determining Optic Sign

Determining optic sign is done with acute bisectrix and optic axis figures. Although optic axis figures are easiest to obtain and are used most commonly for routine work, it is easiest to illustrate the process of determining optic sign by using an acute bisectrix figure with modest 2*V*.

Of the two rays of light that propagate along the acute bisectrix and emerge in the center of an acute bisectrix figure, one vibrates parallel to the *Y* axis (the optic normal)

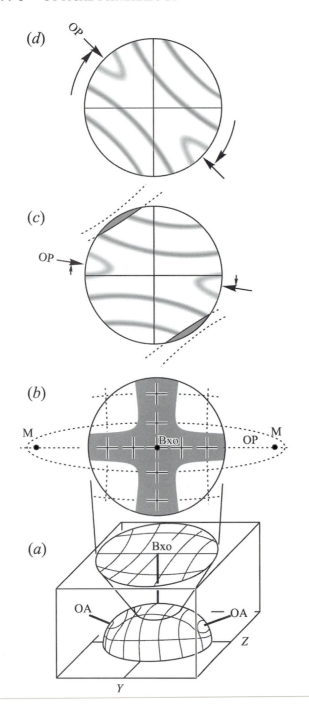

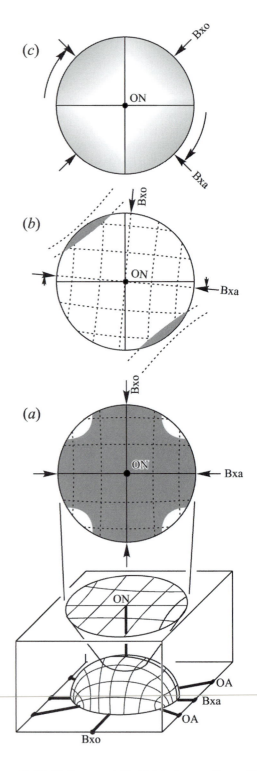

Figure 7.45 Obtuse bisectrix interference figure. (*a*) Pattern of vibration directions in the interference figure. (*b*) With the optic plane E–W, the isogyres form a broad cross. Melatopes are well out of the field of view. Compare with the acute bisectrix figure in Figure 7.41. (*c*) Less than ~15° of stage rotation causes the isogyres to exit the field of view. The pattern of isochromes is similar to that formed in an acute bisectrix figure. (*d*) With the optic plane in a 45° position, the field is occupied only by isochromes.

Figure 7.46 Optic normal (flash) interference figure. The optic normal (ON) emerges in the center of the figure. (*a*) With the Bxa and Bxo E–W and N–S, the field is occupied by a broad fuzzy cross. Only the outer edges of the four quadrants are not extinct. (*b*) Isogyres split and exit the field with a few degrees of stage rotation. The isogyres exit from the quadrants into which the trace of the Bxa is being rotated. (*c*). Isochrome pattern. No isogyres are present with traces of the Bxo and Bxa in 45° positions. The interference color in the middle of the figure is the same as that displayed by the grain in orthoscopic illumination. Colors increase outward in all four quadrants

and has index of refraction n_β. The other vibrates parallel to the obtuse bisectrix (Bxo) along the trace of the optic plane and has index of refraction n_{Bxo}. If the mineral is optically positive, the Bxo is the X axis and $n_{Bxo} = n_\alpha$. If

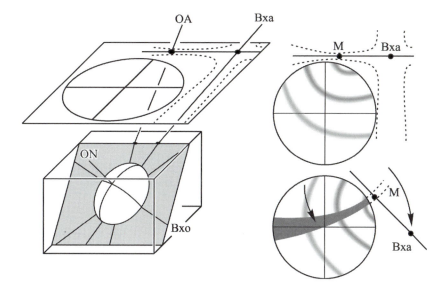

Figure 7.47 Biaxial interference figure for randomly oriented grain. As the stage is rotated, isogyres sweep sequentially across the field but are not parallel to the crosshairs.

the mineral is optically negative, the Bxo is the Z axis and $n_{Bxo} = n_\gamma$. The task, therefore, is to determine whether the ray vibrating parallel to the trace of the optic plane in the center of the field of view is the fast ray or the slow ray. If it is fast, the mineral is optically positive, and if it is slow, the mineral is optically negative. The procedure is as follows (Figure 7.48).

• Obtain an acute bisectrix interference. Finding an appropriately oriented grain is a matter of luck, because many orientations may display the same intermediate retardation.

• Rotate the stage so that the trace of the optic plane is NE–SW. Near the center of the figure the ray vibrating NE–SW vibrates parallel to the Bxo and has index of refraction n_{Bxo}. The ray vibrating NW–SE vibrates parallel to Y and has index n_β. Note the interference color in the center of the figure and determine the retardation Δ_1 that it indicates from the interference color chart (Plate 1).

• Insert the accessory plate with retardation Δ_A (slow NE–SW) and note the change in interference color in the central part of the figure to determine if n_{BXO} is fast or slow compared to n_β. If retardations add, the new interference color will indicate a retardation of $\Delta_1 + \Delta_A$. If retardations subtract, the new interference color will indicate a retardation of $|\Delta_1 - \Delta_A|$.

• Interpretation. If the new interference color between the melatopes indicates that retardations subtract, the ray vibrating parallel to the Bxo must be the fast ray with index n_a, so the Bxo is the X axis, the Bxa must be the Z axis, and the mineral is optically positive (Figure 7.48a). If the new interference color between the melatopes indicates that retardations add, the ray vibrating parallel to the Bxo must be the slow ray with index n_γ, so the Bxo is the Z axis,

the Bxa must be the X axis and the mineral is optically negative (Figure 7.48b). Interference colors outside the melatopes change opposite to the change seen between the melatopes. If the trace of the optic plane is placed NW–SE, the areas of addition and subtraction will be reversed.

Optic axis interference figures are usually easier to obtain because appropriately oriented grains display the lowest order interference color. If $2V$ is small enough that both melatopes are in the field of view, the figure can be treated the same as an acute bisectrix figure. If only one melatope is visible, the figure can be considered to be half of an acute bisectrix figure.

• Start with the isogyre along either the N–S or the E–W crosshair.

• Rotate the stage to place the trace of the optic plane NE–SW with the isogyre convex to the NE. This places the Bxa in the NE quadrant (Figure 7.49). Depending on original grain orientation, either 45°, 135°, 225°, or 315° of clockwise rotation may be required. This figure can now be interpreted just like the SW half of the acute bisectrix figure shown in Figure 7.48. The optic sign can be interpreted with the Bxa in any of the four quadrants by comparison with the appropriate portion of a full acute bisectrix figure with the same optic plane orientation. If $2V$ is nearly 90°, the isogyre will be nearly straight and it may not be possible to determine the position of the Bxa. In this case the optic sign may be recorded as neutral, with $2V$ of around 90°.

Determining 2V

Figure 7.50 shows the paths followed by light waves that propagate parallel to the optic axes in a crystal cut so that

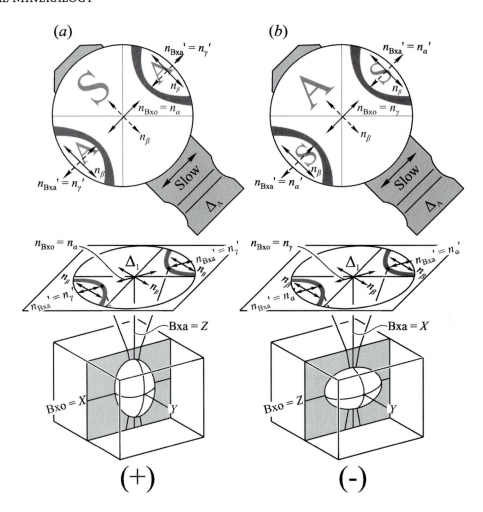

Figure 7.48 Determining optic sign in acute bisectrix interference figures. The trace of the optic plane is NE–SW. In the center of the field of view, the retardation is Δ_1, and the NW–SE ray vibrates parallel to the optic normal Y and has index of refraction n_β. Insert an accessory plate with retardation Δ_A (slow NE–SW) and note the change in interference color. (If the optic plane is NW–SE, the color changes are reversed.) (a) Optically positive. The ray vibrating NE–SW in the center of the figure vibrates parallel to $X = $ Bxo, has index of refraction n_α, and is the fast ray compared to n_β, so retardations subtract to become $|\Delta_1 - \Delta_A|$. Outside the isogyres, the ray vibrating NE–SW has index of refraction n'_γ, which is the slow ray compared to n_β, so retardations add. (b) Optically negative. The ray vibrating NE–SW in the center of the figure vibrates parallel to $Z = $ Bxo, has index of refraction n_γ, and is the slow ray compared to n_β, so retardations add to become $\Delta_1 + \Delta_A$. Outside the isogyres, the ray vibrating NE–SW has index of refraction n'_α, which is the fast ray compared to n_β, so retardations subtract.

the acute bisectrix is vertical. Light following the optic axes has index of refraction n_β and is refracted when it enters the air above the mineral plate. The angle between the refracted optic axes is a larger angle known as **2E**, or the **apparent optic angle**. Assuming that the material between the mineral and the objective lens is air ($n \approx 1$), the relation between 2V and 2E is

$$\sin E = n_\beta \sin V. \qquad (7.17)$$

The spacing between the melatopes in an acute bisectrix figure depends on the value of 2E. For the melatopes

to be at the edge of the field of view, the angular aperture of the objective lens (Figure 7.10) must equal 2E. Hence, for a given objective lens and value of n_β, the largest 2V angle ($2V_{max}$) that still allows the melatopes to be within the field of view (Table 7.1) can be determined by combining Equations 7.6 and 7.17.

$$\text{NA} = n_\beta \sin V_{max} \qquad (7.18)$$

Figure 7.51 shows acute bisectrix interference figures for different values of 2V assuming that $n_\beta = 1.60$ and that the numerical aperture of the objective lens is 0.65.

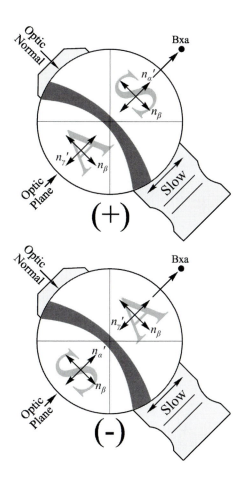

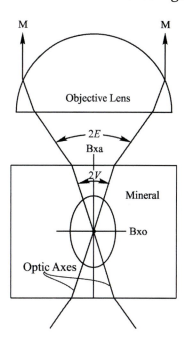

Figure 7.50 *2E* versus *2V*. The angle between the optic axes is *2V* within the mineral. Light following the optic axes is refracted as it exits the mineral to form the apparent optic angle *2E*. The position of the melatopes in an acute bisectrix interference figure depends on the value of *2E*.

Figure 7.49 Determining optic sign with an optic axis figure. The trace of the optic plane is parallel to the slow ray vibration direction in the accessory plate. Retardations subtract (S) on the convex side of the isogyre for optically positive minerals and add (A) for negative. If the optic plane is oriented NW–SE, the color changes are reversed.

Table 7.1 *2V* angle ($2V_{max}$) for Melatopes at the Edge of the Field of View for a Numerical Aperture of 0.65 and Different Values of n_β

n_β	$2V_{max}$
	NA = 0.65
1.40	55°
1.50	51°
1.60	48°
1.70	45°
1.80	42°

Because most biaxial minerals have indices of refraction that fall in the range 1.50–1.70, these figures can be used to estimate *2V* to within 10 or 15°, which is adequate for routine work.

A more convenient method of estimating *2V* uses an optic axis interference figure. This method depends on the observation that the curvature of an isogyre when the trace of the optic plane is in a 45° position depends on the value of *2V* (Figure 7.52). If *2V* is 90°, the isogyre forms a straight line and for smaller values of *2V*, the isogyre is progressively more curved. If *2V* is less than about 25–30°, both melatopes will be in the field of view and an estimate of optic angle can be made based on the distance between the melatopes. For *2V* of less than 5°, the distance between the melatopes is quite small, and the figure looks like a uniaxial figure except for a small gap between the isogyres in the middle. To accurately orient the figure to place the trace of the optic plane in a 45° position, proceed as follows.

• Obtain an optic axis figure. Appropriately oriented grains display low or zero retardation. Off-center figures can be used, but accuracy suffers.

• Rotate the stage to place the isogyre N–S. This places the trace of the optic plane N–S. Unless the melatope is exactly centered, the isogyre will be somewhat to the right or left of the N–S crosshair.

• Rotate the stage 45° clockwise, using the stage goniometer to measure the angle. The trace of the optic plane is now NE–SW and the curvature of the isogyre can be compared with Figure 7.52. The isogyre may be convex to the NE or the SW depending on whether the Bxa is to the NE or SW, respectively. If the isogyre is convex to the SW, 180° of stage rotation places the Bxa in the NE quadrant,

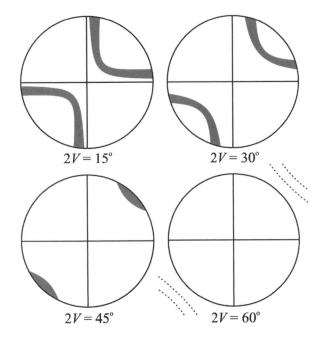

$2V = 15°$ $2V = 30°$

$2V = 45°$ $2V = 60°$

Figure 7.51 Rapid estimation of $2V$ based on separation of isogyres in acute bisectrix interference figures. The positions of the isogyres are shown for $n_\beta = 1.60$ and an objective lens with NA = 0.65.

which duplicates the orientation used to determine optic sign in Figure 7.49.

Dispersion Effects

The values of n_α, n_β, and n_γ vary for different wavelengths of light just as the index of refraction of isotropic minerals varies. Consequently, the values of $2V$ and the orientation of the indicatrix can be expected to vary for different wavelengths of light. Variation in the size of $2V$ is called **optic axis dispersion**, and variation in the orientation of the indicatrix is called **indicatrix dispersion**, or **bisectrix dispersion**. If this dispersion is sufficiently strong, color fringes may be visible along the isogyres in interference figures. The dispersion is termed **weak, moderate,** or **strong,** depending on how noticeable the color fringes are.

Orthorhombic minerals can display only optic axis dispersion because the indicatrix axes are fixed parallel to the crystallographic axes. Although exceptions exist, the orientation of the optic plane in most orthorhombic minerals is the same for all wavelengths. If $2V$ for light at the red end of the spectrum is larger than $2V$ for light at the violet end, the dispersion is described as $r > v$, and if $2V$ for violet light is larger than for red light, the dispersion is described as $v > r$. The formation of the color fringes is shown in Figure 7.53. Note that the red fringe is found where the melatopes for violet light are located and the blue fringe is found where the melatopes for red light are located.

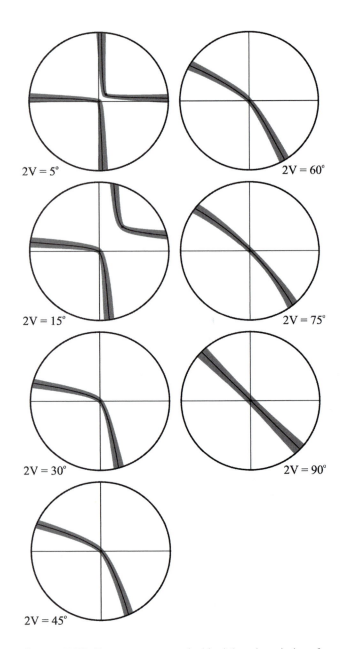

$2V = 5°$ $2V = 60°$

$2V = 15°$ $2V = 75°$

$2V = 30°$ $2V = 90°$

$2V = 45°$

Figure 7.52 Isogyre curvature in biaxial optic axis interference figures with the optic plane aligned NE–SW. The geometry shown here assumes $n_\beta = 1.60$ and NA = 0.65. Compare Figure 7.51.

In addition to optic axis dispersion, monoclinic minerals may display indicatrix dispersion because the indicatrix is free to rotate about the b crystal axis for different wavelengths of light. Three types of indicatrix dispersion are possible, depending on which indicatrix axis coincides with the b axis. **Inclined dispersion** is produced when the optic normal (Y) is parallel to the b axis. The orientation of Y is fixed for all wavelengths, but the X and Z are free to rotate within the {010} plane, which serves as the optic plane for all wavelengths. Color fringes are symmetrical across the trace of the optic plane in acute bisectrix figures. **Horizontal** or **parallel dispersion** is produced

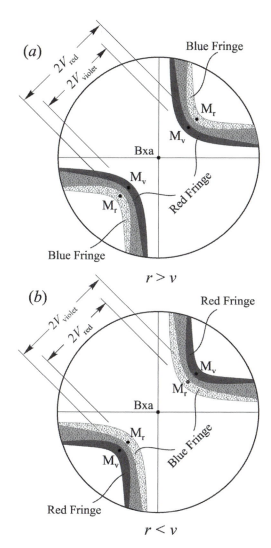

Figure 7.53 Optic axis dispersion: M_v and M_r are the melatopes for violet and red light, respectively. The isogyre for violet light (shaded) marks the area where no violet light comes through. The isogyre for red light (stippled) marks the area where no red light comes through. Red fringes are therefore produced where blue and violet light are lacking, and the blue fringe is produced where red light is lacking. The areas of overlap represent the normal isogyres. (a) $2V$ for red light is greater than $2V$ for violet light ($r > v$). (b) $2V$ for red light is less than $2V$ for violet light ($r < v$). The width of the colored fringes is exaggerated here.

when the obtuse bisectrix is parallel to the b axis. The optic planes for different wavelengths pivot about the b axis so the position of the Bxa for red light emerges at a different point in the {010} plane than the Bxa for violet light. Color fringes are symmetrical across the trace of the optic normal in acute bisectrix figures. **Crossed bisectrix dispersion** is produced when the acute bisectrix is parallel to the b axis. The optic plane pivots about the Bxa, but because the b axis is vertical in this acute bisectrix figure, the color fringes are symmetrical by 2-fold rotation about the Bxa.

REFRACTOMETRY: MEASUREMENT OF INDEX OF REFRACTION

Immersion Method

Refractometry refers to measurement of indices of refraction. The simplest and most convenient method in mineralogy is the **immersion method**. With this method, grain mounts of an unknown transparent mineral are prepared, and the index of refraction of the mineral is identified by comparing it to the known index of refraction of an immersion oil. Immersion oil is commercially available with indices of refraction that cover the range found for most minerals. A typical set consists of a number of small bottles of immersion oil whose indices of refraction range from about 1.4 to 1.8 in increments of 0.002, 0.004, or 0.005. Indices of refraction may be measured to accuracies of about ±0.003 using white light for illumination.

For routine work, relief and the Becke line method provide a means of comparing the index of refraction of mineral and immersion oil. For additional techniques, refer to Nesse (2002), Bloss (1961), and references therein.

Relief

The degree to which mineral grains stand out from the immersion oil or other mounting medium is called **relief** (Figure 7.54). If the indices of oil and mineral are not the same, light is refracted on passing from oil to mineral and the mineral appears to stand out. If the indices of refraction of oil and mineral are the same, light is not refracted at oil–mineral boundaries and the mineral does not appear to stand out. Relief may be described as high, moderate, or low (Figure 7.55). The approximate differences of index of refraction between mineral and immersion oil indicated by these categories are **high relief** > 0.12, **moderate relief** between 0.12 and 0.04, and **low relief** < 0.04. Note that the relief of anisotropic minerals in both grain mount and thin section may change as the stage is rotated in plane light, particularly if the mineral has moderate to high birefringence. This is because the slow and fast rays have different indices of refraction and therefore display different relief, depending on which ray is dominant.

If the index of refraction of the mineral is higher than the oil, the mineral has **positive relief**, and if it is lower, the mineral has **negative relief**. However, in both cases, the mineral will appear to stand out from the oil to the same degree.

Becke Line Method

The **Becke line method** is used to distinguish between positive and negative relief. The Becke line (Figure 7.56)

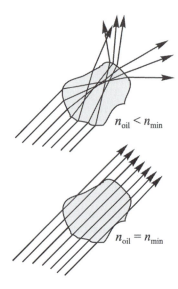

Figure 7.54 Formation of relief in grain mount. If indices of refraction of oil and mineral are different, light is refracted at the mineral–oil boundary and the grain stands out. If the indices of refraction are the same, light passes the mineral–oil boundary without refraction and the mineral grain does not stand out.

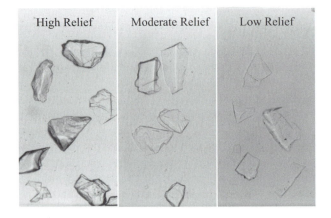

Figure 7.55 Relief. (*Left*) High relief. (*Middle*) Moderate relief. (*Right*) Low relief.

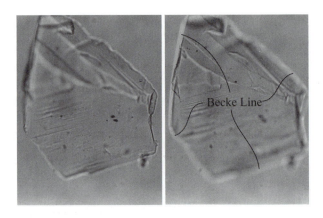

Figure 7.56 The Becke line. (*Left*) Microscope focused on a grain of fluorite that shows moderate relief. (*Right*) Microscope stage lowered. The Becke line is a thin band of light that moves into the medium with the higher index of refraction, which in this case is the immersion oil.

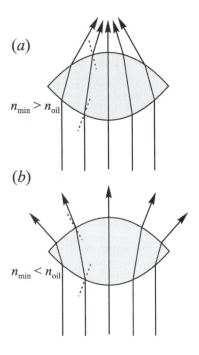

Figure 7.57 Formation of the Becke line by the lens effect. Light is refracted toward the normal on passing into a material with higher index of refraction and away from the normal when passing into a material with lower index of refraction as predicted by Snell's law (Equation 7.3). (*a*) $n_{mineral} > n_{oil}$. The grain acts as a crude converging lens, concentrating light above the mineral. (*b*) $n_{mineral} < n_{oil}$. The grain acts as a crude diverging lens, refracting light away from the center of the grain.

consists of a band or rim of light along the edge of a grain. The Becke line is most easily seen when one is using the medium-power objective if the image is slightly out of focus and the aperture diaphragm is closed somewhat. If the stage of the microscope is lowered so that the distance between the sample and objective lens is increased, the Becke line moves into the material with the higher index of refraction. The formation of the Becke line depends on both the lens effect and the internal reflection effect.

The lens effect depends on the observation that most mineral fragments are thinner on the edges than in the center and act as crude lenses (Figure 7.57). If the mineral has a higher index of refraction than the immersion oil, the grains act as converging lenses to concentrate light toward the center of the grain. If the grains have a lower index of refraction than the immersion oil, they act as diverging lenses and light is concentrated in the immersion oil.

The internal reflection effect depends on the requirement that the edges of grains must be vertical at some point. Moderately converging light from below impinges on the vertical boundary and may be internally reflected or refracted, depending on the angle of incidence and indices of refraction. As can be seen in Figure 7.58, the result of the refraction and internal reflection is to concentrate light into a thin band in the material with the higher index of refraction.

Both the lens and internal reflection effects concentrate light into the material with the higher index of refraction in the area above the mineral grain, in effect forming a cone of light propagating up from the edge of the mineral (Figure 7.59). If the mineral has the higher index of refraction, the cone converges above the mineral, and if the immersion oil has the higher index of refraction, the cone diverges above the mineral. With the microscope crisply focused on a grain, the Becke line is coincident with the edge of the grain, or may disappear. As the stage is lowered, the concentration of light near the top of the grain that forms the Becke line is brought into focus. Hence, **as the stage is lowered, the Becke line appears to move into the material with the higher index of refraction.** The opposite is true if the stage is raised.

Dispersion Effects

The dispersion of immersion oil is greater than the dispersion of most minerals, so it is possible to produce a match of index of refraction for only one wavelength of light. Ideally, the object is to produce a match for light whose wavelength is 589 nm because that is the wavelength for which indices of refraction are usually reported.

If the dispersion curves for mineral and immersion oil intersect in the visible spectrum (Figure 7.60), the oil will have higher indices of refraction for wavelengths shorter than the match, and the mineral will have higher indices of refraction for longer wavelengths. This results in the formation of two Becke lines, one for the shorter wavelengths and one for the longer. When the stage is lowered, the Becke line composed of shorter wavelengths will move into the immersion oil, and the Becke line composed of longer wavelengths will move into the mineral. The color of the two Becke lines depends on the wavelength at which the dispersion curves cross. The relationships are summarized in Table 7.2.

Practical Considerations

Determining the index of refraction of an unknown mineral by using grain mount techniques can be tedious and

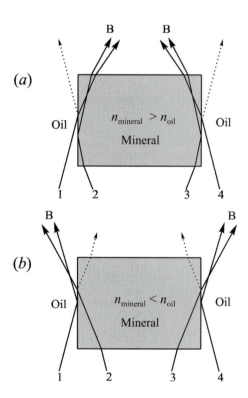

Figure 7.58 Formation of the Becke line by internal reflection at vertical grain boundaries. The Becke line (B) consists of a concentration of light above the mineral–oil boundary in the material with the higher index of refraction. (a) $n_{mineral} > n_{oil}$. Some of rays 1 and 4 is reflected (dotted lines) at the vertical mineral boundary, and some is refracted into the mineral (shaded) even for shallow angles of incidence (compare Figure 7.6c). Rays 2 and 3 strike the mineral–oil interface at greater than the critical angle and are internally reflected in the mineral (compare Figure 7.7). (b) $n_{mineral} < n_{oil}$. Rays 1 and 4 strike the vertical mineral–oil boundary at greater than the critical angle and are totally reflected back into the oil. Some of rays 2 and 3 are refracted into the oil.

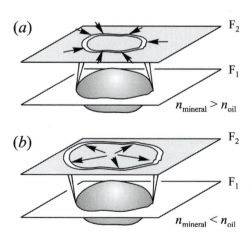

Figure 7.59 Movement of the Becke line as the stage is lowered. The Becke line may be considered to consist of a cone of light that extends upward from the mineral grain. If $n_{mineral} > n_{oil}$, the cone converges upward and if $n_{mineral} < n_{oil}$, the cone diverges upward. If the stage is lowered, the plane of focus goes from F_1 to F_2 and the Becke line appears to move toward the material with the higher index of refraction. (a) $n_{mineral} > n_{oil}$. (b) $n_{mineral} < n_{oil}$.

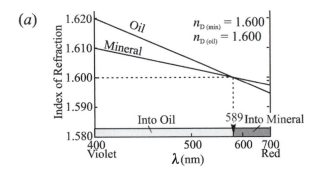

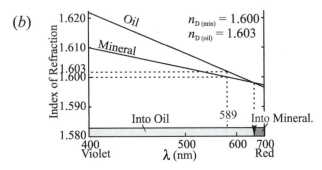

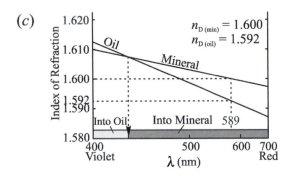

Figure 7.60 Formation of colored Becke lines. Immersion oil typically has higher dispersion than minerals. (*a*) Indices of refraction of oil and mineral matched at light whose wavelength is 589 nm. The mineral has higher index of refraction for longer wavelengths, so these wavelengths, perceived as yellowish orange, go into the mineral as the stage is lowered. The immersion oil has higher index of refraction for shorter wavelengths, and this bluish light forms a Becke line that goes into the oil. (*b*) Indices of oil and mineral are matched near the red end of the spectrum. The mineral has higher index of refraction for red light, and the oil for all other wavelengths, perceived as bluish white. When the stage is lowered, a red Becke line moves into the mineral, and bluish white into the oil. (*c*) Indices of oil and mineral match near the blue end of the spectrum. The oil has higher index of refraction for blue and violet light, the mineral for the rest of the spectrum. When the stage is lowered, a blue-violet Becke line moves into the oil, and yellowish white into the mineral.

frustrating unless it is approached systematically. The method recommended here involves successively bracketing the index of the unknown mineral until a match can be obtained with an immersion oil.

Table 7.2 Becke Line Relationships When the Microscope Stage Is Lowered

Condition	Observation	Interpretation
n_{oil} higher for all wavelengths	White line into oil	$n_{D(oil)} \gg n_{D(mineral)}$
$n_{oil} = n_{min}$ for orange/red light	Red line into mineral, bluish-white line into oil	$n_{D(oil)} > n_{D(mineral)}$
$n_{oil} = n_{min}$ for yellow light (589 nm)	Yellowish-orange line into mineral, pale blue line into oil	$n_{D(oil)} = n_{D(mineral)}$
$n_{oil} = n_{min}$ for blue light	Yellowish-white line into mineral, blue-violet line into oil	$n_{D(oil)} < n_{D(mineral)}$
n_{oil} lower for all wavelengths	White line into mineral	$n_{D(oil)} \ll n_{D(mineral)}$

• Prepare a grain mount as described at the beginning of the chapter, using an immersion oil in the middle of the available range. Examine the grains with the microscope to determine the relief. Use the Becke line method to determine whether the mineral has a higher or lower index of refraction than the immersion oil. If the index of the mineral is higher than the oil, all lower indices of refraction can be eliminated from consideration. If the index of the mineral is lower than the oil, all higher indices of refraction can be eliminated.

• Prepare a second grain mount using an immersion oil somewhere in the remaining range of possible indices of refraction. Use the relief of the mineral in the first grain mount as a guide in selecting the second oil. The object is to bracket the index of refraction of the mineral so that one oil is higher and the other lower than the index of refraction of the mineral. Examine the relief of the grains in the second grain mount and use the Becke line method to determine whether the mineral has an index of refraction higher or lower than the oil.

• Prepare a third grain mount using an immersion oil somewhere in the middle of the now much-reduced range of possible indices, using relief seen in the first two grain mounts as a guide in selecting the oil whose index of refraction is close to that of the mineral. If the first two oils did not bracket the index of the mineral and relief was high in the second oil, it may be advisable to use an oil at the extreme end of the available range to determine whether the mineral actually falls within the range of available oil.

• Make additional grain mounts, in each case selecting an oil to split the range of possible indices, until a match is obtained. With practice, a match can be obtained with as few as four or five grain mounts.

Refractometry in Thin Section

It is not practical to accurately determine indices of refraction of minerals in thin section with a petrographic microscope, but it is possible to make estimates or establish limits. The index of an unknown mineral can be compared to the index of refraction of the cement or to other known minerals in the thin section. The index of refraction of the cement depends on the manufacturer, but is usually about 1.540. Relief provides an estimate of how close the index of a mineral is to the cement. Becke lines formed at mineral–cement and mineral–mineral boundaries can be interpreted as described earlier to provide rough estimates of the index of refraction of an unknown mineral. Plate 2 (on the back side of Plate 1) organizes minerals based on relief in thin section, optical character, optic sign, and birefringence.

Isotropic Minerals

Identification of isotropic minerals requires measurement of one index of refraction. This is done by means of the technique described earlier.

Uniaxial Minerals

Identification of uniaxial minerals may be done by measuring both n_ω and n_ϵ. From these values the birefringence may be calculated ($\delta = |n_\omega - n_\epsilon|$). Measurement of the two indices of refraction requires that grains in specific orientations be selected so that each index of refraction can be measured independently. It is generally most convenient to measure n_ω first and then measure n_ϵ. To measure n_ω and n_ϵ in grain mount, proceed as follows.

- Measure n_ω. Select a grain oriented to pass all light as ordinary ray. Two techniques are available:
  a. Cross the polarizers and search for a grain of the mineral whose optic axis is vertical. A grain in this orientation will display the lowest-order interference color and ideally will remain uniformly dark with stage rotation. All the light propagating along the optic axis passes as ordinary ray with index of refraction n_ω. To confirm that the optic axis is vertical, obtain an interference figure. The melatope should be near the center of the field of view. Determine optic sign if that has not already been done. If grains in this orientation cannot be found, use the second method.
  b. Pick any grain. As long as the mineral is uniaxial, one ray must pass with index of refraction n_ω. The trick is to rotate the stage of the microscope to place a grain in one of its extinction positions so that the ordinary ray vibration direction is parallel to the lower polarizer. Use an accessory plate to identify the slow and fast ray vibration directions (Figure 7.21) in the grain. The appropriate vibration direction is then placed parallel to the lower polarizer (slow if optically negative or fast if optically positive) so that all light passes as ordinary ray.

Once the ordinary ray vibration direction has been placed parallel to the lower polarizer, n_ω can be compared with the index of refraction of the immersion oil using relief and the Becke line. Repeat with new grain mounts by using different immersion oils until a match is obtained.

- Measure n_ϵ. Accurate measurement of n_ϵ requires grains oriented so that the optic axis is horizontal. Grains in random orientations can yield only values for n_ϵ', which is intermediate between n_ω and n_ϵ. To find appropriately oriented grains, scan the grain mount between crossed polarizers to find a grain of the unknown that displays the highest interference color. This grain should have its optic axis approximately parallel to the microscope stage. A symmetrical optic normal (flash) interference figure confirms that the optic axis is horizontal. If the mineral is optically positive, the ϵ ray is the slow ray, and if the mineral is negative, the ϵ ray is the fast ray. Identify the ϵ ray vibration direction and place it parallel to the lower polarizer with the aid of the accessory plate as follows.
  a. From an extinction position, rotate the stage 45° clockwise. Insert an accessory plate to determine which of the two rays is slow and which is fast as described earlier (Figure 7.21).
  b. Rotate the stage 45° clockwise or counterclockwise as required to place the slow ray (for positive) or fast ray (for negative) vibration direction parallel to the lower polarizer (lower polarizer E–W).

Once the stage has been rotated so that the grain passes only ϵ rays, the value of n_ϵ can be compared with the index of refraction of the immersion oil by using relief and Becke line methods. Prepare additional grain mounts with different immersion oils as needed to obtain a match.

Biaxial Minerals

Three indices of refraction (n_α, n_β and n_γ) must be measured for biaxial minerals. Measurement of n_β requires that grains with an optic axis vertical be found, and measurement of n_α and n_γ requires that grains with the optic normal vertical be found. The procedure is as follows.

- Measure n_β
  a. With orthoscopic illumination and crossed polarizers, scan the grain mount to find a grain of the unknown that displays the lowest order interference color. This grain should have the optic axis vertical, so that all light passes with index n_β. Confirm the orientation by obtaining an interference figure, and determine optic sign and $2V$ if not already done.
  b. Remove the upper polarizer and use relief and the Becke line to compare n_β with the immersion oil. Repeat, using the bracketing technique described earlier, until a match is obtained.

- Measure n_α
  a. With orthoscopic illumination and crossed polarizers, scan the grain mount to find a grain of the unknown that displays the highest interference color. This grain should have the optic normal vertical, so that the two rays pass with indices or n_α and n_γ. Correctly oriented grains yield optic normal (flash) figures.
  b. In orthoscopic illumination, determine which vibration direction is the fast ray with index of refraction n_α by using an accessory plate (Figure 7.21). Rotate the fast ray vibration direction to be parallel to the lower polarizer vibration direction so that all light passes with index of refraction n_α.
  c. Remove the upper polarizer and compare n_α with the immersion oil.
  d. Repeat, using the bracketing technique described earlier, until a match is obtained.

- Measure n_γ
  a. With orthoscopic illumination and crossed polarizers, scan the grain mount to find a grain of the unknown that displays the highest interference color. This grain should have the optic normal vertical, so that the two rays pass with indices or n_α and n_γ. Correctly oriented grains yield optic normal (flash) figures.
  b. In orthoscopic illumination, determine which vibration direction is the slow ray with index of refraction n_γ by using an accessory plate (Figure 7.21). Rotate the slow ray vibration direction to be parallel to the lower polarizer vibration direction so that all light passes with index of refraction n_γ.

Time can be saved by comparing all three indices of refraction with each immersion oil. Once n_β has been determined, oils needed to match n_α and n_γ can be selected based on knowledge of the birefringence, $2V$, and comparisons made in earlier oils. Note that grains used to compare n_α can be rotated 90° to compare n_γ.

REFLECTED-LIGHT OPTICS

Opaque minerals can be identified by optical methods only with the use of a reflected light microscope and polished samples. Only a brief summary of some basic material is presented here. Interested readers are encouraged to pursue the subject with books such as Ineson (1989), Craig and Ixer (1990), and Vaughn (1994).

Samples for reflected-light work may be thin sections or rock/mineral samples mounted in an epoxy plug. The surface is polished to a mirror shine with progressively finer abrasives and no coverslip is used. Polished thin sections are convenient because they can be used for transmitted

light microscopy, reflected-light microscopy, and electron microprobe analysis.

One common geometry used for reflected-light microscopes is to adapt a conventional petrographic microscope by the addition of an epi-illuminator (Figure 7.9). This illuminator is mounted above the objective lenses and is equipped with optical elements so that a beam of polarized light passes down through the objective onto the top surface of the sample. Reflected light passes back up through the objective to the observer. The upper polarizer (analyzer) can be inserted into the optical path to determine whether the polarization of the incident light is affected by reflection on the mineral.

Observation in Plane-Polarized Light

(The analyzer is removed from the optical path.)

Color and Reflection Pleochroism

The color seen in plane-polarized light depends on the specific colors of light that are absorbed and reflected by the mineral. Opaque minerals in the isometric crystal system display the same color regardless of sample orientation. All others may display reflection pleochroism, which is a color change observed as the stage is rotated. Tetragonal and hexagonal minerals whose c axis is vertical will not display reflection pleochroism. It is important to note the similarity to the behavior of light in transparent minerals.

Reflectance and Bireflectance

Reflectance (R) is a measure of the percentage of the incident light that is reflected from a mineral's surface. If some wavelengths are more strongly reflected than others, the mineral will be colored. If the mineral is anisotropic (not in the isometric crystal system) the reflectance may vary depending on the orientation of the plane-polarized light relative to crystallographic directions. These bireflectant minerals change brightness with rotation of the microscope stage in plane light. Bireflectant minerals also may display reflection pleochroism.

Observation with Crossed Polarizers

(The analyzer is inserted into the optical path.)

Isotropism versus Anisotropism

Isotropic opaque minerals crystallize in the isometric system. They remain uniformly dark or black with rotation when viewed between crossed polarizers regardless of crystal orientation. Hexagonal and tetragonal minerals whose c axis is vertical also will show the same effect. Plane-polarized light impinging on the surface of an anisotropic opaque mineral with be reflected as two plane-polarized rays that vibrate at right angles to each other, analogous to the way light is split into two rays on passing through a transparent anisotropic mineral. When this light reaches the analyzer, it can be resolved into components that pass through and reach the observer. Anisotropic minerals (all but isometric system) therefore may show variations in the intensity of color or illumination with rotation for most orientations. The strength of the anisotropy is described as ranging from very weak to very strong.

Polarization Colors

The polarization color is the color observed between crossed polarizers; in a crude way, it is analogous to interference colors seen with transparent minerals.

Internal Reflections

If the mineral is not entirely opaque, light may pass some distance into the mineral and then be reflected back to the surface by a crack or other imperfection. Internal reflections therefore show as small areas that are somewhat brighter than their surroundings. Only minerals with some degree of transparency may show them. Normally transparent minerals (carbonates, silicates, etc.) usually display internal reflections but low reflectance.

Selected properties of opaque minerals are listed in Appendix B and are included with the mineral descriptions in Chapters 18 through 20.

TACTICS FOR MINERAL IDENTIFICATION

Rapid identification of minerals with a petrographic microscope requires a systematic approach tempered with common sense and familiarity with a variety of common rocks and minerals. With practice, optical data can be rapidly measured, but because of small grain size, lack of grains in appropriate orientations, alteration, or other complications, it may not be possible to measure all the optical properties that might be desired. For routine work, however, identification often can be made with incomplete data, and for many minerals only selected data are needed to confirm a tentative identification.

To identify and describe minerals optically, proceed as follows.

- Examine the hand sample of the mineral to determine as many of the following characteristics as possible: color, luster, streak, hardness, cleavage/fracture, specific gravity, mineral habit. Provide a tentative identification or list of possibilities based on this information.
- Based on the identity of associated minerals, rock type, or type of mineral deposit, modify the tentative list of minerals that the unknown might be.

Appendix C lists common mineral associations in different rock types and mineral deposits.
- Prepare a thin section, grain mount, or polished section.

Thin Section Identification

Thin sections are most valuable when rock textures and mineral intergrowth relations are of interest. Indices of refraction can only be estimated based on relief, and measurement of $2V$ is approximate. A thin section takes substantially more time to prepare than grain mounts, but the time is often well spent. Proceed as follows.

- Scan the slide to examine different grains of the unknown mineral. Color, relief, twinning, crystal shape, textures, and alteration usually provide the basis for distinguishing different minerals. Cross and uncross polarizers, and rotate the stage as needed. Record the following information.
  a. Color and pleochroism (if any)
  b. Relief relative to cement
  c. Mineral habit, textures, and alteration
  d. Whether the mineral is isotropic or anisotropic
  e. Nature of twinning, if present
  f. Nature of cleavage and/or fracture
- If isotropic, go to the identification tables (Appendix B) and mineral descriptions.
- If anisotropic:
  a. Scan the slide to find a grain of the unknown displaying the lowest interference color. Obtain an interference figure.
     1. If uniaxial: (a) Determine optic sign. (b) Return to orthoscopic illumination (plane light) and record the color associated with ω and the relief associated with n_ω.
     2. If biaxial: (a) Determine optic sign, $2V$, and dispersion characteristics, if any. (b) Rotate the stage to place the optic plane at right angles to the lower polarizer direction. Return to orthoscopic illumination (plane light) and observe the color associated with Y and record the relief of the mineral associated with n_β. Check the Becke line to determine if n_β is greater or less than the index of refraction of cement.
  b. Scan the slide to find a grain of the unknown displaying the highest interference color.
     1. Determine maximum birefringence based on interference color and thin section thickness.
     2. If the mineral is elongate or has cleavage, measure the extinction angle and determine sign of elongation.
     3. Record the color and relief associated with n_ϵ (uniaxial) or n_α and n_γ (biaxial). Use the accessory plate to distinguish the appropriate vibration directions.

4. If crystallographic directions can be recognized, determine optic orientation of biaxial minerals.

c. Go to the identification tables and mineral descriptions to determine the identity of the mineral. For minerals in thin section, the chart printed on the back side of the interference color chart (Plate 2) provides a convenient starting point. Use the other figures and tables in Appendix B as needed to help refine the possibilities. Note that Plate 2, Table B.11, and the mineral descriptions in the latter part of the text all refer to relief in thin section. It is assumed that the cement used in thin section has an index of refraction of 1.540. Some cements may have an index of refraction either higher or lower than 1.540, so relief seen with these cements may be different than reported here. If in doubt about a cement's index of refraction, consult the manufacturer, or prepare a grain mount of fragments of cured cement and measure it by using the immersion method.

Grain Mount Identification

Grain mounts of an unknown mineral are useful because they are quicker to prepare than a thin section and they provide accurate numerical values of indices of refraction, birefringence, and related optical variables. From these data, it may be possible to estimate chemical composition or confirm an identity that was uncertain in thin section or hand sample. It is generally necessary to separate grains of the unknown mineral from others that may be in the sample prior to preparing the mount. Refer to Chapter 10 for a summary of mineral separation procedures. In handling grain mounts, be careful to avoid getting immersion oil on lenses and working parts of the microscope. To identify an unknown mineral in grain mount, proceed as follows.

- Scan the slide to observe as many of the following properties about the unknown mineral as possible: relief relative to immersion oil, whether isotropic or anisotropic, nature of twinning (if present), nature of cleavage and fracture, and alteration.

- If isotropic, use the Becke line to compare the indices of refraction of immersion oil and mineral. Use different immersion oils to prepare additional grain mounts until an index of refraction match is obtained between the mineral and oil. Use the bracketing technique described in detail earlier in this chapter.

- If anisotropic:
  a. Scan the slide to find a grain of the unknown with the lowest interference color. Obtain an interference figure.
    1. If uniaxial: (a) Determine optic sign. (b) Return to orthoscopic illumination (plane light) and record the color associated with ω and the relief associated with n_ω. Check the Becke line to determine if n_ω is greater or less than the index of refraction of the oil.
    2. If biaxial: (a) Determine optic sign, $2V$, and dispersion characteristics, if any. (b) Rotate the stage to place the optic plane at right angles to the lower polarizer direction. Return to orthoscopic illumination (plane light) and observe the color associated with Y and record the relief of the mineral associated with n_β. Check the Becke line to determine if n_β is greater or less than the index of refraction of oil.
  b. Scan the slide to find a grain of the unknown displaying the highest interference color.
    1. If the mineral forms elongate fragments because of cleavage, measure the extinction angle and determine sign of elongation.
    2. Record the color and relief associated with n_ϵ (uniaxial) or n_α and n_γ (biaxial). Use the accessory plate to distinguish the appropriate vibration directions.
  c. Prepare additional grain mounts to find index of refraction matches for n_ω and n_ϵ (uniaxial) or n_α, n_β, and n_γ (biaxial) following the procedures described in detail earlier.
  d. Go to the identification tables (Appendix B) and mineral descriptions to identify the unknown.

Polished Section Identification

Proficiency in identification of minerals with a standard reflecting light microscope depends on gaining experience based on having looked at many samples. Unlike the situation with transmitted-light optics, in which it is possible to make quantitative measurements, most of the items to be observed in reflective light are qualitative. Begin by scanning the polished section both in plane light and with crossed polarizers to note intergrowth textures, colors, alteration, or other features that distinguish among different opaque minerals in the sample. Then proceed systematically to observe the following for each mineral.

- In plane light (analyzer removed):
  a. Color of the mineral
  b. Reflectance. Try to distinguish among very low (like quartz or epoxy), low (like magnetite), moderate (like galena), and high (like pyrite)
  c. Bireflectance and reflectance pleochroism
- With crossed polarizers (analyzer inserted):
  a. Isotropism versus anisotropism
  b. Polarization colors
  c. Internal reflections

Based on these observations, refer to the tables in Appendix B and the mineral descriptions to work out the identification.

REFERENCES CITED AND SUGGESTIONS FOR ADDITIONAL READING

Bloss, F. D. 1961. *An Introduction to the Methods of Optical Crystallography.* Holt, Rinehart & Winston, New York, 294 p.

Bloss, F. D. 1981. *The Spindle Stage: Principles and Practice.* Cambridge University Press, Cambridge, 340 p.

Craig, J. R., and Vaughn, J. D. 1994. *Ore Microscopy and Ore Petrography,* 2nd ed. John Wiley & Sons, New York, 434 p.

Dyer, M. D., Gunter, M. E., and Tasa, D. 2008. *Mineralogy and Optical Mineralogy.* Mineralogical Society of America, Chantilly, VA, 708 p.

Humphries, D. W. 1992. *The Preparation of Thin Sections of Rocks, Minerals, and Ceramics.* Oxford University Press, Oxford, 83 p.

Ineson, P. R. 1989. *Introduction to Practical Ore Microscopy.* Longman Scientific & Technical, Essex, England, 181 p.

Ixer, R. A. 1990. *Atlas of Opaque and Ore Minerals in Their Associations.* Van Nostrand Reinhold, New York, 208 p.

Mertie, J. B., Jr. 1942. Nomograms of optic angle formulae. American Mineralogist 27, 538–551.

Nesse, W. D. 2002. *Introduction to Optical Mineralogy*, 3rd ed. Oxford University Press, New York, 348 p.

Schulze, D. J. 2004. *An Atlas of Minerals in Thin Section.* Oxford University Press, New York. CD-ROM.

CHAPTER 8

Introduction to X-Ray Crystallography

INTRODUCTION

The application of X-rays to the study of minerals has probably been the single most important technological advance in mineralogy. The chemical, physical, and optical properties of minerals indicate that mineral structures consist of regular and repeating arrays of atoms, but do not elucidate what those structures actually are. Study of the interaction of X-rays with mineral structures can, however, allow the detailed arrangements of atoms/ions to be determined. From that knowledge, coupled with knowledge of the chemical composition, has flowed an ever-expanding understanding of the physical and chemical behaviors of minerals in the geological environment, and therefore of the geological processes that have shaped the Earth.

X-ray diffractometers (Figure 8.1) are widely used in the routine analysis and identification of mineral samples. The diffractometer consists of an X-ray tube, a sample holder, and an X-ray detector. With proper instrumentation and techniques, this instrument can be used to identify small samples of an unknown mineral and may be used with mineral samples such as clays and zeolites that are much too fine grained to be examined with hand sample or optical microscopic techniques.

Identification of minerals is routinely accomplished with the **powder method** that is described in this chapter. Sample preparation can be accomplished rapidly and a typical X-ray diffraction run can be done in 20 minutes or less.

Crystal structure determinations require techniques beyond the scope of this book, such as using well-crystallized single crystals and various sample holders and detector systems that allow precise orientation of the sample. Interested readers are encouraged to peruse the references at the end of the chapter for more information on the techniques of crystal structure determination.

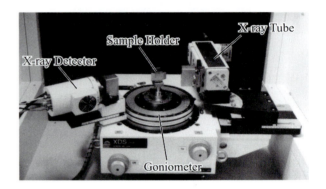

Figure 8.1 X-ray diffractometer. The principal components are an X-ray tube, sample holder, and an X-ray detector. Sample and detector are mounted on the goniometer, which allows both to be synchronously rotated through angles θ and 2θ respectively. Additional components include power supplies, controls, and electronics to allow the output of the detector to be recorded and displayed. Because X-rays can be very hazardous, an enclosure, shields, and other safety devices (not shown) are provided to prevent the operator from being exposed to X-rays. Photo courtesy of Scintag, Inc.

X-RAYS

X-rays are part of the continuous electromagnetic spectrum (Figure 6.6) whose wavelengths are between 0.1 and 10 angstroms (Å; 1 Å = 0.1 nm = 10^{-10} m). X-rays are generated when a stream of high-energy electrons strikes a material. Both the energy of the electrons and the nature of the material that they strike determine the wavelength(s) of X-rays that are emitted.

X-Ray Generation

In a conventional X-ray diffractometer, X-rays are generated in a cathode ray tube (Figure 8.2). Free electrons are produced by heating a filament called the cathode (similar to the filament in a conventional incandescent electric light bulb) to a high temperature. These negatively charged

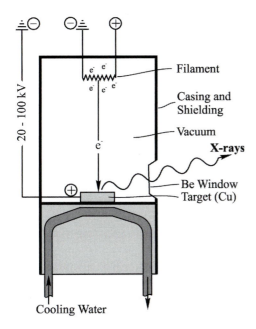

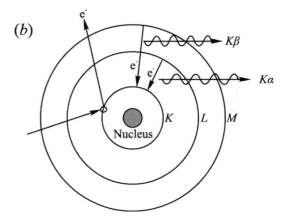

Figure 8.2 X-ray tube. The cathode filament is heated so that it boils off electrons. A voltage of 20 to 100 kV is maintained between the filament and the target (the anode) to accelerate the electrons toward the target. When these high-energy electrons strike atoms in the target, X-rays are generated and exit through a thin beryllium window in the tube. The target is most commonly copper or molybdenum.

electrons are accelerated to high energy by applying a voltage of several tens of kilovolts between the filament and the target or anode. The energy (eV) of each electron in the stream is its charge e times the accelerating voltage V. If the electrons have sufficient energy, both continuous and characteristic spectra of X-rays (Figure 8.3) are produced when the electrons strike atoms of the target.

Continuous Spectrum

The **continuous spectrum** of X-rays is produced when the electrons strike the target and are slowed or halted without changing the electron configuration of atoms in the target. As an electron comes to a halt through a series of one or more collisions, the energy it contained is released as electromagnetic radiation that includes a wide range of wavelengths and a substantial amount of heat. The accelerating voltage (V, kilovolts) determines the shortest wavelength (λ_s, angstroms) and highest-energy X-rays that can be produced:

$$\lambda_s = \frac{12.4}{V}. \tag{8.1}$$

The continuous spectrum is equivalent to background noise and must be removed by filtering or other means.

Characteristic Spectrum

The **characteristic spectrum** is produced when incoming high-energy electrons dislodge a tightly held electron in the

Figure 8.3 X-ray spectra. (*a*) Intensity of X-rays emitted by a copper target operated at 50 kV. The characteristic spectrum (peaks) is superimposed on the continuous spectrum. (*b*) The characteristic spectrum is produced when electrons are dislodged from the *K* shell and electrons from outer shells drop in to occupy the vacancy.

innermost *K* shell of the target atoms. Almost immediately, an electron from an outer, higher energy shell "drops" in to replace the missing *K*-shell electron, with the emission of radiation whose energy is equal to the energy difference ΔE between the outer shell and the *K* shell. The wavelength (in angstroms) of the emitted X-rays is

$$\lambda = \frac{ch}{\Delta E} = \frac{12.4}{\Delta E} \tag{8.2}$$

where c is the speed of light and h is Planck's constant. As Figure 8.3b shows, electrons most commonly drop in from the *L* and *M* shells to produce characteristic wavelengths known as *Kα* and *Kβ*. By Equation 8.2, *Kβ* radiation must have a shorter wavelength than *Kα* because ΔE for the *M–K* transition is greater than for the *L–K* transition. The *Kα* peak actually consists of six slightly different components but, for practical purposes, can be considered to be composed of two different wavelengths, $K\alpha_1$ and $K\alpha_2$, because two slightly different energy levels are present in the L shell from which electrons drop into K. $K\alpha_1$ has a slightly shorter wavelength than $K\alpha_2$ and about twice the intensity. The combined $K\alpha$ peak is typically about ten times as intense as $K\beta$.

Monochromatic X-rays

Effective use of X-rays in an X-ray diffractometer requires that approximately monochromatic (single-wavelength) X-rays be used. Because the $K\alpha$ peak has the greatest intensity, it is most commonly used, and $K\beta$ and much of the continuous spectrum must be removed. This is traditionally done either by filtering or by passage through a monochrometer. Filters are pieces of metal foil that absorb some wavelengths of X-rays and allow others to pass. A piece of nickel foil is usually used with copper radiation to eliminate the $K\beta$ peak and much of the continuous spectrum. A monochrometer is a carefully oriented crystal plate whose interatomic spacing is appropriate to diffract and pass along the $K\alpha$ radiation but not other wavelengths. (See later discussion of diffraction.) A third strategy is to use a solid-state detector (described shortly) that is capable of distinguishing the different wavelengths and setting the device so that it records only a narrow window of energy/wavelengths that includes the desired $K\alpha$ radiation.

The X-rays being used in a diffractometer are never strictly monochromatic. The $K\alpha$ radiation coming from the target actually consists of two closely spaced wavelengths, $K\alpha_1$ and $K\alpha_2$ as just described, and each has a range of wavelengths that overlap, making it difficult to isolate just the $K\alpha_1$ peak, even with solid-state detectors. Further, it is important to have a reasonably high intensity of X-rays. Even if it were possible to make purely monochromatic X-rays by selecting only one wavelength from the spectrum coming from the X-ray tube and excluding all the wavelengths on either side, the intensity would be far too low to be used.

The characteristic spectrum, as the name implies, is characteristic to each different element. Of the elements employed as the target in X-ray tubes used for X-ray diffractometry (Table 8.1), copper is the most common. The weighted average wavelength of the two peaks $K\bar{\alpha}$ is typically used in calculations. The weighted average wavelength of the $CuK\alpha_1$ and $CuK\alpha_2$ radiation is $CuK\bar{\alpha} = 1.5418$ Å.

X-Ray Detection

X-rays may be detected either photographically or electronically. In the early and middle years of the twentieth century, photographic film was widely used to document the diffraction of X-rays by minerals. These photographic technologies are now largely obsolete, having

Table 8.1 Characteristic Wavelengths (Å) of Metals Commonly Used as Targets in X-Ray Tubes[a]

	Metal				
	Mo	**Cu**	**Co**	**Fe**	**Cr**
$K\beta$	0.63225	1.38217	1.62073	1.75653	2.08479
$K\alpha_1$	0.70926	1.54051	1.78892	1.93597	2.28962
$K\alpha_2$	0.71354	1.54433	1.79279	1.93991	2.29351
$K\bar{\alpha}$	0.7107	1.5418	1.7902	1.9373	2.2909

[a] $K\bar{\alpha}$ is the weighted average of $K\alpha_1$ and $K\alpha_2$.

been almost entirely replaced by electronic detection technologies.

The commonly used electronic detectors in X-ray diffractometers are scintillation counters, gas-proportional counters, and solid state devices. In a scintillation counter, a crystal plate called a scintillator emits a brief flash of light when struck by an X-ray photon. This flash of light is electronically detected and yields a brief spike of electrical current that is recorded as a count. Gas-proportional counters consist of a hollow metal tube filled with an inert gas. A voltage of ~1.5 kV is maintained between the tube and a thin wire along its axis. X-ray photons passing through the detector produce ions of the inert gas by dislodging electrons. The electrons move to the positively charged wire while the ions move to the case. The resulting pulse of electric current is recorded as a count. Solid-state detectors, which are based on semiconductor technology, convert the energy of X-ray photons that strike the detector directly to an electrical pulse. Their operation requires very cold temperatures, provided either by liquid nitrogen or a solid-state Peltier device. All these detectors are very sensitive and are designed to make it easy to feed their output directly to a computer for immediate analysis and plotting.

Neither photographic nor electronic X-ray detectors provide a direct image showing the position of individual atoms in a mineral. The atoms are far too small for this to be possible. Rather, the regularly spaced atoms act to diffract the X-rays and produce a characteristic distribution of X-rays after passing through the sample.

X-RAY DIFFRACTION

The wavelength of X-rays used in X-ray diffractometers is around 1 or 2 Å, which is similar to the spacing of atoms in the structure of most minerals. The similarity in dimensions means that X-rays are diffracted by the regularly spaced atoms that comprise a crystal. Based on an analysis of this diffraction in three dimensions, it can be shown that the layers of atoms produce diffraction maxima that effectively "reflect" the incident X-rays if the angle of incidence θ is appropriate. Consider Figure 8.4. Rays 1 and 2 are incident

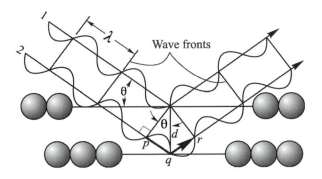

Figure 8.4 Diffraction from parallel planes of atoms (not shown where the X-ray rays are drawn). Rays 1 and 2 impinge on layers of atoms that are distance d apart. Constructive interference ("reflection") of the X-rays diffracted from these layers is possible only if the angle of incidence θ is such that the distance pqr is an integer number of wavelengths.

to the layers of atoms at angle θ. Note that the angle θ is defined differently than in the discussion of reflection and refraction of light in Chapters 6 and 7. The planes of atoms are spaced distance d apart. Diffracted X-rays from each plane of atoms will be in phase only if the angle θ is such that the additional distance pqr traveled by wave 2 is equal to an integer number of wavelengths

$$pqr = n\lambda \tag{8.3}$$

where n is an integer and λ is the wavelength of the X-rays. The distance pqr is twice distance pq, which is related to the interplanar spacing d by the equation

$$pq = d\sin\theta \tag{8.4}$$

or

$$pqr = 2pq = 2d\sin\theta. \tag{8.5}$$

Combining Equations 8.3 and 8.5 yields

$$n\lambda = 2d\sin\theta. \tag{8.6}$$

This relationship is the **Bragg equation**. The references given at the end of the chapter offer a more complete derivation and explanation of diffraction.

In Figure 8.4, the distance pqr, which produces constructive interference between rays 1 and 2, is equal to one wavelength. However, constructive interference also will occur when pqr is equal to any integer number of wavelengths. For this reason, the integer n is included on the left side of the Bragg equation.

For example, the {111} planes of atoms in halite (NaCl) have an interplanar spacing $d_{111} = 3.2555$ Å. If CuKα radiation is used, $\lambda = 1.5418$ Å. Solution of the Bragg equation for $n = 1$ yields a reflection at $\theta = 13.70°$. This is a **first-order reflection** because $n = 1$. For $n = 2$, a **second-order reflection** is produced at $28.27°$. **Third- and fourth-order reflections** for $n = 3$ and $n = 4$ are at $45.27°$ and $71.30°$, respectively. Reflections higher than the fourth order are not possible for {111} because θ cannot exceed $90°$.

Reflection of X-rays from planes of atoms as a result of diffraction in a crystal structure is possible only if the grain happens to be correctly oriented. For an unknown mineral whose d-spacings are unknown, the chances of placing a single grain in the X-ray beam in an orientation to yield a reflection are about the same as winning the lottery. Further, each mineral has many different atomic planes ({100}, {010}, {001}, {111}, {110}, etc.) that potentially can diffract X-rays. A grain oriented to reflect X-rays from one set of atomic planes cannot also reflect X-rays from any of the other planes that can diffract the X-rays. To measure all the different d-spacings in a mineral with a single crystal, it would be necessary to systematically rotate the sample and detector so that a wide range of orientations is sampled. Apparatuses to do this are available, but the procedure involved is complex and exacting. A much simpler solution for routine work is to provide many grains in a myriad of different orientations, which is the basis of the powder method.

POWDER METHOD

The powder method is a powerful and convenient technique for identifying minerals. It is not routinely used to determine the details of crystal structures; single crystal methods are typically better. However, the **Rietveld refinement method** provides detailed structural information on minerals such as clays and zeolites that usually occur in very-fine-grained aggregates and for which single crystals of sufficient size are not available.

Sample Preparation

The sample used with the powder method consists of a finely ground powder prepared by crushing a few tenths of a gram or more of the unknown mineral in a mortar. Relatively pure samples of the unknown are preferred, but mixtures of minerals also may be analyzed. Grains that pass through a 200-mesh (< 0.074 mm) or 400-mesh (< 0.037 mm) sieve are adhered to a glass microscope slide, packed into a well in a plastic or metal sample holder, or otherwise mounted on a holder of similar size (Figure 8.5). Grains in the sample should be as randomly oriented as possible for most identification purposes. For certain analyses of clay minerals, however, the single prominent cleavage needs to be aligned parallel to the surface of the sample holder.

The sample is mounted in the X-ray diffractometer on a holder that pivots relative to the X-ray tube to allow the angle of incidence θ of the X-ray beam to be varied from zero up to nearly $90°$. The X-ray detector is mounted on a concentric goniometer that moves at twice the angular speed between sample and X-ray tube, so that the detector is at the angle 2θ required to record reflections.

Instrumental Output

For routine identification work using CuKα radiation, the detector is run through a 2θ angle of between $5°$ and

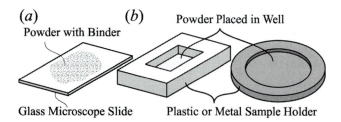

(a) Powder with Binder — Glass Microscope Slide

(b) Powder Placed in Well — Plastic or Metal Sample Holder

Figure 8.5 Sample holders for routine powder diffraction work. (*a*) Mineral powder mounted on 26 × 47 mm glass microscope slide. A small amount of acetate dissolved in acetone helps the powder adhere to the slide. Other materials also may be used to stick the powder to the slide. (*b*) Powder is placed into a well in a plastic or metal mount. Sample holders of other types may be used, depending on the design of the instrument. Some instruments have an apparatus that allows multiple samples to be loaded and run sequentially.

around 70°. This allows the instrument to detect reflections from atomic planes in a mineral whose *d*-spacing ranges from about 17.7 to 1.34 Å. Larger 2θ angles generally pick up only higher-order reflections or minor reflections from atomic planes with Miller indices containing relatively large integers. Smaller angles may be used if desired, but diffractometers should not be set to 0° because the intense beam coming directly from the X-ray tube may damage some detectors.

The sample and detector are rotated through the desired range of θ and 2θ angles, respectively (Figure 8.6a), and the intensity of the reflected X-rays is continuously recorded electronically in a computer. A peak in the intensity of the diffracted X-rays indicates that the mineral possesses atomic planes whose *d*-spacing is appropriate to reflect X-rays for that particular angle θ by Equation 8.6. Because

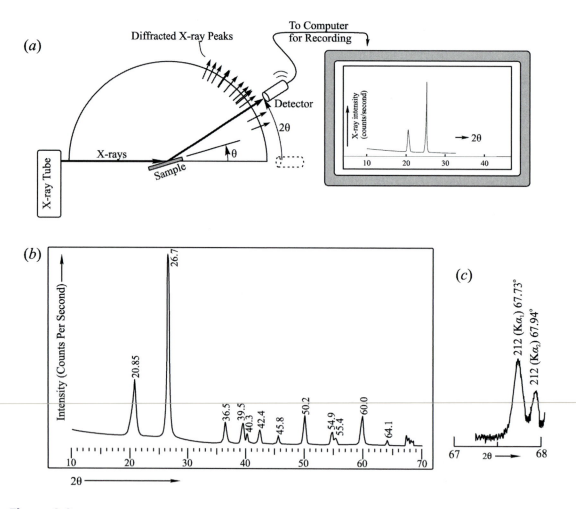

Figure 8.6 Running a powder diffraction record of quartz. (*a*) As the detector moves through its range of 2θ, the sample moves through angle θ, and intensity of the diffracted X-rays is recorded as a time series in a computer and displayed on a screen. Each reflection from the mineral is recorded as a peak on the screen at its 2θ position. (*b*) On the output, the horizontal axis is the 2θ angle and the vertical scale is the intensity of diffracted X-rays, usually in terms of counts per second. (*c*) Because *Kα* radiation consists of two different peaks, $K\alpha_1$ and $K\alpha_2$, each atomic plane actually produces two reflections. The peak for the $K\alpha_1$ wavelength has higher intensity than for $K\alpha_2$. At small values of 2θ, the $K\alpha_2$ reflection may show as a slight bulge on the side of the larger $K\alpha_1$ peak. At larger 2θ, the peaks can be distinguished. The $K\alpha_2$ peak will be on the larger 2θ side of the $K\alpha_1$ peak, as is shown for reflections from the {212} plane in quartz.

the powdered sample contains grains in all orientations, each of the different sets of atomic planes that are capable of diffracting X-rays will (with luck) produce its X-ray peak at the appropriate angle as the detector is moved through its range of 2θ. The intensity of the diffracted X-rays for each peak also is controlled by mineral structure. Some atomic planes are more effective than others at reflecting X-rays.

A diffractometer record for quartz is shown in Figure 8.6b. The peak for $K\alpha_2$ may show as a slight bulge on the side of larger peaks for small and intermediate values of 2θ; otherwise, however, it is not distinguished because the $K\alpha_1$ and $K\alpha_2$ wavelengths are quite close. At large values of 2θ (Figure 8.6c), two peaks may be readily identified, one for $K\alpha_1$ and another at slightly higher 2θ and about half the intensity for $K\alpha_2$. With quartz, for example, the {100} peak occurs at $26.64°$ for $CuK\alpha_1$ and $26.70°$ for $CuK\alpha_2$, a difference of only $0.06°$. However, the peak for {212} occurs at $67.73°$ for $CuK\alpha_1$ and $67.94°$ for $CuK\alpha_2$, a more readily observed difference of $0.21°$. A well-aligned and properly instrumented X-ray diffractometer will be able to distinguish these peaks. As long as the two peaks show a substantial degree of overlap, the combined peak is treated as a single entity and the d-spacing is calculated by using the weighted average of the $K\alpha$ radiation (Table 8.1).

Data Reduction

The data recorded from diffractometer record, are the 2θ angle and the intensity of each peak (Table 8.2). The 2θ angle is conventionally chosen to be the arithmetic center of the peak at a point either one-half or two-thirds of the way up between its base and top (Figure 8.7). The intensity measure (I) may be either **peak height intensity** or **integrated intensity**. Peak height intensity is simply the height of the peak above the background level. Integrated intensity is the area of the peak, equivalent to the total number of counts that produced it. For manual methods, the area of a peak can be approximated by assuming that it is an isosceles triangle. Computer systems can account for the actual peak shape and can calculate areas with considerable precision.

From these data, d-spacings and relative intensity values are calculated. The d-spacing for each peak is calculated from Equation 8.6 assuming $n = 1$. Relative intensity is the ratio of the intensity of the peak (I) divided by the intensity of the highest peak (I_1).

$$\text{Relative intensity} = I / I_1 \times 100 \qquad (8.7)$$

Because each mineral has its own structure and unit cell dimensions, each has its own set of atomic planes at specific d-spacings that is capable of diffracting X-rays. Identification of an unknown mineral is therefore based on comparison of the set of d-spacings and the reflection

Table 8.2 Data Reduction from X-Ray Diffraction Record[a]

2θ	I	I/I_1	d(Å)
20.9	2.8	31	4.25
26.7	9.0	100	3.34
36.5	1.0	11	2.46
39.5	1.0	11	2.28
40.3	0.4	4	2.24
42.4	0.7	8	2.13
45.8	0.4	4	1.98
50.2	1.4	16	1.82
54.9	0.6	7	1.67
55.4	0.3	4	1.66
60.0	1.4	16	1.54
64.1	.3	4	1.45

[a] The values of 2θ and I are recorded for each peak from the X-ray diffraction record. The value of d is calculated from Equation 8.6 and the value of I/I_1 from Equation 8.7, where I_1 is the intensity of the highest peak.

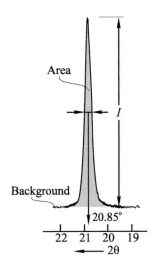

Figure 8.7 Recording data from the diffractometer record: 2θ is taken as the center of the peak measured halfway from the base to the peak. Peak height (I) is the height of the peak above the background level. Integrated peak height is the area of the peak (shaded).

intensities of the unknown sample with d-spacing and intensity data for known minerals.

Powder Diffraction File

X-ray diffraction data for minerals and other compounds are compiled by the International Centre for Diffraction Data (12 Campus Boulevard, Newtown Square, PA 19073–3273, USA). This compilation, known as the **Powder Diffraction File** (PDF), is available in most university and industrial laboratories involved with X-ray diffraction and has been published in different formats. The

5-0490 MINOR CORRECTION

d	3.34	4.26	1.82	4.26	SiO₂	
I/I₁	100	35	17	35	Silicon Oxide	Quartz, Low

Rad. CuKα, λ 1.5404 Filter Ni				
Dia. Cut off Coll				
I/I, G.C. Diffractometer d corr. abs?				
Ref. Swanson and Fuyat, NBS Circular 539, Vol. III (1953)				

Sys. Hexagonal   S.G. D₃⁴ - P3,21
a₀ 4.913   b₀   c₀ 5.405 A   C 1.10
α   β   γ   Z 3
Ref. Ibid.

nα   nβ 1.5448 nγ 1.533   Sign +
2V   D₃ 2.647 mp   Color
Ref. Ibid.

Sample from Lake Toxaway, N.C. Spect. Anal.:
<0.01% Al; <0.001% Ca,Cu,Fe,Mg.
X-ray pattern at 25°C
3-0427, 3-0444
Replaces 1-0649, 2-0458, 2-0459, 2-0471, 3-0419

d Å	I/I₁	hkl	d Å	I/I₁	hkl
4.26	35	100	1.228	2	220
3.343	100	101	1.1997	5	213
2.458	12	110	1.1973	2	221
2.282	12	102	1.1838	4	114
2.237	6	111	1.1802	4	310
2.128	9	200	1.1530	2	311
1.980	6	201	1.1408	<1	204
1.817	17	112	1.1144	<1	303
1.801	<1	003	1.0816	4	312
1.672	7	202	1.0636	1	400
1.659	3	103	1.0477	2	105
1.608	<1	210	1.0437	2	401
1.541	15	211	1.0346	2	214
1.453	3	113	1.0149	2	223
1.418	<1	300	0.9896	2	402,115
1.382	7	212	.9872	2	313
1.375	11	203	.9781	<1	304
1.372	9	301	.9762	1	320
1.288	3	104	.9607	2	321
1.256	4	302	.9285	<1	410

Figure 8.8 Powder Diffraction File. The first three *d* values at the top are for the three most intense peaks; the fourth is the largest *d*-spacing. Below this on the left side are the reference(s) from which the data was taken, crystallographic data, optical and physical property data, and notes. The columns on the right include the *d*-spacing (in angstroms), relative intensity ($I/I_0 \times 100$), and Bragg reflection index for the crystal plane producing the reflection. These data are listed in order of decreasing *d*-spacing. © Joint Committee on Powder Diffraction Standards—International Centre for Diffraction Data.

simplest consists of a collection of paper 3 × 5 inch cards (Figure 8.8), or equivalent images on microfiche. On each card the *d*-spacing data are systematically presented along with selected crystallographic, physical property, and optical data. Books containing facsimile images of the same cards also have been published. Because computer-based systems are widely used with X-ray diffraction, the same compilation of data is readily available in digital form, can be purchased on CD-ROM and other media, and is updated regularly. Data from the PDF are usually included as part of commercially available software used for mineral identification and analysis.

Bragg Reflection Indices

The values shown in the hkl column on the PDF cards are **Bragg reflection indices**. They are related to the actual Miller index (*hkl*) of the reflecting surface in the mineral by the value *n* in the Bragg equation (Equation 8.6).

$$\text{Bragg reflection index} = nh \; nk \; nl \qquad (8.8)$$

In calculation of the *d*-spacings from the experimental data, it is assumed that *n* = 1. This is both reasonable and necessary. The first-order reflections are typically strongest, and until the calculations have been completed and the mineral has been identified, it cannot be known which reflections are actually first order and which are higher order. It is therefore assumed that all reflections are first order, and we accept the fact that *d*-spacings calculated this way may not actually be present in the mineral.

For example, in quartz (Figure 8.8) the {100} planes are 4.26 Å apart. If CuKα radiation is used, this plane produces a first-order reflection at $2\theta = 26.85°$, a second-order reflection at $2\theta = 42.48°$, and a third-order reflection at $2\theta = 65.87°$. If it is assumed that the latter two reflections are actually first-order reflections, then Equation 8.6 yields hypothetical *d*-spacings of 2.128 and 1.418 Å, respectively. These *d*-spacings are one-half and one-third of the actual {100} spacings that are producing the reflections and are equivalent to hypothetical {200} and {300} planes (recall that Miller indices are inversely related to the actual dimensions).

The *d*-spacings of the first three orders of reflections from the {100} plane in quartz are therefore identified as d_{100}, d_{200}, and d_{300}. The Bragg reflection indices are shown without { } or ( ) to avoid confusing them with Miller indices. The common factor, 1, 2, and 3, respectively, in the index indicates the order of the reflection. For example, d_{202} actually represents a second-order reflection from the {101} plane. If, in calculating *d*-spacings from an experimental pattern, two or more *d* values are related by factors of 2, 3, or 4, it is likely that they represent several orders of reflections from the same lattice plane in the mineral.

A complicating factor is that some minerals may have real *d*-spacings such as {200} and {300} that are one-half and one-third of the {100} *d*-spacing. These interatomic planes will produce first-order reflections at the same 2θ angle as the second- and third-order reflections from {100}.

Mineral Identification

Finding a match between the X-ray diffraction pattern of an unknown mineral and one of the thousands of minerals in the PDF file would take an intolerably long time without a systematic approach. Use of a search manual in which d-spacing data for minerals are systematically arranged greatly simplifies the process. Manuals can be obtained from the International Centre for Diffraction Data. Search manuals based on the **Hanawalt** and **Fink methods** organize powder diffraction data by using the d-spacing of peaks with highest intensities.

Hanawalt Method

With the Hanawalt method, minerals are arranged into 87 groups based on the d-spacing of the most intense peak (d_1). Within each group, minerals are arranged in descending order based on the d-spacing of the second most intense peak (d_2). Minerals with the same d_2 are ranked based on the d-spacing of the third most intense peak (d_3). Because intensity values may vary depending on sample preparation and preferred orientation, minerals are listed three times, once as just described, and the others based on ordering the three highest peaks d_2, d_1, d_3 and d_3, d_1, d_2.

Fink Method

The Fink method selects the eight peaks with highest intensity and orders them d_1 through d_8 in order of decreasing d-spacing, not intensity. In earlier versions of the method, eight entries were made in the index, one for each peak in which the d-spacings were ordered according to the following scheme

$$d_1\ d_2\ d_3\ d_4\ d_5\ d_6\ d_7\ d_8$$
$$d_2\ d_3\ d_4\ d_5\ d_6\ d_7\ d_8\ d_1$$
$$d_3\ d_4\ d_5\ d_6\ d_7\ d_8\ d_1\ d_2$$
$$\dots$$
$$d_8\ d_1\ d_2\ d_3\ d_4\ d_5\ d_6\ d_7$$

without regard to the relative intensities. In more recent versions of the Fink index, the number of entries is reduced to four by including only those entries in which the first d-spacing is for one of the four strongest peaks displayed by the mineral.

The entries in the index are broken into as many as 101 different groups, based on size of the first d-spacing in the entry. Within each group, entries are ordered based on the size of the second d-spacing in the entry, and according to the third d-spacing if two or more have the same second value.

Computer-based methods greatly speed the search for the identity of an unknown mineral, particularly if more than one mineral is in the sample. They also provide sophisticated statistical treatment of peak intensity and 2θ positions and, as a consequence can yield greater precision and accuracy.

A certain amount of caution is required in the use of any computer-based search technology. Despite expectations of some based on television and the movies, computer searches rarely provide an unambiguous identity for an unknown mineral. Rather, they often yield a list of possibilities from which to choose. That list of possibilities is strongly influenced by details of the search parameters that the operator specifies. Knowledge of the possible identity of an unknown based on physical and optical properties, rock type, associated minerals, and so forth will help greatly in determining the correct identity. Computers are extremely fast and can search through large quantities of data; but they are no better than the quality of the data provided to them, and the results they provide must be evaluated with sound geologic and scientific judgment.

Sources of Error

X-ray diffraction by means of the powder method is fairly straightforward but is subject to certain errors, including the following.

• Instrumental errors. Such errors include misalignment of the instrument and problems with sample orientation and position in the instrument. If the data is recorded on a paper strip chart rather than with a computer, human errors in placing the pen and labeling the output at the beginning of a run are always a possibility. Computer-based systems can introduce errors as a result of improper calibration and the other digital gremlins that sometimes confound their operators. Instrumental errors may, at least in part, be identified by including as part of the sample a standard such as quartz, fluorite, or another mineral whose d-spacings are accurately known.

• Sample preferred orientations. Minerals with a prominent cleavage tend to acquire a preferred orientation when mounted on or in the sample holder. Intensity values may, therefore, be significantly different than if a random grain orientation were obtained. Techniques to maximize randomness are described in references at the end of this chapter. Intensity also is greatly affected if the samples are not ground small enough; below 0.010 mm is best.

• Recording errors. Multiple runs on the same sample inevitably yield slightly different peak heights and 2θ positions. These values may not be accurately read and recorded, whether electronically or manually.

The consequences of errors at small values of 2θ are substantially greater than for large values. A 0.4° error in reading 2θ at 5° produces a 1.31 Å error in the value of d. A similar error at 50° produces only a 0.014 Å error.

An additional complication is that not all the peaks for the d-spacings shown in the PDF for a mineral will necessarily appear in the record of an X-ray diffraction run. The reasons include variations in instrumental setup and grain size of the sample, and problems with preferred orientation. The missing peaks are typically those with low relative-intensity values, but their absence can add considerable ambiguity to the identification process. A further complication is that many minerals display significant compositional variation that can influence d-spacing and intensity values. If the PDF file is for a sample whose composition is significantly different from the sample being studied, d-spacings may be different enough to prevent easy identification. Prior knowledge of the likely identity of the unknown can do a great deal to minimize these problems. Note, however, that variation in d-spacing can be used in some cases to help estimate mineral composition.

Mixed Samples

X-ray diffraction methods make it possible to identify the individual minerals in mixed rock or mineral samples. Details of the method are outlined in the search manuals referred to earlier. The diffraction pattern produced with mixed samples includes peaks for all the different minerals, with intensities that are roughly proportional to the amounts of the minerals present. The procedure is to identify one of the minerals, then remove from the list of d-spacings all that belong to that mineral. Intensity values for the remaining peaks are recalculated and used to identify subsequent minerals.

This procedure is complicated because peaks for different minerals may overlap, and weak diffraction peaks from one mineral may be masked by nearby larger peaks from another. Manual search systems generally do not work well if more than three minerals are present, but computer-based systems may be able to handle more complex mixtures. Prior knowledge of the probable mineralogy of the sample can obviously greatly simplify the procedure.

Estimation of Relative Mineral Abundance

Estimates of the relative abundance of minerals in a mixture can be based on the intensity of X-ray diffraction peaks for the different minerals. For minerals such as the clays, which typically occur in various mixtures that cannot be separated, this may provide the only means of determining the relative amounts of different species. Calibration can be difficult because X-ray intensity can depend fairly strongly on the details of sample preparation and the absorption properties of the constituent minerals. A starting point is preparation of a series of mixtures of known composition to provide a calibration based on the intensity values of one or more peaks from each mineral. If the calibration samples and the unknown are prepared under the same set of conditions, reasonable accuracy may be obtained.

Table 8.3 Relation between d_{hkl} and Unit Cell Parameters[a]

Crystal System	$\dfrac{1}{d_{hkl}^2} =$	Number of Independent d_{hkl} Measurements Needed
Isometric	$\dfrac{h^2 + k^2 + l^2}{a^2}$	1
Tetragonal	$\dfrac{h^2 + k^2}{a^2} + \dfrac{l^2}{c^2}$	2
Hexagonal	$\dfrac{4}{3a^2}(h^2 + hk + k^2) + \dfrac{l^2}{c^2}$	2
Orthorhombic	$\dfrac{h^2}{a^2} + \dfrac{k^2}{b^2} + \dfrac{l^2}{c^2}$	3
Monoclinic	$\dfrac{\dfrac{h^2}{a^2} + \dfrac{l^2}{c^2} - \dfrac{2hl\cos\beta}{ac}}{\sin\beta} + \dfrac{k^2}{b^2}$	4
Triclinic	$\dfrac{\dfrac{h^2}{a^2}\sin^2\alpha + \dfrac{k^2}{b^2}\sin^2\beta + \dfrac{l^2}{c^2}\sin^2\gamma + \dfrac{2hk}{ab}(\cos\alpha\cos\beta - \cos\gamma) + \dfrac{2kl}{bc}(\cos\beta\cos\gamma - \cos\alpha) + \dfrac{2hl}{ca}(\cos\gamma\cos\alpha - \cos\beta)}{[1 - \cos^2\alpha - \cos^2\beta - \cos^2\gamma + 2\cos\alpha\cos\beta\cos\gamma]}$	6

[a]The values h, k, and l are reflection indices; a, b, and c are unit cell dimensions; α, β, and γ are the angles between crystal axes.

Estimation of Composition

Many minerals display significant solid solution. Because the constituent ions that substitute for each other in the structure typically differ in size, the unit cell dimensions and the *d*-spacing between various atomic planes necessarily vary. Diagrams useful for estimating composition based on *d*-space variation are included with the description of selected minerals in the latter part of this book.

DETERMINING UNIT CELL PARAMETERS

The mathematical relationships between unit cell parameters and the d_{hkl} values are shown in Table 8.3. The number of independent *d*-spacing determinations required to calculate the unit cell parameters is equal to the number of unit cell parameters. With orthorhombic minerals, for example, three different *d*-spacing measurements such as d_{100}, d_{111}, and d_{010} are required to determine the *a*, *b*, and *c* unit cell dimensions, by simultaneous solution of the equation for orthorhombic minerals for each of the three different *d*-spacings and *hkl* values. Triclinic minerals have six

different unit cell parameters (*a*, *b*, *c*, *α*, *β*, *γ*) and require six different *d*-spacings. Finding the one set of *a*, *b*, *c*, *α*, *β*, *γ* values that best fits the actual diffraction data with minimum error can be a time-consuming task and is one for which computers are well suited.

REFERENCES CITED AND SUGGESTIONS FOR ADDITIONAL READING

Azaroff, L. V. 1968. *Elements of X-Ray Crystallography.* McGraw-Hill, New York, 610 p.

Bish, D. L., and Post, J. E. (eds.). 1989. *Modern Powder Diffraction. Reviews in Mineralogy*, Volume 20. Mineralogical Society of America, Washington, DC, 369 p.

Klug, H. P., and Alexander, L. E. 1974. *X-Ray Diffraction Procedures*, 2nd ed. John Wiley & Sons, New York, 966 p.

Moore, D. M., and Reynolds, R. C., Jr. 1997. *X-Ray Diffraction and the Identification of Clay Minerals.* Oxford University Press, Oxford, 378 p.

Pecharsky, V. K., and Zavalij, P. Y. 2009. *Fundamentals of Powder Diffraction and Structural Characterization of Materials.* Springer, New York. 741 p.

Suryanarayana, C., and Norton, M. G. 1998. *X-ray Diffraction, a Practical Approach.* Plenum Press, New York. 273 p.

CHAPTER 9

Chemical Analysis of Minerals

INTRODUCTION

To understand any mineral, it is essential to know the chemical composition. The common analytical tools to determine composition are wet chemical methods, electron probe microanalysis, scanning electron microscopy, X-ray fluorescence (XRF) spectroscopy, and mass spectrometry. An ever-increasing list of analytical tools also includes atomic absorption spectrometry, Raman spectroscopy, nuclear magnetic resonance spectroscopy, neutron activation analysis, neutron scattering analysis, optical emission spectroscopy, plasma emission spectroscopy, and synchrotron radiation analysis. For additional information, consult the references at the end of this chapter. Many commercial and research laboratories are equipped to provide chemical analyses of rock and mineral samples at remarkably modest costs.

This chapter summarizes a few of the analytical techniques that are of particular relevance to mineralogy and describes how raw chemical analyses are converted into the structural formulas that we normally associate with each mineral.

ANALYTICAL METHODS

Wet Chemical

Most science students get some exposure to wet chemical techniques as part of introductory chemistry laboratory activities in high school and college. The analytic techniques are diverse, but all begin with dissolving or digesting a sample in a solvent, which for geological materials may be an acid such as HCl, HNO_4, or HF. Once the sample has been dissolved, various reagents are added to allow gravimetric, colorimetric, titration, and related analyses of the different elements. Jeffery (1975) describes many standard techniques. Gravimetric techniques involve precipitating elements as part of insoluble compounds, so that their mass can be measured. With colorimetry, reagents are added that combine with the element being analyzed to produce compounds that color the solution. The abundance

of the element in question is determined by comparing the color of the sample solution with the color of solutions with known composition. Titration determines the abundance of a substance in the solution by carefully measuring how much reagent is required to produce complete reaction between the reagent and the substance being analyzed.

Wet chemical procedures are generally slow and require great skill to produce high-quality results. Further, the analysis represents an average of all of the material digested. Distinguishing the composition of different parts of a zoned mineral grain is not possible, for example, nor do wet chemical techniques differentiate between a mineral and any fine inclusions that it may contain.

Electron Probe Microanalysis

Electron probe microanalysis is performed with an instrument conventionally called an **electron microprobe**. It is one of the most widely used instruments for determining the chemical composition of minerals. It requires a very small sample and can determine the composition of specific spots within mineral grains. Inclusions of other minerals can be avoided, and the composition variation within a mineral grain can be studied. Unfortunately, the apparatus is expensive, and it requires substantial maintenance support.

Various sample configurations can be used, but one of the most common was shown earlier: a conventional petrographic thin section (Figure 7.1a) prepared without a coverslip. The surface of the sample is carefully polished and then coated with a conductive material, usually a carbon or gold, so electrons that strike the sample are conducted away.

The geometry of an electron microprobe is shown in Figure 9.1. It is basically a cathode ray tube that is evacuated after the sample has been inserted. A tungsten filament is strongly heated and boils off loose electrons. These electrons are accelerated toward the anode plate by a voltage of several tens of kilovolts. A hole is provided in the anode plate so that some of the high-energy electrons pass through and strike the sample. Lenses focus the electrons into a fine beam, using electromagnetic fields produced by electrical coils strategically mounted in the instrument. The precise spot on the sample being struck by the electron beam can be observed with the aid of an optical microscope built into the instrument, and the sample can be moved beneath the beam to accurately position it. Many modern instruments are equipped to handle multiple samples. Multiple spots on each sample can be selected by the operator, and the machine can then be programmed to conduct the analyses automatically—overnight, for example. This capability greatly speeds data acquisition and analysis and removes much of the labor that used to be involved with these instruments.

When the electron beam strikes the sample, X-rays fluoresce in a manner exactly analogous to the generation of X-rays in an X-ray tube (Figures 8.2 and 8.3). When the beam of electrons strikes the sample, some electrons lose their energy by striking atoms in the sample and produce the continuous spectrum. Other incident electrons strike and dislodge electrons in the K shells of atoms in the sample. Electrons from outer shells immediately drop into the vacant K shell sites, with emission of the characteristic X-ray spectrum as described in Chapter 8.

Each of the different elements in the sample emits its own characteristic spectrum with an intensity proportional to the amount of the element present. The abundance of an element in a mineral sample is, therefore, determined based on the intensity of the characteristic X-ray spectrum for that element. The instrumental response is calibrated by using minerals or other materials of known composition. X-rays are detected by using either wavelength-dispersive or energy-dispersive techniques.

A wavelength-dispersive spectrometer, which is able to detect the characteristic $K\alpha$ wavelength for a single element at a time, has a geometry similar to an X-ray diffractometer. X-rays coming from the sample are diffracted by a diffracting crystal whose d-spacing is known. The diffracting crystal is arranged at angle θ (Equation 8.6) required to diffract the $K\alpha$ wavelength of the element being detected. The detector is positioned at 2θ. The detector, usually a scintillation counter or solid-state detector like that used with X-ray diffraction, records the intensity of the characteristic wavelength and sends the output to a computer for processing and display. To detect all the different elements in the sample, the diffracting crystal and detector must be reset for each different $K\alpha$ wavelength. Instruments usually are equipped with several wavelength-dispersive spectrometers to allow analysis of several elements at once.

The energy-dispersive technique uses a single solid-state detector that records all the X-rays coming from the sample, but distinguishes among the different energy levels (wavelengths) present. A computer parses these data and distinguishes the characteristic spectra of each of the different elements present in the sample. The energy-dispersion technique is much faster and more convenient than the wavelength-dispersion technique but not quite as accurate.

A limitation of the electron microprobe is that it works best for elements whose atomic numbers are 11 (sodium) and higher, although some instruments are capable of detecting elements as low as fluorine ($Z = 9$). It is not particularly good at detecting trace amounts of elements and does not distinguish between different oxidation states of the same element such as Fe^{2+} and Fe^{3+}. It does, however, do an admirable job on the major elements that comprise most minerals and routinely detects elements down to concentrations on the order of 0.1 to 0.01 weight percent.

Scanning Electron Microscopy

A scanning electron microscope (SEM) is functionally very much like an electron microprobe in its design and function, and except for a different detector, the major components are shown in Figure 9.1. A sample can be a small fragment of rock or mineral. A coating of carbon or gold is applied to carry away electrons from the electron beam that strikes the sample. The detector in an SEM records electrons emitted from the sample rather than the X-rays that are detected with an electron microprobe.

In addition to the X-rays generated when the electron beam strikes a sample, electrons are emitted from the surface. Some of these are relatively high-energy

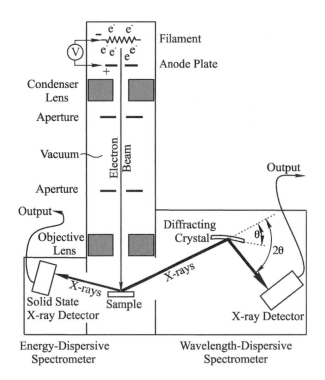

Figure 9.1 Schematic diagram showing major components of an electron microprobe. The output of both wavelength-dispersive and energy-dispersive spectrometers is sent to a computer for data reduction and analysis. See text for additional discussion.

backscattered electrons from the electron beam that are deflected by their interaction with atoms in the sample and come back out of the surface of the sample. Other electrons are relatively low-energy **secondary electrons** that are dislodged from atoms in the sample. Electrons of both types can be detected with an electron detector that operates the same as the scintillation detector used with X-rays. An electron that strikes the detector produces a brief spark of light. This light produces an electrical pulse that is recorded as a count by the instrument's electronics. Since backscattered or secondary electrons have different energy levels, The detectors can be set to detect either kind.

Scanning electron microscope images are produced by systematically sweeping the electron beam back and forth across the sample; the beam stops at each point that will become a pixel in the image. The detector records the output for each pixel, and the electronics package plots the pixel on a computer monitor and records it in a digital image file.

The flux of backscattered and secondary electrons coming from the sample depends both on the orientation of the surface and on the chemical composition. Secondary electrons are most commonly used to generate images that show details of shape and surface topography, and backscatter electrons are usually used to map composition. Unlike a light microscope, SEM images (Figure 9.2) have great depth of field and show amazing detail at magnifications that range from 5× to 300,000× for conventional instruments. By comparison, the magnification of a typical petrographic microscope (Chapter 7) is from 25× to 400×.

The backscatter electrons respond to the composition of the sample because high atomic weight elements tend to scatter a higher fraction of the electron beam back out of the sample. Pyrite (FeS_2), for example, would scatter more electrons and would be brighter in a backscatter image than an adjacent piece of quartz (SiO_2) because of the higher atomic weight of both Fe and S.

Because the SEM and the electron microprobe are very similar, it should be no surprise that they share functionality. SEMs are routinely equipped with energy-dispersive X-ray detectors to record the characteristic spectrum of the elements in a sample and provide semiquantitative chemical analyses. The accuracy is not as good as with an electron microprobe because the surface is not flat and polished; but the elements present in a mineral sample can be identified, thereby confirming an identification. Electron microprobes are similarly equipped so that the beam can be scanned back and forth, and they have backscatter electron detectors to allow quick examination of a sample to distinguish different minerals. The scanning capability also allows the electron microprobe to scan across mineral grains to create detailed maps of chemical composition by means of either the wavelength- or energy-dispersive X-ray detectors.

X-Ray Fluorescence

The **X-ray fluorescence spectrometer** (XRF) (Figure 9.3) is functionally similar to the electron microprobe, but with two significant differences. The first is that the characteristic X-ray spectra of the elements in the sample are excited by the high-energy continuous spectrum of an X-ray tube rather than by an electron beam. This minimizes the amount of continuous spectrum produced by the sample. The second is that heterogeneous samples such as rocks must be either finely ground or fused before being analyzed. This is necessary because the X-rays

Figure 9.2 Secondary electron SEM image of pyrite grains from alluvium of the Canadian River, Oklahoma. Most of the crystals are octahedrons or octahedrons modified by cubes. Photo courtesy of G. Breit, U.S. Geological Survey.

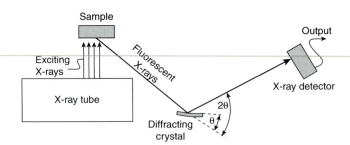

Figure 9.3 X-ray fluorescence spectrometer. The continuous spectrum from the X-ray tube causes elements in the sample to fluoresce their characteristic X-ray spectra. The intensity of the characteristic spectra of the different elements in the sample are detected by using either wavelength- or energy-dispersive techniques to determine the abundance of the elements. A wavelength-dispersive spectrometer is shown.

from the X-ray tube cannot be focused like the electron beam, so the entire surface of the sample fluoresces in secondary X-rays. Fusing the sample or grinding it finely ensures that every point in the sample is representative of the whole. X-ray fluorescence, therefore, yields an average bulk analysis of the sample.

The characteristic X-ray spectra of the different elements in the sample are detected by using the same wavelength- or energy-dispersive techniques employed with the electron microprobe. The instrument is calibrated with standards of known composition. Limits of detection are substantially improved compared to the microprobe because the background noise of the continuous spectra of the elements is minimized. XRF techniques can detect as little as 1 ppm (0.0001 wt %) of some elements, but they are usually limited to sodium and heavier elements.

Mass Spectrometry

The geometry of a **mass spectrometer** is schematically shown in Figure 9.4. The sample is vaporized and introduced into an evacuated ionization chamber where atoms in the sample are ionized. In this case electrons zipping from the hot filament to the anode dislodge electrons from the gaseous atoms. The ionized atoms are accelerated toward the slit in plate B by a voltage maintained between plates A and B. When the ions reach the magnet, their paths are deflected by the magnetic field to follow the curvature of the analyzer tube. The amount that the ions are deflected depends on their mass. Light elements are deflected more than heavy elements. By carefully adjusting the strength of the magnetic field and the accelerating voltage between A and B, different atomic weight ions can be brought to strike the detector at the far end of the instrument. Because the ions striking the detector represent a flow of electricity, detection is basically accomplished by sensing that minute electric current. The abundance of the element striking the detector is interpreted based on the strength of the electric current.

Because mass spectrometers measure mass, they are widely used to measure the relative amounts of different isotopes of elements in minerals. This makes them particularly valuable for radiometric dating of geological materials and for other isotope studies.

Conventions in Reporting Chemical Analyses

Major Elements

Chemical analyses showing the major elemental composition of minerals are typically reported in terms of weight percent of the constituents. An analysis of arsenopyrite, for example, might be reported as 34.30% Fe, 46.01% As, and 19.69% S. For the majority of minerals that contain oxygen as the principal anion, major element analyses are reported in terms of the weight percent of metal elements expressed in oxide form, water, and additional anions.

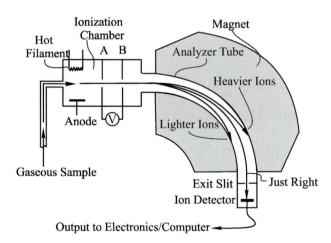

Figure 9.4 Mass spectrometer. The sample is vaporized and injected into the ionization chamber, where the atoms are ionized by interacting with a stream of electrons moving from the hot filament to the anode. A voltage between A and B accelerates the ions through slits and into the curved analyzer tube within the magnet. The magnetic field deflects the path of the elements. Light ions are deflected more than heavy ions. Adjusting the strength of the magnetic field and the voltage between A and B allows each element in the sample to strike the detector in turn.

Table 9.1 Typical Chemical Analysis of a Biotite Sample

Sample	Weight Percent
SiO_2	39.14
Al_2O_3	13.10
TiO_2	4.27
FeO	5.05
Fe_2O_3	12.94
MgO	12.75
MnO	0.14
CaO	1.64
Na_2O	0.70
K_2O	6.55
F	1.11
H_2O^+	2.41
H_2O^-	0.58
	100.38
$O \equiv F, Cl$	0.46
Total	99.92

Table 9.1 shows an analysis for a sample of biotite. In this analysis, and in essentially all others that report results in terms of the oxides, the amount of oxygen in the sample is not actually measured. Rather, weight percent of the cations is measured, and then the amount of oxygen required by the stoichiometry of the oxide species is added in. Not all of this oxygen may actually be present if other anion species such as F and Cl are in the sample; a correction is discussed shortly. It should not be inferred that separate oxide species actually exist in the mineral as structural entities.

Water may be reported as H_2O^- and H_2O^+. The value reported as H_2O^- represents actual molecular water in the sample that is adsorbed onto the surface of mineral grains or occupies small voids. It is not part of the crystal structure. H_2O^- is typically detected by determining the weight loss caused by heating the sample to $100-110°C$ for a period of time. H_2O^+ is inferred to represent hydrogen that is part of a mineral structure, but not necessarily in the form of water molecules. Water analyses are not usually very reliable and are typically available only with wet chemical techniques.

F and Cl are the additional anions usually included in chemical analyses. Their presence requires a correction to the amount of oxygen presumed to be in the sample as part of the metal oxide species. For every two F^- or Cl^- atoms actually in the sample, one O^{2-} must be deleted. The correction is therefore

$$O \equiv F, Cl = \left(wt\% \, Cl \frac{aw \, O}{2aw \, Cl} + wt\% \, F \frac{aw \, O}{2aw \, F} \right) \quad (9.1)$$
$$= (0.226 \, wt\% \, Cl + 0.421 \, wt\% \, F)$$

where wt% Cl and wt% F are the weight percent of Cl and F in the analysis, and aw refers to the atomic weight of the element.

The total of the different constituents in a chemical analysis rarely sums to exactly 100%. The degree to which the total differs from 100% is an indication of analytical errors and, if the total is less than 100%, may indicate that some elements present were not detected. Even if the total comes to nearly 100%, the possibility of compensating errors must be considered. Standards, which consist of carefully analyzed rock and mineral samples, are shared among laboratories to aid with instrumental calibration and to ensure that results from one laboratory can be compared with others.

Trace Elements

Trace element analyses of minerals and rocks are usually reported in terms of parts per million (ppm) of elements, or in some cases parts per million of the oxide of the element. Trace elements either substitute in small amounts for one of the major cations in the structure or are in sites

Table 9.2 Typical Trace Element Analysis of a Pyroxene

Element	Parts per Million (ppm)
Sr	5.67
Y	10.8
Zr	20.4
La	0.554
Ce	2.48
Nd	2.45
Sm	0.856
Eu	0.317
Dy	1.60
Er	0.835
Yb	0.791

Source: Data from Papike and others (1995).

not normally occupied in the ideal crystal structure of the mineral. They form substitutional or interstitial defects (Chapter 4). A typical analysis is shown in Table 9.2.

A common tactic to allow easy comparison among different data sets is to normalize trace element data by using a common standard. The rare earth elements (REE), for example, are commonly normalized by dividing the observed analysis with the elemental composition of an average chondrite (Anders and Grevesse, 1989). A chondrite is a type of primitive meteorite that is inferred, in some sense, to be representative of the bulk chemistry of the solar system as a whole.

CONVERSION OF CHEMICAL ANALYSES TO STRUCTURAL FORMULAS

Converting a chemical analysis into a mineral's structural formula is a relatively straightforward procedure that can be done manually or with the aid of a spreadsheet program in a personal computer. Similar calculations are routinely done as part of all introductory chemistry classes. Table 9.3 illustrates the calculation for a pyroxene whose general formula is XYZ_2O_6. Columns 1 and 2 present the chemical analysis. Note that the analysis sums to 100.12 wt%, suggesting a small analytical error. For purposes of calculation, we assume that the sample weighs 100.12 g; the weight percent values can then be considered equal to the number of grams of each oxide species. Column 3 shows the molecular weight (grams per mole) of the oxide species based on the atomic weight of the cation and oxygen. Table 9.4 lists the molecular weight of the common oxides. The calculations in Table 9.3 are accomplished as follows,

Table 9.3 Recalculation of a Pyroxene Analysis

1	2	3	4	5	6	7	8		
	Weight Percent	Molecular Weight	Moles Oxide	Moles Oxygen	Moles Cation	Cations on Basis of Six O	Structural Allocation		
SiO_2	53.8	60.08	0.89547	1.79095	0.89547	1.961	1.961	2.00	^{IV}Z
Al_2O_3	1.4	101.96	0.01373	0.04119	0.02746	0.060	0.039		
							0.021	1.053	^{VI}Y
TiO_2	0.18	79.90	0.00225	0.00451	0.00225	0.005	0.005		
Cr_2O_3	0.9	151.99	0.00592	0.01776	0.01184	0.026	0.026		
MgO	16.7	40.30	0.41439	0.41439	0.41439	0.907	0.907		
FeO	3.0	71.84	0.04176	0.04176	0.04176	0.091	0.091		
MnO	0.07	70.94	0.00099	0.00099	0.00099	0.002	0.002		
CaO	23.8	56.08	0.42439	0.42439	0.42439	0.929	0.929	0.948	^{VIII}X
Na_2O	0.27	61.98	0.00436	0.00436	0.00871	0.019	0.019		
Total	100.12			2.74030	Factor = 2.18954				

Table 9.4 Molecular Weight of Common Oxide Species

Oxide	Molecular Weight	Oxide	Molecular Weight
H_2O	18.02	K_2O	94.20
CO_2	44.01	CaO	56.08
Na_2O	61.98	Cr_2O_3	151.99
MgO	40.30	MnO	70.94
Al_2O_3	101.96	FeO	71.84
SiO_2	60.08	Fe_2O_3	159.69

taking care to avoid rounding the figures until all calculations are complete

COLUMN 4: Calculate the number of moles of each oxide species by dividing the weight percent value (column 2) by the molecular weight of the oxide (column 3).

COLUMN 5: Calculate moles of oxygen in each oxide species based on the stoichiometry. For example, moles of oxygen from SiO_2 is twice the moles of oxide (column 4) because SiO_2 has two oxygen atoms. The total moles of oxygen in the sample (2.74029 in this case) is summed at the bottom of column 5.

COLUMN 6: Calculate moles of cation in each oxide species based on the stoichiometry. For example, the number of moles of Si is the same as the moles of SiO_2, whereas the number of moles of Al is twice

the number of moles of Al_2O_3. At this point the chemical formula can be written (rounded to three digits to the right of the decimal):

$$Na_{0.009}Ca_{0.424}Mn_{0.001}Fe_{0.042}Mg_{0.414}Cr_{0.012}Ti_{0.002}Al_{0.027}Si_{0.895}O_{2.740}$$

COLUMN 7: In this column the formula is recalculated so that it is based on the number of oxygen anions used in the structural formula. For pyroxene this is six. First calculate the *factor* by dividing the number of oxygens per formula by the moles of oxygen at the bottom of column 5. In this case the *factor* is 6/2.74030 = 2.18954. Cations on the basis of six O are then calculated by multiplying the moles of cations (column 6) by the *factor*.

COLUMN 8: Cations are segregated into different structural sites based on knowledge of the mineral structure. In this pyroxene, three sites are available: one for tetrahedrally coordinated cations (^{IV}Z), one for 6-fold-coordinated cations (^{VI}Y), and one for cations in 8-fold coordination (^{VIII}X). In general for silicates, all the Si and some of the Al get apportioned to the tetrahedral (Z) site, and the remaining Al and the other cations get apportioned to the other site(s). In this case, the tetrahedral site is assigned 1.961 Si and 0.039 Al to make a total of 2.000. The remaining 2.001 cations must occupy the 6- and 8-fold sites (Y and X in the formula). As a first approximation, we will assume that Na and Ca, which are the largest cations, occupy the X site and the remaining cations occupy Y. This yields 1.053

cations in the Y site and 0.948 cation in X, which is very close to the site occupancy in the ideal formula. Based on this assumption, the formula can now be written:

$$\underset{\substack{\text{VIII} \\ X_{0.948}}}{(Na_{0.019}Ca_{0.929})}\underset{\substack{\text{VI} \\ Y_{1.053}}}{(Mn_{0.002}Fe_{0.091}Mg_{0.907}Cr_{0.026}Tr_{0.005}Al_{0.021})}\underset{\substack{\text{IV} \\ Z_2}}{(Al_{0.039}Si_{1.961})}\underset{O_6}{O_6}$$

It is probable that about 0.05 of the cations assigned to the 6-fold Y site are actually in the 8-fold X site. All but the Al are sufficiently large to occupy an 8-fold site (Figure 3.16, Table 4.2, and Appendix A). Without detailed structural information, however, it is not possible to know how much of which element to assign to the 8-fold site.

These data can be further manipulated to show the relative amounts of the end-member compositions for minerals that display significant solid solution. For pyroxenes, the standard end members are wollastonite (wo, $Ca_2Si_2O_6$), enstatite (en, $Mg_2Si_2O_6$), and ferrosilite (fs, $Fe_2Si_2O_6$). The relative amounts of the end members are therefore equivalent to the relative amounts of Ca, Mg, and Fe. Hence,

$$\% \ wo = Ca/(Ca + Mg + Fe) = 0.929/1.927 = 48.2 \ \%$$

$$\% \ en = Mg/(Ca + Mg + Fe) = 0.907/1.927 = 47.1 \ \%$$

$$\% \ fs = Fe/(Ca + Mg + Fe) = 0.091/1.927 = 4.7 \ \%$$

The remaining cations (Cr, Ti, Mn, etc.) are ignored in this calculation.

Each unit cell contains an integer number of **formula units (Z)** where a formula unit is the number of atoms shown in the chemical formula as conventionally written.

The formula for halite is NaCl, so a formula unit consists of one Na and one Cl. Because the unit cell contains four Na and four Cl, $Z = 4$.

REFERENCES CITED AND SUGGESTIONS FOR ADDITIONAL READING

Anders, E., and Grevesse, N. 1989. Abundances of the elements: Meteorite and solar. Geochimica et Cosmochimica Acta 53, 197–214.

Fenter, P., Rivers, M., Sturchio, M., and Sutton, W. (eds.). 2002. *Applications of Synchrotron Radiation in Low-Temperature Geochemistry and Environmental Science.* Reviews in Mineralogy 49, 579 p.

Hawthorne, F. C. (ed.). 1988. *Spectroscopic Methods in Mineralogy and Geology. Reviews in Mineralogy*, Volume 18. Mineralogical Society of America, Washington, DC, 698 p.

Jeffery, P.G. 1975. *Chemical Methods of Rock Analysis*, 2nd ed. Pergamon Press, Oxford, 533 p.

Johnson, W.M., and Maxwell. J. A. 1981. *Rock and Mineral Analysis*, 2nd ed. John Wiley & Sons, New York, 489 p.

Papike, J. J., Spilde, M. N., Fowler, G. W., and McCallum I. S. 1995. SIMS studies of planetary cumulates: Orthopyroxene from the Stillwater Complex, Montana. American Mineralogist 80, 1208–1221.

Reed, S. J. B. 2005. *Electron Microprobe Analysis and Scanning Electron Microscopy in Geology.* Cambridge University Press, Cambridge, 189 p.

Reeves, R. D., and Brooks, R. R. 1978. *Trace Element Analysis of Geological Materials.* John Wiley & Sons, New York, 421 p.

Skoog, D. A., and Leary, J. J. 1992. *Principles of Instrumental Analysis.* Saunders College Publishing, Fort Worth, TX, 700 p.

Wenk, H.-R. (ed.). 2006. *Neutron Scattering in Earth Sciences.* Reviews in Mineralogy & Geochemistry 63, 471 p.

CHAPTER 10

Strategies for Study

INTRODUCTION

The preceding four chapters describe a variety of techniques by which minerals may be studied. The objective of this chapter is to provide systematic guidance to help direct an inquiry. In most cases, the first problem faced is the prospect of identifying an unknown mineral, because without an identity, further study is impractical. The guidance provided here, therefore, is principally intended to lead to an identification. Detailed description of the procedures is omitted, but reference will be made to the relevant parts of the previous four chapters as needed. Additional information about many laboratory procedures can be found in Hutchison (1974), Zussman (1967), and Humphries (1992).

It is common in many introductory mineralogy courses to provide large, relatively pure samples of minerals for study. This makes it easy to observe and measure the normal hand-sample properties and come up with an identification. It is then relatively easy to extend these skills to the extent that identification of the major minerals in coarse-grained rock and mineral deposits becomes routine. These skills are essential for any practicing Earth scientist, particularly anyone doing geological field mapping; but they are just the starting point in the investigation of geological materials. Field examination and identification are often followed by microscopic study of thin and polished sections to confirm hand-sample identification and to provide a wealth of information about textural relations among minerals. X-ray powder diffraction often supplements hand-sample and microscopic work. Chemical analysis of the bulk rock sample and the constituent minerals is the next step. The chemical data provide the basis for a variety of petrological studies that help expand and refine the foundation of knowledge provided by field study.

PRESENTATION OF MINERAL DATA

Descriptions and data on more than 100 common minerals are provided in Chapters 12 through 20. These chapters are organized based on the chemical classification of the minerals. Silicates are presented first because they are by far the most abundant group of minerals. Data and descriptive material for each mineral are organized systematically, so that it is easy to find specific information.

The material for tremolite–ferro-actinolite (Chapter 14) (Figure 10.1) is typical of how the data are presented. Each mineral description begins with a summary of numerical data including the chemical formula, crystallography (crystal system, crystal class, unit cell dimensions), hardness (H), and specific gravity (G). This is followed by optical data including optical class (isotropic, uniaxial, biaxial), optic sign, indices of refraction, birefringence (δ), and 2V for biaxial minerals.

One or more line drawings of a typical crystal of the mineral are provided. Miller indices are shown for some faces, the traces of cleavages are shown with fine lines, and crystal axes are shown with dot–dash lines. For biaxial minerals, the trace of the optic plane is shown with a heavy dashed line, optic axes are shown with a solid line with a ball at the end, and optical indicatrix axes are solid lines labeled X, Y, and Z.

In many cases a drawing of a section through the crystal is provided, showing the relationships among crystal outline, crystal axes, cleavage, and indicatrix axes. This section helps visualize extinction angle relationships as seen in thin section. In the example shown here, the extinction angle between the slow ray (parallel to Z) and the trace of cleavage in sections cut parallel to (010) is to 12° to 34°. In this case the (010) section will display maximum birefringence because the X and Z indicatrix axes are horizontal.

The written material on each mineral presents information on structure, composition, form and twinning, cleavage, color, optical characteristics, alteration, distinguishing features, occurrence, and uses. Various diagrams and photographs may also be included.

MINERAL IDENTIFICATION TACTICS

Mineral Separation

It is commonly necessary to separate individual mineral grains from a rock for detailed hand-sample analysis,

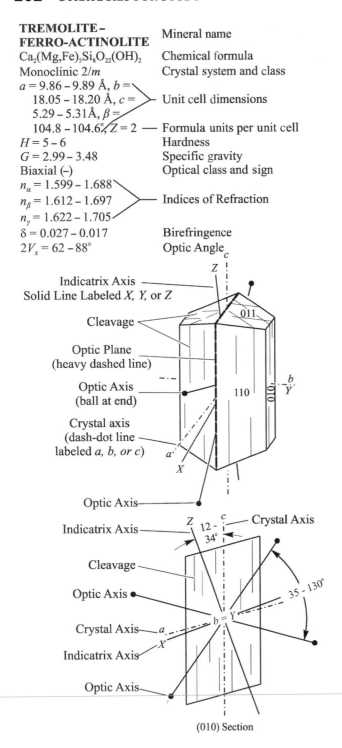

TREMOLITE–
FERRO-ACTINOLITE   Mineral name

$Ca_2(Mg,Fe)_5Si_8O_{22}(OH)_2$ Chemical formula

Monoclinic $2/m$ Crystal system and class

$a = 9.86 – 9.89$ Å, $b =$
18.05 – 18.20 Å, $c =$ Unit cell dimensions
5.29 – 5.31Å, $\beta =$
104.8 – 104.6°, $Z = 2$ Formula units per unit cell

$H = 5 – 6$ Hardness

$G = 2.99 – 3.48$ Specific gravity

Biaxial (–)   Optical class and sign

$n_\alpha = 1.599 – 1.688$
$n_\beta = 1.612 – 1.697$ Indices of Refraction
$n_\gamma = 1.622 – 1.705$

$\delta = 0.027 – 0.017$ Birefringence

$2V_x = 62 – 88°$ Optic Angle

Indicatrix Axis
Solid Line Labeled X, Y, or Z

Cleavage

Optic Plane
(heavy dashed line)

Optic Axis
(ball at end)

Crystal axis
(dash-dot line
labeled a, b, or c)

Optic Axis

Indicatrix Axis

Cleavage

Optic Axis

Crystal Axis

Indicatrix Axis

Optic Axis

(010) Section

Figure 10.1 Presentation of mineral data.

optical study using oil immersion techniques, X-ray diffraction, or chemical analysis. The techniques employed depend on the amount of material needed and its purity. The routine methods include hand extraction, heavy liquid separation, and magnetic separation. In many cases it is necessary to crush the sample to allow individual minerals to be segregated.

Hand Extraction

In most cases, hand extraction methods involve breaking a rock with hammer and chisel to free the desired mineral. A sledge hammer, or hydraulic press designed to break rocks, may be used to deal with large samples. Eye protection should always be used. If grains are small, a dental pick or similar probe can be used to break out and select the desired grains with the aid of a binocular microscope or hand lens.

Crushing

A variety of apparatuses (Figure 10.2) are available to crush rock and mineral samples so that the individual minerals are separated from each other. Jaw crushers can reduce rocks to particles as small as a few millimeters, and rotary mills can produce substantially finer grains. Samples can be manually crushed with a piston-type percussion mortar or a steel mortar and pestle. The crusher must be cleaned between uses to avoid contamination. As a general rule, crushing igneous and metamorphic rocks should produce fragments between one-fourth and one-tenth the average grain size of the rock to ensure that most grains are monomineralic.

Crushing clastic sedimentary rocks requires some care, because the shape of detrital grains may need to be preserved. As long as the rock is not strongly indurated, disaggregation can be accomplished in a steel mortar and pestle, or even by hand. In some cases, calcareous cement can be removed or weakened by soaking in dilute acid. Some fine-grained sedimentary rocks such as shale and mudstone may be mixed with water and disaggregated in a kitchen blender.

Once crushed, the sample is usually sieved to select grains with an appropriate size for examination. Table 10.1 shows common sieve sizes. Particles that pass a 100-mesh sieve and collect on the 170-mesh sieve, or that pass a 140-mesh sieve and collect on the 200-mesh sieve, are commonly used. Examine these grains with a binocular microscope to confirm that each grain is monomineralic. If not, a finer size fraction may be required.

The crushed and sieved material must now be cleaned to remove fine dust from the particle surfaces. This can be accomplished by placing the selected grain size fraction in a beaker. Tap water (or distilled water) is squirted into the beaker to thoroughly agitate the grains and then decanted to remove the fines. This procedure is repeated several times until all of the fines are removed. Wet grains are transferred to a watch glass to dry. Acetone may be used instead of water if time is not available to allow the water to dry.

Density Separation

Heavy liquids are routinely used to separate minerals based on their density. The commonly used liquids are **bromoform**

Figure 10.2 Crushing tools. (*a*) Laboratory jaw crusher. (*b*) Rotary mill: the sample is fed into the chute and is crushed between rotating ceramic jaws. (*c*) Piston-type percussion mortar.

Table 10.1 Sieve Openings

ASTM[a] Number	Openings (mm)
80	0.180
100	0.150
120	0.125
140	0.106
170	0.090
200	0.075
230	0.063
270	0.053
325	0.045
400	0.038
450	0.032

[a] American Society for Testing and Materials.

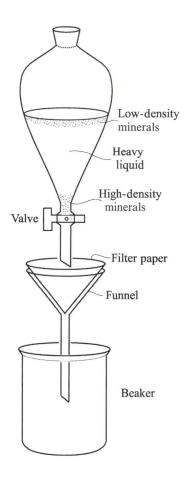

Figure 10.3 Separatory funnel setup.

($G = 2.89$) and **methylene iodide (diiodomethane:** $G = 3.31$); both are toxic and must be used in a chemical fume hood. **Lithium metatungstate** ($G = 3.00$), **ammonium metatungstate** ($G = 2.4$), and **sodium polytungstate** ($G = 3.1$) are less toxic but are more expensive, and their high viscosity makes them more difficult to use. A related product known as **LST** ($G = 2.9$) has a lower viscosity than the other tungstate-based liquids, and its density is strongly temperature dependent. Refer to material safety data sheets for appropriate safety precautions to be employed with all of these materials.

Mineral separations are accomplished with the aid of a separatory funnel (Figure 10.3) in which the heavy liquid and sample are placed. Low- and high-density minerals float and sink, respectively; periodic stirring helps produce a good separation. When separation is complete, the valve at the bottom of the separatory funnel is opened to expel the high-density minerals and part of the liquid. The high-density minerals are retained on a filter paper in a funnel and the liquid passes into a beaker for reuse. The low-density minerals are subsequently collected on a fresh filter paper.

The density of the heavy liquid can be adjusted by diluting with a solvent so that it lies between the densities of the minerals to be separated. Both bromoform and methylene iodide may be diluted with acetone or certain other organic solvents; the tungstate products may be diluted with water. The density of the diluted heavy liquid may be determined by weighing a known volume of the liquid or by comparison with **aerometers**, which are spheres whose density is known.

Grains on the filter papers are washed with the appropriate solvent to remove traces of the heavy liquid. Additional separation among minerals in each density fraction may be accomplished with different heavy liquids.

Because heavy liquids are expensive, they are recovered from mixtures and from waste produced by washing grains and the apparatus. For the tungstate products, simple evaporation in an open container will remove the water. More complex reverse-osmosis apparatus may be used if large volumes are involved. Bromoform and methylene iodide are "washed" with distilled water several times. Because acetone is soluble in water, but bromoform and methylene iodide are not, water extracts acetone from the heavy liquid. The water, which floats on the heavy liquid, is decanted after each washing.

Magnetic Separation

A very simple separation technique uses an ordinary magnet wrapped in a piece of plastic wrap to segregate grains such as magnetite that are ferro- or ferrimagnetic. The crushed and sieved sample is spread on a piece of paper and the magnet is moved over or through the grains to pull out the ferro/ferrimagnetic particles. These are retrieved by stripping the plastic wrap from the magnet.

A **Frantz Isodynamic Magnetic Separator**® (Figure 10.4) is available in many laboratories and can be used to separate paramagnetic minerals based on their magnetic susceptibilities and to segregate paramagnetic from diamagnetic minerals. The instrument consists of a chute positioned between the poles of a strong electromagnet. Mineral grains are fed onto the inclined chute, which is tilted to the side, and gravity pulls them down the chute and to the low side. The magnetic field causes paramagnetic minerals to be pulled to the high side of the chute. A splitting ridge divides the minerals into two populations that are collected in separate containers. Different mineral fractions are separated by running the sample through the apparatus using progressively higher magnetic field strengths and by adjusting the inclination and tilt of the chute. The mineral fraction left after the paramagnetic minerals have been removed usually consists of diamagnetic quartz and feldspar. The closely related **Frantz Magnetic Barrier Separator**® is a significant refinement and allows finer separations among paramagnetic minerals and also among diamagnetic and weakly paramagnetic minerals.

Hand-Sample Identification

Hand-sample identification of an unknown mineral depends on systematically identifying the physical properties described in Chapter 6: color, luster, streak, specific gravity, hardness, cleavage/fracture, and so forth. Although identification based on these relatively crude and imprecise properties may seem haphazard, they yield remarkably good results when applied systematically, but success is by no means guaranteed. Proceed as follows.

• Observe and record crystal habit or grain shape. If crystals are well developed, determine the crystal system to which they belong based on symmetry and the geometry of the crystals.

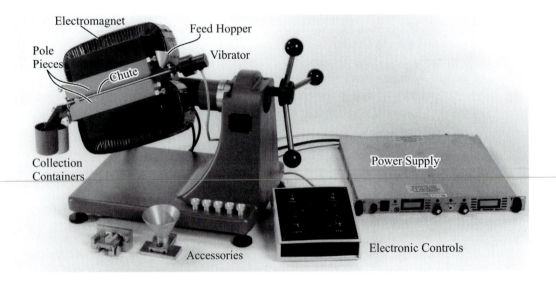

Figure 10.4 Frantz Isodynamic Magnetic Separator®. The crushed sample is placed in the feed hopper and fed onto the chute with the aid of a vibrator. Gravity pulls the grains down the chute between the pole pieces of the electromagnet. The electromagnet provides a magnetic field that deflects the path of mineral grains depending on their magnetic susceptibilities. A splitting ridge divides the lower part of the chute into two channels that guide the separated fractions into separate collection containers. The strength of the magnetic field, and the downward and lateral tilts of the chute, may be adjusted to allow separation of minerals of different susceptibilities. The closely related Frantz Magnetic Barrier Laboratory Separator® provides improved capabilities. Photo courtesy of S. G. Frantz Company.

• Observe and record color, luster, and streak. Use a streak plate if it is available; otherwise just crush a bit of the sample.

• Measure specific gravity, using a Jolly balance, pychnometer, or heavy liquids if sample size and purity allow. An estimate of specific gravity can sometimes be made by simply hefting the sample, if it is large enough.

• Measure the hardness by scratching with a fingernail ($H = 2$–$2\frac{1}{2}$), copper penny ($H = 3$), knife blade ($H = 5$), glass ($H = 5\frac{1}{2}$), and/or piece of quartz ($H = 7$). Scratch the mineral with the scratch tester, and vice versa, to confirm hardness. Minerals that have suffered some alteration may be softer than normal.

• Carefully examine the mineral to determine if it has cleavage, and if so, how many different directions and the angles between them. A hand lens or binocular microscope is a great aid here. Fine grains may reveal cleavage by a flash or sparkle when moved. Be alert to the possibility that what appears to be a cleavage may actually be a parting. On noncleavage surfaces, observe and record the nature of the fracture.

This provides the basic data that should lead to an identity for most of the common minerals. Proceed by looking in Tables B.1, B.2, or B.3 in Appendix B to get a list of possibilities. Table B.1 is for minerals with nonmetallic luster and white, gray, or pale-colored streak. Table B.2 is for minerals with nonmetallic luster and distinctly colored streak. Table B.3 is for minerals with metallic or submetallic luster. Within each table, minerals are grouped based on hardness, whether cleavage is prominent, and color. Within Table B.1, cleavable minerals are further separated into groups based on the number of cleavages. Relatively common minerals are indicated with bold type.

Begin with the appropriate table based on luster and streak. The combination of hardness, cleavage, and color will yield a relatively short list of possible minerals. Combine the lists for one, two, and three or more cleavages on Table B.1 if the number of cleavages cannot be determined with certainty. The list of possible minerals may be shortened by reference to Table B.4, which lists minerals in order of increasing specific gravity. The objective here is to acquire a list of possibilities; identification should await examination of the more detailed information contained in the descriptions of these possible minerals in Chapters 12 through 20. Refer to the mineral index to find the page numbers on which each of the possible minerals is described. Some prior knowledge of the likely mineralogy of the sample based on association can greatly expedite the process.

Many laboratories will be equipped with computer systems that search through a data set of minerals based on physical properties input by the worker to provide an identity. Remember that the identity provided by a computer system can be no better than the quality of the data; garbage in—garbage out. Confirm the computer-suggested identification with a review of the detailed mineral descriptions in Chapters 12 through 20.

Unfortunately, mineral identification is often not the highly organized step-by-step process implied here. Physical properties show substantial variability and, because of grain size, or other difficulties, it may not be possible to determine some properties. Not infrequently, these difficulties make it impossible to identify a sample using hand-sample techniques, and X-ray diffraction, optical, or other means must be used.

In practice, the apparent ease with which experienced mineralogists and mineral collectors recognize minerals depends on a subjective summation of the physical properties and a variety of other subtle cues. Experience with the rocks being examined also helps. Once familiarity has been gained with a mineral, identification may be possible on sight or with the application of one or two diagnostic tests that distinguish among similar minerals. Regardless, if correct identification is critical, most mineralogists confirm hand-sample identifications with optical, X-ray, chemical, or other techniques.

Thin Section Identification

Examination of thin sections provides one of the most productive means of studying rocks. Thin sections are slices of rock about 0.03 mm thick mounted on glass microscope slides (Figure 7.1). Consult Humphries (1992) and Murphy (1986) for additional information concerning specific preparation techniques. While various machines are available to speed the process and improve accuracy and consistency, the basic preparation process is as follows.

• A chip of rock slightly smaller than a microscope slide is cut from a hand sample. Standard petrographic slides are 27 mm × 46 mm.

• If the sample is porous and friable, it should be impregnated with an epoxy. The sample is immersed in a small container of liquid epoxy and placed in a vacuum chamber from which the air is then evacuated. This removes air from pores in the sample. When the vacuum is broken, epoxy is forced into the pores.

• The side of the rock chip to be glued to the microscope slide is ground smooth and flat, beginning with coarse abrasive and finishing with 600 grit abrasive for routine preparations.

• The rock chip is glued to a clean microscope slide by using an epoxy, UV-setting adhesive, or other cement.

• When the cement is cured, surplus rock chip is cut off with a diamond trim saw and the chip is ground to its final thickness, usually 0.03 mm. A number of manufacturers make machines that speed this process. Final grinding to just the right thickness is usually done by hand. The interference color of a mineral in the sample such as quartz is used to determine thickness.

• If desired, the thin section may be stained for K-feldspar, plagioclase, carbonate minerals, and others. Hutchison (1974), Bailey and Stevens (1960), Boone and Wheeler (1968), Houghton (1980), and Ruperto and others (1964) provide instructions.

• The sample is completed either by cementing a cover-slip in place or by polishing the surface to a mirror finish.

Tactics to measure optical properties and identify minerals in thin section are described at the end of Chapter 7.

Grain Mount Identification

Grain mounts are prepared and examined as follows.

• Separate, crush, sieve, and clean grains as just described. The −100/+170 or −140/+200 mesh ranges work well.

• Sprinkle a few dozen grains on a microscope slide and cover with a piece of coverslip. To immerse the grains, place a small drop of immersion oil next to the coverslip; it will be drawn under by capillary action.

Tactics to measure optical properties and identify minerals in grain mount are described at the end of Chapter 7.

Polished Section Identification

Polished sections may either be thin sections whose surface has been polished (no coverslip) or rock chips that are mounted with epoxy in small holders. Polishing is accomplished by progressively grinding the surface with finer and finer abrasives. Grinding through 600 grit abrasive is commonly done on a rotating lap wheel with abrasive disks or powdered abrasive. Finer abrasives, including aluminum oxide and diamond pastes, are usually applied to a fabric disk mounted on a rotating lap. Care must be taken to avoid contaminating one lap with coarser abrasive from another. Details of procedures depend on the equipment that is available.

Procedures for observing optical properties with reflected light are described in Chapter 7.

X-Ray Diffraction

X-ray diffraction is routinely used to confirm an identity made with hand-sample or optical techniques or to identify minerals whose grain size is too fine for either of these methods. The material used for X-ray diffraction may be the fine-grained material left over from grain mount preparation or other material that has been selected. An absolutely pure sample is not needed, but excessive contamination can lead to confusion.

The sample consists of a finely ground powder of the unknown mineral mounted as described in Chapter 8. See that chapter for discussion of procedures.

MINERAL ASSOCIATION

Mineral associations are never random. Some mineral combinations are very common and form certain rock types or mineral deposits, and other mineral combinations are rarely found. It follows, therefore, that knowledge of the mineralogy of common rock types and mineral deposits can aid in making educated guesses about the possible identity of an unknown mineral. Further, knowledge of mineral associations can suggest the presence of a mineral that otherwise might be overlooked.

A certain amount of caution is required, however, because use of association contains a trap. An unthinking application of association may blind the observer to the possibility of an unusual mineral in a rock or mineral deposit. It is not uncommon for the serendipitous discovery of an unusual mineral in a suite of rocks to have implications concerning the geological story that those rocks have to tell.

The lists of mineral associations in Appendix C show the minerals likely to be found in a variety of common rock and mineral deposits. The list is not intended to be exhaustive and does not contain the unusual associations that can confound and delight petrologists, nor does it contain any of the myriad products of alteration and weathering that may be present.

PROBLEMS IN PARADISE

It would be nice if everything worked as it should, but difficulties routinely arise. Some of the common problems include the following.

ALTERATION: Alteration, either from weathering or hydrothermal processes, may affect mineral color, luster, specific gravity, and hardness. In many cases, alteration converts the primary mineral to various fine-grained phyllosilicate and oxide minerals whose presence may be indicated by a cloudy or chalky appearance, red staining, or anomalous softness. In thin section and grain mount, the altering minerals may appear as "grunge" or may be sufficiently coarse grained to be individually identified.

MIXTURES: It is not uncommon for two or more minerals to be intimately intergrown, making it impossible to obtain relatively pure material for hand-sample work. The bulk physical properties of a sample may therefore not be those of any of the constituent minerals. The nature of the intergrowth is generally evident in thin section, and individual constituents may be identified by using microscopic or X-ray techniques.

INCONSISTENCIES IN CRYSTALLOGRAPHIC SETTINGS: In orthorhombic, monoclinic, and triclinic minerals, crystal axes may be assigned in several different ways. The convention adopted in this book is to assign axes based on external morphology. However, in some X-ray structural studies, it is convenient to assign axes differently. This has led to a substantial amount of confusion in the literature, where optical and physical properties are described based on one crystal axis setting and the structure is described using another setting. Even authoritative sources such as the series of reference works by Deer and others (1992–2009) and Anthony and others (1990–2003) are not immune to this problem. A concerted effort has been made to use a consistent crystal-axis setting for all properties in the mineral descriptions that follow, but the reader is cautioned that problems may persist and that care must be exercised when one is comparing data from different sources.

POOR DATA: Most of the optical and physical properties reported in this and other standard sources about minerals are based on observations and reports from over a century ago. These data are uncritically repeated in each new text or reference work, with a few additions or modifications based on new measurements gleaned from the literature. Old errors are thereby perpetuated and new, often unrepresentative data added. As a general rule, data from the reference works by Deer and others (1992–2009) and/or Anthony and others (1990–2003) are used here, but other sources were used where it seemed appropriate. Because the time required to systematically verify and update the data on each mineral is very great, it has not been attempted here. Reader beware.

REFERENCES CITED AND SUGGESTIONS FOR ADDITIONAL READING

Anthony, J. W., Bideaux, R. A., Bladh, K. W., and Nichols, M. C. 1990–2003. *Handbook of Mineralogy* (5 volumes). Mineral Data Publishing, Tucson, Az.

Bailey, E. H., and Stevens, R. E. 1960. Selective staining of K-feldspar and plagioclase on rock slabs and thin sections. American Mineralogist 45, 1020–1025.

Boone, G. M., and Wheeler, E. P. II. 1968. Staining for cordierite and feldspars in thin section. American Mineralogist 53, 327–331.

Deer, W. A., Howie, R. A., and Zussman, J. 1992–2009. *Rock-Forming Minerals* (5 volumes). Geological Society, London.

Fleischer, M., Wilcox, R. E., and Matzko, J. J. 1984. *Microscopic Determination of the Nonopaque Minerals.* U.S. Geological Survey Bulletin 1627, 453 p.

Houghton, H. F. 1980. Refined techniques for staining plagioclase and alkali-feldspars in thin section. Journal of Sedimentary Petrology 50, 629–631.

Humphries, D. W. 1992. *The Preparation of Thin Sections of Rocks, Minerals, and Ceramics.* Oxford University Press, New York, 83 p.

Hutchison, C. S. 1974. *Laboratory Handbook of Petrographic Techniques.* John Wiley & Sons, New York, 527 p.

Murphy, C. P. 1986. *Thin Section Preparation of Soils and Sediments.* A B Academic Publishers, Berkhamsted, England, 149 p.

Nesse, W. D. 2002. *Introduction to Optical Mineralogy*, 3rd ed. Oxford University Press, New York, 348 p.

Phillips, W. R., and Griffen, D. T. 1981. *Optical Mineralogy, the Nonopaque Minerals.* W. H. Freeman, San Francisco, 677 p.

Ruperto, V. L., Stevens, R. E., and Normans, M. B. 1964. Staining of plagioclase feldspars and other minerals with F.D. and C. No. 2. U.S. Geological Survey Professional Paper 501B, p. B152–B153.

Troger, W. E. 1979. *Optical Determination of Rock-Forming Minerals, Part I: Determinative Tables.* English edition of the 4th German edition by Bambaur, H. U., Taborgzky, F., and Trochim, H. D. Schwiezerbart'sche Verlagsbuchhandlung, Stuttgart, 1988 p.

Zussman, J. 1967. *Physical Methods in Determinative Mineralogy.* Academic Press, London, 514 p.

PART THREE

Mineral Descriptions

CHAPTER 11

Silicates

INTRODUCTION

The silicate minerals are by far the most abundant group of minerals in the Earth's crust and require extensive consideration. They are the minerals in which much of the Earth's crustal history is chemically recorded.

Some, including quartz, form exquisite crystals that appeal to mineral collectors, children, and others who appreciate beauty. A few silicates are used as precious or semiprecious gemstones. Most silicates, however, are drab and commonplace. Their ubiquity makes them easy to ignore. Few silicates cause the heart of the dollar-driven to race like a few flecks of gold tend to do. Regardless of this perception, silicates are very important economic resources. A few, such as quartz, beryl, and zircon, are mined or quarried as the source for specific elements (Si, Be, and Zr, respectively). Many are used as industrial minerals, valued not for their elemental content but for their physical or chemical properties. They make industry possible. Silicates also make agriculture possible. The soil in which everything from artichokes to zucchini is grown is composed dominantly of silicate minerals, mostly quartz and clays.

This chapter provides an introduction to the structural classification of the silicates and summarizes some of the major features of the important rock groups that are composed dominantly of silicate minerals. The coverage here is necessarily superficial. More detailed descriptions of silicate structure and classification can be found in succeeding chapters. The sources listed at the end of the chapter provide additional information on petrography and petrology of igneous, sedimentary, and metamorphic rocks.

SILICATE STRUCTURE AND CLASSIFICATION

The basic building block of all silicate minerals is the silicon tetrahedron (Figure 11.1). Si^{4+} is a relatively small cation and conveniently fits in tetrahedral coordination with O^{2-} (refer to the discussion on coordination in Chapter 4). It consists of four O^{2-} anions arranged to occupy the corners of a tetrahedron with a Si^{4+} in the center. An isolated silicon tetrahedron therefore has a net −4 charge. This has profound consequences for the structure of silicate minerals. Recall from the discussion of Pauling's second rule

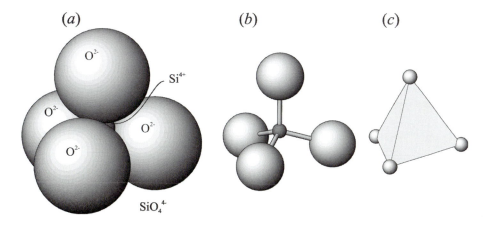

Figure 11.1 A silicon tetrahedron consists of four O^{2-} forming a tetrahedral coordination polyhedron around a Si^{4+}. (*a*) Silicon and oxygen drawn to scale. (*b*) Stick-and-ball representation with silicon and oxygen sizes reduced by half. (*c*) Oxygen anions shown as small spheres at corners of tetrahedron.

in Chapter 4 that the silicon tetrahedron has mesodesmic bonds. What that means is that each O^{2-} has one of its two valence charges satisfied by bonding with the Si^{4+} at the center of the tetrahedron. The remaining -1 charge is available to bond with a Si^{4+} in the center of another tetrahedron. In effect, silicon tetrahedra may polymerize. The degree to which the oxygen anions are shared between adjacent tetrahedra provides the basis for the structural classification of the silicate minerals.

Table 11.1 lists the essential characteristics of the different groups of silicate structures and Figure 11.2 shows the basic configuration of the silicon tetrahedra in those structures. The classification used here is conventional. Consult Liebau (1985) for detailed discussion of other schemes that can and have been used to classify silicates.

The least degree of polymerization is found in the **orthosilicates**—no oxygen anions are shared between adjacent tetrahedra. The silicon tetrahedra form isolated structural entities whose net negative charge is balanced by bonding with other cations such as Mg^{2+}, Fe^{2+}, and Al^{3+}. Sharing of a single O^{2-} between two silicon tetrahedra produces the **disilicates**. In the **ring silicates**, the tetrahedra share two O^{2-} each and form rings, usually with six members. Two O^{2-} per tetrahedron are shared in the single-chain group of the **chain silicates**; in the double-chain group, some tetrahedra share two O^{2-} and others share three. Three O^{2-} per tetrahedron are shared to form continuous sheets in the **sheet silicates**. In the **framework silicates**, all four O^{2-} on each tetrahedron are shared with adjacent tetrahedra to form a three-dimensional framework. The following alternate terminology is used in some sources.

Orthosilicates = Nesosilicates

Disilicates = Sorosilicates

Ring silicates = Cyclosilicates

Chain silicates = Inosilicates

Sheet silicates = Phyllosilicates

Framework silicates = Tectosilicates

Table 11.1 Silicate Classification[a]

	Number of		
Silicate Class	**O^{2-} Shared per Tetrahedron**	**Z:O Ratio**	**Structural Configuration**
Orthosilicates	0	1:4	Isolated tetrahedra
Disilicates	1	2:7	Double tetrahedra
Ring silicates	2	1:3	Rings of tetrahedra
Chain silicates			Chains of tetrahedra
Single chain	2	1:3	
Double chain	2 or 3	4:11	
Sheet silicates	3	2:5	Sheets of tetrahedra
Framework silicates	4	1:2	Framework of tetrahedra

[a] Z refers to the cation(s), usually Si^{4+}, and also Al^{3+}, that occupy the tetrahedral sites.

Table 11.2 Common Silicate Minerals

Mineral or Mineral Group	Chemical Formula
Orthosilicates	
Olivine	$(Mg,Fe)_2SiO_4$
Zircon	$ZrSiO_4$
Garnet	$X_3Y_2(SiO_4)_3$
Aluminum silicates	$AlAlOSiO_4$
Staurolite	$Fe_2Al_9O_6[(Si,Al)O_4]_4(OH)_2$
Chloritoid	$(Fe^{2+},Mg,Mn)_2(Al,Fe^{3+})$ $Al_3O_2(SiO_4)_2(OH)_4$
Topaz	$Al_2(SiO_4)(F,OH)_2$
Titanite	$CaTiOSiO_4$
Disilicates	
Epidote group	$Ca_2Al_2(Al,Fe^{3+})OOH[SiO_4][Si_2O_7]$
Lawsonite	$CaAl_2(Si_2O_7)(OH)_2 \cdot H_2O$
Pumpellyite	$Ca_2MgAl_2[SiO_4][Si_2O_7](OH)_2 \cdot H_2O$
Ring silicates	
Tourmaline	$Na(Mg,Fe,Li,Al)_3Al_6[Si_6O_{18}]$ $(BO_3)_3(O,OH,F)_4$
Beryl[a]	$Al_2Be_3Si_6O_{18}$
Cordierite[a]	$Mg_2Al_3[AlSi_5]O_{18}$
Chain silicates	
Orthopyroxene	$(Mg,Fe)_2Si_2O_6$
Ca-Clinopyroxene	$Ca(Mg,Fe)Si_2O_6$
Orthoamphibole	$(Mg,Fe)_7Si_8O_{22}(OH)_2$
Ca-Clinoamphibole (hornblende)	$Ca_2(Mg,Fe)_5Si_8O_{22}(OH)_2$
Sheet silicates	
Clay minerals	Various
Serpentine	$Mg_3Si_2O_5(OH)_4$
Pyrophyllite	$Al_2Si_4O_{10}(OH)_2$
Talc	$Mg_3Si_4O_{10}(OH)_2$
Muscovite	$KAl_2(AlSi_3O_{10})(OH)_2$
Biotite	$K(Mg,Fe)_3(AlSi_3O_{10})(OH)_2$
Margarite	$CaAl_2(Al_2Si_2O_{10})(OH)_2$
Clintonite	$CaMg_2Al(Al_3SiO_{10})(OH)_2$
Chlorite group	$(Mg,Fe,Al)_3(Si,Al)_4O_{10}(OH)_2 \cdot$ $(Mg,Fe,Al)_3(OH)_6$
Framework silicates	
Silica group (quartz, tridymite, cristobalite)	SiO_2
Feldspars	
Potassium feldspar	$KAlSi_3O_8$
Plagioclase	$NaAlSi_3O_8$–$CaAl_2Si_2O_8$
Feldspathoids	
Leucite	$KAlSi_2O_6$
Nepheline	$Na_3K(Al_4Si_4O_{16})$
Sodalite	$Na_8[Al_6Si_6O_{24}]Cl_2$
Scapolite	$Na_4Al_3Si_9O_{24}Cl$–$Ca_4Al_6Si_6O_{24}CO_3$
Zeolite group	$M_xD_y(Al_{x+2y}Si_{n-x-2y}O_{2n})mH_2O$

[a] Beryl and cordierite also may be considered framework silicates.

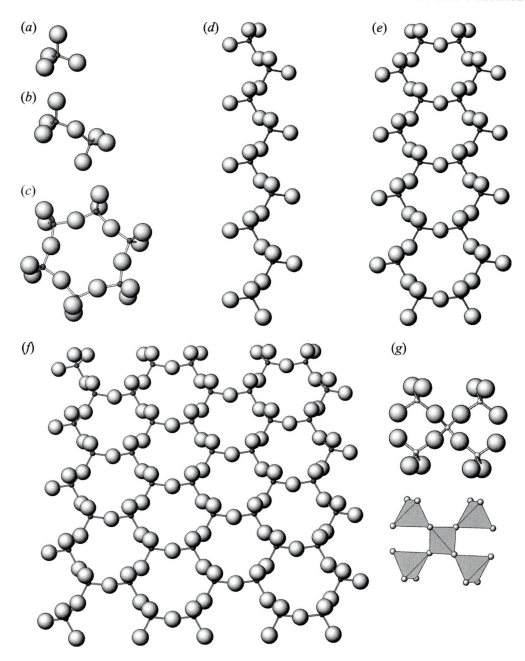

Figure 11.2 Silicate structures. (*a*) Orthosilicates. The basic building block is a single tetrahedron. (*b*) Disilicates. A single oxygen is shared between two tetrahedra. (*c*) Ring silicates. Tetrahedra share two oxygen each to form rings. (*d*) Single-chain silicates. Two oxygen are shared per tetrahedron. (*e*) Double-chain silicates. Tetrahedra share either two and or three oxygens. (*f*) Sheet silicates. Three oxygen per tetrahedron are shared to form continuous sheets. (*g*) Framework silicates. All four oxygens on the central tetrahedron are shared with adjacent tetrahedra as shown in the stick/ball and polyhedra diagrams. While not shown, the oxygens in the outer four tetrahedra are shared with additional tetrahedra to form a three-dimensional framework.

Except for quartz (SiO_2) and its polymorphs, the structures produced by the tetrahedra in their various configurations have a net negative charge. Cations are therefore required to balance charges and to effectively serve as the "mortar" that holds the silicate tetrahedron structures (single tetrahedra, chains, rings, etc.) together. Table 11.2 lists the common minerals in each class of silicate. The structural details of each group of silicates are described in the chapters that follow.

As a general rule, a high degree of polymerization is possible only if silicon is relatively abundant. Framework silicates such as quartz and K-feldspar are therefore found only in rocks containing fairly abundant Si. Orthosilicates such as olivine are restricted to rocks relatively low in silicon and rich in other elements, including iron and magnesium. An important complicating factor to this generalization is that Al may occupy some tetrahedral sites in silicate minerals. Hence, framework silicates such as plagioclase and nepheline are found in some relatively low-silicon rocks (gabbro and syenite, for example), but only if Al is available to occupy some of the tetrahedral sites.

MAFIC VERSUS FELSIC

For purposes of general discussion it is convenient to separate the common silicate minerals into two broad groups, mafic and felsic.

Mafic silicates are those that contain magnesium and/ or iron as major constituents. The name is derived from these elements—*ma* from *ma*gnesium and *f* from the Latin word for iron, *ferrum*. The common mafic silicates include biotite, amphiboles, pyroxenes, and olivine. All tend to be fairly dark colored, with some exceptions. Rocks in which they are abundant are therefore usually dark colored.

The felsic minerals include those that lack Mg and Fe as major constituents. They include the feldspars (from which the name was derived), quartz, muscovite, and feldspathoids.

IGNEOUS ROCKS

It should be no surprise that igneous rocks are composed dominantly of silicate minerals. Oxygen and silicon are, after all, the most abundant elements in the Earth's crust and igneous rocks comprise a large portion of that crust. The silicate minerals found in abundance in igneous rocks include a relatively small selection: quartz, K-feldspar, plagioclase, muscovite, biotite, Ca-clinoamphibole (e.g., hornblende), Ca-clinopyroxene (e.g., augite), orthopyroxene, and olivine. The relative amounts of these minerals, plus one additional group—the feldspathoids—provide the mineralogical framework within which igneous rocks are classified.

Figure 11.3 shows the conventional classification scheme for igneous rocks, which is based on the relative amounts of quartz, K-feldspar, plagioclase, and feldspathoids. Note that quartz and feldspathoid minerals are mutually exclusive; they are not normally found together. This classification is not universally accepted and has significant weaknesses, but it serves as a common point of reference. The classification distinguishes between plutonic rocks (usually coarse grained) and volcanic rocks (usually fine grained). That distinction must first be made based on field relations or other information.

The classification in Figure 11.3 is based on the **modal mineralogy**, that is, the volume percent of the rock comprised by each mineral. This is normally determined by "point counting" the rock. Point counting is done on a thin section with a petrographic microscope. The slide is moved in increments beneath the crosshairs with the aid of a mechanical stage (Figure 7.9) that holds the slide and allows it to be moved systematically by turning small knobs. The mineral beneath the crosshair after each movement is identified and tallied.

The fine grain size of volcanic rocks usually precludes the accurate determination of their modal mineralogy so Figure 11.3b cannot easily be used in practice. Various classification schemes based on the bulk chemical analysis of volcanic rocks are in common use and are described in Philpotts and Agee (2009) and most other texts on igneous petrology. Figure 11.4 shows one of these classifications based on the relative amount of SiO_2 and alkalis (Na_2O + K_2O). Another common method of dealing with fine-grained volcanic rocks is to calculate the **normative mineralogy** based on the chemical analysis. The normative mineralogy is the collection of minerals that could potentially be created from the bulk chemistry given certain assumptions concerning how to allocate the different elements. Refer to Philpotts and Agee (2009) or other igneous petrology books for the details of normative calculations.

Note that a number of the fields on the right side of Figure 11.3a contain two or more names—quartz monzodiorite and quartz monzogabbro, or quartz diorite, quartz gabbro, and quartz anorthosite for example. Table 11.3 shows criteria to distinguish among the diorite and gabbro varieties shown on the diagram. The anorthosite varieties are distinguished from the others because they are composed of more than 90% plagioclase.

Using Figure 11.3 to classify mafic and ultramafic rocks is not practical because these rocks lack quartz and K-feldspar, and in some cases plagioclase. Figure 11.5 shows one means of classifying these rocks based on the relative amounts of plagioclase, orthopyroxene, clinopyroxene, and olivine. This diagram includes members of the gabbro clan (gabbro, norite, olivine gabbro, olivine norite), pyroxenites (orthopyroxenite, clinopyroxenite, websterite, olivine orthopyroxenite, olivine clinopyroxenite), peridotites (lherzolite, wehrlite, harzbergite, dunite), anorthosite, and troctolite.

For purposes of general discussion, it is convenient to group igneous rocks into four broad categories based on the amount of SiO_2 in the bulk chemical analysis.

FELSIC: > 65% SiO_2

INTERMEDIATE: 55–65% SiO_2

MAFIC: 45–55% SiO_2

ULTRAMAFIC: < 45% SiO_2

Some sources place the breaks at 52 and 63% SiO_2.

The felsic rocks include granite, granodiorite, rhyolite, rhyodacite, and similar types. Quartz, K-feldspar,

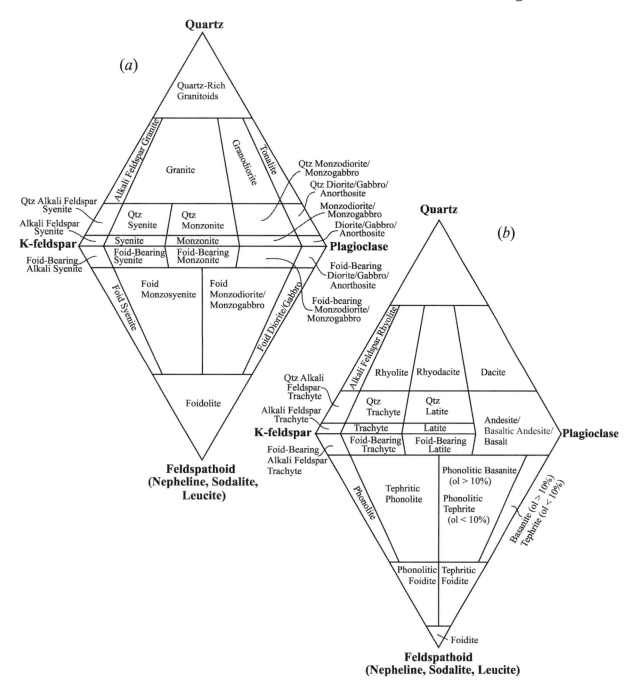

Figure 11.3 Classification of igneous rocks. Classification is based on the relative amounts (volume percentages) of quartz, K-feldspar, plagioclase, and feldspathoids (nepheline, sodalite, leucite). Other minerals are ignored. (*a*) Plutonic rocks. (*b*) Volcanic rocks. Use of this diagram for volcanic rocks is greatly complicated by the fact that fine grain sizes preclude the accurate determination of the amounts of each mineral. Adapted from Streckheisen (1976) and Le Maitre and others (2002).

and plagioclase are all present in substantial amounts. Note that felsic minerals are abundant in felsic rocks. Additional common minerals include biotite, muscovite, and hornblende.

The intermediate group includes diorite, quartz diorite, and related intrusives, and andesitic volcanic rocks. Plagioclase is abundant, and K-feldspar and quartz are fairly minor or absent. Hornblende is the major mafic mineral in intrusive rocks, although biotite and/or some pyroxene also may be present. Andesitic volcanics may contain either pyroxene, hornblende, or biotite as the major mafic mineral.

The mafic group is most abundantly represented by basalt among volcanic rocks and by the gabbro clan among intrusives. The major minerals include plagioclase, clinopyroxene, and orthopyroxene, with or without olivine. Mafic minerals are abundant in mafic rocks.

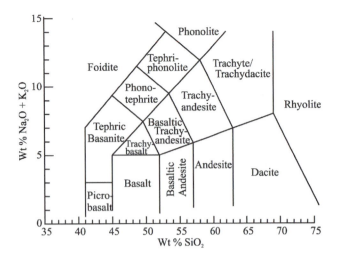

Figure 11.4 Chemical classification of volcanic rocks based on weight percent of SiO_2 and $Na_2O + K_2O$. Based upon Le Bas and others (1986).

Table 11.3 Criteria to Distinguish between Members of the Diorite and Gabbro Clans Shown in Figure 11.3[a]

Criteria	Diorite	Gabbro
Plagioclase composition	< 50% Ca (< An_{50})	> 50% Ca (> An_{50})
Dominant mafic mineral	Amphibole	Pyroxene
Associated rocks	Granodiorite/granite	Other mafic/ultramafic varieties
Wt % SiO2	> 55%	< 55%

[a] Figure 11.5 shows the classification of mafic, ultramafic, and related rocks including members of the gabbro clan.

Ultramafic igneous rocks are not abundant, either in continental or oceanic environments. The major minerals include orthopyroxene, Ca-clinopyroxene, olivine, and, in some cases, hornblende. Rocks containing mostly pyroxene and olivine are known as peridotite. The Earth's mantle has an ultramafic composition, and it is not uncommon to find ultramafic xenoliths of mantle rock carried to the surface in a basaltic lava. An important ultramafic rock is kimberlite, which is a mantle-derived rock consisting of olivine, Mg-rich biotite, and pyroxene. Kimberlite also may contain diamonds.

Rocks that do not fit conveniently in the felsic/intermediate/mafic/ultramafic groupings are not abundant. However, two groups are worthy of mention here—alkaline igneous rocks and anorthosites.

The alkaline igneous rocks include syenite, monzonite, trachyte, latite, phonolite, and feldspathoid-bearing rocks shown in Figure 11.3. Their compositional characteristic is an abundance of K and Na (the alkali elements). Because Si is not sufficiently abundant to allow quartz to form, these rocks are often referred to as silicon

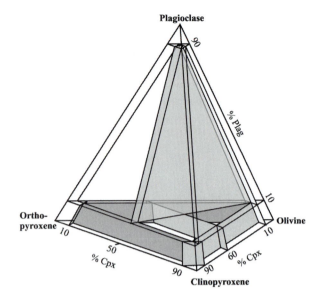

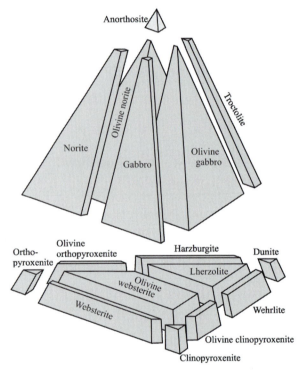

Figure 11.5 Classification of plutonic igneous rocks lacking significant amounts of quartz, K-feldspar, or feldspathoids. Adapted from Streckheisen (1976) and Le Maitre and others (2002).

undersaturated. Common minerals include K-feldspar, plagioclase, and one or more feldspathoids (nepheline, sodalite, leucite). Mafic minerals include amphiboles, pyroxenes, and biotite.

Anorthosite most commonly occurs in batholith-sized bodies, most of which were emplaced in the Proterozoic crust around roughly 1.45 to 1.1 billion years ago. This rock type is composed of plagioclase with minor amounts of pyroxene and olivine. It is common to find large bodies of orthopyroxene syenite or related alkali-rich rock associated with anorthosite. The nature and source of the magma from which anorthosite crystallized remain poorly understood.

Pegmatites are worthy of special mention. They are very-coarse-grained igneous rocks most commonly associated with granitic intrusions. They are interpreted to be produced from the last residual magma in a crystallizing granitic pluton in which water and a variety of less common elements (Li, B, Be, F, U, etc.) have been concentrated. The major minerals are quartz, K-feldspar, Na-rich plagioclase, muscovite, and biotite; so mineralogically, they are like granite. Other common minerals include tourmaline, beryl, garnet, spodumene, and lepidolite. Additional minerals are listed in Table C.1 in Appendix C. Pegmatites are economically important because the minerals found in them are the primary source of beryllium, lithium, rare earth elements, and other elements that are important in many industries.

Pegmatite bodies (Figure 11.6) have a wide range of sizes and shapes, though roughly dike-shaped bodies less than a meter to tens of meters thick are quite common. Most are compositionally zoned, often with a quartz-rich core. The individual mineral grains/crystals in common pegmatites range in size from a few millimeters up to a meter across. In extreme cases, individual crystals up to 7 or 8 meters long have been found.

Introductory geology students often infer that the extremely coarse grain size means that the magma from which pegmatites crystallized must have cooled very, very slowly. However, this is probably not the case. Pegmatites can have cooled no more slowly than associated igneous rocks, and they often are small igneous bodies that cannot have retained their heat for long. The explanation for the very coarse grain size probably has to do with the very water-rich nature of the liquid from which they crystallized.

Water in the pegmatite fluid greatly reduces the viscosity by preventing short-range polymerization of silicon tetrahedra. Even in a magma, silicon bonds to oxygen to form silicon tetrahedra, and those tetrahedra may share oxygen anions (polymerize) to form short-range chains, rings, and so forth. The addition of water prevents that polymerization because some of the water dissociates to H^+ and OH^-. If an H^+ cation bonds to an oxygen anion on a silicon tetrahedra, that oxygen is no longer available to be shared with another silicon tetrahedron. Similarly, the OH^- can occupy a corner of a silicon tetrahedron with the same result.

The reduction in the size of the short-range chains, rings, and so forth of silicon tetrahedra greatly reduces viscosity and allows cations and anions to migrate freely over larger distances to get to a growing mineral nucleii. Once a few mineral nucleii have become established, by either homogeneous or heteorogeneous nucleation, the low fluid viscosity allows the nuclei to grow rapidly and extract

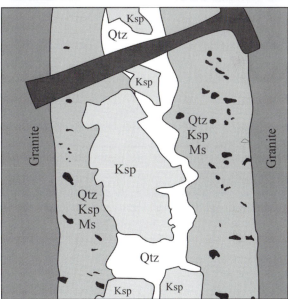

Figure 11.6 This small pegmatite dike (*top*) that has intruded Precambrian granite is compositionally zoned. The outer zones consist of intergrown quartz (Qtz), K-feldspar (Ksp), and muscovite (Ms) with scattered tourmaline grains (black). The central zone consists of large K-feldspar grains, some with crystal faces, and large masses of quartz. The sketch at the bottom shows the major components.

from the liquid the chemical constituents needed for their growth. That rapidly reduces the degree of supersaturation of those constituents in the liquid and effectively shuts down additional mineral nucleation. Because only a few mineral nuclei get established, the crystals that grow from them must be very large. Recall from Chapter 5 that an igneous rock's grain size is fundamentally controlled by the abundance of mineral nuclei that are produced.

Magmatic Processes

Because magma is derived by melting preexisting rocks, it follows that the composition of the rock being melted must have exerted major control on the composition of the magma. For example, felsic magmas are derived by melting relatively felsic continental crust, and mafic magmas are derived by melting the ultramafic magma.

A complicating factor is that most rocks melt incongruently. That means that melting begins at one temperature and is not complete until a higher temperature has been reached. It also means that a partial melt and the residual rock have different compositions, though the sum of melt plus residual rock must equal the composition of the original rock. This provides a mechanism to produce melts of different composition from the same parent rock if melting occurs in stages and the melt is removed after each stage. In general, when many silicate rocks are partially melted, the melt has a higher silica content than the parent rock and the residual rock is depleted in silica. Progressive melting may therefore yield a relatively silica-rich magma first, followed by a series of progressively less silica-rich magmas.

It is possible for a magma, once formed, to be emplaced or erupted without further modification and to subsequently crystallize in place. Often, however, a variety of processes operate on the magma to yield a great diversity of resulting rocks. These magmatic processes include crystal settling, magma mixing, and assimilation. For additional information about these and other magmatic processes, consult Winter (2010) or other books on igneous petrology.

Magmas do not crystallize at a single temperature. Rather, they crystallize over a substantial temperature range as first one mineral begins to grow, followed by another at lower temperature, and so forth. Because the crystals growing from a magma rarely have the same composition as the magma, the composition of the residual magma changes or evolves as crystallization progresses. The crystallization of plagioclase described in Chapter 5 provides a good example. If the crystals stay with the magma from which they are growing, the resulting igneous rock will have the same composition as the parent magma. If, however, the crystals can be removed from the magma as the crystals grow, a mechanism is available to produce magmas with a range of composition.

Consider a magma chamber at some depth, from which crystals are growing. If the crystals are more dense than the magma (as they usually are), they will tend to settle through the magma and accumulate on the magma chamber floor. Magma near the top of the magma chamber will, over time, become depleted in the elements being taken up by the settling crystals and enriched in the remainder. If batches of magma are periodically removed from the magma chamber, to be erupted from a volcano for example, the composition of each batch of magma should be different and the rock formed from each will have differences in mineralogy.

Sequences of compositionally diverse magmas produced by this process are called a **fractionation series**, and the process is called **magmatic differentiation**. The accumulation of crystals on the magma chamber floor becomes a separate igneous rock called a **cumulate**. Fractionation series usually produce progressively more Si-rich magmas with time.

Magmas with different compositions can and do mix with each other. Scenarios in which this can happen include the following.

- Basaltic magma produced in the mantle may pool in the more Si-rich lower crust. Heat from the crystallizing basalt melts the crustal rocks to produce a Si-rich magma that may, under some circumstances, mingle with the basalt.

- Fresh magma may be injected into a magma chamber that has already been crystallizing and differentiating. The fresh and old magma may then mingle.

If the two magmas mix, the resultant will have an intermediate composition and mineralogy.

Magmas cannot exist in isolation from the rocks in which they are contained. They are derived by melting preexisting rocks, travel from their source to final destination through other rocks, and ultimately crystallize within or on those rocks. At every step along the way, the potential exists for chemical interaction between magma and surrounding rocks. The nature of the interaction may be complex and can include direct melting, dissolution, or chemical exchange with minerals of surrounding rocks. The nature of the chemical change in the magma depends on details of the interaction.

Igneous Environments

It was recognized fairly early in the study of igneous rocks that certain rock types are commonly associated and that others rarely or never occur together. The plate tectonic paradigm has provided a useful means of placing these observations within a tectonic framework.

Oceanic Associations

Mid-ocean ridges are the sites of the most prolific magma generation on Earth. Apparently, mantle rocks rise beneath the ridges as part of large-scale convective overturn of the mantle associated with sea floor spreading. Melting triggered by this upwelling yields basaltic magma. This magma erupts onto the ocean floor along the mid-ocean ridge or solidifies in shallow magma chambers to form the large majority of oceanic crust. The oceanic crust spreads laterally and ultimately returns to the mantle in subduction zones. Almost all the rocks formed at the mid-ocean ridge are basaltic and mineralogically consist of plagioclase, pyroxene, and usually a small amount of olivine.

An additional common oceanic association is at intra-plate volcanic centers, like Hawaii. Not all form volcanic edifices large enough to project above sea level. Most are inferred to be positioned over "hot spots" within the mantle, which are capable of generating substantial quantities of basaltic magma.

Convergent Plate Boundaries

Convergent plate boundaries are also prolific sites for magma generation. The abundant volcanic centers that girdle the Pacific Ocean basin, for example, are testimony to the ability to generate magmas in the upper mantle and crust above subduction zones. The nature of the magmas depends significantly on the local conditions associated with the subduction zone. It is clear that most of the magmas, either directly or indirectly, are the result of melting the wedge of upper mantle above the downgoing slab of oceanic crust. The basaltic rocks of the oceanic crust may become altered and hydrated. This water is evidently released from the downgoing slab as the result of metamorphism and can move into the overlying mantle where it promotes melting.

Volcanic rocks found above subduction zones are characteristically andesitic, although the eruptions from each volcanic center almost always are part of a continuum of compositions. The earliest erupted lavas typically are basaltic. With time, the erupted lavas become progressively more Si rich and form abundant andesite and, in some cases, dacite and rhyolitic rocks. The abundance of Si-rich rocks is usually quite restricted in island arc environments where continental crust is absent. Where the subduction zone extends beneath continental crust, dacite and rhyolite may make up a substantial amount of the volcanic rocks. These relations suggest a substantial contribution of continental crustal material to the magmas erupted in continental environments, either as a result of assimilation of crustal material or generation of more Si-rich magmas from the continental crust.

Intrusive rocks in convergent environments are typically more diverse than the volcanic rocks and, on average, tend to be more Si rich. Granodiorite is the most abundant intrusive rock, at least in continental environments where geologists have the most opportunity to make observations. Most granodiorite magma is probably produced when more mafic mantle-derived magmas pool in or beneath the lower continental crust and release heat while crystallizing. That heat then melts portions of the lower crust. Some granodiorite magma is also undoubtedly produced by crystallization and differentiation of mafic magmas in the same environment. In either case, when a sufficient volume of relatively low-density, Si-rich magma is produced, it rises diapirically through the overlying crust to be emplaced at shallower levels, or to be erupted.

Intracontinental

Continental flood basalts are found in many areas of the world. Some are Precambrian; others are Tertiary or Quaternary. Examples include the Miocene Columbia River basalts in Washington, Idaho, and Oregon, U.S.A.; latest Cretaceous Deccan basalts in India, and Precambrian Keweenawan basalts in Michigan, U.S.A. and adjacent areas. All appear to be produced in areas where the continent was being rifted or pulled apart. Great quantities of basaltic magma from the mantle may be erupted over very short periods of geologic time in these environments. Individual lava flows may be a few meters to over 100 meters thick and may extend for over 100 km.

Layered igneous intrusives are large mafic bodies found within continental crust. The bulk composition of these intrusives is about the same as a typical basalt. Sequential crystallization of olivine, pyroxene, plagioclase, and other minerals results in the formation of mineralogically distinct layers as first one group of crystals and then another settles on the magma chamber floor. Layers range in thickness from millimeters to hundreds of meters within bodies that can be several kilometers to hundreds of kilometers in long dimension. Most large, layered igneous intrusives are Precambrian. Many appear to have been formed within a rifting environment where magma formed by melting the upper mantle was able to rise to intermediate or higher levels within the crust. Some, such as the Sudbury body in Ontario, Canada, are located within large meteorite impact craters.

Other intracontinental igneous associations are not voluminous and tend to be quite diverse. One group includes alkaline igneous rocks associated with intracontinental rift systems. The magmas appear to be derived by melting the upper mantle, but complex crustal interaction and contamination also may be inferred. These rocks include syenite, phonolite, and other feldspathoid-bearing rocks.

TERRIGENOUS SEDIMENTARY ROCKS

Terrigenous sedimentary rocks are formed by the accumulation and subsequent lithification of mineral and rock particles derived by weathering and eroding preexisting rocks exposed at the land's surface. Other terms applied to terrigenous sediments are clastic, siliclastic, detrital, and fragmental. Whatever the term used for them, these rocks are always stratified or layered. Mineralogically, terrigenous sedimentary rocks are composed dominantly of silicate minerals. A simplified classification of terrigenous sedimentary rocks, based upon grain size, is shown in Table 11.4.

Conglomerates and breccias are differentiated by the degree to which the pebbles, cobbles, and boulders are rounded by transport processes—conglomerates are composed of rounded clasts; breccias are angular. They can be further subdivided based on grain size and the nature of the particles—quartz pebble conglomerate, chert cobble breccia, and so forth.

Among the sandstones, a distinction can be made between framework grains and matrix. Framework grains are sand particles—usually quartz, feldspar, or lithic fragments. Lithic fragments are recognizable pieces of rock consisting of multiple mineral grains. Matrix consists of clay or other fine-grained minerals between the sand grains. Sandstones with less than 15% matrix are **arenites** (Figure 11.7a); those with 15 to 75% matrix are **wackes** (Figure 11.7b).

Silt- and clay-sized particles collectively are called mud, in reference to the nice oozy material that can be made from silt and clay mixed with water. **Mud rock** is a generic term for all sedimentary rocks composed of over 75% mud-sized particles. The term **siltstone** is used if the particles are dominantly silt sized (distinctly gritty if chewed), **claystone** if the particles are dominantly clay sized (smooth if chewed), **shale** if the rock is fissile (breaks approximately parallel to bedding), and **mudstone** if the rock is not fissile. The terminology for mud rocks has repeatedly been redefined and modified, and consensus has not developed around the details of a single nomenclature.

One source of confusion in mud rock nomenclature is that the term *clay* is used in two ways. The first refers to a size—clay-sized particles are less than 0.004 mm across. The second refers to a group of hydrous sheet silicate minerals collectively called clay. Much of the clay-sized fraction of sedimentary particles is composed of clay minerals, but not all. Quartz and other minerals are usually present, and zeolites may be abundant in some situations. Similarly, clay minerals may be part of the silt-sized fraction of some rocks. As used in this book, the term *clay* refers to the minerals. The fine-grained size fraction of sediments will be referred to as clay sized.

Table 11.4 Simplified Classification of Terrigenous Sedimentary Rocks

Particle Name		Millimeters	Rock Type
Gravel	Boulder	256	Conglomerate (if particles are rounded) Breccia (if particles are angular)
	Cobble	64	
	Pebble	4	
	Granule	2	
Sand	Very coarse sand	1	Sandstone (see Figure 11.7)
	Coarse sand	0.5	
	Medium sand	0.25	
	Fine sand	0.125	
	Very fine sand	0.062	
Mud	Coarse silt	0.031	Mudstone (if lacks fissility) Shale (if fissile) Siltstone (if dominantly silt sized) Claystone (if dominantly clay sized
	Medium silt	0.016	
	Fine silt	0.008	
	Very fine silt	0.004	
	Clay sized		

Sedimentary Processes

The mineralogy of terrigenous sedimentary rocks depends both on the mineralogy of the rocks found in the source

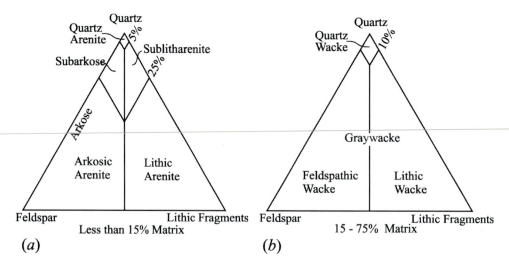

(a)   (b)

Figure 11.7 Classification of sandstones based on quartz, feldspar (K-feldspar + plagioclase), and lithic fragment content of the sand-sized particles. Lithic fragments are polymineralic pieces of rock. (*a*) *Arenites* have less than 15% matrix material (clay, fine quartz, etc.). *Arkose* is a common term for rocks with mostly quartz and feldspar grains. (*b*) *Wackes* have 15% or more matrix material. *Graywacke* is a more generic term applied to wackes with substantial amounts of quartz, feldspar, and lithic fragments. Modified from Dott (1964).

area of the sediment and on the processes of weathering, transport, deposition, and diagenesis.

Mechanical weathering serves to break rocks into smaller and smaller particles. The processes include frost wedging, biological activity (including geologists with rock hammers), exfoliation, and so forth. Were mechanical weathering the only process operating, the mineralogical composition of sediment would be the same as the mineralogy of the rocks being weathered.

Chemical weathering produces mineralogic changes that can be summarized as

Silicates + water + H$^+$
    = clay minerals + quartz + ions in solution.   (11.1)

The water in most cases is meteoric, meaning that it is ultimately precipitation (rain, snow, etc.) that has percolated into the ground to become part of the groundwater supply. The H$^+$ is mostly from carbonic acid (H_2CO_3), produced as a consequence of dissolving atmospheric CO_2 in the water. Some H$^+$ also comes from organic acids produced by microbial activity and the decomposition of organic material in the soil, or sulfuric acid (H_2SO_4) produced from sulfur emitted into the atmosphere by volcanos or from industrial processes.

Typical reactions that can produce clays such as kaolinite and montmorillonite involve the breakdown of minerals such as K-feldspar and pyroxene.

$2KAlSi_3O_8 + 2H^+ + 9H_2O$
K-feldspar
    $= Al_2Si_2O_5(OH)_4 + 4H_4SiO_4 + 2K^+$ (11.2)
    kaolinite

$pCa(Mg,Fe,Al)(Si,Al)_2O_6 + qH_2O + rH^+$
    pyroxene
    $= s(Ca)_{0.5}(Al_5Mg)(Si_4O_{10})_3(OH)_6 \cdot nH_2O$
        montmorillonite
    $+ tFeO \cdot OH + uCa^{2+} + vH_4SiO_4$ (11.3)
       goethite

The H_4SiO_4 represents silica in solution that, in many cases, may be incorporated into the structure of clay minerals produced by weathering other minerals in the same rock. The values of p, q, r, etc. in Equation 11.3 depend on the exact composition of the pyroxene. Some silicate minerals are more susceptible to chemical weathering than others (Figure 11.8).

Microorganisms are intimately involved in many weathering processes (Konhauser, 2007). Freshly exposed rock is rapidly colonized by bacteria, algae, fungi, and lichens, because the minerals in the rocks provide a rich source of bioessential elements. Fungi that colonize rock surfaces send filaments into cracks, cleavages, and along grain boundaries. These both pry mineral grains apart and etch or pit mineral surfaces. Bacteria also coat mineral surfaces with extracellular polymers (polysaccharides

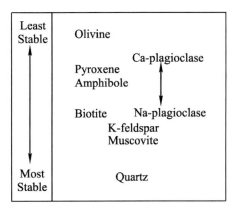

Figure 11.8 Chemical weathering susceptibility of common silicate minerals.

and proteins—think slime) that retain water and increase the time for water to participate in chemical reactions with mineral surfaces. Microbes accelerate dissolution of minerals by production of organic acids that are by-products of metabolism or whose purpose is to release nutrients from the minerals. The cellular surfaces of microbes may also serve as sites where clay and other minerals nucleate. These biologic processes work together with the other processes of mechanical and chemical weathering to accelerate weathering of rocks and the formation of soils.

Clay minerals (see Chapter 13) have many different structures and compositions. The nature of the silicate minerals being weathered and details of the chemical environment control which clay mineral is produced. Time also plays a significant role because weathering is often progressive—one group of clay minerals forms first but, with time, the members of that group may be progressively altered to form other clay minerals.

The iron from mafic minerals may be incorporated into the structure of some clay minerals. Iron may also form iron oxide minerals that stain the weathered rock various shades of red or brown. The goethite formed by weathering of pyroxene (Equation 11.3) is an example.

Transport, most commonly by streams, but also by wind, waves, and other processes, produces both continued chemical weathering and progressive reduction of grain size. Near the source, sediment derived from a plutonic igneous/metamorphic terrane may contain quartz and fairly abundant plagioclase, K-feldspar, muscovite, biotite, and hornblende. With transport, these latter minerals progressively break down to yield clay, as described earlier, so that the sediment issuing from an extensive river system into the ocean typically consists dominantly of quartz and clay. Continued weathering also promotes a reduction of grain size because weathered minerals lose strength and hardness. Similarly, the mechanical abrasion that occurs during transport promotes chemical weathering because chemical reaction rates increase as grain size decreases.

An additional result of transport is segregation of minerals based on grain size, shape, and specific gravity. An obvious result in marine environments is that clay-rich mud can be deposited only in areas in which the water is not vigorously agitated—offshore, for example. Sand, composed of quartz, with or without feldspars and lithic fragments, may be deposited in higher energy environments such as a beach. In fluvial environments, clay-rich mud is deposited in overbank areas, and the quartz-rich sand is deposited in active channels.

The term **compositional maturity** refers to the degree to which sediment lacks both feldspar and lithic fragments or other easily weathered materials. A compositionally mature sandstone, for example, consists almost exclusively of quartz grains, and a compositionally mature mud rock consists dominantly of quartz and clay. A compositionally immature sediment, on the other hand, contains substantial amounts of feldspar and/or lithic fragments. In general, intense weathering in the source area and/or extensive transport produce compositionally mature sediment.

Quartz, feldspars, and clay are the most abundant minerals found in sediment, but not the only ones. Other minerals may be found in small amounts. Because many of these minerals have specific gravities greater than either quartz, clay, or the feldspars, they have collectively been referred to as **heavy minerals**. Some may be major constituents of rocks of the source area; others are present only as accessories. Common heavy minerals and their relative susceptibility to weathering are listed in Table 11.5.

The net result of all these processes is a progressive reduction of grain size, weathering to produce clay and quartz, and winnowing of minerals based on grain size, specific gravity, and shape. It should be no surprise that the most abundant sedimentary rocks are mud rocks, composed dominantly of clay and quartz, with minor amounts of feldspar. The clay is produced by the breakdown of other minerals, and the quartz is largely derived from the parent rocks. Feldspars and lithic fragments are usually a major constituent of clastic sedimentary rocks only if the sediment is close to its source and/or has been derived from an area subjected to rapid erosion so that weathering is limited.

Once deposited, the mineralogy of sediments continues to evolve during the processes collectively referred to as **diagenesis**. Mineralogic changes that may take place include the following.

- Alteration: Primary feldspars and other minerals in the sediment may break down to form clays or zeolites. This process may release silica, which can precipitate to form a cementing agent. The fine-grained clay and zeolite may also serve as a binder holding larger grains together.

- Dissolution: Some minerals may dissolve into the pore fluid over time.

- Recrystallization: Small grains may combine by recrystallization to form larger grains. Some minerals may be dissolved at one location and reprecipitated elsewhere.

- Precipitation: Ions in solution can precipitate either as new minerals or as additions to existing minerals. The silica that cements many arenites typically forms overgrowths on existing quartz grains. Calcite cement precipitates from pore water. Other minerals that may form by diagenetic processes include the feldspars, gypsum, anhydrite, iron oxides, and pyrite.

- Changes in clay mineralogy: The clay minerals evolve and (re)crystallize with time and burial. The detailed clay mineralogy of a sediment after many millions of years may be quite different than when the sediment was originally deposited. These processes are described in more detail in Chapter 13.

Sedimentary Environments

The depositional environments of terrigenous sediment can be broadly grouped into terrestrial, transitional, and marine (Table 11.6). The large majority of preserved sedimentary rocks have been deposited in marine or transitional environments, although most of the sediment undoubtedly spent time as deposits in, or passed through, terrestrial environments.

Sandstones are among the most useful sedimentary rocks when it comes to recording the nature of the source

Table 11.5 Heavy Minerals Commonly Found in Terrigenous Sediments[a]

Stability	Minerals
Extremely stable	**Zircon, tourmaline**, rutile
Stable	**Garnet (low-Fe), staurolite, biotite,** magnetite, ilmenite, apatite, monazite
Moderately stable	**Epidote, kyanite, garnet (Fe), sillimanite, titanite, zoisite**
Unstable	**Amphibole, pyroxene, andalusite**
Very unstable	**Olivine**

[a] Silicates are in bold type.
Source: Adapted from Pettijohn and others (1987).

Table 11.6 Sedimentary Environments

Terrestrial	Transitional	Marine
Alluvial fan and fan delta	Deltaic	Sublittoral (continental shelf)
Fluvial (stream)	Littoral (beach)	Bathyal marine fan (continental slope)
Lacustrine (lake)	Tidal	
Glacial	Estuarian	Abyssal plain
Eolian (wind)		

area, which is referred to as the **provenance** of the sediment. Figure 11.9 shows, in a general way, the mineralogical makeup of sandstones derived from different tectonic environments. The three corners of the triangular diagram are for the relative amounts of framework grains: Q is for quartz, including polycrystalline grains of quartzite and chert, F is for feldspar grains (K-feldspar + plagioclase), and L is for igneous, metamorphic, and sedimentary polycrystalline lithic grains exclusive of quartzite and chert. The diagram distinguishes sand derived from continental craton, magmatic arcs, and recycled orogens (mountain belts).

Three sediment sources are recognized within continental environments in which plutonic igneous and metamorphic rocks are exposed. The most quartz-rich sediment is from continental craton interiors with low relief. Erosion is slow relative to weathering, so that feldspars and other less stable minerals are destroyed. The feldspar-rich sediments are derived from areas of substantial basement uplift where erosion is fast relative to weathering, so that feldspars are not extensively weathered. The transitional continental environment is intermediate between these two extremes.

Within magmatic arcs, the source rocks are dominantly andesitic volcanic rocks, with intermediate to felsic intrusives at depth. Sand derived from this environment is dominated by feldspars and lithic fragments; quartz is not particularly abundant. Sands with the most abundant lithic fragments are derived from volcanic arcs that have not experienced extensive erosional dissection. Deeply dissected arcs expose the felsic intrusives at depth and tend to produce sands with

more feldspar and quartz. Transitional arc environments are intermediate between these extremes.

Recycled orogenic sand is derived from sedimentary rocks exposed in orogenic belts such as the Cordilleran fold-and-thrust belt that extends from northwestern Canada and through the western United States. These rocks may be metamorphosed to some degree and some volcanic rocks may be present, but felsic intrusives are not exposed in large amounts at the surface. Because igneous rocks are not major contributors to the sediment, feldspars are not abundant.

The reader should recognize that the boundaries drawn on Figure 11.9 are in reality gradational and that the sand deposited in any particular environment may be derived from a variety of areas with different tectonic characteristics and rock types. However, the diagram makes the important point that geological setting and tectonics can play a major role in determining the mineralogy of sediment.

METAMORPHIC ROCKS

Metamorphic rocks are those whose mineralogy, texture, and/or composition have been changed, usually as the result of heat, pressure, chemically reactive fluids, and/or deformation. The classification of common metamorphic rocks (Table 11.7) is fairly straightforward by comparison to the names applied to igneous rocks. The nomenclature is based on texture (grain size and presence or absence of foliation) and/or mineralogy. For rock types defined based on texture, additional modifiers are usually applied to specify some aspect of the mineralogy.

Muscovite biotite garnet schist

Quartz microcline plagioclase biotite gneiss

Andalusite cordierite hornfels.

Additionally, if the identity of the parent rock is evident, a metamorphic rock may be referred to by attaching the term "meta-" to the parent rock name: meta-rhyolite, meta-graywacke, and so forth.

Foliation can have two separate but related forms. Mineralogical foliation is produced by the alignment of inequant minerals. Micas, because they form platy crystals, can produce a well-developed foliation known as **schistosity** if coarse grained or **slaty cleavage** if fine. Rocks with schistosity or slaty cleavage break easily along the foliation because it is parallel to the perfect cleavage of the micas. A foliation also can be represented by alternating lighter- and darker-colored layers in the rock. Because it is common in gneiss, compositional layering is also known as **gneissic foliation**.

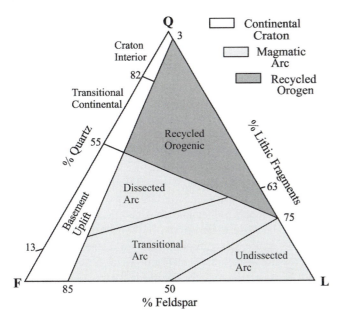

Figure 11.9 Mineralogy of framework grains in sandstone derived from different tectonic settings. Adapted from Dickinson and others (1983). See text for additional discussion.

Metamorphic Variables

The principal factors controlling metamorphism are changes in temperature and pressure, deformation, and migration of

Table 11.7 Classification of Common Metamorphic Rocks

Criteria	Rock Type	Comments
Defined based on texture		
Foliated	Slate	Very fine grained with dull luster on well-developed foliation; sheet silicates usually abundant
	Phyllite	Fine grained with silky luster on well-developed foliation; sheet silicates usually abundant
	Schist	Coarse grained with abundant inequant minerals such as micas that are aligned to produce a foliation
	Gneiss	Coarse grained with layering of light- and dark-colored minerals; roughly equant minerals such as quartz and feldspar are dominant
Not foliated	Hornfels	Fairly fine grained and massive; inequant minerals like sheet silicates may be abundant but are not foliated
	Granofels	Fine to coarse grained with a granular texture dominated by equant minerals such as quartz and feldspars
Defined based on mineralogy	Quartzite	Primarily quartz; derived from quartzose sandstone
	Amphibolite	Abundant amphibole such as hornblende; derived from mafic igneous rocks or closely related sediment; if well foliated, may be referred to as hornblende schist
	Marble	Primarily calcite or dolomite; derived from limestone or dolostone
	Skarn	Abundant Ca-, Mg-, Fe-, and Si-bearing minerals such as amphibole, pyroxene, olivine, garnet, wollastonite, calcite, and dolomite; magnetite may be abundant in some cases; derived from limestone/dolostone adjacent to felsic intrusive; if foliated, may be referred to as a calc-silicate gneiss
	Argillite	Fine to very fine grained, weakly metamorphosed mud rock without well-developed foliation; sheet silicates are abundant
	Serpentinite	Serpentine minerals (antigorite, chrysotile) abundant; derived from mafic or ultramafic igneous rocks
	Greenstone	Weakly metamorphosed mafic meta-igneous rock whose green color is derived from chlorite, amphibole, and/or epidote, other common minerals include titanite, calcite, dolomite, pumpellyite, and plagioclase; may be distinctly foliated

fluids. These may be considered external agents that act on rocks and produce metamorphic changes. One of the primary objectives in studying metamorphic rocks is to determine the conditions to which the rocks were subjected by these external agents, and from that information, deduce something about the tectonic history of the rocks. The bulk composition of the rocks being metamorphosed also is of primary importance in determining the nature of the metamorphosed rock. Shale and a limestone, if subjected to the same conditions of metamorphism, will produce distinctly different mineralogies, for example.

Temperature is perhaps the most important variable in metamorphism because most metamorphic reactions are driven by changes in temperature. The source of the thermal energy may be a nearby igneous intrusion, or more extensive regional sources, such as heat from the mantle or from breakdown of radioactive elements in the rocks. Among other things, high temperature promotes mineral reactions in which volatile components such as H_2O and CO_2 are lost.

The lower temperature limit of metamorphism is generally, and somewhat arbitrarily, taken to be 200°C. Mineralogical processes taking place at lower temperatures are considered to be part of sedimentary diagenesis, although diagenesis and metamorphism are clearly part of a continuum of changes that rocks experience as temperatures increase.

The upper temperature limit of metamorphism is also poorly defined. It should presumably be defined based on the temperature at which rocks are substantially molten. However, different rocks begin melting at different temperatures. For example, in the presence of water in many metamorphic environments, granitic rocks begin melting at 600–700°C, and mafic rocks begin melting at 800–900°C. Also, it may be difficult to determine the extent to which a rock was melted based on the textures and mineralogy preserved after the rock has cooled. An alternative, and also arbitrary, distinction is to reserve the term *igneous* for molten material (magma) that has been moved from where the melting occurred. If forced to pick a temperature, some petrologists would say that between 900 and 1000°C is the upper limit of metamorphism. Others argue that the upper limit goes much higher and that mantle rocks are metamorphic because their mineralogy and textures are produced by processes going on in the solid state.

Pressure also controls metamorphism. Lithostatic pressure (P) due to the mass of overlying rock is given by

$$P = \rho g h \qquad (11.4)$$

where ρ (kg/m^3) is the density of the overlying rock, g (9.80 m/sec^2) is the acceleration of gravity, and h (m) is the depth. Units of pressure are pascals (1 Pa): 1 Pa = 1 kg m^{-1} sec^{-2}. For geological purposes, pressure is often reported in

Table 11.8 Lithostatic Pressure Produced by Different Rock Types

Rock Type	Density (kg/m³)	kbar/km	km/kbar
Granite	2700	0.264	3.8
Basalt	3000	0.294	3.4
Peridotite	3300	0.323	3.1

gigapascals (GPa = 10^9 Pa) or kilobars (kbar): 1 kbar = 0.1 × 10^9 Pa. Table 11.8 shows the lithostatic pressure produced by granite, basalt, and peridotite. Pressure at the base of continental crust 40 km thick is roughly 10–11 kbar. High pressure favors the formation of minerals with high density because cation coordination numbers tend to increase as the pressure forces the anions into more tightly packed arrangements.

Lithostatic pressure is generally considered to be hydrostatic, meaning that it is equal in all directions. In active orogenic belts, an additional deviatoric stress is usually present. The deviatoric stress is the measure to which the pressure is not equal in all directions. The magnitude is generally not large—probably in the range of a few bars to no more than 100 bars—so it does not significantly affect the pressure stability of minerals in metamorphic rocks. However, the deviatoric stress is responsible for the extensive deformation that is found in many metamorphic terranes, and it controls the formation of the foliation and related fabrics found in metamorphic rocks. Detailed study of the deformation fabrics of metamorphic rocks provides a crucial link to understanding the tectonic environment in which the rocks were metamorphosed.

Prior to metamorphism, most rocks contain an aqueous fluid in the pore space and along grain boundaries. In addition, the clays, micas, and amphiboles are hydrous (contain H_2O), and carbonates contain CO_2 in their crystal structure. In general, fluids are lost during progressive metamorphism as the pore fluid is expelled and hydrous and carbonate minerals break down. The temperature stability of hydrous and carbonate minerals is strongly influenced by the composition of the pore fluid. For example, the temperature stability of hydrous minerals such as the micas is reduced if the aqueous pore fluid is diluted with CO_2, CH_4, or other chemical species. Similarly, carbonate minerals can be stable at high temperatures only if the pore fluid contains high levels of CO_2. Note also that the composition of the pore fluid will be affected by the breakdown of hydrous or carbonate minerals.

Metamorphic Processes

During metamorphism, a variety of changes may occur in a body of rock. These changes include recrystallization, new mineral growth, textural changes, and compositional changes.

Recrystallization involves changes in the size and shape of existing mineral grains in a rock. Many small grains of a mineral have much higher surface area than a single large grain with the same mass. Minerals will tend to recrystallize to produce a smaller number of large grains to eliminate the high-energy surface area. Rocks subjected to high temperature, which allows easy migration of ions, tend to recrystallize and become coarser grained.

If minerals are intensely deformed, for example in a shear zone, recrystallization may initially produce a reduction in grain size. Mineral grains with strongly deformed crystal lattices segment into numerous smaller grains whose lattices are unstrained as a means of reducing the strain energy. Subsequent recrystallization may allow the fine grains to coalesce to form larger grains. The grain size of rocks being ductilely deformed is therefore a balance between a reduction caused by deformation and an increase caused by subsequent recrystallization.

New mineral growth involves replacement of less stable minerals with minerals that are more stable under the temperature and pressure conditions to which the rocks are being subjected. For example, the introduction of staurolite with increasing temperature in the metamorphism of a mica schist may be accomplished by the following reaction.

$$17(Mg,Fe)_3Si_4O_{10}(OH)_2 \cdot (Mg,Fe)_3(OH)_6$$
$$+ 30KAl_2AlSi_3O_{10}(OH)_2$$
$$= 6Fe_2Al_9O_6(AlO_4)(SiO_4)_3(OH)_2$$
$$+ 30K(Mg,Fe)_3AlSi_3O_{10}(OH)_2 + 50SiO_2 + 62H_2O$$

17 Chlorite + 30 muscovite
= 6 staurolite + 30 biotite + 50 quartz + 62 H_2O (11.5)

The reaction will progress until either the chlorite or the muscovite is exhausted.

In many metamorphic terranes, the mineralogy of the rocks goes through progressive changes with increasing temperature. The key idea is that the mineralogy of a rock is an indicator of the temperature and pressure to which the rock was subjected. In many cases, the bulk composition of the rocks being metamorphosed does not change except for the loss of fluids. Hence, introduction of a new mineral means that previously existing minerals must be consumed.

A **mineral assemblage** is the collection of minerals found in a particular rock. The term is also used to refer to the reactant mineral assemblage and product mineral assemblage of a particular mineral reaction. In Reaction 11.5, chlorite + muscovite is the reactant mineral assemblage and is stable at lower temperature. Staurolite + biotite + quartz is the product mineral assemblage that is stable at higher temperature. Note that muscovite may be stable at higher temperatures with staurolite + biotite + quartz, but not with chlorite, and that quartz and biotite may be stable with chlorite and muscovite at lower temperatures, but not with staurolite.

The most obvious textural changes that accompany progressive metamorphism include an increase in grain size and formation of foliations. The increase in grain size can be the result of both recrystallization and new mineral growth. Development of foliations generally requires that the rock be deformed during metamorphism. Platy minerals like the micas can be rotated into parallel alignment by the deformation or may (re)crystallize in preferred orientations relative to the deviatoric stress.

Fluids derived from external sources may change the bulk composition of a rock in a process called **metasomatism**, by both dissolving some components and contributing others. For example, skarns deposits consisting of pyroxene, amphibole, garnet, magnetite, and other related minerals may be derived from carbonate-rich sedimentary rocks that have been heated and invaded by fluids from a nearby felsic intrusive.

Metamorphic Grade, Facies, Mineral Zone Boundaries, and Isograds

The term *metamorphic grade* is used in a general sense to describe the intensity of metamorphism, usually with reference to temperature. Figure 11.10 shows one division of temperature-pressure (T–P) space into very-low-, low-, medium-, and high-grade metamorphism. While the different grades have been identified with specific mineralogies in pelitic and other rocks, the term *grade* is probably best used in a generic sense to refer to T–P conditions.

A **metamorphic facies**, properly defined, refers to a set of metamorphic mineral assemblages, repeatedly associated in space and time, such that there is a constant and predictable relation between mineral composition and bulk rock chemical composition (Turner, 1981). The facies concept is firmly grounded in physical and chemical principles because rock mineralogy must change in response to temperature and pressure. However, application of facies terminology has been difficult. All the rocks of different composition in an outcrop are members of the same facies, which is defined based on the specific mineralogy of each rock type. If, in another outcrop, one of the lithologies, by

virtue of a metamorphic mineral reaction, has a different mineral assemblage, all those rocks must be in a different facies because the mineral assemblages are different. Because the number of possible rock compositions is large and the number of metamorphic mineral reactions that can produce different mineral assemblages also is large, the possible number of facies becomes completely unmanageable. Further, rocks subjected to the same temperature–pressure condition may have very different mineralogies depending on the nature of the fluid phase. For these reasons, facies have traditionally been defined based on the mineralogy of mafic rocks, and other rock types are largely ignored. Facies terminology has gradually degenerated in common usage to have meanings not substantially different than the grade terminology outlined above. Figure 11.11 shows the commonly defined facies in T–P space.

Geologists mapping metamorphic terranes routinely find that they can, within a particular group of rocks, identify boundaries in the field that define the first appearance of specific minerals, called **index minerals**, in the direction of increasing metamorphic grade (Figure 11.12). These mapped features are called **mineral zone boundaries** and are named based on the index mineral being introduced (e.g., biotite-in or garnet-in). The term **isograd** is also applied to these boundaries. The rocks between successive mineral zone boundaries are in a **mineral zone**. Similar boundaries may define the last appearance of minerals and are named based on the mineral being lost (e.g., chlorite-out or andalusite-out).

The introduction of an index mineral such as staurolite in pelitic rocks (see next section) must be produced by a mineral reaction such as 11.5. By careful mapping and petrographic study it is often possible to identify the reactant and product mineral assemblages on opposite sides of a mineral zone boundary. Mineral-in boundaries are often associated with mineral-out boundaries—for example, with the clear implication that the mineral being lost is involved in the reaction producing the new mineral. In Reaction 11.5, a chlorite-out boundary could be associated with the staurolite-in boundary.

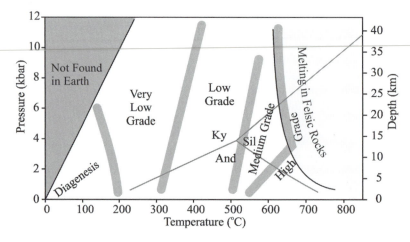

Figure 11.10 Metamorphic grade boundaries. Adapted from Winkler (1976). Stability fields of the aluminum silicates [AlAlOSiO$_4$: andalusite (And), sillimanite (Sil), and kyanite (Ky)] are shown for reference.

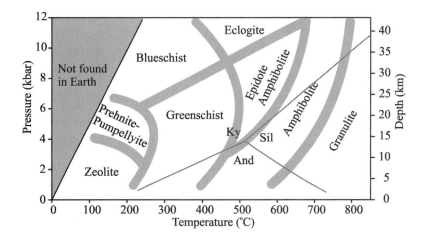

Figure 11.11 Approximate temperature–pressure fields of the eight principal metamorphic facies. Adapted from Spear (1993).

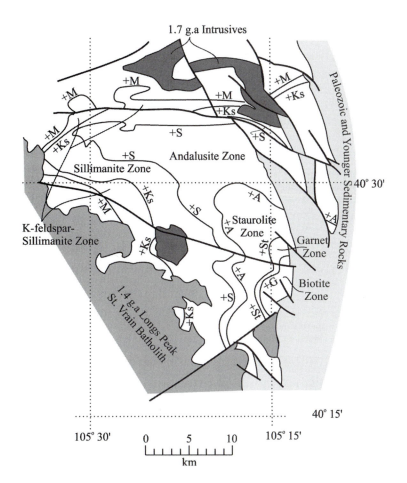

Figure 11.12 Metamorphic zones in Proterozoic pelitic metamorphic rocks (no shading) of the northeastern Front Range, Colorado. The least-metamorphosed, biotite-zone rocks are in the southeastern part of the area. Chlorite-zone rocks are not exposed. Mapped mineral zone boundaries include garnet-in (+G), staurolite-in (+St), andalusite-in (+A), sillimanite-in (+S), and K-feldspar + sillimanite-in (+Ks). Rocks on the high-grade side of the +M boundary were partially melted during metamorphism. The sillimanite boundary is associated with the loss of andalusite, and the K-feldspar boundary is associated with the loss of primary muscovite. Faults and shear zones are shown with heavy lines.

Major Compositional Groups of Metamorphic Rocks

Pelites

Pelitic metamorphic rocks are derived from mud rocks. Because mud rocks contain Al-rich clay minerals, pelitic rocks commonly contain Al-bearing minerals such as muscovite, biotite, garnet, staurolite, cordierite, and the aluminum silicates, in addition to quartz and plagioclase (Table C.3 in Appendix C). K-feldspar may be present in high-grade rocks. Details of which minerals are present at any particular temperature-pressure condition depend on the bulk composition of the rock. A common sequence of mineral zone boundaries found in pelitic rocks is shown in Table 11.9 and Figure 11.12. If foliated, pelites form slate, phyllite, and schist, and if not foliated, argillite and hornfels.

Mafic Rocks

Mafic metamorphic rocks are derived from basaltic volcanics and gabbroic intrusives or closely related immature sediments. Progressive metamorphism of these rocks produces mineralogical changes that depend on the conditions of metamorphism (Table 11.10). Low-grade rocks are greenschists with abundant chlorite, Na-rich plagioclase (albite), and epidote/zoisite, ± Mg-rich calcic amphibole (actinolite). With increasing temperature, chlorite is lost, the amphibole is converted to more Fe-rich hornblende,

Table 11.9 Typical Mineral Zones in Pelitic Rocks[a]

Grade	Index Mineral	Mineral Zone	Common Mineral Assemblage
Very low	Chlorite	Chlorite	Chlorite, muscovite, albite, quartz
	Biotite		
Low		Biotite	Biotite, muscovite, plagioclase, quartz, ± chloritoid, chlorite
	Garnet	Garnet	Biotite, muscovite, garnet, plagioclase, quartz, ± cordierite, garnet
	Staurolite		
Medium	Andalusite or Kyanite	Staurolite	Biotite, muscovite, staurolite, plagioclase, quartz, ± cordierite, garnet
		Andalusite/ Kyanite	Biotite, muscovite, andalusite or kyanite, plagioclase, quartz, ± cordierite, garnet
	Sillimanite	Sillimanite	Biotite, muscovite, sillimanite, plagioclase, quartz, ± garnet
	K-feldspar		
High		K-feldspar-Sillimanite	Biotite, K-feldspar, sillimanite, plagioclase, quartz

[a]The mineral assemblage may vary depending on the bulk composition of the rocks involved.

the Ca content of the plagioclase increases, and the epidote/zoisite is lost. With very-high-grade metamorphism, amphiboles are replaced by pyroxenes.

Blueschist and Related Rocks

The blueschist and related rocks are found in high-pressure/low-temperature metamorphic environments usually associated with subduction zones. The parent rocks are typically graywacke and related immature sediments derived from an island arc. In terms of bulk chemistry these rocks are often broadly mafic, and their mineralogy defines several of the high-P/low-T facies. Zeolite minerals are common in the least metamorphosed of these rocks (Table 11.11). With increasing pressure and temperature, prehnite and pumpellyite may become abundant, followed by sodic amphibole (glaucophane) and lawsonite. The blue color of the sodic amphiboles is the reason these rocks are referred to as blueschists. Other minerals found in these rocks are listed in Table C.3 in Appendix C.

Calc-Silicate Rocks

Many sedimentary rocks contain minor to substantial amounts of carbonate minerals. The original mineralogy usually consists of quartz, clay, calcite ($CaCO_3$), and/or dolomite [$CaMg(CO_3)_2$]. When metamorphosed, these rocks undergo mineral reactions to produce minerals containing both Ca and Si, hence the term calc-silicate. An entire spectrum of mineralogical compositions is possible, depending on the relative amounts of carbonates and silicates in the parent rock, temperature, pressure, fluid composition, and whether these rocks have been metasomatized. Common minerals found in these rocks are listed in Table C.3 in Appendix C. Calcite and dolomite are found in most calc-silicate rocks. Low-grade rocks may contain talc and Mg-biotite; increasing temperature usually involves introduction of amphiboles, pyroxenes, Ca-garnet, scapolite, and perhaps Mg-rich olivine. Members of the epidote group are common.

Metamorphic Environments

A variety of different and somewhat overlapping terminologies have been developed to describe the different environments in which rocks may be metamorphosed. These include terminology based on the local geological setting, plate tectonic environment, and the balance between heating and deformation.

Table 11.10 Mineralogy of Mafic Rocks in Selected Facies

Facies	Mineralogy
Greenschist facies	Chlorite, albite, epidote/zoisite ± actinolite
Epidote-amphibolite facies	Plagioclase, hornblende, epidote ± garnet
Amphibolite facies	Plagioclase, hornblende, ± garnet
Granulite facies	Orthopyroxene, plagioclase ± Ca-clinopyroxene, hornblende, garnet

Table 11.11 Mineralogy of Blueschist and Related Rocks in Selected Facies

Facies	Characteristic Minerals
Zeolite	Zeolites, usually laumontite, wairakite, or analcime
Prehnite-pumpellyite	Prehnite, pumpellyite
Blueschist	Glaucophane, lawsonite, epidote

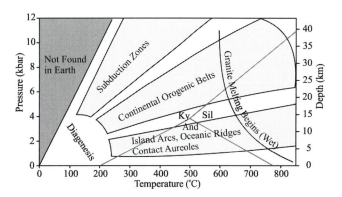

Figure 11.13 T–P conditions found in different tectonic environments. Adapted from Spear (1993).

Terminology based on the local geological setting is perhaps the least obscure, because the nature of the metamorphism usually can be determined with a relatively modest amount of field work. The groupings are as follows.

- Contact metamorphism: Metamorphism in the immediate vicinity of an igneous intrusion that serves as the source of the heat. The metamorphosed rocks typically form an aureole, ranging in thickness from a few millimeters to several kilometers, around the intrusive, with the grade of metamorphism decreasing outward. Metamorphism is commonly not associated with extensive deformation.

- Regional metamorphism: Metamorphism involves large volumes of rocks, usually in orogenic belts. The heat may be derived from the mantle. Igneous intrusives may be common, but metamorphism is not restricted to aureoles around them. Metamorphism and deformation are usually closely related both in time and space.

- Burial metamorphism: Metamorphism involves substantial volumes of rocks, but usually within a depositional basin where sedimentary and volcanic rocks accumulate. Heating is closely related to burial, not to extensive tectonic activity, magmatism, or deformation. Burial metamorphism is essentially a continuation of sedimentary diagenesis.

- Hydrothermal metamorphism: Metamorphism produced by the flow of hot aqueous solutions into and through rocks. The source of the heat is usually an igneous intrusion. The flow of water commonly is concentrated along fault and fracture zones, or in the volume of rocks adjacent to and above an intrusive. Hydrothermal metamorphism is commonly associated with the formation of hydrothermal mineral deposits.

- Shock metamorphism: Metamorphism produced by the impact of a meteorite. New minerals (coesite, stishovite) or melt may be formed, but much of the energy is converted to shattering and displacing the rocks being struck.

- High-strain metamorphism: Intense ductile shearing can produce mylonitic rocks. The energy comes from tectonic sources. The major change is a dramatic reduction in grain size with or without production of fine-grained sheet silicates.

Metamorphic environments also can be described in terms of their plate tectonic environment (Figure 11.13). High-pressure/low-temperature metamorphism is typically produced in subduction zones or other areas where rapid overthrusting moves rocks rapidly to substantial depth. Low-pressure/high-temperature metamorphism is typical of areas of high heat flow, such as island arcs or oceanic ridges, and also in contact metamorphic aureoles. The broad range of intermediate conditions is typically found in continental orogenic belts adjacent to subduction zones.

The terms that refer to the balance between heating and deformation are **thermal, dynamothermal**, and **dynamic metamorphism**. Thermal metamorphism involves simply heating the rocks, without significant deformation. Mineralogical change is dominant; textural change relatively minor. Both contact and burial metamorphism are types of thermal metamorphism. The other extreme is dynamic metamorphism, characterized by intense deformation, but without substantial heating. Textural change is dominant. The high-strain metamorphism associated with faulting, whether brittle or ductile, provides a good example. Dynamothermal metamorphism involves both deformation and substantial temperature changes so the rocks are subjected to both textural and mineralogical changes. Regional metamorphism is, in most cases, dynomothermal.

REFERENCES CITED AND SUGGESTIONS FOR ADDITIONAL READING

Best, M. G. 1995. *Igneous and Metamorphic Petrology.* Blackwell Science, Cambridge, MA, 630 p.

Boggs, S. 2009. *Petrology of Sedimentary Rocks,* 2nd ed. Cambridge University Press, New York, 599 p.

Bucher, K., Frey, J., and Bucher-Nurminen, K. 2002. *Petrogenesis of Metamorphic Rocks,* 6th ed. Springer-Verlag, New York, 341 p.

Dickinson, W. R., Beard, L. S., Brakenridge, G. R., Erjavec, J. L., Ferguson, R. C., Inman, K. F., Knepp, R. A., Lindberg, F. A., and Rybert P. T. 1983. Provenance of North American Phanerozoic sandstones in relation to tectonic setting. Geological Society of America Bulletin 94, 222–235.

Dott, R. L., Jr. 1964. Wacke, graywacke, and matrix—What approach to immature sandstone classification? Journal of Sedimentary Petrology 34, 625–632.

Gill, R. 2010. *Igneous Rocks and Processes.* Wiley-Blackwell, Chichester, 428 p.

Griffen, D. T. 1992. *Silicate Crystal Chemistry.* Oxford University Press, New York, 442p.

Hibbard, M. J. 1995. *Petrography to Petrogenesis.* Prentice-Hall, Englewood Cliffs, NJ, 587 p.

Kerrick, D. M. 1990. *The Al_2SiO_5 Polymorphs.* Reviews in Mineralogy 22, 406 p.

Konhauser, K. 2007. *Introduction to Geomicrobiology.* Blackwell, Malden, MA, 425 p.

Kretz, R. 1994. *Metamorphic Crystallization.* John Wiley & Sons, New York, 507 p.

Le Bas, M. J., Le Maitre, R. W., Streckheisen, A. L., and Zanettin, B. 1986. Chemical classification of volcanic rocks based on the total alkali–silica diagram. Journal of Petrology 27, 745–750.

Le Maitre, R. W., Streckheisen, A., Zanettin, B., Le Bas, M. J., Bonin, B., Bateman, P., Bellieni, G., Dudek, A., Efremova, S., Keller, J., Lameyre, J., Sabine, P. A., Schmid, R., Sørensen, H., and Wooley, A. R. (eds.). 2002. *Igneous Rocks: A Classification and Glossary of Terms,* 2nd ed., Cambridge University Press, Cambridge, 236 p.

Liebau, F. 1985. *Structural Chemistry of Silicates, Structure, Bonding and Classification.* Springer-Verlag, Berlin, 347 p.

Pettijohn, F. J., Potter, P. E., and Siever, R. 1987. *Sand and Sandstone,* 2nd ed. Springer-Verlag, New York, 553 p.

Philpotts, A. R., and Agee, J. J. 2009. *Principles of Igneous and Metamorphic Petrology,* 2nd ed. Cambridge University Press, Cambridge, 667 p.

Ribbe, P. H. (ed.). 1982. *Orthosilicates,* 2nd ed. Reviews in Mineralogy 5, 450 p.

Spear, F. S. 1993. *Metamorphic Phase Equilibria and Pressure–Temperature–Time Paths.* Mineralogical Society of America, Washington, D C, 799 p.

Streckheisen, A. L. 1976. To each plutonic rock its proper name. Earth Science Reviews 12, 1–33.

Tucker, M. E. 2001. *Sedimentary Petrology: An Introduction to the Origin of Sedimentary Rocks,* 3rd ed. Blackwell, Malden, MA, 262 p.

Turner, F. J. 1981. *Metamorphic Petrology,* 2nd ed. McGraw-Hill, New York, 524 p.

Vernon, R. H. and Clarke, G. L. 2008. *Principles of Metamorphic Petrology.* Cambridge University Press, Cambridge, 446 p.

Williams, H., Turner, F. J., and Gilbert, C. M. 1982. *Petrography,* 2nd ed. W. H. Freeman, San Francisco, 626 p.

Winkler, S. J. F. 1976. *Petrogenesis of Metamorphic Rocks,* 4th ed. Springer-Verlag, New York, 334 p.

Winter, J. D. 2010. *Principles of Igneous and Metamorphic Petrology,* 2nd ed. Prentice Hall, Upper Saddle River, NJ, 702 p.

CHAPTER 12

Framework Silicates

INTRODUCTION

More than two-thirds of the Earth's crust is composed of framework silicate minerals. The common framework silicates are listed in Table 12.1. Of these, quartz, plagioclase, and the alkali feldspars are by far the most abundant.

The structure of all these minerals is based on a framework of TO_4 tetrahedra in which T is Si^{4+} or Al^{3+} and each of the four oxygen anions is shared with another tetrahedron. The fact that every O^{2-} is shared between two tetrahedra, coupled with the mutual repulsion of the high-charged cations in those tetrahedra, ensures that the structure of the framework silicates is fairly open. This leads to two important consequences—one compositional and the other having to do with physical properties.

The compositional consequence is that the open structure of framework silicates can easily accommodate large cations such as Ca^{2+}, Na^+, and K^+. These large cations do not readily fit into structures based on close-packed anions in which the largest coordination site is octahedral (6-fold). To accommodate the positive charge of these large cations, the usual practice is to substitute Al^{3+} for Si^{4+} in the tetrahedral (T) sites. Only in the silica group of minerals are all the tetrahedra occupied by Si^{4+}. We will see later in the discussion of the zeolites that the very open structure of these minerals leads to some interesting and useful chemical properties.

The physical property consequence is that the specific gravity for most framework silicates is significantly lower than for minerals whose anion arrangement can be approximated by close packing. Quartz (SiO_2), for example, has a specific gravity of 2.65, whereas forsterite olivine (Mg_2SiO_4), whose structure is based on cubic close packing of the oxygen anions, has a specific gravity of 3.27, even though Mg has a lower atomic mass than Si. Because the structure is fairly open, the framework silicates tend not to be stable at high pressures. This means that framework silicates are minerals generally restricted to the Earth's crust, and often the shallow crust.

SILICA GROUP

The silica group (Table 12.2) includes about a dozen natural and synthetic polymorphic varieties of SiO_2. Of these, **quartz, tridymite**, and **cristobalite** are the common polymorphs, and they are the only ones described here. **Stishovite** and **coesite** are quite rare, as might be inferred from the stability fields shown in Figure 12.1. They are found only where extremely high pressures have been obtained, most notably in rocks that were shocked by meteorite impact.

Quartz, tridymite, and cristobalite have their own distinct structures (described later), and conversion from one to the other requires that chemical bonds be broken—they are reconstructive polymorphs. In addition, the structures of quartz, tridymite, and cristobalite are each represented by α (low) and β (high) varieties. The α and β varieties of each mineral are displacive polymorphs, related by a distortion in the crystal lattice. In each case the β polymorph has higher symmetry and is stable at higher temperature than the α polymorph.

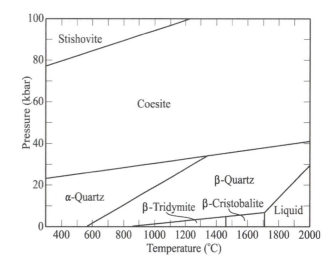

Figure 12.1 Stability fields of the silica minerals. Adapted from Griffen (1992).

Table 12.1 Common Framework Silicate Minerals[a]

Mineral	Chemical Formula	Igneous				Sedimentary	
		Felsic	Intermediate	Mafic	Alkaline	Clastic	Chemical/biologic
Quartz	SiO_2	M	M			M	A
Plagioclase	$Ca_xNa_{1-x}Al_{1+x}Si_{3-x}O_8$	M	M	M	M	M	
Alkali feldspar	$KAlSi_3O_8$	M	M		M	M	
Nepheline	$Na_3K(Al_4Si_4O_{16})$				M		
Sodalite	$Na_8(AlSiO_4)_6Cl_2$				M		
Leucite	$KAlSi_2O_6$				M		
Zeolite group	$M_xD_y(Al_{x+2y}Si_{n-x-2y}O_{2n})\cdot mH_2O$				A		
Scapolite	$Na_4Al_3Si_9O_{24}Cl — Ca_4Al_6Si_6O_{24}CO_3$						

[a] M, major or common; A, accessory or minor mineral, or heavy mineral fraction of clastic sediments.

Table 12.2 Silica Polymorphs

Polymorph	Crystal System	Specific Gravity
α-Quartz	Hexagonal (rhombohedral)	2.65
β-Quartz	Hexagonal (hexagonal)	2.53
α-Tridymite	Orthorhombic	2.26
β-Tridymite	Hexagonal	2.22
α-Cristobalite	Tetragonal	2.32
β-Cristobalite	Isometric	2.2
Coesite	Monoclinic	3.01
Stishovite	Tetragonal	4.35
Moganite	Monoclinic	2.62

The forms of quartz, tridymite, and cristobalite at room temperature are all the α varieties. Quartz may grow as either the α or β variety depending on the temperature and pressure as shown in Figure 12.1. With cooling, original β-quartz will convert by a displacive polymorphic transition to α-quartz as described in Chapter 5. Tridymite and cristobalite grow originally only as the β varieties. Primary α-tridymite and α-cristobalite do not normally grow because there is no T–P field where they are the lowest-energy polymorph. With cooling, both β-tridymite and β-cristobalite convert by a displacive polymorphic transition to lower energy α-tridymite and α-cristobalite that can be preserved because the reconstructive polymorphic transition to the even lower-energy quartz is very sluggish. For a more complete survey of the silica minerals, refer to Heaney and others (1994).

QUARTZ

SiO_2
Hexagonal, trigonal
  division (32)
$a = 4.914$ Å,
  $c = 5.405$ Å, $Z = 3$
$H = 7$
$G = 2.65$
Uniaxial (+)
$n_\omega = 1.544$
$n_\epsilon = 1.553$
$\delta = 0.009$

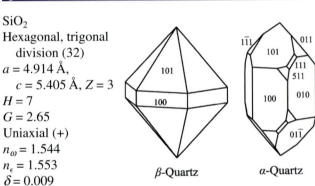

β-Quartz α-Quartz

Structure Figure 12.2 shows the structure of quartz. Silicon tetrahedra are arranged to form interlocking spirals parallel to the c axis. Because the spirals may be either right or left handed, quartz forms entantiomorphous right- and left-handed crystals (Chapter 2). In β-quartz the spirals produce 6-fold rotational symmetry (actually a 6-fold screw axis) reflected in the hexagonal symmetry of the crystal shown above. On cooling through the transition temperature (573°C at 1 atm pressure), the β-quartz structure kinks so that the spirals are reduced to 3-fold symmetry and the point group symmetry is reduced to 32, forming α-quartz.

Moganite is a variety of silica found in microcrystalline quartz and was first discovered in 1984. Its structure consists of alternating sheets of right- and left-handed quartz at the unit cell level (Miehe and Graetsch, 1992). It has now been recognized as a mineral by the Commission on New

Metamorphic			Other		
Mafic	**Calc-silicate**	**Blueschist**	**Hydrothermal**	**Weathering/Alteration**	**Miscellaneous**
		M	M	M	
M		M			
			M		
M		M	M	M	Vessicle filling
M	M				

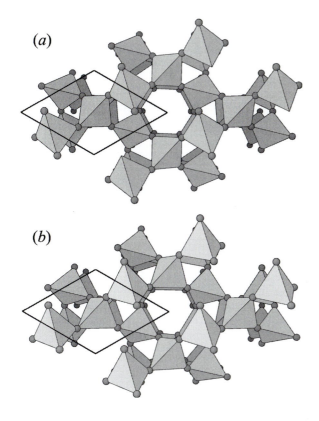

Figure 12.2 Quartz structure viewed down the c axis. A unit cell outline is shown. Silica tetrahedra link together to form spirals that extend parallel to the c axis (Figure 4.12). (*a*) β-Quartz. The spirals produce hexagonal symmetry. (*b*) α-Quartz. The structure is distorted, reducing symmetry to 3-fold.

Minerals, Nomenclature and Classification. It differs from quartz in having a hardness of 6 and lower specific gravity. Because it is fine grained, identification typically requires careful X-ray diffraction study.

Composition Most quartz is relatively pure SiO_2, although trace amounts of other elements may be present. The most common is substitution of small amounts of Al^{3+} and Fe^{3+} for Si^{4+}. Charge neutrality is maintained by inserting appropriate amounts of Fe^{3+}, Na^+, Li^+, or K^+ interstitially, or adding H^+, which bonds to O^{2-} to form OH^-.

Form and Twinning Quartz commonly forms as anhedral grains in both igneous and metamorphic rocks (Figure 12.3). Detrital grains are more or less equant. Crystals of α-quartz are usually prismatic {100} terminated with the positive {101} and negative {1$\bar{1}$1} rhombohedrons, one of which is usually more prominent than the other. Phenocrysts that originally grow as β-quartz crystals are typically stubby prisms terminated with a {101} dipyramid. On cooling, the β-quartz inverts to α-quartz, but the crystals retain the external habit of the original β-quartz. Deformed quartz may display a pattern of **undulatory extinction** seen in thin section with the petrographic microscope. As the stage is rotated, areas of extinction appear to sweep across deformed grains in an irregular or wavy manner.

Quartz also occurs in microcrystalline aggregates known as **chert** and **chalcedony**. Chert and chalcedony may appear quite similar in hand sample, but they are readily distinguished in thin section. Chert is fine granular microcrystalline quartz commonly found as nodules or irregular beds in limestone. Chert may contain substantial amounts of moganite.

Chalcedony is the name given to intergrowths of fibrous microcrystalline quartz and moganite. Up to 80% of the SiO_2 in microcrystalline chalcedony may be moganite. Most chemical analyses show the presence of water. The water is either trapped in submicroscopic pore spaces between fibers or takes the form of hydroxyl, which replaces oxygen in the structure. The specific gravity (2.57–2.64) is lower than that of quartz in proportion to the amount of pore space between

Figure 12.3 Quartz (*a*) Rounded quartz (left) phenocryst in rhyodacite. The other phenocrysts are altered plagio-clase. Plane light. (*b*) Undulatory anhedral quartz in deformed granite. Crossed polarizers. (*c*) Feathery-appearing chal-cedony cementing quartz and microcline grains in a sandstone. Crossed polarizers. (*d*) Quartz crystals. Field of view is about 10 cm wide. (*e*) β-Quartz crystals (now α-quartz).

the fibers, and the indices of refraction vary with the specific gravity (Figure 12.4). Hardness ($6\frac{1}{2}-7$) is slightly less than coarsely crystalline quartz. The fine fibrous nature tends to produce a luster that is dull vitreous to greasy. Individual quartz fibers are elongate along one of the *a* axes and twist along their length so that the *c* axis sweeps out a helix (like a spiral staircase). As seen in thin section (Figure 12.3c) between crossed polarizers, fibers are dark where the *c* axis is vertical and birefringence is zero, and light where *c* is horizontal and birefringence is maximum. The bands of light and dark along the length of fibers produce the feath-ery appearance characteristic of chalcedony. Moganite also forms fibrous grains, but its fibers are length slow, whereas

quartz fibers are length fast. It is possible that moganite represents a rediscovery of length-slow fibrous silica in chalcedony, identified in the old petrographic literature with the now-discredited term **lutecite**.

The various intergrowths of quartz and the feldspars are described in the feldspar section below.

Dauphene twins (rotation on *c*) and Brazil twins (reflec-tion on {110}) are common but not easily detected. In hand sample, the faces needed to recognize these twins are not usually present and in thin section the *c* axis orientations of the twin segments are the same, so optical properties are also the same. Japan twins, which form an elbow shape with {111} as the composition plane, are uncommon.

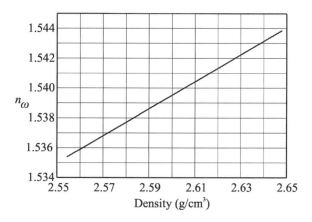

Figure 12.4 Approximate variation of n_ω of chalcedony with density, which is a function of pore space between the fibers. After Frondel (1982).

Cleavage None. Conchoidal fracture. Brittle.

Color Properties

Luster: Vitreous.

Color: Colorless or gray is most common, but almost any color is possible. For a survey of the complex causes of color in quartz, refer to Mossman (1994). Names applied to colored varieties of coarsely crystalline quartz include the following:

ROCK CRYSTAL: Clear crystals

AMETHYST: Violet or purple (related to trace amounts of Fe)

ROSE QUARTZ: Pink or rose red [cause is uncertain, but some rose quartz contains very fine inclusions of pink **dumortierite** ($Al_{27}O_6(BO_3)_4(SiO_4)_{12}(OH)_3$)]

CITRINE: Yellow (related to Fe content, irradiation, or some combination)

MILKY QUARTZ: Milky white, due to abundant minute fluid inclusions

SMOKY QUARTZ: Smoky brown to black (probably from color centers produced by irradiation of quartz containing trace amounts of Al)

Numerous names have been applied to microcrystalline varieties of quartz used for lapidary work. A few of the common terms include

JASPER: Red or brown microcrystalline quartz (usually chert)

FLINT: Black or dark gray microcrystalline quartz (usually chert)

AGATE: Microcrystalline quartz (often chalcedony) with color bands or irregular color patches; often translucent (Figure 2.51d)

Streak: White.

Color in Thin Section: Colorless.

Pleochroism: None.

Optical Characteristics

Relief in Thin Section: Low positive.

Optic Orientation: Crystals are length slow, fibers in chalcedony are length fast. Deformed quartz shows undulatory extinction (Figure 12.3b).

Indices of Refraction and Birefringence: Almost no variation for coarsely crystalline varieties.

Interference Figure: Basal sections yield a uniaxial positive figure. Undulatory quartz and some amethyst and smoky quartz may yield a biaxial figure with a small $2V$ angle.

Alteration Not normally altered. Very stable in the weathering environment.

Distinguishing Features

Hand Sample: Coarsely crystalline samples are distinguished by hardness, conchoidal fracture, and luster. Crystals are quite distinctive. Beryl has a higher specific gravity and an imperfect {001} cleavage; it is usually a shade of blue or green. Microcrystalline quartz is distinguished from other minerals by its hardness; opal is softer and has a lower specific gravity than microcrystalline quartz.

Thin Section/Grain Mount: Recognized by low relief, low birefringence, and lack of cleavage or twinning. Orthoclase and sanidine are biaxial, have cleavage, and may show a single Carlsbad twin plane. Plagioclase is biaxial, has cleavage, and is usually polysynthetically twinned. If not authigenic, microcline shows its distinctive tartan plaid twinning. Cordierite closely resembles quartz but is biaxial and may show distinctive pinite alteration. Beryl is optically negative and has higher indices of refraction. The feldspars and cordierite may be distinguished from quartz by a variety of staining techniques that can be used with thin sections or rock slabs (Chapter 8).

Microcrystalline quartz may have a cloudy appearance in thin section or grain mount. The feathery appearance of chalcedony is distinctive.

Occurrence Quartz is a very common mineral found in a wide variety of geological environments. It is abundant in felsic to intermediate intrusive and extrusive igneous rocks such as granitic pegmatite, granite, granodiorite, quartz diorite, rhyolite, rhyodacite, and dacite, and may be found in small amounts in diorite, gabbro, syenite, and volcanic equivalents. In metamorphic rocks it is abundant in slate, phyllite, schist, gneiss, and quartzite of various types. Because it is stable in the weathering environment, quartz is a major constituent in many clastic sedimentary rocks and may serve as a cementing agent, often in the form of chalcedony. While quartz is not itself a biomineral, it may crystallize from opal precipitated by diatoms and other organisms. Chert layers found in limestone are commonly derived from biogenic amorphous silica, either by direct crystallization or indirectly, by dissolution of the amorphous silica into the pore water of sediment and subsequent precipitation of quartz from the pore water. Chert commonly replaces the carbonate minerals of the host limestone. Hydrothermal vein and replacement deposits typically contain quartz as a gangue mineral, often in the form of beautiful crystals, or as microcrystalline and massive varieties.

Use Quartz is widely used in many industries. Relatively pure quartz sandstone is quarried to provide the raw materials for making glass and to serve as the source of elemental silicon. The applications for silicon include the manufacture of semiconductors for the electronics industry; silicone rubber caulking compound is used to seal around doors, windows, and bathtubs in residential homes. Quartz also is widely used as an abrasive or polish, a filler in plastics, paints, and related materials, and, in the form of sand, as aggregate in asphalt, mortar, and concrete. Synthetic quartz is used in the electronics industry to make electronic oscillators and pressure-sensing devices.

Quartz also is widely used as a semiprecious gemstone in the jewelry industry. Followers of "New Age spiritualism" may ascribe mystical, spiritual, and healing properties to quartz crystals, which are worn as pendants, earrings, or other ornaments. Historically, quartz was used in cures for gastrointestinal, skin, nervous system, insect bite, and other problems, particularly in the Middle Ages. Amethyst derives its name in Greek (*amethystos*) from the belief that it prevented drunkenness, perhaps because its color is similar to the color of red wine. Unfortunately, none of these therapeutic claims is supported by scientific evidence.

Comments Although a very common and widespread mineral, quartz can pose a health risk if present as dust in substantial quantities. Chronic inhalation of quartz dust can lead to silicosis, a pathology that, among other things, involves scarring of lung tissue and impaired lung function. Industrial health and safety regulations in most Western countries include requirements for dust suppression in industries such as mining and quarrying, where exposure to quartz dust is likely. Quartz (crystalline silica) also is listed as a human carcinogen by the International Agency for Research on Cancer. This seems remarkable given that life on Earth has coexisted with this mineral for literally billions of years.

TRIDYMITE

SiO_2
Orthorhombic
   (pseudohexagonal)
   (222)
$a = 9.88$ Å, $b = 17.1$ Å,
   $c = 16.3$ Å, $Z = 64$
$H = 7$
$G = 2.27$
Biaxial (+)
$n_\alpha = 1.468–1.482$
$n_\beta = 1.470–1.484$
$n_\gamma = 1.474–1.486$
$\delta = 0.002–0.004$
$2V_z = 40–90°$

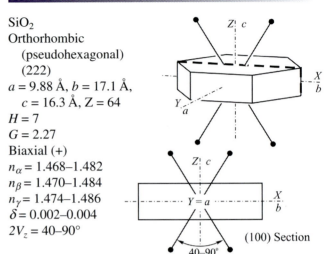

(100) Section

Structure The structure consists of stacked sheets of silicon tetrahedra. The tetrahedra within each sheet form six-member rings with the tetrahedra pointing alternately up and down. Sheets are joined by sharing oxygen anions between the tetrahedra pointing up in one sheet and the tetrahedra pointing down on the overlying sheet. As viewed down the c axis, the rings of tetrahedra within each sheet line up one above the other. The result is the formation of a three-dimensional framework of tetrahedra in which every oxygen anion is shared between two tetrahedra.

The high-temperature β-tridymite structure is hexagonal. With cooling, the structure distorts slightly and the hexagonal symmetry is lost. Low-temperature tridymite varieties with orthorhombic, monoclinic, and triclinic symmetry have been described. All tridymite at room temperature is one of these polymorphic varieties.

Composition Ideally the composition is SiO_2, but the fairly open structure can accommodate varying amounts of Na^+, K^+, and Ca^{2+}, with electrostatic charge balance maintained by substituting Al^{3+} for Si^{4+}.

Form and Twinning Crystals are typically pseudohexagonal platelets dominated by the {100} prism and {001} pinacoid. Tridymite also may form radiating or spherical aggregates or be anhedral granular.

The process of converting from hexagonal β-tridymite to lower symmetry typically causes crystals to be twinned on {106}. Tridymite derives its name from the fact that twins often are composed of three wedge-shaped segments.

Cleavage None. Conchoidal fracture. Brittle.

Color Properties

Luster: Vitreous.
Color: Colorless, gray, white.
Streak: White.
Color in Thin Section: Colorless.
Pleochroism: None.

Optical Characteristics

Relief in Thin Section: Moderate negative.

Optic Orientation: X = b, Y = a, Z = c.

Indices of Refraction and Birefringence: Indices of pure synthetic tridymite are $n_\alpha = 1.469$, $n_\beta = 1.469$, $n_\gamma = 1.473$. Indices of refraction increase modestly with increasing Ca, Na, and K content. Birefringence is low (0.002–0.004), so interference colors in standard thin section are no higher than first-order gray.

Interference Figure: Biaxial positive with 2V between 40° and 90° but usually around 70°. 2V may be different in different segments of a twinned crystal.

Alteration Tridymite at room temperature is always a pseudomorph (paramorph) after β-tridymite. Tridymite also may be pseudomorphically replaced by quartz, either as single crystals or as aggregates. The *c* axis orientation of the replacing quartz generally does not coincide with the *c* axis of the original tridymite.

Distinguishing Features

Hand Sample: The crystal habit is characteristic, but it is not possible to distinguish tridymite from a quartz pseudomorph after tridymite in hand sample.

Thin Section/Grain Mount: Crystal habit, low birefringence, low indices of refraction, and twinning are characteristic. Quartz has higher birefringence and indices of refraction, and is uniaxial. Cristobalite is uniaxial or essentially isotropic.

Occurrence Tridymite is found in felsic to intermediate extrusive and shallow intrusive igneous rocks. It may form phenocrysts, be part of the groundmass, or form small crystals lining vesicles or other voids. It also is found in vesicles in basaltic lavas and in the contact metamorphic zones adjacent to basaltic intrusives; it has been reported in stony meteorites.

Use None.

CRISTOBALITE

SiO_2
Tetragonal
   (pseudoisometric)
   (422)
$a = 4.971$ Å,
   $c = 6.928$ Å, Z = 4
$H = 6.5$
$G = 2.32$–2.36
Uniaxial (−)
$n_\omega = 1.486$–1.488
$n_\epsilon = 1.482$–1.484
$\delta = 0.002$–0.004

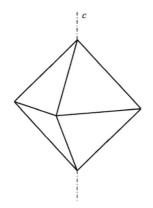

Structure Like tridymite, the structure of cristobalite consists of sheets of silicon tetrahedra arranged in 6-fold rings whose tetrahedra point alternately up and down. Sheets are tied together by sharing oxygen anions between the up tetrahedra on one sheet with the down tetrahedra on the overlying sheet. The cristobalite structure differs in that the rings in adjacent sheets do not line up with each other but are offset, so that the up tetrahedra in one sheet are over the centers of the rings in the sheet below. If the arrangement of tetrahedra in tridymite is equivalent to hexagonal close packing with sheets parallel to {001}, then the tetrahedral arrangement in cristobalite is equivalent to cubic close packing with the sheets parallel to {111}. At high temperatures, the structure of β-cristobalite is isometric, but cooling below 268°C allows a slight distortion of the lattice and conversion to the tetragonal symmetry of α-cristobalite. This conversion is accompanied by twinning.

Composition The ideal composition is SiO_2, but the open structure allows interstitial substitution of Na^+, Ca^{2+}, K^+, or other cations whose electrostatic charge is balanced by substituting Al^{3+} for Si^{4+}.

Form and Twinning Crystals are typically octahedrons, rarely cubes. Also dendritic and skeletal, or intergrown with K-feldspar forming a radial pattern in spherulites produced by devitrifying volcanic glass. All cristobalite at room temperature is actually a pseudomorph of α-cristobalite after β-cristobalite. A fibrous variety known as **lussatite** resembles chalcedony and may be intergrown with it.

    Multiple twins on {111} are produced during the conversion from β- to α-cristobalite. The *c* axis in the different lamellae in the tetragonal structure follows one of the three original *a* axes in the high-temperature isometric structure.

Cleavage None. Brittle.

Color Properties

Luster: Vitreous.
Color: Colorless to white or pale yellowish white.
Streak: White.
Color in Thin Section: Colorless.
Pleochroism: None.

Optical Characteristics

Relief in Thin Section: Moderate negative.

Optic Orientation: One of the *a* axes in β-cristobalite becomes the *c* axis in α-cristobalite after inversion and is therefore the optic axis. Fibers are usually length slow, less commonly length fast, and display parallel extinction in most cases.

Indices of Refraction and Birefringence: Indices of refraction vary within a small range, and birefringence is quite low

(0.002–0.004), so interference colors in standard thin section are no higher than lower first-order gray, and for many orientations, cristobalite may appear nearly isotropic.

Interference Figure: Fine grain size and low birefringence make good figures difficult to obtain. Most cristobalite is uniaxial negative, but biaxial varieties with small $2V$ have been reported.

Alteration Cristobalite may be pseudomorphically replaced by quartz or may pseudomorphically replace tridymite.

Distinguishing Features

Hand Sample: Difficult to identify with certainty in hand sample because of fine grain size and complications related to pseudomorphic relations. It resembles quartz and tridymite, but with a different crystal habit.

Thin Section/Grain Mount: Crystal habit, indices of refraction, and low birefringence are characteristic, as is its presence in spherulites. Tridymite is biaxial and has somewhat higher negative relief. Quartz has higher birefringence and is uniaxial positive. X-ray techniques are often needed to identify fine-grained material.

Occurrence It may form a phenocryst or groundmass mineral in some felsic lavas. Cristobalite is more common as grains lining cavities and vesicles in volcanic rocks and shallow intrusives in association with topaz, garnet, tridymite, and other minerals. It also is a common constituent of the material routinely identified as opal. Cristobalite has been found in meteorites and in sandstone that has been fused by contact with high-temperature magma or by underground coal fires. The fibrous variety lussatite is found in serpentinites and in other environments associated with chalcedony.

Use None.

OPAL

$SiO_2 \cdot nH_2O$
Amorphous
$H = 5.5–6.5$
$G = 2.0–2.25$
Isotropic
$n = 1.43–1.46$

Structure Opal is a mineraloid that consists of a mixture of amorphous and crystalline silica. The amorphous silica may consist of extremely small (0.1–0.5 nm) spherical masses that are closely packed. The crystalline material usually consists of extremely small crystals of tridymite or cristobalite. The amount of crystalline material may be quite variable. If the silica spheres have a fairly uniform size, they can pack together in a regular manner to produce a diffraction grating that yields the beautiful play of color seen in precious opal.

Composition Variable amounts of water may be present, either in the void space between the spheres, or incorporated in some manner in the amorphous silica. Heating or long exposure to air may cause dehydration and cracking.

Form and Twinning Opal forms colloform and encrusting masses, vein fillings, or irregular masses. It may be the cementing agent in sandstone or may replace organic material, as in petrified wood. Opal also forms the skeletons of diatoms, radiolarians, and silicoflagellates, which may be abundant in lacustrine and marine waters.

Cleavage None. Conchoidal fracture. Brittle.

Color Properties

Luster: Vitreous to waxy.

Color: Colorless, gray, white, blue, green, red, and so forth. Gem material may display a beautiful play of colors; fluorescence is common.

Streak: White.

Color in Thin Section: Colorless.

Pleochroism: None.

Optical Characteristics

Relief in Thin Section: Moderate negative.

Indices of Refraction: Index of refraction decreases with increasing water content.

Alteration With dehydration and crystallization, opal converts to microcrystalline quartz, such as chert. The opal shells of microorganisms may be replaced by pyrite.

Distinguishing Features

Hand Sample: Opal resembles microcrystalline quartz but has inferior hardness and lower specific gravity. The play of colors, if present, is distinctive.

Thin Section/Grain Mount: Opal is isotropic and may be cloudy. Analcime and volcanic glass have higher indices of refraction. Fluorite has distinct cleavage.

Occurrence Opal is common in voids in volcanic rocks and shallow intrusives, and may replace part of the rock. It is precipitated both from groundwater and from hydrothermal solutions. Opal is found in sedimentary rocks as a cementing agent and as a replacement of shells, wood fiber, or other organic material. Opal also is deposited around geysers and hot springs as geyserite or siliceous sinter.

Opal is a biomineraloid, being precipitated by microorganisms including diatoms, silicoflagellates, and radiolaria to form elaborate shells. Sponges also form siliceous spicules. Numerous species are involved, including over 1800 modern diatoms. In almost all cases silica is undersaturated in the water, so these organisms must expend cellular energy to concentrate silica to the point of being able to precipitate. When the organisms die, the siliceous shells settle to the bottom, where they accumulate as siliceous ooze, which may later crystallize to form chert. In favorable conditions, such as freshwater lakes in volcanic terranes, diatoms may accumulate in sufficient volumes that they can later be mined as diatomaceous earth.

Use Opal with a play of color is used as a gemstone, and colored varieties are commonly used for lapidary purposes. Diatomaceous earth has numerous uses, including as a polishing compound, a filtration agent in water treatment plants, a sorbent to clean up hazardous chemical spills, and an insecticide.

FELDSPAR GROUP

The feldspars are the most abundant minerals in the Earth's crust. They are found in nearly all igneous rocks, most metamorphic rocks, and are an important constituent of many sedimentary rocks and sediments. Additional information about these important minerals can be found in Smith and Brown (1988) and Ribbe (1983).

Composition

The feldspars include three principal compositional end members: K-feldspar ($KAlSi_3O_8$), albite ($NaAlSi_3O_8$), and anorthite ($CaAl_2Si_2O_8$), whose abbreviations are Ks, Ab, and An, respectively. The abbreviation Or is commonly used to indicate K-feldspar, but this usage is not followed here to avoid the unwanted inference that Or refers to orthoclase, which is a specific variety of K-feldspar. As is shown on Figure 12.5, only compositions between albite and anorthite, and between K-feldspar and albite, are found. The former feldspars are known as **plagioclase** and the latter as **alkali feldspars**.

As was discussed at some length in Chapter 4, the plagioclase feldspars represent a continuous solid solution series accomplished by a coupled substitution, at least at high temperatures. Albite is at one end of the series and anorthite at the other. Ca^{2+} can substitute for Na^+ because they are about the same size; charge neutrality is maintained by substituting Al^{3+} for Si^{4+}. Because relatively little K^+ is included in plagioclase, the composition is often described in terms of the percentage of the anorthite (An) end member, with the presumption that the remainder is dominantly albite (Ab) so that An + Ab = 100. The composition of plagioclase with 35 mol % Ca^{2+} and 65 mol % Na^+ is An_{35}, for example. The plagioclase feldspars are

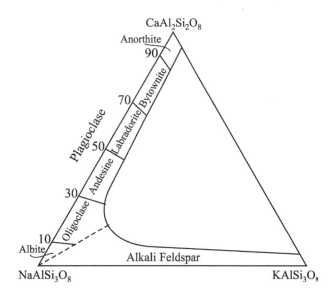

Figure 12.5 Composition range of common feldspars.

conventionally separated into six different named composition ranges.

Albite	$An_0–An_{10}$
Oligoclase	$An_{10}–An_{30}$
Andesine	$An_{30}–An_{50}$
Labradorite	$An_{50}–An_{70}$
Bytownite	$An_{70}–An_{90}$
Anorthite	$An_{90}–An_{100}$

Continuous solid solution in the alkali feldspar series is possible because K^+ and Na^+ have the same charge but is limited to high temperatures because the sizes of these cations are significantly different. The lack of compositions intermediate between K-feldspar and anorthite under any geological conditions follows from the observation that Ca^{2+} and K^+ have substantially different sizes and also have different charges.

Structure

The structure of the feldspars consists of a framework of corner-sharing Si and Al tetrahedra. The tetrahedra are arranged to form four-member rings (Figure 12.6a), with the apex of two of the tetrahedra pointing "up" and two pointing "down," parallel to the a axis. One of the "up" and one of the "down" tetrahedra are T_1 sites; the others are T_2 sites. The rings are joined to other rings by sharing oxygen anions to form crankshaft-like chains that extend parallel to a (Figure 12.6b). These crankshaft-like chains are in turn joined laterally to other chains by sharing the remaining oxygen anions (Figure 12.6c). This results in the formation of sites between chains that can contain large cations such as Ca^{2+}, Na^+, and K^+. These cations are coordinated with nine oxygen anions.

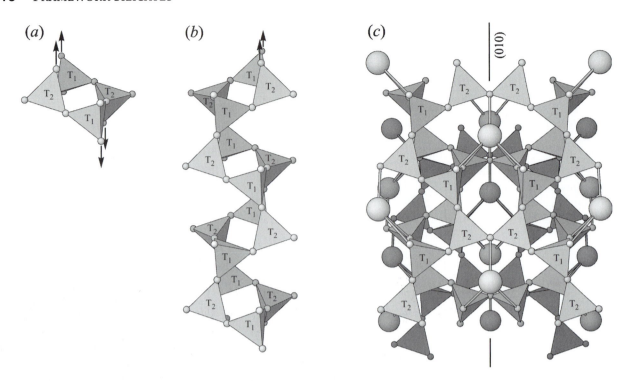

Figure 12.6 Idealized feldspar structure: T_1 and T_2 refer to different tetrahedral sites. View is perpendicular to the (001) plane; the same view is used in Figure 12.7. (*a*) Four-member ring of silicon tetrahedra containing two nonequivalent tetrahedral sites, T_1 and T_2. (*b*) The four-member rings share O^{2-} anions to form a crankshaft-like chain of silicon tetrahedra extending parallel to *a*. (*c*) Chains are cross-linked by sharing the remaining oxygen anions on each tetrahedron to form large coordination sites occupied by Ca^{2+}, Na^+, or K^+ (large spheres).

Table 12.3 Modes of Al:Si Ordering in Tetrahedral Sites in the Feldspars[a]

Ordering Mode	Al:Si Ratio	Description	$t_1o + t_1m$
Mode I	1:3 Alkali feldspar (Figure 12.7b)	Complete disorder; Al is equally likely to be in any of the four tetrahedral sites; on average, 25% of each of the four tetrahedral sites (T_1o, T_1m, T_2o, T_2m) is occupied by Al and 75% occupied by Si	0.5
Mode II		All of the Al is in the two T_1 sites; none is in the T_2 sites; the T_1o and T_1m, sites are, on average, 50% occupied by Al and 50% occupied by Si; the T_2 sites contain only Si	1.0
Mode III		All of the Al is concentrated in just the T_1o site, so its occupancy is 100% Al; the other three sites (T_1m, T_2o, T_2m) contain only Si; this is the highest degree of ordering available	1.0
Mode IV	2:2 Anorthite (Figure 12.7c)	The only way to avoid having Al in adjacent tetrahedra is to have a systematic alternation of Al and Si in the tetrahedral sites: Al is therefore in one-half of each of the T_1o, T_1m, T_2o, and T_2m sites	N/A

[a] The fractions of the T_1o and T_1m sites occupied by Al^{3+} are t_1o and t_1m, respectively.

Al/Si Order–Disorder

The way in which Si^{4+} and Al^{3+} are distributed in the T_1 and T_2 tetrahedral sites leads to the distinction among different order–disorder polymorphs (Chapter 4) of the feldspars and has significant consequences for the exsolution structures of plagioclase. Four different modes of ordering are possible (Table 12.3, Figure 12.7). Modes I, II, and III are available to K-feldspar and albite, which have an Al:Si ratio of 1:3, and Mode IV is found in anorthite, which has an Al:Si ratio of 2:2:

MODE I: Al and Si randomly distributed in tetrahedral sites

MODE II: Al preferentially located in half the T_1 sites

MODE III: Al preferentially located in just one T_1 site

MODE IV: Al and Si occupy alternating tetrahedra

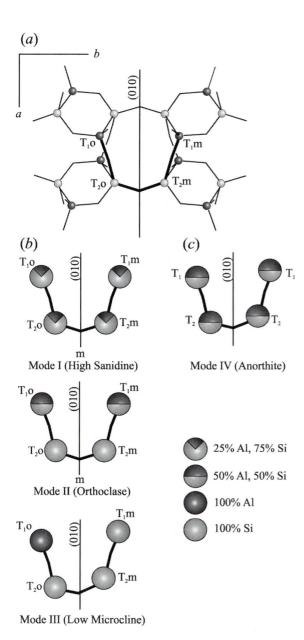

Figure 12.7 Ordering of Al and Si in tetrahedral sites in the feldspars. (*a*) Sketch showing the location of the T_1 (dark spheres) and T_2 (light spheres) cations within tetrahedra. The "o" and "m" identify sites on opposite sides of the (010) plane. Compare with Figure 12.6c. (*b*) Alkali feldspars, which have an Al:Si ratio of 1:3. In Mode I ordering, Al and Si are randomly distributed in the T_1 and T_2 sites so 25% of each site is occupied by Al^{3+} and a mirror plane is possible parallel to {010}. The "o" and "m" sites are structurally identical. In Mode II ordering each T_1 site is half-occupied by Al and by Si. The T_2 sites contain only Si. This ordering preserves the {010} mirror, so the "o" and "m" sites are structurally identical and orthoclase is ideally monoclinic. In Mode III ordering, all of the Al is in the T_1o site, and the T_1m, T_2o, and T_2m sites contain only Si^{4+}. The (010) mirror is destroyed, and the "o" and "m" sites are structurally different. The structure distorts slightly, and symmetry is reduced to triclinic. (*c*) Anorthite, in which the Al:Si ratio is 2:2. Al and Si occupy alternating tetrahedral sites, so T_1 and T_2 sites are occupied, on average, with half Al and half Si. See text for additional discussion.

K-Feldspar

Consider first the K-feldspars, which have an Al:Si ratio of 1:3, and occur in three polymorphs: **sanidine, orthoclase**, and **microcline**. Al and Si may order in the tetrahedral sites by Modes I, II, and III.

Mode I ordering, that is, complete disorder, is displayed in the idealized high sanidine structure (Figure 12.7b), which is approached only at temperatures greater than roughly 1000°C. Because, on average, the two T_1 and two T_2 sites have the same occupancy, a plane of symmetry parallel to {010} is present between adjacent crankshaft-like chains. The structure is therefore monoclinic and the *c* axis is inclined at roughly 116° to the {001} plane. Most K-feldspar, except that grown at relatively low temperature in hydrothermal systems or as a result of diagenesis in sediments (adularia), originally crystallizes with a sanidine structure and a high degree of disorder in the distribution of Al and Si in the tetrahedral sites.

Mode III ordering, which is the most ordered Al/Si distribution possible in the alkali feldspars, is approached in the low or maximum microcline structure. This highly ordered structure can be produced by slow cooling in plutonic igneous or deep metamorphic conditions. The Al and Si, which originally were more or less randomly distributed in the tetrahedral sites, diffuse through the lattice to preferentially place all or most of the Al^{3+} in just the T_1o site (Figure 12.7b). The T_1m or the T_2 sites contain only Si^{4+}. This destroys the mirror plane and results in a slight distortion of the lattice so microcline is triclinic.

The triclinic structure of microcline can be related to the original monoclinic structure of sanidine in one of four different but equally likely ways, depending on exactly which complement of T_1 sites in the crystal structure happens to accumulate the Al. When progressive ordering of Al and Si triggers the conversion from monoclinic to the slightly distorted triclinic microcline structure at around roughly 450°C, different parts of the crystal adopt one or another of the four possible orientations. Two of the orientations are related to each other by reflection on the {010} plane, and the other two are related by rotation about the *b* axis. Recall that different segments of crystals related by a symmetry operation are twins (Chapter 4). In the feldspars, twinning with {010} as a twin plane is called **albite twinning**, and twinning with the *b* axis as a rotation axis is called **pericline twinning**. Microcline that has been produced by ordering of an originally monoclinic K-feldspar always displays a combination of polysynthetic albite and pericline twinning. Twinning is discussed in more detail later.

Ideally, orthoclase would be ordered according to Mode II, in which the Al preferentially is located in the two T_1 sites (Figure 12.7b). Reality, however, is more complicated. Real orthoclase displays a combination of Mode II and Mode III ordering and can be considered to represent an intermediate degree of ordering between sanidine and microcline. The transition with cooling from sanidine to orthoclase is accompanied by a systematic increase in Mode II ordering as Al^{3+} diffuses out of the T_2 sites and into the T_1 sites. However, the distribution is not even between the T_1o and T_1m sites; the T_1o site is favored. This raises a difficulty because any ordering according to Mode III should produce a triclinic structure and orthoclase is monoclinic, at least macroscopically. The resolution to this problem is based on the recognition that most orthoclase is actually triclinic at the atomic scale, but polysynthetically twinned like microcline. The individual twin domains are submicroscopic, so the structure averages out to be macroscopically monoclinic.

The practical significance of this discussion of Al/Si order–disorder in K-feldspar is that it can inform geologists about the thermal history of the rocks containing K-feldspar. Sanidine-bearing rocks must have quenched rapidly from high temperature to prevent Al/Si ordering from occurring. Rocks containing low microcline with grid twinning must have been subjected to an extended period of slow cooling to allow the extensive Al/Si ordering, and so forth. A quick means of estimating the degree of Al/Si order in K-feldspar is provided by measuring the $2V$ angle (see below). X-ray diffraction techniques are described by Kroll and Ribbe (1983).

Plagioclase

The pattern of ordering of Al and Si in the tetrahedral sites in plagioclase depends on composition and leads to some complications not immediately evident.

Na-rich plagioclase displays Modes I, II, and III order–disorder characteristics the same as K-feldspar. **Monalbite** is the high-temperature Na-plagioclase. It is monoclinic, like sanidine, and has a high degree of Al/Si disorder (Mode I ordering) in the tetrahedral sites. On cooling, it polymorphically converts to a triclinic structure at a temperature of around 980°C. Two different processes are involved.

The first mechanism that leads to triclinic symmetry is a collapse or distortion of the structure (Figure 12.8). Na^+ is a loose fit in its structural site (which easily holds the larger K^+ in K-feldspar). At very high temperatures, the energetic vibration of the Na^+ is sufficient to effectively prop the structural site open to preserve monoclinic symmetry. With cooling, however, some of the oxygen anions collapse in toward the Na^+ and the structure distorts from monoclinic to triclinic. Triclinic Na-plagioclase with a high degree of Al/Si disorder like sanidine is known as **analbite**, but it is not normally found in nature.

The second process that causes the structure to become triclinic with cooling is the preferential ordering of Al into the T_1o sites according to the Mode III ordering mechanism described earlier. While some Mode II ordering occurs, Mode III is dominant. Triclinic Na-plagioclase with only a modest amount of Al/Si ordering is called **high albite**, and if highly ordered it is called **low albite**.

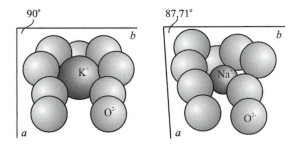

Figure 12.8 Arrangement of oxygen anions about K^+ in sanidine (left) and the distorted or collapsed arrangement around Na^+ in albite (right). The angle between the a and b axes in sanidine is 90°, allowing for monoclinic symmetry. In the albite structure $a \wedge b = 87.71°$, so symmetry is reduced to triclinic. Only at high temperatures is the vibrational energy of Na^+ in albite sufficient to keep the "cage" in which it resides from collapsing.

The Al/Si ordering associated with converting from high albite to low albite apparently progresses fairly easily so that the high albite structure is preserved only in volcanic rocks. Samples from both shallow and deep-seated intrusives and metamorphic rocks are typically well-ordered low albite. Samples with an intermediate degree of order are uncommon.

Ca-plagioclase (anorthite) follows Mode IV ordering (Figure 12.7c) because it contains equal numbers of Al and Si in the tetrahedral sites. The structure is apparently triclinic for all geologically normal conditions because Ca^{2+} is too small to keep its structural site propped open to maintain the monoclinic symmetry of the idealized feldspar structure. With Mode IV ordering, Al and Si tend to distribute themselves to avoid having Al in adjacent tetrahedra. This is known as the "**aluminum avoidance principle.**" Within a given crankshaft-like chain (Figure 12.6b), placing the Al and Si in alternate tetrahedra will, for example, ensure that all the T_1 sites contain Al and all T_2 sites contain Si. In adjacent chains, however, the T_1 and T_2 site occupancy must be reversed because T_2 tetrahedra bond to T_2 tetrahedra across the (010) plane (look in the center of Figure 12.6c). The net result is that T_1 and T_2 sites, on average, contain half Al and half Si. Mode IV ordering is possible only in feldspars with near-equal amounts of Al and Si.

Plagioclase with an intermediate Na–Ca composition poses a problem because it is not possible to have a linear variation in Al/Si order between low albite (Mode III with all the Al in one site) and anorthite (Mode IV with Al and Si in alternating sites) without violating the aluminum avoidance principle. What apparently happens in most cases is that ordering triggers the formation of submicroscopic domains, usually a few tens of angstroms across, alternating between relatively Na-rich plagioclase and Ca-rich plagioclase. The Al/Si tetrahedral framework is continuous, but the Na-rich domains have Mode III ordering and the Ca-rich domains have Mode IV ordering. Most plagioclase with compositions between ~An_{20} and An_{70} has this sub-microscopic structure and is known as **"e" plagioclase**.

Exsolution in the Feldspars

Both alkali feldspar and plagioclase may undergo a process of exsolution (Chapter 5) as they cool after crystallization. The driving force of the exsolution and the textures that result are somewhat different in the two types of feldspar.

Alkali Feldspar

Exsolution in the alkali feldspars was described in Chapter 5. Exsolution occurs because Na^+ and K^+ have significantly different ionic radii (1.32 vs. 1.65 Å). On cooling after crystallization, both the amount of Na^+ permissible in the K-feldspar structure and the amount of K^+ permissible in albite decrease. This is shown in Figure 5.27a. As the result of exsolution, stringers of albite are formed within K-feldspar crystals, yielding what is known as a **perthitic** texture (Figure 12.9). Albite with exsolved K-feldspar lamellae is known as **antiperthite**.

Development of exsolution lamellae depends, in large part, on the rate at which the alkali feldspar cooled after crystallization. Even sanidine and orthoclase in volcanic rocks undergo some exsolution, but rapid cooling typically ensures that the lamellae are submicroscopic. K-feldspar with submicroscopic exsolution lamellae is referred to as **cryptoperthitic**. These samples appear homogeneous in both hand sample and thin section. Plutonic igneous and metamorphic rocks cool at a sufficiently slow rate to allow development of larger exsolution lamellae that may be visible microscopically, or, in some cases, even in hand sample. Microcline is much more prone to developing visible exsolution lamellae than either orthoclase or sanidine. In detail, the exsolution lamellae of albite may form a braided, netlike, vein, patchy, or irregular pattern in the host K-feldspar. Alkali feldspar known as **moonstone** displays a diffuse milky white to pale blue iridescence produced by interaction of light with submicroscopic exsolution lamellae.

Plagioclase

The exsolution textures in plagioclase are more subtle and depend on order–disorder considerations, not the ionic radii of the cations. Exsolution occurs because of incompatibilities in the Al/Si ordering modes in albite and anorthite described earlier. Depending on the bulk plagioclase composition, **peristerite, Bøggild,** or **Huttenlocher** intergrowths may be produced. In most cases these intergrowths are not visible either in hand sample or microscopically. However, the thickness of the repeating lamellae may be about the same as the wavelength of visible light (i.e., several hundred nanometers), so the alternating lamellae may serve as a diffraction grating. It is therefore quite common for these submicroscopic intergrowth structures to cause

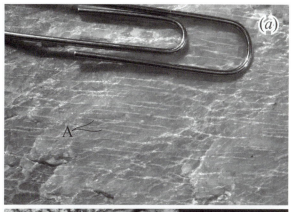

Figure 12.9 Perthitic exsolution. (*a*) Hand sample. Thin stringers of albite (A) in microcline extend roughly parallel to the paper clip. (*b*) Thin section. Light stringers of albite extending NNE are visible in microcline (crossed polarizers). The sample also displays typical twinning, with twin lamellae with diffuse borders extending NW–SE and NE–SW (cf. Figure 12.20b).

plagioclase to be **chatoyant** or **iridescent**, a feature sometimes referred to as **labradorescence** or **schiller**. The term *moonstone* is sometimes also applied to these samples.

Peristerite intergrowths are produced in plagioclase whose bulk composition falls in the range An_2 to ~An_{17}. One set of lamellae consists of nearly pure albite (An_{0-2}) ordered with all Al in just one of the T_1 sites (i.e., Mode III ordering). The other set of lamellae consists of an "e" plagioclase whose composition is ~An_{20-25} (oligoclase).

Bøggild intergrowths are produced in bulk compositions between about An_{46} and An_{60} and consist of lamellae of andesine (~An_{44}) and labradorite (~An_{58}). Both sets of lamellae consist of "e" plagioclase.

Huttenlocher intergrowths are produced in anorthite-rich plagioclase whose bulk composition falls in the approximate range An_{65}–An_{85} and consist of lamellae of labradorite (~An_{68}) and bytownite (~An_{88}). The labradorite consists of "e" plagioclase, and the bytownite is ordered according to the anorthite model (Mode IV).

Other Feldspar Intergrowths

In addition to the exsolution textures just described, the feldspars are commonly involved in a number of intergrowth

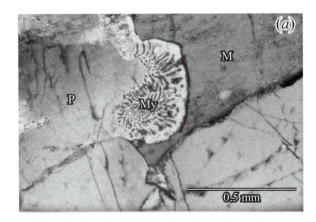

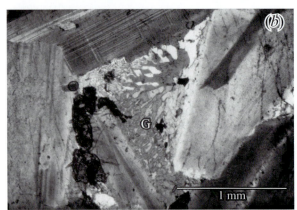

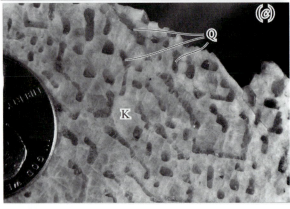

Figure 12.10 Quartz–feldspar intergrowths. (*a*) Myrmekite (My) consists of irregular quartz grains (dark) in plagioclase (light) seen at the contact between plagioclase (P) and microcline (M). Crossed polarizers. (*b*) Granophyre (G) in the space between plagioclase laths in norite. Crossed polarizers. (*c*) Graphic intergrowth of quartz (Q) and microcline (M) from a pegmatite.

relations. These include myrmekite, granophyre, graphic intergrowths, and (anti)rapakivi textures.

Myrmekite (Figure 12.10a) is the intergrowth of vermicular (wormlike) quartz in sodic plagioclase. It is visible only in thin section and is most common at the contact between K-feldspar and plagioclase in plutonic igneous rocks. Development of myrmekite evidently takes place after magmatic crystallization is complete, but the details of the mechanism are subject to some debate.

Granophyric intergrowths (Figure 12.10b) involve irregular, often vermicular quartz intergrown with feldspar (usually K-feldspar or sodic plagioclase). These intergrowths are most common in the interstices between larger, earlier-formed grains, and evidently represent simultaneous crystallization of quartz and feldspar from the last residual melt in an igneous rock.

Graphic intergrowths (Figure 12.10c) are found most commonly in granitic pegmatite and are sometimes referred to as graphic granite. In this intergrowth, large crystals of perthitic K-feldspar enclose smaller volumes of quartz whose cross sections have the appearance of cuneiform, hieroglyphic, or runic writing. In three dimensions, the quartz may form an irregular branching network. Simultaneous crystallization of quartz and K-feldspar is responsible for the intergrowth.

A **rapakivi** texture is produced when rims of sodic plagioclase mantle a phenocryst of K-feldspar, usually in felsic intrusive rocks. An **antirapakivi** texture is produced when a rim of K-feldspar crystallizes on a phenocryst of plagioclase.

Twinning

About 20 different twin laws have been identified in feldspars but only albite, pericline, Carlsbad, Baveno, and Manebach twins are at all common. Albite and pericline twins are polysynthetic and the others are single twins.

Polysynthetic albite and pericline twins are nearly ubiquitous in triclinic feldspars, but are not possible in the monoclinic feldspars. Albite twins (Figure 12.11a) are produced by reflection on {010} and composition planes are parallel to that crystal plane. Pericline twins (Figure 12.11b) are produced by 2-fold rotatation on [010] (the *b* axis). The position of the composition plane in most plagioclase is roughly parallel to {001}, but it varies significantly with composition, degree of Al/Si order, and temperature and pressure at the time of formation. In microcline, the composition plane gets to within about 15° of being parallel to {100}. Both twins may be produced by growth and by deformation. In addition, both twins are readily produced as part of the monoclinic–triclinic transformation that accompanies cooling and ordering of Al and Si in tetrahedral sites, as described earlier. The absence of these twins in adularia is taken to indicate that this feldspar originally grew with triclinic symmetry below the temperature of the monoclinic–triclinic transformation.

Carlsbad, Baveno, and Manebach twins are all single twins. Carlsbad twins (Figure 12.11c) are by far the most common and are produced by 2-fold rotation on [001] (the *c* axis) to yield penetration twins with a slightly irregular composition plane parallel to {010}. Manebach and Baveno twins (Figure 12.11d and e) are produced by reflection on {001} and {021}, respectively. All three twins are possible in both monoclinic and triclinic crystals and are produced only by growth.

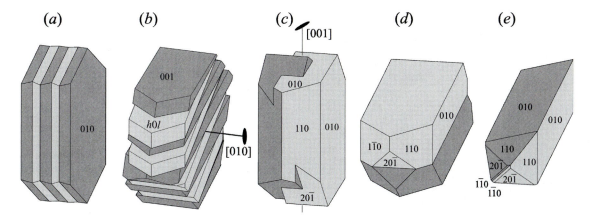

Figure 12.11 Common twins in the feldspars. (*a*) Albite twins on {010}. (*b*) Pericline twins produced by rotation on [010]. (*c*) Carlsbad twins produced by rotation on [001]. (*d*) Manebach twins by reflection on {001}. (*e*) Baveno twins produced by reflection on {021}. In the triclinic varieties, twinning is also possible on {0$\bar{2}$1}.

PLAGIOCLASE

$NaAlSi_3O_8$–$CaAl_2Si_2O_8$

Triclinic $\bar{1}$

Low albite: a = 8.137 Å, b = 12.785 Å, c = 7.158 Å, α = 94.26°, β = 116.60°, γ = 87.71°, Z = 4

Anorthite: a = 8.177 Å, b = 12.877 Å, c = 14.169 Å, α = 93.17°, β = 115.85°, γ = 92.22°, Z = 8

H = 6–6.5

G = 2.60 (Ab)–2.76 (An)

Biaxial (+ or −)

n_α = 1.527–1.577

n_β = 1.531–1.585

n_γ = 1.534–1.590

δ = 0.007–0.013

$2V_x$ = 45–102° (high plagioclase)

$2V_x$ = 75–102° (low plagioclase)

Form and Twinning Plagioclase occurs commonly as both euhedral and anhedral grains (Figure 12.12). Crystals are tabular parallel to {010} and elongate parallel to the *c* or *a* axis. Cross sections are more or less rectangular. **Cleavelandite** is a variety of albite that forms very platy crystals parallel to {010}.

Chemical zoning is common, particularly in volcanic and hypabyssal intrusive rocks, and is expressed as a variation in the extinction angle from one zone to another. Zoning is considered normal if it changes smoothly from a more calcic core to a more sodic rim; the core usually has a larger extinction angle. Reverse zoning is the opposite. Oscillatory zoning involves a rhythmic alternation between calcic and sodic zones usually superimposed on an overall trend from a more calcic core to a more sodic rim.

Twinning is very common. Polysynthetic albite {010} and pericline [010] twins are present in most grains, provided they are sufficiently large. In hand sample the

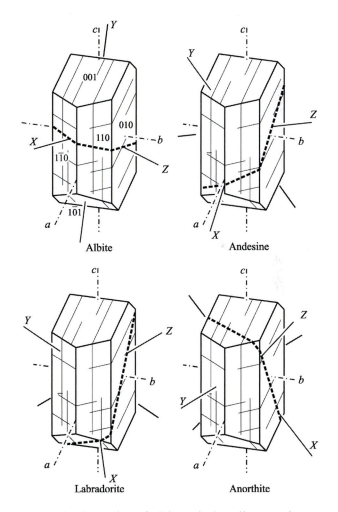

Albite

Andesine

Labradorite

Anorthite

systematic alternation of albite twin lamellae may be recognized by what appear to be fine striations on basal {001} cleavage surfaces. In thin section, albite twins are recognized by parallel lamellae that may be cut at intervals with pericline twin lamellae at a high angle. Carlsbad twins [001] are also quite common; Manebach {001} and Baveno {021} twins are less common.

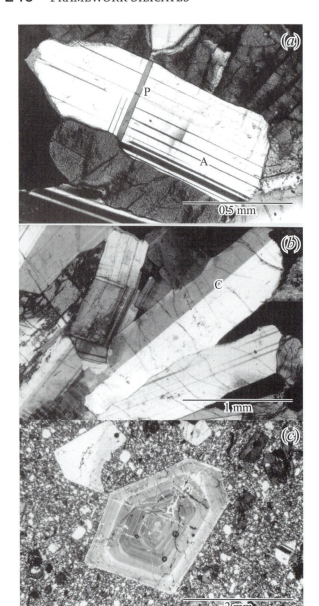

Figure 12.12 Plagioclase in thin section. (*a*) Albite twin lamellae (A) extend NW–SE, and pericline twin lamellae (P) extend NE–SW. (*b*) Large grain in the center displays a Carlsbad twin (C) that divides the grain into right and left portions. (*c*) Phenocryst with oscillatory zoning.

Cleavage Basal {001} cleavage is perfect and {010} cleavage is good. They intersect at an angle of 93–94°. A poor cleavage on {110} is not normally seen either in hand sample or thin section. Cleavage tends to control fragment orientation in grain mounts. In thin section, the cleavage may not be obvious unless the aperture diaphragm is stopped down to emphasize what little relief there is, and is most noticeable along the sample's edge where stress associated with slide preparation is more likely to fracture grains. Conchoidal to uneven fracture. Brittle.

Color Properties

Luster: Vitreous; chatoyant varieties may be pearly.

Color: White and gray are most common, also bluish, greenish, yellowish, or reddish. Varieties called moonstone display a beautiful chatoyance or iridescence related to peristerite, Bøggild, or Huttenlocher intergrowths.

Streak: White.

Color in Thin Section: Colorless in thin section and grain mount. Some may be clouded due to partial alteration to sericite or clay or may appear pale reddish, brownish, or gray due to fine Fe–Ti oxide inclusions.

Pleochroism: None.

Optical Characteristics

Relief in Thin Section: Low positive or negative.

Optic Orientation: Optic orientation varies systematically with composition and provides the basis for several techniques of determining the composition. Indicatrix axes and crystal axes do not coincide, so extinction is inclined to crystal faces and cleavage traces in most orientations.

Indices of Refraction and Birefringence: Indices of refraction increase systematically with Ca content (Figure 12.13), and if measured accurately, can provide a useful estimate of composition. Birefringence varies between 0.007 and 0.013, but the change is too small to be of use in estimating composition. Maximum interference colors in thin section are usually first-order gray or white. Only for very calcic plagioclase will first-order yellow colors be seen.

Interference Figure: 2V varies systematically with composition (Figure 12.13), but has a distinctly different pattern for high and low varieties. The 2V angle by itself cannot be used to estimate composition, but it may be used to resolve ambiguity in other techniques. Fine lamellar twinning or exsolution may make usable figures difficult to obtain in some cases. Optic axis dispersion is weak, usually $v > r$, but also $r > v$.

Alteration Plagioclase is commonly partly altered to sericite (fine-grained white mica), clay, or zeolites. Alteration may be uniformly distributed or may be concentrated along twin lamellae or in the core of grains. Other alteration products include scapolite, calcite, pyrophyllite, and **sausserite** (a fine-grained aggregate of epidote group minerals, albite, sericite, and other minerals).

Distinguishing Features

Hand Sample: Hardness and cleavage distinguish feldspars from other silicates. The presence of parallel striae on cleavage surfaces distinguishes plagioclase from K-feldspar. In the common igneous rocks that contain both plagioclase and K-feldspar, the K-feldspar is more commonly pinkish and the plagioclase is more commonly gray.

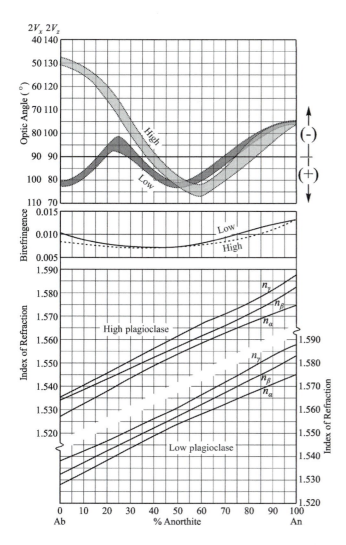

Figure 12.13 Optical properties of low (plutonic) and high (volcanic) plagioclase. Adapted from Smith (1958) and Burri and others (1967).

Table 12.4 Plagioclase Composition in Common Igneous Rock Types

Rock Type	Plagioclase Composition
Felsic (granitic pegmatite/granite/rhyolite/ granodiorite/dacite/etc.)	Albite or oligoclase
Intermediate (andesite/diorite and related)	Andesine
Mafic (basalt/gabbro and related)	Labradorite

and others, 1964; Boone and Wheeler, 1968; and Haughton, 1980.

Determining Composition It is not possible in hand sample to determine composition, but plausible guesses can be made based on the rock type in which the plagioclase is found (Table 12.4). Several optical techniques have been developed to allow composition to be estimated to within a few percent anorthite. If knowledge of composition is important, optical techniques should be supplemented with chemical and/or X-ray techniques.

Grain Mount: The most accurate method depends on measurement of n_α, n_β, and n_γ and comparison with Figure 12.13. However, cleavage tends to control fragment orientation, making it difficult to find grains in appropriate orientations. This difficulty is turned to an advantage with the **Tsuboi method**, which is used for plutonic plagioclase. To use the Tsuboi method, proceed as follows.

1. Prepare a grain mount of crushed plagioclase (Chapter 10).
2. By examination, select a fragment lying on either cleavage surface; it does not matter which.
3. Cross the polarizers and rotate the grain to extinction with the fast ray parallel to the lower polarizer vibration direction. Use the accessory plate to determine which ray is fast (Chapter 7).
4. In plane light, compare the index of refraction n'_α with the index of refraction of the oil.
5. Repeat, using different immersion oils, until a match is obtained.
6. Use Figure 12.14 to determine the composition. If n'_α is measured to an accuracy of ±0.001, accuracy of ±1% An content is possible in the range An_{20} to An_{60}, and ±3% An outside that range. The accuracy for volcanic rocks is probably no better than ±5% An. A significant complication is that plagioclase often is chemically zoned, so grains from a crushed sample of a single crystal may contain a range of compositions.

Thin Section: The **Michel-Lévy** and **Carlsbad–albite** methods are used to determine the composition of plagioclase in thin section. They depend on the fact that optical orientation of plagioclase varies systematically with composition. Both methods involve measurement of extinction

Thin Section/Grain Mount: Low relief, lack of color, biaxial character, and polysynthetic twinning distinguish plagioclase from most other silicates. Cordierite may display similar twinning, but it alters to pinite, and the twins may display a radial pattern in certain orientations. Cordierite also commonly contains pleochroic halos around radioactive inclusions such as zircon, and it has less obvious cleavage. Untwinned plagioclase, particularly in low-grade pelitic metamorphic rocks, may resemble quartz and can easily be overlooked. However, quartz is uniaxial, lacks cleavage, and usually lacks the incipient alteration or cloudiness common in plagioclase. Microcline twins according to the same twin laws, but the albite and pericline twin lamellae are irregular, fuzzy, and tend to pinch and swell, forming a "tartan plaid" pattern (see later: Figure 12.20b).

A number of staining techniques have been developed to aid with rapid hand-sample and thin section identification of plagioclase, K-feldspar, and other minerals with similar appearance (e.g., Bailey and Stevens 1960; Ruperto angles to the trace of the {010} composition plane of albite twin lamellae, and require selection of grains oriented so that the {010} crystallographic plane is vertical (i.e., the *b* crystal axis is horizontal).

The Michel-Lévy method requires measurement of a half-dozen or more grains in the sample to find a maximum value of the extinction angle between {010} and the trace of the fast ray vibration direction. A number of grains must be measured because the extinction angle varies depending on grain orientation and it is necessary to find a grain in the correct orientation to yield the maximum extinction angle. To use the Michel-Lévy method (Figure 12.15), proceed as follows.

1. Scan the slide to find a grain with the {010} composition plane between albite twin lamellae vertical. Usable grains have the following characteristics.

   • Composition planes between lamellae are crisp and sharp. If {010} is inclined significantly, the lamellae will be fuzzy and the position of the boundaries will shift laterally when focus is raised and lowered while viewed with the high-power objective.

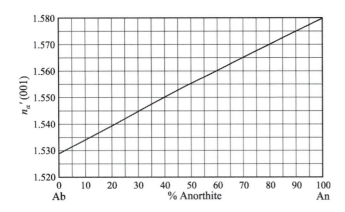

Figure 12.14 Index of refraction of the fast ray ($n_{\alpha'}$) for plagioclase fragments lying on cleavages. The diagram is constructed for fragments on (001) cleavage surfaces but also may be used for fragments on (010) cleavage surfaces. After Morse (1968).

• All twin lamellae have essentially the same interference color between crossed polarizers when placed parallel to the N–S and in the 45° position. If the two sets of lamellae have different interference colors, either {010} is not vertical or the twins are pericline twins. If the grain is divided in half so lamellae on one side are all one color and lamellae on the other half are a different color, the boundary between the two sides is a Carlsbad twin and the Carlsbad-albite method should be used.

2. Between crossed polarizers, start with the composition plane between twin lamellae parallel to the N–S. Rotate clockwise to bring one set of lamellae to extinction and record the extinction angle a_1. Return the composition plane to N–S, then rotate counterclockwise to bring the other set of lamellae to extinction and record the extinction angle a_2. If the two extinction angles differ by 4° or less, calculate their average. If the two extinction angles differ by more than 4°, discard the readings and find a new grain to measure because {010} is not vertical. In most cases, the extinction angle between the trace of {010} and the fast ray vibration direction is less than 45°, but for calcic plagioclase, it may be larger. To identify the fast ray vibration direction within one set of lamellae, begin by rotating the grain to bring those lamellae to extinction. Then rotate 45° clockwise from the extinction position and insert the gypsum plate (slow NE–SW). If the retardations in the lamellae in question subtract (usually to produce first-order yellow), then the fast ray in those lamellae was correctly placed N–S at extinction. If the colors increase (usually to second-order blue), the slow ray was N–S in the extinction position, and rotation in the opposite direction is needed to bring the fast ray vibration direction in the set of lamellae N–S in the extinction position.

3. Repeat for a half-dozen or more grains (more is better). Because only the maximum average extinction angle is useful, time can be saved by quickly moving on if the

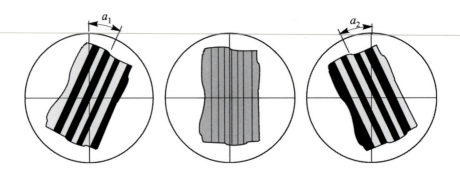

Figure 12.15 Michel-Lévy method. Extinction angles a_1 and a_2 are measured to the fast ray vibration direction in their respective sets of albite twin lamellae. The two angles are averaged. A half-dozen or more grains are measured, and the largest average extinction angle is used to estimate the plagioclase composition from Figure 12.16.

first angle measured on a grain is less than the largest average measurement taken on other grains.

4. The composition is determined from Figure 12.16. The maximum average extinction angle is plotted on the vertical axis; composition is plotted along the horizontal axis. Different curves are provided for high (volcanic) and low (plutonic) plagioclase. For angles less than about 18°, the curves indicate two possible compositions: one less than An_{20} (plutonic) or An_{12} (volcanic) and the other higher. The distinction between the two possibilities can be made based on indices of refraction and optic sign (Figure 12.13).

An_0–An_{20}: Optically (+), $n_\alpha < 1.538$

An_{20}–An_{35}: Optically (–), $n_\alpha > 1.538$

Note that 1.538 is about the same as the index of refraction of many, but not all, commonly used thin section cements. Provided the index of refraction of the cement is known to be about 1.538, the two possibilities can be distinguished by comparing n_α with n_{cement}. To do this, find a plagioclase grain with maximum birefringence whose edge is in contact with cement along a hole in the slide or along the edge. Rotate the stage to place the portion of the grain in contact with cement to an extinction position with the fast ray (n_α) vibration direction parallel to the vibration direction of the lower polarizer. Use the accessory plate to identify fast and slow ray vibration directions. Remove the upper polarizer and compare n_α with n_{cement} using the Becke line method.

Accuracy of the method for low plagioclase is roughly ±5% anorthite content up to about An_{60} and worse at higher An content. The accuracy is probably significantly worse with volcanic plagioclase.

The Carlsbad-albite method is closely related to the Michel–Lévy method, but only one properly oriented grain is needed to obtain a composition. The procedure is as follows (Figure 12.17).

1. Select a grain with both albite and Carlsbad twinning that is oriented with the {010} composition planes between albite twins vertical. Appropriately oriented grains have the following characteristics.
   • The albite twin lamellae are crisp and sharp, indicating that {010} is vertical (see Michel-Lévy method).
   • When the twin lamellae are placed in the N–S and 45° positions, the Carlsbad twin separates the grain into two segments with different interference colors, but the albite lamellae within each segment have essentially the same colors. The composition plane of the Carlsbad twin may be slightly irregular.
2. On the left half of the Carlsbad twin, measure the extinction angles from the albite twin lamellae to the fast ray vibration direction in each albite lamellae set as described in the Michel-Lévy method. Average the two readings provided they differ by less than 4°. If they differ by more, discard the readings and find another grain.

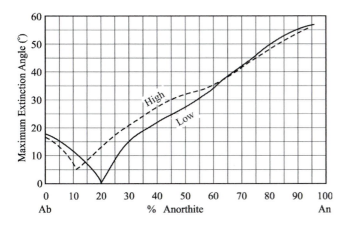

Figure 12.16 Diagram for use with the Michel-Lévy method. The maximum extinction angle to albite twin lamellae in sections cut perpendicular to {010} is plotted on the vertical axis; composition is the horizontal axis. The dashed curve is for high (volcanic) plagioclase, and the solid curve for low (plutonic) plagioclase. See text for additional discussion. Based on curves of Tobi and Kroll (1975).

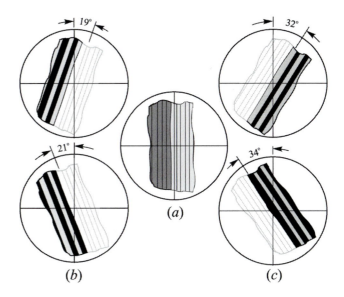

Figure 12.17 Carlsbad–albite method. (a) Grain with twin lamellae N–S between crossed polarizers. The two sides of the Carlsbad twin show different colors, but the albite lamellae within each side are uniform. (b) Extinction angles measured to the fast ray vibration direction in the albite lamellae on the left side of the Carlsbad twin. The average extinction angle in this case is 20°. (c) Extinction angles measured to the fast ray vibration direction in the albite lamellae on the right side of the Carlsbad twin. The average extinction angle in this case is 33°. The composition of the plagioclase (An_{60}) is determined from Figure 12.18a.

3. On the right half of the Carlsbad twin, measure the extinction angles from the albite twin lamellae to the fast ray vibration direction in each albite lamellae set in the same manner and average the readings. The 4° rule still applies.

4. The composition is obtained from Figure 12.18a for low plagioclase and Figure 12.18b for high plagioclase. The solid lines are for the larger of the two average extinction angles found in steps 2 and 3; the dashed line is for the smaller. The point where the two curves intersect indicates the composition, which is read from the bottom of the diagram. The vertical axis indicates the inclination of the c axis to the stage of the microscope.

For extinction angles less than about 20°, there are two or three sets of curves, so the same set of readings may produce several intersections, indicating several possible compositions. For example, with high plagioclase, if the smaller average extinction angle is 10° and the larger is 15°, intersections can be found at An_1 (upper left), An_{29} (top at left center), and An_{66} (bottom at right center), but only one is correct. For relatively albite-rich compositions, the ambiguity is resolved in the same way described for the Michel-Lévy method. For more calcic possibilities, additional grains can be measured to determine which is correct.

Occurrence   Plagioclase is the most common and abundant mineral group in the Earth's crust. As described in Chapter 11, it is abundant in many igneous and metamorphic rocks and is common as a detrital mineral in many sedimentary rocks. The plagioclase compositions found in common igneous rocks are listed in Table 12.4. The plagioclase in pelitic metamorphic rocks is usually albite if subjected to relatively low-grade metamorphism and oligoclase or andesine in medium- and high-grade environments. Fairly pure anorthite may be found in metamorphosed carbonates and amphibolite.

In sedimentary rocks plagioclase occurs as detrital grains, but it is susceptible to normal weathering and diagenetic processes.

Use   The industrial commodity that most commonly contains plagioclase is aggregate, the ordinary sand and gravel used to make concrete, asphalt-based pavement, and so forth. Plagioclase also is a major constituent in many of the rocks that are cut and polished to make countertops, facing stone for buildings, and headstones. Plagioclase may also be used in powdered form as a filler in paint, plastic, and rubber, and in the manufacture of glass and ceramics. Although not currently economic, plagioclase, in the form of large intrusive bodies of anorthosite, has the potential to be a significant source of aluminum.

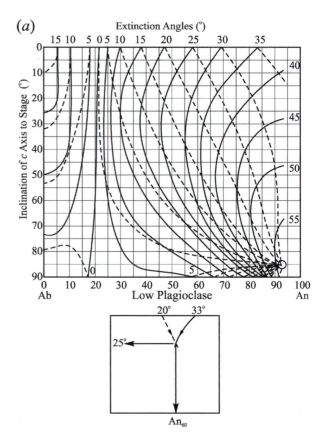

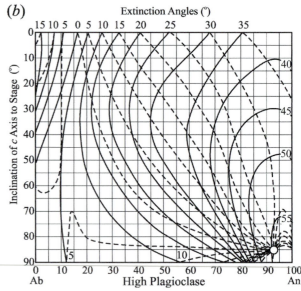

Figure 12.18 Diagrams for use with the Carlsbad–albite method. Solid lines are for the larger extinction angle, and dashed lines for the smaller extinction angle (Figure 12.17). See text for additional discussion. (a) Low plagioclase. A composition of An_{60} is indicated for a smaller extinction angle of 20° and a larger angle of 33°, and the c axis of the grain is inclined 25° from the microscope stage. (b) High plagioclase. From Tobi and Kroll (1975).

ALKALI FELDSPAR

$(K,Na)AlSi_3O_8$
$H = 6–6.5$
$G = 2.55–2.63$

Sanidine, Orthoclase, Microcline, Adularia
Biaxial (−)
$n_\alpha = 1.514–1.526$
$n_\beta = 1.518–1.530$
$n_\gamma = 1.521–1.533$
$\delta = 0.005–0.008$
$2V_x =$ See individual descriptions

Anorthoclase
Biaxial (−)
$n_\alpha = 1.519–1.529$
$n_\beta = 1.524–1.534$
$n_\gamma = 1.527–1.536$
$\delta = 0.005–0.008$
$2V_x = 42–52°$

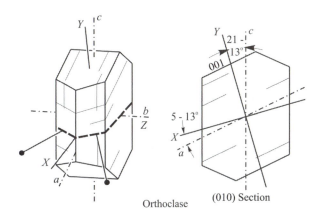

Orthoclase   (010) Section

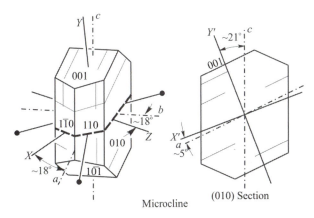

Microcline   (010) Section

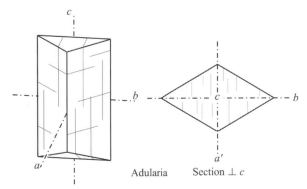

Adularia   Section ⊥ c

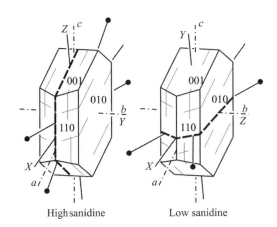

High sanidine   Low sanidine

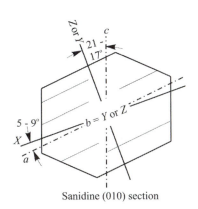

Sanidine (010) section

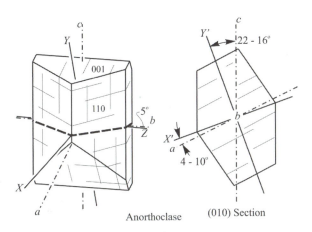

Anorthoclase   (010) Section

The terminology of the alkali feldspars has evolved over several centuries and is not as well defined as might be desired. The problem is that the four varieties of K-feldspar (microcline, orthoclase, sanidine, and adularia) were originally recognized based on physical and optical properties and occurrences. These distinctions were not always well founded on a structural or chemical basis. We now know, for example, that much of the material traditionally identified as monoclinic orthoclase is, as has been described, triclinic, but is submicroscopically twinned so that it appears morphologically and optically monoclinic. Regardless, nothing is likely to replace the old terminology. For current purposes, the distinctions among the alkali feldspars are based on optical and physical properties as follows:

MICROCLINE: Optically triclinic K-feldspar characterized by its distinctive cross-hatched twin pattern. The optic plane is nearly perpendicular to (010) and $2V_x$ is typically greater than 65°, though smaller values are possible.

ORTHOCLASE: Optically monoclinic K-feldspar with $2V_x$ greater than 40° and the optic plane perpendicular to (010).

LOW SANIDINE: Optically monoclinic K-feldspar with $2V_x$ between 0° and 40° and the optic plane perpendicular to (010).

HIGH SANIDINE: Optically monoclinic K-feldspar with $2V_x$ between 0° and 40° and the optic plane parallel to (010).

ADULARIA: K-feldspar produced by authigenic or low-temperature hydrothermal processes in which the {110} prism is dominant. It is structurally similar to microcline but without the grid twinning produced by conversion from monoclinic to triclinic symmetry.

ANORTHOCLASE: Optically triclinic sodic alkali feldspar characterized by grid twinning similar to microcline but on a finer scale, and with $2V_x$ less than 52°; found in volcanic and hypabyssal intrusive rocks. The composition typically lies between $Ab_{63}Or_{27}$ and $Ab_{90}Or_{10}$.

It turns out that $2V$ and indices of refraction in K-feldspar vary systematically with the degree of order of Al and Si in tetrahedral sites (Figure 12.19). A convenient means of expressing that order is to sum the occupancy of Al in the T_1o and T_1m structural sites, which are two of the four available tetrahedral structural sites (Figure 12.7 and Table 12.3). In completely disordered high sanidine, Al is equally likely to be in the T_1o and T_1m sites. Because K-feldspar has one Al for every three Si in the tetrahedral sites, each tetrahedral site will be, on average, one-quarter

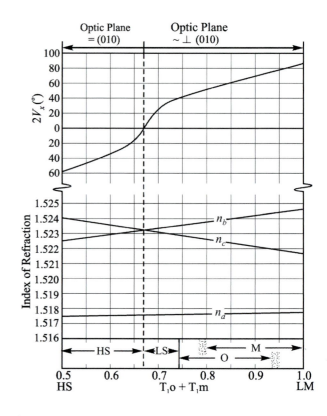

Figure 12.19 Variation of $2V_x$ and indices of refraction for common K-feldspar as a function of the degree of Si–Al order. The occupancies of Al in the T_1o and T_1m tetrahedral sites are t_1o and t_1m, respectively. If Al and Si are randomly distributed among the tetrahedral sites, as in high sanidine (HS), then one-quarter of the T_1o and one-quarter of the T_1m sites are filled with Al: $T_1o + T_1m = 0.50$. If all Al is in the T_1o site, then $T_1o + T_1m = 1.00$ as in low microcline (LM). The principal indices of refraction n_a, n_b, and n_c are for light vibrating most nearly parallel to the a, b, and c crystal axes, respectively. The approximate ranges of high sanidine (HS), low sanidine (LS), orthoclase (O), and microcline (M) are indicated. After Su and others (1984).

occupied by Al, so $T_1o + T_1m = 0.25 + 0.25 = 0.50$. In completely ordered low microcline, all of the Al is in the T_1o site and none is in the T_1m site, so $T_1o + T_1m = 1.0 + 0 = 1.0$. Similarly, ordering according to the orthoclase model (Mode II ordering) puts all of the Al equally in the T_1o and T_1m structural sites; each is half occupied by Al, so $T_1o + T_1m = 0.5 + 0.5 = 1.0$. This accounts for the fact that orthoclase and microcline have substantial overlap on the diagram and emphasizes the point that the nature of the ordering mechanism cannot be determined by optical and physical properties.

As a practical matter, the different varieties cannot reliably be distinguished in hand sample and it is best to use the generic term *K-feldspar* for all varieties until optical, X-ray, or other techniques provide additional information. However, the geological occurrence of a sample can provide guidance as to which variety is most likely.

Use K-feldspar, like plagioclase, is a major mineral in many of the rock types used as aggregate and dimension stone. Its industrial uses are in glassmaking and in the manufacture of ceramics including dinnerware, floor tiles, plumbing fixtures, and electrical insulators. K-feldspar also is used, in powdered form, as a filler in products including paints, plastic, and rubber. Its only legitimate medical use is in dental ceramics; historically, however, K-feldspar has been used to treat bladder and kidney problems.

MICROCLINE

Triclinic $\bar{1}$
$a = 8.578$ Å, $b = 12.960$ Å, $c = 7.211$ Å, $\alpha = 90.65°$,
  $\beta = 115.96°$, $\gamma = 87.65°$, $Z = 4$
$2V_x = {\sim}65–88°$

Form and Twinning (Figure 12.20) Microcline is common as anhedral and euhedral grains in many igneous and metamorphic rocks. Crystals tend to be tabular parallel to {010} and elongate parallel to the c or a axis. Perthitic exsolution is common.

Polysynthetic twins according to the albite and pericline laws are characteristic of microcline that has inverted from monoclinic symmetry and produces a distinctive cross-hatched or "tartan plaid" pattern seen in thin section. The contacts between lamellae are diffuse, and the lamellae appear to pinch and swell, and are discontinuous. The composition plane for the albite twins is {010}. The composition plane for the pericline twins contains the b axis and is commonly inclined about 35° to {010} and may be as close as 12°. Carlsbad twins are also common, particularly in phenocrysts from igneous rocks. Baveno, Manebach, and other twins also may be found. The polysynthetic twins are not evident in hand sample.

Cleavage Perfect on {001} and good on {010} at an angle of ~90.6°. Uneven fracture. Brittle.

Color Properties

Luster: Vitreous.

Color: White, pink, and salmon red are most common, also pale yellow. Blue and green microcline, which contains trace amounts of Pb, is known as **amazonite**.

Streak: White.

Color in Thin Section: Colorless.

Pleochroism: None.

Optical Characteristics

Relief in Thin Section: Low negative.

Optic Orientation: $X \wedge a \approx 18°, Y \wedge c \approx 18°, Z \wedge b \approx 18°$.

Indices of Refraction and Birefringence: Fully ordered low microcline has indices of refraction $n_\alpha = 1.518$, $n_\beta = 1.522$,

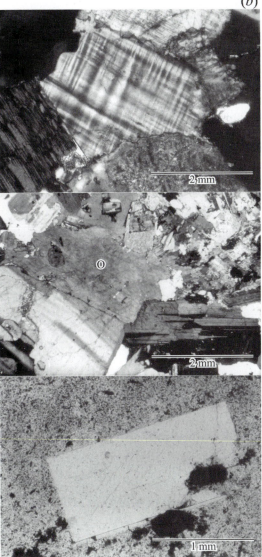

Figure 12.20 K-feldspar. (*a*) Microcline crystals with quartz (dark). (*b*) Photomicrographs. (*Top*) Thin section of granite showing microcline with "tartan plaid" albite and pericline twinning. Crossed polarizers. (*Middle*) Orthoclase (O) in granite. Crossed polarizers. (*Bottom*) Sanidine phenocryst in rhyolite. Plane light.

$n_\gamma = 1.525$, and the indices appear to vary systematically with degree of Al/Si order (Figure 12.19). Birefringence is usually about 0.007, so interference colors are no higher than first-order white in standard thin sections.

Interference Figure: Biaxial negative with 2V usually greater than about 65°, most often 70–85°, but as small as 35° is possible. Optic axis dispersion is weak, r > v. Twinning may make good interference figures difficult to obtain. Fragments on cleavage surfaces yield strongly off-center figures.

Alteration Microcline is commonly altered to sericite and clay.

Distinguishing Features

Hand Sample: Microcline cannot reliably be distinguished from sanidine and orthoclase in hand sample, but the presence of visible exsolution lamellae is almost always associated with microcline. Sanidine is typically colorless, whereas orthoclase and microcline are more commonly, but not always, colored. Blue or green amazonite is almost always microcline.

Thin Section/Grain Mount: The tartan plaid pattern of twins is characteristic. Unlike plagioclase, the twin lamellae typically have fuzzy boundaries and tend to pinch and swell irregularly.

Occurrence Microcline is commonly found in granite, granodiorite, syenite, granitic pegmatite and related plutonic igneous rocks, and in pelitic metamorphic rocks, usually of fairly high grade. It is not found in volcanic rocks except as part of a xenolith.

Because microcline is fairly stable in the weathering environment, it is a common constituent of detrital sedimentary rocks. It is not uncommon for overgrowths of authigenic K-feldspar to develop on detrital grains.

ORTHOCLASE

Monoclinic 2/m
$a = 8.563$ Å, $b = 12.963$ Å, $c = 7.299$ Å,
$\beta = 116.07°$, Z = 4
$2V_x = 40$–~70°

Form and Twinning Orthoclase is common as anhedral and euhedral grains in felsic igneous rocks (Figure 12.20). Crystals are typically elongate parallel to c or a, and the {010} form is usually prominent. Some orthoclase may be microscopically perthitic, and most is perthitic at a submicroscopic scale.

Carlsbad twins are common, particularly in phenocrysts found in igneous rocks. Baveno and Manebach twins are less common. Polysynthetic albite and pericline twins are not found.

Cleavage Perfect cleavage on {001} and good cleavage on {010} intersect at 90°. Uneven to conchoidal fracture. Brittle.

Color Properties

Luster: Vitreous.
Color: White, gray, pink, pale yellow; may display schiller or opalescence.
Streak: White.
Color in Thin Section: Colorless.
Pleochroism: None.

Optical Characteristics

Relief in Thin Section: Low negative.
Optic Orientation: $X \wedge a = +5°$ to $+13°$, $Y \wedge c = +21°$ to $+13°$, $Z = b$.
Indices of Refraction and Birefringence: Indices of refraction vary in a small range with degree of Al/Si order (Figure 12.19). An increase in Na content produces a slight increase in indices of refraction, but the degree of Al/Si ordering is more important. Birefringence is low (~0.007), so interference colors in standard thin section are no higher than first-order white.

Interference Figure: $2V_x$ varies systematically with degree of Al/Si order (Figure 12.19) and increases slightly with Na content. The convention adopted here is to use $2V_x = 40°$ as the dividing line between orthoclase and low sanidine. Orthoclase may have a 2V angle as large as 85°, but it is usually less than 70°. Fragments lying on {001} and {010} cleavages yield off-center obtuse bisectrix and flash figures, respectively. Optic axis dispersion is r > v.

Comments: Orthoclase commonly displays a slight cloudiness, evidently produced by minor alteration.

Alteration The common alteration products are sericite and clay.

Distinguishing Features

Hand Sample: Right-angle cleavage, hardness, and color distinguish the feldspars from other silicates. Orthoclase cannot reliably be distinguished from microcline or sanidine in hand sample, but sanidine is more commonly clear and glassy, and microcline is more commonly strongly colored and may have visible exsolution lamellae. Plagioclase may display parallel striations on the basal cleavage related to albite twinning.

Thin Section/Grain Mount: Lack of polysynthetic twins distinguishes orthoclase from plagioclase and microcline, and the 2V angle and optic orientation distinguish it from sanidine. Anhedral orthoclase is readily mistaken for quartz, but orthoclase has negative relief in standard thin section, has cleavage, is biaxial, and often displays a slight cloudiness.

Occurrence Orthoclase is very common in granite, granodiorite, syenite, and related felsic rocks, particularly in shallow intrusives of Phanerozoic age. Sanidine is more common in volcanic rocks, and microcline is more common in deep-seated intrusives. Orthoclase also is found in contact and regional metamorphic rocks, and as detrital grains in clastic sediments.

SANIDINE

Monoclinic 2/m
$a = 8.603$ Å, $b = 13.036$ Å, $c = 7.174$ Å,
   $\beta = 116.03°$, $Z = 4$
$2V_x = 0–40°$ [OP $\perp$ (010)]
   $0–47°$ [OP $\parallel$ (010)]

Form and Twinning Sanidine is common as phenocrysts that are tabular parallel to {010} and somewhat elongate parallel to a (Figure 12.20). Cross sections tend to be square, rectangular, or six-sided. Acicular crystals are common in felsic volcanic glass and in spherulites derived from that glass. Zoning, expressed as variation in birefringence, is fairly common. Twinning according to the Carlsbad law, which divides crystals into two segments, is common; Baveno and Manebach twins are less common.

Cleavage Perfect cleavage on {001} and good cleavage on {010} intersect at right angles. Fragments tend to lie on cleavage surfaces. Uneven to conchoidal fracture. Brittle.

Color Properties

Luster: Vitreous.
Color: Colorless to white.
Streak: White.
Color in Thin Section: Colorless.
Pleochroism: None.

Optical Characteristics

Relief in Thin Section: Low negative.

Optic Orientation: Two optic orientations are possible. In high sanidine, the optic plane is parallel to {010} with $X \wedge a = +5°$, $Y = b$, and $Z \wedge c = +21°$. Low sanidine is oriented like orthoclase with the optic plane perpendicular to {010} and $X \wedge a = +5°$ to $+9°$, $Y \wedge c = +21°$ to $+17°$, and $Z = b$. Grains elongate parallel to a are length fast.

Indices of Refraction and Birefringence: Indices of refraction vary systematically with the degree of Al/Si order (Figure 12.19). Indices of refraction increase moderately with Na content, but the variation is not substantial for common compositions. Birefringence is low (0.005–0.008), so interference colors in standard thin section are no higher than first-order white.

Interference Figure: $2V_x$ for low sanidine decreases with decreasing Al/Si order from 40° to 0° with the optic plane

perpendicular to (010). In high sanidine, the optic plane is parallel to (010) and $2V_x$ increases to as much as 47° with decreasing order. The low sanidine orientation is more common. Provided the trace of (010) can be recognized, either as the composition plane of a Carlsbad twin or based on crystal habit and cleavage, low and high sanidine can be distinguished in thin section. Find a grain that yields a figure (e.g., an acute bisectrix figure) from which the trace of the optic plane can be recognized. Return to plane light and observe whether the trace of the optic plane is parallel or at right angles to the trace of (010). Optic axis dispersion is $r > v$ for low sanidine and $r < v$ for high sanidine.

Alteration Sericite and clay are the common alteration products.

Distinguishing Features

Hand Sample: Sanidine cannot be reliably distinguished from orthoclase and microcline in hand sample. However, sanidine is most commonly colorless or white, whereas the other K-feldspars are more commonly colored. Plagioclase usually displays parallel striae related to albite twinning on the basal cleavage surface.

Thin Section/Grain Mount: Lack of polysynthetic twinning distinguishes sanidine from microcline and plagioclase. The distinction between orthoclase, low sanidine, and high sanidine is based solely on $2V$ angle and optic orientation. Quartz has a similar appearance, but has positive relief in standard thin section and is uniaxial.

Occurrence Sanidine is the common K-feldspar in silicic volcanic rocks such as rhyolite, rhyodacite, phonolite, and trachyte, and in dikes and other shallow intrusives of the same composition. Most sanidine has between 20 and 60% Na and is usually cryptoperthitic. Compositions with more than about 60% Na are triclinic anorthoclase. Terrigenous sediments may contain sanidine, but microcline and orthoclase are more common.

ADULARIA

This variety of K-feldspar is not strictly defined based on its physical or structural properties but on the basis of habit and occurrence. Crystals are elongate parallel to c and are dominated by the {110} prism. In most cases adularia is crystallized at low temperatures (< roughly 450°C) in hydrothermal systems in veins or as a replacement after other minerals, or is produced authigenically during diagenesis of sediments. It is commonly triclinic but also may be monoclinic, and some crystals may contain domains of both symmetries. The degree of Al/Si order is variable. Indices of refraction, $2V$, and related optical properties are similar to microcline and orthoclase but can be quite variable. Triclinic varieties lack the grid twinning produced by the polymorphic transition from monoclinic to triclinic, presumably because they grew originally with triclinic symmetry.

ANORTHOCLASE

Triclinic $\bar{1}$
$a = 8.287$ Å, $b = 12.972$ Å, $c = 7.156$ Å, $\alpha = 91.05°$,
    $\beta = 116.26°$, $\gamma = 90.15°$, $Z = 4$
$2V_x = 42–52°$

Form and Twinning Crystals tend to be prismatic and to be elongate parallel to the c axis; anhedral grains and microlites are common. Complex twinning according to the albite {010} and pericline {h01} laws are characteristic.

Cleavage Perfect {001} and good {010} cleavages, like the other feldspars, intersect at an angle of about 91°.

Color Properties

Luster: Vitreous.
Color: Colorless, also white, pale yellow, red, green.
Streak: White.
Color in Thin Section: Colorless.
Pleochroism: None.

Optical Characteristics

Relief in Thin Section: Low negative.

Optic Orientation: $X \wedge a \approx +4°$ to $+10°$, $Y \wedge c \approx +16°$ to $+22°$, $Z \wedge b \approx 5°$

Indices of Refraction and Birefringence: Indices of refraction increase with Na content. Birefringence is low (0.005–0.008) like the other alkali feldspars.

Interference Figure: Anorthoclase is biaxial negative with $2V_x$ less than about 52°. Complex twining may make good figures difficult to obtain.

Alteration Alters to sericite and clay like the other feldspars.

Distinguishing Features

Hand Sample: Cannot be reliably distinguished from the other feldspars in hand sample. The colorless crystals most closely resemble sanidine. Orthoclase and microcline are more typically colored. Striae on cleavage surfaces from albite twinning may be visible but are often too closely spaced to be reliably seen.

Thin Section: The combined albite and pericline twins most closely resemble microcline, but the lamellae are usually finer and the 2V is smaller. The occurrence is different.

Occurrence Anorthoclase is found in Na-rich felsic volcanics such as alkali rhyolite, trachyte, phonolite, and equivalent hypabyssal intrusives.

FELDSPATHOIDS

As the name implies, the feldspathoids are similar to the feldspars in both structure and physical properties. The common feldspathoid minerals include nepheline, leucite, and the sodalite group. Nepheline is by far the most abundant. Chemically the feldspathoids are distinguished from the feldspars by having less silica relative to the amount of Na and K. As a consequence, they are rarely, if ever, found in primary association with quartz. The principal occurrence of these minerals is in alkali-rich, silica-poor igneous rocks. Because magmas of these compositions are uncommon, the feldspathoids also are uncommon.

The structure of the feldspathoids is similar to the feldspars and consists of four- and six-member rings linked laterally to form a three-dimensional tetrahedral framework. The presence of six-member rings requires that the structures be somewhat more open than the feldspars, so the specific gravity of these minerals is less. The cavities created in the framework of tetrahedra are occupied by Na^+ and/or K^+, and in the sodalite group by anions and anionic groups including Cl^-, SO_4^{2-}, and S^{2-}.

NEPHELINE

Hexagonal (6)
$Na_3K(Al_4Si_4O_{16})$
$a = 9.993$ Å, $c = 8.374$ Å, $Z = 8$
$H = 5.5–6$
$G = 2.55–2.67$
Uniaxial (–)
$n_\omega = 1.529–1.546$
$n_\epsilon = 1.526–1.544$
$\delta = 0.003–0.005$

Structure The structure is similar to tridymite and consists of alternating layers of Si/Al tetrahedra arranged in rings with tetrahedra alternately pointing up and down. The "up" tetrahedra contain Si^{4+} and the "down" tetrahedra contain Al^{3+}. The "up" tetrahedra share oxygen anions with the "down" tetrahedra from the sheet above. This avoids placing Al^{3+} in adjacent tetrahedra according to the aluminum avoidance principle (see discussion in feldspar section). The Na^+ and K^+ cations are located in the middle of the 6-fold tetrahedral rings. The framework is distorted somewhat compared to the ideal tridymite structure, so that of four sites in the middle of the 6-fold rings, three are smaller and one larger. The three smaller sites contain Na^+ and the larger site contains K^+.

Composition At high temperature, there is complete solid solution to **kalsalite** ($K_4Al_4Si_4O_{16}$), but at most magmatic conditions, intermediate compositions are not possible and nepheline typically has a 3:1 Na:K ratio. Kalsalite exsolution lamellae may be present. Small amounts of Ca may substitute in the structure.

Form and Twinning Uncommon crystals form blocky hexagonal prisms terminated with a basal pinacoid. More common as anhedral masses, particularly in plutonic rocks (Figure 12.21).

Twins on {010}, {010}, {112}, and {335} are possible but not normally seen either in hand sample or thin section.

Cleavage Poor cleavage on {100} and {001}. Subconchoidal fracture. Brittle.

Color Properties

Luster: Subvitreous to somewhat greasy.
Color: White or gray.
Streak: White.
Color in Thin Section: Colorless.
Pleochroism: None.

Optical Characteristics

Relief in Thin Section: Low positive most common, also low negative.

Optic Orientation: Uniaxial, so the optic axis is *c*. Longitudinal sections through elongate crystals are length fast with parallel extinction.

Indices of Refraction and Birefringence: Substitution of K^+ for Na^+ produces a slight increase in refractive indices. Most nepheline has indices of refraction above 1.54, so it shows low positive relief in standard thin section. Birefringence is low, so the maximum interference color seen in thin section is middle first-order gray. Kalsalite has indices of refraction $n_\omega = 1.540$, $n_\epsilon = 1.535$.

Interference Figure: Uniaxial negative, but slight separation of isogyres may be seen in some samples. Owing to low birefringence, isogyres tend to be fairly broad and fuzzy in standard thin section. Ca-rich samples may be optically positive.

Alteration Nepheline may alter to clay minerals, analcime, sodalite, calcite, and cancrinite.

Figure 12.21 Photomicrograph of nepheline (N) in thin section with plagioclase (P) and microcline (M) in nepheline syenite. Crossed polarizers.

Distinguishing Features

Hand Sample: Resembles quartz but is softer; has a somewhat greasy luster and poor cleavage.

Thin Section/Grain Mount: Resembles the feldspars but is uniaxial, lacks good cleavage and twinning, and has lower birefringence. Quartz is uniaxial positive and lacks the clouding common in nepheline.

Occurrence Nepheline is a common mineral in alkali-rich, silicon-poor plutonic and volcanic igneous rocks such as nepheline syenite, foidite, and phonolite. Associated minerals often include K-feldspar, Na-rich plagioclase, biotite, and sodic and sodic–calcic amphibole and/or pyroxene. It also may be associated with leucite and sodalite. **Cancrinite** [$Na_6Ca_2Al_6Si_6O_{24}(CO_3)_2$] is a zeolite-like mineral, colorless, white or pale yellow in hand sample, that is commonly associated with nepheline. Cancrinite is uniaxial (−), $n_\omega = 1.507–1.528$, $n_\epsilon = 1.495–1.503$, $\delta = 0.012–0.025$, $H = 5–6$, $G = 2.42–2.51$ with perfect {100} cleavage. Nepheline almost never occurs as a primary mineral in quartz-bearing rocks. Nepheline also may be found in some contact metamorphosed rocks adjacent to alkali-rich intrusives, presumably as a result of sodium metasomatism.

Use Nepheline, generally as a constituent mineral in nepheline syenite, has fairly extensive industrial uses. Nepheline syenite, which typically also contains K-feldspar and plagioclase as major minerals, is crushed, and Fe–Mg-bearing minerals are magnetically removed. The resulting material is pulverized and used in the manufacture of glass and ceramics. It also may be used as an extender or filler in the manufacture of paints, plastic, and foam rubber. The high aluminum content makes nepheline a potential source of metallic aluminum, but the availability of more economic bauxite has precluded significant development for this purpose except for a few locations in Russia.

LEUCITE

KAlSi$_2$O$_6$
Tetragonal (pseudoisometric) 4/*m*
$a = 13.05$ Å, $c = 13.75$ Å, $Z = 16$
$H = 5.5–6$
$G = 2.45–2.50$
Uniaxial (+)
$n_\omega = 1.508–1.511$
$n_\epsilon = 1.509–1.511$
$\delta = 0.000–0.001$

Structure The structure is an open framework of four- and six-member rings of Si/Al tetrahedra with K^+ cations occupying large openings within this framework. Above about 625°C the structure is isometric, but below that temperature it distorts to tetragonal, with the *c* axis of the tetragonal structure following one of the three crystal axes in the original isometric structure.

Composition Most leucite is relatively pure $KAlSi_2O_6$, but small amounts of Na^+ may replace K^+ and Fe^{3+} may replace Al^{3+}.

Form and Twinning Leucite commonly occurs as equant trapezohedral crystals that show eight-sided to nearly round cross sections (Figure 5.24). It also occurs as small microlites and skeletal or anhedral grains in the groundmass of volcanic rocks. Inclusions of other minerals are common. The transformation from isometric to tetragonal symmetry that accompanies cooling results in multiple twins on {110}. Twins are not visible in hand sample but appear as a complex pattern of concentric lamellae or lamellae that intersect at roughly 60° in thin section.

Cleavage Poor pseudododecahedral cleavage {110} is not normally seen and does not tend to control fragment orientation. Conchoidal fracture. Brittle.

Color Properties

Luster: Vitreous.
Color: White or gray.
Streak: White.
Color in Thin Section: Colorless.
Pleochroism: None.

Optical Characteristics

Relief in Thin Section: Low negative.

Optic Orientation: The optic axis coincides with one of the original *a* axes in the isometric structure.

Indices of Refraction and Birefringence: Indices of refraction show little variation. Birefringence is very low, and small grains appear nearly isotropic in thin section.

Interference Figure: Complex twinning and low birefringence make interference figures difficult to obtain and interpret in thin section. The optic sign is usually positive.

Alteration Leucite may alter to analcime or **pseudoleucite**, which is a mixture of nepheline and K-feldspar.

Distinguishing Features

Hand Sample: Crystal habit and occurrence are characteristic. Analcime is similar but typically occurs as a cavity-filling mineral, not as a phenocryst.

Thin Section/Grain Mount: Low birefringence, crystal habit, and complex twinning are diagnostic. Analcime and members of the sodalite group have higher indices of refraction, are isotropic, and lack the complex twinning.

Occurrence Leucite is an uncommon mineral found in potassium-rich mafic lavas and associated shallow intrusive rock bodies. It is commonly associated with plagioclase, nepheline, sanidine, clinopyroxene, and sodic or sodic–calcic amphiboles. Leucite weathers readily and is not commonly preserved in sediments.

Use Leucite is used in the manufacture of ceramics used in dental prostheses.

SODALITE

Isometric $\overline{4}3m$
$Na_8Al_6Si_6O_{24}Cl_2$
$a = 8.870–8.882$ Å, $Z = 1$
$H = 5.5–6$
$G = 2.27–2.33$
Isotropic
$n = 1.483–1.487$

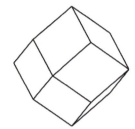

Structure The structure consists of an open tetrahedral framework, similar to the zeolites, that contains fairly large open cavities. The tetrahedra contain the Al^{3+} and Si^{4+} in an alternating pattern. The Cl^- anions, each surrounded by four Na^+ cations, occupy the large cavities.

Composition Sodalite is a member of the sodalite group, which also includes **nosean** $(Na_8Al_6Si_6O_{24}SO_4)$ and **haüyne** $[(Na,Ca)_{4-8}Al_6Si_6(O,S)_{24}(SO_4)_{1-2}]$. **Lazurite** is a variety of haüyne in which Cl^- and S^{2-} substitute for SO_4^{2-}. Most sodalite is relatively close to its ideal composition.

Form and Twinning Crystals, usually dodecahedrons, are rare. Usually massive or as anhedral grains. Twins on {111} are common but are recognized only if betrayed by the external crystal shape, in which case crystals may form pseudohexagonal prisms elongate on [111].

Cleavage {110} poor. Uneven to conchoidal fracture. Brittle.

Color Properties

Luster: Vitreous.

Color: Commonly light to dark blue; also gray, colorless, white, yellow, or green; may fluoresce.

Streak: White. The streak of lazurite may be blue.

Color in Thin Section: Colorless or very pale blue; color may be patchy. Haüyne may be more distinctly blue.

Pleochroism: None.

Optical Characteristics

Relief in Thin Section: Moderate negative.

Indices of Refraction and Birefringence: The index of refraction of nosean is 1.470–1.495 and of haüyne is 1.494–1.510. Sodalite may be weakly birefringent around inclusions.

Alteration Members of the sodalite group may alter to zeolites (often fibrous), clay, calcite, or cancrinite.

Distinguishing Features

Hand Sample: Distinguished by the blue color. If not blue, sodalite may be difficult to distinguish from analcime and leucite. Members of the sodalite group are not readily distinguished from each other in hand sample.

Thin Section/Grain Mount: Members of the sodalite group are distinguished by their isotropic character, low index of refraction, and occurrence. Leucite is anisotropic and has distinctive twinning. Garnet has much higher indices of refraction. Fluorite has an even lower index of refraction and good cleavage. Analcime has poor cubic cleavage and is not blue, but is otherwise similar. The members of the sodalite group may be difficult to distinguish based on optical properties, but sodalite is colorless, lacks abundant inclusions, and is the only member usually found in plutonic rocks. Nosean and haüyne commonly contain inclusions, and haüyne is usually bluish in thin section.

Occurrence Members of the sodalite group are found in silicon-deficient, alkali-rich igneous rocks. Plutonic igneous rocks such as syenite are more likely to contain sodalite. Volcanic rocks such as phonolite, alkali basalt, and feldspathoidal basalt may contain sodalite, nosean, or haüyne. Both sodalite and lazurite may be found in contact metamorphosed carbonate rocks. **Lapis lazuli**, which is derived by contact metamorphism of limestone, consists of lazurite, pyroxene, calcite, and other silicates. It owes its blue color to the lazurite. When polished, lapis lazuli can be strikingly beautiful, particularly if it contains disseminated pyrite.

Use The blue color of sodalite-bearing rocks makes it valuable as a dimension stone used to face buildings and for other decorative purposes. It has been used as a folk medicine treatment for diabetes.

ZEOLITE GROUP

The zeolite group contains over 80 naturally occurring members and roughly 600 additional synthetic zeolites with no natural counterpart. Zeolites are the largest single group of silicate minerals and constitute major rock-forming minerals in some environments. Prior to the 1960s, zeolites were considered to be a relatively minor mineral group, represented mostly by crystals occupying vesicles and other voids in basaltic and andesitic volcanic rocks. However, the introduction of the scanning electron microscope and sophisticated X-ray diffraction techniques demonstrated that zeolites can comprise a major part of certain sedimentary and metamorphic rocks, and are abundant in hydrothermally altered volcanic rocks. The fine grain size in these rocks, however, makes identification and characterization difficult, and in the past much of this material was dismissed as "clay." Occurrences in vesicles and

vugs in volcanic rocks are best known to mineralogy students because only in these occurrences are crystals large enough to be examined in hand sample or microscopically. For more information on this fascinating group of minerals refer to Gottardi and Galli (1985), Bish and Ming (2001), or Tschernich (1992). Some of the more common zeolites are listed in Table 12.5.

Composition and Structure Zeolites are hydrated framework silicates with a general formula

$$M_x D_y (Al_{x+2y} Si_{n-x-2y} O_{2n}) \cdot m H_2O$$

where the ratio of Si to Al varies from 1:1 to 6:1. The M and D cations make up the charge deficiency associated with having Al^{3+} rather than Si^{4+} in the tetrahedral sites. M is usually monovalent Na or K and D is Ca, Mg, or other divalent cations. The M and D stoichiometric amounts are therefore dependent on the amount of Al^{3+}. Water content is variable.

The structures of zeolites consist of relatively stable open frameworks of Al/Si tetrahedra (Figure 12.22) that link together to form open channels and voids whose geometry is different in the different species. The water molecules and the mono- and divalent cations occupy the voids in the framework structure and are weakly bonded. This leads to some interesting and useful properties described next.

Physical Properties In their most common occurrences, zeolites are too fine grained to allow identification by physical properties. Only in cases in which they form large

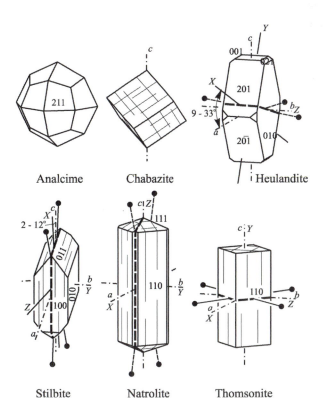

Analcime   Chabazite   Heulandite

Stilbite   Natrolite   Thomsonite

Table 12.5 Physical Properties of Selected Zeolite Minerals

Mineral	Formula	Crystal System	Habit	*H*	*G*	Cleavage	Color
Analcime	$NaAlSi_2O_6 \cdot H_2O$	Isometric $a = 13.72$	Trapezoidal crystals	5–5.5	2.25	{001} poor	White, colorless, gray, pink, greenish, yellowish
Chabazite	$(Ca, Na_2, K_2)_2Al_4Si_8O_{24} \cdot 12H_2O$	Triclinic (pseudohexagonal) $a \cong b \cong c =$ 9.40–9.45 $\alpha = 94.18$–$94.33°$ $\beta = 94.08$–$94.36°$ $\gamma = 94.07$–$94.45°$	Rhombs that are nearly cubic	4.5	2.05–2.20	{101} distinct	White, yellow, pink red, colorless
Clinoptilolite	$(Ca, Na_2, K_2)_3Al_6Si_{30}O_{72} \cdot 21H_2O$	Monoclinic $a = 17.63$ $b = 17.94$ $c = 7.40$ $\beta = 116.39°$	Tabular, radiating aggregates	3.5–4	2.16	{010} perfect	Colorless, white
Erionite	$(K_2,Na_2,Ca,Mg)_4Al_8Si_{28}O_{72} \cdot 28H_2O$	Hexagonal $a = 13.21$ $c = 15.04$	Fibrous	3.5–4	2.08	Poor prismatic?	White
Heulandite	$NaCa_4Al_9Si_{27}O_{72} \cdot 24H_2O$	Monoclinic $a = 7.955$ $b = 17.95$ $c = 7.435$ $\beta = 91.74°$	Tabular to equant	3.5–4	2.20	{010} perfect	Colorless, white, gray, yellow, red, pink, orange, brown, black
Laumontite	$CaAl_2Si_4O_{12} \cdot 4H_2O$	Monoclinic $a = 14.72$ $b = 13.08$ $c = 7.76$ $\beta = 112.01°$	Prismatic	3.5–4	2.3	{010}, {110} perfect, {011} imperfect	White, gray, pink, yellowish, brownish
Mesolite	$Na_2Ca_2Al_6Si_9O_{30} \cdot 8H_2O$	Orthorhombic $a = 18.40$ $b = 56.66$ $c = 6.54$	Prismatic to fibrous, radiating aggregates	5	2.26	{110} perfect	Colorless, white, gray, yellowish
Mordenite	$(Ca, Na_2, K_2)Al_2Si_{10}O_{24} \cdot 7H_2O$	Orthorhombic $a = 18.05$ $b = 20.40$ $c = 7.50$	Fibrous, radiating or felted	3–4	2.13	{100} perfect. {010} distinct	Colorless, white. yellowish, pinkish
Natrolite	$Na_2Al_2Si_3O_{10} \cdot 2H_2O$	Orthorhombic $a = 18.27$ $b = 18.61$ $c = 6.59$	Prismatic to acicular, radiating aggregates	5	2.25	{110} perfect	White, colorless, gray, bluish, yellowish, pink
Phillipsite	$K_2(Ca,Na_2)_2Al_6Si_{10}O_{32} \cdot 12H_2O$	Monoclinic $a = 9.87$ $b = 14.30$ $c = 8.67$ $\beta = 124.2°$	Prismatic, pseudo-orthorhombic	4–4.5	2.20	{010}, {100} distinct	Colorless, white, reddish, yellowish
Scolecite	$CaAl_2Si_3O_{10} \cdot 3H_2O$	Monoclinic $a = 18.51$ $b = 18.98$ $c = 6.53$ $\beta = 90.64°$	Prismatic, radiating aggregates	5	2.26	{110} perfect	Colorless, white
Stilbite	$Na_2Ca_4Al_{10}Si_{26}O_{72} \cdot 30H_2O$	Monoclinic $a = 13.6$ $b = 18.2$ $c = 22.4$ $\beta = 128.2°$	Tabular, sheathlike aggregates	3.5–4	2.19	{010} perfect	White, yellowish, gray, pink, orange, light to dark brown
Thomsonite	$NaCa_2Al_5Si_5O_{20} \cdot 6H_2O$	Orthorhombic $a = 13.09$ $b = 13.05$ $c = 13.23$	Prismatic or lamellar rosettes or radiating aggregates	5	2.35	{100} perfect. {010} good	White, yellowish, pink brown, greenish
Wairakite	$CaAl_2Si_4O_{12} \cdot 2H_2O$	Monoclinic (pseudocubic) $a = 13.69$ $b = 13.64$ $c = 13.56$ $\beta = 90.5°$	Trapezoidal crystals	5.5–6	2.27	?	Colorless, white

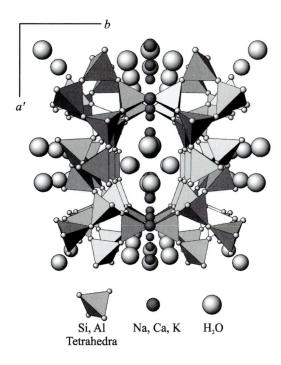

Si, Al   Na, Ca, K   H₂O
Tetrahedra

Figure 12.22 Typical zeolite structure. Heulandite viewed down the *c* axis. The open channels provide space for water molecules and mono- and divalent cations. Structural data from Gunter and others (1994).

crystals in vesicles or other voids in mafic lava flows are zeolites sufficiently large to be identified without X-ray or related techniques. The physical properties of selected zeolite minerals are listed in Table 12.5. Of these, chabazite, natrolite, heulandite, analcime, thomsonite, and stilbite are relatively common as large crystals. These minerals have a vitreous luster, white streak, and are typically white or colorless; or they may display pastel shades of yellow, pink, brown, or green. Specific gravity and hardness also vary within relatively narrow ranges. Identification is therefore made principally on the basis of habit.

ANALCIME: Equant trapezohedral crystals

CHABAZITE: Rhombohedral (pseudocubic) crystals

HEULANDITE: Platy crystals with perfect {010} cleavage; often with a coffin-shaped cross section

NATROLITE: Slender prismatic or acicular crystals that may form beautiful radiating aggregates, also fibrous

STILBITE: Platy crystals tabular on {010} with good cleavage; often forms sheaf-like aggregates

THOMSONITE: Acicular, prismatic, or bladed crystals, often in spherical, botryoidal, or columnar aggregates and masses that may be color zoned

Positive identification in most cases requires X-ray diffraction study.

Optical Characteristics Table 12.6 shows optical properties for selected zeolites. Note that the optical properties display substantial overlap and may vary, depending on the degree of hydration. All have low indices of refraction usually less than the index of refraction of cement in thin section, low birefringence, and all are colorless in thin section and grain mount. It may be difficult to distinguish among different species based solely on optical properties.

Alteration Zeolite minerals may alter to clay or to other zeolites.

Occurrence Zeolites are widely but irregularly distributed. In the large majority of occurrences, zeolites are secondary, meaning that they are produced by alteration of some precursor mineral or minerals. Only analcime occurs as a primary magmatic mineral, in which occurrence it typically forms late-crystallizing grains occupying the interstices between other minerals in basalts, phonolites, and related rocks.

The environments in which zeolites are produced by alteration include the following, ordered roughly by depth of burial.

Weathering under high pH conditions

Sedimentary diagenesis, especially of volcanic ash in saline lakes or certain deep marine environments

Alteration associated with percolating groundwater

Hydrothermal alteration associated with igneous activity

Contact metamorphism

Burial and low-grade regional metamorphism

The most extensive zeolite deposits are found in altered felsic volcanic ash flow sheets and related volcanic rocks, most commonly of late-Mesozoic and Cenozoic age. Zeolitization may be accomplished by percolating groundwater, a hydrothermal circulation system, or contact metamorphism. Volcanic ash deposited in saline lakes also is particularly prone to zeolite alteration. The zeolites commonly formed by alteration include clinoptilolite, chabazite, mordenite, erionite, and phillipsite. The grain size is almost always submicroscopic.

The zeolites that occur most commonly as vesicle and other void-space filling in basaltic volcanics include heulandite, chabazite, natrolite, and stilbite. These crystals may be large and they appear to have crystallized from circulating groundwater, whether driven by a thermal system or not.

Burial and low-grade regional metamorphism of graywacke and related sedimentary rocks derived from andesitic to basaltic sources may produce heulandite and analcime at lower temperatures. Laumontite appears at greater depth

Table 12.6 Optical Properties of Selected Zeolites

Mineral	Optical Class & Sign	Index of Refraction	Birefringence	2V(°)	Optical Orientation
Analcime	Isotropic	$n = 1.479-1.493$			
Chabazite	Uniaxial or biaxial (−) or (+)	$n_\alpha = 1.460-1.513$ $n_\gamma = 1.462-1.515$	0.002–0.010	0–~30	The Bxa is ~ c in biaxial varieties
Clinoptilolite	Biaxial (−) or (+)	$n_\alpha = 1.476-1.491$ $n_\beta = 1.471-1.493$ $n_\gamma = 1.479-1.497$	0.003–0.006	$2V_x$ = small to large	$Y = b$, $Z \wedge a = +30°$ to $+45°$
Heulandite	Biaxial (+)	$n_\alpha = 1.487-1.505$ $n_\beta = 1.487-1.507$ $n_\gamma = 1.488-1.515$	0.001–0.011	0–70, usually ~30	$X \wedge a = +9°$ to $+33°$, $Y \wedge c = -8°$ to $-32°$, $Z = b$; the cleavage trace is length fast
Laumontite	Biaxial (−)	$n_\alpha = 1.502-1.514$ $n_\beta = 1.512-1.524$ $n_\gamma = 1.514-1.525$	0.008–0.016	25–47	$X \wedge a = +30°$ to $+62°$, $Y \wedge b$, $Z \wedge c = -8°$ to $-40°$
Natrolite	Biaxial (+)	$n_\alpha = 1.473-1.490$ $n_\beta = 1.476-1.491$ $n_\gamma = 1.485-1.502$	0.012–0.013	0–64	$X = a$, $Y = b$, $Z = c$; fibers and cleavage fragments are length slow
Stilbite	Biaxial (−)	$n_\alpha = 1.482-1.500$ $n_\beta = 1.489-1.507$ $n_\gamma = 1.493-1.513$	0.006–0.014	30–49	$X \wedge c = +2°$ to $+12°$, $Y \cong b$, $Z \wedge a = +26°$ to $+36°$; elongate grains are length fast
Thomsonite	Biaxial (+)	$n_\alpha = 1.497-1.530$ $n_\beta = 1.513-1.533$ $n_\gamma = 1.518-1.544$	0.006–0.021	41–75	$X = a$, $Y = c$, $Z = b$; fibers and cleavage fragments may be length fast or slow

and higher temperatures. Similar metamorphism of more felsic rocks typically produces a mineralogical progression in which clinoptilolite and/or mordenite appear first, followed by analcime, which in turn is replaced by albite with increasing depth of burial. The grain size of these rocks is almost always very fine.

Normal chemical weathering of felsic volcanic rocks is reported to produce zeolites only when the pH is quite high. One such environment is in saline alkali-rich soils, which are usually found only in arid or semiarid environments.

To a significant degree, zeolite composition is controlled by the chemistry of the original rock. Zeolites that develop in mafic rocks with relatively low silicon content tend to have low Si:Al ratios, and include thomsonite and scolecite. High-silicon zeolites such as mordenite and clinoptilolite tend to occur in altered felsic rocks. Zeolites with an intermediate Si:Al composition such as chabazite, analcime, and phillipsite may be found in rocks of diverse compositions.

Use Zeolites are widely used in many industrial and other processes. Most of these uses depend on the unique structural and chemical behavior of these minerals. Many zeolite-like materials have been synthesized to take advantage of these properties and use of synthetic materials is very extensive.

At room temperature, the structural voids in zeolites can be entirely occupied by water molecules and the M

and D cations. If the temperature is increased, however, the water is released. Above 300 or 400°C, zeolites will have lost most of their water. In many zeolites, moderate heating does not destroy the framework structure. When returned to a lower temperature, the voids can fill back up with water. This property makes zeolites useful as desiccants. They are heated to drive off the water, then used to remove water vapor from gasses such as carbon dioxide, refrigerator Freon, and various organic chemicals used for industrial purposes.

The fact that zeolites have large open frameworks also makes them useful as molecular sieves. They are usually prepared by heating to drive off the water. Molecules small enough to enter the channels are adsorbed into the zeolite and larger molecules are excluded. This provides the basis for separating different constituents in industrial processes. Mordenite, for example, can be used to selectively remove nitrogen from air to produce a gas that is up to 95% oxygen.

The M and D cations in the voids and channels are loosely bound and can be exchanged for other cations under appropriate conditions. This allows zeolites to be used as cation-exchange agents, which is one of their most important industrial uses. One of the earliest cation-exchange applications was to treat water to reduce its "hardness." Hard water contains substantial quantities of Ca^{2+} cations (usually in the form of dissolved $CaCO_3$) that prevent soaps and detergents from lathering properly and leave clothes dingy after being washed. A synthetic Na-bearing zeolite similar to natrolite is exposed to the hard water and exchanges $2Na^+$ for Ca^{2+}. The Ca^{2+} in the water is replaced with Na+, which does not prevent soap from lathering. Once the zeolite has been saturated with Ca^{2+}, it is exposed to a strong NaCl solution. The Na^+ replaces the Ca^{2+} and prepares the zeolite to be used again.

Many other cation-exchange applications have been developed, including removing heavy metals from mine drainage and industrial waste, and dangerous isotopes from radioactive waste. Zeolites may be used to treat contaminated soils to scavenge undesirable metals (Cd, Cu, Pb, Zn) and reduce their availability to plants. One of the more important ion-exchange applications involves the ability of clinoptilolite and mordenite to adsorb ammonium (NH_4^+) ions. Because ammonia is generated in substantial quantities from animal and human waste, these zeolites are used to treat wastewater from sewage treatment plants and from agricultural operations. They also are used to control ammonia in the water used in fish farms, and as an odor-control agent in cat litter.

Zeolites have a variety of agricultural uses. They are used as a soil conditioner to improve soil moisture retention, increase cation-exchange capability, and buffer acidic soils. Their adsorptive capability makes them useful as a slow-release medium for nitrogen fertilizer and as a carrier for herbicides, pesticides, and fungicides. Clinoptilolite has been added to feed for chickens, turkeys, pigs, and cattle to improve feed efficiency and to reduce diarrhea and related health problems. It has the additional benefit of producing a less odoriferous excrement by adsorbing ammonia. The manure thus produced may have improved nitrogen release properties when used as a fertilizer.

Zeolites, or their synthetic analogs, are widely used as catalysts in chemical processes, particularly in petroleum refining, as a sorbent to clean oil spills, and as a filler in some paper products.

Despite wide application, industrial use of natural zeolites is limited by the availability of other, more established commodities, and by the use of synthetic zeolites tailored to specific tasks.

Zeolites have been advocated as treatments for a wide variety of medical conditions including cancer, gastrointestinal problems, diabetes, and detoxification, and it is possible that some uses may prove to be valid. A notable success that has received approval from the U.S. Food and Drug Administration is the use of a synthetic zeolite as a blood-clotting agent to prevent blood loss on the battlefield and in trauma centers.

Although most zeolite minerals appear to be environmentally benign, there are exceptions. The most notable is erionite, which is fibrous. It appears to pose a significant risk of mesothelioma, which is a cancer of the lining of the pleural cavity in which the lungs reside, and other respiratory pathologies, very similar to asbestiform minerals (Ross and others, 1993).

OTHER FRAMEWORK SILICATES

SCAPOLITE

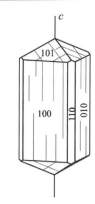

Tetragonal $4/m$
Marialite: $Na_4Al_3Si_9O_{24}Cl$
$a = 12.06$ Å, $c = 7.57$ Å, $Z = 2$
Meionite: $Ca_4Al_6Si_6O_{24}CO_3$
$a = 12.18$ Å, $c = 7.57$ Å, $Z = 2$
$H = 5–6$
$G = 2.50–2.78$
Uniaxial $(-)$
$n_\omega = 1.532–1.607$
$n_\epsilon = 1.522–1.571$
$\delta = 0.004–0.038$

Structure The structure consists of four-member rings of Si/Al tetrahedra that are linked together to form chains parallel to the c axis. These chains are in turn linked laterally through other four-member tetrahedral rings. The Ca^{2+}, Na^+, Cl^-, and CO_3^{2-} occupy voids formed in this open structure.

Composition Scapolite forms a solid solution series extending from meionite to marialite, although most samples have intermediate compositions. The substitution scheme is coupled with $Na^+ + Si^{4+}$ substituting for $Ca^{2+} + Al^{3+}$, similar to plagioclase, or $Na^+ + Cl^-$ substituting for $Ca^{2+} + CO_3^{2-}$. SO_4^{2-} or OH^- may substitute for CO_3^{2-} or Cl^-.

Form and Twinning Crystals are prismatic, usually with flat pyramidal terminations. Commonly as anhedral grains, granular clusters, or columnar aggregates. Some grains may be poikilitic, meaning that they are relatively large and enclose grains of associated minerals. Twinning is not reported.

Cleavage Good cleavages parallel to the {110} faces intersect at 90°. A second set of fair cleavages is parallel to {100} and 45° to the {110} cleavages. Fragment shape tends to be controlled by cleavage. Uneven to conchoidal fracture. Brittle.

Color Properties

Luster: Vitreous.

Color: White, gray, or greenish; less commonly, shades of pink, yellow, and blue. Commonly fluoresces orange to bright yellow in UV light

Streak: White.

Color in Thin Section: Colorless.

Pleochroism: None.

Optical Characteristics

Relief in Thin Section: Low to moderate positive.

Optic Orientation: Because scapolite is tetragonal, the *c* axis is the optic axis. The traces of cleavage in longitudinal sections and cleavage fragments both show parallel extinction and are length fast.

Indices of Refraction and Birefringence: Indices of refraction and birefringence increase systematically with increasing meionite content (Figure 12.23).

Interference Figure: Scapolite is uniaxial negative, but biaxial samples with $2V < 10°$ have been reported.

Alteration Scapolite may be replaced by aggregates containing sericite, calcite, chlorite, epidote, zeolites, or other minerals.

Distinguishing Features

Hand Sample: Characterized by prismatic crystals with square cross sections and cleavages at 45°. Resembles feldspars, but the cleavages have a fibrous appearance.
Thin Section/Grain Mount: Distinguished from feldspar by uniaxial character, and from quartz by optic sign and higher birefringence. Cordierite is biaxial. Nepheline has lower birefringence and less well-developed cleavage.

Occurrence Scapolite is found in both contact and regional metamorphic rocks derived from calcareous sediments, or from gabbroic or similar rocks. In skarn deposits and marble derived from calcareous sediment, scapolite is often associated with calcite and calc-silicate minerals such as titanite, garnet, diopside, and actinolite. In amphibolite and related rocks derived from gabbroic parents, scapolite is associated with hornblende, calcic clinopyroxene, epidote, and titanite. It may replace plagioclase in hydrothermally

altered mafic igneous rocks. Scapolite also is found in granulites, and less commonly in pelitic and psammitic regional metamorphic rocks. It infrequently is found in veins cutting regional metamorphic rocks or epizonal intrusives, in some pegmatites, and ejected volcanic blocks.

Use Scapolite has no industrial uses, but colorless crystals have been used as gemstones.

REFERENCES CITED AND SUGGESTIONS FOR ADDITIONAL READING

Bailey E. H., and Stevens, R. E. 1960. Selective staining of K-feldspar and plagioclase on rock slabs and thin sections. American Mineralogist 45, 1020–1025.

Bish, D. L., and Ming, D. W. 2001. *Natural Zeolites: Occurrence, Properties, Applications.* Reviews in Mineralogy and Geochemistry 45, 654 p.

Boone, G. M., and Wheeler, E. P., II. 1968. Staining for cordierite and feldspars in thin section. American Mineralogist 53, 327–331.

Burri, C., Parker, R. L., and Wenk, E. 1967. *Die optische Orientierung der Plagioklase.* Birkhauser Verlag, Basel, 333 p.

Deer, W. A., Howie, R. A., and Zussman, J. 2001. *Rock-Forming Minerals,* Volume 4A: *Feldspars,* 2nd ed. The Geological Society, London, 992 p.

Frondel, C. 1982. Structural hydroxyl in chalcedony (type B quartz). American Mineralogist 67, 1248–1257.

Gottardi, T., and Galli, E. 1985. *Natural Zeolites.* Springer-Verlag, Berlin, 409 p.

Graziani, G., and Lucchesi, S. 1982. Thermal behavior of scapolites. American Mineralogist 67, 1229–1241.

Griffen, D. T. 1992. *Silicate Crystal Chemistry.* Oxford University Press, New York, 442 p.

Gunter, M. E., Armbruster, T., Kohler, T., and Knowles, C. R. 1994. Crystal structure and optical properties of Na- and Pb-exchanged heulandite-group zeolites. American Mineralogist 79, 675–682.

Guthrie, G. D., Jr., and Mossman, B. T. 1993. *Health Effects of Mineral Dust.* Reviews in Mineralogy 28, 584 p.

Haughton, H. F. 1980. Refined techniques for staining plagioclase and alkali-feldspars in thin section. Journal of Sedimentary Petrology 50, 629–631.

Heaney, P. J., Prewitt, C. T., and Gibbs, G. V. 1994. *Silica, Physical Behavior, Geochemical and Material Applications.* Reviews in Mineralogy 29, 606 p.

Howie, R. A., Deer, W. A., Wise, W. S., and Zussman, J. 2004. *Rock-Forming Minerals,* Volume 4B: *Silica Minerals,* 2nd ed. The Geological Society, London, 982 p.

Kroll, H., and Ribbe, P.H. 1983. Lattice parameters, composition and Al,Si order in alkali feldspar. Reviews in Mineralogy 2, 2nd ed. 57–100.

Miehe, G., and Graetsch, H. 1992. Crystal structure of moganite: A new structure type for silica. European Journal of Mineralogy 4, 693–706.

Morse, S.A. 1968. Revised dispersion method for low plagioclase. American Mineralogist 53, 105–115.

Mossman, G. R. 1994. Colored varieties of the silica minerals. Reviews in Mineralogy 29, 433–467.

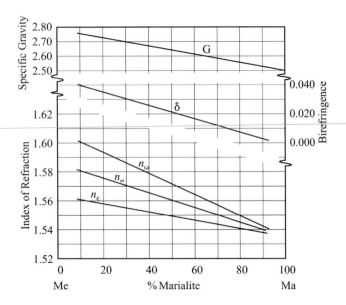

Figure 12.23 Optical properties and specific gravity of scapolite. After Shaw (1960), Ulbricht (1973), and Graziani and Lucchesi (1982).

Ribbe, P. 1983. *Feldspar Mineralogy, 2nd ed.* Reviews in Mineralogy 2, 362 p.

Ross, M., Nolan, R. P., Langer, A. M., and Cooper, W. C. 1993. Health effects of mineral dusts other than asbestos. Reviews in Mineralogy 28, 361–407.

Ruperto, V. L., Stevens, R. E., and Normans, M. B. 1964. Staining of plagioclase feldspars and other minerals with F.D. and C. Red No. 2. U.S. Geological Survey Professional Paper 501B, pp. B152–153.

Sand, L. B., and Mumpton, F. A. (eds.). 1978. *Natural Zeolites: Occurrence, Properties, Use.* Pergamon Press, Oxford, 546 p.

Shaw, D. M. 1960. The geochemistry of scapolite, Part I: Previous work and general mineralogy. Journal of Petrology 1, 218–260.

Smith, J. R. 1958. The optical properties of heated plagioclase. American Mineralogist 43, 1179–1194.

Smith, J. V., and Brown, W. L. 1988. *Feldspar Minerals*, Volume 1. Springer-Verlag, Berlin, 828 p.

Smith, P., and Parsons, I. 1974. The alkali feldspar solvus at 1 kilobar water vapour pressure. Mineralogical Magazine 39, 747–767.

Su, S. C., Bloss, F. D., Ribbe, P. H., and Stewart, D. B. 1984. Optic axis angle, a precise measure of Al, Si ordering in T_1 tetrahedral sites of K-rich alkali feldspars. American Mineralogist 69, 440–448.

Tobi, A. C., and Kroll, H. 1975. Optical determination of the An-content of plagioclases twinned by the Carlsbad law: A revised chart. American Journal of Science 275, 731–736.

Tschernich, R. W. 1992. *Zeolites of the World*. Geoscience Press, Phoenix, AZ, 563 p.

Ulbricht, H. H. 1973. Crystallographic data and refractive indices of scapolites. American Mineralogist 58, 81–92.

CHAPTER 13

Sheet Silicates

INTRODUCTION

Sheet silicates are abundant and important minerals, particularly in geological environments at or within roughly 20 km of the Earth's surface. They are commonly found in intermediate and felsic igneous rocks and in many metamorphic rocks, and are abundant in many sediments and sedimentary rocks, particularly if fine grained. All are hydrous, meaning that hydrogen is included in the structure, usually bonded to oxygen to form OH^-, not molecular water. Sheet silicates are also called **phyllosilicates**, a term derived from the Greek *phyllon*, which means leaf, in allusion to the observation that nearly all members of the group have a flaky or platy habit. They also are referred to as "layer silicates." For additional information about the sheet silicates, please refer to Bailey (1984, 1988), Velde (1995), and Mottana and others (2002).

STRUCTURE AND CLASSIFICATION

Sheet silicates are constructed of sheets of two different types: *O* or octahedral sheets, and *T* or tetrahedral sheets. The octahedral sheets are referred to as *M* sheets in some references. Octahedral and tetrahedral sheets are joined together to form layers. The layers are in turn stacked one atop another and bonded together to form the repeating unit structure of the mineral. The volume between adjacent layers may be vacant, or it may contain interlayer cations and/or water. The single perfect cleavage that is characteristic of most sheet silicates is present because bonding between adjacent layers is typically quite weak. Note that the terms *sheet, layer,* and *unit structure* are used with quite specific meanings outlined here and are not interchangeable.

Octahedral sheets (Figure 13.1a) consist of two planes of OH^- anionic groups. The octahedral sites formed between the OH^- anionic groups are occupied in most cases by either divalent cations such as Fe^{2+} and Mg^{2+}, or trivalent cations such as Al^{3+} or Fe^{3+}. If the cations are divalent, three (= tri) out of three octahedral sites are filled, forming a **trioctahedral** sheet (Figure 13.2a). The ideal formula is often given

as $Mg_3(OH)_6$, the same as the mineral **brucite**, whose structure consists of trioctahedal sheets bonded together with van der Waals bonds. If the cations are trivalent, charge balance requires that only two (= di) out of three sites is occupied, leaving one vacant, forming a **dioctahedral** sheet (Figure 13.2b). The ideal dioctahedral sheet has a formula $Al_2(OH)_6$. This is the same as the mineral **gibbsite**. The net cation charge of +6 is the same in both cases, and the net charge on both dioctahedral and trioctahedral sheets is zero. Note that the "di" and "tri" refer to the occupancy of the octahedral sites, not the charge of the cations.

As should be expected, site occupancy influences unit cell dimensions of the minerals containing these octahedral sheets. In the dioctahedral sheet silicates, the *b* unit cell dimension is usually about 9 Å, whereas in the trioctahedral sheet silicates *b* is usually about 9.2 Å. This can be measured by observing the *d*-spacing of 060 in X-ray powder diffraction work. Reflections are produced for *d*-spacings that are one-sixth of the *b* dimension or about 1.5 and 1.53 Å, respectively. Optical properties also are affected. In most dioctahedral sheet silicates, the optic plane is parallel to (100) and in trioctahedral sheet silicates the optic plane is usually parallel to (010).

Tetrahedral sheets (Figure 13.1b) consist of sheets of tetrahedrally coordinated cations whose composition can be represented T_2O_5, where T represents cations, usually Si^{4+}, Al^{3+}, or less commonly Fe^{3+}. The tetrahedra are arranged in a mesh of 6-fold rings so that three O^{2-} on each tetrahedron are shared with adjacent tetrahedra (Figure 11.2). The three shared oxygen anions are known as the basal oxygens and the fourth, unshared, O^{2-} on each tetrahedra is known as the apical oxygen. A tetrahedal layer is therefore two oxygen atoms thick. In its basic form, a tetrahedral sheet has the chemical formula $Si_2O_5^{2-}$. In many sheet silicates Al^{3+} (or Fe^{3+}) substitutes for Si^{4+}, increasing the net negative charge on the tetrahedral sheet. The symmetry represented by the mesh of 6-fold rings is typically reflected in the nearly hexagonal cross sections of most sheet silicate minerals.

Tetrahedral sheets are always joined with an octahedral sheet. Conceptually this is accomplished by having the apical oxygen anions of the tetrahedral sheet form part of an immediately adjacent octahedral sheet (Figure 13.1c,d). To accomplish this, an OH^- from the octahedral sheet must be

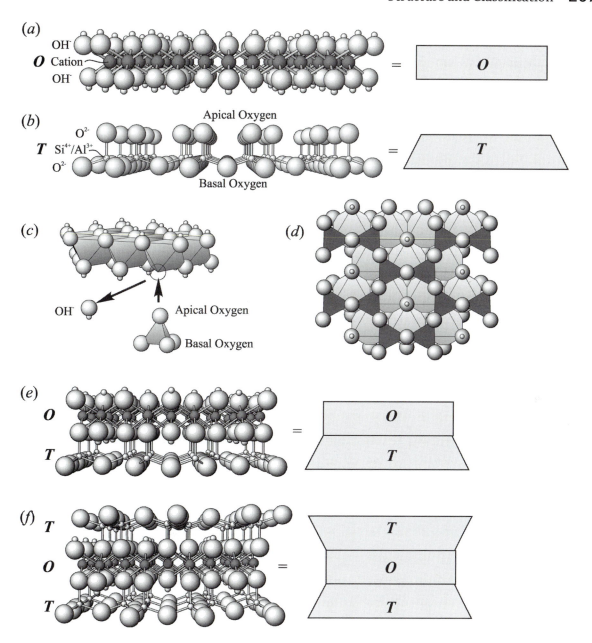

Figure 13.1 Sheet silicate sheets and layers. (*a*) An octahedral sheet consists of two close-packed planes of OH^- with di- or trivalent cations occupying octahedral sites in between. The side view of an octahedral sheet is schematically represented by a rectangle. (*b*) A tetrahedral sheet consists of Si/Al tetrahedra, each of which shares three oxygen atoms with its neighbors to form a continuous hexagonal mesh (cf. Figure 11.2f). The side view of a tetrahedral sheet is schematically represented by a trapezoid. (*c*) Combining a tetrahedral sheet with an octahedral sheet requires that an OH^- be removed from the octahedral sheet to make way for the apical oxygen on each tetrahedron. (*d*) Face-on view of a tetrahedral sheet combined with an octahedral sheet. The apical oxygens on the tetrahedra (dark shading) point away from the viewer and substitute for OH^- within the top of the octahedral sheet. The octahedra are shown with light shading. Note the openings formed in the center of the hexagonal rings of tetrahedra. (*e*) The apical oxygen atoms of the tetrahedral sheet substitute for OH^- on one side of an octahedral sheet to allow the tetrahedral and octahedral sheets to bond to each other, forming a 1:1 or *TO* layer that is three layers of O^{2-}/OH^- thick. (*f*) A second tetrahedral sheet may bond to the other side of an octahedral sheet to form a 2:1 or *TOT* layer that is four layers of O^{2-}/OH^- thick.

removed to make room for each apical oxygen anion from the tetrahedral sheet. Tetrahedral and octahedral sheets are combined to form *TO* and *TOT* layers. These also are known as 1:1 and 2:1 layers, in reference to the ratio of *T* to *O* sheets.

The combination of one tetrahedral and an octahedral sheet is called a *TO* or 1:1 layer (Figure 13.1e) and consists of three planes of anions. On one side are the basal oxygen anions of the tetrahedral sheet and on the other are

OH⁻-anionic groups of the octahedral sheet. The middle plane contains both OH⁻ from the original octahedral sheet and the apical O^{2-} from the tetrahedral sheet. The combination of a T and an O sheet can be represented as follows to form a dioctahedral TO layer.

$$1T \text{ sheet} + 1O \text{ sheet} = TO \text{ layer} + \text{hydroxyl}$$
$$Si_2O_5^{2-} + Al_2(OH)_6 = Al_2Si_2O_5(OH)_4 + 2(OH)^-$$

The equivalent trioctahedral TO layer is $Mg_3Si_2O_5(OH)_4$.

A TOT or 2:1 layer is formed by joining a tetrahedral sheet to both sides of an octahedral sheet (Figure 13.1f). These layers consist of four planes of anions. The outer two planes are the basal oxygen atoms of the tetrahedral sheets. The middle two planes contain both OH⁻ from the original octahedral layer and O^{2-} that are the apical oxygen anions on the tetrahedra. The combination of two T and one O layers to form a dioctahedral TOT layer can be represented:

$$2T \text{ sheets} + 1O \text{ sheet} = TOT \text{ layer} + \text{hydroxyl}$$
$$2(Si_2O_5^{2-}) + Al_2(OH)_6 = Al_2Si_4O_{10}(OH)_2 + 4(OH)^-.$$

The equivalent trioctahedral layer is $Mg_3Si_4O_{10}(OH)_2$.

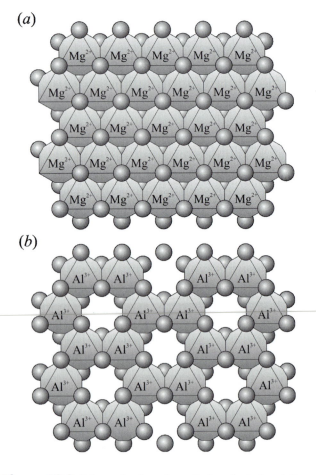

(a)

(b)

Figure 13.2 Schematic plan view of trioctahedral and dioctahedral sheets. (a) Trioctahedral. Divalent cations (usually Mg^{2+} or Fe^{2+}) occupy all (three out of three) octahedral sites. (b) Dioctahedral. Trivalent cations (usually Al^{3+}) occupy two-thirds of the potential octahedral sites.

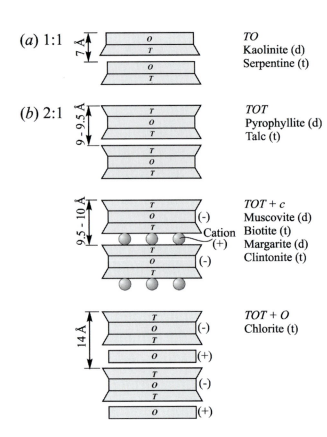

Figure 13.3 Schematic layer silicate unit structure. Dioctahedral minerals are indicated with (d) and trioctahedral with (t). (a) 1:1 layer silicates. TO layers are bonded to each other with weak electrostatic bonds. (b) 2:1 layer silicates. Successive uncharged TOT layers are bonded together with weak electrostatic bonds (TOT, top). TOT layers with a net negative charge are bonded through interlayer cations ($TOT + c$, middle) or through O sheets with a net positive charge ($TOT + O$, bottom).

1:1 Layer Silicates

In the 1:1 group, the unit structure consists of repeating TO layers (Figure 13.3a, Table 13.1). The dioctahedral version is kaolinite $[Al_2Si_2O_5(OH)_4]$, and the trioctahedral version is serpentine $[Mg_3Si_2O_5(OH)_4]$. Because 1:1 layers are electrically neutral, bonding between layers depends on weak electrostatic bonds, including van der Waals and hydrogen bonds. This accounts for the fact that all 1:1 layer silicates are soft. The repeat distance from one layer to the next is typically about 7 Å, so the c unit cell dimension is usually about 7 or 14 Å, depending on whether the unit cell is one or two layers thick.

2:1 Layer Silicates

The 2:1 layer silicates are constructed of repeating TOT layers. In the form just described, dioctahedral and trioctahedral TOT layers are electrically neutral. Recall, however, that cations may substitute in octahedral and tetrahedral sites and change the net layer charge. The most important is substitution of Al^{3+} for Si^{4+} in tetrahedral sites so that

Table 13.1 Common Sheet Silicate Minerals

Group	Structure	Octahedral Sheet	Layer Composition	Interlayer	Unit Structure	Mineral
1:1	*TO*	Dioctahedral	$Al_2Si_2O_5(OH)_4$	None	$Al_2Si_2O_5(OH)_4$	Kaolinite
		Trioctahedral	$Mg_3Si_2O_5(OH)_4$	None	$Mg_3Si_2O_5(OH)_4$	Serpentine
2:1	*TOT*	Dioctahedral	$Al_2Si_4O_{10}(OH)_2$	None	$Al_2Si_4O_{10}(OH)_2$	Pyrophyllite
		Trioctahedral	$Mg_3Si_4O_{10}(OH)_2$	None	$Mg_3Si_4O_{10}(OH)_2$	Talc
	TOT + c	Dioctahedral	$Al_2AlSi_3O_{10}(OH)_2^{1-}$	K^+	$KAl_2AlSi_3O_{10}(OH)_2$	Muscovite
		Trioctahedral	$(Mg,Fe)_3AlSi_3O_{10}(OH)_2^{1-}$	K^+	$K(Mg,Fe)_3AlSi_3O_{10}(OH)_2$	Biotite
	TOT + c	Dioctahedral	$Al_2Al_2Si_2O_{10}(OH)_2^{2-}$	Ca^{2+}	$CaAl_2Al_2Si_2O_{10}(OH)_2$	Margarite
		Trioctahedral	$Mg_2Al(Al_3SiO_{10})(OH)_2^{2-}$	Ca^{2+}	$CaMg_2Al(Al_3SiO_{10})(OH)_2$	Clintonite
	TOT + O	Dioctahedral	$Mg_3(Si,Al)_4O_{10}(OH)_2^{1-}$	$(Mg,Al)_3(OH)_6^{1+}$	$Mg_3(Si,Al)_4O_{10}(OH)_2 \cdot (Mg,Al)_3(OH)_6$	Chlorite

TOT layers have a net negative charge. The varieties of 2:1 structures depend on whether the *TOT* layers have a net negative charge, and if so, what is placed in the interlayer sites to balance the charge (Figure 13.3b).

TOT Structure

Talc and pyrophyllite (Table 13.1) are constructed of trioctahedral and dioctahedral *TOT* layers, respectively. Si^{4+} occupies all tetrahedral sites, so the *TOT* layers are electrically neutral, and bonding between adjacent *TOT* layers depends on van der Waals and hydrogen bonds. The unit structure therefore consists only of successive *TOT* layers. No additional cations are required to maintain charge balance. The weak interlayer bonding is responsible for the softness ($H = 1$) and waxy or greasy feel of both prophyllite and talc. The repeat distance of *TOT* layers in the unit structure is typically between 9 and 9.5 Å. The c unit cell dimension is therefore between 9 and 9.5 Å, or twice that dimension if the unit cell happens to be two layers thick.

TOT + c Structure

Minerals with the *TOT + c* structure include the micas and a less common group of minerals called the brittle micas. The structure consists of *TOT* layers in which some tetrahedral sites are occupied by Al^{3+} rather than Si^{4+}. In the micas, the ratio of Al^{3+} to Si^{4+} in tetrahedral sites is 1:3. Dioctahedral *TOT* layers have the formula $Al_2(AlSi_3O_{10})(OH)_2^{1-}$; trioctahedral *TOT* layers have the formula $Mg_3(AlSi_3O_{10})(OH)_2^{1-}$ (Table 13.1). The net negative charge of these *TOT* layers is balanced by inserting large monovalent cations such as K^+ between the layers to form muscovite.

TOT layer + interlayer cation (c) = unit structure
$$Al_2(AlSi_3O_{10})(OH)_{2^-} + K^+ =$$
$$KAl_2(AlSi_3O_{10})(OH)_2 \text{ (muscovite)}$$

The equivalent trioctahedral mica is $KMg_3(AlSi_3O_{10})(OH)_2$, which is phlogopite. The interlayer cations are situated in the centers of the tetrahedra rings (Figures 13.1d and 11.2f) in the tetrahedral sheets. Because bonding between adjacent *TOT* layers includes both van der Waals bonds and stronger ionic bonds involving the interlayer cations, the hardness of micas is usually between 2 and 3. The repeat distance of layers in this structure is usually between 9.5 and 10 Å, slightly larger than for the *TOT* structure because of the addition of interlayer cations.

The brittle micas are constructed the same as the micas except that more tetrahedra are occupied by Al^{3+} rather than Si^{4+}. In margarite, half the tetrahedral sites are occupied by Al^{3+} so the *TOT* layers are $Al_2(Al_2Si_2O_{10})(OH)_2^{2-}$. To make up the net charge deficiency of −2, divalent cations, usually Ca^{2+}, occupy the interlayer sites.

TOT layer + interlayer cation (c) = *TOT* unit structure
$$Al_2(Al_2Si_2O_{10})(OH)_2^{2-} + Ca^{2+} =$$
$$CaAl_2(Al_2Si_2O_{10})(OH)_2$$

The equivalent trioctahedral brittle mica would be $CaMg_3(Al_2Si_2O_{10})(OH)_2$, but this composition is not found. The most common trioctahedral brittle mica is clintonite, $\sim CaMg_2Al(Al_3SiO_{10})(OH)_2$. Roughly one-third of the octahedral sites contain Al^{3+}, whose charge excess is balanced by substitution of additional Al^{3+} for Si^{4+} in tetrahedral sites.

TOT + O Structures

The *TOT + O* layer silicates are also known as 2:1 + 1 layer silicates. The most common representatives are members of the chlorite group. The structure can conceptually be derived from talc with the addition of an octahedral sheet (a brucite-like sheet) between the *TOT* layers.

Talc-like layer + brucite-like sheet = chlorite unit structure

$$Mg_3Si_4O_{10}(OH)_2 + Mg_3(OH)_6 =$$
$$Mg_3Si_4O_{10}(OH)_2 \cdot Mg_3(OH)_6$$

The contents of the *TOT* layer and the *O* sheet are written separately in the formula to emphasize that the two units are structurally distinct. The thickness of the combined *TOT + O* unit is about 14 Å, or about 9.5 Å for the *TOT* layer and about 4.5 Å for the *O* sheet.

In practice, the *TOT* layers in chlorite have a net negative charge because Al^{3+} substitutes for about a third of the Si^{4+} in the tetrahedral sites within the *TOT* layers, similar to the micas. In contrast, the interlayer octahedral sheet has a net positive charge produced by substitution of Al^{3+}

and some Fe^{3+} for octahedrally-coordinated divalent cations. The net negative charge on the *TOT* layer is balanced by the net positive charge on the interlayer *O* sheet. This results in somewhat stronger bonding between layers than would be anticipated if the layers were electrically neutral and bonded only with van der Waals bonds. The hardness of chlorite is 2–3, about the same as the micas.

Polytypism

The layer silicates have various polytypes related by the pattern in which adjacent layers are stacked atop each other. The micas provide a convenient illustration. Within a *TOT* layer, the top tetrahedral sheet must be offset relative to the bottom tetrahedral sheet so that the apical oxygen atoms can fit into sites provided within the planes of OH^- that produce the central octahedral sheet (Figure 13.4a). This offset can be directed along any of the three pseudohexagonal axes of the tetrahedral sheet. Adjacent *TOT* layers are keyed to each other by the interlayer cations that fit in the middle of the hexagonal rings in the tetrahedral sheets. If the offset direction of successive *TOT* layers repeats itself in a systematic way, the periodicity along the *c* axis direction will be some multiple of the ~9.5° to 10 Å layer thickness. It turns out that six different polytypes are possible in the micas, with repeat distances along the *c* axis that range from ~10 Å (one *TOT* layer) to ~60 Å (six *TOT* layers). Of these, the most common are the *1M* and *2M₁* polytypes. In the *1M* polytype, successive layers show the same direction of offset and the unit cell is one *TOT* layer thick (Figure 13.4b). In the *2M₁* polytype, the successive layers have offset vectors at 120° and the unit cell is two *TOT* layers thick. Both have monoclinic symmetry.

TO STRUCTURES (1:1)

SERPENTINE

(ANTIGORITE, CHRYSOTILE, LIZARDITE)

$Mg_3Si_2O_5(OH)_4$
Monoclinic 2/*m*
(also triclinic, orthorhombic, or hexagonal)
$a = \sim 5.3$ Å, $b = 9.25$ Å, $c = \sim 7.3$ Å, $\beta = 90°$ to 93°, $Z = 2$
$H = \sim 2.5 - 3.5$
$G = 2.55 - 2.65$
Biaxial (−)
$n_\alpha = 1.529 - 1.595$
$n_\beta = 1.530 - 1.603$
$n_\gamma = 1.537 - 1.604$
$\delta = 0.001 - 0.010$ (rarely higher)
$2V_x = 20 - 50°$, highly variable

Structure The three varieties of serpentine are all trioctahedral *TO* sheet silicates. A problem with the ideal dioctahedral *TO* layer structure is that the lateral dimensions of the tetrahedral and octahedral sheets differ somewhat. The *a* and *b* dimensions of the trioctahedral sheets are about 5.4° and 9.3 Å, respectively, compared to about 5.0° and 8.7 Å for the tetrahedral sheets. The different varieties of serpentine correspond to the different ways of dealing with this mismatch.

CHRYSOTILE: The mismatch is accommodated by allowing the layers to curve (Figure 13.5a). As the layers grow, they roll themselves into hollow fibers elongate parallel to *a*.

ANTIGORITE: The tetrahedral sheets are continuous, but periodically there is a reversal in the direction that the apical oxygens face and the side on which the octahedral sheet is mated (Figure 13.5b). This yields a periodic reversal in the sense of curvature of the *TO* layers so that they remain essentially flat rather than curling up.

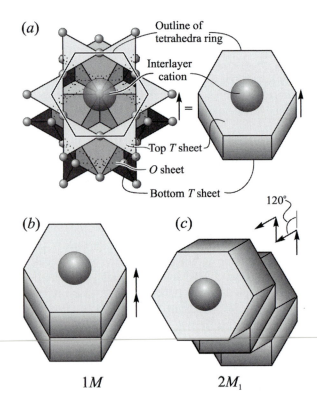

Figure 13.4 Polytypism in the micas. (*a*) In a *TOT* layer, the top *T* sheet must be offset relative to the bottom *T* sheet. Four octahedra from the octahedral sheet (*O*) are shown with a single ring of tetrahedra from the tetrahedral (*T*) sheets on the top (light shading) and bottom (dark shading). Adjacent *TOT* layers are keyed to each other with a cation that fits in the center of the tetrahedra rings. The shape on the right schematically represents position of the tetrahedra rings on bottom and top. (*b*) *1M* polytype. Adjacent *TOT* layers have the same direction of offset. (*c*) *2M₁* polytype. The offset direction in adjacent *TOT* layers is at 120°, and the repeat distance is two layers.

LIZARDITE: The pattern of tetrahedra in the *T* sheet is distorted to allow a match to be achieved with the *O* sheet (Figure 13.5c). In addition, substitution of Al^{3+} for Si^{4+} in tetrahedral sites allows the tetrahedral sheet to expand, and substitution of Al^{3+} for Mg^{2+} in octahedral sites allows the octahedral sheet to shrink to better accommodate a match. Note that these coupled substitutions retain electric neutrality for the *TO* layer.

Composition Most serpentine is close to the ideal composition. Some substitution of Al^{3+} for Si^{4+}, or Al^{3+}, Fe^{3+}, or Fe^{2+} for Mg^{2+} may occur.

Form and Twinning Figure 13.6. Chrysotile forms fine silky fibers in veins or matted masses. Fibers are elongate parallel to *a*. Individual fibers are submicroscopic, with diameters of 220° to 270 Å. A hollow tube down the center of fibers, most commonly around 70 Å across, may be empty or may contain amorphous material. What may appear to be individual fibers teased from a chrysotile vein are actually parallel aggregates of many fibers. Lizardite is the most common serpentine variety and usually forms fine-grained greenish masses. Antigorite may be more or less micaceous and can form scaly or foliated masses. Different varieties may be intergrown. Habit is not a good guide for distinguishing between antigorite and lizardite.

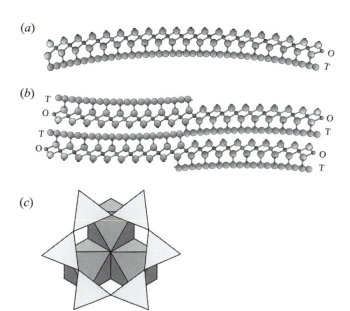

Figure 13.5 Serpentine structures. All are based on dioctahedral *TO* layers. (*a*) Chrysotile. Mismatch in the lateral dimensions of *T* and *O* sheets is accommodated by curving the *TO* layers so that layers roll up into fibers as they grow. (*b*) Antigorite. Continuous *TO* sheets contain reversals in the facing of the tetrahedra. (*c*) Lizardite. The tetrahedra mesh (light shading) is distorted from its ideal hexagonal geometry to match the dimensions of the octahedral sheet. *TO* layers are more or less uncurved.

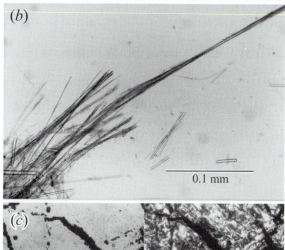

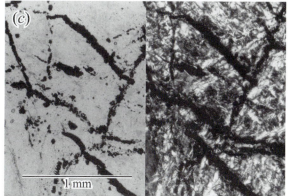

Figure 13.6 Serpentine. (*a*) Hand sample in which veins of cross-fiber chrysotile (C) alternate with antigorite/lizardite (A/L). (*b*) Fibers of chrysotile in grain mount. (*c*) Thin section of antigorite/lizardite. Plane light (*left*) and crossed polarizers (*right*). Note the abundance of opaque magnetite inclusions derived from Fe in the olivine and pyroxene that were altered to form the serpentine.

Each variety may occur in different polytypes. Identifiable twins are rare.

Cleavage Perfect {001} cleavage is visible only in antigorite. Chrysotile fibers are easily separated and are flexible.

Color Properties

Luster: Greasy or waxy in massive varieties, silky in chrysotile.

Color: Shades of green. Chrysotile is most commonly yellowish green, white, or gray. Lizardite and antigorite are commonly light to medium green but variable, owing in part to associated minerals. Magnetite may color samples gray, brown, or black, and hematite may yield shades of brown or red.

Streak: White.

Color in Thin Section: Colorless to pale green.

Pleochroism: Weak, from colorless to pale green or yellow green. Fine-grained varieties may not be discernibly pleochroic because of random grain orientation. Chrysotile fibers are usually darker when aligned parallel to the vibration direction of the lower polarizer.

Optical Characteristics

Relief in Thin Section: Generally low to moderate positive, also low negative.

Optic Orientation: The slow ray vibration direction is generally parallel to fiber length in chrysotile, which shows parallel extinction. In antigorite $X \approx c$, $Y \approx b$, $Z \approx a$, also $X \approx c$, $Y \approx a$, $Z \approx b$, so the trace of cleavage is always length slow. The optic plane may be parallel to either {010} or {100}.

Indices of Refraction and Birefringence: Indices of refraction increase with Fe and with Al content. Birefringence is low, and maximum interference color in thin section is rarely above first-order gray or white. The interference color may be anomalous. The mean index of refraction of lizardite is usually 1.54–1.55 and antigorite is slightly higher but with substantial overlap. The indices of refraction of chrysotile are generally $n_\alpha = 1.532$–1.549 and $n_\gamma = 1.545$–1.556.

Interference Figure: Usually difficult to obtain because of grain size. Basal sections of antigorite yield biaxial negative figures with $2V_x$ between roughly 20° and 50°. Optic axis dispersion may be either $r > v$ or $r < v$.

Comments: Chrysotile is distinctive because of its fibrous nature, but antigorite and lizardite cannot reliably be distinguished optically.

Alteration Oxidation of associated magnetite may stain serpentinite brown or yellow. Serpentine may be altered to chlorite or replaced by quartz.

Distinguishing Features

Hand Sample: The fibrous nature of chrysotile is distinctive. The fibrous amphiboles have different occurrences and are usually either blue or brown. They also grind to a powder in a mortar, whereas chrysotile forms a matted aggregate. Lizardite and antigorite usually form fine-grained green masses that cannot be distinguished from each other except with X-ray diffraction techniques. Chlorite also forms foliated green masses, but is more likely to be coarse grained and is commonly found as a constituent of metamorphosed sediments, whereas the serpentine minerals are found in altered mafic and ultramafic rocks.

Thin Section/Grain Mount: The fibrous amphiboles are distinguished from chrysotile by their higher birefringence and indices of refraction. Chlorite is more strongly pleochroic than antigorite, and brucite is uniaxial.

Occurrence Serpentine is commonly formed by hydrothermal alteration of mafic and ultramafic rocks, such as peridotite and pyroxenite, that contain olivine and pyroxene. The serpentine-rich rock known as serpentinite consists of masses of green antigorite/lizardite that may be cut by veins of chrysotile. Because serpentine contains relatively little iron, the iron content of the original olivine and pyroxene usually goes into the formation of fine-grained magnetite disseminated through the serpentinite. The associated minerals may include talc, calcite, brucite, chlorite, magnetite, and chromite. Serpentine also may be found in contact metamorphosed carbonate rocks in association with calcite, dolomite, forsterite, magnesite, and calc-silicate minerals.

Use The fibrous nature, low thermal conductivity, and incombustibility of chrysotile have made it a valuable industrial mineral, and it is the most common of the minerals collectively known as asbestos. Chrysotile has been widely used in the automotive industry for the manufacture of brake shoes and clutch facings. When mixed with a binder, chrysotile has been used as a fireproof insulation on boilers and heating pipes, and it has been used as a reinforcing fiber in floor tiles, acoustic ceiling tiles, plastics, roofing materials, adhesives, and a wide variety of other industrial and construction materials. It also can be made into a felt or woven into a fabric. Chrysotile accounts for about 98% of the current world production of asbestos, fibrous amphiboles (chapter 14) for the remainder.

Comments A great deal of attention has been given to the health risks associated with inhaling or ingesting asbestos minerals. For a review and additional references, consult Guthrie and Mossman (1993) and Gunter and others (2007). It is important to recognize that asbestos is an industrial product made from fibrous minerals. Asbestos minerals are, of course, naturally occurring, so people also are exposed to them to varying degrees from soil and rock.

That asbestos minerals can pose health risks has been established based on studies of individuals who have had chronic exposure to high levels of asbestos mineral dust in occupational settings such as asbestos mining and milling. Chronic nonoccupational exposure from mine tailings and related settings also can pose health risks. The pathologies to which asbestos has been linked include the following.

Lung cancer, particularly among smokers

Mesothelioma, an otherwise rare cancer of the pleural lining of the lung cavity

Asbestosis, a scarring of the lung tissue associated with asbestos fibers

Pleural plaques (a localized scarring of the tissue lining the lung cavity),

A slight increase in the risk of cancer of the larynx, stomach, colon, and lymphoid tissues

Although the risks associated with chronic occupational exposure are clearly very real, it is unlikely that these risks can be extrapolated in a linear fashion to the low-level exposure that might be found in homes, school rooms, or office buildings where asbestos fibers are part of ceiling tiles, electrical fixtures, or insulation on the plumbing. In fact, we are continuously exposed to low levels of natural mineral fibers in the air and water as the result of normal weathering processes of rocks. It is extremely unlikely that biological defenses would not have developed some level of protection against the risks associated with these natural fibers in the billions of years that life has been evolving on Earth. Regardless, government mandates and lawsuits have prompted the expenditure of billions of dollars to remove asbestos from homes, schools, factories, and office buildings in an attempt to mitigate what, in reality, is a low risk. Most of this money has simply been wasted.

A complicating factor is that regulatory actions and reports in the public media often fail to distinguish among the different varieties of asbestos that may pose significantly different health risks. For example, crocidolite, which is an amphibole asbestos, poses a much greater health risk than chrysotile, particularly with regard to mesothelioma. It is inappropriate to judge the risks posed by chrysotile, which is used in the large majority of all products containing asbestos in North America, based on the behavior of crocidolite.

KAOLINITE

Kaolinite is described in a later section on the clay mineral group.

TOT STRUCTURES (2:1)

TALC

$Mg_3Si_4O_{10}(OH)_2$
Triclinic or monoclinic,
$\bar{1}$, 1, or $2/m$
$a = 5.290$ Å, $b = 9.173$ Å,
$c = 9.460$ Å, $\alpha = 90.46°$,
$\beta = 98.68°$, $\gamma = 90.09°$, $Z = 2$
$H = 1$
$G = 2.58–2.83$
Biaxial (–)
$n_\alpha = 1.538–1.554$
$n_\beta = 1.575–1.599$
$n_\gamma = 1.575–1.602$
$\delta = 0.03–0.05$
$2V_x = 0–30°$

(010) Section

Composition The composition of most talc is relatively close to the ideal formula with minor substitution of Fe, Mn, or other cations for Mg in the octahedral sites and of Al for Si in the tetrahedral sites. The Fe-bearing analog of talc is **minnesotaite** [$Fe_3Si_4O_{10}(OH)_2$], which is found in metamorphic iron formations.

Form and Twinning Talc forms foliated, radiating, or randomly oriented aggregates of irregular flakes or fibers resembling the micas. Talc has a waxy feel and is sectile. Individual flakes are easily bent but are not elastic. Fine-grained compact and massive varieties are sometimes referred to as **steatite**.

Cleavage {001} perfect. Sectile. Thin fragments flexible but not elastic.

Color Properties

Luster: Pearly to somewhat greasy.
Color: White, light to dark green, brown.
Streak: White.
Color in Thin Section: Colorless.
Pleochroism: None.

Optical Characteristics

Relief in Thin Section: Low to moderate positive.

Optic Orientation: $X \wedge c \approx +10°, Y \approx b, Z \approx a$, optic plane = (010).

Indices of Refraction and Birefringence: Indices of refraction increase with increasing Fe content. Birefringence is strong, and the maximum interference color seen in thin section is usually third order. Basal sections and fragments lying on the cleavage display low birefringence because n_β and n_γ are nearly the same. Minnesotaite has indices of refraction in the vicinity of $n_\alpha = 1.592$, $n_\beta = 1.622$, and $n_\gamma = 1.623$.

Interference figure: Cleavage flakes and basal sections yield essentially centered acute bisectrix figures with $2V_x = 0–30°$. Optic axis dispersion is weak to moderate, $r > v$.

Alteration Talc may alter to chlorite.

Distinguishing Features

Hand Sample: Characterized by softness, micaceous habit, and waxy feel. X-ray diffraction may be required to distinguish talc from pyrophyllite.

Thin Section/Grain Mount: Similar to muscovite and pyrophyllite, but with smaller 2V. Phlogopite has higher indices of refraction, stronger relief in thin section, and is usually pale brown. Brucite is uniaxial positive and often shows anomalous interference colors and lower birefringence. Minnesotaite has higher indices of refraction, is weakly colored in shades of pale yellow and green, has a small $2V$, and is typically fibrous.

Occurrence Talc is found in hydrothermally altered mafic and ultramafic rocks in association with serpentine, magnesite, and relict grains of olivine and pyroxene. It also forms in the metamorphism of siliceous dolomite, where it may be associated with calcite, dolomite, tremolite, and related calc-silicate minerals. Talc is a major constituent of talc schist derived by metamorphism of mafic igneous rocks or mafic volcaniclastic sedimentary rocks. These rocks also may contain magnetite, tremolite, chlorite, anthophyllite, or serpentine.

Use Talc is a versatile mineral used in a wide variety of industrial and consumer products. Industrial applications include use as a filler and colorant in paint, plastics, rubber, paper, and various caulking and roofing compounds, and as a lubricant. Talc is sprinkled on asphalt roofing shingles to keep the shingles from sticking together during transport and storage. It also is an ingredient in certain ceramic products. Consumer products that contain talc include cosmetics products such as lipstick, facial products, antiperspirant sticks, and various creams and lotions. Talc (talcum powder) is a major ingredient in a wide variety of baby, body, and foot powders available at any drug store. Some studies have claimed correlations between chronic talc exposure and cancer, though the risk may have been from associated asbestiform minerals, not the talc. Talc is approved by the U.S. Food and Drug Administration for use in preventing the recurrence of pleural effusions, an abnormal collection of fluid between the thin layers of tissue (pleura) lining the lung and the wall of the chest cavity.

PYROPHYLLITE

$Al_2(Si_4O_{10})(OH)_2$
Monoclinic ($2/m$)
$a = {\sim}5.17$ Å, $b =$
    ${\sim}8.96$ Å, $c =$
    ${\sim}18.68$ Å, $\beta =$
    ${\sim}100°$, $Z = 4$
Triclinic ($\bar{1}$ or 1)
$a = {\sim}5.16$ Å, $b =$
    ${\sim}8.96$ Å, $c = {\sim}9.35$ Å,
    $\alpha = {\sim}91°$, $\beta =$
    ${\sim}100°$, $\gamma = {\sim}90°$, $Z = 2$
$H = 1–2$
$G = 2.65–2.90$
Biaxial (–)
$n_\alpha = 1.552–1.556$
$n_\beta = 1.586–1.589$
$n_\gamma = 1.596–1.601$
$\delta = 0.043–{\sim}0.05$
$2V_x = 53–62°$

(010) Section

Structure The most common structure is monoclinic and the unit cell is two *TOT* layers thick. The stacking sequence of *TOT* layers in the triclinic polytype is somewhat different and results in a unit cell that is one *TOT* layer thick.

Composition The composition is usually close to the ideal formula, with the most common substitutions being Al^{3+} for Si^{4+} and Mg^{2+}, Fe^{2+}, Fe^{3+}, or Ti^{4+} for Al^{3+}. Minor Ca^{2+}, Na^+, or K^+ may be present, presumably in interlayer sites.

Form and Twinning Pyrophyllite is commonly found as foliated, radiating, columnar, or massive aggregates of mica-like flakes. Also fibrous. Individual grains may be easily bent. Twins are not generally observed.

Cleavage As with the micas, perfect {001} cleavage controls fragment orientation. Folia are flexible but not elastic. Masses are somewhat sectile.

Color Properties

Luster: Pearly to dull.
Color: White, pale blue, yellow, grayish, or brownish green.
Streak: White.
Color in Thin Section: Colorless.
Pleochroism: None.

Optical Characteristics

Relief in Thin Section: Low to moderate positive.
Optic Orientation: $X \wedge c \approx +10°, Y = a, Z = b$.
Indices of Refraction and Birefringence: Indices of refraction vary within a small range and birefringence is high (~0.05).
Interference Figure: Cleavage fragments and basal sections yield nearly centered acute bisectrix figures with $2V_x = 53–62°$. Optic axis dispersion is weak, $r > v$.

Alteration Not readily altered.

Distinguishing Features

Hand Sample: Distinguished from muscovite by being softer, having a greasy or waxy luster and feel, and because cleavage flakes are not elastic. Pyrophyllite is difficult to distinguish from talc in hand sample and may require either optical or X-ray tests to confirm the identity.

Thin Section/Grain Mount: Distinguished from muscovite and talc by higher $2V$ and from kaolinite and other clay minerals by higher birefringence.

Occurrence Pyrophyllite is a relatively uncommon mineral that occurs in moderately low-grade, Al-rich metamorphic rocks such as metapelites, metabauxite, or metaquartzite, or as a result of hydrothermal alteration of aluminous minerals such as the feldspars, muscovite, aluminum silicates, corundum, or topaz.

Use Pyrophyllite's most common use is in the manufacture of refractory materials used to line kilns, furnaces, and ladles in foundries and related industries where resistance to high temperatures is required. It also is used in the manufacture of ceramics, including wall and floor tile, pottery, china, and porcelain plumbing fixtures. Finely ground pyrophyllite is used as a carrier for pesticides or as a blending agent in animal feeds or fertilizers. Commercial applications include use as a filler in paint, plastics, cements, and construction materials such as gypsum wallboard. Pyrophyllite also is used in cosmetic formulations.

TOT + c STRUCTURES: MICA MINERALS (2:1)

Micas comprise about 4% of the Earth's crust and are found in igneous, sedimentary, and igneous rocks. The current recommended classification (Rieder and others, 1998) identifies 11 dioctahedral micas and 16 trioctahedral micas. Of the dioctahedral micas, only muscovite and paragonite are at all common. Of the trioctahedral micas, only the biotite series and lepidolite are common. Both glauconite and illite can be considered to be interlayer-cation-deficient dioctahedral micas. Glauconite is described with the micas, and illite is described with the clay minerals in a later section.

MUSCOVITE

$KAl_2(AlSi_3O_{10})(OH)_2$
Monoclinic (2/m)
$a = 5.19$ Å, $b = 9.04$ Å,
  $c = 20.08$ Å,
  $\beta = 95.8°$, $Z = 4$
$H = 2.5$ on {001} cleavage
  surfaces, ~ 4 at right
  angles to (001)
$G = 2.77–2.88$
Biaxial (−)
$n_\alpha = 1.552–1.576$
$n_\beta = 1.582–1.615$
$n_\gamma = 1.587–1.618$
$\delta = 0.036–0.049$
$2V_x = 28–47°$

Structure The large majority of muscovite has a unit cell that is two *TOT* layers thick, consisting of the $2M_1$ polytype (Figure 13.4c). Some fine-grained muscovite from low-grade metamorphic rocks or sediments has a unit cell that is only one *TOT* layer thick and is known as a $1M$ polytype (Figure 13.4b).

Composition The common substitutions include the following.

For K	Na, Rb, Cs, Ca, Ba
For [VI]Al	Mg, Fe^{2+}, Fe^{3+}, Mn, Li, Cr, Ti, V
For (OH)	F, Cl
Tetrahedral site	$AlSi_3$ to $Al_{0.5}Si_{3.5}$

Paragonite $[NaAl_2(AlSi_3O_{10})(OH)_2]$ is the sodium analog of muscovite; but because of differences in the sizes of Na and K cations, solid solution is limited to no more than about 10% Na in muscovite and 20% K in paragonite. In X-ray powder diffraction patterns of muscovite, d_{002} decreases with increasing Na content, reflecting its smaller size, and the d_{060} dimension increases with substitution of Al for Si in tetrahedral sites. The term **phengite** is applied to muscovites in which the Si:Al ratio is greater than 3:1 and Mg^{2+} or Fe^{2+} substitutes for Al^{3+} to compensate for the higher charge of the Si. Fluorine commonly replaces some of the OH, but usually in amounts less than about 20%. Chlorine is restricted to less than about 1% of the OH sites.

Form and Twinning Figure 13.7. Well-formed tabular crystals with pseudohexagonal outlines are common in granitic pegmatite and may be referred to as "books" of muscovite in allusion to the fact that thin cleavage sheets can be readily peeled from the crystals. Muscovite is more

Figure 13.7 Muscovite. (*a*) Muscovite "book" from a granitic pegmatite. (*b*) Photomicrograph of muscovite (M) with dark biotite (B) and clear quartz in a mica schist.

commonly found as micaceous flakes or tablets with irregular outlines, tabular parallel to {001}. **Sericite** is a name given to fine-grained white mica, usually muscovite or paragonite, commonly produced by the hydrothermal alteration of feldspars or other K- and/or Al-bearing minerals.

Twins by rotation on [310] with a {001} composition plane are not commonly seen except in large well-formed crystals from granitic pegmatites.

Cleavage Perfect cleavage on {001} strongly controls fragment orientation. Individual cleavage folia are flexible and elastic.

Color Properties

Luster: Vitreous.

Color: Colorless or light shades of green, red, or brown. The edges of muscovite books may appear quite dark, particularly when grains are viewed as part of the rock in which they are found.

Streak: White.

Color in Thin Section: Colorless.

Pleochroism: None.

Optical Characteristics

Relief in Thin Section: Modest positive, changes somewhat with rotation.

Optic Orientation: $X \wedge c = +1° - +4°, Y \wedge a = +1° - +3°$, $Z = b$.

Indices of Refraction and Birefringence: Relatively pure muscovite has indices of refraction at the lower end of the range given above. Substitution of Fe, Mg, Ti, Mn, Cr, and V causes indices of refraction to increase. Substitution of Li causes indices of refraction to decrease. Birefringence is high; interference colors in thin section may be as high as third order, and vivid colors of the second order are typical. Paragonite has indices of refraction $n_\alpha = 1.564–1.580$, $n_\beta = 1.594–1.609$, $n_\gamma = 1.594–1.618$, $\delta = 0.028–0.038$.

Interference Figure: $2V_x$ is between 28° and 47°; relatively pure muscovite has the upper end of the range. Cleavage fragments yield nearly centered acute bisectrix figures. In thin section, good figures are yielded by grains that do not display cleavage and have lower to middle first-order colors. To the uninitiated, these grains do not look particularly like muscovite. Optic axis dispersion is noticeable with $r > v$. Paragonite is biaxial negative with $2V_x = 0–40°$.

Alteration Not readily altered, but may be converted to clay minerals by weathering.

Distinguishing Features

Hand Sample: Habit and perfect cleavage producing elastic folia distinguish the micas from most other minerals.

Biotite, unless very rich in Mg (phlogopite), is usually much darker colored. Pyrophyllite and talc have a greasy luster and feel, and are softer. Chlorite is distinctly green and has inelastic cleavage folia.

Thin Section/Grain Mount: Muscovite is distinguished from most biotite by lack of color. Mg-rich biotite (phlogopite) may be nearly colorless but has smaller 2V. Talc and pyrophyllite are distinguished by having smaller and larger 2V, respectively. A distinctive feature of muscovite and biotite is a pebbly appearance seen near extinction that is called "birds-eye" extinction in allusion to a similar pattern seen in sugar maple wood (*Acer saccharum*) cut parallel to the grain. Paragonite has very similar optical properties, and X-ray or staining techniques (Laduron, 1971) may be needed to distinguish them.

Occurrence Muscovite is a very common mineral found in igneous, metamorphic, and sedimentary rocks. In igneous rocks it is a common constituent of granitic pegmatite, granite, granodiorite, aplite, and related felsic rocks. It is somewhat less common in felsic volcanic rocks. Sericite is widespread in many igneous rocks, produced by the hydrothermal or late-stage magmatic alteration of feldspars and other minerals. The alteration may be selective, replacing only the cores of plagioclase grains or selected twin lamellae. Muscovite is a constituent of a wide variety of metamorphic rocks including slate, phyllite, schist, gneiss, hornfels, and quartzite that are produced by metamorphism of common sedimentary rocks. Terrigenous sediments derived from crystalline terranes and not subjected to extensive weathering or transport often contain muscovite. It is therefore a common mineral in arkosic sandstone and related siliclastic sedimentary rocks.

Phengite and paragonite are usually found in relatively low-grade regional metamorphic rocks. Paragonite is probably more abundant than has often been supposed and has commonly been misidentified as muscovite.

Use One of the earliest uses for muscovite was as a substitute for glass because thin cleaved sheets are transparent. It is still used, though infrequently, for viewing windows in industrial furnaces and ovens. It is now widely used in electronics and industrial applications. Muscovite sheets and ground muscovite are used in the electronics industry to make components as diverse as capacitors, transistors, insulators, and the windows on microwave tubes used in microwave ovens. Industrial applications include use as a filler in plastic, paint, and wallboard cement, coatings on wallpaper to produce a silky luster, mold release agents in the manufacture of automobile tires, and as a constituent of drilling mud used in the drilling for oil and gas. Consumer products that contain muscovite include nail polish, lipstick, and eye shadow. The subtle luster seen in many colored cosmetic creams is there because of the presence of pulverized muscovite. The use of muscovite as a treatment for colitis and other problems associated with the digestive tract has been studied in rats, particularly in China.

BIOTITE

K(Fe,Mg)$_3$AlSi$_3$O$_{10}$(OH)$_2$
Monoclinic 2/m
a = ~5.3 Å, b = ~9.2 Å,
c = ~10.3 Å,
β = ~100°, Z = 2
H = 2–3
G = 2.7–3.3
Biaxial (–)
n_α = 1.522–1.625
n_β = 1.548–1.696
n_γ = 1.549–1.696
δ = 0.03–0.07
$2V_x$ = 0–25°

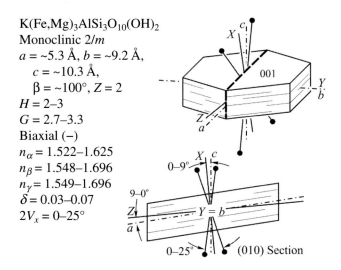

Structure Most biotite consists of the *1M* polytype (Figure 13.4b), meaning that the unit cell is one *TOT* layer thick. Polytypes that are two or three *TOT* layers thick are known.

Composition The composition variation can be represented by four end members (Figure 13.8):

KFe$_3$(AlSi$_3$O$_{10}$)(OH)$_2$: Annite

KMg$_3$(AlSi$_3$O$_{10}$)(OH)$_2$: Phlogopite

KFe$_2$Al(Al$_2$Si$_2$O$_{10}$)(OH)$_2$: Siderophyllite

KMg$_2$Al(Al$_2$Si$_2$O$_{10}$)(OH)$_2$:Eastonite

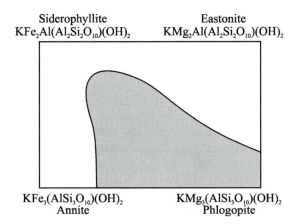

Figure 13.8 Compositional range of common biotite (shaded). Adapted from Deer and others (1992).

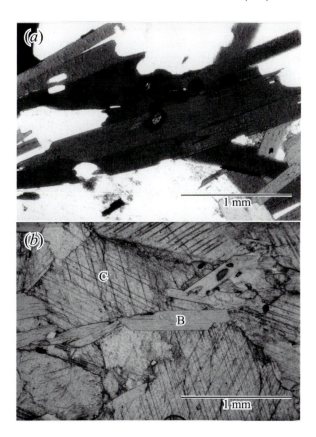

Figure 13.9 Biotite. (*a*) Foliated biotite grains in mica schist. Note the dark pleochroic halos around radioactive inclusions. (*b*) Pale laths of phlogopite-rich biotite (B) with calcite (C) in marble.

Other substitutions include Na and to a lesser degree Ca, Ba, Rb, and Cs for K, and substantial amounts of F and lesser Cl for OH. Samples with substantial amounts of Fe^{3+} substituting in the octahedral sites balanced by substituting O for OH in the hydroxyl sites are known as **oxybiotite**.

Form and Twinning Figure 13.9. Crystals are usually tabular with a pseudohexagonal outline and are relatively common, particularly in felsic volcanic rocks. Also found as micaceous or tabular grains or grains with irregular shapes. Twins with {001} composition planes are possible but not usually observed.

Cleavage Perfect {001} cleavage produces thin elastic folia.

Color Properties

Luster: Vitreous.
Color: Most commonly black or dark brown, sometimes greenish or reddish. Phlogopite-rich biotite may be light brown to nearly colorless.
Streak: White or gray.

Color in Thin Section: Typically brown, brownish green, or reddish brown and strongly pleochroic. Phlogopite-rich biotite may be pale brown to nearly colorless.

Pleochroism: Usually strongly pleochroic with $Z \approx Y > X$, so grains are darker when the trace of cleavage is parallel to the lower polarizer vibration direction. The common colors are X = colorless, light tan, pale greenish brown, pale green; $Y \approx Z$ = brown, olive brown, dark green, dark red-brown. The intensity of color generally increases with increasing iron content. Cleavage flakes and basal sections yield darker colors with little pleochroism. Pleochroic halos are common around minerals, such as zircon, containing radioactive elements (Figure 5.28).

Optical Characteristics

Relief in Thin Section: Moderate to moderately high positive, low if Mg rich.

Optic Orientation: $X \wedge c = 0° - +9°, Y = b, Z = a$.

Indices of Refraction and Birefringence: Indices of refraction increase with iron content (Figure 13.10). Birefringence is strong and yields maximum interference colors in the third or even fourth order, but these colors are usually masked by the mineral color. Cleavage flakes and sections cut parallel to {001} show low birefringence.

Interference Figure: Cleavage flakes and sections cut parallel to {001} yield nearly centered acute bisectrix figures with several orders of isochromes and small 2V. Some biotite may have 2V of 0°, meaning that it is sensibly uniaxial. Optic axis dispersion is weak, usually $r < v$, less commonly $r > v$ for Mg-rich varieties.

Alteration Biotite commonly alters to chlorite, which may mantle grains or be interleaved along cleavage traces (Figure 13.13). Biotite also may alter to clay minerals,

including vermiculite and sericite, Fe–Ti oxides, and epidote.

Distinguishing Features

Hand Sample: Dark color and micaceous habit with one perfect cleavage distinguish biotite. Phlogopite may be light brown or tan.

Thin Section/Grain Mount: Dark color, strong pleochroism, high birefringence, and micaceous habit distinguish biotite. Light-colored phlogopite resembles muscovite but has a smaller 2V. Chlorite is more distinctly green, has lower birefringence, and usually displays anomalous interference colors. Stilpnomelane is similar but lacks the "birds-eye" texture of micas seen near extinction and has a second cleavage at nearly right angles to the basal cleavage.

Occurrence Biotite is a very common mineral. In igneous rocks it is characteristic of silicic and alkalic rocks such as granite, granodiorite, quartz diotite, pegmatite, syenite, nepheline syenite, rhyolite, rhyodacite, dacite, and phonolite. It also is found as a late-stage magmatic product in more mafic rocks including diorite, gabbro, norite, and anorthosite. Mg-rich biotite (phlogopite) is found in peridotite and other ultramafic varieties.

In metamorphic rocks, biotite is very common in a wide variety of hornfels, phyllites, schists, and gneisses and may persist from greenschist facies through strongly migmatitic rocks. Mg-rich biotite is also found in marble and related metamorphosed carbonate-rich rocks.

Biotite also is a relatively common detrital mineral, particularly in immature sediments, but it yields to clay minerals with extended weathering and transport.

Use Biotite has few commercial uses, but vermiculite, most of which is a hydrated alteration product of biotite, has numerous applications. Alteration, accomplished either by weathering or hydrothermal processes, results in leaching of interlayer K cations and replacement with Ca, Mg, and H_2O, with ion exchanges in other sites as needed to maintain electric neutrality. As a result of adding the interlayer water, vermiculite is prone to dramatic expansion when heated. It owes its name to the observation that books of vermiculite, when heated, expand into wormlike shapes. For most applications, the vermiculite is heated to force it to expand, producing a low-density product that looks like dirty fluffed-up biotite. Expanded vermiculite is used as an insulation material, a filler in gypsum wallboard or other construction materials, and in a variety of other industrial applications. The most frequently encountered use for most people is as an additive in potting soil used to grow house plants.

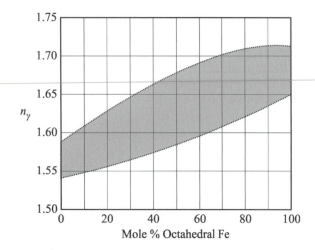

Figure 13.10 Approximate range of n_γ (shaded) as a function of mole fraction octahedral Fe in biotite.

LEPIDOLITE

$K(Li,Al)_3(Si,Al)_4O_{10}(F,OH)_2$
Monoclinic ($2/m$)
$a = 5.209$ Å, $b = 9.011$ Å,
   $c = 10.149$ Å,
   $\beta = 100.77°$, $Z = 2$
$H = 2.5$ on {001} cleavage
   surfaces, ~3 to 4 at
   right angles
$G = 2.8–2.9$
Biaxial (−)
$n_\alpha = 1.524–1.548$
$n_\beta = 1.543–1.585$
$n_\gamma = 1.545–1.587$
$\delta = 0.018–0.039$
$2V_x = 0–58°$

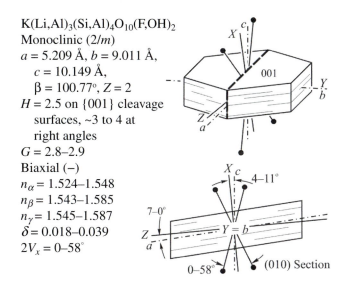

Structure Lepidolite is considered a trioctahedral mica in which the octahedral sites are occupied by Li^{1+} and Al^{3+}. The common polytypes are the $1M$ and $2M_2$ (similar to the $2M_1$ in Figure 13.4) except the offset vector is at 60° rather than 120°.

Composition The ideal end-member compositions are **polylithionite** [$KLi_2AlSi_4O_{10}(F)_2$] and **trilithionite** [$KLi_{1.5}Al_{1.5}AlSi_3O_{10}(F)_2$]. Most lepidolite has five out of six of the octahedral sites occupied, but compositions intermediate between lepidolite and muscovite are not uncommon (Figure 13.11). Interlayered mixtures of trioctahedral lepidolite and dioctahedral muscovite may account for compositions intermediate between lepidolite and muscovite. Considerable amounts of Na, Rb, and Cs may substitute for K, and Fe, Mg, and Mn may enter octahedral sites. OH and F readily substitute for each other.

Form and Twinning Lepidolite is micaceous and commonly forms scaly–granular masses, less commonly as large cleavable plates. Twins by rotation on [310] with a {001} composition plane are uncommon and not generally observed in hand sample.

Cleavage Perfect cleavage on {001} like the other micas. Individual cleavage folia are flexible and elastic.

Color Properties

Luster: Vitreous.

Color: Usually some shade of lilac or pink, also purple, rose-red, violet-gray, yellowish, white, or colorless.

Streak: White.

Color in Thin Section: Usually colorless, less commonly pale pink.

Pleochroism: If present, pleochroism is X = almost colorless, $Y = Z$ = pink, pale violet.

Optical Characteristics

Relief in Thin Section: Low to moderate positive, which may change with stage rotation. The value of n_α is less than 1.540 for some samples so relief may be low negative for some grain orientations.

Optic Orientation: $X \wedge a = 90°$ to 87°, $Y = b$, $Z \wedge c = 0°$ to 7°.

Indices of Refraction and Birefringence: Optical properties vary mostly depending on the amount of Mn and Fe^{3+} rather than with Li content. In general higher indices of refraction are associated with higher iron content; most lepidolite have indices of refraction in the middle of the ranges given above. Birefringence is moderately high, and bright colors of the second order are typical in thin section.

Interference Figure: Cleavage flakes yield nearly centered acute bisectrix interference figures with $2V_x$ usually between 30° and 50°. Optic axis dispersion is $r > v$.

Alteration Not readily altered.

Distinguishing Features

Hand Sample: The micaceous habit and pink color are usually diagnostic. However, rose-colored muscovite is very similar, and X-ray diffraction or chemical tests may be required to distinguish between the two minerals. Unfortunately, standard electron microprobe and X-ray fluorescence analyses are not capable of detecting Li because of its low atomic weight.

Thin Section: Lepidolite closely resembles muscovite but has lower indices of refraction and lower birefringence. If present, the pale pink pleochroism is a useful diagnostic property.

Occurrence Lepidolite is the most common Li-bearing mineral and is usually found in Li-bearing granitic pegmatites associated with other Li-bearing minerals such as spodumene and amblygonite. Other associated minerals include quartz, feldspars, tourmaline, beryl, and topaz. Lepidolite may be formed by direct crystallization from Li-rich pegmatitic fluids or by metasomatic replacement of previously crystallized biotite or muscovite, or reaction between K-feldspar and Li-rich pegmatitic fluids. It has infrequently been found in high-temperature hydrothermal deposits, and in granitic igneous rocks.

Use Lepidolite has been mined as a source of lithium, although most lithium is derived from spodumene or from Li-bearing brines. Lepidolite has a number of historical uses as a pharmaceutical and is currently used in the preparation of various homeopathic remedies, but it does not appear to be approved for any use by the U.S. Food and Drug Administration.

GLAUCONITE

$(K, Ca, Na)_{1.6}(Fe^{3+},$
$\quad Al, Mg, Fe^{2+})_4$
$\quad Si_{7.3}Al_{0.7}O_{20}(OH)_4$
Monoclinic, $2/m$
$a = {\sim}5.23$ Å, $b = {\sim}9.06$ Å,
$\quad c = {\sim}10.15$ Å, $\beta = {\sim}101°$,
$\quad Z = 1$
$H = 2$
$G = 2.4$–2.95
Biaxial (–)
$n_\alpha = 1.59$–1.61
$n_\beta = 1.61$–1.64
$n_\gamma = 1.61$–1.64
$\delta = {\sim}0.02$–0.03
$2V_x = 0$–$20°$

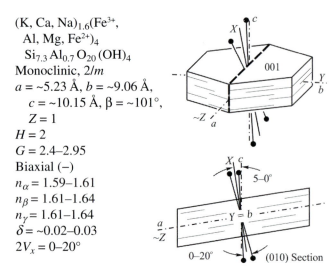

Structure and Composition Glauconite is a dioctahedral layer silicate whose structure is similar to that of muscovite. It differs in that the octahedral sites contain Fe^{3+} as the most abundant cation, along with some Mg^{2+}, Fe^{2+}, and Al^{3+}. The Al^{3+} content of the tetrahedral sites is lower than in muscovite to compensate for divalent cations in octahedral sites. Glauconite can be considered a clay mineral similar to illite (see later discussion of clay minerals), but its distinctive common occurrence warrants a separate description. Some glauconite contains layers of expanding clay structure. **Celadonite** is a similar mineral with less octahedral Fe^{3+} and Al^{3+}, more octahedral Mg^{2+} and Fe^{2+}, and less tetrahedral Al^{3+}.

Form and Twinning Figure 13.12. Glauconite most commonly forms small green pellets or granules in sedimentary rocks, or forms casts of foraminifera or other microfossils. In most cases, these are all composed of

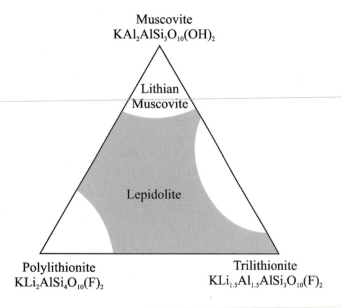

Figure 13.11 Lepidolite composition. The common range is shown with the shaded area.

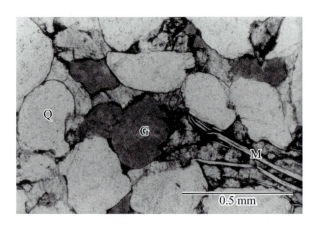

Figure 13.12 Glauconite pellets (G) with rounded quartz (Q) grains and elongate detrital muscovite (M). Calcite forms the cement between grains, and the opaque mineral is hematite.

aggregates of fine, irregular flakes. Celadonite is usually found as radiating masses in vesicles and other voids in basaltic igneous rocks. No twinning is reported.

Cleavage Perfect {001} cleavage, like the other micas.

Color Properties

Luster: Earthy or dull.

Color: Green or blue-green, sometimes with reddish limonite stains.

Streak: Dull green.

Color in Thin Section: Usually green.

Pleochroism: X = yellowish green to green; $Y = Z$ = green, bluish green, darker yellowish green.

Optical Characteristics

Relief in Thin Section: Moderate positive.

Optic Orientation: $X \wedge a = +5° - +10°, Y = b, Z \sim c$.

Indices of Refraction and Birefringence: Indices of refraction increase with Fe content and decrease with increasing fraction of expanding clay structure. Maximum second-order interference colors are usually masked by the mineral color.

Interference Figure: Fine grain size usually precludes obtaining good interference figures. Basal sections and cleavage fragments should yield approximately centered optically negative acute bisectrix figures with small $2V_x$, $r > v$.

Alteration Reddish or yellowish limonitic staining due to weathering is common in glauconite.

Distinguishing Features

Hand Sample: The distinctive occurrence and green color are characteristic, but green pellets in sedimentary rocks are not necessarily glauconite. Chlorite forms similar pellets.

Thin Section/Grain Mount: Occurrence and green color are distinctive. Chlorite in similar pellets has lower birefringence.

Occurrence Glauconite characteristically forms small rounded pellets in clastic sediments deposited in marine conditions. If abundant, these pellets give sediments a green color, hence the term "greensand." Limestone and marls also may contain glauconite pellets. Celadonite fills vesicles and other voids in basaltic volcanics.

Use In the past, glauconite was used as a fertilizer to provide K for forage crops, but this use dwindled in the late 1800s with the development of potash from other sources. During the 1940s and 1950s, glauconite was reintroduced as a soil amendment, and it has, at times, been used in the manufacture of bricks and as a colorant to make green glass. Its principal current uses are in water treatment and as a soil amendment. In water treatment it is used to remove soluble iron and manganese salts, and hydrogen sulfide. For this purpose, glauconite pellets are extracted and processed to give them a manganese dioxide coating. As a soil amendment, glauconite releases potassium slowly to aid plant growth, loosens clayey soil, and improves water retention.

TOT + c STRUCTURES: BRITTLE MICAS (2:1)

MARGARITE

$CaAl_2(Al_2Si_2O_{10})(OH)_2$
Monoclinic 2/m
$a = 5.10$ Å, $b = 8.83$ Å,
  $c = 19.15$ Å, $\beta = 95.5°$,
  Z = 4
H = 3.5–4.5
G = 2.99–3.08
Biaxial (−)
$n_\alpha = 1.630–1.638$
$n_\beta = 1.642–1.648$
$n_\gamma = 1.644–1.650$
$\delta = 0.010–0.014$
$2V_x = 40–67°$

Composition Up to 30% Na^+ may substitute for Ca^{2+} with charge balance maintained principally by increasing the Si:Al ratio in tetrahedral sites.

Form and Twinning Margarite has a micaceous habit, forming platy or scaly aggregates and foliated masses. Intergrowths with muscovite or paragonite are not uncommon.

Cleavage Perfect {001}. Cleavage folia are brittle.

Color Properties

Luster: Vitreous to pearly.
Color: Grayish pink, pale yellow, or pale green.
Streak: White.

Color in Thin Section: Colorless.
Pleochroism: None.

Optical Characteristics

Relief in Thin Section: Moderately high positive.
Optic Orientation: $X \wedge c = +11°$ to $+13°$, $Y \wedge a = −6°$ to $−8°$, $Z = b$. Maximum extinction angle to the trace of cleavage is therefore 6–8° and the cleavage trace is length slow.
Indices of Refraction and Birefringence: Indices of refraction decrease with substitution of Na for Ca. Birefringence is relatively low, so maximum interference color in standard thin section is usually no higher than first-order yellow.
Interference Figure: Cleavage flakes and sections parallel to {001} yield nearly centered acute bisectrix figures with strong optic axis dispersion, r < v. 2V decreases with substitution of Na for Ca.

Alteration Margarite may alter to clay minerals.

Distinguishing Features

Hand Sample: Greater hardness and brittle character distinguish margarite from muscovite and other similar sheet silicates.
Thin Section/Grain Mount: Similar to muscovite but with higher indices of refraction and lower birefringence. Chlorite and chloritoid are green and pleochroic.

Occurrence Most commonly as a prograde metamorphic mineral in aluminous pelitic rocks and metamarls, and also in metamorphosed basalts and anorthosites, where the Ca and Al are derived from original plagioclase. Margarite also may pseudomorphically replace aluminous minerals such as andalusite and corundum in metamorphic pelites.

Use None.

CLINTONITE

$CaMg_2Al(SiAl_3O_{10})(OH)_2$
Monoclinic 2/m
$a = 5.204$ Å $b = 9.026$ Å,
  $c = 9.812$ Å,
  $\beta = 100.3°$, Z = 2
H = 3.5 on {001},
  6⊥{001}
G = 3.0–3.1
Biaxial (−)
$n_\alpha = 1.643–1.648$
$n_\beta = 1.655–1.662$
$n_\gamma = 1.655–1.663$
$\delta = 0.012 - 0.015$
$2V_x = 0–33°$

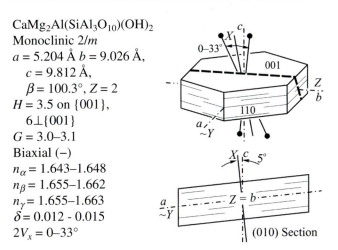
(010) Section

Composition Most clintonite has a composition fairly close to the formula given here; some shows a higher Si:Al ratio in the tetrahedral sites, with correspondingly less Al in octahedral sites. Some Fe may substitute for Mg, and F may substitute in substantial quantities for OH.

Form and Twinning Clintonite has a typical micaceous habit, forming irregular platy grains or platy aggregates. Twins with {001} composition planes are not common.

Cleavage Perfect {001} cleavage, but not as good as muscovite or biotite. Folia are brittle.

Color Properties

Luster: Vitreous, pearly to submetallic.

Color: Colorless, yellow, green, reddish brown.

Streak: White.

Color in Thin Section: Colorless or pale brown.

Pleochroism: Moderate with X = colorless, light orange, yellow, or reddish brown; $Y \approx Z$ = pale brown, pale green, $X < Y \approx Z$. Sections are darker when the trace of cleavage is parallel to the lower polarizer vibration direction.

Optical Characteristics

Relief in Thin Section: Moderately high positive.

Optic Orientation: $X \wedge c = +5°$, $Y \approx a$, $Z = b$, also $X \wedge c = +10°$, $Y = b$, $Z \approx a$.

Indices of Refraction and Birefringence: The range of indices of refraction is small, with Fe-bearing samples having higher values. Maximum interference color in thin section is first-order yellow.

Interference Figure: Cleavage fragments and sections parallel to {001} show no cleavage and low birefringence, and yield centered acute bisectrix figures with small $2V$. Optic axis dispersion is weak, $r < v$.

Alteration Clintonite may alter to clay minerals.

Distinguishing Features

Hand Sample: Resembles muscovite and phlogopite, but both are softer and cleavage folia are not brittle.

Thin Section/Grain Mount: Has lower birefringence than either muscovite or phlogopite, and chloritoid has larger $2V$ and is usually optically positive and more distinctly green. Chlorite is more distinctly green, has lower relief, lower birefringence, and may be optically positive. Margarite has a larger $2V$.

Occurrence Clintonite is found in chlorite or talc schist and in metamorphosed carbonate-rich rocks.

Use None.

TOT + O STRUCTURE

CHLORITE

$(Mg,Fe,Al)_3 (Si,Al)_4 O_{10} (OH)_2 \cdot$
$\quad (Mg,Fe,Al)_3 (OH)_6$
Monoclinic 2/*m* (also triclinic)
a = 5.3 Å, b = 9.2 Å,
$\quad c$ = 14.3 Å, β = 97°, Z = 2
H = 2–3
G = 2.6–3.3
Biaxial (+ or −)
n_α = 1.55–1.67
n_β = 1.55–1.69
n_γ = 1.55–1.69
δ = 0.00–0.015
$2V_x$ = 0–40°
$2V_z$ = 0–60°

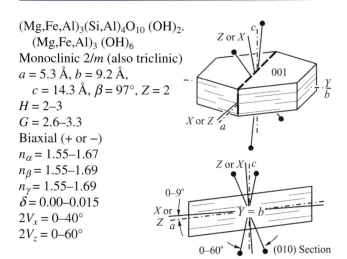

Structure Chlorite has the *TOT + O* structure described earlier. In common chlorite, both octahedral sheets are trioctahedral, meaning that all octahedral sites are occupied, usually by Mg, Fe, or Al. However, in some varieties either or both of the octahedral sheets may be dioctahedral. These uncommon varieties are not considered here.

Composition The principal substitutions are of Mg, Fe^{2+}, Fe^{3+}, and Al in octahedral sites, and Si and Al in tetrahedral sites as described earlier. Continuous solid solution extends from Mg-rich chlorite (clinochlore) to Fe-rich chlorite (chamosite). Substitution of minor amounts of Mn and Cr in octahedral sites also is common.

Form and Twinning Figure 13.13. Most commonly it occurs as fine-grained scaly or foliated massive aggregates, or isolated micaceous grains. Well-formed platy crystals with hexagonal outline are uncommon. Chlorite also may partially or completely replace other Mg/Fe-bearing minerals. Chlorite is a common constituent of the clay fraction of soil and clastic sediments, and may form oolitic or spherulitic balls similar to glauconite in some sedimentary rocks. Twins with {001} composition planes are not uncommon but are difficult to recognize.

Cleavage Perfect {001}. Folia are flexible but inelastic.

Color Properties

Luster: Vitreous to somewhat pearly, waxy, or dull.

Color: Various shades of green, infrequently yellow, white, pink, or rose-red.

Streak: Greenish white to white.

Color in Thin Section: Usually very pale to medium green.

Pleochroism: Moderate in shades of green. For optically positive varieties $X \approx Y > Z$: $X \approx Y$ = pale green, green,

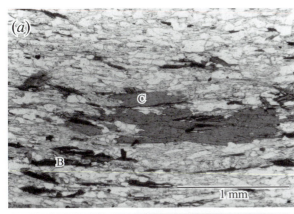

Figure 13.13 Chlorite. (*a*) Foliated chlorite grains (C) with biotite (B) in a thin section of moderately low-grade pelitic schist. The finer-grained matrix includes scaly chlorite and muscovite, and clear quartz. (*b*) Chlorite (C) alteration along cleavage traces in biotite (B) in a thin section of granodiorite.

brownish green; Z = colorless, pale green, pale yellowish green. For optically negative varieties X < Y ≈ Z: X = colorless, pale green, pale yellowish green; Y ≈ Z = pale green, green, brownish green.

Optical Characteristics

Relief in Thin Section: Moderate to moderately high positive.

Optic Orientation: The optic plane is normally parallel to (010) and the acute bisectrix is nearly perpendicular to (001). For optically positive varieties X ≈ a, Y = b, Z ≈ c, and the trace of cleavage is length fast. For optically negative varieties X ≈ c, Y = b, Z ≈ a, and the trace of cleavage is length slow.

Indices of Refraction and Birefringence: Indices of refraction increase with Fe + Mn + Cr content (Figure 13.14). Birefringence is usually low, so interference colors are rarely higher than first-order white or yellow in standard thin section. Anomalous brownish interference colors are common in optically positive varieties, and anomalous bluish or purplish interference colors are common in optically negative varieties.

Interference Figure: Cleavage flakes and sections parallel to {001} yield centered acute bisectrix figures. Optic sign can be used to distinguish between low and high

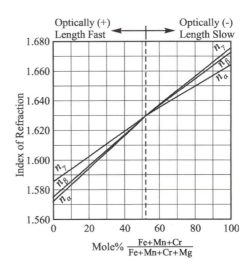

Figure 13.14 Approximate variation of chlorite optical properties as a function of composition. After Albee (1962).

Fe + Mn + Cr, but there are many exceptions. Some may be essentially uniaxial. Optic axis dispersion is usually $r < v$. As noted earlier, the sign of elongation may be used to establish optic sign.

Alteration Oxidation may produce iron stains, and chlorite may weather to clay minerals, but otherwise it is fairly resistant to alteration.

Distinguishing Features

Hand Sample: Green color, inelastic folia, and hardness distinguish chlorite from other sheet silicates.

Thin Section/Grain Mount: Distinguished from muscovite and biotite by green color, pleochroism, and weak birefringence. Serpentine usually has lower refractive indices and is less pleochroic. In soils and sediments, chlorite is indistinguishable from other clay minerals and must be identified with X-ray diffraction techniques.

Occurrence Chlorite is a common and widespread mineral. It is a major rock-forming mineral in low- and medium-grade pelitic and mafic metamorphic rocks; the green in greenschist facies rocks comes from chlorite. It is also common in many igneous rocks, usually derived by alteration of primary Mg/Fe-bearing minerals such as biotite and hornblende. Chlorite also is a common product of weathering and is widespread in soil, clay-bearing sediments, and sedimentary rocks.

Use By itself, chlorite does not have significant industrial uses. Rocks such as chlorite schist may occasionally be used as construction stone or for decorative purposes. However, chlorite is a common constituent of clay and therefore appears in a wide variety of industrial and consumer applications described in the following section on clay.

CLAY MINERALS

The term *clay* has been used in different ways. Originally the term was applied to the fine-grained fraction of sediment and sedimentary rocks, variously defined as particles whose dimension is less than 1/256 mm (~0.004 mm) or less than 0.002 mm. The size was chosen because particles this small or smaller cannot be identified with conventional optical or physical methods. Sediment made dominantly of these fine particles was conventionally called clay, so the constituent particles were referred to as clay-sized or just clay. Once X-ray diffraction techniques were applied, it was discovered that most clay-sized sediment is composed of a collection of sheet silicate minerals, so the term *clay mineral* has come to refer to fine-grained (<0.002 mm) sheet silicate minerals. The terminology used here is as follows.

CLAY: Sediment or rock (claystone) composed dominantly of particles less than 0.002 mm

CLAY-SIZED: Particles whose dimension is less than 0.002 mm.

CLAY MINERAL: Sheet silicate minerals that occur in the clay-sized fraction of soils, sediments, sedimentary rocks, and weathered or altered rocks.

ARGILLACEOUS: Rock or sediment containing significant amounts of clay minerals:

This leads to two potential sources of confusion. The first is that clay-sized sediment includes not only clay minerals but a variety of others, most commonly quartz, but also carbonates, zeolites, iron oxides, and others. The second is that the definition of clay mineral contains a size requirement, which is a nonmineralogical criterion. Several minerals, such as chlorite, talc, pyrophyllite, and serpentine, occur both as clay minerals and as more coarse-grained constituents of igneous and metamorphic rocks.

The following minerals included in this section occur as clay minerals.

KAOLINITE: $Al_2SiO_2O_5(OH)_4$

SMECTITE: $\sim Ca_{0.17}(Al, Mg, Fe)_2 (Si, Al)_4O_{10}(OH)_2 \cdot nH_2O$

ILLITE: $\sim K_{0.8}Al_2(Al_{0.8}Si_{3.2})(OH)_2$

VERMICULITE: $\sim(Mg, Ca)_{0.3} (Mg, Fe^{2+}, Fe^{3+}, Al)_3 (Si, Al)_4O_{10} (OH)_2$

CHLORITE: $(Mg, Fe, Al)_3 (Si, Al)_4O_{10}(OH)_2, (Mg, Fe, Al)_3 (OH)_6$

Physical and optical properties are listed in Table 13.2.

Form and Twinning Generally too fine grained to allow recognition of grains in hand sample or with light microscopes. Scanning electron microscopes reveal that most clay particles are platy, though some are lath-shaped or fibrous.

Cleavage Perfect {001}.

Color Properties

Luster: Earthy.
Color: White, gray, tan, brown.
Streak: White.
Color in Thin Section: Typically very-fine-grained gray or brown earthy aggregates.
Pleochroism: None.

Optical Characteristics (Table 13.2)

Relief in Thin Section: Low to moderate positive or negative.

Comments: Clay minerals are too fine grained to allow optical properties to be measured. Coarse-grained samples of essentially the same material include chlorite,

	Kaolinite	Illite	Smectite	Vermiculite	Chlorite
a (Å)	5.15	5.2	5.2	5.3	5.3
b (Å)	8.95	9.0	9.1	9.25	9.2
c (Å)	7.4	10.0	~10–15	14.4	14.3
β (°)	104.8	~90	~90	~97	97
Z	2	2	2	2	2
Point group	1	2/m	2/m	2/m	2/m

Table 13.2 Properties of Clay Minerals

Mineral	H	G	Class and Sign	$2V_x°$	n_α	n_β	n_γ	δ
Kaolinite	2–2.5	2.61–2.68	Biaxial (−)	24–50	1.553–1.565	1.559–1.596	1.560–1.570	0.005–0.007
Smectite	1–2	~2–3	Biaxial (−)	Small	1.48–1.61	1.50–1.64	1.50–1.64	0.01–0.04
Illite	1–2	2.6–2.9	Biaxial (−)	< 10	1.54–1.57	1.57–1.61	1.57–1.61	~0.03
Vermiculite	1–2	~2.3	Biaxial (−)	0–18	1.520–1.564	1.530–1.583	1.530–1.583	0.02–0.03
Chlorite	2–3	2.6–3.3	Biaxial (−) or (+)	$2V_x = 0$–40, $2V_z = 0$–60	1.55–1.67	1.55–1.69	1.55–1.69	0.000–0.015

serpentine, talc, pyrophyllite, and glauconite, which were described separately earlier, and vermiculite, which is briefly described in the section on biotite.

Structure and Classification

The clay minerals (Figure 13.15) are layer silicates with structures described in Figure 13.3. The commonly used classification divides clays into 1:1 clays constructed of repeating *TO* layers usually with interlayer spacing of about 7 Å, and 2:1 clays constructed of repeating *TOT* layers with interlayer spacings of either about 10 or 14 Å. An additional group, called mixed-layer clays, may include combinations of any of these components interlayered in a variety of ways. Interlayer spacing is measured by means of X-ray diffraction techniques, which are outlined shortly.

1:1 Clay Minerals

The structure of 1:1 clays consists of repeating *TO* layers bonded with weak electrostatic bonds (Figure 13.15a). This group includes dioctahedral kaolinite $Al_2Si_2O_5(OH)_4$ (Figure 13.3) and trioctahedral serpentine, described earlier. Dickite and nacrite are polymorphs of kaolinite related by different repeating patterns for stacking the layers. The interlayer spacing of the kaolinite group is typically about 7 Å, although halloysite $[Al_2Si_2O_5(OH)_4 \cdot 2H_2O]$ has an ~10 Å interlayer spacing because a single layer of water molecules occupies the interlayer position. Halloysite layers have a strong tendency to roll up into cylinders and typically adopt a fibrous habit. Interlayer water is readily lost on heating.

2:1 Clay Minerals

The 2:1 clays are constructed of repeating *TOT* layers. The *TOT* layers have a net negative charge that is less than one per formula unit, which requires that cations be placed between layers (the interlayer) to maintain charge balance. Two groups of 2:1 clays, with interlayer spacing of about 10 and 14 Å, are distinguished by the nature of the interlayer and correspond to the *TOT* + *c* and *TOT* + *O* structures described earlier.

In the 10 Å clays (Figure 13.15b), *TOT* layer charge is balanced with interlayer cations, usually K, Na, or Ca; they have a *TOT* + *c* structure and are similar to the micas. The 10 Å clays can be separated into low-charge and high-charge groups depending on the net charge of the *TOT* layers and the substitution mechanism that produces the charge.

The net negative charge of the *TOT* layers in the low-charge group is usually between about 0.2 and 0.6 per formula unit and commonly around 0.33. Ca and Na are the usual interlayer cations. The most abundant group of these minerals, collectively known as **smectites**, may be either dioctahedral or trioctahedral. In dioctahedral smectites, the TOT-layer charge is the result of either substitution of Al^{3+} for Si^{4+} in tetrahedral sites or substitution of divalent

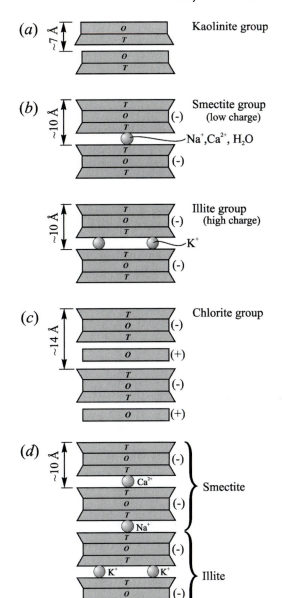

Figure 13.15 Clay mineral structure. Refer also to Figures 13.2 and 13.3. (*a*) 7 Å clays have 1:1 or *TO* structures. (*b*) 10 Å clays have 2:1 or *TOT* + *c* structures. The *TOT* layers in the smectite group have a relatively low charge and require fewer Na^+ or Ca^{2+} cations in interlayer sites. In the illite group, *TOT* layers have a higher negative charge and require more cations in interlayer sites to maintain charge balance. (*c*) 14 Å clays have the chlorite structure in which octahedral sheets with a positive charge occupy the interlayer between negatively charged *TOT* layers. (*d*) A common mixed-layer clay contains both smectite and illite layers.

cations (usually Mg^{2+}) for Al^{3+} in octahedral sites. In the less common trioctahedral varieties, the *TOT*-layer charge is the result of substitution of Al^{3+} for Si^{4+} in tetrahedral sites. Relatively low layer charge of smectites coupled with the presence of cations in only about a third of the interlayer sites allows water to easily move into the interlayers, causing the structure to expand. Smectites with a single layer of water molecules between *TOT* layers have an interlayer spacing of about 12.5 Å, and with two layers of water molecules, about 15.2 Å. The water can be introduced and removed reversibly from smectite at room temperature in response to changing moisture content in the soil or rock. Only when the smectite is thoroughly dried is the interlayer spacing approximately 10 Å.

In the high-charge group, the *TOT* layers have a net negative charge of between 0.8 and 1.0 per formula unit, mostly as a result of substitution of Al^{3+} for Si^{4+} in tetrahedral sites. The clay minerals in this group include illite and glauconite–celadonite. All are dioctahedral, and the interlayer cation is dominantly K^+. Because structure and composition are nearly the same as those of muscovite, these clays have been dubbed "mica-like." Collectively they are known as the **illite group**. The presence of between 0.8 and 1.0 K^+ cation per formula unit in interlayer sites ensures that the layers are bonded to each other fairly strongly. Water molecules do not easily enter the interlayer because most interlayer cation sites are occupied and interlayer bonding is stronger, so members of the illite group do not swell when moistened.

A third group of 10 Å clays is called **vermiculite**. It is usually produced by alteration of biotite, or less commonly chlorite. Vermiculite is an expanding clay, like the smectite group, but has a higher negative layer charge, usually around 0.6 per formula unit. The changes that accompany alteration of biotite to vermiculite are oxidation of some Fe^{2+} to Fe^{3+} in octahedral sites, which decreases the net negative charge of the *TOT* layer from 1.0 to about 0.6 per formula unit. To compensate for the reduction of negative charge in the *TOT* layer, some K^+ cations are stripped from interlayer sites. Additional K^+ is exchanged for Mg^{2+} or Ca^{2+} and water is introduced into the interlayer, often structurally bonded to the interlayer cations. The water in the interlayer expands the structure, so that at room conditions the interlayer spacing is about 14.4 Å. Only when heated to drive off the water does the structure collapse to about 10 Å interlayer spacing.

The 14 Å clays (Figure 13.15c) have the chlorite structure (*TOT* + *O*) and are called chlorite. The interlayer is an octahedral sheet (a plane of cations bonded in octahedral coordination with two planes of OH) with a net positive charge. The chlorite that occurs in clays may be either tri- or dioctahedral, unlike most coarsely crystalline chlorite, which is trioctahedral.

Mixed-Layer Clay Minerals

Natural clays rarely correspond fully to the end members just described. Most contain, to one degree or another, combinations of kaolinite-like, illite-like, smectite-like, vermiculite-like, and/or chlorite-like layers. In some cases the different layer types alternate in a regular manner; in others, they appear random (Figure 13.15d). The reader should not infer that mixed-layer clays are mechanical

Table 13.3 Properties of Palygorskite and Sepiolite

Palygorskite	Sepiolite
Earthy to waxy: White, gray, yellowish, gray-green; colorless in thin section	Earthy to waxy: Gray, white, yellowish; colorless or gray in thin section
$H = 2$–2.5	$H = 2$–2.5
$G = 1.0$–2.6	$G > 2$
$n_\alpha = 1.522$–1.528	$n_\alpha = 1.515$–1.520
$n_\beta = 1.530$–1.546	$n_\beta = $ nd
$n_\gamma = 1.533$–1.548	$n_\gamma = 1.525$–1.529
$\delta = 0.011$–0.020	$\delta = 0.009$–0.010
Biaxial (−), $2V_x = 30$–$61°$	Biaxial (−) $2V_x = 0$–$50°$

mixtures of two or more types of clay. The interlayering occurs at the unit cell level. Mixed-layer clays are classified by combining the names of the constituent layers. Illite/smectite contains layers of both illite and smectite, for example, and is probably the most common variety. Chlorite/smectite also appears to be common.

Additional Clay Minerals

Palygorskite [$(Mg,Al)_2Si_4O_{10}(OH) \cdot 4H_2O$], and sepiolite [$(Mg_4Si_6O_{15}(OH_2) \cdot 6H_2O$] (Table 13.3) are additional clay minerals with complex structures that form fibrous grains. Palygorskite may form fibrous mats known as mountain leather, and sepiolite may form fine-grained masses, known as meerschaum, that can be carved. They are formed by weathering and alteration of Mg-bearing rocks, and are common constituents of soil and sediment.

Geology of Clay

Clays form as the result of the interaction of silicate rock and water under conditions of relatively low temperature.

Rock + water + ions in solution
= clay ± other minerals + ions in solution

The hydrolysis of K-feldspar is a simple example of this process.

K-feldspar + hydrogen ion + water
= kaolinite + potassium ion + quartz
$$2KAlSi_3O_8 + 2H^+ + H_2O$$
$$= Al_2Si_2O_5(OH)_4 + 2K^+ + 4SiO_2$$

The environments in which clays are formed are all near the Earth's surface and include weathering and soil formation, sediment transport and deposition, sediment compaction and diagenesis, and hydrothermal alteration. Clay is therefore abundant in weathered rock, soil, fine-grained clastic sediment and sedimentary rocks, and hydrothermally altered rocks. The clay minerals that form in any particular environment depend on the nature of the parent rock, temperature, availability and chemistry of the water, and

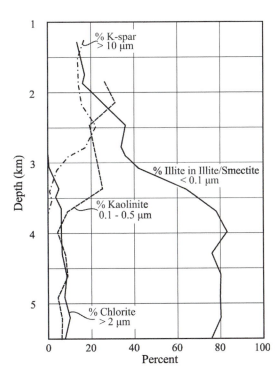

Figure 13.16 Clay mineralogy in selected size fractions of Miocene–Oligocene sediment in the Gulf Coast of the United States. See text for additional discussion. Adapted from Hower and others (1976).

time. For a summary of these considerations, consult Velde (1995) or Weaver (1989). Above ~200°C (400°C in some hydrothermal conditions), clay minerals are converted by metamorphic processes into coarser-grained chlorite, micas, and other minerals.

Once formed, clays change or recrystallize in response to changing environmental conditions. A common example is the conversion of smectite into illite during diagenesis. The clay produced by normal weathering of common silicate rocks and subsequent transport tends to be fairly rich in smectite. Hence, most clay-rich sediment, when first deposited, is rich in smectite and has relatively little illite. With burial, however, the amount of illite layers in the clay structure systematically increases (Figure 13.16) as the smectite recrystallizes. This pattern is found in many sedimentary basins. The temperature within basins at which illite layers begin to appear ranges from about 50 to 100°C and the conversion progresses over a temperature interval of 35 to 100°C. Note that conversion of smectite to illite is possible only if a source of K is available. In many cases, fine-grained detrital K-feldspar in the sediment is consumed to provide the necessary K, and the extent of conversion of smectite to illite is limited by the availability of this K-feldspar. Other changes that occur with increasing depth include an increase in the amount of chlorite and a decrease in the amount of kaolinite.

Soil mineralogy also is influenced by climate. Intense weathering in wet climates tends to favor formation of kaolinite plus oxide minerals such as gibbsite, whereas dry temperate climates favor smectite. With progressive weathering, smectite may convert to illite/smectite and then to illite.

Bacteria are capable of playing a role in the production of clay minerals. It appears to begin with precipitation of iron oxide minerals on the cellular surface, which serve as kinetically favorable sites for precipitation of material of variable clay composition, morphology, and structure. Because bacteria are ubiquitous in weathering, transport, and diagenetic environments, their role in facilitating the mineralogic processes in those environments deserves study.

Identification

Because most clay minerals occur in submicroscopic grains, identification is usually not possible with conventional hand sample or microscopic techniques, and the physical and optical properties just listed are not particularly useful. Plasticity of wet soil or sediment is an indication that clay is present, as is the earthy odor that we associate with moist soil. Lightly breathing on an argillaceous sample to add a bit of warm moisture often releases a clay odor.

In most cases, identification depends on X-ray diffraction techniques. The basic outline of the procedure is sketched next. For more details, consult Moore and Reynolds (1997) or Brindley (1980).

1. Sample preparation.

DISAGGREGATE THE ROCK/SOIL: A variety of equipment and techniques can be used depending on the nature of the sample and the equipment available in the laboratory. Samples should be crushed, not ground.

SEPARATE THE CLAY FROM OTHER MINERALS: A common technique is to thoroughly mix the disaggregated sample with water in a kitchen-type blender, and then decant the supernate containing the clay.

CHEMICAL PRETREATMENT: Iron oxides and organic matter can be removed by chemical treatment, if needed, and clays can be saturated with selected cations (e.g., Mg) by soaking in an appropriate solution.

PARTICLE-SIZE SEPARATION: Usually accomplished by centrifuging the clay–water mixture. The size fraction remaining in the water can be controlled by varying the centrifuge run time. The supernate is the material used for X-ray analysis.

SLIDE PREPARATION: Much routine work involves using a sample in which (ideally) all clay grains are oriented with {001} parallel to the surface of the slide. The simplest oriented samples are prepared by placing the clay–water supernate on a glass slide and allowing it to dry.

GLYCOLATING: Ethylene glycol (used in automobile antifreeze) readily occupies the interlayer sites in smectites and expands the interlayer spacing to about 16.9 Å. Samples to be glycolated are placed in a vapor of ethylene glycol at 60°C for at least 8 hours.

2. Sample analysis. In most cases, samples are analyzed in an X-ray diffractometer in three conditions: air-dried, solvated with ethylene glycol, and after heating to collapse expandable layers. The d-spacing of the $\{00l\}$ and other interplanar spacings are measured in each of the three conditions (Table 13.4), and the identity of the mineral(s) in the sample can, with luck and experience, be deduced from that information. Mixed-layer clays pose additional complexity that is beyond the scope of this presentation.

Uses

Clay has been used since before recorded history and continues to be used in a wide variety of industrial and commercial products. Uses include the following.

PAPER COATING AND FILLER: The paper on which this is printed contains kaolinite.

DRILLING MUD ADDITIVE: Used for drilling oil/gas wells.

CERAMICS: Probably the earliest use of clay, and an enduring one.

FILLER: Used in rubber, plastics, caulking compound, paint, and related products.

COSMETICS: Both as a filler and colorant.

REFRACTORY PRODUCTS: Including firebrick and foundry sand.

BUILDING PRODUCTS: Including bricks and adobe.

PORTLAND CEMENT: Clay in pulverized shale is an essential component.

ABSORBENTS: Used to absorb oil and other chemicals in industrial settings and as cat litter.

FOOD: Certain types of clay are approved food additives.

PHARMACEUTICALS: Clays have been used for medical purposes since ancient times. Many uses continue today, and new applications are still being found. Williams and others (2009) provide a useful review.

Clay in the Environment

Geophagy is the practice of deliberately eating Earth materials. In modern society, geophagy may be considered to be of aberrant behavior, and it can certainly be a manifestation of mental illness in some cases. However, geophagy has been a common practice since ancient times and continues today in many parts of the world, including the southeastern United States. A substantial variety of minerals are involved in geophagy (Limpitlaw, 2010), but the clay minerals are perhaps the most commonly eaten. The clay may be powdered, or formed into tablets, disks, or other shapes. An Internet search will yield numerous sources for edible clays. An element of hucksterism is apparent in many of these sources, but clay is also routinely sold in grocery stores, pharmacies, and markets in many parts of the world. The adverse health consequences of eating clay can include blockage of the lower intestine, nutrient deficiency, nutrient excesses, damage to the teeth and gums, poisoning, and parasitic invasions. However, geophagy also can be beneficial, and the clay may serve as a source of nutrients (e.g., K, Na, Fe,), calm upset stomachs, and control diarrhea. Unraveling the beneficial from the adverse consequences of eating Earth materials will require the cooperation of medical personnel and geoscientists.

Pelotherapy is the application of muds (peloids) to the skin to treat rheumatic disorders, osteoarthritis, infections, sciatica, skin disorders, and a variety of other conditions. A wide range of claims are made regarding the efficacy of these treatments, but the fact that people pay large sums of money for the pelotherapy in health and beauty spas in many parts of the world suggests that there is a strong perception of a benefit. Clay is also used in a number of folk remedies. Readers may be familiar with the practice of applying a daub of mud or clay to a bee sting to (presumably) absorb some of the venom. No claims are made here about whether it works, and the author is not inclined to conduct a test.

Table 13.4 Clay Mineral Identification[a]

Clay Mineral Group	Air-Dried	Heated	Glycolated
Kaolinite group	7.16	7.16	7.16
Smectite group	~15.0 (variable)	~10	16.9
Illite group	10.1	10.1	10.1
Vermiculite	14.4	~10	14.4
Chlorite	14.2	14.2	14.2

[a] The values are the d-spacing in angstroms for $\{00l\}$ planes measured with X-ray diffraction techniques.

It has recently been recognized that some clays have antibacterial properties. Certain clays mixed with water are capable of killing bacteria by what appears to be a chemical exchange, either by providing something toxic to the bacteria or by depriving the bacteria of nutrients essential for metabolism, although details of the process are poorly understood. All the clays that display antibacterial properties contain smectite as a major constituent and consist of very small mineral particles.

Conventional applications of clay minerals in medications include use both as an inert filler and as active ingredients. Kaopectate® is is a well-known antidiarrhea medication that originally was a mixture of kaolinite and pectin. The kaolinite was replaced with atapulgite (palygorskite) in the 1980s, and the product is now formulated with bismuth subsalicylate. Kaolinite-based preparations are still available in some parts of the world and are routinely used in veterinary medicine.

Clays are minerals with major environmental significance. They provide enormous benefits but also pose significant risks.

Perhaps the most obvious benefit is that clay minerals are a major constituent of the soils. Plants have evolved to take advantage of the chemical behavior of clay minerals. These include the ability to adsorb/absorb and release various molecules, including water and nitrogen compounds, that plants can use. A good case can be made that clay minerals provide the interface between the solid Earth and all life.

Clay-rich rocks and soil also pose substantial engineering problems. Soils rich in smectite are capable of shrinking when dried and swelling when wet. These swelling soils can do serious damage to the foundations of homes and other buildings, and to roadways and related improvements. The clays shrink during dry periods and swell during wet periods. Some of the most serious problems are posed by sediment containing volcanic ash that is deposited in marine conditions. This volcanic ash, which may alter to a clay rich in smectite, can have a very high potential for shrinking and swelling. Relict-altered ash beds are referred to as bentonite beds and can be a particular problem. The term *bentonite* has been applied to rocks containing abundant swelling clays. Clay-rich rock and soil also tends to be more susceptible to slope stability problems.

Clays have been used extensively in the environment protection and remediation industry. A typical use is the installation of a layer of clay a meter or so thick beneath a new waste dump or containment pond to prevent migration of contaminated water into the groundwater supply. Similar clay caps may be installed over existing dumps to prevent infiltration of water into the refuse, where the water could become contaminated. Clays may also be used to absorb contaminants from industrial effluent or from other contaminated water supplies.

OTHER SHEET SILICATES

STILPNOMELANE

$K_{0.6}$ (Fe, Mg)$_6$Si$_8$Al (O, OH)$_{27}$·2–4H$_2$O
Triclinic $\bar{1}$
a = 21.8–22.1 Å, b = 21.8–22.1 Å, c = 17.6–17.7 Å, α = ~125°, β = ~96°, γ = 120°, Z = 8
H = 3–4
G = 2.59–2.96
Biaxial (–)
n_α = 1.543–1.634
n_β = 1.576–1.745
n_γ = 1.576–1.745
δ = 0.030–0.110
$2V_x$ = ~0°

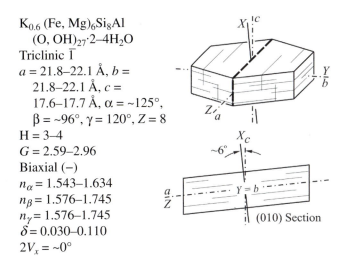

Structure In the common sheet silicates, the apical oxygen anions in the tetrahedral T sheets point inward toward, and are shared with, a continuous octahedral O sheet. In stilpnomelane, groups of the tetrahedra within the tetrahedral sheet point away from the octahedral sheet at the center of the TOT layer of which they are a part (Figure 13.17). These outward-pointing apical oxygen anions are shared with similar outward-pointing tetrahedra on adjacent TOT layers and provide relatively strong bonding between layers.

Composition Stilpnomelane displays substantial compositional variation, principally in the relative amount of Fe^{2+}, Fe^{3+}, and Mg in octahedral sites. Substitution of Fe^{3+} is coupled with replacing OH$^-$ with O^{2-} to maintain charge balance.

Form and Twinning Figure 13.18. It is micaceous, like most sheet silicates. Grains may be arranged in foliated, sheaf-like, or radiating aggregates, or interlayered with chlorite or other micaceous minerals. No twinning is reported.

Cleavage An excellent cleavage on {001} is not as good as the micas. Fair cleavage on {010} intersects {001} at 90°.

Color Properties

Luster: Vitreous to pearly on cleavage surfaces; may be submetallic.

Color: Golden brown, reddish brown, dark green, black.

Streak: White to tan.

Color in Thin Section: Pale yellow, brown, or green.

Pleochroism: Distinct with $Z \approx Y > X$: X = pale yellow to golden yellow; $Y \approx Z$ = brown, reddish brown, greenish brown, black. Sections are darker when the trace of cleavage is parallel to the vibration direction of the lower polarizer.

Figure 13.17 Idealized structure of stilpnomelane. Some tetrahedra point outward from the *TOT* layers and share oxygen anions with similar tetrahedra from adjacent *TOT* layers. Adapted from Guggenheim and Eggleton (1987).

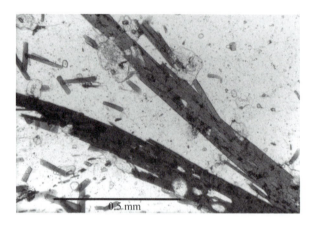

Figure 13.18 Stilpnomelane grains in a stilpnomelane schist.

Optical Characteristics

Relief in Thin Section: High positive, may change with rotation.

Optic Orientation: $X \wedge c = \sim +6°, Y = b, Z \wedge a = \sim 0°$, the trace of cleavage is length slow.

Indices of Refraction and Birefringence: Indices of refraction and birefringence increase with Fe^{3+} content.

Interference Figure: Stilpnomelane is effectively uniaxial negative. Cleavage fragments yield nearly centered optic axis figures.

Alteration Stilpnomelane may alter to chlorite, iron oxides, and clay.

Distinguishing Features

Hand Sample: Similar to biotite, but the basal cleavage is less perfect and stilpnomelane has a second cleavage at 90°.

Thin Section/Grain Mount: Resembles biotite, but biotite has $2V$ up to 25° and lacks the second cleavage. Also, stilpnomelane lacks the distinctive mottling or "birds-eye" appearance seen near extinction in biotite. Birefringence of Fe^{3+}-rich stilpnomelane is higher than that of biotite.

Occurrence Fairly common in low-grade regional metamorphic rocks (greenschists) along with chlorite, albite, and muscovite. Also in blueschist regional metamorphic rocks associated with glaucophane, calcite, epidote-group minerals, and pumpellyite. Stilpnomelane also may be found in metamorphosed iron formations.

Use None.

PREHNITE

$Ca_2Al(AlSi_3O_{10})(OH)_2$
Orthorhombic *mm*2
$a = 4.646$ Å, $b = 5.491$ Å,
　$c = 18.52$ Å, $Z = 2$
$H = 6–6.5$
$G = 2.80–2.95$
Biaxial (+)
$n_\alpha = 1.610–1.637$
$n_\beta = 1.615–1.647$
$n_\gamma = 1.632–1.670$
$\delta = 0.020–0.035$
$2V_z = 64–70°$

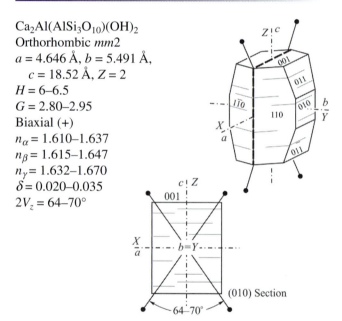

Structure The structure consists of complex sheets of Si^{4+} and Al^{3+} tetrahedra parallel to {001}. Some tetrahedra share all four oxygen ions, and some share only two. Ca^{2+} in 7-fold coordination and Al^{3+} in 6-fold coordination with oxygen from the tetrahedra and additional OH^- between the layers serve to bind adjacent sheets together.

Composition Some Fe^{3+} may substitute for Al^{3+}; otherwise the composition is usually fairly close to the ideal formula.

Form and Twinning Commonly occurs as lamellar or columnar masses and also as radiating, globular, fanlike or sheaflike aggregates. A pattern resembling a bow tie is common in which adjacent grains are joined on surfaces nearly parallel to {001}. Crystals are uncommon and are either columnar parallel to *c* or tabular parallel to {001}. Fine lamellar twinning is reported, but not common.

Cleavage Good {001} cleavage and poor cleavage parallel to {110} faces. The basal cleavage tends to control fragment orientation. Uneven fracture. Brittle.

Color Properties

Luster: Vitreous, somewhat pearly on {001}.
Color: Light to dark green, white, yellow, gray, pink; may slowly fade on exposure to light.
Streak: White.
Color in Thin Section: Colorless.
Pleochroism: None.

Optical Characteristics

Relief in Thin Section: Moderately high positive.
Optic Orientation: $X = a$, $Y = b$, $Z = c$.
Indices of Refraction and Birefringence: Indices of refraction and birefringence increase with Fe^{3+} content. Low-Fe samples are most common and have values near the lower end of the range given above. Interference colors may be anomalous.
Interference Figure: Biaxial (+) with a fairly large $2V_z$ (64–70°); the optic axis dispersion is weak, $r > v$.

Alteration Prehnite may alter to chlorite or zeolite minerals.

Distinguishing Features

Hand Sample: The green color and single good cleavage aid identification. The sheaflike or globular/radiating aggregates are fairly distinctive.
Thin Section/Grain Mount: Lawsonite and pumpellyite have higher indices of refraction and lower birefringence. Epidote, with birefringence similar to that of prehnite, typically has a greenish color in thin section and has higher indices of refraction. The bow-tie structure, if present, is distinctive.

Occurrence Prehnite is found in amygdules, veins, or other cavities in mafic to intermediate igneous rocks, often in association with zeolites and calcite. Contact-metamorphosed impure limestones and marls may also contain prehnite associated with calcite, wollastonite, garnet, tremolite, or other calc-silicate minerals. The most abundant occurrence, however, is as a constituent in low-grade regional metamorphic rocks derived from graywacke, basalt, or other mafic to intermediate rocks. Associated minerals include lawsonite, pumpellyite, albite, epidote-group minerals, chlorite, and zeolites. Small masses of prehnite have been found along the cleavage in biotite from granitic rocks.

Use Prehnite has been used as an ornamental material or semiprecious gem.

APOPHYLLITE

$KCa_4Si_8O_{20}(F,OH) \cdot 8H_2O$
Tetragonal (4/m 2/m 2/m)
$a = 8.963$ Å, $c = 15.804$ Å, $Z = 2$
$H = 4.5–5$
$G = 2.33–2.37$
Uniaxial (+) or (−)
$n_\omega = 1.531–1.542$
$n_\epsilon = 1.533–1.543$
$\delta = 0.000–0.003$

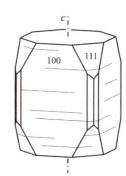

Structure The structure consists of 4-fold rings of Si tetrahedra linked together to form kinked sheets parallel to (001). The K, Ca, F, OH, and H_2O are located between these sheets.

Composition Compositions extend from **hydroxyapophyllite** to **fluorapophyllite**, which contain all OH and F, respectively. Most apophyllite is relatively fluorine rich. **Natroapophyllite** contains Na instead of K and is orthorhombic.

Form and Twinning Crystals are square in cross section and may resemble cubes modified by octahedrons; also platy parallel to (001) or granular massive. Rare twins on {111}.

Cleavage Perfect cleavage on {001} and imperfect on {110}.

Color Properties

Luster: Vitreous, pearly on {001} cleavage surfaces.
Color: Colorless, white, gray or a pale shade of pink, green, or yellow.
Streak: White.
Color in Thin Section: Colorless.
Pleochroism: None.

Optical Characteristics

Relief in Thin Section: Low negative or positive.
Optic Orientation: Apophyllite is uniaxial, so the trace of the {001} cleavage always shows parallel extinction and is length fast for optically positive samples and length slow for optically negative samples.
Indices of Refraction and Birefringence: Indices of refraction decrease with increasing substitution of F for OH. Birefringence is low, and interference colors are often anomalous.
Interference Figure: Apophyllite is usually uniaxial negative, less commonly positive. Biaxial varieties with $2V$ up to 60° have been reported.

Alteration Apophyllite may alter to clay minerals, calcite, opal, or quartz.

Distinguishing Features

Hand Sample: Distinguished from associated zeolites by crystal habit, basal cleavage, and pearly luster on cleavage surfaces.

Thin Section: Apophyllite has higher indices of refraction than the zeolites, with which it may be associated, and has a perfect basal cleavage. Interference colors are commonly anomalous. Natroapophyllite is biaxial positive with $2V_z = 32°$, $n_\alpha = 1.536$, $n_\beta = 1.538$, and $n_\gamma = 1.544$, $\delta = 0.008$.

Occurrence Apophyllite most commonly occurs in amygdules, cavities, and veins in basaltic rocks in association with zeolites, prehnite, pectolite, and calcite. It is also found in the contact aureole around granitic intrusions and in metamorphosed limestone or other calc-silicate rock. It may form as an alteration of wollastonite.

Use None.

CHRYSOCOLLA

$(Cu,Al)_2H_2Si_2O_5 \cdot nH_2O$
Orthorhombic (?)
$a = 5.72$ Å to 5.92 Å,
$\quad b = 17.7$ Å to 18.0 Å
$\quad c = 8.00$ Å to 8.28 Å,
$\quad Z = ?$
$H = {\sim}2\text{–}4$
$G = 1.93\text{–}2.4$
Biaxial (–)?
$n_\alpha = 1.575\text{–}1.585$
$n_\beta = 1.597$
$n_\gamma = 1.598\text{–}1.635$
$\delta = 0.023\text{–}0.050$
$2V = ?$

Chrysocolla coating breccia
1 cm

Structure The structure is poorly defined and is often amorphous. At least some contains silicon tetrahedral sheets and is therefore grouped here with the sheet silicates.

Composition Cu and Al are approximately in the ratio of 3:1.

Form and Twinning Typically forms botryoidal and encrusting masses; often cryptogranular. Rare crystals are acicular.

Cleavage Conchoidal fracture.

Color Properties

Luster: Usually earthy, also vitreous or porcelaneous.
Color: Blue, blue-green or green; brown or black if impure.

Streak: White or pale greenish if relatively pure.
Color in Thin Section: If not too fine grained to observe, colorless to pale blue-green.
Pleochroism: Reported colorless to pale blue-green.

Optical Characteristics

Relief in Thin Section: Moderate positive.
Optic Orientation: ??
Indices of Refraction and Birefringence: Generally so fine grained that measurement is impractical.
Interference Figure: Too fine grained to allow interference figures to be observed. Based on the reported indices of refraction, $2V_z$ should be 24°.

Alteration No consistent alteration.

Distinguishing Features

Hand Sample: Resembles turquoise but has lower hardness. Most samples will adhere to the tip of the tongue, whereas turquoise will not.

Thin Section: Amorphous/microgranular nature makes thin section identification impractical.

Occurrence Chrysocolla occurs in the oxidized portion of Cu-bearing sulfide mineral deposits.

Use Minor ore of copper. May be used as a semiprecious gem, particularly if intergrown with chalcedony to increase hardness.

REFERENCES CITED AND SUGGESTIONS FOR ADDITIONAL READING

Albee, A. L. 1962. Relationships between the mineral association, chemical composition and physical properties of the chlorite series. American Mineralogist 47, 851–870.

Bailey, S. W. (ed.). 1984. *Micas.* Reviews in Mineralogy 13, 584 p.

Bailey, S. W. (ed.). 1988. *Hydrous Sheet Silicates (Exclusive of Micas).* Reviews in Mineralogy 19, 755 p.

Brindley, G. W. 1980. Quantitative X-ray analysis of clays, in G. W. Brindley and G. Brown (eds.), *Crystal Structures of Clay Minerals and Their X-Ray Identification.* Monograph No. 5, pp. 411–438. Mineralogical Society, London.

Deer, W. A., Howie, R. A., and Zussman, J. 1992. *The Rock Forming Minerals,* 2nd ed., Longman Group Ltd., London, 696 p.

Deer, W. A., Howie, R. A., and Zussman, J. 2009. *Rock-Forming Minerals,* Volume 3B: *Layered Silicates: Excluding Micas and Clay Minerals,* 2nd ed. The Geological Society, London, 320 p.

Fleet, M. E. and Howie, R. A. 2004. *Rock-Forming Minerals,* Volume 3A: *Micas,* 2nd ed. The Geological Society, London, 780 p.

Guggenheim, S., and Eggleton, R. A. 1987. Modulated 2:1 layer silicates: Review, systematics, and predictions. American Mineralogist 72, 724–738.

Gunter, M. E., Belluso, E., and Mottana, A. 2007. Amphiboles: Environmental and health concerns. Reviews in Mineralogy and Geochemistry 67, 453–516.

Guthrie, G. D., Jr., and Mossman, B. T. 1993. Health effects of mineral dusts. Reviews in Mineralogy 28, 584 p.

Hower, J., Eslinger, E. V., Hower, M. E., and Perry, E. A. 1976. Mechanism of burial metamorphism of argillaceous sediment: 1. Mineralogical and chemical evidence. Geological Society of America Bulletin 87, 725–737.

Laduron, D. M. 1971. A staining method for distinguishing paragonite from muscovite in thin section. American Mineralogist 56, 1117–1119.

Limpitlaw, U. G. 2010. Ingestion of Earth materials for health by humans and animals. International Geology Review 52, 726–744.

Moore, D. M., and Reynolds, R. C., Jr. 1997. *X-Ray Diffraction and the Identification and Analysis of Clay Minerals*, 2nd ed. Oxford University Press, New York, 378 p.

Mottana, A., Sassi, F. P., Thompson, J. B., Jr., and Guggenheim, S. (eds.). 2002. *Micas: Crystal Chemistry and Metamorphic Petrology*. Reviews in Mineralogy and Geochemistry 46, 499 p.

Rieder, M., Cavazzini, G., D'Yakonov, Y. S., Frank-Kamenetskii, V. A., Gottard, G., Guggenheim, S., Koval, P. V., Müller, G., Neiva, A. M. R., Radoslovich, E. W., Robert, J. L., Sassi, F. P., Hiroshi, T., Weiss, Z, and Wones, D. R. 1998. Nomenclature of the micas. The Canadian Mineralogist 36, 41–48.

Skinner, H. C. W., Ross, M., and Frondel, C. 1988. *Asbestos and Other Fibrous Minerals: Mineralogy, Crystal Chemistry and Health Effects*. Oxford University Press, New York, 204 p.

Velde, B. (ed.). 1995. *Origin and Mineralogy of Clays*. Springer-Verlag, Berlin, 334 p.

Weaver, C. E. 1989. *Clays, Muds and Shales*. Elsevier Science Publishers, Amsterdam, 819 p.

Williams, L. B., Haydel, S. E., and Ferrell, R. E., Jr. 2009. Bentonite, Bandaids, and borborygmi. Elements 5, 99–104.

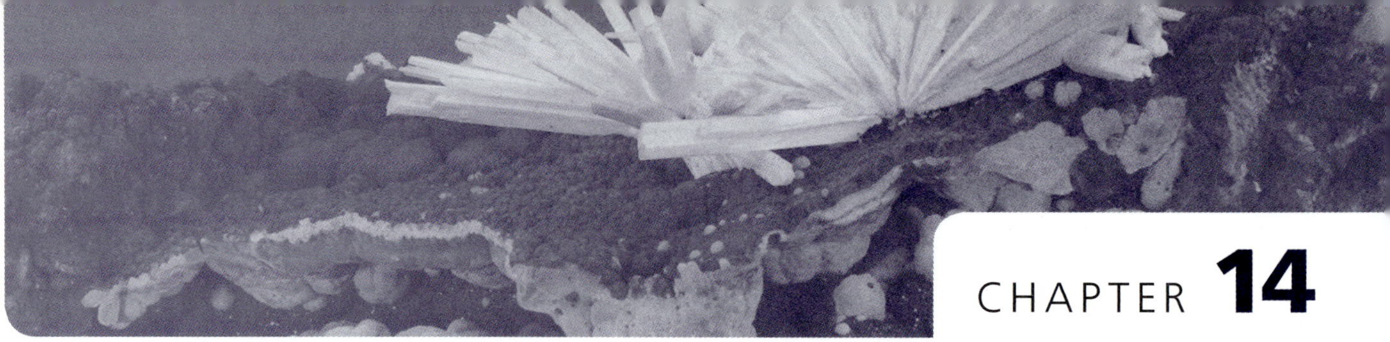

CHAPTER **14**

Chain Silicates

INTRODUCTION

The chain silicates are common Fe/Mg-bearing silicates represented by two principal groups, the pyroxenes and the amphiboles. The pyroxenes and the amphiboles are constructed, respectively, of single and double chains of silicon tetrahedra (see Chapter 11). In terms of abundance, the pyroxenes are among the most common minerals, given the abundance of pyroxene in the basaltic and related rocks that make up the oceanic crust. Although not as abundant as the pyroxenes, the amphiboles are very common minerals in a wide variety of igneous and metamorphic rocks exposed on the continents. The pyroxenoids are relatively uncommon minerals with structures that are closely related to the pyroxene group.

PYROXENE GROUP

Structure and Classification

Pyroxenes, with a general formula of XYZ_2O_6, are constructed of single chains of tetrahedra (Figure 14.1) that extend parallel to the *c* axis. The Z cations in the tetrahedra are most commonly Si^{4+}, although some may be Al^{3+}. The chains are stacked atop each other in the *a* direction in an alternating fashion so that bases of tetrahedra face bases, and unshared apical oxygens face unshared oxygens. The X cations occupy the M2 sites between the bases of the tetrahedra in somewhat distorted 6- or 8-fold coordination, depending on how the chains are stacked and the size of the cation that occupies the site. Y cations occupy the 6-fold M1 structural sites between the apical oxygens on the points of adjacent chains.

It is convenient to treat a pair of tetrahedal chains whose apical oxygens face each other and the associated strip of octahedrally coordinated cations in the M1 sites as a structural unit that looks something like the cross section of a steel I-beam. Note the relationship to the *TOT* layers in the sheet silicates shown earlier (Figure 13.1). These *TOT* I-beams are then stacked atop each other in an interlocking fashion to form the structure of the mineral.

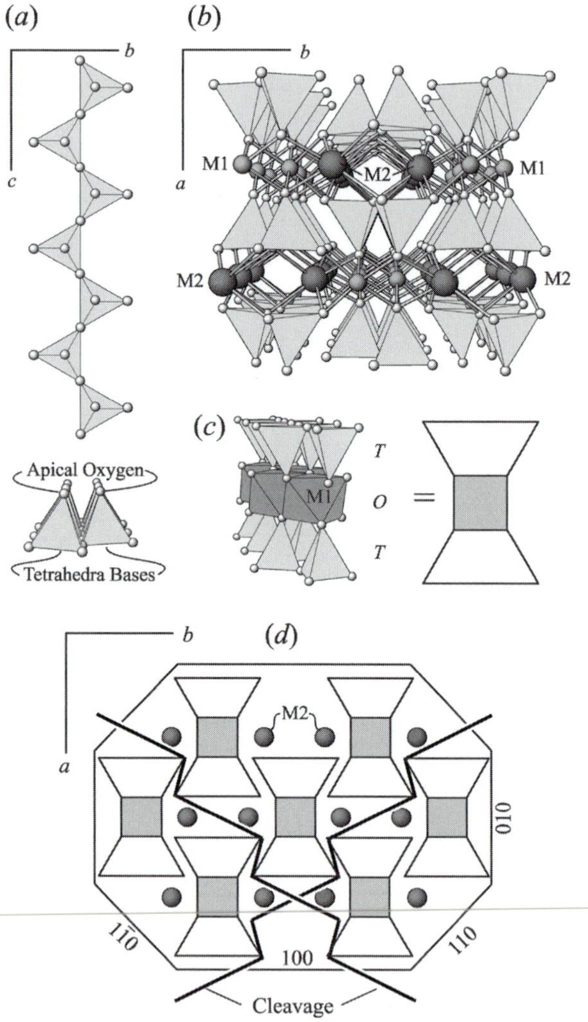

Figure 14.1 Pyroxene structure. (*a*) Top and end views of a single chain of silicon tetrahedra. Chains may be kinked somewhat. (*b*) Chains are stacked in alternating fashion to form 6-fold coordinated M1 sites between the unshared apical oxygens of the tetrahedra and 6- to 8-fold coordinated M2 sites between the bases. (*c*) Two chains and the intervening octahedral M1 sites form a structural *TOT* "I-beam." (*d*) View similar to (*b*) but with I-beams shown. Cleavage at about 87° and 93° between the I-beams is parallel to the {110} faces in the monoclinic unit cell.

Whether the M2 sites between the bases of the tetrahedral chains have 6- or 8-fold coordination is determined by exactly how the adjacent I-beams line up with each other. If they line up to produce 6-fold coordination for the M2 site, the structure acquires orthorhombic symmetry, and if the alignment yields 8-fold coordination sites, the symmetry is monoclinic.

Crystals produced with this structure are typically blocky prisms elongate along c with square or approximately octagonal cross sections. Cleavage splits between the I-beams and produces two directions at about 87° and 93°, which is characteristic of the pyroxenes.

Pyroxene classification is based on the occupancy of the M2 site and symmetry (Table 14.1). The composition of most common pyroxenes can be shown on a triangular diagram (Figure 14.2a) whose vertices are $Ca_2Si_2O_6$ (wollastonite), $Mg_2Si_2O_6$ (enstatite), and $Fe_2Si_2O_6$ (ferrosilite). Wollastonite, enstatite, and ferrosilite are abbreviated wo, en, and fs, respectively. The three groups of common pyroxenes are orthopyroxene (opx), low-Ca-clinopyroxene (pigeonite), and Ca-clinopyroxene (cpx). While wollastonite is one of the vertices on the diagram, its structure is somewhat different and it is classified as a pyroxenoid, not a pyroxene.

Orthopyroxene (opx) can be considered a simple solid solution series between enstatite ($Mg_2Si_2O_6$) and ferrosilite ($Fe_2Si_2O_6$). As the name implies, orthopyroxene is orthorhombic. Both the M1 and M2 sites are octahedral and both can contain Mg and Fe, but the larger Fe cations are concentrated in the somewhat larger M2 sites.

Low-Ca-clinopyroxene (pigeonite) typically has somewhat more Ca than orthopyroxene, but solid solution does not extend to Ca-clinopyroxene. Most M2 structural sites between the bases of the tetrahedra are occupied by Fe and Mg; some are occupied by Ca. The M1 site is occupied by Mg and Fe.

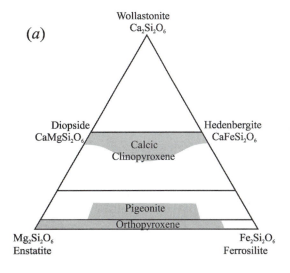

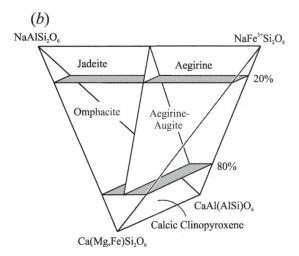

Figure 14.2 Pyroxene classification. (*a*) Calcic and iron–magnesium pyroxenes. The common compositions are shaded. (*b*) Sodic and sodic–calcic pyroxenes.

Table 14.1 Pyroxene Classification[a]

Group	X	Y	Z_2O_6	Mineral(s)	Symmetry	Comment
Magnesium–iron	Mg, Fe	Mg, Fe	Si_2O_6	Enstatite–Ferrosilite	Orthorhombic	Orthopyroxene (opx)
	Mg, Fe, Ca	Mg, Fe	Si_2O_6	Pigeonite	Monoclinic	Low-Ca clinopyroxene
Calcium	Ca	Mg, Fe	Si_2O_6	Diopside–Hedenbergite		Ca clinopyroxene (Ca-cpx or cpx)
		Mg, Fe, Al	$(Si,Al)_2O_6$	Augite		
Calcium–sodium	Ca, Na	Mg, Fe^{2+}, Al, Fe^{3+}	Si_2O_6	Omphacite		Sodic–Calcic clinopyroxene
		Mg, Fe^{2+}, Fe^{3+}	Si_2O_6	Aegirine–Augite		
Sodium	Na	Al	Si_2O_6	Jadeite		Sodic clinopyroxene
		Fe^{3+}	Si_2O_6	Aegirine		
Lithium	Li	Al	Si_2O_6	Spodumene		Lithium clinopyroxene

[a] Based on the general formula XYZ_2O_6 where X cations occupy M2 sites, Y cations occupy M1 sites, and Z cations occupy tetrahedral sites.

Ca-clinopyroxene (cpx) includes compositions that extend from diopside ($CaMgSi_2O_6$) to hedenbergite ($CaFeSi_2O_6$). The M2 sites contain mostly Ca and the M1 sites contain mostly Mg and Fe. The most abundant Ca-clinopyroxene is augite, in which some Al substitutes both in the M1 sites and for Si in tetrahedral sites, and some Na, Fe, or Mg substitutes for Ca in the M2 site.

The composition space for the sodium-bearing clinopyroxenes is shown in Figure 14.2b. Extensive solid solution is possible between the sodic, sodic–calcic, and calcic clinopyroxenes. The common compositions fall in two ranges. One includes aegirine and aegirine–augite, and the second includes jadeite and omphacite. Those groupings are used in the mineral descriptions that follow.

Pyroxene species can be distinguished from each other only with difficulty in hand sample, and then based mostly on color and occurrence. Orthopyroxenes and pigeonite are most commonly brown; pigeonite is usually restricted to volcanic rocks. Ca-clinopyroxenes and sodic pyroxenes are both commonly shades of dark green or black. Light green pyroxene in metamorphosed carbonate is usually diopside. Optical properties can help distinguish among the various species, but if knowledge of composition is important, chemical analysis is essential. Figure 14.3 shows photomicrographs of pyroxenes in thin section.

Geology of Pyroxenes

Igneous

The most common pyroxenes in igneous rocks are augite, pigeonite, and orthopyroxene. The rocks in which they are found are intermediate to mafic (andesite, diorite, basalt, gabbro, etc.), and the associated minerals include plagioclase, olivine, and hornblende. It is common to find both augite and either pigeonite or orthopyroxene in the same rock.

Augite is the most abundant of these minerals and is common in both volcanic and intrusive rocks, usually basalt, gabbro, or other related varieties. It typically forms blocky prismatic crystals or roughly equant grains, depending on when in the crystallization sequence it has grown. Crystals are usually enriched in Mg relative to the melt from which they are growing. The residual magma from which augite is growing therefore becomes enriched in Fe as the pyroxene grows. This commonly results in formation of zoned crystals with more Mg-rich cores and Fe-rich rims.

With cooling, the amount of Mg/Fe that can substitute for Ca in the M2 site of augite becomes progressively more limited. Exsolution lamellae of low-Ca pyroxene (Figure 14.4a) form in the augite and the augite becomes

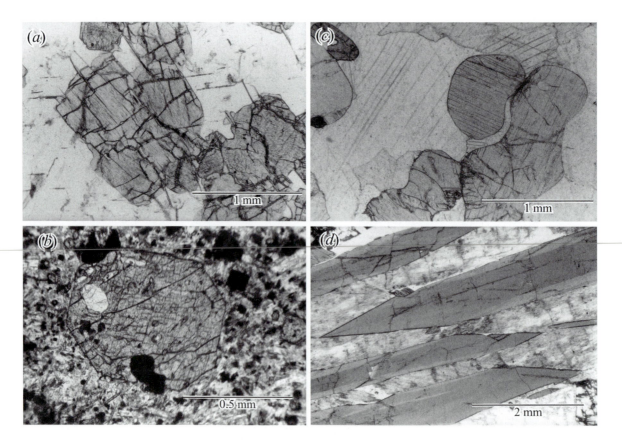

Figure 14.3 Photomicrographs of pyroxene in thin section. (a) Anhedral orthopyroxene in norite. (b) Augite phenocryst in basalt. (c) Rounded diopside in marble. (d) Aegirine in syenite.

progressively richer in Ca in a manner analogous to the way that albite lamellae exsolve from K-feldspar with cooling (Figure 5.27). These lamellae may be either pigeonite, usually oriented roughly parallel to (001) of the host augite, or orthopyroxene, oriented parallel to (100). In most cases the pigeonite converts to orthopyroxene on cooling, as described shortly.

Orthopyroxene may crystallize from fairly typical gabbroic/basaltic magmas and also from more felsic varieties such as dacite, diorite, granodiorite, tonalite, and granite. It is a common phenocryst in andesitic volcanics. At high temperatures, orthopyroxene may accommodate more Ca in the M2 sites than at low temperatures. With slow cooling, the excess Ca is expelled to form exsolution lamellae of augite parallel to (100) (Figure 14.4b). Orthopyroxene crystallized directly from magma is known as orthopyroxene of the Bushveld type, and it contains augite lamellae parallel to only (100). The name is derived from the Bushveld Complex in South Africa. Orthopyroxene also may form by conversion from pigeonite.

Pigeonite also is a common mineral to grow from mafic magmas and may have up to about 10% Ca in the M2 sites

at high temperatures. With cooling, some of this Ca is expelled from the M2 sites and forms exsolution lamellae of Ca-clinopyroxene (augite) roughly parallel to {001} of the host pigeonite (Figure 14.4c). If cooling is not too fast, pigeonite converts through a reconstructive transformation to form orthopyroxene at temperatures below ~1100°C for Mg-rich pigeonite or ~950°C for Fe-rich pigeonite. Pigeonite is preserved only in volcanic rocks where cooling is fast.

The conversion from pigeonite to orthopyroxene may trigger the formation of a second set of exsolution lamellae because orthopyroxene cannot contain as much Ca as pigeonite can. To take care of the excess Ca, lamellae of augite exsolve parallel to the {100} planes of the new orthopyroxene (Figure 14.4d). Orthopyroxene that originally grew as pigeonite crystals may therefore contain two different sets of exsolution lamellae of augite, one roughly parallel to (001) of the original pigeonite and another parallel to (100) of the orthopyroxene. This is known as orthopyroxene of the Stillwater type; the name is derived from the Stillwater Complex in Montana.

Aegirine and aegirine–augite are less common and are found most often in alkali-rich igneous rocks such as

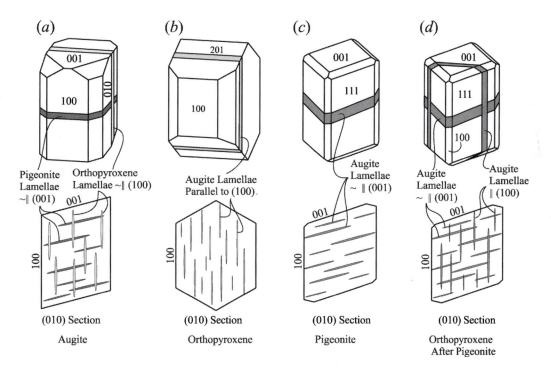

Figure 14.4 Exsolution lamellae may be responsible for partings in pyroxene. (*a*) Augite. Pigeonite may exsolve as lamellae roughly parallel to (001); the exact orientation depends on composition and temperature. The pigeonite converts to orthopyroxene with slow cooling. Primary orthopyroxene lamellae may exsolve parallel to (100). (*b*) Primary orthopyroxene (Bushveld type). Lamellae of augite develop parallel to (100) of the orthopyroxene. (*c*) Pigeonite. Augite lamellae may exsolve roughly parallel to (001). The exact orientation depends on temperature and composition. (*d*) Orthopyroxene converted from pigeonite (Stillwater type). Unless cooled rapidly, pigeonite converts to orthopyroxene. The "(001)" lamellae of augite in the original pigeonite are preserved, and additional lamellae of augite develop parallel to (100) of the orthopyroxene.

syenite, nepheline syenite, and phonolite. Growth of these minerals requires the combination of high Na and modest to low Si found in these rocks and not found in more common granite, granodiorite, diorite, and gabbro or their volcanic equivalents.

Spodumene, with its high Li content, is restricted to Li-bearing granitic pegmatites. Gigantic crystals of spodumene, up to 10 m long, have been mined for their Li content from pegmatites in the Black Hills of South Dakota.

Metamorphic

Calcic clinopyroxenes are found both in metamorphosed carbonate rocks and in metamorphosed-mafic rocks. The clinopyroxene in metamorphosed carbonates usually is fairly diopside rich, reflecting the abundance of Ca and Mg derived from calcite and dolomite. Associated minerals include calcite, dolomite, tremolite/actinolite, members of the epidote group, and olivine. Metamorphism of rocks such as basalt in the presence of some water often produces amphibolite, a rock consisting of hornblende, plagioclase, and other minerals. With sufficiently high temperature, the hornblende, which is hydrous, may break down to produce Ca-clinopyroxene.

The sodic and sodic-calcic clinopyroxenes have fairly restricted occurrences in metamorphic rocks. Jadeite is restricted to metagraywacke, blueschists, and related rocks that have been subjected to high pressures but only modest temperatures. This combination of conditions is most often found in subduction zone environments. Omphacite, with its higher Ca content, may be found in the higher temperature–pressure parts of this environment. Omphacite is also found in eclogite, a rock of roughly basaltic composition formed under very high pressure conditions in the lower crust or upper mantle.

Orthopyroxene also makes an appearance in granulite-facies metamorphic rocks. These rocks have been heated enough to eliminate most water, through various dehydration reactions. The Mg and Fe originally contained in biotite or amphiboles, which are hydrous, often get relegated to orthopyroxene.

Sedimentary

Pyroxenes, being anhydrous, are not particularly stable in the weathering environment and readily alter to clay minerals. Pyroxene is therefore a major constituent of sediments derived from pyroxene-bearing rocks only if erosion is rapid and transport is limited. Once deposited, normal diagenetic processes usually convert pyroxene in sediment to clay minerals. Any pyroxene that is preserved in the heavy mineral fraction of sediment, however, may be useful in determining the provenance of the sediment.

ORTHOPYROXENE

$(Mg,Fe)_2Si_2O_6$
Orthorhombic $2/m\ 2$
$a = 18.22–18.43$ Å,
$\quad b = 8.81–9.08$ Å,
$\quad c = 5.17–5.24$ Å, z
$H = 5–6$
$G = 3.21–3.96$
Biaxial (+ or −)
$n_\alpha = 1.649–1.768$
$n_\beta = 1.653–1.770$
$n_\gamma = 1.657–1.788$
$\delta = 0.007–0.020$
$2V_x = 50–132°$

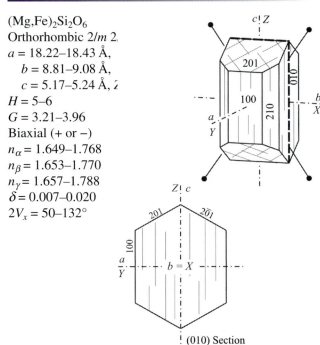

(010) Section

Composition The principal chemical variation is substitution of Mg and Fe in the M1 and M2 sites (Figure 14.5). Most orthopyroxene falls in the enstatite range. Up to 10 or 15% of other cations including Al, Mn, Ti, Cr, and Ca may be present in some cases. The abbreviation for orthopyroxene is commonly taken to be "opx." In the past orthopyroxene compositions have been divided into additional named ranges that included enstatite (en_{100-88}), bronzite (en_{88-70}), hypersthene (en_{70-50}), ferrohypersthene (en_{50-30}), eulite (en_{30-12}), and orthoferrosilite (en_{12-0}). Use of these names is no longer recommended by the Commission on New Minerals, Nomenclature and Classification of the International Mineralogical Association.

Form and Twinning Figure 14.3a. Euhedral crystals are stubby prisms with four- or eight-sided cross sections. Longitudinal sections are usually roughly rectangular. Anhedra and irregular grains also are common. Late-growing orthopyroxene may form highly irregular grains that occupy the interstices among earlier-formed grains or entirely enclose them. Exsolution lamellae of augite (Figure 14.4) may be thin and tabular; or they may pinch and swell, or form irregular rows of blebs. Exsolution lamellae are not usually seen in hand sample.

Cleavage Good cleavage on {210}, like the other pyroxenes, intersects at angles of 88° and 92°. A parting that bisects the cleavages may be present parallel to {100}. For orthopyroxene that has converted from pigeonite, a second parting approximately parallel to {001} of the

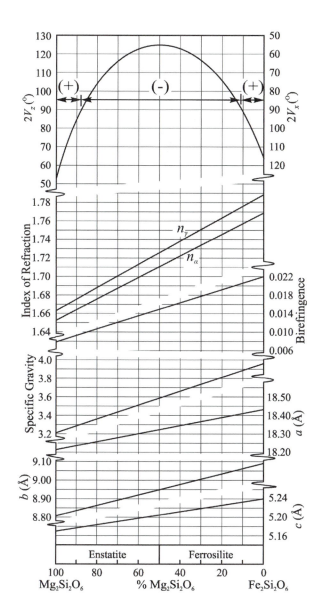

Figure 14.5 Optical properties, specific gravity, and unit cell dimensions of orthopyroxene. Adapted from Leake (1968), Jaffe and others (1975, 1978), and Deer and others (1978).

original pigeonite also may be present. Fracture uneven. Brittle.

Color Properties

Luster: Vitreous, bronzite may be submetallic.

Color: Mg-rich varieties most often brown or greenish brown; also white, tan, or green. Fe-rich varieties green, dark brown, or greenish black.

Streak: White to gray.

Color in Thin Section: Varies with composition. Mg-rich enstatite is colorless or with very pale pinkish to greenish pleochroism; more Fe-rich varieties are darker colored with reddish to greenish pleochroism.

Pleochroism: $X < Y < Z$, stronger with higher Fe content: X = colorless, pink, brownish pink, pale yellow; Y = colorless, light brown, yellow, pinkish yellow, greenish yellow; Z = colorless, light green, gray green, green, bluish green.

Optical Characteristics

Relief in Thin Section: Moderately high to high positive.

Optic Orientation: $X = b$, $Y = a$, $Z = c$. Extinction is parallel in longitudinal sections and symmetrical in basal sections. Crystals are length slow.

Indices of Refraction and Birefringence: Indices of refraction and birefringence vary systematically (Figure 14.5).

Interference Figure: 2V varies systematically with composition (Figure 14.5). Mg-rich enstatite and Fe-rich ferrosilite are both optically positive; intermediate compositions are negative. Optic axis dispersion measured about the X axis is $r > v$, weak to strong.

Comments: The value of 2V measured in thin section can be used to estimate composition to within 5 or 10 mol % Mg. Measurement of indices of refraction in grain mount provides greater accuracy.

Alteration Opx readily alters to serpentine, talc, fine-grained pale amphibole, or other silicates.

Distinguishing Features

Hand Sample: Nearly right-angle cleavage and stubby or nearly equant crystals distinguish pyroxenes from amphiboles. The {100} parting that intersects the cleavages at about 45° may be confusing to students new to the mineral. Brown color distinguishes orthopyroxene from clinopyroxene. The submetallic luster of bronzite is quite distinctive. High-iron varieties may be indistinguishable from Ca-clinopyroxene.

Thin Section/Grain Mount: Opx is distinguished from pigeonite and Ca-clinopyroxene by having lower birefringence, parallel extinction in longitudinal sections, and pale pink to green pleochroism. Andalusite may appear similar to enstatite but lacks the ~90° cleavage and is optically negative. The occurrences for andalusite and enstatite are distinctly different.

Occurrence Common in mafic and ultramafic igneous rocks (gabbro, norite, peridotite, pyroxenite, etc.) associated with Ca-clinopyroxene, plagioclase, and olivine. Opx with intermediate or higher iron content may be found in diorite, or even some syenite or granite. Basalt and andesite often contain orthopyroxene phenocrysts. Very-high-grade metamorphic rocks of the granulite facies may also contain orthopyroxene. Associated minerals may include feldspars, Ca-clinopyroxene, hornblende, biotite, and garnet.

Use No significant economic uses. Rocks containing orthopyroxene are sometimes used for dimension stone or decorative purposes. Occasionally used as a minor gemstone.

PIGEONITE

$(Mg,Fe^{2+},Ca)_2Si_2O_6$
Monoclinic $2/m$
$a = 9.71$ Å, $b = 8.95$ Å,
$\quad c = 5.24$ Å, $\beta = 108.5°$,
$\quad Z = 4$
$H = 6$
$G = 3.17–3.46$
Biaxial (+)
$n_\alpha = 1.682–1.732$
$n_\beta = 1.684–1.732$
$n_\gamma = 1.705–1.757$
$\delta = 0.023–0.029$
$2V_z = 0–32°$

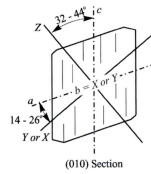

(010) Section

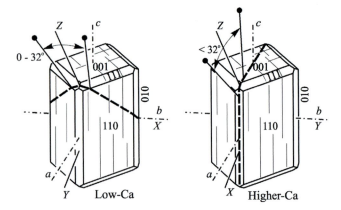

Low-Ca   Higher-Ca

Composition Pigeonite is the common low-calcium clinopyroxene. Typically between 10 and 30% of the M2 sites are filled with Ca; the balance is filled with Fe and Mg. The M1 site is filled almost exclusively with Mg and Fe. Most pigeonite has a Mg:Fe ratio of between 70:30 and 30:70. Minor Na, Ti, Cr, Al, and Fe^{3+} may be present. High temperature of formation favors higher Ca content. Slow cooling allows Ca to migrate out of the structure to form exsolution lamellae of Ca-clinopyroxene (Figure 14.4).

Form and Twinning Crystals are prismatic and somewhat elongate parallel to c. Cross sections are square or eight sided. Longitudinal sections are rectangular. In mafic and intermediate volcanic rocks it may form groundmass grains. Lamellae of exsolved augite (Figure 14.4) are common roughly parallel to (001) in pigeonite preserved in shallow intrusive rocks but are not usually seen in hand sample. Compositional zoning may be expressed in thin section by variations in color, extinction angle, or birefringence. Pigeonite may be mantled by augite.

Cleavage Good {110} like the other pyroxenes at 87° and 93°. Partings on {100} and ~{001}. Fracture uneven. Brittle.

Color Properties

Luster: Vitreous.

Color: Brown, greenish brown, black.

Streak: Light to dark gray.

Color in Thin Section: Colorless, pale brownish green, pale yellowish green.

Pleochroism: Absent or weak: X = colorless, pale greenish brown, yellow; Y = pale brown, greenish brown; Z = colorless, pale yellow, pale green.

Optical Characteristics

Relief in Thin Section: High positive.

Optic Orientation: Low-Ca pigeonite: $X = b, Y \wedge a = -14°$ to $-26°$, $Z \wedge c = +32°$ to $+44°$, optic plane normal to (010). With increasing Ca content, the $2V$ angle closes on Z and then opens in the (010) plane so that $X \wedge a = -22°$ to $-26°$, $Y = b, Z \wedge c = +40°$ to $+44°$, to optic plane = (010). The extinction angle to the trace of cleavage in (010) sections is between 32° and 44°.

Indices of Refraction and Birefringence: Indices of refraction and birefringence increase with increasing Fe content. Interference colors in standard thin section range up to lower or middle second order.

Interference Figure: Biaxial negative with a small $2V$. Optic axis dispersion is weak, either $r > v$ or $r < v$.

Alteration Most pigeonite in plutonic rocks converts to orthopyroxene on cooling. Alteration products include fine-grained amphibole, serpentine, talc, or chlorite.

Distinguishing Features

Hand Sample: Color, cleavage, and habit distinguish pigeonite as a pyroxene. Distinguishing among pigeonite, orthopyroxene, and augite phenocrysts in volcanic rocks is not really practical in hand sample.

Thin Section/Grain Mount: Distinguished from the other pyroxenes by a smaller optic angle. Olivine has higher birefringence and no cleavage.

Occurrence Pigeonite is found in dacitic, andesitic, and basaltic volcanic rocks as groundmass grains or phenocrysts. It commonly crystallizes in mafic to intermediate intrusive rocks, but converts by reconstructive processes to orthopyroxene on cooling. Only shallow intrusives are likely to preserve pigeonite. Metamorphism of iron formations at sufficiently high temperatures may yield pigeonite, but most subsequently converts to orthopyroxene.

Use None.

CALCIC CLINOPYROXENE

Diopside CaMgSi$_2$O$_6$
$a = 9.75$ Å, $b = 8.90$ Å,
    $c = 5.25$ Å, $\beta = 105.6°$,
    $Z = 4$
Hedenbergite CaFeSi$_2$O$_6$
$a = 9.85$ Å, $b = 9.03$ Å,
    $c = 5.24$ Å,
    $\beta = 104.8°$, $Z = 4$
Augite (Ca,Mg,Fe^{2+},Fe^{3+},Al)$_2$
    (Si,Al)$_2$O$_6$
$a = {\sim}9.7$ Å,
    $b = {\sim}8.8$ Å, $c = {\sim}5.3$ Å,
    $\beta = 106.9°$, $Z = 4$

Monoclinic 2/m
$H = 5.5$–6
$G = 3.19$–3.56
Biaxial (+)
$n_\alpha = 1.664$–1.745
$n_\beta = 1.672$–1.753
$n_\gamma = 1.694$–1.771
$\delta = 0.018$–0.034
$2V_z = 25$–70°

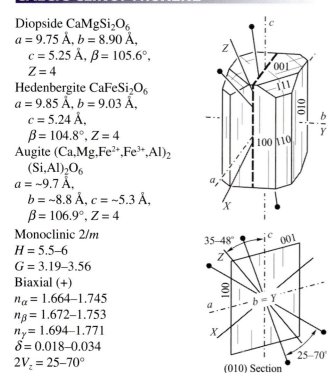

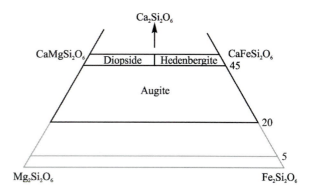

Figure 14.6 Commonly used composition classification of the Ca-clinopyroxenes (Morimoto, 1988).

Composition There is complete solid solution among the Ca-rich clinopyroxenes. Pyroxene along the diopside–hedenbergite join (Figure 14.2a), whose M2 site contains nearly all Ca, is generally restricted to metamorphic rocks. Most igneous Ca-clinopyroxene has somewhat less Ca and more Fe/Mg in the M2 site and contains significant Al, both in tetrahedral and octahedral (M1) sites. This pyroxene is conventionally called augite. Various schemes have been applied to divide augite into compositional ranges; Figure 14.6 shows a simple one. However, given the diversity of composition and the inability to accurately deduce composition based on physical or optical properties, the terms calcic clinopyroxene (Ca-cpx or cpx) may be used for all varieties of Ca-rich clinopyroxene. Extensive solid solution toward the Na-rich clinopyroxenes is also possible and most augite contains some Na substituting for Ca. Ti-bearing augite is called **titanaugite**.

The amount of Ca in the ~8-fold M2 sites is, in part, a function of temperature. High temperature allows more extensive substitution of Fe/Mg for Ca. With cooling, lamellae of pigeonite or orthopyroxene are exsolved to allow the Ca-cpx to become more Ca rich. Pigeonite lamellae form roughly parallel to (001) of the augite and orthopyroxene lamellae are parallel to (100). Figure 14.4. The exact orientation of the "(001)" lamellae depends on temperature. Exsolution lamallae are not usually obvious in hand sample.

Form and Twinning Figure 14.3b. Crystals are usually stubby prisms elongate parallel to c with four- or eight-sided cross sections. Also as granular masses and anhedral grains that range from equant to highly irregular and may enclose associated minerals. Overgrowths of hornblende are relatively common. Compositional zoning may be visible in thin section and is expressed by variations in color or birefringence (Figure 5.13).

Single and multiple twins with {100} composition planes are fairly common as are multiple twins on {001}. Although visible in thin section, twins are not usually evident in hand sample. The combination of twinning and exsolution lamellae may produce a herringbone pattern in certain orientations in thin sections.

Cleavage Fair to good cleavage on {110} intersects at 87° and 93°. {100} and {001} partings may be distinct. Fracture uneven. Brittle.

Color Properties

Luster: Vitreous.

Color: Pale green (Mg rich) to black (Fe rich), usually dark green, greenish black, or brownish black.

Streak: White or gray.

Color in Thin Section: Colorless, gray, pale green, pale brown, or brownish green. Color zoning is fairly common, particularly in Fe- and Ti-rich varieties, and may form a distinctive "hourglass" pattern consisting of four triangular segments radiating from the center of the crystal.

Pleochroism: Fe-rich samples may be weakly pleochroic: X = pale green, bluish green; Y = pale greenish brown, green, bluish green; Z = pale brownish green, green, yellowish green. Titanaugite is more distinctly colored in shades of brown and violet.

Optical Characteristics

Relief in Thin Section: High positive.

Optic Orientation: $X \wedge a = -20°$ to $-33°$, $Y = b$, $Z \wedge c = +35°$ to $+48°$.

Indices of Refraction and Birefringence: Indices of refraction increase with Fe + Mn content but have too much variability to be used to estimate composition. The trend is shown in Figure 14.7. Birefringence generally increases with increasing Fe content.

Interference Figure: The $2V_z$ angle increases with increasing Ca content (Figure 14.7). The common range of $2V_z$ is $50° ± 10°$. Optic axis dispersion is weak to strong, usually $r > v$.

Alteration Calcic clinopyroxenes commonly alter to fine-grained, light-colored amphibole (uralite) and to chlorite, serpentine, biotite, carbonates, or clay.

Distinguishing Features

Hand Sample: Dark color, approximately right-angle cleavage, and hardness distinguish pyroxene from most other silicates. Beware the partings, however, because they may be more distinct than the cleavage. Hornblende has better cleavage at 56° and 124°. Diopside is usually light green and is distinguished from olivine in metamorphosed carbonates by having cleavage.

Thin Section/Grain Mount: Distinguished from opx by higher birefringence, inclined extinction in (010) sections, and smaller 2V. Pigeonite has a smaller 2V, and jadeite has a larger 2V. The other Na-rich clinopyroxenes (aegirine, aegirine–augite, omphacite) are usually more distinctly green and have larger 2V. Olivine lacks cleavage and has higher birefringence. Wollastonite has lower indices of refraction and birefringence, is colorless, and has better and more cleavages.

Occurrence Augite is common in intrusive mafic igneous rocks such as gabbro, pyroxenite, anorthosite, norite, and peridotite. Fe-rich augite may be found in more silicic rocks such as diorite and granodiorite. Basaltic and andesitic volcanics also commonly contain augite. Cr-rich diopside is common in ultramafic rocks such as kimberlite.

In metamorphic rocks, augite is found in amphibolite, hornblende gneiss, and granulites. Skarns, marble, and other metamorphosed carbonate rocks commonly contain fairly diopside-rich compositions in association with tremolite–actinolite, grossular garnet, epidote, wollastonite, forsterite, calcite, and dolomite.

Use Diopside is a component of some ceramic products, including those used for dental applications. It also has been used as a gemstone.

AEGIRINE, AEGIRINE–AUGITE

Aegirine: $NaFe^{3+}Si_2O_6$
Aegirine–Augite
   $(Na,Ca)(Fe^{3+},Fe^{2+},Mg)Si_2O_6$
Monoclinic 2/m
$a = 9.7$ Å, $b = 8.8$ Å,
  $c = 5.3$ Å, $\beta = 105–107.5°$,
  $Z = 4$
$H = 6$
$G = 3.40–3.60$
Biaxial (+ or −)
$n_\alpha = 1.700–1.776$
$n_\beta = 1.710–1.820$
$n_\gamma = 1.730–1.836$
$\delta = 0.028–0.060$
$2V_x = 60–110°$

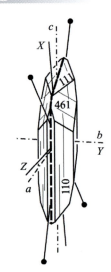

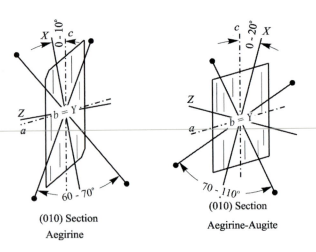

(010) Section
Aegirine

(010) Section
Aegirine-Augite

Composition The principal substitutions are $NaFe^{3+}$ for $Ca(Mg,Fe^{2+})$ in the M1 and M2 structural sites. Solid solution extends to the Ca-clinopyroxenes. Limited amounts of Al may substitute in octahedral M1 sites and almost no Al substitutes for Si in tetrahedral sites. The division between

Figure 14.7 Variation of $2V_z$ (solid lines) and n_β (dashed lines) for calcic clinopyroxene. Adapted from Deer and others (1992).

aegirine and aegirine–augite is taken here (Figure 14.2) to be 80 mol % Fe^{3+} in the octahedral M1 site, and the boundary between aegirine-augite and augite is taken at 20 mol % Fe^{3+} in the octahedral M1 site. In the past, the term **acmite** was used interchangably for aegirine. Acmite also has been used in a more restricted sense for brownish samples of sodic pyroxene that have sharply pointed crystals; the term aegirine was used for stubby crystals of green to black sodic pyroxene. The current convention is to abandon the term *acmite* and to use *aegirine* for all.

Form and Twinning Aegirine may have sharply pointed terminations (Figure 14.3d) or blocky prismatic crystals; aegirine–augite usually forms stubby prismatic crystals similar to augite. Also fibrous and as anhedral grains. Spongy intergrowths with riebeckite may be found. Simple and lamellar twinning on {100} is common

Cleavage Good {110} cleavage at 87° and 93°, also a {100} parting. Uneven fracture. Brittle.

Color Properties

Luster: Vitreous.

Color: Dark green to greenish black or reddish brown, less commonly brown or yellowish green.

Streak: White or gray.

Color in Thin Section: Pale to dark green or yellowish green, darker with higher Fe^{3+} content. Color zoning may be evident with crystal rims darker than cores.

Pleochroism: Distinct with X = emerald green, dark green, bright green; Y = grass green, yellowish green, yellow; Z = brownish green, green, yellowish brown, yellow.

Optical Characteristics

Relief in Thin Section: High positive.

Optic Orientation: $X \wedge c$ = +10° to –20°, $Y = b$, $Z \wedge a$ = +7° to +36°.

Indices of Refraction and Birefringence: Indices of refraction and birefringence increase with Fe^{3+} content (Figure 14.8). The scatter in the data is probably due to variations in Mg:Fe^{2+}. High interference colors may be masked by mineral color.

Interference Figure: $2V$ varies systematically with Fe^{3+} content. Samples with more than roughly 50 to 55% Fe^{3+} are optically negative; less Fe^{3+}-rich compositions are positive. Optic axis dispersion is moderate to strong, $r > v$.

Alteration The common alteration is fine-grained amphibole (uralite) or chlorite.

Distinguishing Features

Hand Sample: Distinguished as a pyroxene by color, cleavage, and hardness. Long slender crystals can be

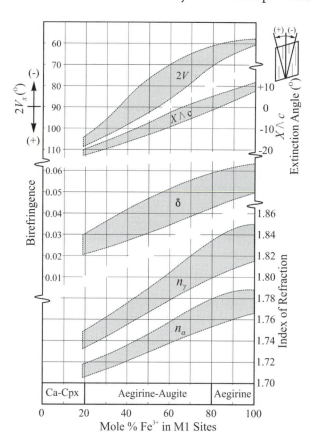

Figure 14.8 Variation of optical properties of aegirine and aegirine–augite with Fe^{3+} content. Data from Larsen (1942), Sabine (1950), and Nolan (1969), and Deer and others (1978).

mistaken for hornblende, but hornblende has better cleavage at 56° and 124°. Aegirine and aegirine–augite cannot reliably be distinguished from other pyroxenes in hand sample.

Thin Section/Grain Mount: Aegirine and aegirine–augite are distinguished from other pyroxenes by color and pleochroism, smaller extinction angle in (010) sections, length fast characteristic, and higher indices of refraction and birefringence. The $2V$ angle is larger than for most other pyroxenes, and samples may be either optically positive or negative. The distinction between aegirine and aegirine–augite can be made based on a combination of indices of refraction, extinction angle in (010) section, birefringence, and $2V$. Aegirine is generally darker than aegirine–augite. Amphiboles have cleavage at 56° and 124° and are length slow. Epidote shows only one good cleavage and generally is lighter colored.

Occurrence Aegirine and aegirine-augite are typical minerals of alkali-rich magmas, meaning magmas with substantial amounts of Na and K. Typical occurrences include alkali granite, syenite, nepheline syenite, and phonolite. Associated minerals include K-feldspar,

Na-amphiboles, and either feldspathoids or quartz. Carbonatites and their associated alkali-rich silicate magmas also may crystallize aegirine or aegirine–augite, and Na-rich fluids derived from carbonatites may replace biotite or hornblende in the country rock with these pyroxenes as a consequence of a process called **fenitization**. Occurrences in metamorphic rocks are generally restricted to blueschists and related rocks in alpine metamorphic belts.

Use None.

JADEITE

$NaAlSi_2O_6$
Monoclinic $2/m$
$a = 9.4$ Å, $b = 8.6$ Å,
$\quad c = 5.2$ Å, $\beta = 107.6°$, $Z = 4$
$H = 6$
$G = 3.24–3.43$
Biaxial (+) or (−)
$n_\alpha = 1.640–1.681$
$n_\beta = 1.645–1.684$
$n_\gamma = 1.652–1.692$
$\delta = 0.006–0.021$
$2V_z = 60–96°$

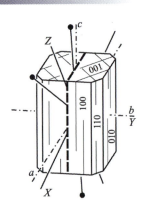

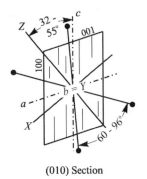

(010) Section

Composition Most jadeite is fairly close to the ideal formula. The principal variation is substitution of Ca^{2+} for up to 20% of the Na^+ in the M2 sites between the bases of the tetrahedra. Charge balance is maintained by substituting a divalent cation such as Mg^{2+} or Fe^{2+} for Al^{3+} in the M1 sites. Some Fe^{3+} also may substitute for Al^{3+}. Omphacite is compositionally intermediate between jadeite and augite, but the three species are not part of a solid solution series. Omphacite has a slightly different structure than either jadeite or augite.

Form and Twinning Crystals are stubby to elongate prisms with four- or eight-sided cross sections. Jadeite is most common as anhedral granules and as aggregates of acicular to fibrous grains. Simple and lamellar twins on {001} and {100} are sometimes seen in thin section but are not evident in hand sample.

Cleavage Good cleavage on {110} like the other pyroxenes intersects at 87° and 93°. A parting on {100} is reported but not common. Splintery fracture. Fibrous masses are very tough.

Color Properties

Luster: Vitreous.

Color: Light to medium green; also white, colorless, greenish blue. Rarely blue or violet.

Streak: White.

Color in Thin Section: Colorless. Grains in grain mount may be pale green.

Pleochroism: None in thin section, weak pale yellow to green pleochroism in grain mount.

Optical Characteristics

Relief in Thin Section: Moderately high positive.

Optic Orientation: $X \wedge a = -14°$ to $-32°$, $Y = b$, $Z \wedge c = +32°$ to $+55°$.

Indices of Refraction and Birefringence: Indices of refraction increase with Fe^{3+}, Ca, and Mg. Birefringence is usually in the range 0.012 to 0.018, so maximum interference colors in thin section are first-order yellow or red. High Fe^{3+} may cause anomalous interference colors and lower birefringence.

Interference Figure: $2V_z$ averages around 70°, larger than for many pyroxenes. Optic axis dispersion is weak to very strong, $r > v$.

Alteration Jadeite may be replaced by amphibole, or less commonly, by analcime or nepheline.

Distinguishing Features

Hand Sample: Massive varieties are distinguished by green color and toughness. Polished surfaces of nephrite have a somewhat oily luster; jadeite is vitreous. Granular material in metagraywacke may be difficult to recognize and distinguish in hand sample.

Thin Section/Grain Mount: Jadeite has lower indices of refraction than other pyroxenes. Omphacite has higher birefringence and aegirine and aegirine–augite have higher birefringence and are more distinctly green. Jadeite with anomalous interference colors resembles zoisite, but zoisite has higher indices of refraction and parallel extinction in principal sections. Tremolite has a smaller extinction angle in (010) sections and has typical amphibole cleavage. The approximate composition of the sodic pyroxenes can be estimated based on the value of n_β and the (211) d-spacing obtained from X-ray diffraction study (Figure 14.9).

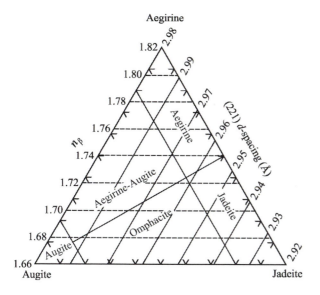

Figure 14.9 Composition of sodic and sodic–calcic amphiboles from glaucophane schist may be estimated based upon the value of n_β, measured optically, and the (221) *d*-spacing measured with X-ray diffraction (Essene and Fyfe, 1967).

OMPHACITE

$(Ca,Na)(Mg,Fe^{2+},Fe^{3+},Al)Si_2O_6$
Monoclinic $2/m$
$a = 9.45\text{–}9.68$ Å,
   $b = 8.57\text{–}8.90$ Å,
   $c = 5.23\text{–}5.28$ Å,
   $\beta = 105\text{–}108°$, $Z = 4$
$H = 5\text{–}6$
$G = 3.16\text{–}3.43$
Biaxial (+)
$n_\alpha = 1.662\text{–}1.701$
$n_\beta = 1.670\text{–}1.712$
$n_\gamma = 1.685\text{–}1.723$
$\delta = 0.012\text{–}0.028$
$2V_z = 56\text{–}84°$

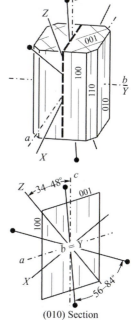

(010) Section

Occurrence

Jadeite is restricted to metamorphic rocks subjected to relatively high pressures and moderate temperatures. Typical occurrences are in glaucophane schist, metagraywacke, and related rocks in alpine metamorphic belts. It may occur as nearly monomineralic pods or veins or as disseminated grains. The commonly associated minerals are albite, glaucophane, lawsonite, quartz, chlorite, garnet, zoisite, tremolite, calcite, aragonite, and micas. The following typical metamorphic reaction may produce jadeite at the expense of plagioclase.

Plagioclase + water
  = jadeite + lawsonite + quartz
$2Na_{0.5}Ca_{0.5}Al_{1.5}Si_{2.5}O_8 + H_2O$
  $= NaAlSi_2O_6 + CaAl_2Si_2O_7(OH)_2 \cdot H_2O + SiO_2$

With increasing temperature, more Ca may substitute for Na in the M2 site, so in equivalent higher grade rocks, the pyroxene may be omphacite.

Use

Jadeite is the principal mineral in **jade**, which is used for decorative purposes and jewelry. The interlocking crystals of jadeite in jade make the material quite tough. The term *jade* also is used for a very similar-looking material composed of felted amphibole crystals. Amphibole jade is also called nephrite. The material sold as jade may be either material, or in some cases, almost anything green. Historically, jadeite has been used to treat a wide range of human maladies and it continues to be used in Chinese folk medicine.

Composition

Omphacite is compositionally intermediate between augite and jadeite (Figure 14.2) but has slightly different structure. The M2 site is occupied by between 20 and 80% Na; the balance is occupied by Ca. The occupancy of the M1 sites by trivalent cations is about equal to the Na content in M2 sites to maintain charge balance, and Al^{3+} generally exceeds Fe^{3+}. The balance of the M1 sites is occupied by Mg and Fe^{2+}.

Form and Twinning

Crystals are stubby prisms with four- or eight-sided cross sections. More common as anhedral grains. Simple and lamellar twinning on {100} is common.

Cleavage

Good cleavage on {110} like the other pyroxenes intersects at 87° and 93°. Parting on {100}. Uneven to conchoidal fracture. Brittle.

Color Properties

Luster: Vitreous to silky.

Color: Green or dark green.

Streak: Gray.

Color in Thin Section: Colorless to pale green.

Pleochroism: Weakly pleochroic: X = colorless, Y = very pale green, Z = very pale green to blue green.

Optical Characteristics

Relief in Thin Section: High positive.

Optic Orientation: $X \wedge a = -18°$ to $-23°$, $Y = b$, $Z \wedge c = +34°$ to $+48°$.

Indices of Refraction and Birefringence: Indices of refraction generally increase with Fe^{3+} content, but with considerable variability. Birefringence is moderate, so maximum interference colors in thin section are usually upper first to lower second order.

Interference Figure: The normal range for $2V_z$ is between 60° and 75°, although the full range is larger. Optic axis dispersion is moderate, $r > v$ or $r < v$.

Alteration Alteration to fine-grained fibrous green amphibole is common. Omphacite may break down to a symplectic intergrowth of diopside and plagioclase.

Distinguishing Features

Hand Sample: Recognized as a pyroxene based on color, cleavage, and hardness. Amphiboles have better cleavage at 56° and 124°. Omphacite cannot reliably be distinguished from other pyroxenes in hand sample.

Thin Section/Grain Mount: Distinguished from the Ca-clinopyroxenes by larger $2V$ and from jadeite by darker color, and higher indices of refraction and birefringence. Aegirine and aegirine–augite are more distinctly colored, and have higher indices of refraction and smaller extinction angles in (010) section. The composition of omphacite from glaucophane schist may be estimated based on the value of n_β and the (221) d-spacing (Figure 14.9).

Occurrence Omphacite is a characteristic mineral of rocks formed in high-pressure conditions such as those found in the upper mantle or lower crust. Omphacite and garnet are major minerals in eclogite, the high-pressure metamorphic/igneous equivalent of basaltic/gabbroic rocks. Omphacite also is less commonly found in alpine metamorphic belts in blueschist, metagraywacke, and related rocks; also may be found in granulites.

Use None.

SPODUMENE

LiAlSi$_2$O$_6$
Monoclinic 2/*m*
$a = 9.45$ Å, $b = 8.39$ Å,
　$c = 5.215$ Å,
　$\beta = 110°$, $Z = 4$
$H = 6.5–7$
$G = 3.03–3.23$
Biaxial (+)
$n_\alpha = 1.648–1.668$
$n_\beta = 1.655–1.671$
$n_\gamma = 1.662–1.682$
$\delta = 0.014–0.027$
$2V_z = 58–68°$

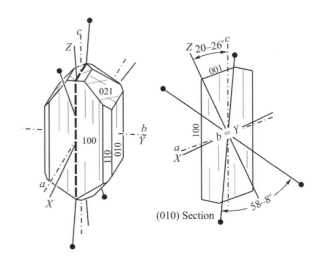

(010) Section

Composition Spodumene shows little compositional variation. Minor amounts of Na^+ and K^+ may substitute for Li^+ in the M2 sites, and Fe^{3+} may substitute for Al^{3+} in M1 sites. Trace amounts of other elements are common.

Form and Twinning Crystals are prismatic with roughly octagonal cross sections. Also as lath-like acicular crystals and cleavable masses. Crystals in some pegmatites may be very large.

Cleavage Good cleavage on {100} at 87° and 93° like the other pyroxenes. {100} parting. Uneven to subconchoidal fracture. Brittle.

Color Properties

Luster: Vitreous.
Color: White or grayish white, also pale shades of blue, green, and yellow.
Streak: White.
Color in Thin Section: Colorless.
Pleochroism: None.
　Fluoresces yellow, orange, or pink under both longwave and shortwave ultraviolet light.

Optical Characteristics

Relief in Thin Section: Moderately high positive.
Optic Orientation: $X \wedge a = 0°$ to $-6°$, $Y = b$, $Z \wedge c = +20°$ to $+26°$.

Indices of Refraction and Birefringence: Indices of refraction vary within a small range. Substitution of Na for Li causes indices of refraction to decrease; substitution of Fe^{3+} for Al increases both indices of refraction and birefringence.

Interference Figure: $2V_z$ varies in a small range (58°–68°). Optic axis dispersion is $r < v$.

Alteration Spodumene is readily altered to Li-mica, albite, and **eucryptite** (LiAlSiO$_4$), which is uniaxial (+), colorless, granular, or prismatic, with $n_\omega = 1.572$ and $n_\epsilon = 1.586$.

Distinguishing Features

Hand Sample: Light color, prismatic crystals, hardness, and cleavage distinguish it from other likely minerals in pegmatites. The combination of cleavage and parting may make spodumene resemble tremolite, but tremolite has only two cleavages and they are at 56° and 124°.

Thin Section/Grain Mount: Occurrence in pegmatites and small extinction angle distinguish it from other pyroxenes. Aegirine has a similar extinction angle, but has a very different occurrence and is colored. Higher indices of refraction quickly distinguish spodumene from feldspars.

Occurrence Spodumene is restricted to Li-bearing pegmatite. It is commonly associated with quartz, K-feldspar, plagioclase, tourmaline, beryl, and lepidolite (Li-mica).

Use Mined as a source of lithium, which has a wide range of uses including pharmaceuticals, ceramics, batteries, and lubricants. Among the historical medical uses of spodumene is treatment of mental health issues, perhaps suggesting that spodumene may have had some level of efficacy because modern medicine uses Li-based preparations as treatments for bipolar disorder and depression.

PYROXENOIDS

Introduction

The pyroxenoid group differs from the pyroxenes principally in that the chains of silicon tetrahedra are distorted or twisted (Figure 14.10). In the pyroxenes (Figure 14.10a), the repeat distance along the chains is two tetrahedra, or about 5.2 Å, which is the c unit cell dimension. In the pyroxenoids, the repeat distance is three or more tetrahedra. Cations in distorted octahedral coordination tie the chains together as in the pyroxenes.

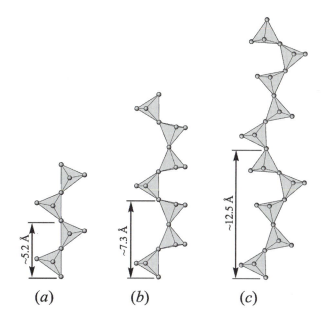

Figure 14.10 Chain structure of pyroxenes and pyroxenoids. Chains of silicon tetrahedra in the pyroxenoids are twisted or kinked and have repeat distances of three or more tetrahedra. Two repeats are shown. Cations, in distorted octahedral coordination, tie the chains together in the pyroxenoids, similar to the pyroxenes. (*a*) Pyroxene. Repeat equals two tetrahedra. (*b*) Wollastonite. Repeat equals three tetrahedra. (*c*) Rhodenite. Repeat equals five tetrahedra.

Structure Wollastonite occurs in three structural varieties: wollastonite-Tc, wollastonite-2M, and pseudowollastonite. All are constructed of tetrahedral chains with a repeat distance of three tetrahedra (Figure 14.10b) and differ in details of how the chains are positioned relative to each other. Wollastonite-Tc, which is triclinic, is the most common and is described here.

Composition Most wollastonite is relatively pure $CaSiO_3$, but both Mn and Fe^{2+} may substitute for Ca.

Form and Twinning Wollastonite is usually bladed, columnar, or fibrous with grains elongate parallel to b and flattened parallel to {100} or {001}. Twins with {100} composition planes are common, but are not visible in hand sample.

Cleavage A single perfect cleavage on {100} and good cleavages on {001} and {$\bar{1}$02} produce splintery cleavage fragments. The angles between cleavages are {100} ∧ {001} = 84.5° and {100} ∧ {$\bar{1}$02} = 70° Uneven fracture. Brittle.

Color Properties

Luster: Vitreous.

Color: White, gray, or pale green. Some may be fluorescent.

WOLLASTONITE

$CaSiO_3$
Triclinic $\bar{1}$
a = 7.94 Å,
   b = 7.32 Å,
   c = 7.07 Å,
   α = 90.03°, β = 95.37°,
   γ = 103.43°, Z = 6
H = 4.5–5
G = 2.86–3.09
Biaxial (–)
n_α = 1.616–1.645
n_β = 1.628–1.652
n_γ = 1.631–1.656
δ = 0.013–0.017
$2V_x$ = 36–60°

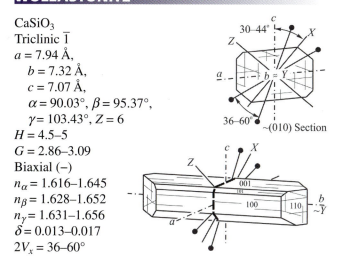

Streak: White.

Color in Thin Section: Colorless.

Pleochroism: None.

Optical Characteristics

Relief in Thin Section: Moderate to moderately high positive.

Optic Orientation: $X \wedge c = -30°$ to $-44°$, $Y \wedge b \approx 0°$, $Z \wedge a = +35°$ to $+49°$, optic plane ~ (010). Cleavage fragments and the trace of cleavage are length slow or length fast, depending on orientation, because Y is parallel to fiber length.

Indices of Refraction and Birefringence: Pure wollastonite has indices of refraction near the lower end of the range given above. Indices of refraction increase with Fe and Mn content. Maximum interference color in thin section is upper first order.

Interference Figure: Most wollastonite has $2V$ of about 40°; larger values are for Fe-bearing samples. Optic axis dispersion is weak to moderate, $r > v$.

Alteration Pectolite, calcite, or other Ca-rich minerals may replace wollastonite.

Distinguishing Features

Hand Sample: Resembles tremolite, with which it is often associated, but has different cleavage. The different varieties of wollastonite can be distinguished only with X-ray diffraction techniques.

Thin Section/Grain Mount: Resembles tremolite and pectolite, but both have higher birefringence and are consistently length slow. Pyroxene such as diopside has higher relief, is optically positive, and rarely displays a bladed or elongate habit.

Occurrence Wollastonite is a common mineral in metamorphosed limestone and dolomite. Associated minerals include calcite, dolomite, tremolite, members of the epidote group, grossular, diopside, and other Ca-Mg silicates. Occurrences in some alkalic igneous rocks have been reported.

Use Wollastonite, because it has an acicular habit when crushed, has replaced chrysotile asbestos in some industrial applications such as fire-resistant interior and exterior construction board, siding, roofing tile, and insulation. Paints and other coatings may contain wollastonite to improve workability and color brightness, and to increase film strength. The plastics industry uses wollastonite as a filler and to improve the working property of plastic during manufacturing processes. It also is used in ceramics including bioceramics, and as a flux in metallurgical processes.

Comments Because processing wollastonite produces splintery fibers, the mineral has come under scrutiny as a possible health risk, similar to the asbestos minerals. The available studies suggest that wollastonite, if inhaled, does not pose an unusual health hazard.

RHODONITE

$MnSiO_3$
Triclinic $\bar{1}$
$a = 9.758$ Å,
$\quad b = 10.499$ Å,
$\quad c = 12.205$ Å,
$\quad \alpha = 108.98°$,
$\quad \beta = 102.92°$,
$\quad \gamma = 82.52°$,
$\quad Z = 20$
$H = 5.5–6.5$
$G = 3.57–3.76$
Biaxial (+)
$n_\alpha = 1.711–1.734$
$n_\beta = 1.715–1.739$
$n_\gamma = 1.724–1.748$
$\delta = 0.011–0.017$
$2V_z = 63–87°$

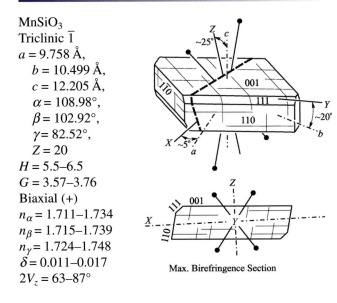

Max. Birefringence Section

Structure The structure consists of distorted chains of silicon tetrahedra (Figure 14.10c) tied laterally through Mn in octahedral coordination.

Composition Up to 20% of the Mn may be replaced by Ca and/or Mg/Fe.

Form and Twinning Rhodonite usually occurs in massive granular aggregates or anhedral grains. Crystals are tabular parallel to (001). Lamellar twins with (010) composition planes are not common and not evident in hand sample.

Cleavage Two perfect cleavages on {110} and {$\bar{1}$10}, and a fair cleavage on {001}. The angle between the perfect cleavages is 92.5°. Conchoidal to uneven fracture. Brittle.

Color Properties

Luster: Vitreous.

Color: Pink, rose red, or red brown.

Streak: White.

Color in Thin Section: Colorless to pale pink.

Pleochroism: Weak: X = yellowish red, Y = pinkish red, Z = pale yellowish red.

Optical Characteristics

Relief in Thin Section: High positive.

Optic Orientation: $X \wedge a \approx 5°$, $Y \wedge b \approx 20°$, $Z \wedge c \approx 25°$, the optic plane is roughly parallel to (010).

Indices of Refraction and Birefringence: Indices of refraction decrease with increasing Ca content. The maximum interference color in thin section is upper first-order yellow or red.

Interference Figure: 2V increases slightly with increased Ca content. Optic axis dispersion is weak, r < v, crossed bisectrix dispersion may be distinct.

Alteration Rhodonite may alter to rhodochrosite or Mn oxides and hydroxides.

Distinguishing Features

Hand Sample: Characterized by pink color and nearly right-angle cleavage. Dark alteration/weathering products that include Mn oxides may mask the mineral color. Rhodochrosite is softer, has rhombohedral cleavage, and reacts with acid.

Thin Section/Grain Mount: Pale pink color, if present. **Bustamite** [(Ca,Mn)SiO$_3$] and pyroxmangite [(Mn,Fe)SiO$_3$] are similar minerals. Pyroxmangite has higher birefringence, and bustamite is optically negative with 2V less than 60°.

Occurrence Generally restricted to Mn-bearing hydrothermal systems and skarn deposits. It commonly is associated with other Mn minerals, such as rhodochrosite and pyrolusite, as well as quartz, other carbonates, and a variety of sulfide minerals.

Use The pink to red color makes it useful as a decorative stone and in the lapidary arts.

PECTOLITE

Ca$_2$NaH(SiO$_3$)$_3$
Triclinic $\bar{1}$
a = 7.99 Å,
$\quad$ b = 7.04 Å,
$\quad$ c = 7.02 Å,
$\quad$ α = 90.52°,
$\quad$ β = 96.18°,
$\quad$ γ = 102.47°,
$\quad$ Z = 2
H = 4.5–5
G = 2.86–2.90
Biaxial (+)
n_α = 1.592–1.610
n_β = 1.603–1.615
n_γ = 1.630–1.645
δ = 0.026–0.039
$2V_z$ = 50–63°

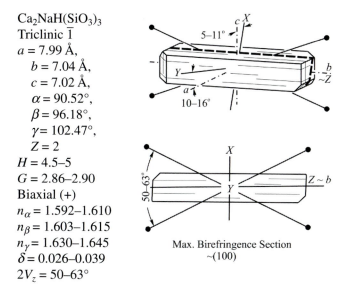

Max. Birefringence Section
~(100)

Structure Distorted chains of silicon tetrahedra are linked laterally through Ca and Na cations.

Composition Most pectolite is near the ideal composition. Mn may substitute for Ca in some samples.

Form and Twinning Figure 14.11. Pectolite commonly forms radiating groups or compact masses of acicular crystals elongate along *b*. Twins with {100} composition planes are rare.

Cleavage Perfect cleavages on {100} and {001} intersect at 95°.

Color Properties

Luster: Vitreous.
Color: White, colorless.
Streak: White.
Color in Thin Section: Colorless.
Pleochroism: None.

Optical Characteristics

Relief in Thin Section: Moderate positive.

Optic Orientation: X $\wedge$ c = –5° to –11°, Y $\wedge$ a = +10° to +16°, Z $\approx$ b, optic plane is nearly parallel to (100). Cleavage fragments and trace of cleavage in thin section are length slow.

Indices of Refraction and Birefringence: Substitution of Mn for Ca causes indices of refraction to increase. Interference colors in thin section are up to upper second order.

Interference Figure: 2V varies in a modest range for most samples, though values as small as 35° have been reported, probably as a consequence of substitution of Mn. Optic axis dispersion is weak, r > v or r < v.

Alteration Pectolite may be replaced by pale pink **stevensite**, which is an Mg-rich clay mineral in the smectite group.

Distinguishing Features

Hand Sample: Resembles wollastonite and acicular zeolites such as natrolite. Wollastonite is not common as radiating

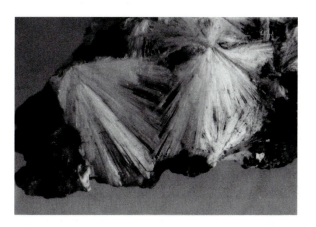

Figure 14.11 Photograph of radiating masses of pectolite. Field of view is about 10 cm wide.

masses in altered volcanics, and natrolite is more likely to form slender radiating crystals, as opposed to radiating masses.

Thin Section/Grain Mount: Pectolite has higher birefringence than most zeolites, with which it is commonly associated. Wollastonite has lower birefringence and displays both length fast and length slow sign of elongation.

Occurrence Pectolite is most common as an amygdule and cavity filling in basalt flows, diabase dikes, and related rocks. Associated minerals include zeolites and calcite. Pectolite may also be found in serpentinites, as a primary mineral in alkalic igneous rocks such as nepheline syenite, and in skarns or other Ca-rich metamorphic rocks.

Use None.

AMPHIBOLE GROUP

Structure and Classification

Amphibole structure and classification are very similar to those of the pyroxenes. The basic building block consists of a double chain of silicon tetrahedra (Figures 11.2e and 14.12a). These chains extend parallel to the *c* crystal axis and are stacked in an alternating fashion, like the pyroxenes. Bases of tetrahedra face bases, and the unshared points (i.e., the apical oxygens) on tetrahedra face points (Figure 14.12b,c,d). This geometry produces five different structural sites between the chains. Octahedral M1, M2, and M3 sites are between the points of the chains; larger M4 and A sites are between the bases.

Two chains with intervening cations in octahedral M1, M2, and M3 sites form a structure analogous to a strip of a *TOT* layer in the layer silicates (Figure 14.12b). These *TOT* strips are stacked together and bonded through cations in the M4 and A sites to form the amphibole structure. Cleavage occurs between the *TOT* strips (Figure 14.12e). Because the *TOT* strips in the amphiboles are wider than the equivalent "I-beams" in the pyroxenes, cleavage is at about 56° and 124°, rather than nearly at right angles. In general, cleavage in the amphiboles is better than in the pyroxenes.

The amphiboles have a general formula

$$A_{0-1}B_2C_5T_8O_{22}W_2$$

where

A = **Na**, □, K, Ca, Li
B = **Na, Ca, Fe²⁺, Mg**, Mn²⁺, Li
C = **Mg, Fe²⁺, Al, Fe³⁺**, Mn²⁺, Mn³⁺, Ti⁴⁺, Li
T = **Si, Al**, Ti⁴⁺
W = **(OH)⁻**, F⁻, Cl⁻, or O²⁻

Bold type indicates the more common occupants, □ indicates a vacancy. The A cations occupy the A structural sites

(Figure 14.12); in some amphiboles, the A site is filled, in others it is partially or entirely vacant. The two B cations are located in the M4 sites. These are analogous to the M2 sites in pyroxenes and may have either 8- or 6-fold coordination depending on exactly how the *TOT* strips are arranged relative to each other. If 8-fold, the B cations are usually Ca^{2+} or Na^+, and if 6-fold, the B cations are Mg^{2+} or Fe^{2+}. The five C cations occupy the octahedral M1, M2, and M3 sites, which are the octahedral sites in the middle of the *TOT* strips. The eight T sites are tetrahedral and usually contain Si and some Al. The W in the formula signifies additional anions, most commonly $(OH)^-$; because amphiboles contain hydroxyl, they are considered hydrous minerals.

Because of the compositional complexity and new discoveries, amphibole classification is revised regularly. The most recent classification by the Commission on New Minerals, Nomenclature and Classification of the International Mineralogical Association came out in 2003 (Leake and others, 2003), and revisions were suggested in 2007 (Hawthorne and Oberti, 2007). A significant problem is that routine electron microprobe (Chapter 9) chemical analyses do not provide enough chemical data to uniquely classify many amphiboles. The microprobe cannot distinguish between Fe^{2+} and Fe^{3+}, nor can it detect chemical elements lighter than fluorine ($Z = 9$). Hydrogen ($Z = 1$) is very important in the structure, and it is now recognized that lithium ($Z = 3$) can be an important chemical constituent in some amphiboles. These elements can be analyzed by using secondary ion and laser ablation mass spectrometry techniques, but the needed instrumentation is not nearly as widely available as the electron microprobe. X-ray diffraction techniques (Chapter 8) can distinguish amphiboles from most other minerals but cannot parse the compositional complexity. An additional issue is that determining the chemical classification realistically requires a computer program, even with a complete chemical analysis, and the complete classification scheme is far too complicated to use for hand sample and thin section work.

The abbreviated classification shown in Table 14.2 includes the majority of common amphiboles. The four main groups are classified primarily on the identity of the X cations in the M4 structural sites (Figure 14.12). Each of these amphibole groups has a structural and compositional analog in the pyroxenes (Table 14.1). Like the pyroxenes, the amphiboles can be divided into those that are orthorhombic—the orthoamphiboles—and those that are monoclinic—the clinoamphiboles. The common ortho- and clinoamphiboles can be shown on a triangular diagram whose vertices are $Mg_7Si_8O_{22}(OH)_2$, $Fe_7Si_8O_{22}(OH)_2$, and $Ca_7Si_8O_{22}(OH)_2$ (Figure 14.13). Note that a mineral with the latter composition does not exist.

The clinoamphiboles are the most common amphiboles, and they display considerable diversity of composition involving both simple substitutions and coupled

Table 14.2 Abbreviated Amphibole Classification Based on the General Formula $A_{0-1}B_2C_5T_8O_{22}W_2$[a]

Group	A	B_2	C_5	$T_8O_{22}(OH)_2$	Mineral	Symmetry	Comment
Iron–magnesium		$(Mg,Fe)_2$	$(Mg,Fe)_5$	$Si_8O_{22}(OH)_2$	Anthophyllite	Orthorhombic	Orthoamphibole
		$(Mg,Fe)_2$	$(Mg,Fe)_3Al_2$	$Al_2Si_6O_{22}(OH)_2$	Gedrite		
		$(Mg,Fe)_2$	$(Mg,Fe,Ca)_5$	$Si_8O_{22}(OH)_2$	Cummingtonite–Grunerite	Monoclinic	Low-Ca clinoamphibole
Calcic		Ca_2	$(Mg,Fe)_5$	$Si_8O_{22}(OH)_2$	Tremolite–Actinolite		Ca clinoamphibole
	$(Na,K)_{0-1}$	Ca_2	$(Mg,Fe^{2+},Fe^{3+},Al)_5$	$(Si,Al)_8O_{22}(OH)_2$	Hornblende		
	Na	Ca_2	$(Mg,Fe)_4Ti$	$Si_6Al_2O_{22}(OH)_2$	Kaersutite		
Sodic–calcic	Na	CaNa	$(Mg,Fe)_5$	$Si_8O_{22}(OH)_2$	Richterite		Na–Ca clinoamphibole
	Na	CaNa	$(Mg,Fe)_4Fe^{3+}$	$Si_7AlO_{22}(OH)_2$	Katophorite		
Sodic		Na_2	$(Mg,Fe^{2+})_3(Al,Fe^{3+})_2$	$Si_8O_{22}(OH)_2$	Glaucophane–Riebeckite		Na clinoamphhibole
	Na	Na_2	$(Mg,Fe^{2+})_4(Al,Fe^{3+})$	$Si_8O_{22}(OH)_2$	Eckermannite–Arfvedsonite		

[a]The A cations occupy the A sites, B cations the M4 sites, C cations the M1, M2, and M3 sites (Figure 14.12); W is assumed to be (OH).
Source: Adapted from Leake and others (2003).

substitutions. Consider the composition of tremolite $[Ca_2Mg_5Si_8O_{22}(OH)_2]$ as a starting point. Simple substitutions that involve cations of the same charge are listed in Table 14.3, where □ = vacancy. Coupled substitutions that involve cations of different charge in two or more sites are listed in Table 14.4. These substitution schemes can be used individually, or in combination. For example, substitution of Al^{3+} for Si^{4+} can be balanced by inserting some Na^+ in the A site combined with substituting some Al^{3+} for Mg^{2+} in octahedral sites.

Given this compositional diversity, it should be obvious that optical and physical properties are not up to the task of providing unambiguous information about the composition of amphibole. Knowledge of the rock type in which the amphibole is found, combined with physical and optical properties, can allow a reasonable guess to be made, but is not infallible. The Mg-Fe amphiboles found in mafic metamorphic rocks usually are brown; anthophyllite is the most common variety. Sodic amphiboles are usually blue; glaucophane is found in blueschists, and riebeckite in alkalic igneous rocks. The calcic and sodic–calcic amphiboles are most commonly dark green to black and occur in a wide variety of rocks; the term **hornblende** is commonly applied to these amphiboles. White or pale green amphibole in metamorphosed carbonates is usually tremolite. Optical properties allow some additional refinement in classification, but the properties of the calcic and sodic–calcic amphiboles overlap substantially. If knowledge of composition is important, the only real recourse is detailed chemical analysis with electron microprobe or other techniques. Figure 14.14 shows some common amphiboles.

Geology of Amphiboles

To understand the geological distribution of the amphiboles, it is first necessary to recognize that they are hydrous. The fact that hydroxyl is an essential constituent of the structure means that amphiboles are not stable in anhydrous environments. We also can expect that very high temperatures will cause dehydration and breakdown to form anhydrous minerals like the pyroxenes. Further, the amphiboles have a higher Si:O ratio (4:11) than either the pyroxenes or olivine, so they should be found in more Si-rich rocks than either of those minerals. These considerations lead to the following generalizations.

• Amphiboles are not normally abundant in mafic and ultramafic igneous rocks. These magmas have a relatively low silicon content, little dissolved water, and they crystallize at high temperatures. If present, amphiboles usually crystallize late in the magmatic history, when the residual melt is enriched both in Si and H_2O, and temperatures are lower. Overgrowths of amphibole on pyroxene in mafic and ultramafic igneous rocks are fairly common.

• Amphiboles, particularly calcic and sodic–calcic varieties, should be common in intermediate to felsic igneous rocks. These magmas typically contain significant amounts of dissolved water and intermediate to high Si contents. Whether amphibole or biotite is the major Mg/Fe-bearing mineral in these rocks is determined, in part, by the abundance of K (for biotite) versus Na/Ca (for amphibole). As a general rule, amphibole is typical in intermediate igneous rocks such as diorite, granodiorite, andesite, and dacite, whereas biotite is more common in felsic varieties.

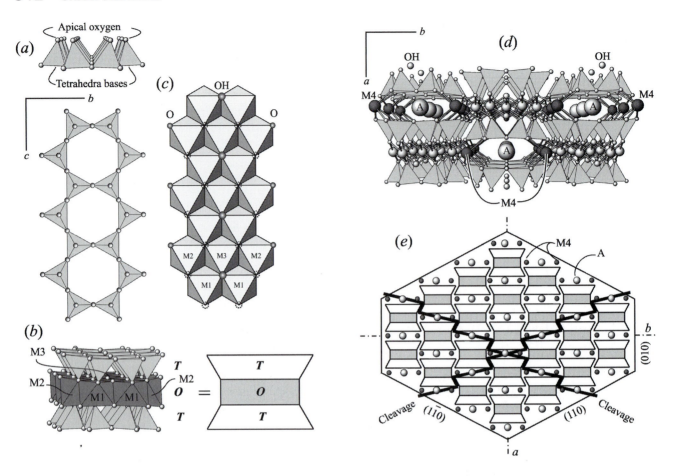

Figure 14.12 Amphibole structure. (*a*) Top and end views of an idealized double chain of tetrahedra. (*b*) Two chains of tetrahedra (*T*) are arranged facing each other forming octahedral (*O*) M1, M2, and M3 sites in between. The cross section shape of this *TOT* strip, or "I-beam," is shown to the right. (*c*) Top view of the octahedral layer showing distribution of M1, M2, and M3 sites. O and OH that are not shared with a tetrahedral chain are shown as filled spheres on the top surface of the octahedrons. O that are shared with a tetrahedral chain are shown as dotted circles. (*d*) *TOT* strips extend parallel to the *c* axis and are stacked atop each other in the *a* direction to produce M4 and A sites between the bases of the tetrahedra. The M1, M2, and M3 cations are shown, rather than their coordination polyhedra. (*e*) View down the *c* axis showing the arrangement of *TOT* strips ("I-beams") and their relation to the typical amphibole cross section. Cleavage between chains parallel to (110) and (1$\bar{1}$0) is at about 56° and 124°.

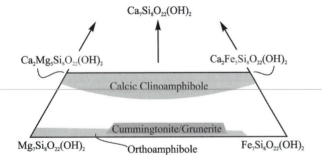

Figure 14.13 Fe–Mg–Ca amphiboles; common compositions are shown with shading. The calcic clinoamphiboles have additional compositional diversity beyond variation in the amount of Fe, Mg, and Ca.

• Amphiboles are common in rocks with intermediate to mafic compositions subjected to medium- to high-grade regional metamorphic conditions. The pore fluid

Table 14.3 Simple Substitutions in the Amphiboles

Formula	Structural Site	Cations
A	A	□, Na, K
B	M4	Ca, Mg, Fe^{2+}
C	M1, M2, M3	Mg, Fe^{2+}
C	M1, M2, M3	Al^{3+}, Fe^{3+}
T	Tetrahedral	Si^{4+}, Ti^{4+}

accompanying metamorphism under these conditions usually is water rich, because water is available from the breakdown of hydrous minerals such as the micas and chlorite. Amphibolite-facies metamorphism gets its name from the abundant amphibole in metamorphosed mafic rocks. At very high temperatures, amphiboles break down by dehydration to produce the pyroxenes typical of granulite facies metamorphism.

Table 14.4 Selected Coupled Substitution Schemes in Clinoamphiboles

A	B	C	T	=	A	B	C	T
		(Fe^{2+},Mg^{2+})	Si^{4+}	=			(Al^{3+},Fe^{3+})	Al^{3+}
			Si^{4+}	=	Na^+			Al^{3+}
	Ca^{2+}			=	Na^+	Na^+		
	Ca_2^{2+}	(Fe^{2+},Mg^{2+})		=	Na^+	Na_2^+	Al^{3+}	

Students are cautioned that these are generalizations, subject to the usual disclaimer that Mother Nature may not have read this book to discover what minerals are supposed to occur in each geological environment. If we discover minerals occurring in unlikely environments, it is not that the minerals are in the wrong place, it is that our understanding of the mineral formations is imperfect.

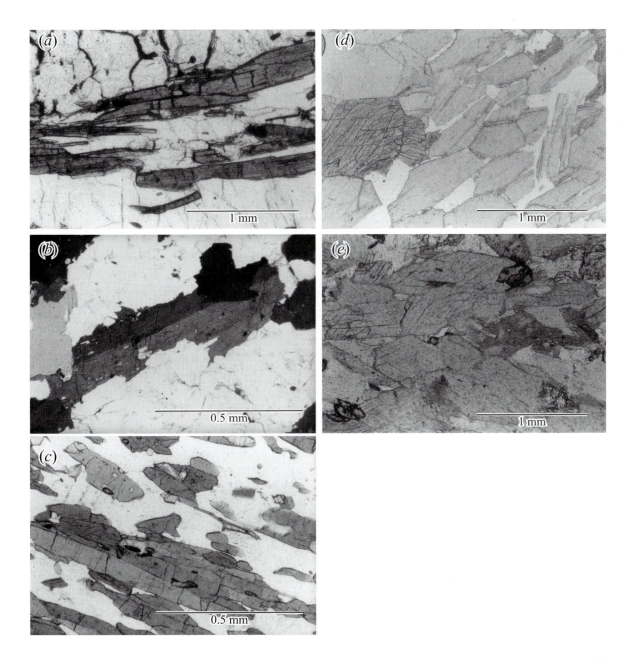

Figure 14.14 Common amphiboles in thin section. (*a*) Anthophyllite with cloudy cordierite (*top*) and quartz. (*b*) Hornblende in granodiorite. Note the twin. (*c*) Hornblende in amphibolite. (*d*). Tremolite in marble. (*e*) Glaucophane in glaucophane schist. The small high-relief grains at the lower left and upper right are titanite.

ORTHOAMPHIBOLE

Anthophyllite:
$(Mg,Fe)_2(Mg,Fe)_5Si_8O_{22}(OH)_2$,
$a = 18.54$ Å, $b = 18.03$ Å,
$\quad c = 5.28$ Å,
$\quad Z = 4$
Gedrite: $(Mg,Fe)_2(Mg,Fe)_3$
$\quad Al_2\,Al_2Si_6O_{22}(OH)_2$,
$a = 18.53$–18.60 Å,
$\quad b = 17.74$–17.89 Å,
$\quad c = 5.25$–5.28 Å, $Z = 4$
Orthorhombic $2/m\ 2/m\ 2/m$
$H = 5.5$–6
$G = 2.85$–3.57
Biaxial (+ or −)
$n\alpha = 1.587$–1.694
$n_\beta = 1.602$–1.710
$n_\gamma = 1.613$–1.722
$\delta = 0.013$–0.028
$2V_x = 65$–122° (anthophyllite)
$2V_x = 72$–133° (gedrite)

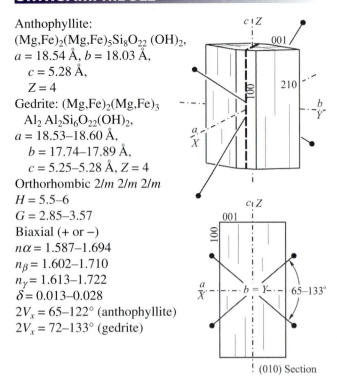

(010) Section

Composition At high temperatures, it appears that anthophyllite and gedrite are part of a continuous solid solution series. At lower temperatures, however, a solvus exists so that lamellae of gedrite may exsolve parallel to (010) and (120) in anthophyllite, and vice versa. Anthophyllite is usually restricted to fairly Mg-rich compositions (>60% Mg), and gedrite has more Fe- and Al-rich compositions. Gedrite may contain some Na in the A site, which remains vacant in anthophyllite.

Form and Twinning Figure 14.14. Anthophyllite and gedrite are common as euhedral to subhedral columnar, bladed, or acicular crystals with the diamond-shaped cross section typical of amphiboles. Also found as coarse to fine radiating aggregates and as an asbestiform variety. Exsolution lamellae of gedrite in anthophyllite, and vice versa, are always present, although the lamellae may be submicroscopic. Twins are not reported.

Cleavage Perfect {210} like all amphiboles (the Miller index is different because the *a* unit cell dimension in the orthoamphiboles is twice that of the clinoamphiboles). Also distinct cleavage on {100} and {010} unlike the clinoamphiboles. Irregular fracture. Brittle. Fibers are elastic.

Color Properties

Luster: Vitreous.

Color: Clove brown, less commonly gray, grayish brown, yellowish brown, brownish green, or white.

Streak: Gray.

Color in Thin Section: Colorless to pale brown, grayish brown or greenish brown; darker with higher Fe content.

Pleochroism: Weak with $Z \geq Y \geq X : X$ = colorless, pale yellow, pale grayish brown, yellowish green; Y = colorless, pale yellow, pale grayish brown, yellowish green; Z = pale grayish brown, purplish brown, grayish green. Grains are darkest when the long dimension (*c* axis) is parallel to the vibration direction of the lower polarizer.

Optical Characteristics

Relief in Thin Section: Moderate to high positive.

Optic Orientation: $X = a$, $Y = b$, $Z = c$, trace of cleavage, and elongate grains are length slow.

Indices of Refraction and Birefringence: Indices of refraction for both minerals increase fairly systematically with Fe content (Figure 14.15). Gedrite has slightly higher birefringence than anthophyllite, but the ranges overlap.

Interference Figure: Both minerals may be either optically positive or negative, and optic angle varies systematically with composition. Anthophyllite is most commonly optically negative, and gedrite is most commonly positive. Optic axis dispersion is weak to moderate, $r > v$ if negative, $r < v$ if positive.

Alteration Fine-grained serpentine, talc, chlorite, or other layer silicates are the usual alteration products.

Distinguishing Features

Hand Sample: Brown color distinguishes anthophyllite and gedrite from hornblende and related greenish-black clinoamphiboles. Cummingtonite, however, often is quite similar in hand sample. Cleavage at 56° and 124° distinguishes the amphiboles from pyroxenes.

Thin Section/Grain Mount: Parallel extinction in longitudinal sections distinguishes anthophyllite and gedrite from all clinoamphiboles, including cummingtonite. Cummingtonite also commonly has lamellar twins. Sillimanite is colorless and has a single cleavage seen in basal sections. Gedrite and anthophyllite are quite similar; but gedrite has higher indices of refraction and is most commonly optically positive, whereas anthophyllite is most commonly optically negative. In addition, exsolution lamellae of gedrite may be very fine and difficult to see in anthophyllite, whereas anthophyllite lamellae in gedrite are more commonly large enough to be visible.

Occurrence Both occur in metamorphosed mafic rocks of medium to high grade and are unknown in igneous rocks. Cordierite is a very commonly associated mineral; others include garnet, aluminum silicates, plagioclase, and sometimes staurolite. The protolith for anthophyllite cordierite gneiss layers that are fairly common in Proterozoic metamorphic terranes is something of an enigma. It is clearly very different than the normal composition range of igneous, sedimentary, or metamorphic rocks.

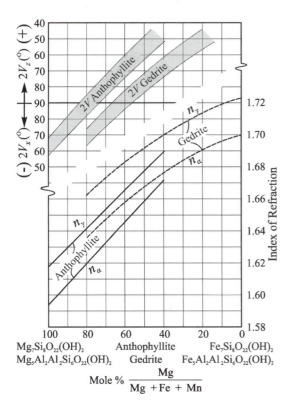

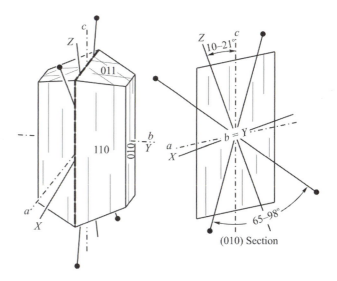

(010) Section

Figure 14.15 Optical properties of orthoamphiboles. Indices of refraction display substantial scatter and only the trends are shown. Data from Winchell (1938), Rabbitt (1948), Deer and others (1963).

Use Fibrous varieties have been used as a form of asbestos, but this use is limited. Fibrous anthophyllite has been implicated in the health risks that are associated with other asbestos minerals. See additional discussion in the section on serpentine (Chapter 13).

CUMMINGTONITE–GRUNERITE

$(Mg,Fe,Ca)_2(Mg,Fe)_5Si_8O_{22}(OH)_2$
Monoclinic $2/m$
$a = 9.52\text{–}9.56$ Å,
$\quad b = 18.14\text{–}8.39$ Å,
$\quad c = 5.31\text{–}5.34$ Å,
$\quad \beta = 102.1°, Z = 2$
$H = 5\text{–}6$
$G = 3.10\text{–}3.60$
Biaxial (+ or −)
$n_\alpha = 1.630\text{–}1.686$
$n_\beta = 1.638\text{–}1.709$
$n_\gamma = 1.655\text{–}1.729$
$\delta = 0.020\text{–}0.045$
$2V_z = 65\text{–}98°$

Composition Minerals in the cummingtonite–grunerite solid solution series are more Fe rich than anthophyllite; compositions with more than 70% Mg are uncommon. Small amounts of Ca usually are present in the B structural sites–somewhat more than in the orthoamphiboles. Grunerite has less than 50% Mg; cummingtonite has more.

Form and Twinning Usually columnar, bladed, or acicular and elongate parallel to c; also parallel or radiating aggregates. Basal sections have the diamond shape of amphiboles. Fibrous variety are called **amosite** or **montasite**. Simple or lamellar twinning on {100} is common.

Cleavage Very good {110} cleavage that intersects at 56° and 124° like the other amphiboles. Brittle.

Color Properties

Luster: Vitreous.

Color: Dark brown to dark green, gray, tan.

Streak: White.

Color in Thin Section: Colorless to pale green or brown, darker with higher Fe content.

Pleochroism: Weak pleochroism in colored varieties: X = colorless, pale yellow; Y = pale yellow, pale brown; Z = pale green, pale brown.

Optical Characteristics

Relief in Thin Section: Moderately high to high positive.

Optic Orientation: $X \wedge a = +2°$ to $-9°$, $Y = b$, $X \wedge c = +10°$ to $+21°$, trace of cleavage and crystal length (= c axis) are length slow. The $Z \wedge c$ extinction angle varies fairly systematically with composition (Figure 14.16).

Indices of Refraction and Birefringence: Indices of refraction decrease systematically with Mg content, as does the birefringence (Figure 14.16).

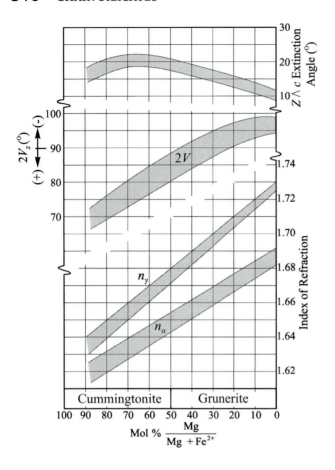

Figure 14.16 Optical properties of cummingtonite–grunerite. Data from Winchell (1938), Deer and others (1963, 1997), and Klein (1964).

Interference Figure: 2V varies systematically with Mg content; grunerite is usually optically negative, cummingtonite is usually optically positive. Optic axis dispersion is weak: r > v for grunerite, r < v for cummingtonite.

Alteration Overgrowths of hornblende are not uncommon, and thin lamellae of hornblende may exsolve under conditions of slow cooling. Alteration to chlorite, talc, or serpentine minerals is possible.

Distinguishing Features

Hand Sample: Distinguished as an amphibole by habit and cleavage. Brown varieties closely resemble orthoamphibole; green varieties closely resemble the calcic clinoamphiboles. A firm identification requires optical or other tests.

Thin Section/Grain Mount: Cummingtonite–grunerite is most similar to orthoamphibole and tremolite/actinolite in thin section. Orthoamphiboles display parallel extinction in (010) sections, whereas cummingtonite–grunerite has inclined extinction, somewhat higher birefringence, and prevalence of fine polysynthetic twinning. A positive optic sign distinguishes cummingtonite from members of the tremolite–actinolite series, and the occurrences of the two amphibole groups are different.

Occurrence Cummingtonite–grunerite occurs in metamorphosed mafic and ultramafic rocks, sometimes with an orthoamphibole with which it is easily confused, particularly in hand sample. Cummingtonite–grunerite also may be found in granitic gneiss, hornfels, and granulite. Associated minerals include orthoamphibole, cordierite, garnet, plagioclase, hornblende, and biotite. Cummingtonite is occasionally found in intermediate volcanic rocks, diorite, gabbro, norite, and in skarns. Grunerite may be found in metamorphosed iron-rich sediments along with magnetite, quartz, and other iron oxides and silicates.

Use The asbestiform variety (amosite/montasite) has been used as asbestos in some industrial applications, but mining has been limited to South Africa, so distribution is not wide. It has been implicated in the health risks associated with asbestos. See the section on serpentine for more information (Chapter 13).

TREMOLITE–FERRO-ACTINOLITE

$Ca_2(Mg,Fe)_5Si_8O_{22}(OH)_2$
Monoclinic 2/m
$a = 9.84–10.02$ Å,
   $b = 18.05–18.20$ Å,
   $c = 5.29–5.31$ Å,
   $\beta = 104.8–104.6°$,
   $Z = 2$
$H = 5–6$
$G = 2.99–3.48$
Biaxial (–)
$n_\alpha = 1.599–1.688$
$n_\beta = 1.612–1.697$
$n_\gamma = 1.622–1.705$
$\delta = 0.027–0.017$
$2V_x = 62–88°$

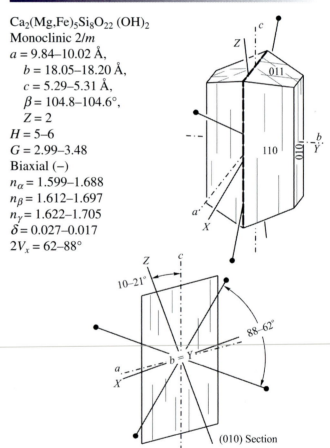

Composition Tremolite and ferro-actinolite represent a Mg–Fe solid solution series, with little substitution of Al^{3+} in either octahedral or tetrahedral sites. Compositions with more than 90% Mg are tremolite, between 90 and 50% Mg are actinolite, and less than 50% Mg are ferro-actinolite. Compositions in the ferro-actinolite range are uncommon. There is some debate about whether solid solution extends continuously between hornblende and tremolite–ferro-actinolite. The facts that both tremolite and hornblende coexist in the same rocks and that exsolution lamellae of one may be found in the other suggest that solid solution is not continuous, at least under some conditions of temperature, pressure, and/or composition.

Form and Twinning Commonly occurs as columnar, bladed, or acicular crystals elongate parallel to the *c* axis; also fibrous or asbestiform. Basal sections are diamond shaped. Simple and lamellar twins with {100} composition planes are common; lamellar twins on {001} are rare. A variety of jade composed of interlocking fibers of actinolite is called **nephrite**.

Cleavage Good cleavage on {110} intersects at 56° and 124°. Parting on {100}. Irregular fracture. Brittle. Fibrous aggregates of nephrite jade can be quite tough.

Color Properties

Luster: Vitreous.

Color: White, gray, pale green, dark green, darker with Fe content.

Streak: White.

Color in Thin Section: Colorless (tremolite) to deep green (ferro-actinolite)

Pleochroism: Tremolite has none. Actinolite and ferro-actinolite: $X \leq Y < Z$; X = colorless, pale yellowish green, pale green; Y = pale yellowish green, pale green; Z = pale green, green, bluish green, dark green.

Optical Characteristics

Relief in Thin Section: Moderate to high positive.

Optic Orientation: $X \wedge a = 5°$ to $-6°$, $Y = b$, $Z \wedge c = +11°$ to $+21°$. The $Z \wedge c$ extinction angle seen in (010) sections decreases with increasing Fe content above about 25% ferro-actinolite.

Indices of Refraction and Birefringence: Indices of refraction increase with Fe content (Figure 14.17), and the maximum birefringence, seen in (010) sections, yields interference colors in the upper first to middle second order.

Interference Figure: The entire series is optically negative, with $2V$ decreasing with increasing Fe content. Optic axis dispersion is weak, $r < v$.

Alteration Chlorite, talc, and carbonates are the common alteration products.

Distinguishing Features

Hand Sample: Cleavage, hardness, and habit indicate that it is an amphibole. The white to light green color of tremolite and some actinolite distinguishes them from hornblende.

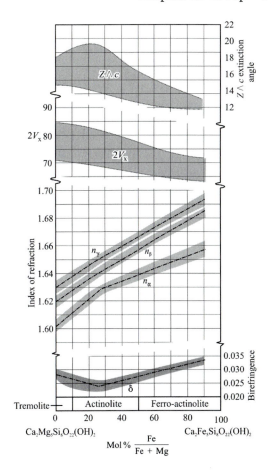

Figure 14.17 Optical properties of the tremolite–ferro-actinolite series. Data from Verkouteren and Wylie (2000).

Dark green actinolite and ferro-actinolite cannot be distinguished from hornblende.

Thin Section/Grain Mount: Pale color distinguishes tremolite and actinolite from hornblende. Colored samples closely resemble hornblende. If the identity is in doubt, use the term *calcic clinoamphibole*. Colorless varieties of richterite also are quite similar.

Occurrence Tremolite and actinolite are common minerals in contact and regionally metamorphosed limestone, dolomite, and other calcareous sediments. The commonly associated minerals include calcite, dolomite, forsterite, garnet, diopside, wollastonite, talc, and members of the epidote group.

Tremolite and actinolite also occur in low-grade regionally metamorphosed mafic and ultramafic igneous rocks. Associated minerals include talc, carbonates, antigorite, epidote group, and chlorite. Metagraywacke, blueschist, and related rocks in alpine metamorphic belts also may contain tremolite or actinolite. Ferro-actinolite is restricted to metamorphosed iron formations where it may be associated with calcite, dolomite, and cummingtonite.

Pyroxene in mafic and ultramafic igneous rocks commonly alters to **uralite**, which consists of fine-grained, light-colored amphibole, usually presumed to be tremolite–actinolite.

Use Compact massive nephrite, which usually consists of finely intergrown acicular crystals, can be used as a decorative stone and may be sold as a variety of jade. Fibrous tremolite may be used as an asbestos mineral and has been implicated, along with chrysotile and other amphibole asbestos minerals, as a health hazard. See the discussion under serpentine (Chapter 13).

HORNBLENDE

$(Na,K)_{0-1}Ca_2(Mg,Fe^{2+},Fe^{3+},Al)_5$
$(Si,Al)_8 O_{22}(OH)_2$
Monoclinic $2/m$
a = ~9.9 Å,
$\quad b$ = ~18.0 Å,
$\quad c$ = ~5.3 Å,
$\quad \beta$ = ~105.5°,
$\quad Z$ = 2
H = 5–6
G = 3.02–3.59
Biaxial (+ or –)
n_α = 1.60–1.70
n_β = 1.61–1.71
n_γ = 1.62–1.73
δ = 0.014–0.034
$2V_x$ = 35–130°

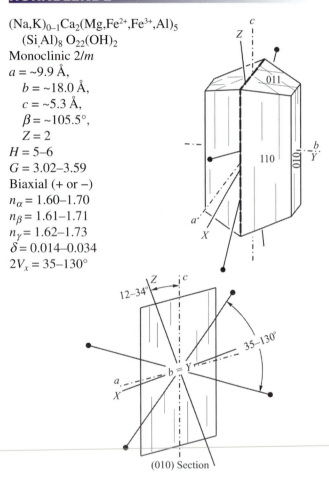

(010) Section

Composition Hornblende displays considerable variation in composition. In comparison to tremolite and actinolite, hornblende contains significant Na and K in the A site, and Fe^{3+} and Al^{3+} in the octahedral M1, M2, and M3 sites. Charge balance is maintained by substituting Al^{3+} for Si^{4+} in tetrahedral sites (Table 14.4) up to about $Al_{2.25}Si_{5.75}$. The material collectively called hornblende here includes members of the following series.

MAGNESIOHORNBLENDE–FERROHORNBLENDE:
$Ca_2(Mg,Fe^{2+})_4AlSi_7AlO_{22}(OH)_2$

TSCHERMAKITE–FERROTSCHERMAKITE:
$Ca_2(Mg,Fe^{2+})_3Al_2Si_6Al_2O_{22}(OH)_2$

EDENITE–FERRO-EDENITE:
$NaCa_2(Mg,Fe^{2+})_5Si_7AlO_{22}(OH)_2$

PARGASITE–FERROPARGASITE:
$NaCa_2(Mg,Fe^{2+})_4AlSi_6Al_2O_{22}(OH)_2$

MAGNESIOHASTINGSITE–HASTINGSITE:
$NaCa_2(Mg,Fe^{2+})_4Fe^{3+}Si_6Al_2O_{22}(OH)_2$

Additionally, Ti, Mn, and Cr may substitute in M1, M2, and M3 sites, and O, F, and Cl may substitute for (OH). **Oxyhornblende** is a variety in which most of the Fe has been oxidized to the trivalent state, with charge balance maintained by substituting O^{2-} for $(OH)^-$. If knowledge of composition is important, chemical analysis, such as with an electron microprobe, is required; optical, physical, and X-ray methods are not up to the task.

Form and Twinning Crystals are usually slender prismatic to bladed, and elongate parallel to c (Figure 14.14b, c). Cross sections are diamond shaped, bounded by the {110} prism and {010} pinacoid. Also in anhedral to highly irregular grains, and in fibrous masses that may mantle or replace pyroxene. {100} simple and lamellar twins are common.

Cleavage Good on {110} intersects at about 56° and 124°. Fragments are usually elongate. Parting on {100}. Irregular fracture. Brittle.

Color Properties

Luster: Vitreous.

Color: Most commonly green, dark green, black, also brown.

Streak: Gray to greenish gray.

Color in Thin Section: Distinctly colored, usually in shades of green, bluish green, or brown. Oxyhornblende is brown.

Pleochroism: Distinct, usually $Z > Y > X$, less commonly $Y > Z > X$. Green varieties are usually X = pale yellow, yellowish green, light blue-green; Y = green, yellowish green, gray green; Z = dark green, dark blue green, dark gray green. Brown varieties are usually X = yellow, greenish yellow, light greenish brown; Y = yellowish brown, brown, reddish brown; Z = grayish brown, brown, reddish brown. Color zoning may be evident in thin section.

Optical Characteristics

Relief in Thin Section: Moderate to high positive.

Optic Orientation: $X \wedge a = +3°$ to $-9°$, $Y = b$, $Z \wedge c = +12°$ to $+34°$ optic plane = (010). The trace of cleavage and cleavage fragments are length slow.

Indices of Refraction and Birefringence: Indices of refraction increase with Fe content (Figure 14.18). Most common hornblende has indices of refraction in the range n_α = 1.655 ± 0.010, n_β = 1.665 ± 0.010, and n_γ = 1.675 ± 0.015 with birefringence between 0.018 and 0.025. Mineral color may mask interference color.

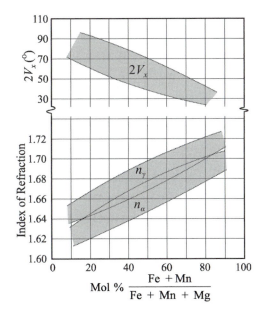

Figure 14.18 Optical properties of hornblende. The bands show the common range of indices of refraction and 2V but do not include the full range of values reported in the literature.

Interference Figure: Common hornblende is optically negative, with $2V_x = 75° \pm 10°$. Optic axis dispersion is moderate, either $r > v$ or $r < v$. Much pargasite, and some tschermakite and edenite, is optically positive with a large 2V.

Alteration Hornblende may be altered to biotite, chlorite, or other Fe–Mg silicates.

Distinguishing Features

Hand Sample: Habit, cleavage, and hardness distinguish hornblende as an amphibole and not pyroxene. The dark green to black color usually distinguishes hornblende from other amphiboles, although some other varieties also may have the same color.

Thin Section/Grain Mount: Green color, distinct pleochroism, cleavage, and inclined extinction in (010) sections distinguish hornblende from most other minerals. Aegirine and aegirine–augite have cleavage at nearly right angles in basal sections and usually have higher birefringence. The properties of actinolite and hornblende overlap. Oxyhornblende may resemble biotite, but has inclined extinction and larger 2V.

Occurrence Hornblende may be found in a wide variety of igneous rocks but is a characteristic mineral of intermediate varieties including diorite, granodiorite, andesite, and dacite. It also may be a late-crystallizing mineral in mafic rocks such as gabbro or norite, and is not uncommon in granitic/rhyolitic rocks, syenite, and phonolite. Oxyhornblende is generally restricted to volcanic rocks. Primary pyroxene in mafic igneous rocks may alter to aggregates of fine amphibole called **uralite**. Uralite more

commonly consists of tremolite-actinolite, but may be hornblende. Alteration usually begins along grain boundaries or cracks. Recrystallization may yield optically continuous masses of amphibole.

Hornblende is common in medium- and high-grade metamorphic terranes in metamorphosed mafic rocks such as amphibolite and hornblende gneiss. In general, the amount of Al in both tetrahedral and octahedral sites, and the Na content of the A site increase with increasing grade of metamorphism. Marble, skarns, and other metamorphosed carbonate rocks also may contain hornblende, but tremolite-actinolite is more common in these rocks.

Use Except as a constituent of decorative or dimension stone, hornblende has no significant industrial use.

GLAUCOPHANE–RIEBECKITE

Glaucophane: $Na_2Mg_3Al_2Si_8O_{22}(OH)_2$
Riebeckite: $Na_2Fe_3^{2+}Fe_2^{3+}Si_8O_{22}(OH)_2$
Monoclinic 2/m
$a = 9.55–9.76$ Å,
  $b = 17.7–18.05$ Å,
  $c = 5.30–5.33$ Å,
  $\beta = 103.6°$,
  $Z = 2$
$H = 5–6$
$G = 3.05–3.50$
Biaxial (+ or −)
$n_\alpha = 1.594–1.701$
$n_\beta = 1.612–1.711$
$n_\gamma = 1.618–1.717$
$\delta = 0.006–0.023$
2V: Glaucophane $2V_x = 10–50°$
Riebeckite: $2V_z = 0–135°$

Composition The composition fields for the sodic amphiboles are shown in Figure 14.19. The principal substitutions are Fe^{2+} for Mg and Fe^{3+} for Al, which are represented by the two axes of the diagram. Compositions between

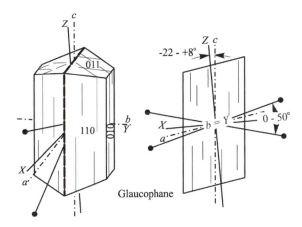

Glaucophane

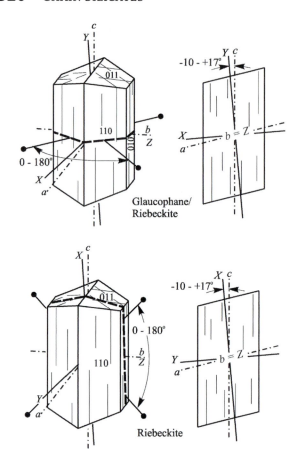

Glaucophane/Riebeckite

Riebeckite

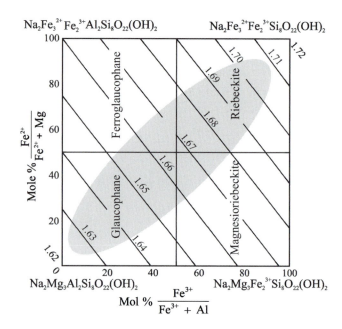

Figure 14.19 Composition and variation of n_γ for the sodic amphiboles. Compositions found in blueschist and related rocks fall in the shaded field extending from glaucophane to riebeckite. Compositions found in metamorphosed iron formations and alkalic igneous rocks usually fall in the riebeckite and magnesioriebeckite fields. Data from Deer and others (1963), Borg (1967), Coleman and Papike (1968), and Ernst and Wai (1970).

30% and 70% Fe^{3+} used to be known as **crossite,** but Leake and others (2003) recommend that the division between glaucophane and riebeckite be placed at 50% Fe^{3+}. Solid solution extends toward the sodic–calcic amphiboles and arfvedsonite.

Form and Twinning Glaucophane commonly forms bladed to slender prismatic crystals and may also be fibrous or granular (Figure 14.14e). Riebeckite commonly forms slender bladed or prismatic crystals, with or without terminations. Grains may be irregular or may enclose associated minerals. Asbestiform riebeckite, known as **crocidolite,** is common in metamorphosed iron formations. Simple and lamellar twinning on {100}.

Cleavage Good cleavage on {110} intersects at 56° and 124°. Parting on {100}. Conchoidal to uneven fracture. Brittle. Fibers are flexible.

Color Properties

Luster: Vitreous.

Color: Lavender blue, blue, dark blue, gray, or black; darker with increasing iron content.

Streak: White to gray.

Color in Thin Section: Glaucophane is pale to medium blue; riebeckite is dark blue.

Pleochroism: Distinct. Glaucophane: X = colorless, pale blue, yellow; Y = lavender blue, bluish green; Z = blue, greenish blue, violet. Riebeckite: X = dark blue; Y = indigo blue, grayish blue; Z = blue, yellowish green, yellowish brown; magnesioriebeckite tends to be lighter. Grains are darkest when the long dimension (c axis) is parallel to the vibration direction of the lower polarizer.

Optical Characteristics

Relief in Thin Section: Moderate to moderately high positive.

Optic Orientation: The position of the indicatrix axes remains nearly fixed for all compositions, with one nearly parallel to c, one parallel to b, and one nearly parallel to a. The relative lengths of the indicatrix axes and the position of the optic plane change fairly systematically with increasing Fe content. In most glaucophane, the optic plane is parallel to (010). With increasing Fe content, the optic angle closes on the axis near a (X), then opens again in a plane nearly parallel to (001), as shown in the glaucophane/riebeckite orientation diagram. With additional Fe, the optic angle closes about the indicatrix axis parallel to b (Z) and then opens again in an optic plane nearly parallel to (100). The latter orientation is typical of most riebeckite. The indicatrix axis closest to c makes an angle with c of +17° to −22°; the extinction angle in (010) sections is therefore less than 22°. The trace of cleavage is length slow in most glaucophane, length fast in most riebeckite, and either length slow or length fast, depending on grain orientation, for intermediate compositions.

Indices of Refraction and Birefringence: Indices of refraction increase with Fe content (Figure 14.19).

Interference Figure: Glaucophane is usually biaxial negative with moderate to small 2V. Intermediate compositions may be either negative or positive, with almost any 2V. Riebeckite is usually optically negative; magnesioriebeckite may be optically positive. Optic axis dispersion in glaucophane is weak, r < v. Riebeckite shows strong optic axis dispersion, r < v or r > v.

Alteration Glaucophane may be altered to other amphiboles, and riebeckite is sometimes altered to iron oxides and carbonates. Fibrous riebeckite partially replaced by quartz is called **tiger eye** and is used in lapidary work.

Distinguishing Features

Hand Sample: Distinguished from the other amphiboles by the distinct blue color. Sodalite is usually massive granular and lacks good cleavage. Crocidolite is distinguished from other asbestos minerals by its blue color.

Thin Section/Grain Mount: The blue pleochroism distinguishes the sodic amphiboles from other amphiboles. Glaucophane is usually length slow; riebeckite is usually length fast. Intermediate compositions, however, could be either length fast or length slow. Blue tourmaline is darkest when the length is perpendicular to the vibration direction of the lower polarizer.

Occurrence Glaucophane is characteristic of high-pressure, low-temperature regional metamorphic rocks, which, because of their color, are called blueschists. Associated minerals include lawsonite, pumpellyite, chlorite, albite, quartz, jadeite, and members of the epidote group. Riebeckite is found in alkali-rich igneous rocks such as syenite, nepheline syenite, and alkali granite. Crocidolite, which is asbestiform riebeckite, is found in metamorphosed iron formations.

Use A few percent of the world's asbestos production is crocidolite, most of which is mined in South Africa. Some of this material, known as "tiger eye," has been partially replaced by quartz. It displays a very attractive chatoyancy when cut and polished.

Comments Crocidolite asbestos appears to be a more significant health risk than the more commonly used chrysotile. Refer to the discussion on health risks of asbestos minerals in the section on serpentine (Chapter 13).

Other Amphiboles

The properties of some of the less common sodic and sodic–calcic amphiboles are summarized in Table 14.5. Note that most are various shades of dark green in hand sample and therefore cannot normally be distinguished

Table 14.5 Properties of Less Common Amphibolesᵃ

Name	Composition	Color	Color in Thin Section	$2V_x°$	n_α	n_β	n_γ	δ	$Z \wedge c$
Kaersutite	$NaCa_2(Mg,Fe)_4TiSi_6Al_2O_{22}(OH)_2$	Brown or black	Yellowish, reddish, or greenish brown	66–84	1.670–1.707	1.690–1.741	1.700–1.772	0.019–0.083	0° to +19°
Richterite	$NaCaNa(Mg,Fe)_5Si_8O_{22}(OH)_2$	Brown, reddish brown, or green	Colorless to yellow or violet	64–87	1.605–1.685	1.615–1.700	1.622–1.712	0.015–0.029	+15° to +40°
Katophorite	$NaCaNa(Mg,Fe)_4Fe^{3+}Si_7AlO_{22}(OH)_2$	Rose red, dark red brown to bluish black	Yellow, reddish brown, bluish green	0–70	1.639–1.681	1.658–1.688	1.660–1.690	0.007–0.021	−36° to −70°
Arfvedsonite–Eckermannite	$NaNa_2(Mg,Fe)_4(Fe^{3+},Al)Si_8O_{22}(OH)_2$	Dark bluish green, greenish black or black	Pale bluish green, yellowish green, brownish green, gray-green, or gray violet	0–100?	1.612–1.700	1.625–1.709	1.630–1.710	0.005–0.020	−30° to +53°

ᵃ All have H = 5–6, G = 2.9–3.5, and are monoclinic.

from hornblende. Optical properties and occurrence provide some aid in identification.

The occurrences of these amphibole varieties are as follows.

KAERSUTITE: Alkalic volcanic rocks such as trachy-basalt and trachyandesite, and in intrusive rocks such as syenite and lamphrophyre dikes.

RICHTERITE: Metamorphosed carbonate rocks such as marble and skarn deposits. It is likely that some material conventionally identified as tremolite is actually richterite.

KATOPHORITE: Alkalic volcanic rocks such as trachyte and phonolite.

ARFVEDSONITE AND ECKERMANNITE: Usually in alkali-rich igneous rocks such as alkali granite, nepheline syenite, syenite, trachyte, and phonolite. Sodic pyroxene may form parallel intergrowths.

REFERENCES CITED AND SUGGESTIONS FOR ADDITIONAL READING

Borg, I. Y. 1967. Optical properties and cell parameters in the glaucophane–riebeckite series. Contributions to Mineralogy and Petrology 15, 67–92.

Coleman, R. G., and Papike, J. J. 1968. Alkali amphibole of the Cazadero, California. Journal of Petrology 9, 105–122.

Deer, W. A., Howie, R. A., and Zussman, J. 1963. *Rock-Forming Minerals*, Volume 2: *Chain Silicates*. Longman, London, 379 p.

Deer, W. A., Howie, R. A., and Zussman, J. 1978. *Rock-Forming Minerals*, Volume 2A: *Single-Chain Silicates*. John Wiley & Sons, New York, 668 p.

Deer, W.A., Howie, R.A., and Zussman, J. 1992. *The Rock Forming Minerals*, 2nd edition. John Wiley & Sons, New York, 696 p.

Deer, W. A., Howie, R. A., and Zussman, J. 1997. *Rock-Forming Minerals,* Volume 2B: *Double-Chain Silicates*, 2nd ed.. The Geological Society, London, 784 p.

Deer, W. A., Howie, R. A., and Zussman, J. 2001. *Rock Forming Minerals*, Volume 2A: *Single Chain Silicates*, 2nd ed., The Geological Society, London, 668 p.

Ernst W. G., and Wai, C. M. 1970. Mossbauer, infrared, X-ray and optical study of cation ordering and dehydrogenation in natural and heat-treated sodic amphiboles. American Mineralogist 55, 1226–1258.

Essene, E. J., and Fyfe, W. S. 1967. Omphacite in Californian metamorphic rocks. Contributions to Mineralogy and Petrology 15, 1–23.

Hawthorne, F. C., and Oberti, R. 2007. Classification of the amphiboles. Reviews in Mineralogy and Geochemistry 67, 55–88.

Hawthorne, F. C., Oberti, R., Ventura, G. D., and Mottana, A. (eds.). 2007. *Amphiboles: Crystal Chemistry, Occurrence, and Health Issues*. Reviews in Mineralogy and Geochemistry 67, 545 p.

Hess, H. H. 1949. Chemical composition and optical properties of common clinopyroxenes. American Mineralogist 34, 621–626.

Jaffe, H. W., Robinson, P., and Tracey R. J. 1975. Orientation of pigeonite exsolution lamellae in metamorphic augite: Correlation with composition and calculated optimal phase boundaries. American Mineralogist 60, 9–28.

Jaffe, H. W., Robinson, P., and Tracey, R. J. 1978. Orthoferrosilite and other iron-rich pyroxenes in microperthite gneiss of the Mount Marcy area, Adirondack Mountains. American Mineralogist 63, 1116–1136.

Klein, C. 1964. Cummingtonite–grunerite series: A chemical, optical and X-ray study. American Mineralogist 49, 963–982.

Larsen, E. S. 1942. Alkali rocks of Iron Hill, Gunnison County, Colorado. U.S. Geological Survey Professional Paper 197a.

Leake, B. E. 1968. Optical properties and composition in the orthopyroxene series. Mineralogical Magazine 36, 745–747.

Leake, B. E., Woolley, A. R., Birch, W. D., Burke, E. A. J., Ferraris, G., Grice, J. D., Hawthorne, F. C., Kisch, H. J., Krivovichev, V. G., Schumacher, J. C., Stephenson, N. C. N., and Whittaker, E. J. W. 2003. Nomenclature of amphiboles: Additions and revisions to the International Mineralogical Association's 1997 recommendations. The Canadian Mineralogist 41, 1355–1362.

Morimoto, N. 1988. Nomenclature of pyroxenes. American Mineralogist 73, 1123–1133.

Nolan, J. 1969. Physical properties of synthetic and natural pyroxenes in the system diopside–hedenbergite–acmite. Mineralogical Magazine 37, 216–229.

Rabbitt, J.C. 1948. A new study of the anthophyllite series. American Mineralogist 33, 263–323.

Sabine, P. A. 1950. The optical properties and composition of the acmitic pyroxenes. Mineralogical Magazine 29, 113–125.

Winchell, A. N. 1938. The anthophyllite and cummingtonite–grunerite series. American Mineralogist 23, 329–333.

Verkouteren, J. R., and Wylie, A. G. 2000. The tremolite–actinolite-ferro-actinolite series: systematic relationships among cell parameters, composition, optical properties, and habit, and evidence of discontinuities. American Mineralogist 85, 1239–1254.

CHAPTER 15

Disilicates and Ring Silicates

DISILICATES

The disilicates are a relatively small group of minerals that, however, are quite widely distributed. Members of the epidote group (zoisite, clinozoisite, epidote, allanite) in particular are very common accessory minerals in a wide variety of igneous and metamorphic rocks, and are a frequent constituent of the heavy mineral fraction of clastic sediments.

Structure and Classification

As described in Chapter 11, the characteristic feature of disilicates is a pair of silicon tetrahedra that share a single oxygen (Figure 15.1a). All disilicates, therefore, contain Si_2O_7 in the formula. Although the basic silicate structure is a double tetrahedron, the structure of most disilicates is more conveniently described in terms of chains of edge-sharing aluminum octahedra. These chains are linked laterally through the double tetrahedra, other cations, and, in some cases, single tetrahedra. They are structurally quite similar to the aluminum silicates, staurolite, and chloritoid, all of which are constructed of chains of edge-sharing aluminum octahedra linked laterally with single silicon tetrahedra, rather than double tetrahedra.

The members of the epidote group (Table 15.1) have a structure consisting of two different chains of octahedra (Figure 15.2a). One is a simple edge-sharing chain whose octahedral sites are designated M2. The H in the chemical formula is attached to O anions in this chain. The second is an edge-sharing chain of M1 octahedra with additional M3 octahedra attached on alternate sides along its length. The M2 sites in the simple chain usually contain only Al. In the M1/M3 chain, the M1 sites are usually occupied by Al, and the M3 sites contain the non-Al cations, usually Fe or Mn plus the remaining Al. The two chains are arranged in an alternating fashion and extend parallel to the b axis. These chains are linked laterally through both single (SiO_4) and double (Si_2O_7) tetrahedra, both of which share oxygen anions with the octahedra (Figure 15.2b). The combination of octahedral chains extending parallel to b and the cross-linking tetrahedra produces cavities (A sites) with distorted 7- to 11-fold coordination that contain

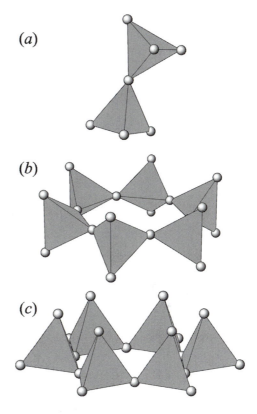

Figure 15.1 Structural elements in disilicates and ring silicates. (*a*) Double silicon tetrahedra. (*b*) Six-member tetrahedral ring found in beryl and cordierite. (*c*) Six-member tetrahedral ring found in tourmaline.

larger cations, usually Ca (Figure 15.2c). It should be no surprise that the cleavage in these minerals is between the chains and that most are elongate parallel to the b axis. The structure of zoisite, which has orthorhombic symmetry, is related to the epidote structure by a twinlike doubling of the unit cell along a (Figure 15.2d).

The other disilicates described here, pumpellyite and lawsonite, have structures similar to the epidote group. Both have octahedral chains that extend parallel to b. In pumpellyite, both single and double tetrahedra provide the cross-links between octahedral chains. In lawsonite, only double tetrahedra are available to do the cross-linking. Properties of some less-common disilicates are listed in Table 15.2.

Table 15.1 Epidote Group Minerals[a]

Mineral	A	M2	M1	M3	Composition	Symmetry
Zoisite	Ca	Al	Al	Al	$Ca_2Al_3OOH[Si_2O_7][SiO_4]$	Orthorhombic
Clinozoisite–epidote	Ca	Al	Al	Al,Fe³⁺	$Ca_2 Al_2 (Al,Fe^{3+}) OOH[Si_2O_7][SiO_4]$	Monoclinic
Allanite	Ca,Mn,Ce,La,Y,Th	Al	A1,Fe³⁺	Fe²⁺,Fe³⁺,Ti	$(Ca,Mn,Ce,La,Y,Th)_2Al\ (Al,Fe^{3+})$ $(Fe^{2+},Fe^{3+},Ti)\ OOH[Si_2O_7][SiO_4]$	Monoclinic

[a] A, Ml, M2, and M3 are structural sites (Figure 15.2).

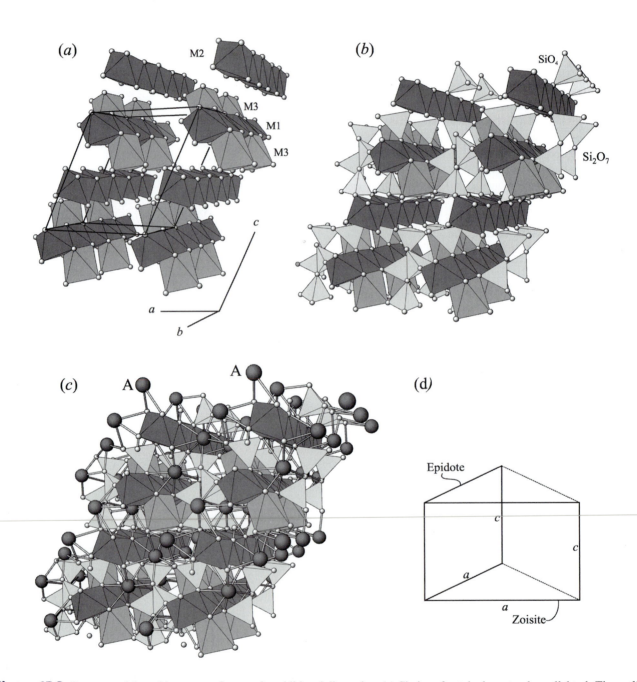

Figure 15.2 Structure of the epidote group. See text for additional discussion. (*a*) Chains of octahedra extend parallel to *b*. The outline of one unit cell is shown. M2 octahedra form simple chains; M1 octahedra form chains with M3 octahedra on alternate sides. (*b*) Single and double silicon tetrahedra cross-link the octahedral chains. (*c*) Additional large cations (A) in 7- to 11-fold coordination occupy interstices between octahedra and tetrahedra. (*d*) The orthorhombic zoisite unit cell is related to the monoclinic epidote unit cell by a twinlike doubling of the epidote unit cell along *a*. View is looking down the *b* axis.

ZOISITE

$Ca_2Al_3OOH[Si_2O_7][SiO_4]$
Orthorhombic $2/m$ $2/m$ $2/m$
$a = 16.19$ Å, $b = 5.55$ $c = 10.03$ Å, $Z = 4$
$H = 6–7$
$G = 3.15–3.37$
Biaxial (+)
$n_\alpha = 1.685–1.705$
$n_\beta = 1.688–1.710$
$n_\gamma = 1.697–1.725$
$\delta = 0.005–0.020$
$2V_z = 0–69°$

Zoisite

(010) Section

Ferrian Zoisite

(100) Section

Composition Most zoisite is relatively close to the composition given here. Less than about 8% of octahedral sites may be occupied by Fe^{3+}; the monoclinic clinozoisite–epidote structure is favored for higher Fe contents. **Thulite** is a pink variety in which up to 2% of octahedral sites may be occupied by Mn. Compositional zoning in Fe content is common.

Form and Twinning It is common as anhedral grains and granular aggregates. Crystals are columnar, prismatic, or acicular parallel to the b axis. Oriented intergrowths with epidote are possible. Twinning is not reported.

Cleavage Perfect {100} and imperfect {001} cleavage. Uneven to conchoidal fracture. Brittle.

Color Properties

Luster: Vitreous.

Color: White, gray, greenish brown, greenish gray, or pink (thulite).

Streak: White.

Color in Thin Section: Colorless to pale pink (thulite).

Pleochroism: Thulite is X = pale pink, pink; Y = colorless or pale pink; Z = pale yellow.

Optical Characteristics

Relief in Thin Section: High positive.

Optic Orientation: Two orientations depend on composition. Low-iron (≤3%) zoisite has $X = a$, $Y = b$, $Z = c$, optic plane = (010). More Fe-rich ferrian zoisite has $X = b$, $Y = a$, $Z = c$, optic plane = (100). The trace of cleavage, cleavage fragments, and length of elongate crystals are length fast in ferrian zoisite, and either length fast or length slow in zoisite.

Indices of Refraction and Birefringence: Both indices of refraction and birefringence increase with Fe and Mn content. Interference colors for all but the most Fe-rich varieties are lower than mid–first order and are usually anomalously blue or greenish yellow. Zoning may be expressed by variation of birefringence.

Interference Figure: Increasing Fe content causes the 2V angle to close from about 50° to 0° on the c axis in the (010) plane for zoisite and open from 0° to over 60° in the (100) plane for ferrian zoisite. Optic axis dispersion is strong, $r > v$ in zoisite, $r < v$ in ferrian zoisite.

Alteration Not commonly altered.

Distinguishing Features

Hand Sample: Grayish color and single good cleavage distinguish zoisite from amphiboles. Clinozoisite and epidote are usually distinctly green.

Thin Section/Grain Mount: Distinguished by high relief, low birefringence, and parallel extinction to the trace of cleavage in sections cut perpendicular to b and in cleavage fragments. Interference colors usually are anomalous. Clinozoisite has inclined extinction in sections cut perpendicular to b. Epidote has higher birefringence, inclined extinction, and usually displays patchy greenish pleochroism. In small grains, however, cleavage may not be evident. Apatite forms elongate crystals with hexagonal cross sections and is uniaxial negative. Pink andalusite resembles thulite, but is biaxial negative and has lower indices of refraction and relief.

Occurrence Zoisite is common in medium-grade metamorphic rocks derived from Ca-rich sedimentary rocks such as calcareous shale, calcareous sandstone, and limestone, and in amphibolite derived from mafic igneous rocks. It is less common in blueschists and eclogite. Hydrothermal alteration of plagioclase may yield a fine-grained material called **saussurite** composed of albite, zoisite or clinozoisite, sericite, and other silicates.

Table 15.2 Properties of Less Common Disilicates and Ring Silicates

	Piemontite	Melilite	Vesuvianite
Composition	$Ca_2(Al,Mn^{3+},Fe^{3+})_3\ O(OH)\ Si_2O_7SiO_4$	$(Ca,Na)_2(Mg,Al)(Si,Al)_2O_7$	$Ca_{19}(Al,Mg)_{13}B_{0-5}\ Si_{18}O_{68}(OH,O,F)_{10}$
Crystal system and habit	Monoclinic: columnar, bladed or acicular crystals; anhedral grains and aggregates	Tetragonal: tabular crystals with square, rectangle, or octagon cross-sections; anhedral grains; often with rodlike inclusions parallel to c	Tetragonal: stubby prisms; also anhedral grains or grains in columnar or radial patterns
Color	Reddish brown, black, light streak; vitreous	Colorless, honey yellow, gray green, brown, green brown, white streak; vitreous to resinous	Yellow, green, brown, light streak; vitreous
Color in thin section	Yellow, pink, violet, red, $X < Y < Z$	Colorless to pale yellow	Colorless to pale yellow or green
Sign and 2V	Biaxial (+) or (−), $2V_z = 64$–$106°$	Uniaxial (−) or (+)	Uniaxial (−)
Indices of refraction	$n_\alpha = 1.730$–1.794 $n_\beta = 1.740$–1.807 $n_\gamma = 1.762$–1.829	$n_\omega = 1.629$–1.669 $n_\epsilon = 1.616$–1.661	$n_\omega = 1.703$–1.752 $n_\epsilon = 1.700$–1.746
δ	0.025–0.073	0.000–0.011	0.001–0.009
Physical	$H = 6$–6.5, $G = 3.38$–3.61	$H = 5$–6, $G = 2.94$–3.07	$H = 6$–7, $G = 3.32$–3.34
Cleavage	{001} perfect	{001} poor	{110} very poor
Occurrence	Epidote group: hydrothermal systems, Mn-bearing metamorphic rocks	Silica-deficient mafic volcanic rocks	Contact-metamorphosed limestone, altered mafic rocks, nepheline syenite

Use Tanzanite is a gem variety of zoisite from Tanzania that contains small amounts of vanadium. The blue color is achieved by heat treatment; untreated material is reddish brown. Thulite also is used as a gem, and notable samples are found in Norway.

CLINOZOISITE–EPIDOTE

$Ca_2Al_2(Al,Fe^{3+})OOH$
$[Si_2O_7]\ [SiO_4]$
Monoclinic $2/m$
$a = 8.88$–8.89 Å,
$b = 5.58$–5.63 Å,
$c = 10.16$–10.15 Å,
$\beta = 115.5$–$115.4°$,
$Z = 2$
$H = 6$–7
$G = 3.21$–3.49
Biaxial (+ or −)
$n_\alpha = 1.703$–1.751
$n_\beta = 1.707$–1.784
$n_\gamma = 1.709$–1.797
$\delta = 0.004$–0.049
$2V_z = 14$–$90°$
$2V_x = 90$–$64°$

(010) Section

The ideal composition of clinozoisite is usually taken as $Ca_2Al_2AlOOH[Si_2O_7][SiO_4]$, and epidote is $Ca_2Al_2Fe^{3+}OOH[Si_2O_7][SiO_4]$; they form a solid solution series. Substitution of Al^{3+} and Fe^{3+} occurs mostly in the M3 site; the M1 and M2 sites contain dominantly Al^{3+}. Pure epidote should ideally have 33.3%. Fe^{3+}—one Fe^{3+} in the M3 site and two Al^{3+} in the M1 and M2 sites. Natural epidote, however, may have up to about 35% Fe^{3+}, indicating that some Fe^{3+} may enter the M1 and/or M2 sites; some Fe^{2+} may substitute for Ca^{2+} in the A site. Defining epidote composition based on the ratio of total Fe^{3+} to total Al^{3+} leads to the result that samples with more than 33.3% Fe^{3+} may be considered to have more than 100% epidote. The division between epidote and clinozoisite has traditionally been taken as the change in optic sign that occurs between 30 and 45% epidote (10–15% octahedral Fe^{3+}). However, the Commission on New Minerals, Nomenclature and Classification of the International Mineralogical Association now recommends that the dividing line be placed at 50% epidote, or 16.7% octahedral Fe^{3+}. Most natural clinozoisite contains more than about 8% octahedral Fe^{3+}; lower iron content favors formation of orthorhombic zoisite. Substitution of Mn^{3+} in octahedral sites yields **piemontite**. Some Fe^{2+} may substitute for Ca^{2+} in the A sites. **Tawmanite** is a rare Cr^{3+}-bearing member of the epidote group.

Form and Twinning Figure 15.3. Epidote and clinozoisite are common as anhedral grains and granular aggregates. Crystals are columnar, bladed, or acicular and elongate parallel to *b*. Columnar and radial aggregates are relatively common. Sections through crystals are usually six sided or rectangular. Detrital grains are rounded or platy parallel to the cleavage. Zoisite may form oriented intergrowths with epidote, and epidote may form overgrowths on brown allanite cores. {100} lamellar twins are not common.

Cleavage {001} perfect. Small grain size may preclude observing it either in hand sample or thin section. Uneven fracture. Brittle.

Color Properties

Luster: Vitreous.

Color: Clinozoisite: pale green to gray; epidote: pistachio green, yellowish green, green, black.

Streak: White.

Color in Thin Section: Clinozoisite: colorless; epidote: yellowish green.

Pleochroism: Y > Z > X, X = colorless, pale green, pale yellow; Y = yellowish green; Z = colorless, pale yellowish green. The color may be patchy or concentric. Even small amounts of Mn may produce the pink colors characteristic of piemontite.

Optical Characteristics

Relief in Thin Section: High positive.

Optic Orientation: Optic orientation varies with composition. In all cases Y = b and the optic plane is (010). The $Z \wedge a$ extinction angle between the slow ray and the trace of cleavage in (010) sections varies from as much as −60° for low-Fe clinozoisite to +40° for high-Fe epidote. Most clinozoisite has $Z \wedge a = 0°$ to +30° and most epidote has $Z \wedge a = +25°$ to +40°.

Indices of Refraction and Birefringence: Indices of refraction and birefringence increase fairly systematically with Fe^{3+} content (Figure 15.4). Clinozoisite usually displays a maximum of first-order colors in thin section; interference colors often are anomalously blue or greenish

Figure 15.3 Epidote. (a) Well-formed 2 cm crystal. (b) Photomicrograph of a thin section of granodiorite showing of epidote (center) enclosed in biotite that is partially replaced by chlorite. Hornblende is at the upper left (plane light).

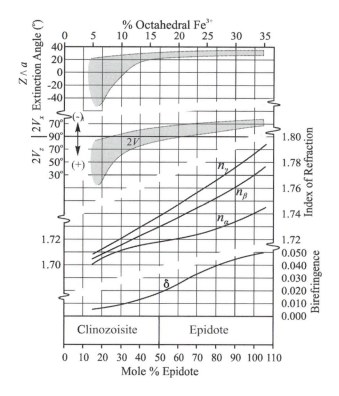

Figure 15.4 Optical properties of clinozoisite–epidote. Compiled from Johnson (1949), Deer and others (1962), Myer (1965), Strens (1966), and Hormann and Raith (1971), with modifications after Franz and Liebscher (2004). Indices of refraction and birefringence have some scatter about the lines shown.

yellow. Epidote displays higher interference colors, and, in grains oriented to display low-order colors, the interference colors also may be anomalously blue or greenish yellow.

Interference Figure: $2V$ varies with Fe^{3+} content but with considerable variability. Clinozoisite $2V_z$ is between 14° and ~100°, usually greater than 65°. Epidote is optically negative with $2V_x$ between ~90° and 64°. Optic axis dispersion is usually strong, $r < v$ for clinozoisite or $r > v$ for epidote. Inclined bisectrix dispersion also may be strong and may reverse the positions of the color fringes expected from the optic axis dispersion.

Alteration No consistent alteration. Clinozoisite and epidote are relatively resistant in the weathering environment so both are common detrital grains in clastic sediments.

Distinguishing Features

Hand Sample: The distinctive green color and single cleavage are characteristic. Students often confuse granular masses of epidote with olivine, but olivine lacks cleavage and, except for metamorphosed carbonates, has different occurrences. Massive epidote also tends to have a more intense green or pistachio green color than olivine.

Thin Section/Grain Mount: Clinozoisite is distinguished from epidote by lower birefringence, different optic sign, and lack of green color. Allanite is usually brown and piemontite pink. Zoisite and pumpellyite both resemble clinozoisite, but zoisite has parallel extinction in (010) sections and colorless pumpellyite has lower indices of refraction and higher birefringence. Hornblende has two cleavages, lower indices of refraction and birefringence, and more fully saturated green color. The small size of many grains may make positive identification difficult in thin section.

Occurrence Epidote and clinozoisite are common accessory minerals in a wide variety of regional and contact metamorphic rocks including pelites, metacarbonates, and felsic to mafic meta-igneous rocks. Clinozoisite is favored in aluminous rocks and epidote in more iron-rich rocks. Many igneous rocks contain epidote or clinozoisite as accessory minerals, and epidote is common in hydrothermal systems or as the result of deuteric alteration, which is produced by late residual water-rich fluids left after a magma is nearly completely crystallized. Alteration along fault and fracture systems often yields epidote. Because epidote and clinozoisite are fairly resistant in the weathering environment, they are common constituents of the heavy mineral fraction of many clastic sediments.

Use Sometimes used as a semiprecious gem.

ALLANITE

$(Ca,Mn,Ce,La,Y,Th)_2Al$
$\quad (Al,Fe^{3+})(Fe^{2+},Fe^{3+},Ti)$
$\quad OOH[Si_2O_7][SiO_4]$
Monoclinic $2/m$
$a = 8.93$ Å, $b = 5.77$ Å, $c = 10.16$ Å, $\beta = 114.7°$, $Z = 2$
$H = 5–6.5$
$G = 3.4–4.2$ (lower if metamict)
Biaxial (+ or −)
$n_\alpha = 1.690–1.813$
$n_\beta = 1.700–1.857$
$n_\gamma = 1.706–1.891$
$\delta = 0.013–0.036$
$2V_x = 40–90°$ (negative)
$2V_z = 90–57°$ (positive)

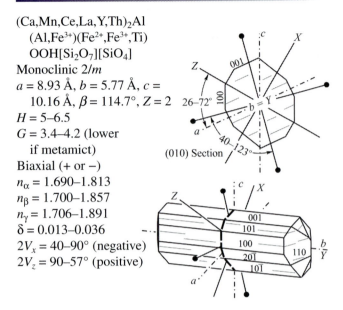

Composition The principal compositional variation is substitution of trivalent Ce, La, Y, and other rare earth elements for Ca^{2+}. Charge balance is maintained by substituting Fe^{2+} for Al^{3+}; allanite is the only member of the epidote group with significant amounts of ferrous iron. Th and U are also usually present in significant amounts, substituting for Ca. The structure may become metamict, as radioactive elements contained in the structure decay over time.

Form and Twinning It often occurs as anhedral or subhedral equant grains. Crystals are columnar, bladed, or acicular and elongate parallel to the b axis. Because allanite contains substantial amounts of radioactive elements, pleochroic halos may develop around allanite grains enclosed in biotite, chlorite, hornblende, or other minerals. Overgrowths of epidote on allanite are common. Twinning on {100} is not common.

Cleavage Imperfect on {001}, poor on {100} and {110}. Uneven to conchoidal fracture. Brittle.

Color Properties

Luster: Vitreous, resinous to submetallic.

Color: Light brown to black.

Streak: Gray.

Color in Thin Section: Usually brownish yellow to brown, also colorless or greenish.

Pleochroism: Distinct in shades of brown or, less commonly, green: X = reddish brown; Y = dark brown, brownish yellow; Z = dark reddish brown, greenish brown. Metamict varieties are nonpleochroic.

Optical Characteristics

Relief in Thin Section: High to very high positive, metamict varieties may be low.

Optic Orientation: Usually $X \wedge c = -1°$ to $-47°$, $Y = b$, $Z \wedge a = +26°$ to $+72°$. (010) sections show highest birefringence and the $Z \wedge a$ extinction angle to the trace of cleavage, if visible.

Indices of Refraction and Birefringence: Indices of refraction increase with Fe and rare earth element content. Strongly metamict varieties may be isotropic, with an index of refraction as low as 1.54. Chemical zoning may be expressed by variation in birefringence, which is usually no higher than mid–second order in standard thin section.

Interference Figure: Allanite is usually biaxial negative with 2V between 40° and 90°; optically positive varieties with 2V between 90° and 57° are less common. Optic axis dispersion is usually strong, $r > v$ or $r < v$. Strong mineral color may make interference figures difficult to interpret.

Alteration Metamict allanite is relatively common. Becoming metamict involves an increase in volume, so surrounding grains may develop cracks that radiate away from the allanite grain.

Distinguishing Features

Hand Sample: Dark color, pitchy or resinous luster, and association with granitic rocks. Firm identification may require other techniques. Strongly metamict varieties will not yield X-ray diffraction peaks because the crystal structure is destroyed.

Thin Section/Grain Mount: Nonmetamict samples are distinguished from other epidote-group minerals by brown color. Metamict samples are optically isotropic and surrounding minerals may contain cracks radiating from the allanite. Garnet has a higher index of refraction than metamict allanite.

Occurrence Allanite is a common accessory mineral in felsic igneous rocks such as granite, granodiorite, monzonite, syenite, and some equivalent volcanic rocks. Granitic pegmatite may contain large masses. It commonly is associated with biotite, hornblende, or other mafic silicate minerals. Allanite also may be found in skarn deposits or other metamorphosed carbonate rocks, and in amphibolite or granitic gneiss.

Use Potentially a source for metallic cerium, though monazite is more commonly used for that purpose. Cerium is used in the "flints" in cigarette lighters and is used in the lining of self-cleaning ovens to improve the oxidation of food residue.

LAWSONITE

$CaAl_2(Si_2O_7)(OH)_2 \cdot H_2O$
Orthorhombic $2/m\ 2/m\ 2/m$
$a = 8.80$ Å, $b = 5.85$ Å,
  $c = 13.1$ Å, $Z = 4$
$H = 6$
$G = 3.05–3.12$
Biaxial (+)
$n_\alpha = 1.663–1.665$
$n_\beta = 1.672–1.675$
$n_\gamma = 1.682–1.686$
$\delta = 0.019–0.021$
$2V_z = 76–87°$

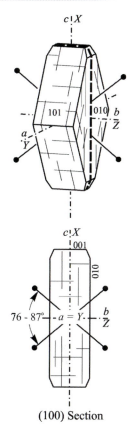

(100) Section

Composition Lawsonite shows little deviation from the ideal formula given above. Minor amounts of Fe, Ti, or Mg may replace Al, and minor amounts of Na may replace Ca.

Form and Twinning Figure 15.5. Crystals are usually tabular on {010} with a prismatic cross section. Also as anhedral grains and granular masses. May be acicular parallel to the b axis. Simple or lamellar twinning with twin planes parallel to the {101} prism faces is common.

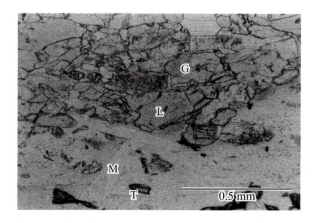

Figure 15.5 Photomicrograph of thin section of glaucophane schist with lawsonite (L), muscovite (M), glaucophane (G), and high-relief titanite (T) in plane light.

Cleavage Perfect cleavages on {100} and {010} intersect at right angles. Imperfect cleavages parallel to the {101} prism faces intersect at 67°. Uneven to conchoidal fracture. Brittle.

Color Properties

Luster: Vitreous to greasy.

Color: Colorless, white, bluish green, bluish gray.

Streak: White.

Color in Thin Section: Colorless, sometimes weakly bluish green.

Pleochroism: Not usually pleochroic. Grain mounts may be weakly pleochroic: X = light blue or brownish yellow; Y = yellow, yellowish green, blue green; Z = colorless or light yellow.

Optical Characteristics

Relief in Thin Section: Moderately high positive.

Optic Orientation: $X = c$, $Y = a$, $Z = b$.

Indices of Refraction and Birefringence: Little variation in either. Interference colors in standard thin sections range up to first-order red.

Interference Figure: Cleavage fragments on the {010} cleavage yield acute bisectrix figures with large $2V$. Optic axis dispersion is very strong, $r > v$.

Alteration Lawsonite may be replaced by pumpellyite as part of normal metamorphic mineral reactions.

Distinguishing Features

Hand Sample: Crystals, if present, are distinctive. Multiple cleavages distinguish lawsonite from zoisite and clinozoisite.

Thin Section/Grain Mount: Lawsonite resembles zoisite and clinozoisite but has two good cleavages, common twinning, and lacks anomalous interference colors. In (010) sections that display maximum birefringence, clinozoisite has inclined extinction and lawsonite should have symmetrical extinction to the {101} cleavages if they happen to be visible. Andalusite has lower birefringence and is usually optically negative.

Occurrence Lawsonite is a common mineral in glaucophane schist and related low-temperature, high-pressure metamorphic rocks. It commonly is associated with glaucophane, pumpellyite, jadeite, chlorite, and albite-rich plagioclase. Lawsonite also is found in metamorphosed gabbro, diabase, and related mafic rocks, and is infrequently found in marble and chlorite schist.

Use None.

PUMPELLYITE

$Ca_2MgAl_2[SiO_4]$
$[Si_2O_7](OH)_2 \cdot H_2O$
Monoclinic $2/m$
$a = 8.83$ Å, $b = 5.90$ Å, $c = 19.17$ Å, $\beta = 97.1°$, $Z = 4$
$H = 5–6$
$G = 3.16–3.25$
Biaxial (+ or −)
$n_\alpha = 1.665–1.710$
$n_\beta = 1.670–1.720$
$n_\gamma = 1.683–1.726$
$\delta = 0.008–0.020$
$2V_z = 7–110°$

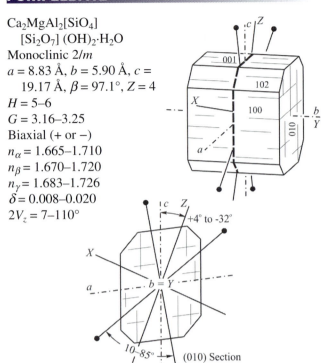

(010) Section

Composition The principal compositional variation is substitution of Fe^{3+} for Al, and Fe^{2+}, Fe^{3+}, and Al for Mg. Some Na or K may also substitute for Ca.

Form and Twinning Crystals are columnar, bladed, or acicular, and elongate parallel to b. Pumpellyite more commonly forms subhedral to anhedral grains or radiating, subparallel, or randomly oriented aggregates of elongate to fibrous grains. Twins with irregular composition planes parallel to {100} and {001} are common and may yield 4-fold sectors in appropriately cut sections.

Cleavage Distinct cleavage on {001} and fair cleavage on {100} intersect at 97°. Brittle.

Color Properties

Luster: Vitreous.

Color: Green, bluish green, brown, or greenish black.

Streak: White or gray.

Color in Thin Section: Colorless, green, yellow, or brown; darker colors are associated with higher Fe content.

Pleochroism: Colored varieties are distinctly pleochroic $Y > Z \geq X$: X = colorless, yellow, brownish, or greenish yellow; Y = light green, green, blue-green, brownish yellow; Z = yellow, brownish yellow, or reddish brown.

Optical Characteristics

Relief in Thin Section: High positive.

Optic Orientation: $X \wedge a = +4°$ to $+32°$, $Y = b$, $Z \wedge c = +4°$ to $−34°$. The trace of cleavage in longitudinal sections (parallel to *b*) and cleavage fragments may be either length fast or length slow, and show parallel extinction. In grains cut perpendicular to *b* that show maximum birefringence, the maximum extinction angle to the {001} cleavage is between 4° and 32°, with the larger values associated with higher Fe content. In some Fe-rich samples the positions of the *Z* and *Y* indicatrix axes are reversed.

Indices of Refraction and Birefringence: Indices of refraction and birefringence increase with Fe content. Interference colors are no higher than upper first order for most samples and are often anomalously blue or yellowish brown.

Interference Figure: Most pumpellyite is optically positive, and 2*V* increases with iron content. Some high-Fe samples are optically negative with large 2*V*. Optic axis dispersion is very strong, *r* < v. Also weak inclined bisectrix dispersion.

Alteration Pumpellyite does not display consistent alteration but may show metamorphic reaction relations with associated minerals.

Distinguishing Features

Hand Sample: Closely resembles members of the epidote group but has two cleavages; some samples have a darker color. X-ray diffraction may be needed for positive identification.

Thin Section/Grain Mount: Weakly colored pumpellyite resembles zoisite and clinozoisite. Zoisite has lower birefringence and shows parallel extinction to cleavage in (010) sections. Clinozoisite has higher indices of refraction and lower birefringence. Colored pumpellyite resembles epidote, but the latter is optically negative. Chlorite may have similar color and anomalous interference colors but has lower birefringence and indices of refraction. Lawsonite has lower indices of refraction, parallel extinction in grains that yield maximum birefringence, better cleavage, and generally is colorless or bluish.

Occurrence Pumpellyite is a common mineral in glaucophane schist and related low-temperature/high-pressure metamorphic rocks found in alpine metamorphic belts. Associated minerals include glaucophane, lawsonite, clinozoisite, epidote, chlorite, actinolite, and calcite. Metamorphosed or hydrothermally altered mafic igneous rocks such as basalt and gabbro may contain pumpellyite and it may be a vesicle filling in basalt. Skarns and related metamorphosed carbonate rocks infrequently contain pumpellyite. It may be a detrital mineral in sediment derived from pumpellyite-bearing rocks, but is easily misidentified as epidote.

Use None.

RING SILICATES

Structure and Classification

The ring silicates are a small group of minerals with only three common species: beryl, cordierite, and tourmaline.

Table 15.3 Properties of Axinite

Composition	$Ca_4(Fe^{2+},Mg,Mn^{2+},Zn)_2$ $Al_4B_2Si_8O_{30}(OH)_2$
Crystal system and habit	Triclinic: wedge-shaped, bladed or tabular crystals; radiating clusters of grains or granular aggregates
Color	Lilac–brown, yellowish, colorless; vitreous
Color in thin section	Colorless, pale yellow, violet; weak pleochroism
Sign and 2*V*	Biaxial (–),63–90°
Indices of refraction	n_α = 1.565–1.694 n_β = 1.660–1.701 n_γ = 1.668–1.706
δ	0.009–0.016
Physical	*H* = 6.5–7, *G* = 3.18–3.43
Cleavage	{100} perfect

Except for slight distortion of the lattice, beryl and cordierite are isostructural. The structure of all three minerals is based on 6-fold rings of silicon and aluminum tetrahedra (Figure 15.1b, c) that are stacked atop each other to form columns. Adjacent columns of 6-fold rings are cross-linked through other cations. Additional cations and/or anions may occupy the center of the columns.

Axinite (Table 15.3) has a structure that does not fall neatly in the conventional silicate classification but is included with the ring silicates based on the presence of rectangular 6-fold rings of tetrahedra. Crystals and masses of axinite may be found in contact-metamorphosed carbonate rocks and also in altered mafic igneous rocks; associated minerals may include prehnite, datolite, zoisite, tourmaline, vesuvianite, calcite, and actinolite. Axinite less commonly occurs in greenschist-facies regional metamorphic rocks associated with pumpellyite.

BERYL

$Al_2Be_3Si_6O_{18}$
Hexagonal 6/*m* 2/*m* 2/*m*
a = 9.20–9.27 Å,
   c = 9.19–9.25 Å, *Z* = 2
H = 7.5–8
G = 2.63–2.97 (increases
   with alkali content)
Uniaxial (–)
n_ω = 1.560–1.610
n_ϵ = 1.557–1.599
δ = 0.003–0.009

(a)

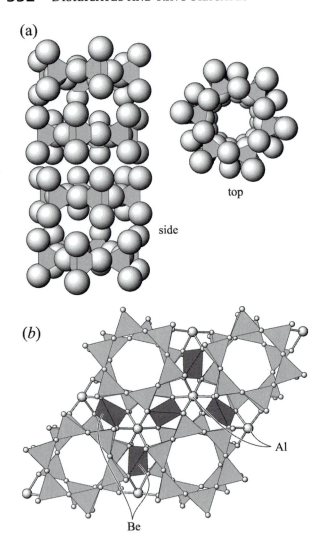

top

side

(b)

Al

Be

Figure 15.6 Structure of beryl. (*a*) Side and top view of column structure. The oxygen anions are drawn at about 70% of full size. (*b*) Columns of tetrahedral rings are cross-linked through Be in tetrahedral coordination (dark tetrahedra) and through Al in octahedral coordination. View down the *c* axis, with oxygen anions drawn as small spheres.

Structure In the beryl structure (Figure 15.6), the tetrahedra in the 6-fold rings are arranged so that the plane of the ring also is a plane of symmetry of the constituent tetrahedra (Figure 15.1b). Tetrahedra rings are stacked to form columns so that successive rings are rotated relative to the rings above and below. This allows the oxygen anions from one tetrahedral ring to nestle in the low spots between oxygen anions in the adjacent rings. Each column of rings therefore consists of two concentric cylinders of oxygen anions. These columns of rings extend parallel to the *c* axis and are stacked next to each other in a hexagonal array. Beryllium occupies distorted 4-fold sites formed by oxygen anions from adjacent columns of rings, and links the rings laterally and vertically. Al occupies distorted 6-fold sites formed by oxygen anions from three adjacent columns of rings.

The combination of tetrahedra rings and cross-linking Be tetrahedra actually forms a three-dimensional framework in which each oxygen anion on every tetrahedra is shared with another tetrahedra. In a strict sense, beryl may therefore be considered a framework silicate. Because the rings are the dominant structural element, beryl is grouped with the ring silicates here.

Composition The primary compositional variation is in the presence of alkali ions including Li, Na, K, and Cs. Li may substitute for either Al or Be, but the other cations are mostly located in the center of the channels in the middle of the stacked rings. Water molecules also may be located in the channels. Charge surplus associated with cations in the channels is probably balanced by substituting Fe^{2+}, Li^+, or Mg^{2+} for Al^{3+} in octahedral sites or by leaving vacancies in the Be sites. Small amounts of Cr and Mn may substitute for Al and are probably responsible for the color of **emerald** (green) and **morganite** (pink), respectively.

Form and Twinning Figure 15.7. Crystals are relatively common and form {100} hexagonal prisms with blunt terminations, usually consisting only of {001} pinacoids. Some crystals taper from one end to the other. Also as tabular hexagonal prisms, and anhedral or interstitial grains. Small needles may form radiating masses. Inclusions may form bands parallel to prism faces.

Cleavage Imperfect {001} cleavage. Conchoidal to uneven fracture. Brittle.

Color Properties

Luster: Vitreous.

Color: Bluish green, blue, green, greenish yellow, yellow, colorless.

Streak: White.

Color in Thin Section: Colorless.

Pleochroism: None.

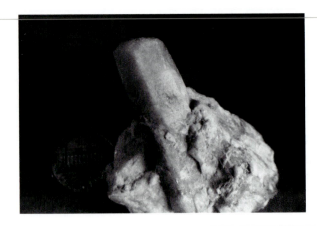

Figure 15.7 Beryl crystal from pegmatite.

Optical Characteristics

Relief in Thin Section: Low to moderate positive.

Optic Orientation: Longitudinal sections parallel to *c* are length fast and show parallel extinction.

Indices of Refraction and Birefringence: Indices of refraction and birefringence show a general increase with alkali content, but with considerable variability. Beryl typically has indices of refraction $n_\omega = 1.580 \pm 0.005$ and $n_\epsilon = 1.575 \pm 0.005$. Chemical zoning may be expressed by variation in the birefringence between the center and edge of grains.

Interference Figure: Basal sections yield centered optic axis figures. Strain of the crystal structure may produce biaxial figures with $2V$ up to $17°$ and variable optic orientation.

Alteration The usual alteration products of beryl are clay minerals or sericite.

Distinguishing Features

Hand Sample: Crystal habit, hardness, and color are distinctive. Apatite is softer, and quartz has lower specific gravity.

Thin Section/Grain Mount: Both quartz and nepheline resemble beryl, but beryl has higher indices of refraction. Quartz is optically positive and length slow. Apatite has higher indices of refraction. Nepheline shows negative relief in standard thin sections ($n_{\text{cement}} = 1.540$) and is not found in the quartz-rich granitic pegmatites and granite in which beryl is most frequently found.

Occurrence Beryl is a common mineral in granitic pegmatite and typically is associated with quartz, K-feldspar, albite, muscovite, biotite, and tourmaline. Beryl is less common in granite and can be found infrequently in nepheline syenite. High-temperature hydrothermal veins may contain beryl associated with tin and tungsten minerals. Contact metasomatism of gneiss, schist, or carbonate rocks also may produce beryl.

Use Beryl is the primary source for metallic beryllium. Beryllium has properties similar to aluminum and is often alloyed with copper to improve strength and hardness. Beryl is also an important gemstone. The more common blue-green variety is known as **aquamarine** and is most commonly found in pegmatites. Green Cr-bearing beryl is known as **emerald**, and most is mined from metamorphosed carbonate rocks. High-quality emerald can be extremely valuable. Other gem varieties include pink to red **morganite** and yellow **golden beryl**. Beryl, probably because of its beauty as a gemstone, had many historical medical uses, particularly in the Middle Ages.

CORDIERITE

$Mg_2Al_3[AlSi_5]O_{18}$
Orthorhombic
  (pseudohexagonal)
   $2/m\ 2/m\ 2/m$
$a = 17.08$ Å,
  $b = 9.73$ Å,
  $c = 9.36$ Å,
$Z = 4$
$H = 7$
$G = 2.53–2.78$
Biaxial (+ or −)
$n_\alpha = 1.527–1.560$
$n_\beta = 1.532–1.574$
$n_\gamma = 1.537–1.578$
$\delta = 0.005–0.017$
$2V_x = 35–106°$

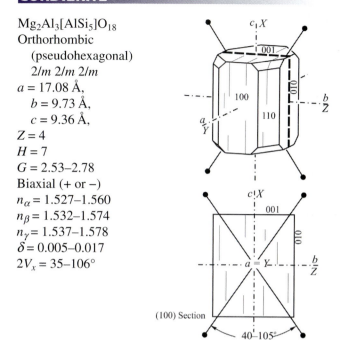

(100) Section

Structure The cordierite structure is essentially the same as beryl (Figure 15.6) with Mg in place of Al in octahedral sites, and Al instead of Be in cross-linking tetrahedral sites. One of six of the tetrahedra in the rings must be occupied by Al to provide charge balance. This distorts the rings and reduces the symmetry to orthorhombic. As in beryl, all tetrahedra are linked by sharing oxygens to form a three-dimensional framework, and cordierite in a strict sense should be classified as a framework silicate. It is included here because the rings are the dominant structural element.

Composition The principal compositional variation is substitution of Fe^{2+} for Mg in octahedral sites, although

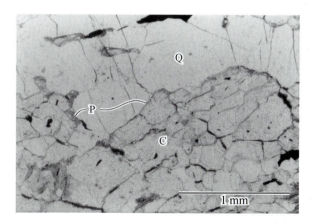

Figure 15.8 Photomicrograph of thin section of anthophyllite–cordierite gneiss showing cordierite (C) and quartz (Q) in plane light. Note that the cordierite has minor pinite (P) alteration at the edge of grains and is slightly cloudy because of fine opaque inclusions.

most cordierite is fairly Mg rich. Fe^{3+} may substitute for Al. The channels in the centers of the rings may contain variable amounts of H_2O and CO_2, and small amounts of large alkali ions, usually Na or K.

Form and Twinning Figure 15.8. Cordierite most commonly occurs as anhedral to highly irregular porphyroblastic grains, often with numerous inclusions of quartz, opaque grains, or other minerals. Less common as hexagonal prismatic crystals. Lamellar and cyclic twins are common, but not visible in hand sample. The twin planes are {110} and {130}. Basal sections in thin sections reveal a radial or cyclic pattern with roughly hexagonal symmetry. Twins in longitudinal section may resemble twins seen in plagioclase. Cordierite from high-grade metamorphic rocks is more likely to be twinned than from medium-grade rocks.

Cleavage Fair cleavage on {010} and poor cleavage on {100} and {001}. Cleavage is usually not obvious in hand sample or thin section. Subconchoidal fracture. Brittle.

Color Properties

Luster: Vitreous.

Color: Gray, grayish blue, blue, or indigo blue.

Streak: White.

Color in Thin Section: Colorless to very pale blue; darker colors are associated with Fe content.

Pleochroism: Usually none in thin section; grain mounts may be distinctly pleochroic. Colored samples are $Y > Z > X : X$ = colorless, pale yellow, pale green; Y = pale blue to violet; Z = pale blue.

Optical Characteristics

Relief in Thin Section: Low negative or positive.

Optic Orientation: $X = c$, $Y = a$, $Z = b$. Elongate crystals are length fast.

Indices of Refraction and Birefringence: Indices of refraction increase with Fe content and also with H_2O content in channels.

Interference Figure: Most cordierite is biaxial negative with $2V$ between 40° and 90°. High water and Na content yield the lower $2V$ angles. Some cordierite is biaxial positive with large $2V$. Optic axis dispersion is weak, usually $r < v$.

Comment Yellowish halos around radioactive minerals, such as zircon, are quite common.

Alteration Cordierite is readily altered to **pinite**, which is a fine-grained greenish or yellowish aggregate of chlorite, muscovite, and other silicates. The alteration may develop along cracks and grain margins, leaving ragged remnants of cordierite in a matrix of pinite.

Distinguishing Features

Hand Sample: Resembles quartz but is more distinctly blue. Corundum is harder. Small anhedral grains in metamorphic rocks may be difficult to distinguish from quartz, particularly if the bluish color is not evident. Porphyroblasts in pelitic metamorphic rocks may stand out on weathered surfaces; but distinguishing among cordierite, andalusite, and staurolite is not always possible, particularly if the minerals are anhedral and significantly altered. Cordierite is much less likely to form euhedral crystals than either staurolite or andalusite. Staurolite is brown and, if euhedral, has a prismatic crystal habit and often displays cross-shaped twins. Euhedral andalusite is typically prismatic, with a roughly square cross section, and the cross-shaped inclusion pattern seen on basal sections of chiastolite is unlike anything seen with cordierite. Unaltered andalusite may be pinkish, whereas cordierite is more typically blue. Pinite alteration of cordierite may significantly reduce the hardness.

Thin Section/Grain Mount: Closely resembles quartz and orthoclase, or if twinned, plagioclase. Quartz is uniaxial, lacks alteration products such as pinite, and in many metamorphic rocks may display undulatory extinction due to strain that is unlike that displayed by cordierite. Orthoclase has more distinct cleavage and has lower indices of refraction than the cements usually used in thin sections (n_{cement} = 1.540). Twinned cordierite resembles plagioclase, but plagioclase has more distinct cleavage and lacks radial twinning. Yellowish halos are common in cordierite around zircon grains or other minerals containing radioactive elements. Similar halos are not found in quartz or feldspars. Cordierite also may contain fine, dust-like inclusions of opaque minerals, or numerous rounded inclusions of quartz. Thin sections may be stained to distinguish cordierite from the feldspars (Boone and Wheeler, 1968).

Occurrence Cordierite is a common mineral in medium- and high-grade pelitic metamorphic rocks. Cordierite is common as porphyroblasts in hornfels found in contact metamorphic zones. In regional metamorphic rocks, cordierite may form anhedral or porphyroblastic grains similar in size and shape to quartz and feldspar. Associated minerals include chlorite, andalusite, sillimanite, kyanite, staurolite, muscovite, biotite, and chloritoid. Cordierite is also common in certain mafic metamorphic rocks associated with orthoamphibole and sometimes garnet. Granite, pegmatite, members of the gabbro clan, and andesite infrequently contain cordierite.

Use A blue gem variety is called **iolite**. The low thermal expansion of cordierite makes it useful as a ceramic. However, ceramics with the composition of cordierite are most commonly made by using talc [$Mg_3Si_4O_{10}(OH)_2$], alumina (Al_2O_3), and SiO_2 as starting materials.

TOURMALINE

Na(Mg,Fe,Li,Al)$_3$Al$_6$
[Si$_6$O$_{18}$] (BO$_3$)$_3$(O,OH,F)$_4$
Hexagonal (trigonal) $3m$
$a = 15.8$–16.0 Å,
   $c = 7.1$–7.25 Å, $Z = 3$
$H = 7$
$G = 2.90$–3.22
Uniaxial (–)
$n_\omega = 1.631$–1.698
$n_\epsilon = 1.610$–1.675
$\delta = 0.015$–0.035

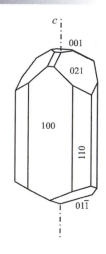

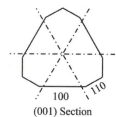

(001) Section

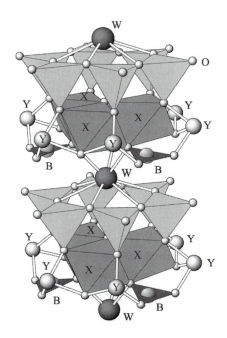

Figure 15.9 Tourmaline structure. Column structure consists of tetrahedral rings, triangular groups of octahedra (X), and large cation sites (W) along the axis of the column. Boron (B) is situated in triangular coordination along the edges of the columns. Columns are stacked adjacent to each other and are cross-linked through octahedrally coordinated Y cations.

Structure The tourmaline structure (Figure 15.9) is somewhat more complicated than that of beryl. The tetrahedra in the rings are arranged so that their bases form a common plane, and one corner on each tetrahedron "points" away from that base (Figure 15.1c). These tetrahedral rings combine with a set of three octahedra (X sites) constructed of O and OH arranged to form a triangle, and a site containing large cations (W sites) to form columns that extend parallel to c. All tetrahedral rings are arranged to point in the same direction. Adjacent columns are stacked next to each other in a hexagonal array. Additional cations in octahedral coordination (Y sites) cross-link between columns. Boron cations are coordinated with three oxygen anions to form triangular groups nestled into the columns of tetrahedra and octahedra. Because the tetrahedra within the rings all point in the same direction, tourmaline lacks center symmetry and may be piezoelectric. Further, the triangular grouping of octahedra within columns reduces the symmetry to trigonal, rather than hexagonal.

Composition Tourmaline exhibits great diversity in its composition. The general formula may be given as WX$_3$Y$_6$ [Si$_6$O$_{18}$](BO$_3$)$_3$(O,OH,F)$_4$. The W cations are usually Na, less commonly Ca or K. The X cations are most commonly Mg and Fe, less commonly Mn, Li, and Al. The Y cations are most commonly Al, less commonly Fe^{3+} or Mg. The normal compositional range lies among the following end members; Schorl-rich tourmaline is most common.

    SCHORL: NaFe$_3$Al$_6$[Si$_6$O$_{18}$](BO$_3$)$_3$(O,OH,F)$_4$

    DRAVITE: NaMg$_3$Al$_6$[Si$_6$O$_{18}$](BO$_3$)$_3$(O,OH,F)$_4$

    ELBAITE: Na(Li,Al)$_3$Al$_6$[Si$_6$O$_{18}$](BO$_3$)$_3$(O,OH,F)$_4$

Chemical zoning is common and is usually expressed by variation in color seen both in hand sample and thin section, and indices of refraction or birefringence seen in thin section.

Form and Twinning Figure 15.10. Crystals are quite common and usually form vertically striated prisms that are stubby to elongate parallel to c. The combination of a dominant {100} trigonal prism with a subordinate {110} hexagonal prism may give crystals a somewhat rounded triangular appearance. Terminations are usually fairly blunt and consist of combinations of trigonal pyramids that are generally different on opposite ends of the crystal because center symmetry is lacking. Fractures roughly perpendicular to crystal length may be filled with quartz, feldspar, or micas. Twinning is rare on {101} and {401}.

Cleavage Cleavages on {110} and {101} are very poor. Fracture is conchoidal. Brittle.

Color Properties

Luster: Vitreous.

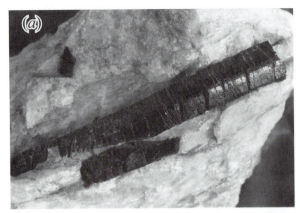

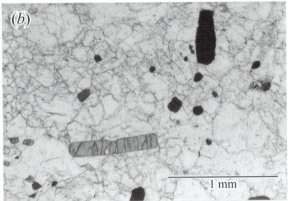

Figure 15.10 Tourmaline. (a) Vertically striated crystal with fractures roughly perpendicular to the length, from pegmatite. The field of view is about 20 cm wide. (b) Thin section showing tourmaline crystals in a metaquartzite in plane light. The ordinary ray is dark and extraordinary ray is light colored. The grain oriented N–S and the basal sections are dark; the grain oriented E–W, which is parallel to the lower polarizer vibration direction, is light.

Color: Schorl is black, dravite tends to be brown, and elbaite-rich tourmaline is brightly colored red, blue, green, or yellow. Almost any color is possible.

Streak: White to gray.

Color in Thin Section: Highly variable and consistently pleochroic. Schorl displays shades of blue, green, and gray, less commonly pink. Dravite-rich samples are light brown, yellow, or colorless. Elbaite-rich samples are light colored to colorless. In general, color intensity increases with Fe content. Concentric color zoning is common.

Pleochroism: Tourmaline is consistently strongly pleochroic with $\omega \gg \epsilon$, so elongate crystals are darkest when the long dimension is perpendicular to the vibration direction of the lower polarizer, and basal sections are dark.

Optical Characteristics

Relief in Thin Section: Moderate to high positive.

Optic Orientation: Longitudinal sections show parallel extinction and are length fast.

Indices of Refraction and Birefringence: Both indices of refraction and birefringence increase with increasing Fe + Mn content, but the wide diversity of composition precludes using optical properties to estimate composition.

Interference Figure: Basal sections that will yield uniaxial negative figures remain uniformly colored with rotation in plane light. The dark color may make figures difficult to interpret.

Alteration Tourmaline may alter to various phyllosilicates, including sericite, chlorite, and lepidolite (Li-mica). It is fairly stable in the weathering environment.

Distinguishing Features

Hand Sample: Columnar crystals with rounded triangular cross sections and vertical striae are characteristic. Common schorl resembles hornblende but tourmaline lacks good cleavage.

Thin Section/Grain Mount: Crystal habit, distinct pleochroism, and moderate birefringence distinguish tourmaline. Biotite and hornblende may display similar colors, but have different crystal habits and are darkest when their long dimension is aligned parallel to the vibration direction of the lower polarizer. Hornblende has inclined extinction. Light-colored tourmaline may resemble topaz, apatite, or corundum, but topaz is biaxial, apatite has lower birefringence, and corundum has higher indices of refraction.

Occurrence Tourmaline, usually schorl, is a characteristic mineral in granitic pegmatites. It is a common accessory mineral in granite, granodiorite, and related felsic igneous rocks. Rocks that have been hydrothermally altered adjacent to pegmatite and related felsic intrusives may contain abundant tourmaline. Tourmaline is also a common accessory mineral in schist, gneiss, quartzite, and phyllite. Dravite-rich tourmaline may be found in metasomatically altered limestone and dolomite in contact metamorphic zones. Because tourmaline is relatively resistant in the weathering environment, it is a common constituent in the heavy mineral fraction of clastic sediments.

Use Tourmaline is used as a gemstone. Green varieties are referred to as tourmaline or **Brazilian emerald**, pink or red stones may be called **rubellite**, and blue varieties may be called **indicolite**. Piezoelectric properties allow tourmaline to be used in the manufacture of pressure gauges and other electronic components. Tourmaline has historically been used for a variety of folk medicine purposes, probably because it forms beautiful crystals. Modern applications include use in skin care preparations that explicitly appeal to tourmaline's identity as a gemstone. It is probably no more or less harmful than any of the other creams and cosmetics that people apply to their skin in the quest for health and beauty.

REFERENCES CITED AND SUGGESTIONS FOR ADDITIONAL READING

Boone, G. M., and Wheeler, E. P. II. 1968. Staining for cordierite and feldspars in thin section. American Mineralogist 53, 327–331.

Deer, W. A., Howie, R. A., and Zussman, J. 1962. *Rock-Forming Minerals*, Volume 1: *Ortho- and Ring Silicates*. Longman, London, 333 p.

Deer, W. A., Howie, R .A., and Zussman, J. 2001. *Rock-Forming Minerals,* Volume 1B: *Disilicates & Ring Silicates,* 2nd ed. The Geological Society, London, 630 p.

Franz, G., and Liebscher, A. 2004. Physical and chemical properties of the epidote minerals—An introduction. Reviews in Mineralogy & Geochemistry 56, 1–82.

Hormann, P. K., and Raith, M. 1971. Optisch Daten, Gitterkonstanten, Dichte and magnetische Suszeptibilität von Al-Fe (III) Epidoten. Neues Jahrbuch für Mineralogie Abhandlungen 116, 41–60.

Johnston, R. W. 1949. Clinozoisite from Camaderry Mountain, Co. Wicklow. Mineralogical Magazine 28, 505–515.

Liebscher, A., and Franz, G. (eds.). 2004. *Epidotes*. Reviews in Mineralogy & Geochemistry 56, 628 p.

Myer, G. H. 1965. X-ray determination curve for epidote. American Journal of Science 263, 78–86

Strens, R. G. J. 1966. Properties of the Al–Fe–Mn epidotes. Mineralogical Magazine 35, 928–944.

CHAPTER 16

Orthosilicates

INTRODUCTION

The common orthosilicates are listed in Table 16.1. Note that in all of the chemical formulas, the Z:O ratio is always 1:4, consistent with the ratio required by the structure. These minerals may be divided into two groups. The first includes olivine, garnet, and zircon, in which all of the oxygen is part of one or another of the silicon tetrahedron. The second includes the aluminum silicate polymorphs, staurolite, chloritoid, topaz, and titanite. In these minerals, cations such as Al^{3+} or Ti^{4+} coordinate with O and/or OH that are not part of a silicon tetrahedron. This accounts for the extra anions in the chemical formulas. A good survey of the orthosilicates is provided by Ribbe (1982) and Deer and others (2001).

The basic structure of all of the first group of orthosilicates consists of isolated silicon tetrahedra bonded laterally through various other cations (Ca, Mg, Fe, Al, etc.) held in octahedral or higher coordination with the oxygen anions of the silicon tetrahedra. The differences in the structures depend on the size and charge for the other cations.

The structures of the second group of orthosilicates are more conveniently described in terms of the polyhedral framework defined by cations other than Si. A high-charged cation, usually Al^{3+} or Ti^{4+}, is a major constituent of these minerals. A typical arrangement is for these cations to form chains or sheets of edge-sharing octahedra, or chains of corner-sharing octahedra. These chains or sheets are linked laterally through silicon in tetrahedral coordination and, if present, other cations. Members of the epidote group, lawsonite, and pumpellyite, described in Chapter 15, all have structures with similar characteristics.

Titanite is the only common member of the second group that is not Al rich. It serves as an important repository for Ti in the wide variety of igneous and metamorphic rocks in which it is found. The abundance of aluminum in others means that they are found only in Al-rich rocks. For the aluminum silicates, staurolite, and chloritoid, this means pelitic metamorphic rocks. For topaz, which is fluorine bearing, this means F-bearing felsic igneous rocks and granitic pegmatites, some hydrothermal environments, and occasionally metamorphic rocks.

OLIVINE

$(Mg,Fe)_2SiO_4$
Forsterite (Fo): Mg_2SiO_4
Fayalite (Fa): Fe_2SiO_4
Orthorhombic ($2/m\ 2/m\ 2/m$)
Fo: $a = 4.75$ Å,
 $b = 10.20$ Å,
 $c = 5.98$ Å, $Z = 4$
Fa: $a = 4.82$ Å,
 $b = 10.48$ Å,
 $c = 6.09$ Å, $Z = 4$
$H = 6.5–7$
$G = 3.22–4.39$
Biaxial (+ or −)
$n_\alpha = 1.635–1.827$
$n_\beta = 1.651–1.869$
$n_\gamma = 1.670–1.879$
$\delta = 0.035–0.052$
$2V_x = 46–98°$

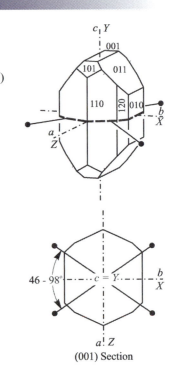

(001) Section

Structure The structure may be visualized as a somewhat distorted hexagonally close-packed array of oxygen with one-eighth of the tetrahedral sites occupied by Si^{4+} and one-half of the octahedral sites occupied by divalent cations (Figures 4.10, 16.1). The Si share oxygen with adjacent octahedra, but not with other tetrahedra. In the ideal structure, all the octahedral sites would be identical, but in practice, two different sites are developed. The M1 sites that form edge-sharing chains parallel to c get stretched out along that axis as though to keep the cations within the octahedra farther apart (Pauling's Rule 3, Chapter 4). The M2 sites that are affixed to the sides of these chains are somewhat less distorted. The net result of this distortion is to decrease the symmetry from the ideal hexagonal to orthorhombic, and crystals tend to be somewhat elongate parallel to the c axis.

Table 16.1 Common Orthosilicate Minerals[a]

Mineral		Formula	Igneous				Terrigenous Sedimentary	Metamorphic				Other	
			Felsic	Intermediate	Mafic	Feldspathoidal		Pelitic	Mafic	Calc-silicate	Blueschist	Hydrothermal	Miscellaneous
Olivine		$(Mg,Fe)_2SiO_4$			M						M		
Zircon		$ZrSiO_4$	A	A	A	A	A	A	A		A		
Garnet	Pyrope	$Mg_3Al_2(SiO_4)_3$											Ultramafic igneous
	Almandine	$Fe_3Al_2(SiO_4)_3$					A	M		M			
	Spessartine	$Mn_3Al_2(SiO_4)_3$	A				A						
	Grossular/ Andradite	$Ca_3(Al, Fe)_2(SiO_4)_3$					A		M				
Al Silicates	Andalusite						A	M					
	Sillimanite	$AlAlOSiO_4$	A	A		A	A	M					
	Kyanite						A	M					
Staurolite		$Fe_2Al_9O_6[(Si,Al)O_4]_4(OH)_2$					A	M					
Chloritoid		$(Fe^{2+},Mg,Mn)_2(Al,Fe^{3+})$ $Al_3O_2(SiO_4)_2(OH)_4$						M					
Topaz		$Al_2(SiO_4)(F,OH)_2$	A				A					A	
Titanite		$CaTiOSiO_4$	A	A	A	A	A		A	A	A		

[a]M = major or common; A = accessory, minor, or heavy mineral fraction of terrigenous sediments.

Composition Complete solid solution exists between the forsterite (Fo) (Mg_2SiO_4) and fayalite (Fa) (Fe_2SiO_4) end members, although most olivine tends to be relatively forsterite rich. Mn also may freely substitute in the octahedral sites to form the **tephroite** (Te) (Mn_2SiO_4) end member, but most common olivine contains little Mn and may be treated as a simple forsterite–fayalite solid solution series. Because Fe and Mg have different ionic radii and atomic masses, many physical and optical properties vary systematically with composition (Figure 16.2) and accurate measurement of one or several of these properties can provide an estimate of composition.

When olivine crystals grow from a magma, the first-formed crystals are more Mg rich than the magma (Figure 5.6). The result is that the residual magma becomes progressively enriched in Fe as olivine crystals grow. Unless equilibrium conditions are maintained, olivine crystals become zoned, with crystal cores being more Mg rich than rims. Refer to Chapter 5 for additional information.

Form and Twinning Figure 16.3. Olivine typically forms anhedral equant grains or aggregates of grains in plutonic

igneous rocks and metamorphosed carbonate-bearing rocks. Euhedral crystals are more common in volcanic rocks and tend to be slightly elongate parallel to c and flattened on a or b. Uncommon twins include simple or multiple twins with diffuse boundaries on {100}, {011}, or {012}.

Cleavage Poor on {010} and {110}, not usually seen. Conchoidal fracture. Brittle.

Color Properties

Luster: Vitreous.

Color: Olive to yellowish green, darker with increasing Fe.

Streak: White.

Color in Thin Section: Colorless, Fe-rich varieties may be yellowish.

Pleochroism: Colored varieties: $X = Z$ = pale yellow, Y = orange-yellow.

Optical Characteristics

Relief in Thin Section: High positive.

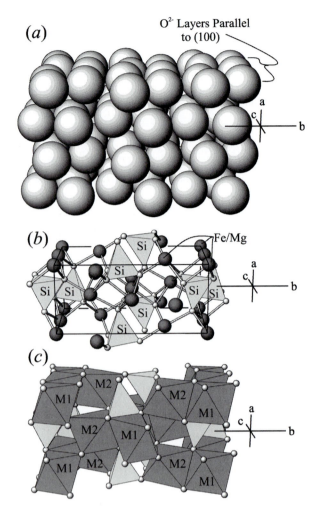

Figure 16.1 Olivine structure. (a) Oxygen anions form a somewhat loose and distorted hexagonal close-packed array with layers parallel to (100). The distortion reduces symmetry to orthorhombic. (b) Silicon tetrahedra point alternately up and down relative to the {100} plane and are tied laterally through Mg/Fe cations in octahedral coordination. Oxygen anions are shown as the smallest spheres; some of the oxygen anions shown in (a) have been omitted. The unit cell is shown. (c) M1 octahedral sites form edge-sharing chains that extend parallel to the c axis. M2 sites are attached to the sides of the M1 chains. All of the oxygen anions shown in (a) appear here as small spheres.

Optic Orientation: X = b, Y = c, Z = a. Elongate crystals show parallel extinction and may be either length fast or slow depending on how they are cut.

Indices of Refraction and Birefringence: Indices of refraction and birefringence vary systematically with Fe content (Figure 16.2).

Interference Figure: 2V varies systematically with Fe content (Figure 16.2). The 2V angle for olivine from volcanic rocks may be somewhat smaller than that for olivine of the same composition from plutonic rocks. Optic axis dispersion is weak with r < v if optically positive and r > v if negative.

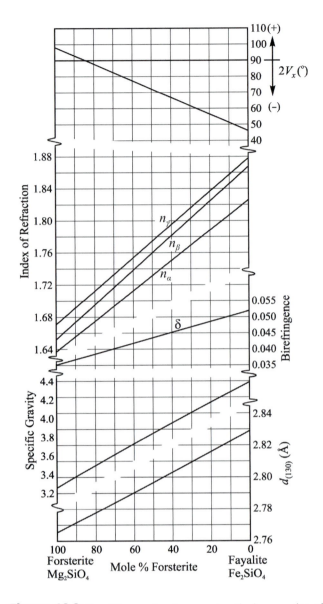

Figure 16.2 Specific gravity, $d_{(130)}$, and optical properties of olivine.

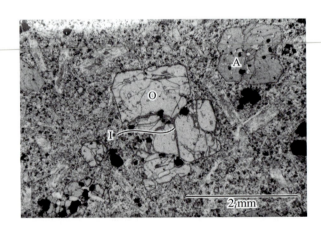

Figure 16.3 Olivine phenocryst (O) in basaltic dike. Note the alteration to iddingsite (I) along grain boundaries and cracks. The phenocryst at the upper right is augite (A). Plane light.

Comments: Chemical zoning (Mg-rich core, more Fe-rich rim) may be recognized by variation in interference color. In sections cut parallel to (001) the rim will have higher interference color than the core, and in sections parallel to (100) the opposite is true.

Alteration Olivine commonly alters to distinctive products visible in thin section. **Iddingsite** and **chlorophaeite** are combinations of various minerals and not subject to rigorous identification. Iddingsite is the term given to fine-grained reddish- to yellowish-brown material that usually contains a combination of goethite, clay, chlorite, quartz, talc, and other minerals. Despite the fact that it is an aggregate, iddingsite may display birefringence of 0.04–0.05 with $n_{low} \approx 1.70$ and $n_{high} \approx 1.75$. The brownish color usually masks the interference color. Chlorophaeite is an essentially isotropic orangish to greenish material composed of some combination of iron oxide/hydroxide, chlorite, and other low-birefringence silicates. Its index of refraction is roughly 1.5–1.6. Alteration to serpentine (chrysotile or antigorite) is also common. Olivine from igneous rocks may be mantled by pyroxene or hornblende as the result of normal magmatic reactions between olivine and the melt.

Distinguishing Features

Hand Sample: Distinguished by green color, vitreous luster, conchoidal fracture, and granular nature. May be confused with epidote, which usually has a richer green or pistachio green color and cleavage. Greenish quartz also may appear similar, but quartz occurs in distinctly different rocks.

Thin Section/Grain Mount: High birefringence, distinctive fracturing, lack of cleavage, and alteration products are distinctive. Pyroxenes have lower birefringence, cleavage, simple or lamellar twinning, and for the monoclinic varieties, inclined extinction in maximum birefringence sections. Epidote has cleavage, inclined extinction, is optically negative, and may display a patchy pistachio green color.

Occurrence Common in mafic and ultramafic igneous rocks associated with Ca-rich plagioclase, Ca-clinopyroxene, and orthopyroxene. Metamorphosed carbonate-bearing rocks may contain relatively pure forsterite in association with calcite, dolomite, diopside, epidote-group minerals, grossular garnet, tremolite, and related Ca-Mg-bearing minerals.

Discussion Under sufficiently high pressures, such as those obtained in the Earth's mantle, olivine is no longer stable relative to two different high-pressure polymorphic structures. Both high-pressure polymorphs are about 6% more dense than olivine and are based on cubic close packing of oxygen rather than hexagonal close packing. They are approximately isostructural with the mineral spinel and are therefore known as *β*-spinel and *γ*-spinel. Si is in tetrahedral coordination and Mg/Fe is octahedral; they differ only in details of site occupancy. The seismic velocity increase at around 400

km depth in the mantle has been interpreted to be caused by the change from the olivine structure to the spinel structure. Further, it has been argued (Green and Burnley, 1989; Iidaka and Suetsuga, 1992) that deep earthquakes in subduction zones are triggered by the reconstructive polymorphic transition of olivine to the spinel structure.

Uses Clear green olivine, known as **peridote** is a minor gemstone; some of the finest material comes from the island of Zebirget in the Red Sea. Presumably because its appeal as a gemstone promoted the placebo effect, peridote has historically been used to treat gastrointestinal problems and other maladies. It is used in industry to make molds for certain types of casting procedures in foundries, and infrequently as an abrasive. Some dental ceramics contain forsterite as a strengthening agent.

GARNET

$X_3Y_2(SiO_4)_3$
Isometric $4/m \bar{3} 2/m$
$a = 11.46–12.05$ Å,
   $Z = 8$
$H = 6.5– 7$
$G = 3.1–4.2$
Isotropic
$n = 1.714–1.890$

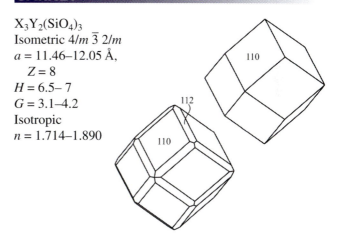

Structure The structure of garnet consists of isolated silicon tetrahedra bonded laterally through cations in 6-fold and 8-fold coordination (Figure 16.4). Si tetrahedra and Y octahedra share single oxygen anions to form chains extending parallel to each of the *a* axes. The spaces between these chains are distorted 8-fold sites occupied by the X cations.

Composition Natural garnets are conventionally divided into two groups (Table 16.2). Those with Al^{3+} in the Y structural site are the **pyralspite** group and include **py**rope, **al**mandine, and **sp**essartine. The **grandite** group has Ca^{2+} in the X site and includes **gr**ossular, **and**radite, and hydrogrossular. With inclusion of the rare Cr-bearing garnet **uvarovite** [$Ca_2Cr_3(SiO_4)_3$], the group can be called the **ugrandite** group. Within the pyralspite group, extensive substitution of Mg, Fe, and Mn is possible in the X site, although most natural pyralspite garnets have compositions that fall between pyrope and almandine or between almandine and spessartine. Continuous solid solution is also present between grossular and andradite in the grandite group. Garnet with a general composition $(Mg,Fe,Ca)_3Al_2(SiO_4)_3$, intermediate between the two groups, is found in high temperature (>700°C) metamorphic rocks. The size difference between Ca^{2+} and Mg^{2+}/Fe^{2+} suggests that exsolution could occur at lower temperatures, forming grossular lamellae within an almandine garnet, for example. However, at the microscopic scale at least, most garnet appears not to develop exsolution textures. Hydrogrossular (also known as **hibschite**) is a garnet variety in which a Si^{4+} is missing from a tetrahedral site. Charge balance is maintained by bonding a H^+ to each of the four oxygens surrounding the vacant site.

The ideal isometric symmetry may be reduced to orthorhombic or triclinic in some garnets, particularly members of the grandite group, by a combination of ordering of Al^{3+} and Fe^{3+} in the octahedral sites and by distortions caused by the hydrogrossular substitution. Heating, which destroys the ordering, causes these garnets to revert to isometric symmetry.

Form and Twinning Figure 16.5. Garnet typically forms euhedral to subhedral crystals comprised of dodecahedral {110} and/or trapezohedral {112} faces. In thin section, these crystals usually yield six- or eight-sided shapes. Granular or irregular masses are not uncommon. Isotropic garnet does not display twinning. Anomalously birefringent varieties often display transformation twins that form a pattern of six or more wedge-shaped sectors radiating from the center or a tweed-like intersecting pattern of twin lamellae.

Cleavage None. Uncommon parting on {110}, irregular conchoidal fracture with lots of angular corners and edges. Brittle.

Color Properties

Luster: Vitreous.

Color: Pyrope, almandine, spessartine: wine red to red-dish brown, also black, orange, dark pink, brown. Grossular and hydrogrossular: green, colorless, pink, brown. Andradite: yellow, green, brown, black.

Streak: White.

Color in Thin Section and Grain Mount: Pale version of the hand-sample color.

Optical Characteristics

Relief in Thin Section: High positive.

Optic Orientation: Typical pyralspite garnet is isotropic, but grandite garnets and some spessartine may be anomalously anisotropic, consistent with their reduced symmetry.

Indices of Refraction: The index of refraction varies systematically with composition in the two groups. The composition of typical pyralspite garnet from medium-grade pelitic metamorphic rocks can be estimated from

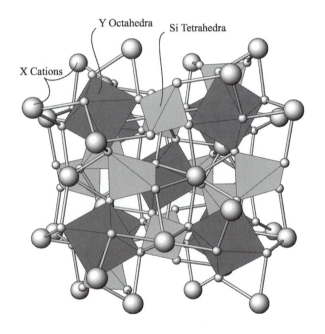

Figure 16.4 Garnet structure. Isolated silicon tetrahedra share single oxygen anions with Y cations in 6-fold coordination. The spaces between are distorted 8-fold sites occupied by the larger X cations.

Table 16.2 Garnet

Group	End Member	Composition	n		G (calc)	a (Å)
			Pure Synthetic	Normal Range		
Pyralspite	Pyrope	$Mg_3Al_2(SiO_4)_3$	1.714	1.720–1.770	3.582	11.46
	Almandine	$Fe_3Al_2(SiO_4)_3$	1.830	1.770–1.820	4.315	11.53
	Spessartine	$Mn_3Al_2(SiO_4)_3$	1.800	1.790–1.810	4.197	11.62
Grandite	Grossular	$Ca_3Al_2(SiO_4)_3$	1.734	1.735–1.770	3.594	11.85
	Andradite	$Ca_3Fe_2^{3+}(SiO_4)_3$	1.877	1.850–1.890	3.859	12.05
	Hydrogrossular	$Ca_3Al_2(SiO_4)_{3-x}(OH)_{4x}$	—	1.675–1.734	3.1–3.6	Varies

knowledge of index of refraction, specific gravity, and/or unit cell dimension (Figure 16.6a). The garnet from most skarns and metamorphosed carbonate-rich rocks is intermediate between grossular and andradite, and its composition can be estimated from Figure 16.6b.

Interference Figure: The anomalously birefringent varieties have variable optical orientation and may be uniaxial or biaxial with variable $2V$.

Alteration   Garnet may alter to chlorite, or occasionally to other minerals such as hornblende, epidote, or iron oxide. Garnet is commonly involved in a variety of mineral reactions in metamorphic rocks, and may display reaction or replacement textures with other minerals.

Distinguishing Features

Hand Sample: Crystal habit, hardness, and color distinguish garnet in hand sample.

Thin Section/Grain Mount: Recognized based on isotropic habit, crystal habit, and high relief. The grandite group from metamorphosed carbonate-bearing rocks may be anomalously birefringent. Spinel is distinctly green or brown, occurs as octahedrons, and often has a somewhat lower index of refraction. Basal sections of apatite in thin section may be confused with garnet, but have lower relief and yield uniaxial interference figures.

Occurrence   The occurrence and identity of associated minerals are fairly reliable guides to the variety of garnet. Pyrope is generally limited to ultramafic igneous rocks and the serpentinites derived from them. Almandine-rich garnet is the typical garnet in mica schist and gneiss. Spessartine and compositions intermediate between spessartine and almandine are found in felsic igneous rocks such as pegmatite, granite, and rhyolite. Grossular and andradite are the typical garnets found in metamorphosed carbonate-rich rocks associated with calcite, dolomite, tremolite, diopside, wollastonite, and epidote. Hydrogrossular is an uncommon variety found in metamorphosed carbonate rocks.

Comments   Garnet is a characteristic mineral of many metamorphic rocks. The chemical composition of the

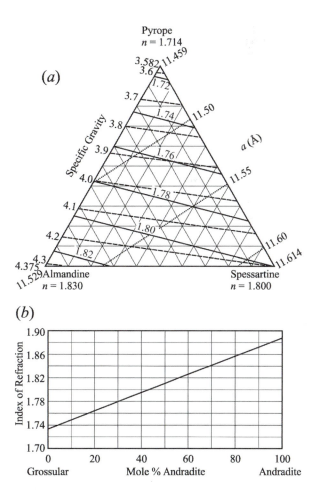

Figure 16.5 Garnet. (*a*) 2 cm garnet crystal in mica schist. (*b*) Photomicrograph of garnet porphyroblast in mica schist. The larger dark grains are biotite, and the finer grains in the groundmass are quartz, plagioclase, muscovite, and chlorite. Plane light.

Figure 16.6 Optical properties of common garnet. (*a*) The pyralspite group. Composition can be estimated if index of refraction and the unit cell dimension *a* and/or specific gravity are known. (*b*) The grandite group. Adapted from Winchell (1958).

almandine-rich garnet found in mica schist is influenced by the temperature, pressure, and related conditions at the time of growth. If garnet growth spanned some period of time during which temperature and pressure changed, it may acquire a chemical zoning that reflects those changing conditions. Detailed microprobe analysis of the chemical zoning in garnet and associated minerals can therefore be used to infer the time–temperature–pressure history of the rocks containing the garnet.

Use Used as a semiprecious gemstone. The irregular fracture makes garnet valuable as an abrasive, particularly for sandpaper. Also used in filters to help purify water in wastewater treatment plants. Probably because of its beautiful, well-formed crystals, garnet found application in early medical practice; in Roman times it was believed that garnet tied with camel hair would "harden the spleen."

ZIRCON

$ZrSiO_4$
Tetragonal $4/m\ 2/m\ 2/m$
$a = 6.61$ Å,
    $c = 5.98$ Å, $Z = 4$
$H = 7.5$
$G = 4.68$
Uniaxial (+)
$n_\omega = 1.920–1.960$
$n_\epsilon = 1.967–2.015$
$\delta = 0.036–0.065$

Structure Zr^{4+} is a relatively large cation (Appendix A) and fits comfortably in 8-fold coordination. The structure can be considered to consist of chains parallel to c in which Si tetrahedra alternate with Zr in 8-fold coordination.

Composition Zircon always contains some hafnium (Hf), and continuous solid solution extends to the rare mineral **hafnon** ($HfSiO_4$). Over 50 elements have been identified in zircon, including the rare earth elements (La through Lu), uranium, and thorium. Radioactive decay of uranium and thorium may cause the structure to become metamict (see Chapter 5).

Form and Twinning Figure 16.7. Most grains are microscopic tetragonal prismatic crystals elongate along the c axis with dipyramidal terminations. Euhedral overgrowths may develop on subhedral or rounded cores. Detrital zircon may be euhedral to rounded, depending on the amount of transport. Zircon is not usually twinned, but twinning on {111} has been reported.

Color Properties

Luster: Subadamantine; greasy if metamict.

Color: Grayish, yellowish, or reddish brown; less commonly colorless, yellow, gray, pink, or bluish green. Thermoluminescent, cathodoluminescent; may fluoresce under UV light.

Streak: White.

Color in Thin Section: Colorless to pale brown, although high relief and small grain size may make color difficult to recognize. Some samples may be cloudy or may show concentric color zoning or patchy color.

Pleochroism: Weak pleochroism ($\omega < \epsilon$) of colored varieties is more likely to be visible in grain mount.

Cleavage Prismatic {110} and pyramidal {111} cleavages are poor and not usually seen either in hand sample or thin section, nor do they tend to control orientation in grain mount. Conchoidal fracture. Brittle.

Optical Characteristics

Relief in Thin Section: Very high positive.

Optic Orientation: Crystals are length slow and show parallel extinction in longitudinal sections or as grains in a grain mount.

Indices of Refraction and Birefringence: Natural zircon that is not metamict has indices of refraction $n_\omega = 1.924–1.934$, $n_\epsilon = 1.970–1.977$, and $\delta = 0.036–0.053$. High-Hf zircon has higher indices. Metamict zircon has lower indices and birefringence and, if strongly metamict, may be isotropic with an index of around 1.80. The interference color of relatively fresh zircon ranges up to third and forth order in thin section.

Interference Figure: Small grain size often precludes obtaining an interference figure. Basal sections yield uniaxial positive optic axis figures with numerous isochromes. Metamict varieties are usually essentially isotropic, but may yield anomalous biaxial figures with small $2V$.

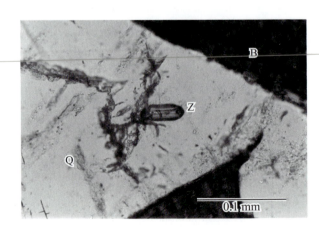

Figure 16.7 Photomicrograph of a thin section of granite showing a small zircon crystal (Z) in quartz (Q). The dark mineral is biotite (B). Plane light.

Alteration Except for becoming metamict owing to radioactive decay of U and Th, zircon is quite stable and neither weathers nor alters to other minerals.

Distinguishing Features

Hand Sample: While present in many different rock types as an accessory mineral, zircon is rarely seen in hand samples because of its small grain size. In uncommon cases of large or abundant grains, it is recognized by its crystal habit and hardness.

Thin Section/Grain Mount: Usually recognized as small, elongate, high-relief grains with bright interference colors included in other minerals. Zircon may produce dark halos in surrounding minerals such as biotite, cordierite, and hornblende owing to radioactive decay of Th, U, and other radioactive elements in the zircon. Rutile is darker colored and has higher indices of refraction and birefringence. Titanite has a different habit, is biaxial, and has higher birefringence. **Xenotime** (YPO_4) is a similar mineral that is isostructural with zircon but has higher birefringence (0.070–0.107) and lower indices of refraction (n_ω = 1.690–1.724, n_ϵ = 1.760–1.827).

Occurrence Zircon is widely distributed in many igneous, sedimentary, and metamorphic rocks, but usually only as microscopic crystals. Because it is durable in the weathering environment, zircon is common in the heavy mineral fraction of sediments. If these sediments are subsequently melted, detrital zircon grains can be preserved as grains in the igneous rocks produced from those melts. These zircon grains that have gone through the full rock cycle will often acquire overgrowths of fresh zircon mantling the old detrital cores.

Use Zircon is unique in that it is both a source of metals and used as an industrial mineral. It is the principal source of metallic zirconium and hafnium, and also a source for rare earth elements. Zirconium is used in nuclear reactors. As an industrial mineral zircon is used in the manufacture of refractory bricks and as a foundry sand used in making castings. Most of the commercially produced zircon comes from placer deposits on modern or ancient beaches. Australia is the major producer, followed by South Africa and the United States. Zircon also may be used as a gemstone. Cubic zirconia, which is widely used in inexpensive jewelry, is a synthetic product made from zirconium extracted from zircon.

Comments From a geological point of view, probably the most important element found in zircon is U^{4+}, which substitutes for Zr^{4+} in abundances of a few hundred parts per million. This uranium provides the basis for radiometric dating using the uranium–lead method. A sophisticated technology has been developed to acquire remarkably precise ages from zircon grains extracted from igneous or other

Table 16.3 Application of Zircon Chemistry to Solution of Geologic Problems

Chemistry	Application
U and Th substitution for Si^{4+}	U–Pb geochronology
He formed by decay of U and Th	Determination of exhumation and landscape development using U–Th–He thermochronometry
Hf^{4+} substitution for Si^{4+}	Investigation of continental growth and the crustal versus mantle source of magmas
Ti^{4+} substitution for Si^{4+}	Ti in zircon thermometry
Y and rare earth elements	Reconstruct magmatic histories and fingerprint magmatic sources
Oxygen isotope composition	Document the contribution of sediments and crust to the sources of magmas; examine crustal recycling
Melt inclusions	Composition of magma from which zircon crystallized

Source: Adapted from Hanchar and Hoskins (2003) and Harley and Kelly (2007).

rocks. In particular, precision dating of zircon grains from Precambrian plutons has become one of the most important techniques used to unravel the complex and often obscure geological story to be told by these rocks. Zircon has considerable value beyond being Earth's timekeeper because of the trace elements that it contains. These trace elements can help document magmatic processes of melting and crystallization of the magmas from which it crystallizes; in addition, they can help constrain models of crustal formation and evolution (Table 16.3).

ALUMINUM SILICATES

The **aluminum silicate polymorphs** andalusite, sillimanite, and kyanite ($AlAlOSiO_4$) are common minerals in pelitic metamorphic rocks. In some sources they are referred to as the **aluminosilicate polymorphs**. The formula is usually given as Al_2SiO_5. However, writing the formula as $AlAlOSiO_4$ clearly shows that the minerals are orthosilicates and that there are two distinct Al structural sites.

Structure Andalusite, sillimanite, and kyanite are reconstructive polymorphs whose structures can be described in terms of edge-sharing chains of aluminum octahedra that are linked laterally through Si and Al. The structures differ in the details of Al coordination, chain geometry, and the manner in which Si and Al link adjacent chains to each other. Note that these structures are quite similar to the structures of epidote, lawsonite, pumpellyite, and staurolite, all of which are based on chains of edge-sharing aluminum octahedra.

In andalusite, edge-sharing chains of Al^{3+} octahedra extend parallel to the *c* axis and are bonded laterally

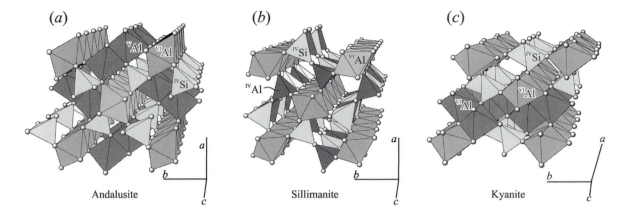

Figure 16.8 Structure of the aluminum silicates. (*a*) Andalusite. Edge-sharing chains of $^{VI}Al^{3+}$ octahedra (medium shading) extend parallel to *c* and are bonded laterally through pairs of $^{V}Al^{3+}$ in irregular 5-fold coordination (dark), and single $^{IV}Si^{4+}$ tetrahedra (light). (*b*) Sillimanite. Edge-sharing chains of $^{VI}Al^{3+}$ octahedra (medium shading) extend parallel to *c* and are bonded laterally through alternating $^{IV}Al^{3+}$ (dark) and $^{IV}Si^{4+}$ (light) tetrahedra. (*c*) Kyanite. Straight chains of $^{VI}Al^{3+}$ octahedra (medium shading) extend parallel to *c* and are bonded laterally through zigzag chains of $^{VI}Al^{3+}$ octahedra (dark) and $^{IV}Si^{4+}$ (light) in tetrahedral coordination.

through pairs of Al^{3+} in distorted 5-fold coordination and Si^{4+} in tetrahedral coordination (Figure 16.8a). The formula can therefore be expressed $^{VI}Al^{V}AlO^{IV}SiO_4$, where the superscripts refer to the coordination of the cations. The structure is relatively open, and andalusite has the lowest specific gravity (3.13–3.16) of the polymorphs. Relatively weak linkages between chains allow cleavage parallel to the {110} faces.

The structure of sillimanite consists of chains of edge-sharing Al^{3+} octahedra parallel to *c* that are linked laterally through Al^{3+} and Si^{4+}, both in tetrahedral coordination (Figure 16.8b). The formula that accounts for coordination can be written $^{VI}Al^{IV}AlO^{IV}SiO_4$. Ideally the Al^{3+} and Si^{4+} occupy alternating tetrahedra (i.e., they are ordered), but some degree of randomness or disorder in the distribution of these elements is possible.

The structure of kyanite can be considered to be derived from a distorted cubic close-packed array of oxygen with Al^{3+} occupying 40% of the octahedral sites and Si^{4+} occupying 10% of the tetrahedral sites. This close-packed structure yields the highest specific gravity of the polymorphs (*G* = 3.53–3.67), so it should be no surprise that kyanite is the polymorph stable at high pressure. Half the aluminum is distributed in the octahedral sites to form zigzag chains parallel to the *c* axis and the other half in octahedra that form straight chains. Silicon in tetrahedral coordination links the chains together (Figure 16.8c). The structural formula can be written $^{VI}Al^{VI}AlO^{IV}SiO_4$. Kyanite is notable because it displays a significant difference in scratch hardness depending on direction. When scratched parallel to the *c* axis, it is significantly softer than when scratched at right angles. This is presumably because bonds within the octahedral chains are stronger than the bonds that tie adjacent chains to each other. When kyanite is scratched parallel to the chains that extend parallel to

c, adjacent chains are easily split apart. When scratched at right angles, the stronger bonds along the length of the chains must be broken.

Occurrence The aluminum silicate polymorphs are characteristic minerals of medium- and high-grade pelitic metamorphic rocks such as mica schist, biotite gneiss, and hornfels. Recall that pelitic metamorphic rocks are derived from sedimentary rocks such as mudstone and shale, which, because they are clay rich, tend to contain substantial amounts of aluminum.

Each polymorph has a specific temperature–pressure range over which it is stable, so these minerals have been used extensively to establish the temperature–pressure conditions to which the rocks containing them have been subjected during the course of metamorphism. Because they are important minerals in metamorphic petrology, the temperature–pressure (T–P) stability of the aluminum silicates has been repeatedly studied and more than 15 different diagrams have been published. Figure 16.9 shows a recent determination that, while probably not the final word, is consistent with field relationships and can be used with reasonable confidence. A good summary of the mineralogy of the aluminum silicate minerals is provided by Kerrick (1990). Problems with establishing the T–P stability fields of the polymorphs include order–disorder in sillimanite and small compositional variations of the minerals. The Al–Si order–disorder in sillimanite is not particularly important, but it appears that small amounts of Fe and/or Mn entering the structure of andalusite may have a significant effect on its stability field.

Andalusite is stable at relatively low temperatures and pressures, and is commonly found in contact metamorphic aureoles of shallow- and intermediate-depth intrusives and

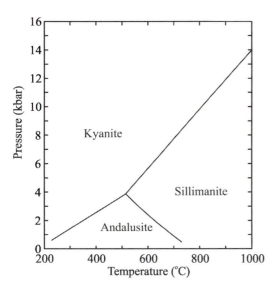

Figure 16.9 Aluminum silicate stability fields, From Hemingway and others (1991).

in rocks subjected to medium-grade metamorphism at relatively low pressures. Sillimanite is found in rocks that have been subjected to higher temperatures, and kyanite is characteristic of rocks subjected to relatively high pressures. These minerals may, depending on grade of metamorphism and bulk chemistry of the rocks, be associated with cordierite, staurolite, garnet, chlorite, muscovite, biotite, chloritoid, K-feldspar, and plagioclase.

All three polymorphs are less commonly found in granitic pegmatites and granite. A considerable literature has developed, debating whether these occurrences represent primary magmatic crystallization. The alternative explanations include crystallization by postmagmatic processes (i.e., metamorphism) or derivation as xenocrysts (foreign crystals). Xenocrysts may be from partially assimilated fragments of metamorphic wall rock through which the magma intruded or could be relicts of the rocks from which the magma was derived by partial melting. One root of this debate derives from the faulty assumption that the aluminum silicates are "metamorphic minerals" and that occurrences in igneous rocks are anomalous. Although each case must be evaluated on its own merits based on field, textural, chemical, and related evidence, there is no compelling reason that these minerals should not crystallize directly from a magma of appropriate composition and under appropriate temperature–pressure conditions. Sillimanite, in particular, appears to be a common accessory mineral in many granitic rocks.

Kyanite infrequently occurs as part of eclogite, which is an ultramafic rock metamorphosed at high pressure. The associated minerals include pyrope-rich garnet and omphacite pyroxene.

All three polymorphs can locally be important detrital minerals, although they are not particularly stable in the weathering environment.

Use Heating the aluminum silicates to high temperatures can cause them to convert to **mullite** [$Al_6Si_2O_{13}$] plus quartz. Mullite has a structure similar to that of sillimanite and is quite refractory. Mullitized aluminum silicate is used to manufacture refractory materials such as high-alumina bricks and related products used to line blast furnaces, ladles, and kilns, and in similar applications requiring resistance to high temperatures. Sillimanite is least commonly used for this purpose because it is usually fine grained, and therefore difficult and expensive to extract from the rocks in which it is found. It also requires higher temperatures (and more money) to mullitize. The choice between andalusite and kyanite depends on availability and on details of the requirements for the specific application. Andalusite, and to a lesser degree kyanite, are used in the manufacture of certain ceramic products, abrasives, and filler materials. The largest commercial deposits of andalusite in the world are in South Africa, where it is extracted from aluminum-rich pelitic metamorphic rocks and from detrital deposits derived from these rocks. Of the three polymorphs, only kyanite has historically been used for medical purposes (stress reduction, throat problems), probably because its bladed blue crystals are more striking and likely to trigger the placebo effect. The chiastolite variety of andalusite, which shows a cross-shaped pattern of inclusions in basal sections, is sometimes used in jewelry and has religious significance to some people.

ANDALUSITE

$^{VI}Al^VAlO^{IV}SiO_4$
Orthorhombic $2/m\ 2/m\ 2/m$
$a = 7.79$ Å, $b = 7.91$ Å,
   $c = 5.56$ Å, $Z = 4$
$H = 6.5 - 7.5$
$G = 3.13–3.16$
Biaxial (−)
$n_\alpha = 1.629–1.640$
$n_\beta = 1.634–1.645$
$n_\gamma = 1.638–1.650$
$\delta = 0.009–0.013$
$2V_x = 71–88°$

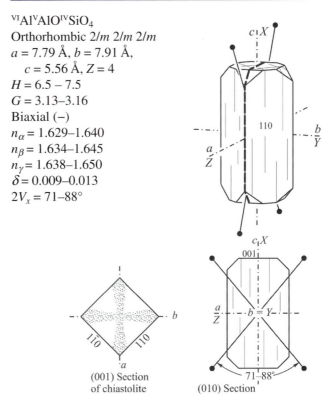

Composition Most andalusite is relatively pure $AlAlOSiO_4$, although significant Mn^{3+} and Fe^{3+} may substitute for Al^{3+} in octahedral sites. Extensive solid solution to **kanonaite** ($MnAlOSiO_4$) is possible.

Form and Twinning Figure 16.10. Andalusite commonly forms porphyroblasts in medium-grade mica schist. These crystals are usually prismatic {110} and elongate parallel to c with a nearly square cross section. A variety called **chiastolite** contains dark carbonaceous inclusions that form a cross along the diagonals of the prism. Andalusite also forms anhedral grains and irregular masses. Numerous inclusions of quartz, fine opaques, or other minerals are common but are usually visible only in thin section. Twinning on {101} is rare.

Cleavage Good {110} prismatic cleavages are nearly at right angles and are parallel to prism faces in cross sections, but may not be evident in hand sample, particularly if the sample is altered. Cleavage on {100} is poor and not usually evident. Uneven to subconchoidal fracture. Brittle.

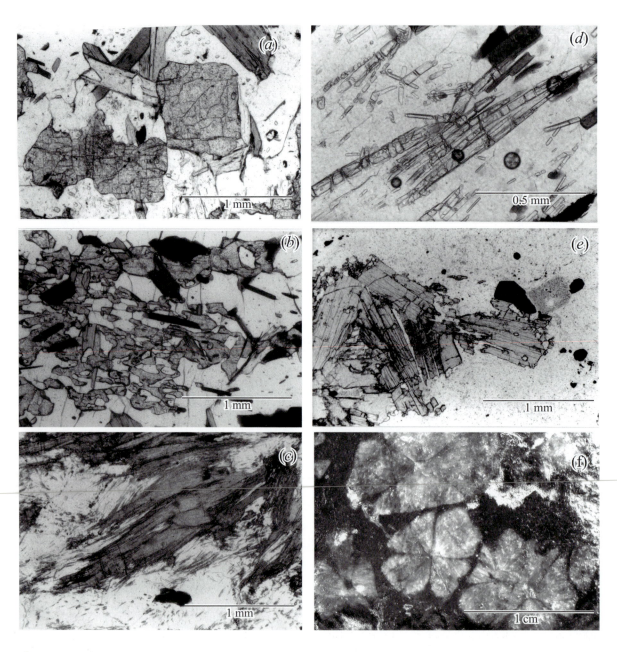

Figure 16.10 Aluminum silicates in mica schist. (*a*) Blocky andalusite crystal in thin section. (*b*) Strongly sieved andalusite porphyroblast (high relief) in thin section. (*c*) Fibrolitic sillimanite mass in thin section. (*d*) Elongate prismatic crystals of sillimanite in thin section. (*e*) Bladed kyanite in thin section. (*f*) Hand sample showing cross-shaped pattern of inclusions in basal sections of chiastolite. All thin sections plane light.

Color Properties

Luster: Vitreous.

Color: Pink, white, pinkish brown, violet or gray; darker colors are associated with high Mn and Fe.

Streak: White.

Color in Thin Section: Usually colorless in thin section or grain mount, less frequently with a pale pink or green cast.

Pleochroism: Weak in colored varieties: X = reddish pink, $Y = Z$ = greenish yellow.

Optical Characteristics

Relief in Thin Section: Moderately high positive.

Optic Orientation: $X = c$, $Y = b$, $Z = a$, optic plane = (010). Length fast with parallel extinction to trace of cleavage and crystal faces in longitudinal sections, symmetrical extinction in basal sections. Samples with $Mn^{3+} + Fe^{3+} > 6.5$ mol % are length slow, $X = a$, $Y = b$, $Z = c$, and are biaxial positive.

Indices of Refraction and Birefringence: Indices of refraction increase with substitution of $Mn^{3+} + Fe^{3+}$ for Al^{3+}. Most andalusite has birefringence in the range 0.009–0.013, which gives it first-order gray and white interference colors in standard thin sections. Birefringence decreases to nearly zero at $Mn^{3+} + Fe^{3+} \approx 6.5$ mol % and increases above that point.

Interference Figure: Most andalusite is biaxial negative with $2V > 80°$. Basal sections yield acute bisectrix figures with the melatopes outside the field of view. Cleavage fragments yield strongly off-center obtuse bisectrix figures. Optic axis dispersion is weak, $r < v$ or occasionally $r > v$. Andalusite with more than ~6.5 mol % $Mn^{3+} + Fe^{3+}$ is optically positive with $2V_z$ between 65° and 85°. In the composition range where the transition from optically positive to negative is made, $2V$ is quite variable.

Alteration Andalusite may alter to sericite (fine-grained white mica) or to chlorite or other sheet silicates. It is commonly involved in metamorphic mineral reactions and may show reaction relations with minerals such as cordierite, staurolite, garnet, sillimanite, and kyanite.

Distinguishing Features

Hand Sample: Porphyroblasts in pelitic metamorphic rocks may be euhedral or may form highly irregular grains riddled with inclusions of quartz. Other porphyroblasts with which andalusite may be confused, particularly if altered, include cordierite, staurolite, and kyanite. The habit of euhedral crystals is distinctive, particularly of the chiastolite variety, as is the pinkish color of some samples. Staurolite is brown and forms cross-shaped twins. Kyanite is bladed and may be

blue, at least in patches, and the difference in hardness across and parallel to the length of crystals is distinctive. Anhedral cordierite with abundant quartz inclusions may appear to be similar but is more typically blue, not pink. The hardness of altered andalusite may be substantially reduced.

Thin Section: Moderately high relief, large $2V$, and length fast character. Sillimanite is biaxial +, length slow, and has higher birefringence and a slender prismatic to fibrous habit. Kyanite has inclined extinction and higher birefringence. Staurolite has a distinctly honey yellow color and higher indices of refraction. Orthopyroxene may have higher birefringence, is length slow, and has a different occurrence. Apatite is uniaxial, and has lower birefringence and slightly anomalous interference colors.

SILLIMANITE

$^{VI}Al^{IV}AlO^{IV}SiO_4$
Orthorhombic $2/m\ 2/m\ 2/m$
a = 7.488 Å,
$\quad b$ = 7.681 Å,
$\quad c$ = 5.777 Å, Z = 4
H = 6.5–7.5
G = 3.23–3.27
Biaxial (+)
n_α = 1.653–1.661
n_β = 1.657–1.662
n_γ = 1.672–1.683
δ = 0.018–0.022
$2V_z$ = 20–30°

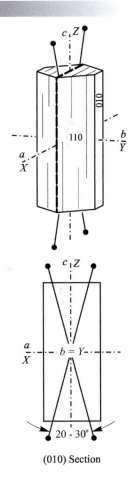

(010) Section

Composition Most sillimanite is relatively pure $AlAlOSiO_4$, although minor amounts of Fe^{3+}, Cr^{3+}, or Ti^{4+} may be present.

Form and Twinning Figure 16.10. Sillimanite commonly forms slender prismatic or fine fibrous crystals. Masses of fine fibrous crystals are called **fibrolite**. These masses may form swirled or matted aggregates a few millimeters to a centimeter across. Cross sections through crystals are usually approximately diamond shaped.

Cleavage Perfect on {010}, apparently between the chains of aluminum octahedra. Uneven fracture. Brittle. Fibrous aggregates may be tough.

Color Properties

Luster: Vitreous.

Color: Colorless or white, less commonly yellow, brown, or blue.

Streak: White.

Color in Thin Section and Grain Mount: Usually colorless, although matted fibrolite masses may be pale brown.

Pleochroism: Not usually pleochroic. Pleochroism for dark-colored varieties and thick fragments is listed in Table 16.4.

Optical Characteristics

Relief in Thin Section: High positive.

Optic Orientation: $X = a$, $Y = b$, $Z = c$, optic plane {010}. Elongate grains show parallel extinction and are length slow.

Indices of Refraction and Birefringence: Indices of refraction show little variation. In thin section, interference colors are up to lower second order. Fine fibrolite fibers show lower interference colors because they are thinner than a standard thin section.

Interference Figure: Basal sections of crystals of sufficient size yield acute bisectrix interference figures with $2V$ between 20° and 30°. Fragments on the {010} cleavage yield flash figures. Optic axis dispersion is strong, $r > v$.

Alteration Sillimanite may alter to sericite (fine-grained white mica) and may show reaction relations with associated minerals.

Distinguishing Features

Hand Sample: Slender crystals or fibrous clots in medium- and high-grade pelitic metamorphic rocks are characteristic. The fibrous nature of fibrolite is visible with a hand lens; it is not uncommon for fibrolite clots to be partially replaced by muscovite. Resembles wollastonite and tremolite but has only one cleavage and a different occurrence.

Table 16.4 Pleochroism in Sillimanite[a]

Hand-Sample Color	X	Y	Z
Yellow	Yellow	Green-yellow	Colorless
Brown	Colorless or pale yellow	Colorless or pale yellow	Violet-brown
Blue	Colorless or pale yellow	Colorless or pale yellow	Blue

[a] Most sillimanite in thin section is not visibly pleochroic.

Thin Section: Distinguished by high relief, moderate bire-fringence, parallel extinction, and slender prismatic or fibrous crystal habit. Kyanite has inclined extinction and is biaxial negative with a large $2V$. Tremolite displays typical amphibole cleavage and inclined extinction. Wollastonite is biaxial negative and has three cleavage directions. Fine fibers of sillimanite found in granitic rocks are often erroneously identified as rutile.

KYANITE

$^{VI}Al^{VI}AlOSiO_4$
Triclinic $\bar{1}$
$a = 7.126$ Å,
  $b = 7.852$ Å, $c = 5.572$ Å,
  $\alpha = 89.99°$, $\beta = 101.11°$,
  $\gamma = 106.03°$, $Z = 4$
$H = {\sim}5$ parallel to c, ${\sim}7$
  at right angles on {100}
  cleavage surfaces
$G = 3.53–3.67$
Biaxial (−)
$n_\alpha = 1.710–1.718$
$n_\beta = 1.719–1.725$
$n_\gamma = 1.724–1.734$
$\delta = 0.012–0.016$
$2V_x = 78–84°$

Composition Most kyanite is relatively pure, although minor substitution of Fe^{3+}, Ti^{4+}, or Cr^{3+} may be present.

Form and Twinning Figure 16.10e. Kyanite commonly forms elongate bladed or columnar crystals that may appear to be bent or twisted. Single or multiple twins with {100} composition planes are common; multiple twins on {001} are less common. Twinning is generally not obvious in hand sample.

Cleavage Perfect on {100}, good on {010}, where {100} ∧ {010} = 79°. Cleavage fragments may be splintery. A basal parting on {001} may be conspicuous, cutting at about 85° to the length of crystals. Splintery fracture. Brittle.

Color Properties

Luster: Vitreous.

Color: Patchy blue, gray, or white, less commonly light green.

Streak: White.

Color in Thin Section: Usually colorless, though some may show patchy light blue colors. Color may be more distinctive in grain mount.

Pleochroism: Weak if colored, $X < Y < Z$: X = colorless, Y = light violet blue, Z = light cobalt blue.

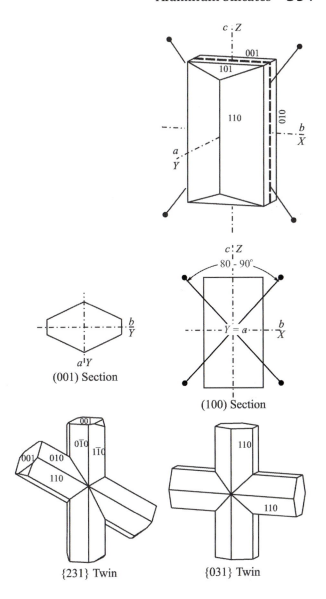

Optical Characteristics

Relief in Thin Section: High positive.

Optic Orientation: The optic plane is nearly at right angles to (100). $Z \wedge c$ and $Y \wedge b$ are both about $+27°$ to $+32°$ and $X \wedge a$ is only a few degrees. The maximum extinction angle of about $30°$ is obtained on grains or cleavage fragments with {100} parallel to the microscope stage. Basal sections yield extinction angles of less than $3°$, and sections parallel to {010} yield extinction angles of around $7°$. Both prismatic crystals and cleavage fragments are length slow.

Indices of Refraction and Birefringence: Indices of refraction vary in a small range. In thin section the 0.012–0.016 birefringence produces first-order interference colors. Fragments on cleavage surfaces in grain mounts yield intermediate birefringence.

Interference Figure: Grains in thin section and fragments with (100) horizontal yield acute bisectrix interference figures with the melatopes outside the field of view. Optic axis dispersion is weak, $r > v$.

Alteration Kyanite may alter to sericite, which is fine-grained white mica, or to chlorite or other phyllosilicates. Reaction textures with other aluminous minerals such as andalusite, sillimanite, staurolite, cordierite, or garnet are not uncommon.

Distinguishing Features

Hand Sample: Distinguished by bladed habit, color, cleavage, and distinct variation of scratch hardness with direction.

Thin Section/Grain Mount: Different habit, higher indices of refraction, inclined extinction, optical character, and sign of elongation distinguish kyanite from andalusite and sillimanite.

STAUROLITE

$Fe_2Al_9O_6[(Si,Al)O_4]_4(OH)_2$
Monoclinic (pseudoorthorhombic) $2/m$
$a = 7.863$–7.871 Å,
  $b = 16.534$–16.613 Å,
  $c = 5.632$–5.663 Å,
  $\beta = 90.00$–$90.45°$, $Z = 2$
$H = 7$–7.5
$G = 3.74$–3.83
Biaxial (+)
$n_{\_} = 1.736$–1.747
$n_{\_} = 1.740$–1.754
$n_{\_} = 1.745$–1.762
$\delta = 0.009$–0.015
$2V_z = 80$–$90°$

Structure The structure of staurolite is based on slightly distorted cubic close-packed oxygen and hydroxyl, and is responsible for the relatively high specific gravity. The aluminum occupies octahedral sites, and both iron and silicon occupy tetrahedral sites. The cations are distributed so that Al-bearing octahedra extend in edge-sharing chains parallel to c, with additional Al octahedra and Si and Fe tetrahedra linking between chains. The octahedra chains are very similar to those found in kyanite, which makes it easy for kyanite to epitaxially nucleate on staurolite; intergrowths of these minerals are common.

Composition The number of Fe per 24 oxygens is usually less than the ideal 2; it commonly ranges from 1.1 to 1.7 because of poorly understood substitutions involving Mg, Al, Ti, Zn, Si, and other cations in the tetrahedral and octahedral sites. Almost all of the iron is Fe^{2+}. The amount of hydroxyl also is variable from about 1.5 to 2.

Form and Twinning Figure 16.11. Crystals are commonly prismatic and elongate parallel to c, with the {110} and

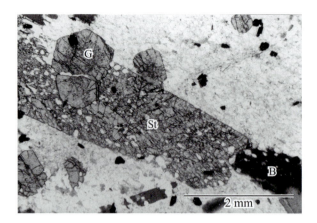

Figure 16.11 Sieved staurolite porphyroblast (St) with porphyroblasts of garnet (G) and biotite (B) in thin section of mica schist. Plane light.

{101} prisms and {010} and {001} pinacoids the dominant forms. The name staurolite is derived from the Greek for *cross* in allusion to the common cross-shaped penetration twins. Twinning on {031} produces a cross at 90°, and on {231} produces a cross at about 60°; neither is obvious in thin section unless fortuitously cut.

Cleavage A relatively poor cleavage on {010} is usually not observed in hand sample or thin section and does not tend to control orientation in grain mount. Subconchoidal fracture is typical, and tenacity is brittle.

Color Properties

Luster: Vitreous.

Color: Brown, with reddish, yellowish, or blackish cast.

Streak: Gray.

Color in Thin Section: Pale honey yellow or tan in thin section; darker in grain mounts. Some may show sector zoning.

Pleochroism: Distinct with X = colorless or pale yellow, Y = pale yellow to yellowish brown, Z = golden yellow to reddish brown.

Optical Characteristics

Relief in Thin Section: High positive.

Optic Orientation: $X = b$, $Y = a$, $Z = c$, optic plane = (100).

Indices of Refraction and Birefringence: Indices of refraction increase with iron content. In thin section the first-order white interference color may appear somewhat yellowish because of mineral color. Infrequently the interference colors may be anomalously bluish.

Interference Figure: $2V_z$ increases with decreasing iron content and in some, $2V_z$ may exceed 90°. Optic axis dispersion is weak to moderate, $r > v$. Basal sections yield acute bisectrix figures with melatopes well out of the field of view.

Alteration Staurolite may alter to fine-grained white mica (sericite) or chlorite. Staurolite is involved in a variety of metamorphic mineral reactions and therefore may display reaction relations with associated minerals such as andalusite, kyanite, garnet, and cordierite.

Distinguishing Features

Hand Sample: Crystal habit, characteristic twins, and brown color are distinctive.

Thin Section/Grain Mount: Color, pleochroism, relief, and habit make staurolite relatively distinctive. It may resemble brown tourmaline, but tourmaline is uniaxial and is darkest when its long axis is perpendicular to the vibration direction of the lower polarizer, whereas staurolite is the opposite.

Occurrence Staurolite is a common mineral in medium-grade pelitic metamorphic rocks and may be associated with garnet, cordierite, chloritoid, aluminum silicates, muscovite, and biotite. Staurolite is common in the heavy mineral fraction of sediments derived from pelitic metamorphic rocks.

Use In industry it is restricted to use as an abrasive, usually in sandblasting applications, where its hardness is an advantage. Well-formed twinned crystals are used to make pendants and earrings that are sold with names such as "fairy stones" or "tears of Christ."

CHLORITOID

$(Fe^{2+},Mg,Mn)_2(Al,Fe^{3+})$
$Al_3O_2(SiO_4)_2(OH)_4$
Monoclinic 2/*m* (also triclinic) pseudohexagonal
$a = 9.50$ Å,
    $b = 5.50$ Å, $c = 18.22$ Å,
    $\beta = 101.95°$, $Z = 4$
$H = 6.5$
$G = 3.46–3.80$
Biaxial (+ or −)
$n_\alpha = 1.705–1.730$
$n_\beta = 1.708–1.734$
$n_\gamma = 1.712–1.740$
$\delta = 0.005–0.022$, usually 0.010–0.012
$2V_z = 36–72°$ (monoclinic)
$2V_x, = 55–88°$ (triclinic)

Structure Chloritoid resembles the micas and, like them, has a layered structure. In chloritoid, however, the sheet-like structure is defined by the distribution of cations in octahedral coordination (Figure 16.12). Two types of octahedral layers are present. One consists of two layers of O

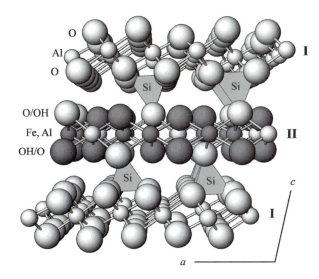

Figure 16.12 Chloritoid structure. Octahedral layer I contains Al^{3+} sandwiched between two sheets of O. Octahedral layer II contains Fe^{2+} (small dark spheres) and Al^{3+} sandwiched between sheets of O and OH (large dark spheres). Si^{4+} occupies isolated tetrahedral sites between alternating layers I and II.

with Al in octahedral coordination in between. The second consists of two layers of (O + OH) with Fe + Al in octahedral coordination in between. These octahedral layers are tied to each other through Si^{4+} in tetrahedral coordination with O from the octahedral layers. Several different polytypes, including triclinic varieties, are possible because the octahedral layers can be stacked in different arrangements.

Composition The ideal formula is $Fe^{2+}_2AlAl_3O_2(SiO_4)_2(OH)_4$, but some Fe^{3+} substitutes for Al^{3+}, and Mg^{2+} and Mn^{2+} typically substitute for Fe^{2+}, although most chloritoid is Fe rich. The Mn-rich variety of chloritoid is called **ottrelite**.

Form and Twinning Figure 16.13. Crystals are usually platy parallel to {001} with a roughly hexagonal basal section. Often foliated like the micas. Random cuts through porphyroblasts are usually rectangular. Lamellar twinning on {001} is common. Chloritoid commonly contains inclusions of associated minerals, particularly quartz.

Cleavage A cleavage on {001} is parallel to the sheets in the structure, but is less perfect than in the micas. Prismatic cleavage on {210} is imperfect to poor. A parting is possible on {010}. The {001} cleavage strongly controls fragment orientation. Cleavage flakes are brittle.

Color Properties

Luster: Somewhat pearly on cleavage surfaces.

Color: Dark gray, greenish gray, greenish black; distinctly green in thin plates.

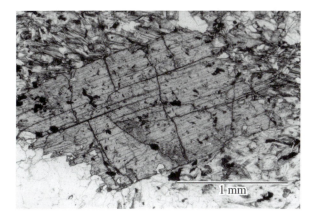

Figure 16.13 Photomicrograph of chloritoid in a thin section of schist. Plane light.

Streak: White, grayish, or slightly greenish.

Color in Thin Section: Green to colorless.

Pleochroism: Distinct: X = green or gray-green; Y = blue-gray, indigo, or blue-green; Z = colorless, yellow, or pale green. Color zoning with an hourglass pattern is common.

Optical Characteristics

Relief in Thin Section: High positive.

Optic Orientation: Variable. Y or $X \wedge a <\sim +20°$, X or $Y = b$, $Z \wedge c = +2°$ to $+30°$ for monoclinic varieties. Triclinic varieties are similar. The trace of cleavage is length fast and extinction is usually inclined between 0° and 20°, depending on the specific optic orientation of the grain and how it happens to be cut.

Indices of Refraction and Birefringence: Mn- and Mg- rich varieties have lower indices of refraction than Fe-rich varieties; substitution of Fe^{3+} for Al^{3+} tends to increase indices of refraction but is not sufficiently systematic to allow compositions to be estimated. Birefringence is usually 0.010–0.012, so maximum interference color in thin section is up to pale first-order yellow. Some varieties have higher birefringence and interference colors are often anomalous.

Interference Figure: Because Z is nearly normal to the basal cleavage, cleavage fragments yield nearly centered acute bisectrix figures if positive or obtuse bisectrix figures if negative. Different samples from the same locality may display different values of $2V$ because of differences in stacking of the octahedral layers. Monoclinic varieties are usually optically positive and triclinic varieties are usually optically negative. Optic axis dispersion is strong, $r > v$, but may be quite variable in different grains from the same sample.

Alteration Chloritoid may alter to chlorite, fine-grained white mica (sericite), or clay. It may display metamorphic mineral reaction relations with the minerals with which it is associated.

Distinguishing Features

Hand Sample: Strongly resembles chlorite and can be difficult to distinguish in hand sample.

Thin Section/Grain Mount: Closely resembles chlorite and may resemble green biotite. Biotite has higher birefringence and nearly parallel extinction. Chlorite has lower indices of refraction and relief, and most chlorite varieties have nearly parallel extinction. If present, the hourglass structure in chloritoid is characteristic.

Occurrence Chloritoid is common as porphyroblasts in low- to medium-grade regionally metamorphosed rocks such as mica, chlorite, and glaucophane schists, phyllite, and quartzite. It often is associated with aluminum silicates, garnet, chlorite, muscovite, and staurolite. Hydrothermal processes may form chloritoid in quartz–carbonate veins and in hydrothermally altered rocks, and it may be associated with corundum in emery deposits.

Use None.

TITANITE

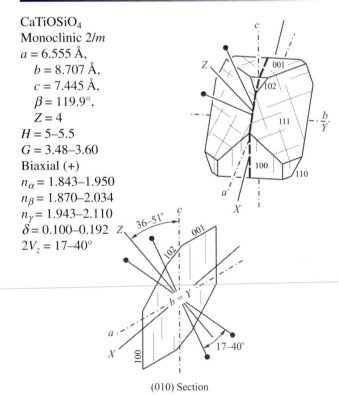

$CaTiOSiO_4$
Monoclinic $2/m$
$a = 6.555$ Å,
$b = 8.707$ Å,
$c = 7.445$ Å,
$\beta = 119.9°$,
$Z = 4$
$H = 5–5.5$
$G = 3.48–3.60$
Biaxial (+)
$n_\alpha = 1.843–1.950$
$n_\beta = 1.870–2.034$
$n_\gamma = 1.943–2.110$
$\delta = 0.100–0.192$
$2V_z = 17–40°$

(010) Section

Structure The structure of titanite, also called **sphene**, consists of kinked chains of corner-sharing TiO_6^{8-} octahedra

that run parallel to [101] (between the *a* and *c* axes). These chains are held together laterally through Si^{4+} in tetrahedral coordination and Ca^{2+} in irregular 7-fold coordination. Note that the crystal setting used here is based on external morphology. Structural studies usually place *a* parallel to the octahedral chains and *c* parallel to −*a* as defined here.

Composition Both Al^{3+} and Fe^{3+} may substitute for Ti^{4+}, usually coupled with substitution of OH^- or F^- for O^{2-} to maintain charge balance. Uranium, thorium, and a variety of rare earth elements may substitute for Ca^{2+} in small amounts.

Form and Twinning Wedge-shaped crystals with {111} and {110} prisms and {100} and {001} pinacoids are common; cross sections in thin section usually are diamond shaped (Figure 16.14). Grains are typically small and often are not visible in hand samples of the rocks in which titanite occurs. Also common are somewhat rounded to irregular grains. Simple twins on {100} are not uncommon and may be expressed in thin section as a composition plane along the long diagonal of diamond-shaped sections. Lamellar twinning on {221} is less common.

Cleavage Fair to good prismatic {110} cleavage is not obvious in thin section but may control fragment orientation. Unlike the other orthosilicates, this cleavage is not parallel to the octahedral chains. Rather, it cuts across the chains in such a way as to avoid cutting Si–O bonds. Partings parallel to the twins may be more prominant than cleavage in hand sample. Uneven fracture. Brittle.

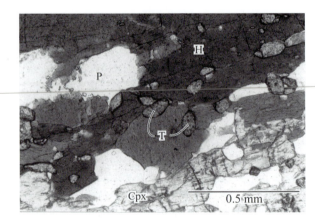

Figure 16.14 Photomicrograph of subhedral to anhedral titanite (T) grains with hornblende (H), plagioclase (P), and Ca-clinopyroxene (Cpx) in a thin section of amphibolite. Plane light.

Color Properties

Luster: Adamantine to resinous.

Color: Brown, gray, green, yellow, red, black.

Streak: White.

Color in Thin Section: Shades of brown or yellowish brown, less commonly colorless.

Pleochroism: Nonpleochroic to weakly pleochroic, $X < Y < Z$: X = colorless to pale yellow; Y = yellowish brown, pink, or greenish yellow; Z = orangish brown, greenish brown, green, or red.

Optical Characteristics

Relief in Thin Section: Very high positive.

Optic Orientation: $X \wedge a = -6°$ to $-21°$, $Y = b$, $Z \wedge c = +36°$ to $+15°$.

Indices of Refraction and Birefringence: Neither appear to vary systematically with composition. Extreme birefringence yields upper-order white interference colors in thin section and grain mount, but interference color may be masked by mineral color. Strong dispersion may prevent complete extinction in white light for some sample orientations.

Interference Figure: Figures are biaxial positive with numerous isochromes and small to moderate $2V$. Optic axis dispersion is very strong, $r > v$; inclined dispersion is weak. Fragments on the {110} cleavage yield off-center figures.

Alteration The usual alteration product of titanite is an aggregate of titanium oxides, quartz, and other minerals known as **leucoxene**, which in hand sample may appear earthy white or yellow white, and in thin section may be nearly opaque. Titanite may be produced by alteration of biotite or clinopyroxene.

Distinguishing Features

Hand Sample: Titanite is characterized by luster and crystal habit in the infrequent occurrences where crystals are large enough to work with. Hardness is less than staurolite and greater than sphalerite.

Thin Section/Grain Mount: Very high relief, extreme birefringence, and habit are characteristic.

Occurrence Titanite is a common accessory mineral in a wide variety of igneous rocks, in which it may be the dominant Ti-bearing mineral. In metamorphic rocks it is most common in relatively mafic varieties such as amphibolite. Titanite is also found in the heavy mineral fraction of clastic sediments.

Use Titanite is a minor ore of titanium.

Comment Titanite may be used for radiometric dating of the rocks in which it is found because it often contains substantial amounts of U and Th.

TOPAZ

$Al_2SiO_4(F,OH)_2$
Orthorhombic $2/m\ 2/m\ 2/m$
$a = 4.65$ Å,
  $b = 8.80$ Å,
   $c = 8.39$ Å, $Z = 4$
$H = 8$
$G = 3.49–3.57$
Biaxial (+)
$n_\alpha = 1.606–1.635$
$n_\beta = 1.609–1.637$
$n_\gamma = 1.616–1.644$
$\delta = 0.008–0.011$
$2V_z = 44–68°$

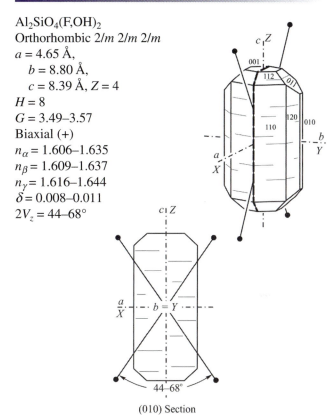

(010) Section

Structure The structure is based on close-packed sheets of O alternating with sheets of $(F,OH)_2O$ parallel to (010). Al^{3+} is distributed in octahedral sites between these sheets to form crankshaft-like chains of edge-sharing octahedra parallel to c that are cross-linked through Si^{4+} in tetrahedral coordination.

Composition The only significant compositional variation is in the relative amounts of F and OH.

Form and Twinning Crystals are usually stubby to elongate prisms parallel to the c axis (Figure 16.15). Basal sections range from nearly square to a rounded diamond shape with eight faces. Anhedral grains and irregular masses are also common. Detrital grains tend to be platy parallel to the basal cleavage. Twins on {010} are rare.

Cleavage The basal cleavage {001} is perfect and is the only plane that can cut through the structure without breaking the relatively strong Si–O bonds. Subconchoidal to uneven fracture. Brittle.

Figure 16.15 Topaz. (a) Topaz crystal. (b) Photomicrograph of basal section of topaz in thin section. Plane light.

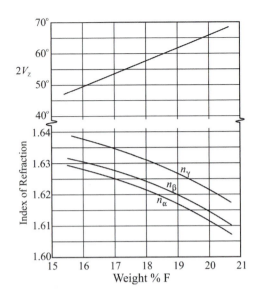

Figure 16.16 Optical properties of topaz. After Ribbe and Rosenberg (1971).

Color Properties

Luster: Vitreous.

Color: Most often colorless or gray but may be yellow, pink, red, brown, green, blue, or violet. Coloration often due to radiation-induced color centers and/or trace amounts of transition metals.

Streak: White.

Color in Thin Section: Colorless. Thick fragments in grain mount may be a pale version of the hand sample color.

Pleochroism: Weak in colored samples: $X = Y$ = shades of yellow and Z = shades of pink.

Optical Characteristics

Relief in Thin Section: Moderate positive.

Optic Orientation: $X = a$, $Y = b$, $Z = c$, optic plane = (010). Trace of cleavage shows parallel extinction in longitudinal sections and is length fast.

Indices of Refraction and Birefringence: Indices of refraction vary systematically with F content (Figure 16.16). Maximum interference color in standard thin section is first-order white with a tinge of yellow. Fragments resting on a cleavage surface display birefringence of about 0.003.

Interference Figure: $2V_z$ increases systematically with increasing F content. Basal sections yield centered acute bisectrix figures. Optic axis dispersion is distinct with $r > v$.

Alteration Relatively stable in the weathering environment. May hydrothermally alter to fine-grained white mica (sericite), clay, or fluorite.

Distinguishing Features

Hand Sample: Resembles quartz but is distinguished by crystal habit, basal cleavage, high specific gravity, and hardness.

Thin Section/Grain Mount: Resembles quartz but has higher relief in thin section, is biaxial, and has cleavage.

Occurrence Occurs most commonly in both volcanic and intrusive felsic igneous rocks. Well-formed crystals may occupy vesicles or other cavities in rhyolitic volcanics and large masses may be found in granitic pegmatites. Topaz also is found in hydrothermal tungsten, tin, molybdenum, or gold mineral deposits, in hydrothermally altered rocks adjacent to granitic intrusions, and infrequently in metamorphic quartzite and schist. Because it is relatively stable in the weathering environment, topaz is not uncommon in the heavy mineral fraction of sediments.

Use Colorless, yellow, brown, or pink topaz may be used as a gemstone. The color of most blue topaz has been induced by radiation. Probably because topaz is both beautiful and valuable, hence likely to produce the placebo effect in patients, it has historical medical uses that include treatment for eye problems, gout, poisons, and gastrointestinal upset.

ADDITIONAL ORTHOSILICATES

Additional orthosilicate minerals include dumortierite, chondrodite, monticellite, and datolite. Table 16.5 lists their properties.

Dumortierite is a relatively uncommon mineral found in granitic pegmatite, aplite, quartz veins, and hydrothermally altered rocks. It also occurs in medium- to high-grade gneiss, schist, and quartzite in association with other aluminous minerals. Fine dumortierite needles may be responsible for the pink color of some rose quartz. It forms acicular or fibrous grains and fibrous masses. The fibrous habit and blue color are distinctive, both in hand sample and in thin section. The blue amphiboles have inclined extinction.

Chondrodite is the most common member of the humite group, whose members are

Norbergite	$Mg_2SiO_4 \cdot XMg\ (F,OH)_2$
Chondrodite	$2(Mg_2SiO_4) \cdot Mg(F,OH)_2$
Humite	$3(Mg)_2SiO_4) \cdot Mg(F,OH)_2$
Clinohumite	$4(Mg_2SiO_4) \cdot Mg(F,OH)_2$

These minerals are found in contact-metamorphosed dolomitic limestone in association with calcite, dolomite, tremolite, grossular, wollastonite, olivine, monticellite, graphite, phlogopite, spinel, and pyrrhotite. The structure of these minerals consists of layers of olivine-like structure parallel to (100) that alternate with layers with brucite-like structure. The olivine-like layers are one, two, three, and four layers thick in the different minerals as is indicated by their formulas. Fe^{2+}, Mn^{2+}, and Ti may substitute for Mg, and Al may substitute for Si in limited amounts. They usually form rounded or irregular grains, or granular masses. Crystals are often platy, but they have diverse habits. Chondrodite may resemble forsterite but is more typically light yellow or red, where forsterite is green. It is distinguished from forsterite in thin section by lower birefringence, smaller 2V, and pale yellow pleochroism.

Monticellite is another mineral found in metamorphosed dolomitic limestone in association with calcite, dolomite, tremolite, diopside, forsterite, wollastonite, humite group, and related calc-silicate minerals. The structure is the same as olivine, with half the octahedral sites occupied by Ca rather than Mg. Some Fe^{2+} may substitute for Mg, and continuous solid solution is possible to **kirschsteinite** ($CaFeSiO_4$). Monticellite typically forms anhedral, equant, or irregular grains. Crystals are elongate on c. Monticellite may resemble both forsterite and the humite group in hand sample but forsterite is typically green, and chondrodite is light yellow to red. In thin section monticellite is distinguished from forsterite by lower birefringence and from diopside by being optically negative with a larger 2V and poorer cleavage. Members of the humite group are all optically positive.

Datolite is most commonly a secondary mineral found in amygdules and other voids in basalt and related rocks. Crystals are typically short prismatic to roughly equant. Datolite also forms botryoidal and globular masses and may be compact massive. It is characterized by its pale green color and crystals with many faces.

Table 16.5 Properties of Less Common Orthosilicates

	Dumortierite	Chondrodite	Monticellite	Datolite
Composition	$Al_7(BO_3)(SiO_4)_3O_3$	$2(Mg_2SiO_4) \cdot Mg(F,OH)_2$	$CaMgSiO_4$	$CaBSiO_4(OH)$
Crystal system	Orthorhombic	Monoclinic	Orthorhombic	Monoclinic
Color	Blue, greenish blue, violet, purple, brown; vitreous	Yellow, brown, red, grayish green; may fluoresce	Colorless, white, pale greenish gray, yellowish gray	Colorless to white, often w/ greenish tinge; commonly fluorescent
Color in thin section	Deep to pale blue; pleochroic	Pale yellow to colorless; pleochroic	Colorless	Colorless
Sign and 2V	Biaxial (−), 13–55°	Biaxial (+), 50–85°	Biaxial (−), 69–88°	Biaxial (−), 72–75°
Indices of refraction	$n_\alpha = 1.659–1.686$ $n_\beta = 1.684–1.722$ $n_\gamma = 1.686–1.723$	$n_\alpha = 1.592–1.617$ $n_\beta = 1.602–1.635$ $n_\gamma = 1.621–1.646$	$n_\alpha = 1.638–1.654$ $n_\beta = 1.648–1.664$ $n_\gamma = 1.650–1.674$	$n_\alpha = 1.622–1.626$ $n_\beta = 1.649–1.654$ $n_\gamma = 1.666–1.670$
δ	0.011–0.037	0.028–0.034	0.012–0.020	0.044–0.046
Physical	$H = 7–8.5\ G = 3.21–3.41$ Cyclic twins on {110}	$H = 6.5\ G = 3.16–3.26$ {001} simple and lamellar twins	$H = 5–5.5\ G = 3.03–3.27$ {031} cyclic twins produce 6-point star	$H = 5–5.5\ G = 2.96–3.00$
Cleavage	{100} good, {110} poor	{100} poor	{010} poor/fair	None; conchoidal to uneven fracture

REFERENCES CITED AND SUGGESTIONS FOR ADDITIONAL READING

Deer, W. A., Howie, R. A., and Zussman, J. 2001. *Rock-Forming Minerals,* Volume 1A*: Orthosilicates*, 2nd ed. The Geological Society, London, 919 p.

Green, H. W. II, and Burnley, P. C. 1989. A new self-organizing mechanism for deep-focus earthquakes. Nature 341, 733–737.

Griffen, D. T. 1992. *Silicate Crystal Chemistry.* Oxford University Press, New York, 442 p.

Hanchar, J. M., and Hoskins, P. W. O. (eds.). 2003. *Zircon.* Reviews in Mineralogy and Geochemistry 53, 500 p.

Harley, S. L., and Kelly, N. M. 2007. Zircon, tiny but timely. Elements 3, 13–18.

Hemingway, B. S., Robie, R. A., Evans, H. T., Jr., and Kerrick, D. M. 1991. Heat capacities and entropies of sillimanite, fibrolite, andalusite, kyanite, and quartz and the Al_2SiO_5 phase diagram. American Mineralogist 76, 1597–1613.

Iidaka, T., and Suetsuga, D. 1992. Seismological evidence for metastable olivine inside a subducting slab. Nature 356, 593–595.

Kerrick, D. M. 1990. *The Al_2SiO_5 Polymorphs.* Reviews in Mineralogy 22, 406 p.

Liebau, F. 1985. *Structural Chemistry of Silicates, Structure, Bonding and Classification.* Springer-Verlag, Berlin, 347 p.

Louisnathan, S. J., and Smith, J. V. 1968. Cell dimensions of olivine. Mineralogical Magazine 36, 1123–1134.

Ribbe, P. H. (ed.). 1982. *Orthosilicates*, 2nd ed., Reviews in Mineralogy 5, 450 p.

Ribbe, P. H., and Rosenberg, P. 1971. Optical and X-ray determinitive methods for fluorine in topaz. American Mineralogist 56, 1812–1821.

Spear, F. S. 1993. *Metamorphic Phase Equilibria and Pressure–Temperature–Time Paths.* Mineralogical Society of America, Washington, DC., 719 p.

Winchell, H. 1958. The composition and physical properties of garnet. American Mineralogist 42, 595–600.

Carbonates, Sulfates, Phosphates, Tungstates, Molybdates, and Borates

STRUCTURE AND CLASSIFICATION

The minerals described in this chapter all have structures based on anionic groups that have a net −2 to −5 charge.

CARBONATES: CO_3^{2-}

SULFATES: SO_4^{2-}

PHOSPHATES: PO_4^{3-}

TUNGSTATES: WO_4^{2-}

MOLYBDATES: MoO_4^{2-}

BORATES: BO_3^{3-}, BO_4^{5-}, and others

For the first five mineral groups (all but the borates), the cation in each anionic group commands more than half of the available charge of the oxygen anions. Minerals constructed using these anionic groups are therefore anisodesmic (Chapter 3); the anionic groups must be treated as discrete structural elements. The carbonate group consists of three oxygens in triangular arrangement around the central carbon. In the others, the anionic group is a tetrahedron with oxygen anions at the corners of a tetrahedron and the cation in the center.

In the borates, which are included in this chapter for convenience, the B^{3+} may coordinate with either three or four O^{2-} or OH^- anions to form a variety of complex structures.

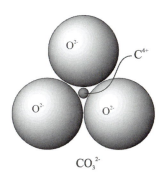

Figure 17.1 CO_3^{2-} anionic groups. A carbon cation is situated at the center of three oxygen anions.

CARBONATES

The essential structural element in all carbonate minerals is the CO_3^{2-} anionic group (Figure 17.1). Carbon, with its +4 charge, commands two-thirds of the net −6 charge of the three oxygen anions with which it coordinates. These oxygen anions are therefore more strongly bonded to the central C^{4+} cation than to any other cations in the mineral structure. Further, an oxygen anion in one CO_3^{2-} group cannot also be shared with another CO_3^{2-} group. The structure of all carbonates must therefore be based on CO_3^{2-} groups that are bonded laterally through various cations. In some cases additional OH^- groups or other anionic components may be present.

The common carbonate minerals can be divided into the calcite group, dolomite group, aragonite group, and a hydrated carbonate group (Table 17.1). The calcite and dolomite groups share essentially the same structure and collectively are called the **rhombohedral carbonates**, because they have rhombohedral symmetry. Additional carbonate minerals are listed in Table 17.2.

Table 17.1 Major Groups of Carbonate Minerals

Calcite Group		Dolomite Group		Aragonite Group		OH-Bearing Group	
Calcite	$CaCO_3$	Dolomite	$CaMg(CO_3)_2$	Aragonite	$CaCO_3$	Azurite	$Cu_3(CO_3)_2(OH)_2$
Magnesite	$MgCO_3$	Ankerite	$Ca(Mg,Fe)(CO_3)_2$	Witherite	$BaCO_3$	Malachite	$Cu_2CO_3(OH)$
Rhodochrosite	$MnCO_3$	Kutnohorite	$CaMn(CO_3)_2$	Strontianite	$SrCO_3$		
Siderite	$FeCO_3$			Cerussite	$PbCO_3$		
Smithsonite	$ZnCO_3$						

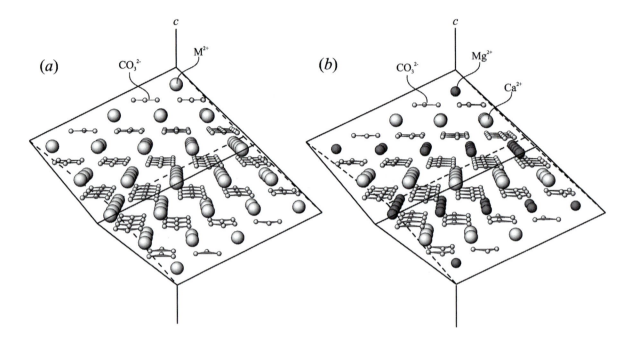

Figure 17.2 Structure of the rhombohedral carbonates. O^{2-} anions at the corners of triangular CO_3^{2-} groups are shown as small spheres. (*a*) In the calcite structure, CO_3 groups form sheets parallel to {001} separated by octahedrally coordinated cations (M^{2+}). Because the position of C cations repeats every third carbonate sheet and the direction that the CO_3 groups "point" alternates every other sheet, only every sixth CO_3 sheet is a duplicate. (*b*) In the dolomite structure, layers of larger Ca cations alternate with layers of smaller Mg, Fe, or Mn cations.

RHOMBOHEDRAL CARBONATES (CALCITE AND DOLOMITE GROUPS)

The structure of the rhombohedral carbonates can be derived from the halite structure, with CO_3 groups in place of Cl, and M^{2+} in place of Na (Figure 17.2). The cube is shortened along one of its 3-fold axes to form a rhombohedron, and the plane of the triangular CO_3 groups is perpendicular to this axis, which becomes the *c* axis. As a result of this geometry, the CO_3 groups form layers perpendicular to the *c* axis. Each layer of CO_3 groups effectively consists of a simple close-packed layer of oxygen anions with C occupying selected triangular sites within the layer. Successive layers are offset so that the position of the C is equivalent to cubic close packing (i.e., their position is repeated in every third layer). Although the position of the C atoms is repeated in every third CO_3 layer, the corners of the triangular CO_3 groups alternate by a mirror plane every

layer. The result is that an exact duplicate layer of CO_3 groups is repeated only every sixth layer. The unit cell must be six CO_3 layers thick or about 15 to 17 Å along the *c* axis.

Divalent cations (M = Ca, Mg, Mn, or Fe) occupy octahedral (6-fold) sites between the layers. Each oxygen anion therefore coordinates with one C cation within its layer, one divalent cation above it, and one below. In the calcite group, the octahedral sites are all occupied by the same cation(s) (Figure 17.2a). The dolomite group must accommodate two distinctly different-sized cations, one of which is Ca. This is accomplished by having layers of Ca alternate with layers of smaller Mg, Fe, or Mn (Figure 17.2b).

Solid solution among the end members of the calcite group is controlled largely by ionic radii (Appendix A). Solid solution between magnesite (Mg, 0.86 Å) and siderite (Fe^{2+}, 0.92 Å) is complete, as is solid solution between siderite and rhodochrosite (Mn, 0.96 Å). Solid solution between magnesite and rhodochrosite is somewhat limited. Ca, however, is substantially larger (1.14 Å) so calcite shows only

Table 17.2 Properties of Less Common Carbonate Minerals

Mineral	Unit Cell	Physical Properties[a]	Optics	Comments
Cerussite $PbCO_3$	Orthorhombic $2/m\ 2/m\ 2/m$ $a = 5.15$ $b = 8.47$ $c = 6.11$ $Z = 4$	H: 3–3.5 G: 6.55 Cl: {110}g, {021}pr Tw: {110}, {130}, cyclic C: Gray, colorless St: White L: Adamantine	$n_\alpha = 1.803$ $n_\beta = 2.074$ $n_\gamma = 2.076$ $\delta = 0.273$ Biaxial (−) $2V = 9°$ $r > \upsilon$, strong Colorless	Crystals variable, often tabular ‖(010) or acicular on c or a; found in oxidized zone of Pb-bearing hydrothermal sulfide deposits; effervesces in dilute HCl
Kutnohorite $CaMn(CO_3)_2$	Hexagonal (rhombohedral) $a = 4.85$ $c = 16.34$ $Z = 3$	H: 3.5–4 G: 3.12 Cl: {101} pf Tw: {001}, {102} C: White, pink St: White L: Vitreous	$n_\omega = 1.727$ $n_\epsilon = 1.535$ $\delta = 0.192$ Uniaxial (−) Colorless	Associated with other Mn minerals in hydrothermal deposits and sediments; effervesces in dilute HCl if powdered
Nahcolite $NaHCO_3$	Monoclinic $2/m$ $a = 7.48$ $b = 9.69$ $c = 3.48$ $Z = 4$ $\beta = 93.4°$ $X \wedge c = +27°, Y = b,$ $Z \wedge a = −24°$	H: 2.5 G: 2.16–2.21 Cl: {101}pf, {111} and {100}g Tw: {101} common C: White, gray St: White L: Vitreous	$n_\alpha = 1.377$ $n_\beta = 1.503$ $n_\gamma = 1.583$ $\delta = 0.206$ Biaxial (−) $2V = 75°$ $r < v$, weak Colorless	Forms grains elongate on c and granular massive aggregates; easily soluble in water; most common in evaporite deposits from saline lakes, as an efflorescence, or in hot spring deposits; effervesces in dilute HCl
Smithsonite $ZnCO_3$	Hexagonal (rhombohedral) $a = 4.653$ $c = 15.026$ $Z = 6$	H: 4–4.5 G: 4.30–4.45 Cl: {101} pf Tw: N/A C: Dirty brown, also colorless, green, blue, pink St: White L: Vitreous to pearly	$n_\omega = 1.850$ $n_\epsilon = 1.625$ $\delta = 0.225$ Uniaxial (−) Colorless	Rhombohedral or scalenohedral crystals; often forms reniform masses, encrustations, or granular masses; found as a secondary mineral in the oxidized portion of Zn-bearing hydrothermal sulfide deposits; effervesces in dilute HCl
Trona $Na_3H(CO_3)_2 \cdot 2H_2O$	Monoclinic $2/m$ $a = 20.11$ $b = 3.49$ $c = 10.31$ $\beta = 103°$ $Z = 4$ $X = b, Y \wedge c = +7°,$ $Z \wedge a = +7°$	H: 2.5–3 G: 2.11–2.13 Cl: (100)pf. {$\bar{1}$11}, {001}pr Tw: None C: Gray, yellowish white, also colorless, white, yellow, light brown St: White L: Vitreous	$n_\alpha = 1.416$ $n_\beta = 1.494$ $n_\gamma = 1.542$ $\delta = 0.126$ Biaxial (−) $2V = 75°$ $r < \upsilon$, distinct Colorless	Fibrous or columnar ‖ b, bladed ‖{001}, massive; found in evaporite deposits from saline lakes and as soil efflorescence; effervesces in dilute HCl, soluble in water, alkaline taste

[a] Cl = cleavage (pf = perfect, g = good pr = poor), Tw = twinning, C = color, St = streak, L = luster.

limited solid solution with Mg, Fe, and Mn carbonates. Ca is the largest cation that routinely fits into the 6-fold site. It also fits in the larger coordination site provided in the aragonite structure. Calcite and aragonite are polymorphs.

Solid solution among the members of the dolomite group is also possible. Dolomite [$CaMg(CO_3)_2$] and ankerite [$Ca(Mg,Fe)(CO_3)_2$] form a solid solution series, and some Mn may substitute for either Fe or Mg. There is little substitution of other cations for the Ca.

Based on the layered structure, it might be reasonable to suspect that cleavage in these carbonates is between the layers of CO_3 groups, i.e., {001} cleavage. That is not the case, however, because bonding from the CO_3 to the layers of divalent cations is substantially stronger than bonds between CO_3 groups within CO_3 layers. Because CO_3 groups are offset from one CO_3 layer to the next, the cleavage is inclined to the c axis, and the three cleavage directions define a rhomb whose surfaces are parallel to the three edges of the triangular carbonate groups.

This discussion leads to a problem with nomenclature in the rhombohedral carbonates. The prominent cleavage has conventionally been considered a unit face with {101} ({10$\bar{1}$1} for those using the Miller–Bravais convention) as an index, consistent with the Laws of Bravais and Haüy (Chapter 2). The problem is that this convention requires a unit cell along c that is one-fourth the dimension actually measured in X-ray diffraction studies ($c = 15$–17 Å), or alternatively, an a dimension that is four times its actual ~5 Å dimension. The thoughtful reader might reasonably suggest that Miller indices should be reassigned based on the correct unit cell to yield {104} as the index for the cleavage. That, however, would lead to confusion in the comparison of data from different sources. The practice followed here is to use the conventional Miller indices and report both the actual unit cell dimensions and unit cell dimensions consistent with the Miller indices. In comparing information from different sources, readers are cautioned to confirm the unit cell convention being used.

CALCITE

CaCO₃
Hexagonal (rhombohedral) $\bar{3}\, 2/m$
$a = 4.990$ Å, $c = 17.060$ Å, $Z = 6$
   ($a = 19.960$ Å, $c = 17.060$ Å, $Z = 96$ to be consistent with Miller indices)
$H = 3$ on cleavage (varies with direction), 2.5 on {001}
$G = 2.71$
Uniaxial (−)
$n_\omega = 1.658$
$n_\epsilon = 1.486$
$\delta = 0.172$

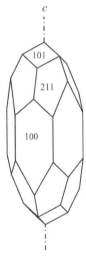

Composition Most calcite, particularly if formed in a sedimentary environment, is nearly pure CaCO₃. Crystallization at elevated temperatures allows some Mg, Fe, Mn, or Zn to substitute for Ca; Mg is the most common. Some larger cations such as Ba or Sr also may substitute for Ca.

Form and Twinning Calcite displays a wide variety of forms, but most crystals display some combinations of scalenohedrons, prisms, and rhombohedrons (Figure 17.3). The most common are the {211} scalenohedron, {100} prism, and {101} rhombohedron. Rhombohedrons, when present, usually modify other more prominent forms and simple rhombohedrons by themselves are not particularly common. Crystals in which the {211} scalenohedron dominate are sometimes referred to as "**dogtooth spar.**" The calcite in limestone usually forms fine anhedral grains. In marble and related metamorphic rocks, calcite is usually anhedral granular. Fossil shells and thin veins may be fibrous or columnar. High-quality clear calcite is referred to as **Iceland spar**.

Lamellar twins on the negative rhombohedron {012} are common and will be parallel to the edge or along the long diagonal between faces of cleavage rhombs (Figure 17.4).

Figure 17.3 Calcite. (*a*) Calcite crystals. Field of view is about 20 cm wide. (*b*) Photomicrograph of calcite in marble. Plane light.

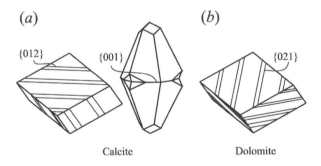

Figure 17.4 Twins in rhombohedral carbonates. (*a*) Lamellar {012} twins may be parallel to rhomb face edges or parallel to the long diagonal. Simple {001} twins may form distinctive crystals. (*b*) Dolomite. Lamellar {021} twins may be parallel to either the short or the long diagonal between rhombohedral cleavage traces.

Etching the face of a cleavage rhomb with dilute acid may reveal the presence of these twins. These twins are commonly produced as a result of deformation, and detailed study of their orientation may indicate the nature of the stresses that produced the deformation. The pressure of a knife blade on one of the obtuse edges of a cleavage rhomb may be sufficient to cause twinning (Figure 5.25). Samples also may be caused to twin as a result of cutting or grinding in the preparation of thin sections or when samples are crushed to prepare grain mounts. Simple twins with {001} composition planes also are common. Twins on {101} are infrequent.

Cleavage {101} Rhombohedral cleavage is perfect. The angle between cleavages is 74.9°. Parting along {012} twin lamellae. Fracture conchoidal but difficult to produce. Brittle.

Color Properties

Luster: Vitreous (earthy if very fine grained).

Color: Colorless, white, or gray most commonly, but almost any color is possible depending on the presence of impurities or fine disseminated inclusions.

Streak: White.

Color in Thin Section: Colorless.

Other: May fluoresce under UV light.

Optical Characteristics

Relief in Thin Section: Moderate negative to high positive, changes with rotation.

Optic Orientation: Extinction is inclined or symmetrical to cleavage traces and parallel to the length of elongate crystals. The fast ray vibrates parallel to the short diagonal of rhombohedral faces and traces of rhombohedral cleavage. The angle between the fast ray (ϵ') and the trace of {012} twin lamellae is greater than about 55° in sections cut so that one of the cleavage surfaces is nearly horizontal.

Indices of Refraction and Birefringence: Indices of refraction vary systematically with composition (Figure 17.5). The index $n_{\epsilon\{101\}}$ for fragments lying on a cleavage surface is 1.566 for pure calcite. Because birefringence is extreme (0.172), interference colors in standard thin section are

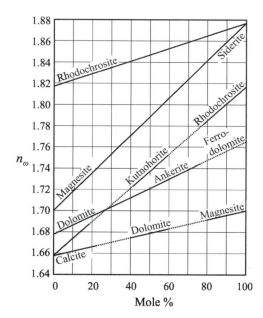

Figure 17.5 Index of refraction n_{ω} in the rhombohedral carbonates. After Kennedy (1947).

usually upper-order creamy white even if the optic axis is close to vertical. Twin lamellae may show as pastel bands of pink or green, and zones in which twin lamellae overlap may not go entirely extinct with stage rotation. Unless the optic axis is close to vertical, grains usually display a distinct change of relief with stage rotation, lower when the short diagonal of the rhomb is parallel to the lower polarizer vibration direction.

Interference Figure: Uniaxial negative with numerous isochromes and thin, well-defined isogyres. Some calcite, particularly from metamorphic rocks, may be biaxial negative with 2*V* up to about 15°.

Alteration Calcite may be replaced by dolomite through diagenetic processes, or replaced by quartz, opal, iron, manganese oxides, or other minerals to yield pseudomorphs. It is soluble in many natural waters and may be removed from rocks by dissolution.

Distinguishing Features

Hand Sample: Recognized by hardness, rhombohedral cleavage, and reaction with effervescence in dilute HCl (1 part concentrated HCl, 9 parts water). Dolomite has a higher specific gravity and does not react vigorously with cold dilute HCl unless powdered. Aragonite, which reacts vigorously with dilute cold HCl, has higher specific gravity than calcite.

Thin Section/Grain Mount: Cleavage, extreme birefringence, and change of relief with rotation distinguish the rhombohedral carbonates from most other minerals. The orthorhombic carbonates (aragonite group) are biaxial and lack rhombohedral cleavage. Index of refraction measurement in grain mount will unambiguously distinguish calcite from dolomite and from the other rhombohedral carbonates. Use the following criteria to help distinguish between calcite and dolomite in common carbonate rocks.

- Calcite is more commonly twinned, and lamellae are not parallel to the short diagonal of the rhomb.
- Dolomite more commonly forms euhedral rhombs in limestone and dolomite; it may be cloudy or stained with iron oxides.
- Dolomite has higher refractive indices.
- Staining techniques (Friedman, 1959; Warne, 1962; Wolfe and others, 1967) can speed identification in both hand sample and thin section.

Occurrence Calcite is a very common and widespread mineral. It is an important constituent in many clastic sedimentary rocks as a cementing agent or as fossil fragments. Calcite forms by chemical precipitation from water in many different environments (marine, lacustrine, groundwater, hydrothermal, etc.) and is an important biomineral (Chapters 1 and 5). It is produced by both biologically controlled and biologically induced processes. Biologically controlled mineralization yields the shells of many marine invertebrates including mollusks, foraminifera, and coccoliths. A good example of biologically induced mineralization is the dramatic "whiting" events seen in subtropical marine waters and saline lakes (Robbins and Blackwelder, 1992). Whitings originate with blooms of plankton, which induce the crystallization of calcite and aragonite crystals on their cellular surface. The abundance of fine carbonate crystals gives the water its white color. Calcite and aragonite produced in this manner may be a major contributor to the mass of limestone deposited in carbonate bank environments. Calcite also is a major mineral in evaporite deposits, which are described later. Calcite is common in metamorphic rocks derived from carbonate-bearing sediments and is the major mineral in marble, where it may be associated with diopside, tremolite, olivine, garnet, epidote, wollastonite, and other calcsilicate minerals. Calcite in igneous rocks is relatively rare, but may be found in silica-poor, alkali-rich igneous rocks containing nepheline or other feldspathoids, as vesicle fillings in volcanic rocks, or as a major mineral in rare calcite-rich intrusions called carbonatites. Hydrothermal mineral deposits commonly contain calcite as a gangue mineral, and it may fill fractures and form beautiful crystals. It is common as a joint coating or fracture filling in almost any rock and may be formed by the alteration of Ca-bearing minerals such as plagioclase.

Use Calcite is a very important industrial mineral. It is an essential raw material in the manufacture of portland cement. Calcite, usually in the form of limestone, is ground and mixed with a pulverized silicon- and aluminum-rich rock (usually shale) that contains some iron. The mixture is intensely heated to form a mixture of Ca silicates and aluminates that is subsequently finely ground to form portland cement. This gray powder is mixed with sand, gravel, and water to form concrete that hardens or cures by chemical reaction.

Intense heating of calcite alone drives off CO_2 and produces CaO, which is known as **quicklime**. Quicklime is mixed with sand and water to form mortar, which sets or hardens by chemical reaction between the CaO and the water to form various Ca hydroxides. Most mortar used now also includes substantial amounts of portland cement. Quicklime and hydrated lime [$Ca(OH)_2$], which also is produced from calcite, are widely used in the steel industry as a flux, in paper manufacture, wastewater treatment, industrial flue gas desulfurization, soil stabilization, asphalt manufacture, sugar refining, and as a soil amendment. Hydrated lime also is used as a mild abrasive in kitchen and bath cleaners.

Calcite is used as a pharmaceutical. Several commercial antacid preparations are simply pulverized calcite, as are some of the calcium supplements that women in particular are encouraged to take to ward off the bone loss associated with osteoporosis. The calcium in vitamin tablets also is commonly calcite. Pulverized calcite is added to cattle feeds to help improve digestion and is included in poultry feed to maintain the quality of egg shells.

Box 17.1 Limestone and Dolostone

Most limestone is deposited in shallow marine environments in locations without a substantial contribution of clastic sediment. Most of the calcite is produced by biologic activity. The foundation of biologic activity is photosynthesis, so most calcite production is limited to the photic zone: that is, it is limited by the depth to which sunlight penetrates the water, which is generally less than about 100 meters. The most abundant photosynthetic activity is within a few tens of meters of the surface, and it is here that calcite-producing organisms thrive. These calcite-producing organisms include mollusks, corals, brachiopods, crinoids, bryozoans, and other marine invertebrates that precipitate carbonate shells. Also included are many single-celled organisms such as coccolithophores, foraminifera, and various bacteria and algae. The identity of the species has, of course, changed over the course of geologic time.

Many limestone layers contain significant amounts of dolomite, and a carbonate consisting of over 75% dolomite is considered a dolostone. Dolomite, however, does not usually precipitate directly from seawater, and most dolostones contain fossils that indicate a normal marine environment of deposition where calcite is expected. In most cases, therefore, the formation of major dolostone layers and dolomitic limestone is the result of postdeposition diagenesis of original calcite in a limestone. Details of the process by which dolomite replaces calcite in limestone beds have been subject to considerable debate. The traditional explanations involve processes by which the pore water in limestone provides Mg^{2+} to produce the recrystallization of calcite to dolomite. The pore water may either be normal seawater or a brine produced by some degree of evaporation of seawater. However, recent studies (e.g., Wright, 1999; Sánchez-Román and others, 2008) show that dolomite growth may be mediated by both sulfate-reducing and aerobic bacteria. This suggests the possibility that dolomitization may be partly or largely a biologically driven process.

Calcite is a major constituent of marble, which is widely used in the creation of monuments and sculpture, and as dimension stone for architectural purposes. Unfortunately, marble, which is easily cut, shaped, and polished with basic tools, is readily degraded when exposed to the elements. Rain, which is naturally acidic because of dissolved CO_2, will dissolve calcite. Surface alteration to softer gypsum is produced by reaction with sulfate pollutants in the atmosphere that are derived in part from burning sulfur-bearing coal or other fuels.

MAGNESITE

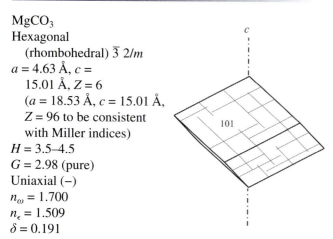

$MgCO_3$
Hexagonal
  (rhombohedral) $\bar{3}$ $2/m$
$a = 4.63$ Å, $c =$
  15.01 Å, $Z = 6$
  ($a = 18.53$ Å, $c = 15.01$ Å,
  $Z = 96$ to be consistent
  with Miller indices)
$H = 3.5–4.5$
$G = 2.98$ (pure)
Uniaxial (–)
$n_\omega = 1.700$
$n_\epsilon = 1.509$
$\delta = 0.191$

Composition Solid solution is complete between magnesite and siderite ($FeCO_3$), and substantial but incomplete between magnesite and rhodochrosite ($MnCO_3$), consistent with what should be expected based on the sizes of the cations (see earlier in this chapter, and Chapter 4). Substitution of Ca for Mg is limited.

Form and Twinning Crystals are usually rhombs, but are uncommon. More commonly as compact granular aggregates or earthy masses, less common as cleavable masses. Some fine-grained masses may contain intergrown siliceous material and look somewhat like porcelain. Twins by translation gliding on {001} in the {100} direction may occur.

Cleavage Perfect {101} cleavage, like the other rhombohedral carbonates. The angle between cleavages is 72.6°. Conchoidal fracture. Brittle.

Color Properties

Luster: Vitreous.

Color: White, gray, or colorless. Fe-bearing samples may be yellow or brown.

Streak: White.

Color in Thin Section: Colorless.

Optical Characteristics

Relief in Thin Section: Low negative to high positive, marked change with rotation.

Optic Orientation: Extinction is inclined or symmetrical to cleavage traces.

Indices of Refraction and Birefringence: Indices of refraction increase with Fe and Mn content (Figure 17.5). Birefringence is extreme, so interference colors in thin section are upper-order creamy white for most grain orientations. The index of refraction $n_{\epsilon'(101)}$ for fragments lying on {101} cleavage surfaces is 1.602 for pure magnesite, higher for Fe-bearing samples.

Interference Figure: Basal sections yield uniaxial negative figures with numerous isochromes.

Alteration Magnesite has no systematic alteration, but Fe-bearing samples may display reddish stains or be altered to iron oxides and hydroxides.

Distinguishing Features

Hand Sample: Reacts vigorously with dilute HCl only if powdered, like dolomite, from which magnesite is distinguished by higher specific gravity. White massive varieties resemble chert but are softer. Intergrown siliceous material (opal, chalcedony) may make the hardness greater than 4.5.

Thin Section/Grain Mount: Resembles calcite and dolomite but lacks twin lamellae and has higher indices of refraction.

Occurrence The most common occurrence is in altered or metamorphosed Mg-rich igneous rocks such as peridotite, pyroxenite, or dunite. It is reported infrequently from evaporite deposits either as a direct precipitate or as the result of later diagenetic alteration of preexisting carbonate minerals. It also may be a gangue mineral in hydrothermal mineral deposits. Some of the major commercial magnesite deposits are hosted in sedimentary dolomite layers and appear to have been produced by hydrothermal replacement of dolomite by magnesite. It also is a relatively uncommon biomineral and may precipitate by a process of biologically induced mineralization on cyanobacteria in waters with sufficient dissolved Mg (Thompson and Ferris, 1990).

Use Magnesite may be used as an ore mineral for the production of metallic Mg and other Mg compounds, but most Mg currently comes from brines and seawater. The major use of magnesite is as a source for MgO (dead-burned magnesite) used in the manufacture of refractory materials. Natural MgO is the mineral periclase. Magnesite is intensely heated to drive off CO_2 to produce MgO.

SIDERITE

FeCO$_3$
Hexagonal
   (rhombohedral) $\bar{3}$ 2/m
$a = 4.69$ Å, $c = 15.38$ Å, $Z = 6$
   ($a = 18.77$ Å, $c = 15.38$ Å,
   $Z = 96$ to be consistent
   with Miller indices)
$H = 4$–4.5
$G = 3.96$ (lower w/ Mg or Mn)
Uniaxial (−)
$n_\omega = 1.875$
$n_\epsilon = 1.635$
$\delta = 0.240$

Composition Complete solid solution extends to magnesite and to rhodochrosite. Substitution of Ca for Fe is limited to no more than 10 or 15% due to difference in cation size.

Form and Twinning Crystals are commonly {101} rhombohedrons, sometimes modified by a {001} basal pinacoid. More common as granular aggregates and cleavable masses. Also found as oolites or as nodular and botyroidal forms consisting of radiating fibers or acicular grains. May form lamellar twins on {012} similar to calcite, or less frequently, single twins by reflection on 001}.

Cleavage Perfect {101} cleavage like the other rhombohedral carbonates. The angle between cleavages is 73°. Uneven to conchoidal fracture. Brittle.

Color Properties

Luster: Vitreous.

Color: Usually some shade of light to dark yellowish, reddish or grayish brown.

Streak: White.

Color in Thin Section: Colorless or pale yellowish brown.

Pleochroism: Colored varieties pleochroic $\epsilon < \omega$, so samples are darker when the long diagonal between cleavages is parallel to the vibration direction of the lower polarizer.

Optical Characteristics

Relief in Thin Section: Moderate to high positive, changes with rotation.

Optic Orientation: Extinction is inclined or symmetrical to cleavage traces and the short diagonal between cleavages is length fast.

Indices of Refraction and Birefringence: Indices of refraction and birefringence decrease with Mg and Mn content (Figure 17.5). The index of refraction $n_{\epsilon'(101)}$ for fragments lying on {101} cleavage surfaces is 1.748 for pure siderite, lower for Mg- and Mn-bearing samples.

Interference Figure: Basal sections yield uniaxial negative figures with numerous isochromes.

Alteration Siderite is commonly altered to iron oxides and hydroxides (goethite, hematite, etc.). Pseudomorphs of these minerals after siderite are relatively common.

Distinguishing Features

Hand Sample: Distinguished from other rhombohedral carbonates by color and high specific gravity. It effervesces vigorously in dilute HCl only if powdered.

Thin Section/Grain Mount: Indices of refraction are higher than the other rhombohedral carbonates, and both ω and ϵ rays have indices of refraction higher than the adhesive commonly used in thin sections (~1.540), unlike calcite, dolomite, or magnesite. Siderite commonly has lamellar twinning that magnesite lacks.

Occurrence Siderite may occur as concretionary or oolitic forms in bedded Fe-rich sedimentary rocks called clay ironstones. It also is a constituent in metamorphosed iron formations. Hydrothermal systems often precipitate siderite as a gangue mineral. Larger masses may be produced by hydrothermal alteration of limestone. It also is a relatively uncommon biomineral (Chapter 5).

Use In some areas it is used as an iron ore. It also can be used as a raw material in the manufacture of iron oxide and hydroxide pigments, which are widely used in paints, cosmetics, and other products where red or brown colors are desired.

RHODOCHROSITE

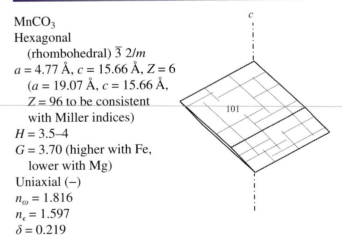

MnCO$_3$
Hexagonal
   (rhombohedral) $\bar{3}$ 2/m
$a = 4.77$ Å, $c = 15.66$ Å, $Z = 6$
   ($a = 19.07$ Å, $c = 15.66$ Å,
   $Z = 96$ to be consistent
   with Miller indices)
$H = 3.5$–4
$G = 3.70$ (higher with Fe,
   lower with Mg)
Uniaxial (−)
$n_\omega = 1.816$
$n_\epsilon = 1.597$
$\delta = 0.219$

Composition Rhodochrosite displays complete solid solution to siderite, and limited solid solution to magnesite. Some Ca may substitute for Mn. Kutnohorite [CaMn(CO$_3$)$_2$], with the dolomite structure, is compositionally intermediate between rhodochrosite and calcite.

Figure 17.6 The Alma King, at 16.5 cm across, is perhaps the largest rhodochrosite crystal in a collection. It is from the Sweethome Mine, Park County, Colorado. The acicular white crystals are quartz, the dark grains are sphalerite, and the equant crystals (lower left) are fluorite. © Denver Museum of Nature and Science.

Form and Twinning Rare crystals are {101} rhombohedrons (Figure 17.6), which may have curved faces similar to dolomite. More commonly as cleavable masses, granular encrustations, or fine-grained masses. Lamellar twinning on {012}, like calcite, is uncommon.

Cleavage Perfect {101} rhombohedral cleavage, like the other rhombohedral carbonates. The angle between cleavages is 73°. Uneven to conchoidal fracture. Brittle.

Color Properties

Luster: Vitreous.

Color: Usually pink or red, also yellow, gray, or brown.

Streak: White.

Color in Thin Section: Colorless or pale pink, some samples may display color zoning.

Pleochroism: Colored varieties are pleochroic in shades of pink with $\omega > \epsilon$, so color is darkest when the long diagonal between cleavage traces is parallel to the vibration direction of the lower polarizer.

Optical Characteristics

Relief in Thin Section: Moderate to high positive, changes with rotation.

Optic Orientation: Extinction is inclined or symmetrical to cleavage traces. The fast ray vibration direction is parallel to the short diagonal between cleavage traces.

Indices of Refraction and Birefringence: Indices of refraction vary systematically with composition (Figure 17.5). The index of refraction $n_{\epsilon(101)}$ for fragments lying on {101} cleavage surfaces is 1.702 for pure rhodochrosite, higher for Fe-bearing samples, and lower for Mg-bearing samples. In zoned samples, darker-colored areas usually have higher indices of refraction. Birefringence is extreme and produces creamy upper-order white interference colors in thin section and grain mount, even if the optic axis is nearly vertical.

Interference Figure: Uniaxial negative optic axis figures display numerous isochromes and isogyres that flare substantially toward the edge of the figure.

Alteration Rhodochrosite may alter to dark-colored Mn oxide and hydroxide minerals or may be replaced by quartz or other minerals to form pseudomorphs.

Distinguishing Features

Hand Sample: Pink to red color and rhombohedral cleavage are characteristic. Effervesces in dilute HCl only if powdered, like dolomite.

Thin Section/Grain Mount: Distinguished from the other rhombohedral carbonates by indices of refraction, pink color, if present, and association with other Mn-bearing minerals.

Occurrence Rhodochrosite is usually restricted to hydrothermal vein and replacement deposits. It is often associated with sulfide minerals, other carbonates, barite, fluorite, rhodonite, and quartz. It also is an uncommon biomineral (Chapter 5).

Use Rhodochrosite is a minor ore of Mn. Good crystals can be highly valued by mineral collectors.

DOLOMITE–ANKERITE

Hexagonal
  (rhombohedral) $\bar{3}$
Uniaxial (−)
Dolomite: $CaMg(CO_3)_2$
$a = 4.80$ Å, $c =$
  15.98 Å, $Z = 3$
  ($a = 19.21$ Å, $c = 15.98$ Å,
  $Z = 48$ to be consistent
  with Miller indices)
$n_\omega = 1.679–1.690$
$n_\epsilon = 1.500–1.510$
$\delta = 0.179–0.182$
Ankerite: $Ca(Mg,Fe)(CO_3)_2$
$a = {\sim}4.82$ Å, $c = {\sim}16.10$ Å, $Z = 3$
  ($a = {\sim}19.28$ Å, $c = {\sim}16.10$ Å,
  $Z = 48$ to be consistent
  with Miller indices)
$n_\omega = 1.690–1.750$
$n_\epsilon = 1.510–1.548$
$\delta = 0.182–0.202$
$H = 3.5–4$
$G = 2.86$ (dolomite) to 3.10 (ankerite)

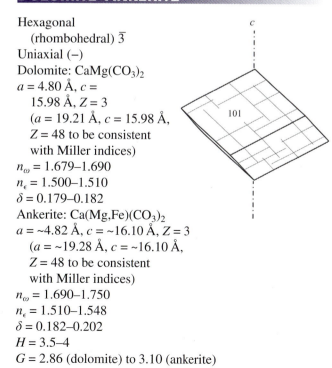

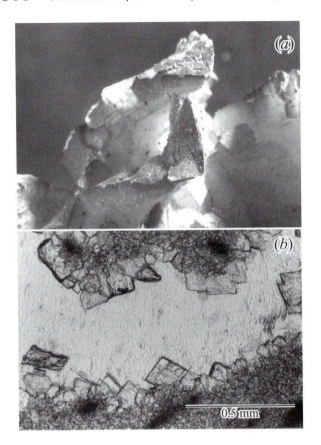

Figure 17.7 Dolomite. (*a*) Dolomite crystals. Note that the crystal faces are curved. Field of view is about 3 cm wide (*b*) Photomicrograph showing dolomite rhombs lining a void in a thin section of limestone.

Composition Dolomite and ankerite are part of a solid solution series whose end members are $CaMg(CO_3)_2$ and $CaFe(CO_3)_2$ (ferrodolomite). Dolomite is taken here to include samples with less than 20 mol % ferrodolomite component. More Fe-rich samples are considered ankerite. Samples with more than about 75% ferrodolomite component have not been found in nature. Mn may substitute for Mg/Fe, and extensive solid solution toward kutnohorite $[CaMn(CO_3)_2]$ is possible.

Form and Twinning Crystals are {101} rhombohedrons that may have curved or saddle-shaped faces (Figure 17.7). More common as coarse- to fine-grained aggregates. Infrequently as fibrous or columnar aggregates, or as oolites. Lamellar twins on {021} are moderately common. Lamellae may be parallel to either the short or long diagonal between rhombohedral cleavage traces (Figure 17.4).

Cleavage Perfect {101} cleavage like the other rhombohedral carbonates. The angle between cleavages is ~74°. Subconchoidal fracture. Brittle.

Color Properties

Luster: Vitreous.

Color: Dolomite is colorless, white, or gray, often tinged with yellow or brown; ankerite may be white, yellow, yellowish brown or brown, rarely gray or blue.

Streak: White.

Color in Thin Section: Colorless.

Optical Characteristics

Relief in Thin Section: Low negative to high positive, changes with rotation.

Optic Orientation: Extinction is symmetrical or inclined to the trace of cleavage. The fast ray vibration direction is parallel to the short diagonal between cleavages.

Indices of Refraction and Birefringence: Indices of refraction increase with Fe content (Figure 17.5). Fragments of pure dolomite lying on a cleavage surface show $n_{\epsilon'(101)} = 1.58$. Birefringence is extreme, so interference colors in both thin section and grain mount are upper-order creamy white, even if the optic axis is nearly vertical.

Interference Figure: Basal sections yield uniaxial negative figures with numerous isochromes.

Alteration Dolomite and ankerite may be pseudomorphically replaced by other carbonates, quartz, pyrite, iron oxides, or other minerals. Iron-rich samples may be brownish due to oxidation or weathering.

Distinguishing Features

Hand Sample: Dolomite closely resembles calcite, from which it is most easily distinguished because dolomite effervesces vigorously in dilute HCl only if powdered. Dolomite and ankerite are also far more likely than calcite to form rhombohedral crystals, and these crystals may have curved faces. A brownish or yellowish color suggests, but does not demonstrate, that the sample is ankerite and not dolomite. Ankerite and siderite can be distinguished based on specific gravity.

Thin Section/Grain Mount: Dolomite closely resembles calcite. See the calcite description for criteria by which they may be distinguished. Index of refraction can be used to estimate composition in dolomite–ankerite solid solutions (Figure 17.5) and to distinguish among the rhombohedral carbonates.

Occurrence Dolomite is quite common and typically is found as a major constituent of sedimentary dolostone, which, unfortunately, also is known as dolomite. Dolostone is generally considered to be formed from limestone as a result of diagenetic alteration of calcite to dolomite. In at least some cases, the precipitation of dolomite is mediated by bacteria (e.g., Sánchez-Román and others, 2008). The replacement of calcite by dolomite generally produces pore space in the original limestone that may allow dolostone beds to serve as the reservoir rock for petroleum deposits.

Dolomite and ankerite are both common gangue minerals in hydrothermal vein and replacement deposits. Ankerite also may be found as veins or joint filling in shale, clay ironstones, and coal beds. Carbonatites, which are unusual igneous bodies composed largely of carbonate minerals, may contain either dolomite or ankerite along with calcite. Metamorphic rocks that include dolomite include marble, calc-silicate gneiss, skarn, and related rocks derived from limestone, dolostone, and other carbonate-rich sediments. Associated minerals include calcite, tremolite, diopside, and garnet. Ankerite may be found in metamorphic iron formations and other metamorphic rocks derived from Fe-rich sediments.

Use Dolomite is a potential ore mineral for metallic Mg, but most Mg currently is extracted from seawater. Industrial uses are numerous and include glassmaking, flue gas desulfurization, wastewater treatment, and metallurgical fluxing. Dolomite is used to manufacture portland cement, and in the production of MgO for use in refractories and CaO for lime. Agricultural applications include use as a soil amendment and as a feed additive for cattle and poultry. Dolostone is commonly crushed and used as aggregate in concrete, asphalt, and similar applications. Cut into blocks and sheets, dolostone is used as dimension stone in the construction industry and to make monuments. Dolomite is sold as a dietary supplement, similar to calcite, and it provides both dietary calcium and magnesium. It also can serve as an antacid.

ARAGONITE GROUP

The orthorhombic carbonates were listed earlier (Table 17.1); aragonite, witherite, and strontianite are described in this section. The properties of cerussite were listed earlier (Table 17.2). The orthorhombic aragonite structure is favored for cations whose radius in octahedral coordination is larger than roughly 1.1 Å; the calcite structure is favored for smaller cations. Because Ca^{2+} has an ionic radius of 1.14 Å in octahedral coordination, it may fit in either structure. To accommodate the larger cations, the aragonite structure provides distorted 9-fold coordination sites arranged between corrugated layers of CO_3 groups that are parallel to {001} (Figure 17.8). The corrugations in the carbonate layers are produced by having rows of CO_3 groups extending parallel to a that are alternately above and below the average plane of the layer. The larger cations are distributed in a manner equivalent to slightly distorted hexagonal close packing, accounting for the fact that members of the group have pseudohexagonal symmetry.

ARAGONITE

$CaCO_3$
Orthorhombic $2/m\ 2/m\ 2/m$
a = 4.96 Å, b = 7.97 Å,
$\quad c$ = 5.74 Å, Z = 4
H = 3.5–4
G = 2.94–2.95
Biaxial (–)
n_α = 1.530
n_β = 1.680
n_γ = 1.685
δ = 0.155
$2V_x$ = 18°

Composition Most aragonite is relatively pure $CaCO_3$. Minor amounts of Sr or Pb may substitute for Ca in some samples. The amount of Sr substituting for Ca in aragonite from fossil scleractinian corals has been used to infer sea-surface temperatures at the time the corals were alive.

Form and Twinning Common as radiating or columnar aggregates of grains elongate along c. Crystals often have penetration twins on {110} to form columnar cyclic twinned crystals with pseudohexagonal outlines and vertical striae (Figure 17.9). Some twinned crystals may

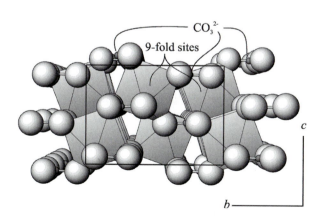

Figure 17.8 Aragonite structure. Triangular CO_3^{2-} groups are arranged in rows parallel to a (into the page). Oxygen anions are shown as larger spheres and carbon cations as small spheres. Large cations such as Ca^{2+}, Ba^{2+}, Sr^{2+}, or Pb^{2+} occupy 9-fold coordination sites between the CO_3^{2-} groups. A unit cell is outlined.

Figure 17.9 Aragonite crystals.

be platy parallel to {001}. {100} twinning also may be polysynthetic. Shells of marine invertebrates may be composed of aragonite and some cave formations (stalactites, stalagmites, etc.) are composed of aragonite.

Cleavage Imperfect {010}. and poor {110} cleavage. Subconchoidal fracture. Brittle.

Color Properties

Luster: Vitreous.

Color: Colorless or white.

Streak: White.

Color in Thin Section: Colorless.

Optical Characteristics

Relief in Thin Section: Low negative to high positive, changes with stage rotation.

Optic Orientation: X = c, Y = a, Z = b, optic plane = (100). Extinction is parallel to crystal elongation (c axis) in principal sections and the trace of cleavage is length fast.

Indices of Refraction and Birefringence: Indices of refraction have little variation. Sr reduces and Pb increases indices of refraction. Birefringence is extreme and produces creamy upper-order colors, even if an optic axis is close to vertical.

Interference Figure: Basal sections yield centered acute bisectrix figures with numerous isochromes and small (18°) 2V. Optic axis dispersion is weak, r < v.

Alteration Aragonite commonly inverts to its polymorph calcite, and pseudomorphs of calcite after aragonite are common. Aragonite also may be replaced by dolomite.

Distinguishing Features

Hand Sample: Easily confused with calcite. Aragonite reacts with vigorous effervescence in dilute HCl, particularly if fine grained, but lacks good rhombohedral cleavage,

has a different crystal habit, and is more dense than calcite. The other members of the aragonite group have higher specific gravity.

Thin Section/Grain Mount: Resembles calcite but lacks rhombohedral cleavage, is biaxial, and has slightly higher indices of refraction. Strained calcite may, however, also be biaxial with small 2V. X-ray diffraction techniques may be required if fine grained. Staining techniques (Friedman, 1959) may also distinguish between calcite and aragonite.

Occurrence Aragonite occurs in carbonate-bearing blueschist facies metamorphic rocks (high pressure, low temperature) associated with glaucophane, lawsonite, pumpellyite, and related minerals. The temperature to pressure stability fields of calcite and aragonite are shown in Figure 17.10. Despite having a high pressure stability field, aragonite often crystallizes metastably under low-pressure, near-surface conditions. Cave and hot spring deposits (stalactites, stalagmites, tuffa, etc.) may be composed of aragonite. It may also precipitate directly from warm seawater as fine needles and form ooliths in calcareous muds. Some shale deposits may contain large well-formed crystals apparently crystallized from pore water. Aragonite is a significant biomineral (Chapter 5), being produced by mineralization that can be either controlled or induced biologically. Biologically controlled mineralization produces the shells of many marine invertebrates, including mollusks and corals. Aragonite also may precipitate on the surfaces of plankton in subtropical marine waters and may contribute to the whiting events described in the section on calcite. Metastable (biologically mediated?) crystallization of aragonite in evaporite deposits is fairly common, but inversion to calcite usually occurs rapidly. In igneous rocks, aragonite may be found as vesicle and cavity fillings in basalt and andesite. Aragonite is reported in the oxidized zone of hydrothermal sulfide deposits and in iron-rich sediments associated with siderite and iron oxides.

Use Aragonite could potentially have any of the same uses as calcite; but lack of large deposits and the abundance of calcite in limestone and related rocks ensure that aragonite

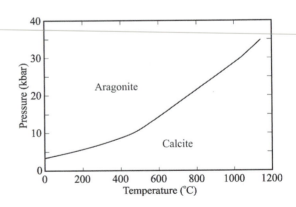

Figure 17.10 Temperature–pressure stability fields of calcite and aragonite. After Carlson (1980).

has little industrial use. Pearls are made, at least in part, of aragonite and have historically been valuable gems. Mother-of-pearl, which also contains aragonite, is used extensively in the decoration of guitars and other stringed instruments. Both pearls and mother-of-pearl have found their way into the medical practice of some cultures, including Chinese traditional medicine and the traditional Ayurvedic medicine of India. King Charles VI (1368–1422) of France reportedly took ground-up pearls as a treatment for his melancholy and episodes of insanity. The evidence that they worked is scant because the monarch suffered from psychoses until his death. Followers of New Age spiritualism may ascribe spiritual, healing, or mystical powers to aragonite crystals.

WITHERITE

$BaCO_3$
Orthorhombic $2/m \; 2/m \; 2/m$
$a = 5.31$ Å $b = 8.90$ Å,
$\quad c = 6.43$ Å, $Z = 4$
$H = 3.5$
$G = 4.29–4.32$
Biaxial (–)
$n_\alpha = 1.529$
$n_\beta = 1.676$
$n_\gamma = 1.677$
$\delta = 0.148$
$2V_x = 16°$

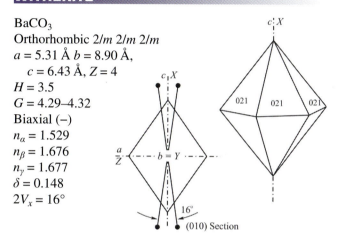

(010) Section

Composition Most witherite is close to pure $BaCO_3$, but minor substitution of Ca, Sr, or Mg for Ba may occur.

Form and Twinning Crystals are stubby pseudohexagonal dipyramids, the result of cyclic twinning on {110} that is always present. Granular, columnar, or fibrous aggregates also occur, and some may have rounded or botryoidal shapes.

Cleavage Distinct cleavage on {010} and poor cleavages on {110} and {012}.

Color Properties

Luster: Vitreous.

Color: Colorless, white, or light grayish or yellowish brown.

Streak: White.

Color in Thin Section: Colorless.

Optical Characteristics

Relief in Thin Section: Low negative to high positive, may change with stage rotation.

Optic Orientation: X = c, Y = b, Z = a, optic plane = (010). Extinction is parallel to {010} cleavage traces in principal sections.

Indices of Refraction and Birefringence: Refractive indices have little variation. Birefringence is extreme, so interference colors in grain mount and thin section are upper-order creamy white for most grain orientations.

Interference Figure: Basal sections yield acute bisectrix figures with numerous isochromes and small (16°) 2V. Optic axis dispersion is weak, r > v.

Alteration Witherite may alter to barite or may be an alteration product after barite.

Distinguishing Features

Hand Sample: Effervesces vigorously in dilute HCl, which distinguishes it from barite. Aragonite has a lower specific gravity. The green color barium gives to a flame is distinctive.

Thin Section/Grain Mount: Distinguished from rhombohedral carbonates by lack of rhombohedral cleavage and biaxial character. Indices of refraction are slightly higher than aragonite and strontianite, and lower than cerussite.

Occurrence This relatively uncommon mineral is most frequently found in hydrothermal sulfide deposits hosted in limestone or other calcareous sediment. It is usually associated with galena and barite.

Use Witherite is used as an ore mineral for barium, which has numerous industrial uses. Economic witherite deposits are far less common than barite deposits, so the latter mineral is a substantially more common source of barium.

STRONTIANITE

$SrCO_3$
Orthorhombic $2/m \; 2/m \; 2/m$
$a = 5.11$ Å, $b = 8.42$ Å,
$\quad c = 6.03$ Å, $Z = 4$
$H = 3.5$
$G = 3.75$
Biaxial (–)
$n_\alpha = 1.517–1.525$
$n_\beta = 1.663–1.686$
$n_\gamma = 1.667–1.690$
$\delta = 0.148–0.165$
$2V_x = 7–10°$

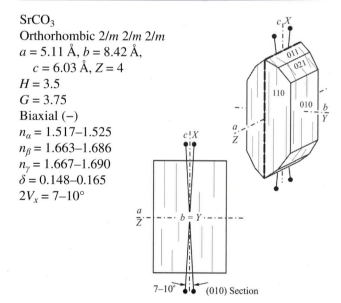

(010) Section

Composition Most strontianite contains up to about 20% Ca substituting for Sr. Some Ba also may substitute for Sr.

Form and Twinning Usually granular massive or as aggregates of columnar, fibrous, or acicular grains. Crystals are elongate parallel to c and often are composed of cyclic twins on {110}, like aragonite, yielding a pseudohexagonal cross section. Twins may be single, cyclic, or polysynthetic.

Cleavage Good {110} prismatic cleavage in two directions intersect at about 64°. Uneven fracture. Brittle.

Color Properties

Luster: Vitreous.
Color: Colorless or white; also yellow, greenish, or brownish.
Streak: White.
Color in Thin Section: Colorless.

Optical Characteristics

Relief in Thin Section: Low negative to high positive, depending on orientation. May change with rotation.

Optic Orientation: $X = c$, $Y = b$, $Z = a$, optic plane = (010). Longitudinal sections and elongate cleavage fragments are length fast with parallel extinction. Extinction in basal sections is symmetrical.

Indices of Refraction and Birefringence: Indices of refraction increase with Ca and Ba content. Birefringence is extreme, so interference colors are high-order creamy white for most grain orientations.

Interference Figure: Basal sections yield centered acute bisectrix figures with small $2V$. Optic axis dispersion is weak, $r < v$.

Alteration Strontianite may be altered to celestine.

Distinguishing Features

Hand Sample: Effervesces vigorously in dilute HCl, which distinguishes it from celestine. High specific gravity distinguishes strontianite from aragonite. The crimson color that Sr produces in a flame is distinctive (dissolve a bit of the sample in HCl, then introduce to a clean flame).

Thin Section/Grain Mount: Resembles the other orthorhombic carbonates. Aragonite, witherite, and cerussite have higher indices of refraction.

Occurrence Most commonly found in veins and masses in limestone deposited by hydrothermal solutions. Also found in hydrothermal sulfide deposits. Strontianite is a biomineral and is found in scleractinian coral along with aragonite.

Use Strontianite is an ore mineral for strontium, but celestine ($SrSO_4$) deposits are more economic to exploit, and little strontium is produced from strontianite. A major use of strontium has been in the faceplate glass on cathode ray tubes, specifically including old style television sets and computer monitors. The strontium is added to provide shielding from the X-rays produced in all cathode ray tubes and to improve the optical quality of the glass. Strontium is used in the manufacture of ceramics and in

permanent ceramic magnets. Fireworks and other pyrotechnics commonly use strontium compounds to provide brilliant red colors. Strontium chloride is used in toothpaste for sensitive teeth, and strontium phosphate is used in the manufacture of fluorescent lights.

OH-BEARING CARBONATES

Azurite and malachite are the only common minerals in this group (Table 17.1), and they are typically restricted to the oxidized portion of copper-bearing hydrothermal sulfide mineral deposits. Both are beautiful and are common in mineral collections. Other hydrated carbonates found in marine and saline lake evaporite deposits were listed earlier (Table 17.2).

MALACHITE

$Cu_2CO_3(OH)_2$
Monoclinic $2/m$
$a = 9.48$ Å, $b = 12.03$ Å,
    $c = 3.21$ Å,
    $\beta = 98.75°$,
    $Z = 4$
$H = 3.5–4$
$G = 3.9–4.05$
Biaxial (−)
$n_\alpha = 1.655$
$n_\beta = 1.875$
$n_\gamma = 1.909$
$\delta = 0.254$
$2V_x = 43°$

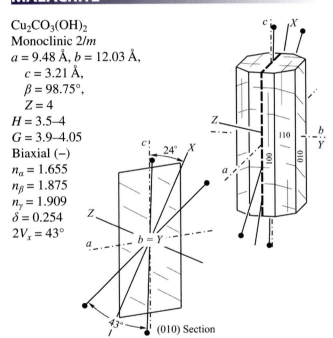

(010) Section

Structure The structure consists of chains of $CuO_2(OH)_4$ and $CuO_4(OH)_2$ octahedra that extend parallel to the c axis. CO_3 groups tie the chains together laterally.

Composition Malachite shows relatively little compositional variation.

Form and Twinning Most commonly as massive encrusting or botryoidal forms with concentric color banding consisting of parallel fibrous grains (front cover). Also granular or earthy. {100} twins are common.

Cleavage Perfect {$\overline{2}$01} cleavage is rarely seen because grain size is typically too small; fair cleavage on {010}. Subconchoidal to uneven fracture. Brittle.

Color Properties

Luster: Adamantine to vitreous on crystals, also silky if fibrous. Fine granular material may be earthy.

Color: Vivid green.

Streak: Pale green.

Color in Thin Section: Green.

Pleochroism: $X < Y < Z$: X = nearly colorless or pale green; Y = yellowish green; Z = dark green.

Optical Characteristics

Relief in Thin Section: High positive.

Optic Orientation: $X \wedge c = -24°$, $Y = b$, $Z \wedge a = +32°$, optic plane = {010}. Elongate grains and fibers have inclined extinction and are length fast.

Indices of Refraction and Birefringence: Birefringence is extreme, but high-order creamy white interference color will be masked by mineral color.

Interference Figure: Fragments on the {$\bar{2}01$} cleavage yield nearly centered acute bisectrix figures with numerous isochromes.

Alteration Malachite may alter to azurite or cuprite.

Distinguishing Features

Hand Sample: Bright green color and vigorous effervescence in dilute HCl are diagnostic.

Thin Section/Grain Mount: Green color and habit are characteristic.

Occurrence Malachite is a common mineral in the near-surface oxidized portion of Cu-bearing hydrothermal sulfide mineral deposits (Chapter 19). It typically is associated with azurite and generally is produced when primary Cu sulfide minerals, such as chalcopyrite, are oxidized and altered by acidic water percolating down from the surface. The copper may be reprecipitated as copper carbonate and copper oxide minerals in the area above the water table. The carbonate component may be acquired from CO_2 dissolved in meteoric water or from carbonate minerals in the host rock for the mineral deposit.

Use Used as a minor ore of copper. The brilliant green color makes it valuable as a decorative stone and for use in jewelry. The use of malachite for medicinal purposes goes back at least 5000 years; it has been used for a wide variety of maladies, including epilepsy, anemia, bleeding, arthritis, influenza, and gastrointestinal disorders. Copper is an essential nutrient, and small amounts of copper compounds—usually $CuSO_4$ (chalcocyanite) or CuO (tenorite)—are included in many multivitamin supplements. While copper is an essential nutrient in small amounts, high levels can pose significant health risks. Being a carbonate, malachite will dissolve in acidic gastric juices, and some portion of the copper can then become available. Ingestion of malachite or other Cu-bearing minerals could, therefore, potentially be hazardous. Malachite is also an antimicrobial agent effective against at least *Staphylococcus aureus*.

AZURITE

$Cu_3(CO_3)_2(OH)$
Monoclinic $2/m$
a = 4.97 Å, b = 5.84 Å,
　c = 10.29 Å,
　β = 92.4°, Z = 2
H = 3.5–4
G = 3.77
Biaxial (+)
n_α = 1.730
n_β = 1.756
n_γ = 1.836
δ = 0.106
$2V_z$ = 68°

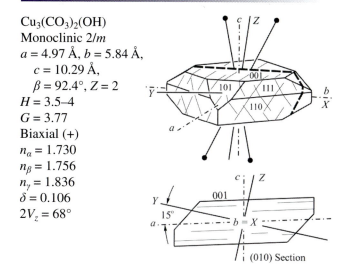

Structure The structure is complex, with Cu coordinating with two O and two OH forming a square. These square groups are linked to form chains parallel to b and cross-linked with CO_3 groups.

Composition Usually relatively close to the composition given by the formula.

Form and Twinning Crystals are varied, often tabular parallel to (001), or short-prismatic parallel to c with many modifying forms. Most common as encrusting or rounded masses that may be color banded with individual bands consisting of radiating or parallel masses of acicular grains (front cover).

Cleavage Perfect {011}, fair {100}, and poor {110} cleavage. Conchoidal fracture. Brittle.

Color Properties

Luster: Vitreous. Fibrous varieties may be silky; earthy if fine grained.

Color: Azure blue.

Streak: Blue, lighter than mineral color.

Color in Thin Section: Blue.

Pleochroism: Weak: X = clear blue, Y = azure blue, Z = dark blue or blue-violet.

Optical Characteristics

Relief in Thin Section: High positive.

Optic Orientation: $X = b$, $Y \wedge a = +15°$, $Z \wedge c = -12°$, optic plane $\perp$ (010). Crystals elongate on c are length slow.

Indices of Refraction and Birefringence: Indices of refraction vary little. Birefringence is extreme, but the upper-order creamy white interference color is masked by mineral color.

Interference Figure: Biaxial positive with moderately large 2V. Optic axis dispersion is $r > v$, also horizontal dispersion.

Alteration Azurite may be replaced by malachite.

Distinguishing Features

Hand Sample: Vivid blue color and effervescence in dilute HCl are characteristic.

Thin Section/Grain Mount: Distinguished by blue color and indices of refraction.

Occurrence   Azurite is a common mineral in the near-surface oxidized portion of Cu-bearing hydrothermal sulfide mineral deposits. It typically is associated with malachite and generally is produced when primary Cu sulfide minerals, such as chalcopyrite, are oxidized and altered by acidic water percolating down from the surface. The copper may be reprecipitated as copper carbonate and copper oxide minerals in the area above the water table. The carbonate component may be acquired from CO_2 dissolved in meteoric water or from carbonate minerals in the host rock of the mineral deposit.

Use   A minor ore of copper. Azurite also may be used for decorative purposes, and when pulverized, has been used as a pigment. Azurite has been used in homeopathic preparations and, probably because of its beauty, has a history of use in folk remedies.

Box 17.2   Evaporites

Evaporite deposits are produced by precipitation of minerals from saturated surface or near-surface waters in response to solar evaporation. Warren (2006) provides an exhaustive review. The minerals in evaporite deposits are all relatively soluble in water. Table 17.3 lists common evaporite minerals. The requirement for formation of an evaporite deposit is a restricted basin from which evaporation can progressively increase the concentration of cations and anions in the water. For marine waters, the most abundant cations are Na^+, Ca^{2+}, Mg^{2+}, and K^+, and the most abundant anions are Cl^-, SO_4^{2-}, and HCO_3^-. The minerals produced by evaporating seawater are, therefore, dominantly chlorides, sulfates, and carbonates. In playas and other saline lakes, the chemistry will depend on the composition of the surrounding bedrock, and any additions from groundwater and hydrothermal systems.

If normal seawater were to simply evaporate, minerals would begin to precipitate in the following sequence.

Calcite
Gypsum
Anhydrite
Glauberite
Halite
Polyhalite
Epsomite
+ Additional sulfates and chlorides

Halite would constitute most of the volume of the minerals because Na^+ and Cl^- are by far the most abundant ions in solution.

In actual marine evaporite systems, the crystallization sequences can be quite complex. Water from open marine sources, rivers, rain/snow, and/or groundwater is continually added, and evaporation waxes and wanes with the seasons and changing climate. Additionally, basins fill with evaporite minerals and a variable flux of clastic sediment, sea level may change, and tectonic activity may change basin geometry. Regardless, most marine evaporite systems are dominated by limestone, gypsum beds, and rock salt, plus variable amounts of terrigenous sediment. In some environments, primary anhydrite may be precipitated rather than gypsum.

The chemistry of saline lakes depends on the chemistry of the water being fed into them from surrounding areas, as well as the biologic and chemical processes going on in the water. As a consequence, the evaporite minerals in those environments may differ significantly from those for marine systems. For example, saline lakes in the Eocene Green River basin in Wyoming had high levels of Na^+ and HCO_3^- and precipitated thick beds of trona and other sodium carbonate minerals. Saline lakes in Death Valley, California, are enriched in boron, presumably from thermal springs and other hydrothermal sources, and precipitate a suite of borate minerals.

Table 17.3   Selected Minerals Formed in Evaporite Deposits

Carbonates		Sulfates		Halides		Borates	
Aragonite	$CaCO_3$	Anhydrite	$CaSO_4$	Carnallite	$KMgCl_3 \cdot 6H_2O$	Borax	$Na_2(BO(OH))_2$ $(BO_2(OH))_2 \cdot 8H_2O$
Calcite	$CaCO_3$	Epsomite	$MgSO_4 \cdot 7H_2O$				
Dolomite	$CaMg(CO_3)_2$	Glauberite	$Na_2Ca(SO_4)_2$	Halite	$NaCl$	Colemanite	$Ca_2B_6O_{11} \cdot 5H_2O$
Magnesite	$MgCO_3$	Gypsum	$CaSO_4 \cdot 2H_2O$	Sylvite	KCl	Kernite	$Na_2B_4O_6(OH)_2 \cdot 3H_2O$
Nahcolite	$NaHCO_3$	Kieserite	$MgSO_4 \cdot H_2O$			Tincalconite	$Na_2B_4O_5(OH)_4 \cdot 3H_2O$
Trona	$Na_3H(CO_3)_2 \cdot 2H_2O$	Thenardite	Na_2SO_4			Ulexite	$NaCaB_5O_6 \cdot 5H_2O$
		Polyhalite	$K_2MgCa_2(SO_4)_4 \cdot 2H_2O$				

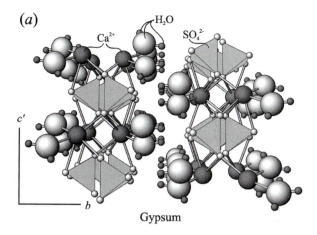

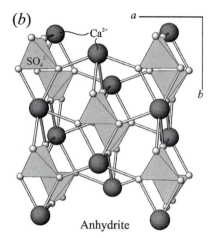

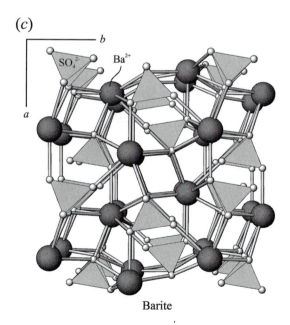

SULFATES

Sulfur is somewhat unusual in that it serves as an anion in the sulfide minerals and as a cation in the sulfates. In the sulfates, sulfur atoms lose all six of their outer electrons, four from the $3p$ and two from the $3s$ subshells (Table 3.3). These small high-charged cations coordinate with four oxygen anions to form SO_4^{2-} groups. Of the available eight negative charges on the four oxygen anions, six are commanded by the sulfur. Sulfur–oxygen bonds are therefore far stronger than bonds between oxygen and any other cations. These structures are anisodesmic, and the SO_4^{2-} groups are discrete structural elements that bond laterally through other cations in the sulfate minerals (Figure 17.11). Only three common sulfate minerals (gypsum, anhydrite, and barite) are described in detail here. Additional sulfate minerals are listed in Table 17.4.

GYPSUM

$CaSO_4 \cdot 2H_2O$
Monoclinic $2/m$
$a = 6.52$ Å, $b = 15.18$ Å,
   $c = 6.29$ Å, $\beta = 127.4°$, $Z = 4$
$H = 2$ on {010} cleavage
$G = 2.30–2.37$
Biaxial (+)
$n_\alpha = 1.519–1.521$
$n_\beta = 1.522–1.526$
$n_\gamma = 1.529–1.531$
$\delta = 0.010$
$2V_z = 58°$

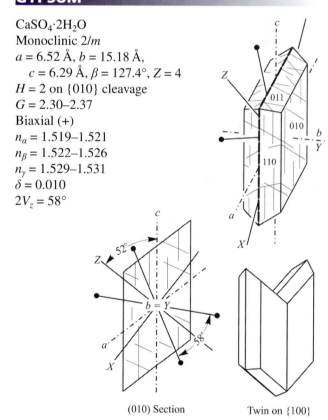

(010) Section   Twin on {100}

Structure The structure consists of strongly bonded layers of $SO_4^{2-} + Ca^{2+}$ parallel to (010) alternating with layers of H_2O molecules (Figure 17.11a). Hydrogen bonding holds the H_2O molecules to the $CaSO_4$ layers and allows for the perfect cleavage on {010}. The strongest linkages within layers are parallel to the c axis, which corresponds to the length of fibers in satin spar. Each Ca^{2+} is coordinated with six O^{2-} from the SO_4^{2-} groups and two O^{2-} from water molecules.

Figure 17.11 Structure of sulfate minerals. SO_4^{2-} groups are shown with oxygen anions as light-colored small spheres. (a) Gypsum viewed down the a axis. Cleavage on {010} splits along the layers of water molecules on the sides of the $CaSO_4$ layers. Hydrogen is shown as small dark spheres. (b) Anhydrite viewed down the c axis. (c) Barite viewed down the c axis.

Table 17.4 Properties of Less Common Sulfate Minerals[a]

Mineral	Unit Cell	Physical Properties[a]	Optical Properties	Comments
Alunite $KAl_3(SO_4)_2(OH)_6$	Hexagonal (rhombohedral) 3m $a = 6.97$ $c = 17.38$ $Z = 3$	H: 3.5–4, G:2.6–2.9 Cl: {001}g, {102}pr Tw: None C: White or gray, also reddish St: White L: Vitreous	n_ω 1.568–1.620 n_ϵ 1.590–1.641 $\delta = 0.020$–0.023 Uniaxial (+) Colorless	Crystals tabular {001} or rhombs; usually fine-grained granular, plumose or flaky aggregates or disseminated grains; massive varieties resemble limestone; common in hydrothermally altered felsic to intermediate volcanic rocks
Antlerite $Cu_3SO_4(OH)_4$	Orthorhombic 2/m 2/m 2/m $a = 8.25$ $b = 12.01$ $c = 6.04$ $Z = 4$ $X = b, Y = a, Z = c$	H: 3.5, G: 3.88–3.94 Cl: {010}pf, {100}pr C: Light to dark green St: Paler green L: Vitreous	$n_\alpha = 1.726$ $n_\beta = 1.738$ $n_\gamma = 1.789$ $\delta = 0.063$ Biaxial (+) $2V = 53°$, $r < v$, v. strong X = yellow-green, Y = blue-green, Z = green	Crystals thick tabular on {010}, equant or short prismatic on c; also cross-fiber grains, acicular, or granular; found in the oxidized portion of Cu-bearing hydrothermal sulfide deposits, usually in arid regions
Celestine $SrSO_4$	Orthorhombic 2/m 2/m 2/m $a = 8.36$ $b = 5.35$ $c = 6.87$ $Z = 4$ $X = c, Y = b, Z = a$	H: 3–3.5, G: 3.96 Cl: {001}pf, {210}g, {010}pr Tw: Rare C: Colorless, white, pale blue, reddish, greenish, brownish St: white L: vitreous	$n_\alpha = 1.621$ $n_\beta = 1.623$ $n_\gamma = 1.630$ $\delta = 0.009$ Biaxial (+) $2V = 50°$, $r < v$ Colorless	Similar to barite; found in cavities and veins and as disseminated grains in limestone or dolostone; also in marine evaporite deposits, hydrothermal veins, and as a biomineral; principal source of Sr
Epsomite $MgSO_4 \cdot 7H_2O$	Orthorhombic 222 $a = 11.876$ $b = 12.002$ $c = 6.859$ $Z = 4$ $X = a, Y = c$ $Z = b$	H: 2–2.5, G: 1.68 Cl: {010}pf, {101}p, conchoidal fracture Tw: {110} rare C: Colorless, white, pale pink, pale green St: White L: Vitreous, silky if fibrous	$n_\alpha = 1.432$ $n_\beta = 1.455$ $n_\gamma = 1.461$ $\delta = 0.029$ Biaxial (–) $2V = 52°$, $r < v$, weak Colorless	Rare as prismatic crystals; typically fibrous crusts, reniform or botryoidal masses, or woolly efflorescences; found in evaporite deposits from saline lakes, as an efflorescence on soil, in mines, and around fumeroles; dehydrates readily in dry air; minor biomineral
Glauberite $Na_2Ca(SO_4)_2$	Monoclinic 2/m $a = 9.99$ $b = 8.19$ $c = 8.41$ $\beta = 112.2°$ $Z = 4$ $X \wedge c = +30°, Y = b$	H: 2.5–3, G: 2.75–2.85 Cl: {001}pf, {110}pr C: Gray, yellow, colorless St: White L: Vitreous	$n_\alpha = 1.515$ $n_\beta = 1.535$ $n_\gamma = 1.536$ $\delta = 0.021$ Biaxial (–) $2V = 7°$, $r > v$, strong Colorless	Variable habit; tabular on {001}, prismatic [110], or dipyramid-like; slight saline taste; found in marine and saline lake evaporite deposits, fumarole deposits, and as amygdule filling
Kieserite $MgSO_4 \cdot H_2O$	Monoclinic 2/m $a = 6.89$ $b = 7.69$ $c = 7.52$ $\beta = 116°$ $Z = 4$ $Z \wedge c = +77°, Y = b$	H: 3.5, G: 2.57 Cl: {110} and{111}pf, {$\bar{1}$11}, {$\bar{1}$01}and {011} fair Tw: {001} rare C: Colorless, white, gray, pale yellow St: White L: Vitreous	$n_\alpha = 1.520$ $n_\beta = 1.533$ $n_\gamma = 1.584$ $\delta = 0.064$ Biaxial (+) $2V = 55°$, $r > v$, moderate Colorless	Massive coarse to fine crystalline; found in marine evaporite deposits with halite, polyhalite, anhydrite, etc.
Polyhalite K_2MgCa_2 $(SO_4)_4 \cdot 2H_2O$	Triclinic $\bar{1}$ $a = 6.95$ $b = 8.89$ $c = 6.95$ $\alpha = 104°$ $\beta = 114° = 101°$ $Z = 1$	H: 3.5, G: 2.78 Cl: {10$\bar{1}$}pf, {010} parting Tw: {010}, {100} common, polysynthetic C: Colorless, white, gray, salmon St: White L: Vitreous	$n_\alpha = 1.547$ $n_\beta = 1.562$ $n_\gamma = 1.567$ $\delta = 0.020$ Biaxial (–) $2V = 62$–70°, $r < v$ Colorless	Rare crystals are tabular on {010} or elongate on c, more commonly massive, also fibrous or foliated; found in marine evaporite deposits with halite and anhydrite, also in fumerole deposits
Thenardite Na_2SO_4	Orthorhombic 2/m 2/m 2/m $a = 9.75$ $b = 12.29$ $c = 5.85$ $Z = 8$ $X = c, Y = b, Z = a$	H: 2.5–3, G: 2.67 CI: {010}pf, {101}g, {100}pr Tw: {100} common cruciform, also {011} C: Colorless, white, gray, yellow, brown, red St: White L: Vitreous, somewhat resinous	$n_\alpha = 1.469$ $n_\beta = 1.475$ $n_\gamma = 1.484$ $\delta = 0.015$ Biaxial (+) $2V = 83°$, $r > v$, weak Colorless	Soluble in water, tastes faintly salty; crystals are {111} dipyramids with additional faces, or tabular parallel to {010}, rarely prismatic on a; found in evaporite deposits from saline lakes, as an efflorescence on soil and around fumeroles

[a] CI = Cleavage (pf = perfect, g = good, pr = poor), C = color, Tw = Twinning, St = Streak, L = Luster.

Several unit cell conventions for gypsum are in common usage. The one used here lets the common prisms be the {011} and {110} prisms that are parallel to the *a* and *c* crystal axes, respectively.

Composition Deviation from the ideal formula is usually restricted to limited substitution of Sr or Ba for Ca.

Form and Twinning Crystals have a variety of habits, but most are tabular parallel to (010), less commonly prismatic or acicular parallel to the *c* crystal axis (Figure 17.12). Large euhedral crystals are known as **selenite**. Granular or foliated masses are quite common. **Satin spar** consists of parallel aggregates of fibrous gypsum and may fill veins. Simple contact twins on {100} are common and may form "swallowtail" twins. Twins on {001} are less common. Prismatic crystals may be curved, and tabular crystals and cleavage sheets may be bent in certain directions with relative ease. Granular-massive rock gypsum is known as **alabaster** if white or light colored.

Cleavage Three cleavage forms in four directions. Perfect {010} cleavage dominates fragment shape. A second cleavage on {100} is good. Two good cleavages parallel to the {$\bar{1}$11} prism intersect at angles of 42° and 138°. Cleavage folia are flexible in certain directions, but not

Figure 17.12 Gypsum. (*a*) Selenite with a swallowtail twin. (*b*) Bladed crystals. Field of view is about 10 cm across.

elastic. Brittle. Massive material may be carved easily with a pocket knife.

Color Properties

Luster: Vitreous, cleavage surfaces may be pearly; satin spar is silky.

Color: Usually colorless or white; less commonly pale shades of gray, red, yellow, brown, or blue.

Streak: White.

Color in Thin Section: Colorless.

Optical Characteristics

Relief in Thin Section: Low negative.

Optic Orientation: $X \wedge a = -15°$, $Y = b$, $Z \wedge c = +52°$, optic plane = (010). The trace of the prominent {010} cleavage is either length fast or length slow depending on grain orientation. Fragments on the {010} cleavage display 38° inclined extinction to the trace of {100} cleavage, with the fast ray closer to the cleavage.

Indices of Refraction and Birefringence: Refractive indices are less than the cement usually used in thin section and interference colors are about the same as quartz (first-order gray or white).

Interference Figure: The 2*V* angle is about 58° at room temperature but decreases to 0° at 91°C. Above that temperature 2*V* opens into an optic plane perpendicular to (010). Optic axis dispersion is strong, *r* > *v*, as is inclined bisectrix dispersion. Fragments on {010} cleavage yield flash figures.

Alteration Replacement of gypsum by quartz, opal, calcite, or celestine may produce pseudomorphs. Gypsum often is produced by hydrating sedimentary beds of anhydrite.

Distinguishing Features

Hand Sample: Characterized by hardness and four different cleavage surfaces. Anhydrite has higher specific gravity, cleavages at 90°, and often has a subtle purplish cast to its color.

Thin Section/Grain Mount: Gypsum may be confused with anhydrite or barite. Both anhydrite and barite have higher birefringence, higher indices of refraction, and either parallel or symmetrical extinction in principal sections.

Occurrence Gypsum is a very common mineral in marine evaporite deposits and may be associated with halite, sylvite, calcite, dolomite, and anhydrite, as well as clay and silicate detrital grains. Biologically induced mineralization of gypsum by cyanobacteria has been documented in alkaline lakes (Thompson and Ferris, 1990) and undoubtedly takes place in evaporite deposits. The stability relationship between gypsum and anhydrite is shown in Figure 17.13.

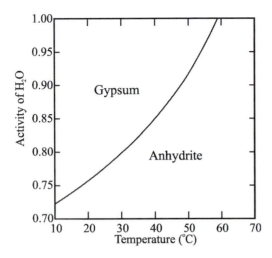

Figure 17.13 Stability of gypsum and anhydrite at one atmosphere pressure. Decreasing activity of water is produced by increasing salinity. Adapted from Hardie (1967).

Gypsum is the stable phase in most evaporite environments, although under high salinity conditions or high temperatures, anhydrite may be the stable phase. Both anhydrite and gypsum may crystallize metastably; kinetic and biomineralization processes may dictate which crystallizes. In many evaporite deposits, gypsum is the primary mineral, but with burial and increase in temperature and pressure, the gypsum commonly dehydrates and recrystallizes as anhydrite at depths of a few meters up to a kilometer. Subsequent uplift and erosion to bring the evaporite material back to the surface may allow the secondary anhydrite to become hydrated and recrystallize back to gypsum. This latter hydration involves an increase in volume, so original bedding may be crumpled or disrupted as a consequence.

Gypsum also crystallizes as an efflorescence in desert soils and is common as an evaporite mineral in sabkah environments. Sabkahs are low-lying, salt-encrusted surfaces adjacent to perennial or ephemeral bodies of water and are not regularly inundated with water. They are found in coastal areas such as the Persian Gulf, southern Australia, and Mexico, and adjacent to playa lakes. Groundwater, supplied either from adjacent terrestrial sources or from the nearby body of water, is brought by capillary action to the surface, where it evaporates to produce a brine in the near-surface environment. Gypsum, anhydrite, and halite are minerals that commonly crystallize from these brines and, over time, may build up significant volumes of evaporite material.

Gypsum may also precipitate around fumeroles or volcanic vents. Infrequently it is found in the near-surface oxidized zone of hydrothermal sulfide deposits.

Use Gypsum is one of the earlier minerals to be exploited by people; its use goes back at least 5000 years. About 70% of the gypsum now mined goes to manufacture gypsum wallboard, also known as sheetrock or drywall. Gypsum wallboard is used to cover interior walls of most houses,

apartments, and offices in North America and in many other parts of the world. To make wallboard, gypsum is calcined (heated) to drive off part of the water and then ground to form a material called stucco (similar material is sold as plaster of paris). Stucco is mixed with water, reinforcing fibers, and other additives to form a thick slurry that is extruded and wrapped with heavy paper to make wallboard. This material sets or hardens by recrystallizing to form gypsum. When cured, it forms a stiff panel that is attached to interior wall framing with screws or nails. Joints between panels are finished with strips of mesh or heavy paper set in a plaster-like compound.

Gypsum also is used in portland cement to control the setting rate, and as a soil amendment to improve soil structure and workability, provide sulfur and calcium to plants, and control the availability of other soil nutrients. Calcined gypsum has medical applications. It is used as a dietary supplement to provide needed sulfur both for people and animals, to make casts to support broken bones, and as a special casting plaster to make dental molds. Additional medical applications of gypsum include treatment of periodontal disease, bone grafting, antibiotic delivery, and oral hygiene. In Roman times, gypsum was mixed with rock salt to combat halitosis.

Alabaster, which is fine-grained white gypsum, is used as an ornamental stone and for sculpture, but its softness restricts its utility.

ANHYDRITE

$CaSO_4$
Orthorhombic
  $2/m \ 2/m \ 2/m$
$a = 6.99$ Å, $b = 7.00$ Å,
  $c = 6.24$ Å, $Z = 4$
$H = 3–3.5$
$G = 2.9–3.0$
Biaxial (+)
$n_\alpha = 1.569–1.574$
$n_\beta = 1.574–1.579$
$n_\gamma = 1.609–1.618$
$\delta = 0.040–0.044$
$2V_z = 42–44°$

Structure The structure (Figure 17.11b) consists of sulfate groups arrayed in a regular fashion defining planes parallel to (100), (010), and (001). Ca^{2+} cations occupy somewhat irregular 8-fold sites among the sulfate groups, and are arrayed so that sulfate and Ca ions alternate parallel to the c axis. Cleavage develops between the planes of sulfate anionic groups.

Composition Most anhydrite is close to the ideal composition, but small amounts of Sr or Ba may substitute for Ca.

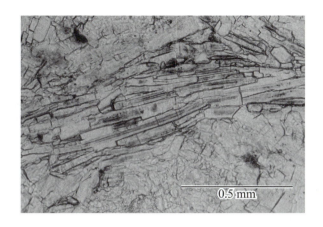

Figure 17.14 Photomicrograph of a thin section of bedded anhydrite.

Form and Twinning Anhydrite forms blocky crystals or crystals tabular parallel to {100}, {010}, or {001} (Figure 17.14). Masses of anhedral grains are much more common; fibrous or radiating aggregates are less so. Simple or repeated twins on {011} are fairly common.

Cleavage Three cleavage directions at right angles: {010} perfect, {100} very good, and {001} good. Uneven to splintery fracture. Brittle.

Color Properties

Luster: Vitreous.

Color: Colorless, white, or gray, sometimes with a slight bluish cast; less commonly blue, red, or brown.

Streak: White.

Color in Thin Section: Colorless.

Optical Characteristics

Relief in Thin Section: Moderate positive.

Optic Orientation: $X = b$, $Y = a$, $Z = c$, optic plane = (100).

Indices of Refraction and Birefringence: Refractive indices show little variation. Maximum birefringence is displayed by (100) sections and interference colors in standard thin section range up to lower third order.

Interference Figure: (001) sections and cleavage fragments yield centered acute bisectrix figures. Optic axis dispersion is distinct, $r < v$.

Alteration Anhydrite commonly becomes hydrated to form gypsum. Alteration is accompanied by a volume increase that may cause original anhydrite beds to crumple or otherwise deform. Anhydrite may be pseudomorphically replaced by quartz, calcite, dolomite, gypsum, and other minerals.

Distinguishing Features

Hand Sample: Most closely resembles gypsum but has three cleavages at right angles as well as greater hardness

and specific gravity; may have a subtle purplish cast to its color. Barite has higher specific gravity, and calcite effervesces vigorously in dilute HCl.

Thin Section/Grain Mount: Higher birefringence and indices of refraction distinguish anhydrite from gypsum. Barite has higher indices of refraction and lower birefringence.

Occurrence Anhydrite most commonly occurs in marine evaporite deposits. Associated minerals include calcite, dolomite, gypsum, and halite. It apparently may either precipitate directly from seawater or be produced by dehydration of gypsum. Anhydrite is also found in hydrothermally altered limestone and dolostone, and in the near-surface, oxidized portion of hydrothermal sulfide deposits. Primary magmatic anhydrite is not common but has been reported in intermediate volcanic rocks. See Box 17.1 on evaporite deposits and the section on gypsum for more information.

Use Anhydrite is used as a soil amendment and is mixed with portland cement to control the rate of curing. Anhydrite has been used as a source of sulfur to make sulfuric acid and other sulfur compounds, but other sources for sulfur, including gypsum, are usually more economic. Calcium sulfate is used as a carrier for antibiotics and for bone grafting, among other medical applications. In many cases, however, it is not clear whether anhydrite is being specified or whether gypsum is the phase actually being used.

BARITE

$BaSO_4$
Orthorhombic
   $2/m\ 2/m\ 2/m$
$a = 8.88$ Å, $b = 5.45$ Å
   $c = 7.15$ Å, $Z = 4$
$H = 2.5–3.5$
$G = 4.5$
Biaxial (+)
$n_\alpha = 1.634–1.637$
$n_\beta = 1.636–1.639$
$n_\gamma = 1.646–1.649$
$\delta = 0.012$
$2V_z = 36–40°$

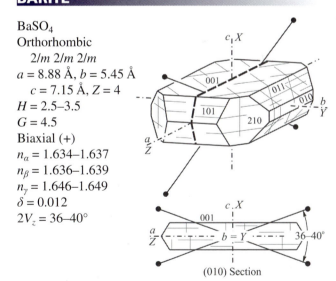

Structure Sulfate anionic groups are arranged in planes parallel to (010) so that oxygens form 12-fold coordination sites for Ba^{2+} cations within the same planes (Figure 17.11c).

Composition Most barite is close to ideal composition, but complete solid solution extends to celestine ($SrSO_4$) and some substitution of Pb and Ca for Ba is possible.

Form and Twinning Crystals are tabular parallel to (001), or less frequently prismatic parallel to *a* or *b*

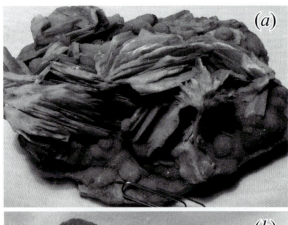

Figure 17.15 Barite. (*a*) Intergrown platy barite crystals with mammillary masses of quartz (near paper clip). (*b*) Two intergrown barite roses that formed as concretionary masses in sedimentary rocks.

crystal axes (Figure 17.15). Crystals often are intergrown to form rosettes or platy aggregates. Barite also forms granular aggregates, cleavable masses, and concretionary masses of radial fibers. Deformation-induced twins on {110} are possible, otherwise it is not usually twinned.

Cleavage Four cleavage surfaces: two very good cleavages on {210} intersect at 78°, a fair to good cleavage on {010}, and a perfect cleavage on {001} that intersects the other three at 90°. Uneven fracture. Brittle.

Color Properties

Luster: Vitreous.

Color: White, pale yellow, gray, pale green, pale blue, red, brown.

Streak: White.

Color in Thin Section: Colorless.

Optical Characteristics

Relief in Thin Section: Moderately high positive.

Optic Orientation: $X = c$, $Y = b$, $Z = a$, optic plane = (010). Sections cut parallel to c show parallel extinction, and basal

sections and fragments on the {001} cleavage show symmetrical extinction to the prismatic cleavage. Fibrous and prismatic crystals are usually elongate parallel to *a* and are length slow. Crystals elongate parallel to *b* may be either length fast or length slow, depending on orientation.

Indices of Refraction and Birefringence: Indices of refraction decrease slightly with Sr content and increase with Pb content. Interference color in standard thin section is no higher than first-order yellow.

Interference Figure: (100) sections yield centered acute bisectrix figures with moderate 2*V*. Optic axis dispersion is weak, *r* < *v*.

Alteration Barite may be altered to witherite ($BaCO_3$) or may be replaced by minerals such as quartz, calcite, dolomite, or pyrite.

Distinguishing Features

Hand Sample: The specific gravity, which is unusually high for a nonmetallic mineral, distinguishes barite from other minerals with similar appearance. Anhydrite has three cleavages at right angles, and the carbonates all react with dilute HCl. Celestine, which is isostructural with barite, has lower specific gravity.

Thin Section/Grain Mount: Resembles gypsum but has higher indices of refraction and inclined extinction in most sections. Celestine has larger 2*V* and lower indices of refraction.

Occurrence Barite is a common gangue mineral in hydrothermal mineral deposits in association with galena, sphalerite, pyrite, quartz, fluorite, and carbonates. It also occurs as concretionary masses and veins in shale, limestone, sandstone, and other sedimentary rocks. Carbonatites may contain barite as a primary magmatic mineral. Barite is an uncommon biomineral (Chapter 5).

Use Barite is the primary ore mineral from which barium is extracted. Most barite, however, is mined as an industrial mineral and used in applications that take advantage of the high specific gravity. The majority is used in the petroleum industry as a component of drilling "mud" in oil and gas wells. Drilling mud is pumped down the center of the drill stem and exits through the drill bit as a well is being drilled. The mud lubricates the bit and also carries rock cuttings from the bottom of the drill hole up to the surface. Pulverized barite is added to the drilling mud to increase its density. This is done to prevent high-pressure gas encountered during drilling from blowing the mud out of the drill hole and venting to the surface, where it could be accidentally ignited.

Barite and barium compounds derived from it have over 2000 industrial uses. These include glassmaking, paper and playing card manufacture, ceramics, as a high-density

Table 17.5 Less Common Phosphate and Vanadate Minerals

Mineral	Unit Cell	Physical Properties[a]	Optical Properties	Comments
Amblygonite LiAl(PO$_4$)F	Triclinic $\bar{1}$ a = 5.18 b = 7.11 c = 5.03 α = 112°, β = 97.8°, γ = 68.3° Z = 2 $X \wedge b$ = 30–40°, $Y \wedge a$ = 15–20°, $Z \wedge c$ = 20°–30°	H: 5.5–6 G: 3.0–3.1 Cl: {100}pf, {110}g {1$\bar{1}$1} fair, {001}pr Tw: {$\bar{1}$$\bar{1}$1} common C: White; less commonly beige, pink, greenish, gray St: While L: Vitreous to greasy	n_α = 1.575–1.595 n_β = 1.587–1.610 n_γ = 1.590–1.622 δ = 0.014–0.027 Biaxial (–) 2V = 50–90° r < v, weak Colorless	Crystals are stubby prisms, often rough, and may be very large; found in Li-bearing granitic pegmatites
Autunite Ca(UO$_2$)$_2$(PO$_4$)$_2$ · 10–12H$_2$O	Tetragonal 4/m 2/m 2/m a = 7.0 c = 20.7 Z = 2	H: 2–2.5 G: 3.1–3.2 Cl: {001}pf, {100}pr Tw:{110} C: Bright yellow St: Yellow L: Vitreous Fluoresces yellowish green in UV light	n_ω = 1.577 n_ϵ = 1.554 δ = 0.023 Uniaxial (–) ω = pale yellow, ϵ = yellow	Crystals thin and tabular on {001}; commonly forms scaly or earthy aggregates in the near-surface oxidized zone of sediment-hosted uranium deposits and U-bearing veins and pegmatites
Carnotite K$_2$(UO$_2$)$_2$ · (VO$_4$)$_2$ 3H$_2$O	Monoclinic 2/m a = 10.47 b = 8.41 c = 6.91 β = 103.7° Z = 2 $X \sim c$, $Y = b$, $Z \sim a$	H: soft G: 4.7–5.0 Cl: {001}pf Tw: ?? C: Bright yellow, lemon yellow St: Yellow L: Earthy; pearly, or silky if coarse	n_α = 1.75–1.78 n_β = 1.90–2.06 n_γ = 1.92–2.08 δ = 0.17–0.30 Biaxial (–) 2V = 43–60° X = colorless, Y = Z = yellow	Usually fine-grained earthy masses or scaly grains; found in sediment-hosted uranium deposits and as an alteration on other U-bearüig minerals
Lithiophilite–triphylite Li(Mn, Fe)PO$_4$	Orthorhombic 2/m 2/m 2/m a = 4.86 b = 10.36 c = 6.01 Z = 4 $X = a$ or c, $Y = c$ or a, $Z = b$	H: 4.5–5 G: 3.34–3.58 Cl: {001}pf,{010}g,{110}pr Tw:N/A C: bluish or greenish gray (triph.), light-dark brown, salmon (lith.), St: White, gray L: Vitreous to subresinous	n_α = 1.670–1.705 n_β = 1.677–1.710 n_γ = 1.684–1.720 δ = 0.007–0.015 Biaxial (+ or –) 2V$_Z$ = 0–180° r < v or r > v strong Colorless to pale pink	Crystals prismatic parallel to c; more common as cleavable masses; found in Li-bearing granitic pegmatites
Pyromorphite Pb$_5$(PO$_4$)$_3$Cl	Hexagonal 6/m a = 9.97 c = 7.32 Z = 2	H: 3.5–4 G: 7.04 Cl: {101}pr Tw: {112} rare C: Green, yellow, brown St: White L: Resinous to subadamantine	n_ω = 2.058 n_ϵ = 2.048 δ = 0.010 Uniaxial (–) Colorless or pale tints of hand-sample color $\omega < \epsilon$	Isostructural with apatite; crystals are hexagonal prisms; often in reniform, globular, fibrous or granular masses; found in the near-surface oxidized portion of Pb-bearing sulfide deposits
Wavellite Al$_3$(OH)$_3$ (PO$_4$)$_2$ · 5H$_2$O	Orthorhombic 2/m 2/m 2/m a = 9.62 b = 17.34 c = 6.99 Z = 4 $X = b$, $Y = a$, $Z = c$	H: 3.5–4 G: 2.32–2.37 Cl: {110}pf, {10l}g, {010}fair Tw: ?? C: White, greenish white, green, yellow, brown St: White L: Vitreous to pearly	n_α = 1.518–1.535 n_β = 1.526–1.543 n_γ = 1.545–1.561 δ = 0.025–0.027 Biaxial (+) 2V = 72° r > v weak Colorless or pale: X = blue, green; Y = yellow-brown; Z = pale brown, pale yellow, colorless	Usually radiating or encrusting aggregates of acicular crystals elongate on c; found as a fairly common secondary mineral deposited by hydrothermal solutions in many rock types including low-grade Al-rich metamorphic rocks, phosphate deposits, massive limonite, and in hydrothermal veins with cassiterite and other phosphates

[a] CI = Cleavage (pf = perfect, g = good, pr = poor), C = color, Tw = Twinning, St = Streak, L = Luster.

filler in plastics, rubber and related materials, and as a pigment. People may swallow substantial quantities of chemically purified BaSO$_4$ (blanc fixe) to allow certain portions of the gastrointestinal tract to be more visible in X-ray images.

PHOSPHATES

The key structural element of phosphate minerals is the presence of tetrahedral PO$_4^{3-}$ anionic groups. The phosphorus ions have a +5 charge, so phosphate groups are anisodesmic; the oxygen anions are more strongly bonded to the P^{5+} than to other cations in the mineral. The PO_4^{3-} groups must be treated as structural units. Only a few phosphate minerals are likely to be encountered routinely. Apatite is a widespread accessory mineral in many rocks, and monazite and xenotime are not uncommon in granitic rocks. Turquoise, although actually fairly rare, is widely used as a decorative stone in jewelry. Less common phosphate and closely related vanadate minerals are listed in Table 17.5.

APATITE

$Ca_5(PO_4)_3(OH,F,Cl)$
Hexagonal $6/m$
$a = 9.32–9.64$ Å,
    $c = 6.78–6.90$ Å, $Z = 2$
$H = 5$
$G = 3.1–3.35$
Uniaxial $(-)$
$n_\omega = 1.629–1.667$
$n_\epsilon = 1.624–1.666$
$\delta = 0.001–0.007$

Structure The structure consists of PO_4^{3-} groups bonded laterally through Ca^{2+} (Figure 17.16). Half the Ca^{2+} is in distorted 6-fold coordination, and the other half is in distorted 7-fold coordination with O from the phosphate groups and additional (OH,F,Cl).

Composition The principal compositional variation is in the relative amounts of OH, F, and Cl. The end members are fluorapatite [$Ca_5(PO_4)_3F$], hydroxylapatite [$Ca_5(PO_4)_3OH$],

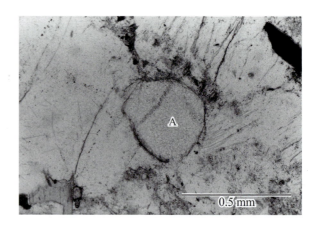

Figure 17.17 Photomicrograph of apatite in thin section.

and chlorapatite [$Ca_5(PO_4)_3Cl$]. Most apatite has compositions that lie between fluorapatite and hydroxylapatite, with more F than OH. A variety called **carbonate–apatite** contains tetrahedral CO_3OH^{3-} groups substituting for PO_4^{3-}. Carbonate–apatite is the dominant biomineral found in bones and teeth. **Dahllite** is carbonate–apatite with dominantly OH in the hydroxyl sites; **francolite** contains dominantly F. Phosphate tetrahedra also may be replaced by SO_4^{2-}, SiO_4^{4-}, or CrO_4^{2-}. Charge balance is maintained by substituting Na^+ for Ca^{2+} or, for example, by substituting a sulfate group and a silicate group for two phosphate groups. Trace amounts of U, V, Mn, and Fe commonly are present. The term **collophane** is given to fine-grained or cryptocrystalline material, such as phosphate rock and fossil bone, that contains apatite and usually a small amount of calcite.

Form and Twinning Crystals are common, but are usually small, and consist of hexagonal prisms terminated with {001} pinacoid and/or {101} dipyramid. Longitudinal sections through crystals are usually elongate-rectangular. Apatite also forms anhedral grains (Figure 17.17) and granular or columnar aggregates. Collophane is usually brown and forms colloform, encrusting, spherulitic, oolitic, and related structures and is a major constituent of some fossil bone. Twins are rare on {111} or {103}.

Cleavage Poor cleavage on {001} is not usually visible in either hand sample or thin section and does not have a strong control on fragment orientation in grain mount. Conchoidal to uneven fracture. Brittle.

Color Properties

Luster: Vitreous.

Color: Most commonly grayish blue-green, also green, yellow, yellowish green, brown, blue, pink, violet, orange, etc. Collophane is brown.

Streak: White.

Color in Thin Section: Colorless. Strongly colored samples may display a pale version of the hand-sample color. Collophane is brown.

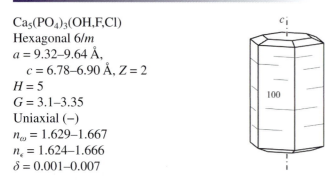

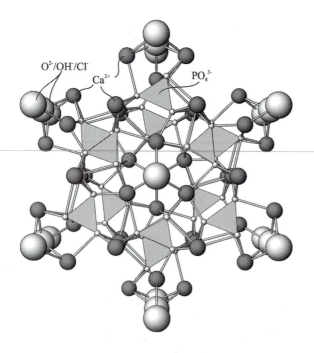

O^{2-}/OH$^-$/Cl$^-$

Ca^{2+}

PO$_4^{3-}$

Figure 17.16 Structure of apatite viewed down the c axis. The hexagonal symmetry is evident.

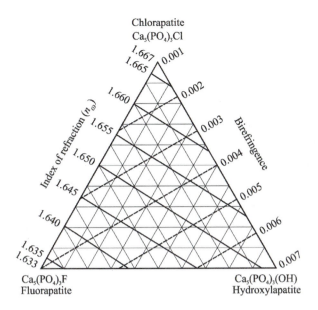

Figure 17.18 Approximate variation in birefringence (dashed lines) and n_ω (solid lines) in fluorapatite, hydroxyapatite, and chlorapatite.

Pleochroism: Colored varieties may be weakly pleochroic $\epsilon > \omega$, so elongate grains are darker when aligned with the vibration direction of the lower polarizer.

Optical Characteristics

Relief in Thin Section: Moderately high positive.

Optic Orientation: Elongate grains are length fast and show parallel extinction.

Indices of Refraction and Birefringence: Indices of refraction vary with composition (Figure 17.18) but with considerable variability. Common apatite has indices of refraction in the range n_ω = 1.633–1.650 and n_ϵ = 1.629–1.647 with δ = 0.003–0.005. Carbonate–apatite has indices of refraction n_ω = 1.603–1.638, n_ϵ = 1.598–1.619 with δ = 0.007–0.017. Collophane may be essentially isotropic.

Interference Figure: Basal sections in thin section yield uniaxial negative interference figures with diffuse isogyres on a first-order gray field. Small grain size may make interference figures difficult to obtain in thin section. Some samples, particularly carbonate–apatite, may be biaxial negative with $2V$ up to 20°.

Alteration Apatite is relatively stable in most geological environments and not readily altered. Pseudomorphs of clay, turquoise, serpentine, or other phosphate minerals have been reported.

Distinguishing Features

Hand Sample: Color, crystal habit, and hardness are diagnostic. Beryl is harder and usually lacks dipyramid terminations.

Thin Section/Grain Mount: Distinguished by moderately high relief and low birefringence. In thin section it usually displays a slightly pebbly surface texture with subtle pastel color highlights in plane light. Garnet is isotropic and topaz, sillimanite, and andalusite have higher birefringence and are biaxial. Zoisite has higher relief and usually displays anomalous interference colors. Beryl has lower refractive indices. Colored varieties may resemble tourmaline, but apatite is darker in plane light when the long dimension of grains is aligned parallel to the lower polarizer vibration direction.

Occurrence Apatite is a very common accessory mineral in a wide variety of igneous and metamorphic rocks and is the most common phosphate mineral. Grains are usually small and visible only in thin section. Only in granitic pegmatites and in skarns, marble, and calc-silicate gneiss derived from carbonate-rich sediment is apatite likely to be coarsely crystalline. Economic quantities of apatite are sometimes found in nepheline syenite and related rocks, and in carbonatites.

Apatite is common as detrital grains in clastic sedimentary rocks, usually as oval to somewhat elongate grains. Chlorapatite and carbonate–apatite are the major constituents of collophane in phosphatic limestone, shale, and ironstone and in some nearly pure phosphate beds. The apatite in these rocks may be derived from the skeletal remains of fish and other vertebrates or may be precipitated from seawater. Bone fragments and fish scales often serve as the nuclei for precipitation of secondary apatite.

Use Apatite, usually as collophane in sedimentary rocks, but also from some igneous sources, is a source of phosphate for use in fertilizer and other industrial purposes. Fluorine is also extracted from apatite, and is used to treat drinking water and for other industrial purposes including aluminum refining. Additional by-products that may be extracted from apatite during processing include uranium and vanadium. In some processes, the calcium is extracted. Apatite is a minor gemstone.

Comments Apatite is a very important biomineral because vertebrate bones and teeth are composed dominantly of carbonate–apatite (see Chapter 5 for more information). For browsing animals, the fact that their teeth are made of apatite with a hardness of 5 poses a significant problem. The dirt and dust that is ingested along with grass and other plants eaten by cattle, horses, rodents, and other herbivores can grind down the teeth because it usually contains quartz whose hardness is 7. Strategies that have developed by evolutionary processes to deal with this problem include replacing teeth as they wear out (elephants), continually growing the teeth (rodents), and providing large teeth with lots of mass to wear away (horses). Modern biomedical applications are numerous and include use in cements, ceramics, coatings on implants and bone grafting.

MONAZITE

$(Ce,La,Th)PO_4$
Monoclinic $2/m$
$a = 6.78$ Å,
$b = 7.00$ Å,
$c = 6.45$ Å,
$\beta = 104°$, $Z = 4$
$H = 5$
$G = 5.0–5.3$
Biaxial (+)
$n_\alpha = 1.772–1.800$
$n_\beta = 1.778–1.801$
$n_\gamma = 1.823–1.860$
$\delta = 0.045–0.075$
$2V_z = 6–26°$

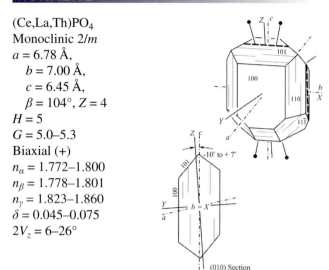

(010) Section

Structure The structure consists of slightly distorted PO_4 tetrahedra bonded laterally through rare earth cations in 9-fold coordination with oxygen from the tetrahedra.

Composition Any of the light rare earth elements may occupy the 9-fold sites, but Ce and La are usually the most abundant. Between 4 and 12% of the 9-fold sites are usually occupied by Th, and some samples have up to about 30%. Minor U is also usually present. Si and S may substitute for some of the P in the tetrahedra. Radioactive decay of Th and other radioactive elements may produce damage to the crystal lattice and potentially cause the mineral to become metamict. However, because monazite anneals the damage at relatively low temperatures, metamict monazite is uncommon.

Form and Twinning Crystals are usually small, nearly equant, flattened parallel to $\{100\}$, or elongate parallel to b. Simple twins on $\{100\}$ are common and lamellar twinning on $\{001\}$ is rare.

Cleavage A single cleavage on $\{100\}$ is distinct, a second at right angles on $\{010\}$ is fair to poor. A parting on $\{001\}$ is probably related to the rare lamellar twinning on the same plane. Altered samples display the cleavage more clearly in thin section. Conchoidal to uneven fracture. Brittle.

Color Properties

Luster: Resinous to waxy.

Color: Yellow, reddish yellow, or reddish brown.

Streak: White.

Color in Thin Section: Yellow or colorless.

Pleochroism: Weak, $Y > X \approx Z$: X = colorless or pale yellow, Y = yellow or dark yellow, Z = greenish yellow.

Optical Characteristics

Relief in Thin Section: High positive.

Optic Orientation: $X = b$, $Y \wedge a = +7°$ to $+24°$, $Z \wedge c = +7°$ to $-10°$, optic plane $\perp$ (010). The extinction angle seen in (010) sections is less than 10° to the trace of $\{100\}$ cleavage, which is length slow. Fragments on the $\{100\}$ cleavage show parallel extinction to the trace of $\{010\}$ cleavage if displayed.

Indices of Refraction and Birefringence: Indices of refraction generally increase with Th content and decrease with alteration. Interference colors in standard thin section range up to third and fourth orders. Fragments on the $\{100\}$ cleavage display maximum birefringence.

Interference Figure: Basal sections yield nearly centered acute bisectrix figures with small $2V$ and numerous isochromes. Optic axis dispersion is weak, $r < v$, or rarely $r > v$. Horizontal bisectrix dispersion is weak.

Alteration Monazite is relatively stable in the weathering environment. Alteration to a brown limonite-like material along cleavages and grain margins may occur and some grains are clouded by fine opaque particles.

Distinguishing Features

Hand Sample: Specimens large enough to deal with in hand sample are uncommon. Monazite is distinguished from zircon by different crystal habit and lower hardness, and from titanite by greater specific gravity. X-ray diffraction or chemical tests are usually required on small grains that may be extracted from host rocks.

Thin Section/Grain Mount: Recognized by high birefringence, high relief in thin section, and pale yellow color. Staurolite is more distinctly pleochroic and has much lower birefringence. Small grains may resemble zircon, titanite, xenotime, or members of the epidote group. Zircon is uniaxial and not usually colored. Epidote has larger $2V$ and birefringence is often lower. Titanite is more strongly colored, has larger $2V$, and extreme optic axis dispersion. Xenotime has higher birefringence and is uniaxial. Because monazite contains radioactive elements, it may produce pleochroic halos in enclosing biotite, hornblende, cordierite, or other minerals, but so may zircon, titanite, xenotime, and allanite. Backscatter images in either an electron micro-probe or an SEM can be used to locate the very small grains that may be present in some samples.

Occurrence Monazite is an accessory mineral in granite, granitic pegmatite, syenite, and carbonatites. Because it is stable in the weathering environment, monazite may be part of the heavy mineral fraction of clastic sediments. It is infrequently found in hydrothermal vein deposits. In metamorphic rocks it is found in metamorphosed dolostone and in mica schists, gneiss, and granulites.

Use Monazite is one of the primary sources for Th, Ce, and other rare earth elements. Most economic deposits are placers, usually on beaches. Cerium is used in lighter flints, glass, automobile catalytic converters, and a variety of other products including the mantles used in gas lanterns. Thorium is also used in the mantles used in gas lanterns and has been used as a fuel in nuclear reactors. Thorium also has a variety of applications in electronic equipment.

XENOTIME

YPO$_4$
Tetragonal
    $4/m\ 2/m\ 2/m$
$a = 6.89$ Å, $c =$
    6.03 Å, $Z = 4$
$H = 4$–5
$G = 4.3$–5.1 (4.28 calc.
    for pure YPO$_4$)
Uniaxial (+)
$n_\omega = 1.690$–1.724
$n_\epsilon = 1.760$–1.827
$\delta = 0.070$–0.107

Structure The structure consists of PO$_4^{3-}$ tetrahedra bonded through Y^{3+} (yttrium) in strongly distorted 8-fold coordination with oxygen anions from the tetrahedra. Xenotime is isostructural with zircon.

Composition Rare earth elements, particularly Ce (cerium) and Er (erbium), may substitute for Y (yttrium) along with small amounts of Ca, Zr, Th, or U. P^{5+} may be replaced by Si^{4+} to a limited extent to maintain charge balance when Zr^{4+} or U^{4+} substitutes for Y^{3+}.

Form and Twinning Crystals are elongate tetragonal prisms resembling zircon; also found as radial aggregates or rosettes. Detrital grains are usually elongate and somewhat rounded.

Cleavage Good prismatic cleavages on {110} intersect at 90° but tend not to have a strong control on fragment orientation. Uneven to splintery fracture. Brittle.

Color Properties

Luster: Vitreous to resinous.

Color: Yellowish to reddish brown, less commonly yellow, gray, salmon pink, or greenish.

Streak: Pale brown, yellowish, or reddish.

Color in Thin Section: Colorless, yellow, or pale brown.

Pleochroism: Weak, ω = pale pink, yellowish brown, or yellow; ϵ = pale brownish yellow, grayish brown, or greenish. Grains are darker with their long dimension parallel to the vibration direction of the lower polarizer.

Optical Characteristics

Relief in Thin Section: High positive.

Optic Orientation: Parallel extinction to crystal length and cleavage traces in longitudinal sections.

Indices of Refraction and Birefringence: Indices of refraction probably increase with substitution of Th for Y. Birefringence is very strong and yields high-order creamy white colors in standard thin section that may be masked by mineral color.

Interference Figure: Small grain size may make figures difficult to obtain. Basal sections yield uniaxial positive figures with numerous isochromes.

Alteration Xenotime is not readily altered.

Distinguishing Features

Hand Sample: Closely resembles zircon but has inferior hardness and cleavage at right angles.

Thin Section/Grain Mount: Small grains in thin section may be difficult to distinguish from zircon, monazite, titanite, and rutile. Zircon has higher indices of refraction and lower birefringence. Monazite is biaxial and has inclined extinction, higher relief, lower birefringence, and different habit. Titanite has a different habit, is biaxial, and has higher relief. Rutile has higher relief and is darker colored. Indices of refraction measured in grain mount will generally resolve ambiguities: zircon, monazite, titanite, and rutile all have higher indices of refraction. An electron microprobe or SEM may be required to identify small grains.

Occurrence Xenotime is a fairly common accessory mineral in granite, granodiorite, syenite, granitic pegmatite, and related rocks. It frequently is misidentified as zircon. In metamorphic rocks it may be found in mica schists and gneiss, or, less commonly, in marble. Xenotime is found as detrital grains in placer deposits, in beach sands, and as a heavy mineral in sandstone and related clastic sedimentary rocks.

Use Placer deposits of xenotime are a source of yttrium, which is used to manufacture phosphors used in fluorescent lights, old-style (cathode ray tube) color television sets and computer monitors, and in zirconia ceramics.

TURQUOISE

CuAl$_6$(PO$_4$)$_4$(OH)$_8$·4H$_2$O
Triclinic $\bar{1}$
$a = 7.48$ Å, $b = 9.95$ Å,
    $c = 7.69$ Å, $\alpha = 111.6°$,
    $\beta = 115.4°$, $\gamma = 69.4°$, $Z = 1$
$H = 5$–6
$G = 2.6$–2.8
Biaxial (+)
$n_\alpha = 1.61$
$n_\beta = 1.62$
$n_\gamma = 1.65$
$\delta = 0.04$
$2V_z = 40°$

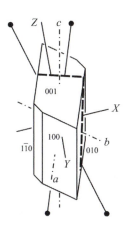

Structure Phosphate groups are linked laterally through Al in 6-fold coordination with O from the phosphate groups and additional OH. Cu occupies relatively open voids in the structure and coordinates with 4 OH and two H_2O molecules.

Composition Substitution of Fe^{3+} for Al^{3+} forms a continuous solid solution series between turquoise and **chalcosiderite** $[CuFe_6(PO_4)_4(OH)_8 \cdot 4H_2O]$. Most samples, however, are near the turquoise end of the series.

Form and Twinning Usually fine grained and massive, filling veins or forming reniform, stalactitic, or encrusting forms. Also as disseminated grains.

Cleavage Perfect {001} and good {010} cleavage is rarely seen due to fine grain size. Conchoidal to smooth fracture on massive material. Brittle.

Color Properties

Luster: Waxy on fractured surfaces, vitreous if polished.

Color: Turquoise green, also sky blue, apple green, or greenish gray.

Streak: White or greenish.

Color in Thin Section: Pale blue or green.

Pleochroism: X ~ colorless, Z = pale blue to pale green.

Optical Characteristics

Relief in Thin Section: Moderately positive.

Optic Orientation: $X \wedge b = 50°$, $Y \wedge a = 41°$, $Z \wedge c = 32°$.

Indices of Refraction and Birefringence: Indices of refraction probably increase with Fe content, but are not documented.

Interference Figure: Fine grain size will generally preclude obtaining interference figures. $2V_z$ is reported as $40°$ with strong optic axis dispersion, $r < v$.

Alteration Being produced in Cu-bearing hydrothermal systems, turquoise is subject to replacement by other minerals as part of normal dissolution/precipitation processes.

Distinguishing Features

Hand Sample: The blue-green color is distinctive. Chrysocolla is similar but is softer (2–4) and has the property of adhering when touched gently with the tongue.

Thin Section/Grain Mount: Color and occurrence are distinctive. Chrysocolla has lower indices of refraction.

Occurrence Turquoise is found in hydrothermally altered volcanic rocks, usually at least distally related to copper mineralization.

Table 17.6 Tungstates and Molybdates

Mineral	Unit Cell	Physical Properties[a]	Optical Properties	Comments
Scheelite $CaWO_4$	Tetragonal $4/m$ $a = 1.92$ $c = 1.94$ $Z = 4$	H: 4.5–5 G: 5.9–6.1 Cl: {101}g Tw: {110} contact and penetration C: White, yellowish, or brownish St: White L: Vitreous Fluoresces bluish white in short-wave UV light	$n_\omega = 1.921$ $n_\epsilon = 1.938$ $\delta = 0.017$ Uniaxial (+) Colorless	Crystals usually elongate tetragonal dipyramids; commonly massive granular or columnar; found in skarn deposits formed by contact metamorphism of limestone adjacent to granitic intrusive; also hydrothermal veins and greisen; an important ore mineral for tungsten
Wolframite $(Fe,Mn)WO_4$	Monoclinic $2/m$ $a = 4.77$ $b = 5.71$ $c = 4.98$ $\beta = 90.5°$ $Z = 2$ $X = b$, $Z \wedge c = 17°$ to $21°$	H: 4–4.5 G: 7.1–7.5 Cl: {010}pf Tw: Common {100} contact C: Brownish black to iron black St: Reddish brown to brownish black L: Submetallic	$n_\alpha = 2.17$–2.31 $n_\beta = 2.22$–2.40 $n_\gamma = 2.30$–2.46 $\delta = 0.13$–0.15 Biaxial (+) $2V = 73°$–79° X = Yellow, orange; Y = Greenish yellow, red-brown; Z = Olive, dark red-brown	Crystals common: short prismatic parallel to c and flattened on {100}; also granular massive, columnar, or lamellar; found in granitic pegmatites, greisen, and hydrothermal vein systems
Wulfenite $PbMoO_4$	Tetragonal 4 $a = 5.44$ $c = 12.11$ $Z = 4$	H: 3 G: 6.5–7.0 Cl: {010}g Tw: {001} C: Yellow, orange, or red; also gray or white St: White L: Vitreous to adamantine	$n_\omega = 2.405$ $n_\epsilon = 2.283$ $\delta = 0.122$ Uniaxial (−) Yellow to pale orange, $\omega > \epsilon$	Crystals tabular on {001}, often very thin, less commonly pyramidal or dipyramidal if twinned; found in the near-surface oxidized portion of Pb- and Mo-bearing hydrothermal sulfide deposits

[a] Cl = cleavage (pf = perfect, g = good), Tw = twinning, C = color, St = streak, L = luster.

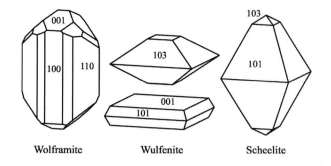

Figure 17.19 Crystal habit of wolframite, wulfenite, and scheelite.

Use Turquoise is used as a decorative stone in jewelry and in a wide variety of other art and craft objects. Material of inferior quality may be stabilized with epoxy or have the color artificially enhanced. Imitation turquoise, made of almost anything with an appropriate color, is also available on the market. Given its use as a gemstone, it is perhaps surprising that turquoise has few historical uses for medical purposes.

TUNGSTATES AND MOLYBDATES

Tungsten and molybdenum cations are readily oxidized to the +6 charge that they have in the tungstate and molybdate minerals. Both the tungstate and molybdate anionic groups consist of four oxygen anions that form a tetrahedral coordination site for the heptavalent cation. However, both W^{6+} and Mo^{6+}, at 0.56 and 0.55 Å, respectively, are somewhat large for a normal tetrahedral site, so the four oxygen anions do not form a regular tetrahedron. Because the W and Mo cations have a high charge, they command more than half of the available anion charge. Tungstates and molybdates are anisodesmic and the WO_4^{2-} and MoO_4^{2-} tetrahedra form discrete anionic groups in mineral structures.

The more common tungstate and molybdate minerals are listed in Table 17.6 and crystal habit is shown in Figure 17.19. Because W^{6+} and Mo^{6+} have the same charge and similar size, they may substitute for each other in some minerals. Scheelite forms a solid solution series with **powellite** ($CaMoO_4$), and wulfenite forms a solid solution series with **stolzite** ($PbWO_4$).

Wolframite displays continuous solid solution between the **hübnerite** ($MnWO_4$) and **ferberite** ($FeWO_4$) end members. Members of the solid solution series are generally restricted to high-temperature hydrothermal vein systems in association with sulfides (pyrrhotite, pyrite, sphalerite, arsenopyrite), cassiterite, scheelite, topaz, magnetite, and/or hematite, or in quartz or pegmatitic vein systems closely associated with granitic intrusive rocks.

Scheelite may be found in carbonate rocks that have been contact metamorphosed by intrusive granitic rocks. Associated minerals include carbonates, diopside, garnet, tremolite, wollastonite, and other calc-silicate minerals. Scheelite also is found, often with wolframite, in high-temperature, quartz-rich hydrothermal vein systems that typically are closely associated with granitic intrusives. Other associated minerals include topaz, fluorite, albite, sulfides (pyrite, pyrrhotite, molybdenite, chalcopyrite, etc.), tourmaline, and apatite.

Wolframite and scheelite together are the principal source for the world's tungsten. This important element is alloyed with iron to make high-strength steel. The filament in conventional incandescent lightbulbs is metallic tungsten. Tungsten carbide, whose hardness is exceeded only by diamond, is used to make durable cutting tools for many industrial applications and is also used as an abrasive.

Wulfenite commonly forms beautiful tabular crystals in the near-surface oxidized zone of Pb-bearing hydrothermal sulfide deposits. It has been used as a minor source of molybdenum, which is used as an alloying agent in high-strength steel and in lubricants.

BORATES

The most common building block of the borates is the triangular BO_3^{3-} group. Each B^{3+} commands one of the available two negative charges on each O. That means that the borates are mesodesmic (Chapter 4); each oxygen may be shared with another borate group to form doublets, chains, rings, and so forth, similar to the silicates. B^{3+} may also form triangular $BO_2(OH)^{2-}$ groups, as in borax. Additional structural elements found in borates are tetrahedral BO_4^{5-}, BO_3OH^{4-}, and $BO_2(OH)_2^{3-}$ groups. The triangular and tetrahedral groups can be linked together to form a variety of complex structures.

The borate minerals are, with few exceptions, found in evaporite deposits from saline lakes. These deposits are typically formed in tectonically active terranes associated with plate boundaries and abundant volcanism. Because boron does not fit in the structure of the silicate minerals in common igneous rocks, it is concentrated in late residual magmatic liquids. Thermal springs and other hydrothermal waters associated with the volcanism presumably carry the boron from these late magmatic liquids to the surface.

Precipitation of borate minerals may be in an apron adjacent to a hot spring or on playa lake beds in response to seasonal evaporation. Borate minerals are typically interbedded with lacustrine sediment and volcanic ash, and may grade laterally into shallow water lacustrine limestone. Because borate minerals are fairly soluble in water, they are not found in wet climates. Notable occurrences include the arid intermontane valleys in southeastern California, Tibet, Turkey, and in the Andes Mountains of Peru, Bolivia, Argentina, and Chile.

Of the 150 or so borates that have been identified, borax, ulexite, colemanite, kernite, and tincalconite are the most common; their properties are listed in Table 17.7. Crystal

Table 17.7 Borate Minerals

Mineral	Unit Cell	Physical Properties[a]	Optical Properties	Comments
Borax $Na_2[BO(OH)]_2$ $[BO_2(OH)]_2 \cdot 8H_2O$	Monoclinic $2/m$ $a = 11.86$ $b = 10.67$ $c = 12.20$ $\beta = 106.6°$ $Z = 4$ $Y \wedge c = +33°$ to $36°$, $X = b$	H: 2–2.5 G: 1.71 Cl: {100}pf, {110}g Tw: {100} rare C: White or gray, may be greenish or bluish St: White L: Vitreous	$n_\alpha = 1.447$ $n_\beta = 1.469$ $n_\gamma = 1.472$ $\delta = 0.025$ Biaxial (−) $2V = 39$–$40°$ $r > v$, strong Colorless	Soluble in water, slightly alkaline taste; crystals stubby prisms with 8-sided cross sections; commonly as granular aggregates; found in evaporite deposits from saline lakes; readily alters by dehydration to crumbly tincalconite on exposure to air
Colemanite $Ca_2B_6O_{11} \cdot 5H_2O$	Monoclinic $2/m$ $a = 8.74$ $b = 11.26$ $c = 6.10$ $\beta = 110.1°$ $Z = 2$ $Y \wedge c = -6°$, $X = b$	H: 4–4.5 G: 2.42 Cl: {010}pf, {100}g Tw: None reported C: White, colorless, gray, or yellowish St: White L: Vitreous	$n_\alpha = 1.586$ $n_\beta = 1.592$ $n_\gamma = 1.614$ $\delta = 0.028$ Biaxial (+) $2V = 56°$ $r < v$, weak Colorless	Decrepitates on heating; crystals stubby prisms; common as granular or cleavable masses; found in evaporite deposits from saline lakes

Kernite
$Na_2B_4O_6(OH)_2 \cdot 3H_2O$
	Monoclinic $2/m$
	$a = 15.52$
	$b = 9.14$
	$c = 6.96$
	$\beta = 108.9°$
	$Z = 4$
	$X \wedge a = +38°$,
	  $Y \wedge c = -19°$, $Z = b$
		H: 2.5–3
		G: 1.91–1.93
		Cl: {100}, {001} pf {$\bar{2}$01} fair
		Tw: {110}
		Cl: White
		St: White
		L: Vitreous, alters to
		  earthy tincalconite
			$n_\alpha = 1.455$
			$n_\beta = 1.472$
			$n_\gamma = 1.487$
			$\delta = 0.032$
			Biaxial (−)
			$2V = 80°$
			$r > v$, distinct
			Colorless
				Crystals nearly equant; common as cleavable and granular masses; may appear fibrous due to intersecting perfect cleavages; found in evaporite deposits from saline lakes

Tincalconite $Na_2B_4O_5(OH)_4 \cdot 3H_2O$	Hexagonal 32 $a = 11.12$ $c = 21.20$ $Z = 6$	H: ?? Soft G: 1.88 Cl: ?? Tw:?? C: White St: White L: Earthy	$n_\omega = 1.461$ $n_\epsilon = 1.473$ $\delta = 0.013$ Uniaxial (+) Colorless	Forms as earthy dehydration product on other borates; found in evaporite deposits from saline lakes

Ulexite $NaCaB_5O_6 \cdot 5H_2O$	Triclinic $\bar{1}$
	$a = 8.71$
	$b = 12.72$
	$c = 6.69$
	$\alpha = 90.3°$, $\beta = 109.1°$,
	  $\gamma = 105.1°$
	$Z = 2$ $X \wedge c = -69°$,
	  $Y \wedge c = +21°$, $Z \sim b$
		H: 2.5
		G: 1.96
		Cl: {010}pf,{1$\bar{1}$0}g, {110}pr
		Tw: {010}, {100} polysynthetic
		C: White, colorless
		St: White
		L: Vitreous, aggregates
		  may be silky
			$n_\alpha = 1.491$–1.496
			$n_\beta = 1.504$–1.506
			$n_\gamma = 1.519$–1.520
			$\delta = 0.026$
			Biaxial (+)
			$2V = 73$–$78°$
			Colorless
				Crystals are acicular to fibrous parallel to c; may form nodular masses, crusts, compact veins, or loose "cotton balls"; found in evaporite deposits from saline lakes

[a] Cl = cleavage (pf = perfect, g = good pr = poor), Tw = twinning, C = color, St = streak, L = luster.

habits of borax and colemanite are shown in Figure 17.20. Minerals that may be associated with borates in saline lake evaporite deposits include halite, sylvite, gypsum, anhydrite, trona, nahcolite, thenardite, glauberite, polyhalite, and kieserite.

Borax, ulexite, and colemanite are the principal sources of boron, which has a variety of industrial uses. The largest use is in the manufacture of glass fiber insulation and borosilicate glass. Borates are also used in detergents and soap (e.g., Boraxo® brand hand soap), as a flux in metallurgical processes, and in ceramics and wood preservatives. The ^{10}B isotope is a good neutron absorber, so boron is used in shielding material in nuclear reactors. Borate minerals have been used in the past as antiseptics and food preservatives. Boron currently finds use in a wide range of products including pharmaceuticals, cosmetics, insecticides,

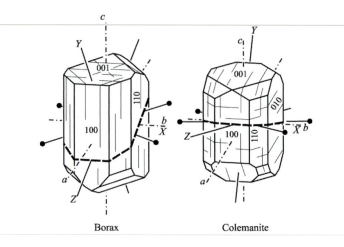

Figure 17.20 Crystal habit of borax and colemanite.

and adhesives. Boron carbide is an excellent abrasive. A relatively new application is boron-fiber-reinforced plastic, which is used in aerospace applications to form high-strength panels, structural members, and wing and fuselage surfaces.

REFERENCES CITED AND SUGGESTIONS FOR ADDITIONAL READING

Alpers, C. N., Jambor, J. L., and Nordstrom, D. K. (eds.). 2000. Sulfate minerals—Crystallography, geochemistry, and environmental significance. Reviews in Mineralogy & Geochemistry 40, 608 p.

Carlson, W. D. 1980. The calcite–aragonite equilibrium: Effects of Sr substitution and anion orientation disorder. American Mineralogist 65, 1252–1262.

Chang, L. L. Y., Howie, R. A., and Zussman, J. 1996. *Rock-Forming Minerals,* Volume 5B: *Nonsilicates: Sulfates, Carbonates, Phosphates, Halides,* 2nd ed. The Geological Society, London, 383 p.

Friedman, G. M. 1959. Identification of carbonate minerals by staining methods. Journal of Sedimentary Petrology 29, 87–97.

Hardie, L. A. 1967. The gypsum–anhydrite equilibrium at one atmosphere pressure. American Mineralogist 52, 171–200.

Kennedy, G. C. 1947. Charts for correlation of optical properties with chemical composition of some common rock-forming minerals. American Mineralogist 32, 561–573.

Reeder, R. J. (ed.). 1983. *Carbonates: Mineralogy and Chemistry.* Reviews in Mineralogy 11, 394 p.

Robbins, L. L., and Blackwelder, P. L. 1992. Biochemical and ultrastructural evidence for the origin of whitings: A biologically-induced calcium carbonate precipitation mechanism. Geology 20, 464–468.

Sánchez-Román, M., Vasconcelos, C., Schmid, T., Dittrich, M., McKenzie, J. A., Zenobi, R., and Rivadeneyra, M. A. 2008. Aerobic microbial dolomite at the nanometer scale: Implications for the geologic record. Geology 36, 879–882.

Spencer, R. J. 2000. Sulfate minerals in evaporite deposits. Reviews in Mineralogy & Geochemistry 40, 173–192.

Thompson, J. B., and Ferris, F. G. 1990. Cyanobacterial precipitation of gypsum, calcite, and magnesite from natural alkine lake water. Geology 18, 995–998.

Warne, W. St. J. 1962. A quick field or laboratory staining scheme for the differentiation of the major carbonate minerals. Journal of Sedimentary Petrology 32, 29–38.

Warren, J. K. 2006. *Evaporites, Sediments, Resources and Hydrocarbons.* Springer-Verlag, Berlin, 1035 p.

Wolfe, K. H., Easton, A. J., and Warne, S. 1967. Techniques of examining and analysing carbonate skeletons, minerals, and rocks, in G. V. Chilingar, H. J. Bissell, and R.W. Fairbridge (eds.), *Carbonate Rocks, Developments in Sedimentology,* 9B. Elsevier, Amsterdam, pp. 253–342.

Wright, D. T. 1999. The role of sulfate-reducing bacteria and cyanobacteria in dolomite formation in distal ephemeral lakes of the Coorong region, South Australia. Sedimentary Geology 126, 147–157.

CHAPTER **18**

Oxides, Hydroxides, and Halides

INTRODUCTION

The minerals described in this chapter are grouped together because they are constructed with simple anions. Oxygen is the anion in the oxides, hydroxyl (OH^-) anionic groups are found in hydroxide minerals, and halogen elements (Cl^-, F^-, and Br^-) are the anions in the halides. Although it can be argued that hydroxyl groups are complex anions, they behave in crystal structures like simple anions, hence hydroxides are described here.

OXIDES

The oxides as a group include compounds in which cations of one or more metals are combined with oxygen. All of these minerals, except ice, are isodesmic, meaning that they do not contain anionic groups. Rather, each cation–oxygen bond commands less than half of the available −2 charge on the oxygen, and the chemical bonding is dominantly ionic. Ice is a molecular crystal.

Most of the minerals in the group have fairly high symmetry, reflecting the fact that their structures are usually based on a systematic packing of the oxygen anions, often in cubic or hexagonal close packing. The cations occupy tetrahedral and octahedral sites between these regularly packed anions. The oxides are conveniently grouped on the basis of the cation to oxygen ratio (Table 18.1).

Many of the oxide minerals described here are widely distributed; except for ice, however, they are often relatively minor constituents of the rocks in which they occur. The abundance of silicon in most geological environments ensures that most of the oxygen is tied up in silicate minerals. Where geologic concentrations of these minerals occur, they may be mined as sources of metals.

Table 18.1 Oxide Minerals

Mineral	Composition	Mineral	Composition
X_2O		**X_2O_3**	
Cuprite	Cu_2O	Hematite	Fe_2O_3
Ice	H_2O	Corundum	Al_2O_3
XO		Ilmenite	$FeTiO_3$
Periclase	MgO		
Zincite	ZnO	**XO_2**	
XY_2O_4		Rutile	TiO_2
Magnetite	$FeFe_2O_4$	Cassiterite	SnO_2
Chromite	$FeCr_2O_4$	Pyrolusite	MnO_2
Spinel	$MgAl_2O_4$	Uraninite	UO_2
Chrysoberyl	$BeAl_2O_4$		

X_2O GROUP

CUPRITE

Cu_2O
Isometric $4/m\,\bar{3}\,2/m$
$a = 4.27$ Å, $Z = 2$
$H = 3.5–4$
$G = 6.14$
Isotropic
$n = 2.849$

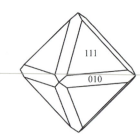

Structure The structure consists of oxygen anions in a simple body-centered cubic arrangement with Cu atoms located in the centers of four of the eight cubelets into which the cube may be divided (Figure 18.1). Each Cu is therefore coordinated with just two oxygen anions. This somewhat unusual geometry is favored because the bonds are fairly covalent and involve one $4s$ and one $4p$ orbital, which hybridize to form sp orbitals that have a linear arrangement.

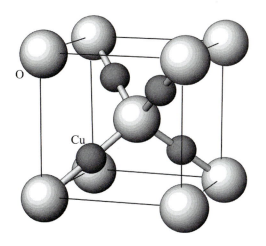

Figure 18.1 Cuprite structure. Oxygen anions occupy the corners and center of the unit cell; Cu occupies four of the eight octants of the cube.

Composition Most samples are probably close to the ideal composition.

Form and Twinning Crystals are octahedrons, sometimes cubes, dodecahedrons, or combinations. Commonly as massive, granular, or earthy aggregates.

Cleavage {111} fair cleavage. Conchoidal to uneven fracture. Brittle.

Color Properties

Luster: Submetallic to adamantine.
Color: Ruby red to deep brownish red or almost black.
Streak: Brownish red.
Color in Thin Section: Red, orange-yellow, or yellow, not pleochroic.

Optical Characteristics

Relief in Thin Section: Very high positive.
Index of Refraction: Isotropic with $n = 2.849$.
Polished Section: Bluish gray. May be anomalously anisotropic with very weak bireflectance and reflection pleochroism, and strong anisotropism with crossed polarizers with blue-gray to olive green polarization colors. Deep red internal reflections are characteristic.

Alteration Cuprite may be altered to malachite, native copper, or other copper minerals.

Distinguishing Features

Hand Sample: Deep red color is fairly distinctive. Hematite is harder (if coarsely crystalline) and has a lighter red streak. Cinnabar is softer and has a scarlet red streak.

Thin Section/Grain Mount: Cinnabar is deep red and uniaxial, and hematite is practically opaque.
Polished Section: Bluish gray color with red internal reflections is characteristic.

Occurrence Cuprite is found in the near-surface oxidized portion of copper-bearing hydrothermal sulfide deposits. It generally is produced by altering and oxidizing primary copper sulfide minerals such as chalcopyrite. Cuprite frequently is associated with iron oxides and hydroxides (limonite), malachite, azurite, native copper, and black **tenorite** (CuO).

Use Cuprite is an ore mineral for copper, which is used extensively for electrical wiring. Copper also is used in a wide variety of other industrial, commercial, and household products. Copper is alloyed with tin to make bronze.

ICE

H_2O
Hexagonal $6/m\ 2/m\ 2/m$
$a = 4.498$ Å,
$\quad c = 7.338$ Å, $Z = 4$
$H = 1.5$
$G = 0.92$
Uniaxial (+)
$n_\omega = 1.309$
$n_\epsilon = 1.311$
$\delta = 0.002$

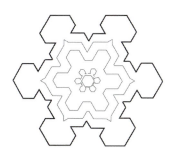

Structure The structure is unusual in that ice consists of H_2O molecules that bond together with hydrogen bonds. Each electrically neutral water molecule can be treated as a sphere with a radius of 1.38 Å. The charge distribution is controlled by the location of the positively charged hydrogen nuclei that are located at two of the corners of a tetrahedron within the sphere; the other two corners have negative charge. The water molecules are arranged so that the hydrogen on one molecule points to one of the negatively charged corners of the charge tetrahedron on an adjacent water molecule to provide the hydrogen bond (Figure 18.2). The result is that each water molecule coordinates in tetrahedral fashion with four adjacent water molecules. The three-dimensional structure produced by repeating this pattern is hexagonal.

Composition Most ice is relatively pure, but ice produced by freezing salt water may contain saline fluid inclusions.

Form and Twinning Ice in the form of snow or frost forms beautiful skeletal crystals with hexagonal symmetry as shown. Glacial ice is a granular aggregate of ice grains, the same as ice cubes from the freezer or the ice on a frozen lake. {001} twins are common and may be produced by deformation.

Cleavage None. Conchoidal fracture. Brittle.

Color Properties

Luster: Vitreous.
Color: Colorless, white.
Streak: White.

Optical Characteristics Conventional thin sections and grain mounts are not practical because ice melts at 0°C.

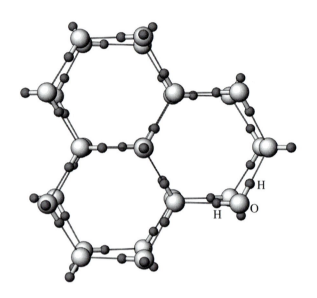

Figure 18.2 Ice structure viewed down the *c* axis. Water molecules consisting of one oxygen (large sphere) and two hydrogens (small spheres) are arranged in a hexagonal array. Hydrogen bonds between water molecules are shown with thinner lines. The positions of the H and O are shown accurately, but the sizes of the atoms are reduced. Compare Figure 3.11.

Alteration Ice melts at 0°C or may sublimate to form water vapor in the atmosphere.

Distinguishing Features

Hand Sample: Not easily mistaken for any other mineral.

Occurrence Ice forms as a sediment (snow) by precipitation from water vapor in the atmosphere. Snow may subsequently be recrystallized to form glacial ice, which, strictly speaking, is a metamorphic rock. Ice also forms by crystallization from liquid water to form ice on lakes, rivers, and other bodies of water. This ice might reasonably be considered an igneous rock. It also acts as a very effective cementing agent in soil in areas in which the average temperature drops below freezing. In areas with permafrost, ice in the soil persists year round.

Use Perhaps its most important use is to cool the drinks that slake the thirst of geologists after long hot days in the field. Ice has been used as a construction material to build dwellings in arctic regions and supports a major recreational industry (skiing and snow boarding) in many parts of the world. Ice fishing and ice skating would not be possible without it. Ice also poses significant environmental risks when it covers roads and bridges, or when icebergs calved from glaciers endanger shipping. Snow avalanches are a significant threat to life and transportation systems in alpine areas. Glacial ice forms a substantial portion of the world's resource of fresh water, although not one that is readily available for extensive human use.

Comments Although not usually included when most people think about minerals, ice is a very common mineral that has great geological, meteorological, economic, and social importance.

Box 18.1 Paleoclimate

Glacial ice contains records of paleoclimate data that go back many thousands of years. Cores of ice taken from glaciers show annual accumulations of snow. As snow compacts and recrystallizes to form glacial ice, it traps small bubbles of air. When the ice is later melted, the small samples of air can be analyzed to determine atmospheric composition when the snow fell. Also preserved in glacial ice are atmospheric dust (e.g., from volcanos) and pollen from vegetation, both of which have climate significance.

Paleotemperatures can be estimated based on the composition of the oxygen isotopes that make up the H_2O molecules in the ice (Allègre, 2008). Water molecules are made with both the heavier ^{18}O isotope and the lighter and more abundant ^{16}O isotope. The relative amounts of the light and heavy water molecules that evaporate from water depend on temperature—the heavier molecules evaporate less easily at lower temperatures. Atmospheric water vapor is derived mostly by evaporation from the ocean, so if sea-surface temperatures are low, the ratio of ^{18}O to ^{16}O in water vapor in the atmosphere will be lower than if sea-surface temperatures are high. Snow crystallizing from water vapor in the atmosphere reflects the oxygen isotope composition of the atmosphere. Changes in the oxygen isotope composition of ice from successive years of snow accumulation can, therefore, provide a record of changing climate.

XO GROUP

The XO group includes the minerals periclase (MgO) and zincite (ZnO), neither of which is common. Their properties are listed in Table 18.2. Periclase is isostructural with NaCl, and zincite has a structure analogous to the wurtzite polymorph of sphalerite (ZnS).

XY$_2$O$_4$ MINERALS

Spinel Group

The spinel group of minerals (Table 18.3) crystallizes in the isometric crystal system. The structure (Figure 18.3) is based on cubic close-packed oxygen with X and Y cations in tetrahedral and octahedral coordination sites among the oxygen anions. The unit cell, which is about 8 Å on a side, contains 8 formula units (Z = 8) or 8 X cations and 16 Y cations. The X cations are divalent, and the Y cations are either all trivalent or half quadrivalent and half divalent. The unit cell contains 32 octahedral sites and 64 tetrahedral sites. Sixteen of the octahedral sites are occupied, and 8 of the tetrahedral sites are occupied. The two types of spinel structures depend on which cations go in which site.

The **normal spinel structure** is displayed by members of the spinel and chromite series. Divalent X cations occupy the 8 tetrahedral (4-fold) coordination sites within the unit cell. Smaller trivalent Y cations occupy the 16 larger octahedral (6-fold) coordination sites. The general formula based on site occupancy is therefore $^{IV}X^{VI}Y_2O_4$, where the roman numerals refer to the coordination number for each cation.

Members of the magnetite series have **inverse spinel structures**. In magnetite, for example, the unit cell contains eight Fe^{2+} (the X cations) in octahedral coordination sites. The 16 Fe^{3+} cations (Y cations) are distributed eight to tetrahedral sites and eight to octahedral sites. Based on site occupancy, the formula could be written $^{VI}(Fe^{2+}Fe^{3+})$ $^{IV}Fe^{3+}O_4$, or for the general case, $^{VI}X^{VI}Y^{IV}YO_4$.

Both normal and inverse spinel structures place large cations in tetrahedral sites that they normally would not occupy based on size considerations alone (Chapter 3). However, the combination of chemical bonds produced by both structures produces a low-energy configuration that outweighs the ion size considerations.

Of the members of the spinel group, only chromite, magnetite, and Fe/Mg spinel are very common, so they are the only species described in detail here.

Many other minerals and chemical compounds crystallize with the spinel structure and are sometimes referred to as spinel. A particular source of confusion to students is the use of the term *spinel* in reference to high-pressure polymorphs of olivine (e.g., γ-Mg$_2$SiO$_4$) that, based on geophysical and experimental studies, should be present in the Earth's mantle. γ-Mg$_2$SiO$_4$ has the spinel structure with Si in tetrahedral coordination and Mg in octahedral coordination. Although the cations have the same coordination as in olivine, a different distribution allows the spinel structure to be significantly more dense than olivine.

Table 18.3 Spinel Group Minerals

Mineral	Composition	Mineral	Composition
Spinel series		Magnetite series	
Spinel	MgAl$_2$O$_4$	Magnetite	FeFe$_2$O$_4$
Hercynite	FeAl$_2$O$_4$	Magnesioferrite	MgFe$_2$O$_4$
Gahnite	ZnAl$_2$O$_4$	Ulvöspinel	FeFeTiO$_4$
Galaxite	MnAl$_2$O$_4$	Franklinite	ZnFe$_2$O$_4$
Chromite series		Jacobsite	MnFe$_2$O$_4$
Chromite	FeCr$_2$O$_4$	Trevorite	NiFe$_2$O$_4$
Magnesiochromite	MgCr$_2$O$_4$	Hausmannite	MnMn$_2$O$_4$

Table 18.2 XO Oxide Minerals

Mineral	Unit Cell	Physical Properties[a]	Optical Properties	Comments
Zincite ZnO	Hexagonal 6*mm* a = 3.249 c = 5.205 Z = 2	H: 4 G: 5.68 Cl: {100}pf, {001} parting Tw: {001} C: Orangish yellow to deep red St: Orangish yellow L: Subadamantine	n_ω = 2.013 n_ϵ = 2.029 δ = 0.016 Uniaxial (+) TS: Deep red to yellow PS: Pinkish brown with strong red/yellowish internal reflections	Crystals consist of {101} pyramid, {100} prism, and {00$\bar{1}$} pedion; usually granular; rare except at Franklin, New Jersey, USA
Periclase MgO	Isometric 4/m $\bar{3}$ 2/m a = 4.213 Z = 4	H: 5.5 G: 3.58 Cl: {100}pf Tw: None C: Colorless, white, or gray; also yellow, brown, or black with inclusions St: White L: Vitreous	n = 1.735–1.756 Isotropic Colorless	Cubic crystals rare, usually rounded or irregular grains; found as disseminated grains in metamorphosed limestone and dolostone

[a] Cl = cleavage (pf = perfect), Tw= twinning, C= color, St = streak, L = luster, TS = thin section, PS = polished section.

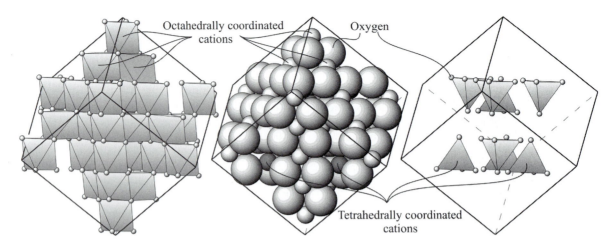

Octahedrally coordinated cations

Oxygen

Tetrahedrally coordinated cations

Figure 18.3 Spinel structure. Oxygen anions form an approximately cubic close-packed arrangement with layers parallel to {111}, which is horizontal in the illustration. Within each unit cell, cations occupy 8 tetrahedral and 16 octahedral sites between the oxygen layers. All the tetrahedral sites are shown; the octahedral sites shown on the left include both those within the unit cell and additional octahedra that are partially within adjacent unit cells. The cube outlines are slightly larger than the unit cell. The corners of the unit cell are the centers of octahedrally coordinated cations at the corners of the cube. See text for additional discussion.

$FeFe_2O_4$
Isometric $4/m\,\bar{3}\,2/m$
$a = 8.396$ Å, $Z = 8$
$H = 5.5–6.5$
$G = 5.18$
Opaque
Ferrimagnetic

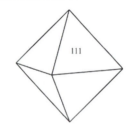

111

Composition Magnetite usually contains some Ti as part of a solid solution series to ulvöspinel, $^{VI}(Fe^{2+}Ti^{4+})^{IV}Fe^{2+}O_4$. The solid solution involves substitution of Ti^{4+} for Fe^{3+} in octahedral sites coupled with substitution of Fe^{2+} for Fe^{3+} in the tetrahedral sites to maintain charge balance. Ti-rich samples are possible only under low-oxidization conditions in which the iron is mostly Fe^{2+}. The availability of oxygen allows the ulvöspinel component of magnetite to be oxidized to form ilmenite and more pure magnetite by the following reaction.

$$3Fe_2^{2+}Ti^{4+}O_4 + 0.5O_2 = 3Fe^{2+}TiO_3 + Fe^{2+}Fe_2^{3+}O_4$$

Ulvöspinel + oxygen = ilmenite + magnetite   (18.1)

At a given temperature, oxidation forces the exsolution of ilmenite out of Ti-bearing magnetite, so that the magnetite contains less Ti. At a given oxidation condition, decreasing the temperature accomplishes the same thing. It is quite common to find that magnetite from igneous and metamorphic rocks contains exsolution lamellae of ilmenite produced either as a consequence of cooling or late oxidation of the magnetite in response to changing environmental conditions.

Ilmenite, however, forms a solid solution series with hematite ($Fe_2^{3+}O_3$). The composition of the ilmenite–

hematite solid solution is influenced by the reaction of magnetite and oxygen to form hematite.

$$2Fe^{2+}Fe_2^{3+}O_4 + 0.5O_2 = 3Fe_2^{3+}O_3$$

Magnetite + oxygen = hematite

(18.2)

All other things being equal, oxidation tends to make the ilmenite–hematite solid solution more hematite rich, provided magnetite is present. This reaction is also temperature sensitive.

The compositions of coexisting magnetite and ilmenite are, therefore, sensitive to both the temperature and the oxidation state. It is therefore possible to analyze coexisting magnetite and ilmenite in an igneous or metamorphic rock to provide information about the chemical environment (oxidation condition) and temperature history of the rock.

Other compositional variation in magnetite includes substitution of Mg^{2+} for Fe^{2+} as part of a continuous solid solution to magnesioferrite ($MgFe_2O_4$) and substitution of some Al^{3+} for Fe^{3+}. Some Mn, Cr, and Ca also may be present.

Form and Twinning Octahedral crystals are common, dodecahedrons are less so. Also granular massive (Figure 18.4) and as isolated anhedral grains. Twinning on {111} (spinel twin) is common, usually forming simple twins, but also multiples.

Cleavage None, but a parting on the {111} twins may be distinct. Uneven fracture. Brittle.

Color Properties

Luster: Dull metallic to splendent.

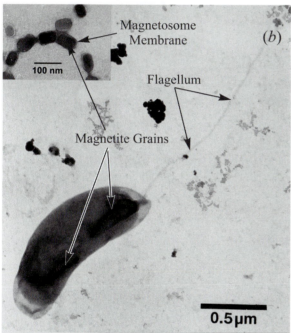

Figure 18.4 Magnetite. (*a*) Massive granular hand sample. (*b*) Transmission electron microscope images (courtesy of D. Bazylinski) of a magnetotactic bacterium (strain MV-1) showing a chain of magnetite grains with a gap in the middle. The cell possesses a single polar flagellum that can be used for propulsion. Inset shows detail of magnetite grains and the magnetosome membrane that surrounds them and anchors them in place.

Color: Black.
Streak: Black.

Optical Characteristics

Thin Section: Opaque.

Polished Section: Brownish gray (but somewhat variable) without bireflectance or reflection pleochroism. Isotropic.

Alteration Magnetite is readily altered to hematite or goethite and related iron oxides and hydroxides in the weathering environment. The term **martite** is applied to hematite pseudomorphs after magnetite.

Distinguishing Features

Hand Sample: Magnetite is readily attracted by a magnet. The variety **lodestone** is naturally magnetic and will attract and hold small objects such as paper clips. Ilmenite is similar but is only weakly magnetic. Chromite has a more resinous luster and is not magnetic.

Thin Section/Grain Mount: Difficult to distinguish from other opaque minerals.

Polished Section: Ilmenite is anisotropic, and sphalerite typically displays internal reflections. Magnetite often contains exsolved lamellae of ilmenite or hematite.

Occurrence Magnetite is a very common accessory mineral found in a wide variety of igneous and metamorphic rocks, usually as small euhedral to anhedral grains. Segregations of massive magnetite occur in some mafic magmatic systems. Contact metamorphism of limestone or dolostone adjacent to granitic intrusives may produce massive replacement deposits of magnetite associated with diopside, tremolite, garnet, calcite, dolomite, and other calc-silicate minerals. Sedimentary and metamorphic iron formations often contain magnetite as a major mineral. Magnetite may be concentrated in the heavy mineral fraction of clastic sediments, particularly if the sediments are derived from areas subject to fairly rapid erosion and limited chemical weathering. Once deposited, detrital magnetite may be oxidized to other iron oxides and hydroxides. Magnetite is a significant biomineral (Figure 18.4), and small magnetite grains are commonly found in the cells of many species, including the brains of humans (Chapter 5). In at least some cases, magnetite appears to have biologic uses in orientation (cyanobacteria) and navigation (birds).

Use Magnetite is mined in many locations as an ore of iron from which cast iron and steel are derived. It is also used as a feedstock in the manufacture of other iron-bearing industrial materials. As a mineral, magnetite has some use as a filter media and black pigment. Crushed magnetite has been used as aggregate to make high-density concrete for applications that have included nuclear reactors. Magnetite appears in some folk medicine and historical medical preparations, where it has been claimed to help chronic wheezing, tinnitus, cancer, epilepsy, and insomnia; it has even been believed to be an aphrodisiac. An interesting medical research application is to attach very small particles of magnetite to the surfaces of individual cells so that the cells can be manipulated and controlled by applying an external magnetic field (Dobson, 2008).

Comments Microscopic grains of magnetite provided some of the most critical evidence that allowed our concepts of continental drift and plate tectonics to be developed. The key observation is that magnetite can acquire and preserve a magnetic polarity under the influence of an external magnetic field. In a volcanic rock, for example, small grains of magnetite will become magnetized parallel to the Earth's magnetic field when the magnetite cools through its Curie temperature (the temperature above which ferrimagnetism is destroyed). Under appropriate conditions, that magnetization will be preserved over great spans of geological time. An instrument called a magnetometer allows the direction and magnitude of this "fossil magnetism" to be measured. If the age of the rock is known, perhaps by radiometric dating, then the Earth's magnetic polarity and the distance and direction to the magnetic poles for that particular time can be deduced. If many samples of different ages are measured, it is possible to determine how the landmass containing the samples moved relative to the Earth's magnetic poles, and also to document the sequence of magnetic polarity reversals through time.

Paleomagnetic information of this type was some of the most convincing evidence that documented continental drift and allowed the development of the current model of seafloor spreading and plate tectonics. This very common and ordinary mineral found in rocks the world over had a very important story to tell.

Box 18.2 Sedimentary Iron Formations

Iron formations are sedimentary rocks (Figure 18.5) that are a major resource for iron. They are characteristic of the Precambrian and most were deposited between 3.5 billion and 1.9 billion years ago with a peak around 2.5 billion years. Trendall (2002) and Klein (2005) provide recent reviews. Iron formations are commonly part of a greenstone belt that has been subjected to both metamorphism and deformation. Layers may be hundreds of meters thick and may extend for hundreds of kilometers. The least metamorphosed units consist of thinly bedded hematite- and magnetite-rich sediment alternating with siltstone, chert, or other material. Other common minerals include siderite, dolomite–ankerite, greenalite, stilpnomelane, riebeckite, and sometimes pyrite. Because modern analogs are lacking, it is difficult to know the primary, prediagenesis and premetamorphism mineralogy. Possibly it consisted of iron hydroxides [$Fe(OH)_2$, $Fe(OH)_3$], SiO_2 gels, carbonate mud, and poorly defined Fe-silicate and hydrous Na-, K-, and Al- silicate gels.

The origin of sedimentary iron formations remains a matter of debate. The metamorphism and deformation to which most have been subjected makes it quite difficult to reconstruct the environment of deposition. A distinction is often made between banded iron formations and granular iron formations. With banded iron formations, the delicate nature of the fine laminations and lack of current-generated sedimentary structures argues for deposition below the maximum depth for wave motion, which is roughly 200 meters. With granular iron formations, whose banding is much coarser and which were mostly deposited after ~2.2 billion years ago, the presence of oolites and other sedimentary structures in some units argues for shallow water. There typically is little contribution of clastic sediment. It is generally agreed that the iron was extracted from anoxic seawater and that the iron probably entered the water from hydrothermal sources. It is probably not a coincidence that the peak of iron formation generation coincides with the Great Oxidation Event. Some authors (e.g., Trendall, 2002;

Figure 18.5 Soudan banded-iron formation in northeastern Minnesota. The dark layers are dominantly hematite, and the light layers are mostly microcrystalline quartz (jasper). Note the pen at right (arrow) for scale. Photo courtesy of Graham Baird.

Kronhouser, 2007) have argued that precipitation of the iron oxides, silica, and other material was mediated by microorganisms. It is now well documented that microorganisms are capable of precipitating iron oxide and hydroxide minerals by both biologically induced and biologically controlled processes (Chapter 5). However, a direct biologic origin for sedimentary iron formations is far from universally accepted, and some authors (e.g., Klein, 2005) argue for a purely chemical origin.

CHROMITE

FeCr$_2$O$_4$
Isometric $4/m\,\overline{3}\,2/m$
$a = 8.38$ Å, $Z = 8$
$H = 5.5–6$
$G = 5.09$ (lower with Mg & Al)
Isotropic
$n = 1.90–2.12$

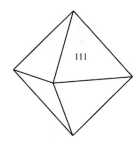

Composition Chromite forms a continuous solid solution series with magnesiochromite (MgCr$_2$O$_4$), and substantial Mg is present in most samples. Al, Fe^{3+}, Mn, and Zn may be present, and continuous solid solution to hercynite (FeAl$_2$O$_4$) is possible.

Form and Twinning Crystals are octahedrons that may be modified by cube faces. Commonly massive granular or as anhedral grains (Figure 2.51). Simple or multiple twins on {111}, according to the spinel law (Figure 5.21a).

Cleavage None. Parting on the {111} twin planes. Uneven fracture. Brittle.

Color Properties

Luster: Metallic to somewhat pitchy.

Color: Black.

Streak: Brown.

Color in Thin Section: Essentially opaque, dark brown on thin edges.

Optical Characteristics

Polished Section: Grayish white to light brown without bireflectance or reflection pleochroism. Usually isotropic, but may show weak anisotropism. Reddish-brown internal reflection common.

Alteration Chromite may alter to limonite and other oxides and hydroxides of Fe, Cr, and Mg, and to chrome-bearing clays.

Distinguishing Features

Hand Sample: Resembles magnetite and ilmenite, but chromite is not magnetic and has a somewhat more resinous luster. Some chromite, however, may have some magnetic character because of intergrown magnetite.

Thin Section/Grain Mount: Difficult to distinguish from other opaque minerals in conventional thin section or grain mount. Some chromite may be transparent and dark brown on thin edges.

Polished Section: Resembles magnetite but is less reflective and has internal reflections.

Occurrence Chromite is found in mafic and ultramafic igneous rocks such as gabbro, peridotite, dunite, and pyroxenite as an accessory mineral. In some cases, masses of chromite may be generated by crystal settling or other magmatic segregation processes. Chromite may be concentrated in detrital sands in some locations.

Use Chromite is the only ore mineral for chromium. About 70% of world production is alloyed with iron to make stainless steel and other high-strength alloys. Metallic chromium is used to plate steel objects to protect them from corrosion and abrasion, and for decorative purposes. Chromium appears in a wide variety of chemical products used for industrial applications as diverse as tanning leather, printing, catalyzing chemical reactions, preserving wood, and in paints. Chromite also is used to make refractory products for foundries.

SPINEL SERIES

Spinel: MgAl$_2$O$_4$
Hercynite: FeAl$_2$O$_4$
Isometric $4/m\,\overline{3}\,2/m$
$a = 8.103$ Å (spinel),
    $a = 8.135$ Å (hercynite),
    $Z = 8$
$H = 7.5–8$
$G = 3.55–4.40$ (higher with Fe)
Isotropic
$n = 1.714–1.835$

Composition Only common varieties that form a solid solution series between spinel and hercynite are described here. The term **pleonaste** has been applied to compositions intermediate between the Mg and Fe end members. Extensive solid solution is possible to galaxite and gahnite (Table 18.3). Cr may also substitute for Al as part of solid solution to chromite; Cr-bearing spinel has been called **picotite**.

Form and Twinning Crystals are octahedrons, sometimes modified by cube and dodecahedron faces. Also common as anhedral grains and granular masses. Simple and multiple twins by the spinel law on {111} are common (Figure 5.21).

Cleavage None, but a parting on the {111} twin planes may be present. Conchoidal, uneven, to splintery fracture. Brittle.

Color Properties

Luster: Vitreous.

Color: Usually green or blue-green, also colorless, blue, red; darker with Fe content.

Streak: White.

Color in Thin Section: Corresponds with hand-sample color. Not pleochroic.

Optical Characteristics

Relief in Thin Section: Very high positive.

Index of Refraction: Index of refraction increases systematically with Fe content, from 1.714 for pure spinel to 1.835 for pure hercynite.

Alteration Spinel is relatively resistant in the weathering environment. Alteration to clay or other layer silicates is possible.

Distinguishing Features

Hand Sample: Crystal habit, hardness, and green color are characteristic. Hercynite-rich samples may resemble magnetite but are not magnetic and have a light streak.

Thin Section/Grain Mount: Isotropic character and strong (usually green or blue green) color are characteristic. Garnet forms dodecahedrons and is lighter colored.

Occurrence Spinel is a common accessory mineral in aluminous metamorphic rocks associated with andalusite, kyanite, sillimanite, corundum or cordierite, in contact and regionally metamorphosed limestone and dolostone, and less commonly in granitic pegmatites and hydrothermal veins. Because it is resistant in the weathering environment, it may be found in detrital sediments.

Use Spinel has been used as a gemstone but does not have significant industrial applications. It has been used in dental ceramics.

CHRYSOBERYL

$BeAl_2O_4$
Orthorhombic $2/m\ 2/m\ 2/m$
$a = 5.49$ Å,
$\quad b = 9.42$ Å, $c = 4.43$ Å,
$Z = 4$
$H = 8.5$
$G = 3.68 – 3.75$
Biaxial (+)
$n_\alpha = 1.732 – 1.747$
$n_\beta = 1.734 – 1.749$
$n_\gamma = 1.741 – 1.758$
$\delta = 0.008 – 0.011$
$2V = \sim 70°$ some reports as
$\quad$ low as 10°

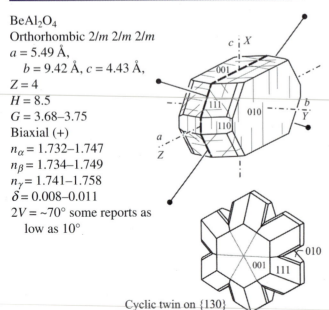

Cyclic twin on {130}

Structure Chrysoberyl is isostructural with olivine (Figure 16.1). Oxygen forms a slightly distorted hexagonal close-packed arrangement within which Be occupies the tetrahedral sites and Al occupies the octahedral sites.

Composition Up to about 5% Fe^{3+} may substitute for Al^{3+} in octahedral sites. The gem variety **alexandrite** contains minor Cr^{3+} substituting for Al^{3+}. Mg, Ti, and Ca are reported in some analyses.

Form and Twinning Crystals are usually tabular on (001) with a striation on (001) parallel to the *a* axis. Twinning on {130} is very common and may form simple twins or pseudohexagonal "sixlings" with or without reentrant angles.

Cleavage {110} good, {010} fair, and {001} poor. Uneven to conchoidal fracture. Brittle.

Color Properties

Luster: Vitreous. Needlelike inclusions and tubelike cavities produce chatoyancy or asterism in the gem variety called **cat's eye**.

Color: Green, greenish white, greenish yellow, white. Alexandrite is red in natural sunlight, green in artificial light.

Streak: White.

Color in Thin Section: Colorless.

Pleochroism: Grain mounts may be weakly colored: X = rose red, Y = orange, Z = green.

Optical Characteristics

Relief in Thin Section: Very high positive.

Optic Orientation: $X = c$, $Y = b$, $Z = a$.

Indices of Refraction and Birefringence: Little variation.

Interference Figure: At room temperature $2V_z$ is between 70° and 10°, $r > v$, weak. With increasing temperature, $2V_z$ gradually decreases to about 0°, $r < v$.

Alteration Chrysoberyl is quite resistant in the weathering environment, so it may appear as detrital grains. It may be altered to layer silicates such as chlorite and clay.

Distinguishing Features

Hand Sample: Great hardness, crystal habit and twinning, and green color are diagnostic.

Thin Section/Grain Mount: Has very high relief and low birefringence. Corundum, beryl, and tourmaline are uniaxial; the latter two minerals have much lower relief in thin section.

Occurrence Chrysoberyl is fairly common in granitic pegmatites, where it occurs with quartz, K-feldspar, albite, muscovite, biotite, tourmaline, and beryl. Chrysoberyl is infrequently found in aluminous mica schists and in skarn deposits derived by contact metamorphism of dolostone. Because of its resistance in the weathering environment and relatively high specific gravity, chrysoberyl is a constituent of some placer deposits.

Use The alexandrite and cat's eye varieties are valued as gems. It is a potential source of beryllium, but is not exploited because beryl is far more abundant. Probably because it has been used as a gemstone and is perceived as valuable, chrysoberyl has had historical uses as a treatment for ailments of the pancreas, spleen, and nervous system.

X₂O₃ GROUP

The hematite group has the general formula X_2O_3. The common minerals in this group are hematite, corundum, and ilmenite, in which X_2 is Fe_2^{3+}, Al_2, and $Fe^{2+}Ti^{4+}$, respectively. All share a common structure based on hexagonally close-packed oxygen anions, with metal cations occupying octahedral sites between the anions (Figure 18.6). For

convenience, these minerals can be considered to be constructed of layers of octahedra parallel to {001}. Within each layer, only two-thirds of the octahedra contain cations and the occupied octahedra share two oxygen anions (share edges) with their neighbors. Successive layers of octahedra share a common layer of oxygen anions and are arranged so that occupied octahedra in adjacent layers share three oxygen anions (share a face). Ilmenite differs from both hematite and corundum in that two different cations (Fe^{2+} and Ti^{4+}) are distributed systematically among the octahedral sites.

A full unit cell is constructed of six successive layers of octahedra and is about 13–14 Å along the *c* axis, with an *a* dimension of about 5 Å. In many references, the Miller indices of crystallographic planes and faces are indexed to a unit cell that has twice the *a* unit cell dimension. In the mineral descriptions that follow, both the unit cell dimensions measured with X-ray diffraction techniques and a unit cell consistent with conventionally assigned Miller indices are provided. As usual, it is important to confirm the crystallographic conventions being used when comparing data from different sources.

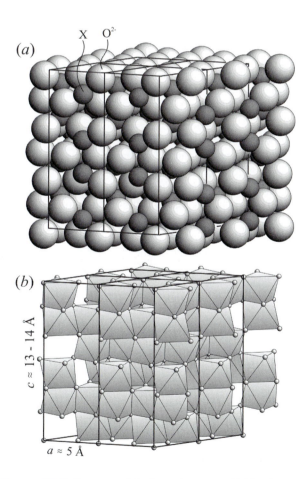

Figure 18.6 Structure of the hematite group. (*a*) X cations and O^{2-} anions drawn approximately to scale. The oxygen anions are in an approximately hexagonal close-packed arrangement. The outline of four unit cells is shown. (*b*) Distribution of X octahedra within the structure.

HEMATITE

Fe_2O_3
Hexagonal (trigonal) $\bar{3}\,2/m$
$a = 5.038$ Å,
   $c = 13.772$ Å,
   $Z = 6$
   ($a = 10.076$ Å,
   $c = 13.772$ Å,
   $Z = 24$ to be consistent
   with Miller indices)
$H = 5$–6 (earthy material may be lower)
$G = 5.25$ if pure, may be less
Opaque

Composition Hematite usually contains some Ti, and above about 1050°C there is complete solid solution between hematite and ilmenite ($FeTiO_3$) that involves substitution of $Fe^{2+}Ti^{4+}$ for $2Fe^{3+}$. Up to about 10% Al can substitute for Fe, as can some Mn.

Form and Twinning Crystals are usually platy with a hexagonal outline and may have triangular markings on the basal pinacoid. Platy grains may form micaceous aggregates. Coarsely crystalline material with metallic luster is known as **specular hematite**. Fine-grained, massive hematite may form oolitic, botryoidal, or rounded shapes with radiating structures. Red earthy hematite may be admixed with clay and other minerals, and may form masses or disseminated grains. Twins on {001} and {101} may be lamellar.

Cleavage None, but parting on {001} and {101} due to twinning. Uneven to subconchoidal fracture. Brittle, elastic in thin laminae.

Color Properties

Luster: Crystalline varieties are metallic. Fine-grained varieties are dull or earthy.

Color: Steel gray for coarse-crystalline varieties; dull to bright brownish red for fine-grained samples.

Streak: Rich red-brown.

Color in Thin Section: Opaque; deep red-brown for very small crystals or thin edges.

Optical Characteristics

Indices of Refraction: Fine grains are uniaxial negative, n_ω = 3.15–3.22, n_ϵ = 2.87–2.94, δ = 0.28.

Polished Section: Grayish white with bluish tint, weak bireflectance and reflection pleochroism. Deep red internal reflections are common. Distinct anisotropism with gray-blue and gray-yellow polarization colors.

Alteration Hematite may be altered to iron hydroxide minerals. It is stable in the weathering environment and is commonly produced by weathering.

Distinguishing Features

Hand Sample: Fine-grained varieties are distinguished from goethite (and limonite) by a red streak. The streak of cinnabar is more carmine red, and its specific gravity is higher. Black metallic varieties are distinguished from magnetite and ilmenite by much weaker or no magnetism and red streak.

Thin Section/Grain Mount: The red-brown color on thin edges distinguishes hematite from magnetite, ilmenite, and other likely opaque minerals.

Polished Section: Grayish-white color with red internal reflections and anisotropic character distinguish hematite from magnetite and ilmenite.

Occurrence Hematite is readily produced by weathering or hydrothermal alteration of iron-bearing minerals in almost any rock type. It is not common as a primary mineral in igneous rocks but may be found in some syenite, trachyte, granite, and rhyolite. Late-stage magmatic processes may alter Fe-bearing minerals or cause the precipitation of hematite from residual fluids. Hematite may be a major mineral in iron formations, almost all of which are of Precambrian age; associated minerals include magnetite, quartz, carbonates, and various Fe-silicates. The red color of many sediments is due to the presence of disseminated hematite and iron hydroxide minerals.

Use Hematite is an important ore of iron and is mined for that purpose in many areas of the world. It also is used as a pigment and polishing abrasive and as black crystals, is cut as a gem. Probably because it is red and can be associated with blood, hematite has numerous historical medical uses, most commonly to treat bleeding and blood disorders. To the degree that the iron from hematite is bioavailable in the gut, hematite could potentially be a source of dietary iron, although dietary supplements typically use ferrous sulfate, gluconate, or fumarate for that purpose.

CORUNDUM

Al_2O_3
Hexagonal (trigonal) = $\bar{3}$ $2/m$
a = 4.75 Å, c = 12.98 Å, Z = 6
   (a = 9.51 Å,
   c = 12.98 Å,
   Z = 24 to be consistent
   with Miller indices)
H = 9
G = 3.98–4.02
Uniaxial (−)
n_ω = 1.766–1.794
n_ϵ = 1.758–1.785
δ = 0.008–0.009

Composition Corundum is usually almost pure Al_2O_3. Blue corundum (sapphire) contains Fe and Ti, red (ruby) contains Cr, and yellow and green varieties contain Fe^{3+} and Fe^{2+}.

Form and Twinning Crystals are usually hexagonal prisms that may taper toward the ends (Figure 18.7). Basal pinacoids may display triangular markings. Crystals may also be tabular on {001}. Also as anhedral grains. Lamellar twins on {101} are common. Simple twins on {101} and {001} are less common.

Cleavage None; parting on the {101} and {001} twins. Uneven to conchoidal fracture. Brittle.

Color Properties

Luster: Vitreous to adamantine.

Color: Most commonly white, gray, or gray-blue, also red (ruby), blue (sapphire), yellow, or green.

Streak: White (hardness makes streak difficult to obtain).

Color in Thin Section: Colorless or a pale version of the hand sample color.

Pleochroism: Colored versions are $\omega > \epsilon$.

Other Star sapphire (Figure 6.12) contains fine needles of rutile that exsolved from original more Ti-rich corundum. The needles align themselves parallel to the three 2-fold rotation axes at right angles to c and produce a distinct asterism when viewed down the c axis.

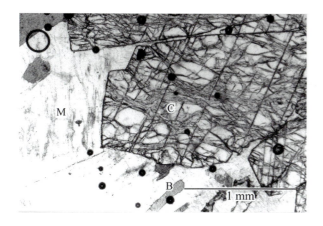

Figure 18.7 Photomicrograph of corundum (C), microcline (M), and small grains of biotite (B) in a thin section of mica schist. Note the rhombohedral parting in the corundum. The small circles are bubbles in the cement used for the thin section. Plane light.

Optical Characteristics

Relief in Thin Section: Very high positive.

Optic Orientation: Crystals are length fast, and extinction is parallel to the basal {001} parting.

Indices of Refraction and Birefringence: Pure corundum has indices of refraction $n_\omega = 1.768$, $n_\epsilon = 1.760$. Indices of refraction increase with Fe and Cr content. Interference colors should be first-order gray and white in standard thin sections; but owing to its great hardness, thin sections containing substantial amounts of corundum are almost always thicker than the standard 0.03 mm, and higher-order colors are encountered.

Interference Figure: Basal sections yield uniaxial negative interference figures. Some corundum may be biaxial with $2V_x$ as high as 50° or 60°, perhaps owing to twinning.

Alteration Corundum may be altered to fine-grained aggregates of aluminous minerals such as muscovite, margarite, andalusite, sillimanite, kyanite, diaspore, gibbsite, or clay.

Distinguishing Features

Hand Sample: Crystal habit and great hardness are characteristic. Alteration may allow samples to be anomalously soft. **Sapphirine** [(Mg,Al)$_8$(Si,Al)$_6$O$_{20}$], which is an uncommon chain silicate found in aluminous metamorphic rocks, may resemble blue corundum. Sapphirine, however, is monoclinic or triclinic, with fair cleavage on {010} and poor cleavage on {001 and {110}, $H = 7.5$, $G = 3.40–3.58$, crystals are tabular; it is usually light to dark blue; also green, white, red or yellow.

Thin Section: High relief, low birefringence, and uniaxial character are diagnostic. Twin lamellae are often visible.

Great hardness usually makes it difficult to grind thin sections to the standard (0.03 mm) thickness. Sapphirine has similar indices of refraction and birefringence ($n_\alpha = 1.701–1.731$, $n_\beta = 1.703–1.741$, $n_\gamma = 1.705–1.745$, $\delta = 0.005–0.012$); but it is biaxial ($2V_x = 47–114°$) and usually has distinct pleochroism in shades of colorless, blue or pink.

Occurrence In igneous rocks, corundum is infrequently found in Al-rich Si-poor varieties such as syenite and associated feldspathoidal pegmatites, and is usually associated with feldspars and feldspathoids, but not quartz. Al-rich pelitic metamorphic rocks may contain corundum associated with aluminum silicate(s), micas, spinel, and other aluminous minerals. Gem-quality corundum usually is derived from metamorphosed limestone or dolostone; the Al presumably is derived from detrital clay minerals. Other occurrences include metamorphosed bauxite deposits (emery), xenoliths in mafic rocks, and Si-poor hornfels.

Use Corundum is an important gemstone, and high-quality ruby can be more valuable than diamond. The color of natural ruby and sapphire can be improved by various processes, although treated samples have less value than naturally colored stones. Gem-quality corundum can be synthesized but is usually recognized because it lacks the imperfections that almost always are present in natural samples.

The gem varieties of corundum have numerous historical uses in medicine—the Middle Ages, in particular, saw many applications. Presumably because it is red, ruby became associated with treatment of blood and bleeding problems. Other applications include plague, eye problems, bladder and kidney problems, and for poisoning. Not surprisingly, there are no modern applications.

Corundum (emery), long used as an abrasive in sandpaper, grinding wheels, and polishing compounds, is being replaced by more uniform synthetic products (silicon carbide, alumina, etc.) in many applications. Emery also is used as a surface-hardening and nonskid agent in concrete, epoxy coatings, and grouting materials. End uses include pedestrian walkways, steel ship decks, roads, bridges, airport runways, and parking ramps.

ILMENITE

FeTiO$_3$
Hexagonal (trigonal) $\overline{3}$
$a = 5.089$ Å,
   $c = 14.09$ Å, $Z = 6$
   ($a = 10.18$ Å, $c = 14.09$ Å,
   $Z = 24$ to be consistent
   with Miller indices)
$H = 5–6$
$G = 4.70–4.79$
Opaque
Weakly magnetic

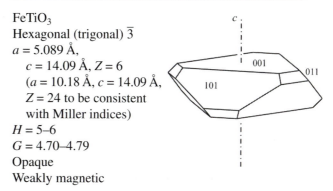

Composition Ilmenite forms a solid solution series with hematite above about 1050°C; most ilmenite contains some Fe^{3+} substituting for $Fe^{2+}Ti^{4+}$. Substantial amounts of Mg and Mn may also substitute for Fe^{2+}.

Form and Twinning Crystals are tabular parallel to {001} with a hexagonal cross section modified by rhombohedral truncations. Also as granular masses and anhedral grains, and common as exsolution lamellae in magnetite. Simple {001} and lamellar {101} twins are common.

Cleavage None, but partings on the {001} and {101} twins may be distinct. Conchoidal to irregular fracture. Brittle.

Color Properties

Luster: Metallic.

Color: Black.

Streak: Black.

Optical Characteristics

Thin Section: Opaque.

Polished Section: Brownish with a pink or violet tint, distinct bireflectance, and reflection pleochroism. Dark brown internal reflections are rare. Anisotropism is strong in crossed polarizers with gray to brownish-gray polarization colors.

Alteration Ilmenite may be altered to leucoxene, a name applied to a fine-grained yellowish, grayish, or brownish mixture of Ti- and Fe-bearing oxides, hydroxides, and other minerals.

Distinguishing Features

Hand Sample: Distinguished from specular hematite by a black streak and from magnetite by weak magnetism. Pure ilmenite is paramagnetic above −218°C, meaning that it does not behave like a magnet. Ilmenite–hematite solid solutions, however, are weakly magnetic. In addition, most ilmenite, particularly if found in large masses, contains intergrown and exsolved magnetite that contributes to the magnetic behavior.

Thin Section/Grain Mount: Entirely opaque. Hematite may be dark red-brown on thin edges or in small grains. Magnetite crystallizes in octahedrons that yield four- or six-sided cross sections rather than the tabular shapes typical of ilmenite.

Polished Section: Somewhat browner color than either magnetite or hematite. Hematite usually displays red internal reflections.

Occurrence Ilmenite is a common accessory mineral in a wide variety of igneous and metamorphic rocks, either as exsolved lamellae in magnetite grains or as separate grains. Larger masses of ilmenite are typically associated with bodies of gabbro, norite, anorthosite, and related mafic and ultramafic rocks. Detrital sands often contain ilmenite as part of the heavy mineral fraction; it also is a biomineral.

Use Ilmenite is a major ore mineral from which titanium is extracted. Much of the world's economic deposits are alluvial or beach placer deposits. Metallic titanium is similar to aluminum, but has higher strength and greater corrosion resistance. Titanium is also alloyed with iron to make high-strength steel. The major use of titanium is to produce brilliant white pigment (TiO_2) used in paints, plastics, and many other products. Ilmenite is also used as a filter agent in water treatment plants, as a sandblast abrasive, as foundry sand, and as a coating on welding rods.

XO$_2$ GROUP

Rutile (TiO_2) and cassiterite (SnO_2) are isostructural and crystallize to form tetragonal crystals. Pyrolusite (MnO_2), which shares the same structure, is included in the later discussion of manganese oxide and hydroxide minerals. The structure can be visualized as chains of edge-sharing octahedra that extend parallel to c along each corner of the unit cell (Figure 18.8). An additional edge-sharing chain of octahedra extends parallel to c in the center of the unit

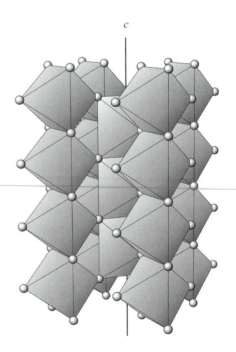

Figure 18.8 Rutile and cassiterite structure. The Ti^{2+} or Sn^{2+} cations occupy octahedra that form edge-sharing chains extending parallel to c along the corners and center of the unit cell.

cell and shares corners with and cross-links the chains on the corners.

Uraninite has the fluorite structure (see later: Figure 18.12) because U is too large to conveniently fit into octahedral coordination. In this structure, which is isometric, the anions are arranged in sheets in which the anions form a square pattern. Successive sheets are stacked directly atop each other to form cubic (8-fold) coordination sites. Alternate 8-fold sites are occupied by the cations. This places U cations at the corners and center of the unit cell.

RUTILE

TiO$_2$
Tetrahedral 4/m 2/m 2/m
$a = 4.593$ Å, $c = 2.959$ Å, $Z = 2$
$H = 6$–6.5
$G = 4.23$–5.5
Uniaxial (+)
$n_\omega = 2.605$–2.613
$n_\epsilon = 2.899$–2.901
$\delta = 0.29$

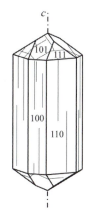

Composition Most rutile is nearly pure TiO$_2$, but some Fe, Ta, Sn, and Nb may substitute for Ti, and some Cr and V may be present.

Form and Twinning Crystals are elongate tetragonal prisms, often with dipyramidal terminations. Also as anhedral grains or slender, acicular, or fibrous crystals. Contact twins on {101} are common and may be elbow shaped or cyclic (Figures 5.20, 5.23). Also twins by gliding on {011} and {092}, and rare contact twins on {031}.

Cleavage {110} good, {010} fair, with a parting on {092} and {011} twins. Conchoidal, subconchoidal to uneven fracture. Brittle.

Color Properties

Luster: Metallic to adamantine.

Color: Reddish brown, also black, violet, yellow, or green.

Streak: White.

Color in Thin Section: Yellowish to reddish brown, darker with Fe content.

Pleochroism: Weak, $\epsilon > \omega$.

Optical Characteristics

Relief in Thin Section: Extreme positive.

Optic Orientation: Elongate crystals show parallel extinction and are length slow, but extreme birefringence makes determining sign of elongation in thin section impractical.

Indices of Refraction and Birefringence: Little variation of refractive indices. Birefringence is extreme and yields upper-order white interference colors that are masked by mineral color.

Interference Figure: Basal sections yield uniaxial positive figures with numerous isochromes. May be anomalously biaxial (+) with $2V_z$ up to 10°.

Polished Section: Gray with bluish tint, distinct bireflectance, and reflection pleochroism. Abundant strongly colored internal reflections (white, brown, violet, green, etc.). Strong crossed-polarizer anisotropy, colors often masked by internal reflections.

Alteration Alteration to leucoxene (gray, yellow, or brown mix of Ti and Fe oxides and hydroxides and other minerals) is not uncommon. Rutile may form as an alteration product after other Ti-bearing minerals.

Distinguishing Features

Hand Sample: Characterized by adamantine to metallic luster and red-brown color.

Thin Section/Grain Mount: Color, extreme birefringence, extreme relief, and habit are characteristic. Hematite is opaque unless quite thin. Cassiterite is similar but has lower birefringence and refractive indices. Fine fibrous sillimanite has much lower birefringence and is not colored. **Anatase** and **brookite** are polymorphs of rutile with similar structure and properties. Anatase is uniaxial negative, $n_\omega = 2.561$, $n_\epsilon = 2.488$; brookite is biaxial positive, $n_\alpha = 2.583$, $n_\beta = 2.584$, $n_\gamma = 2.700$, $2V_z = 0$–30°.

Polished Section: Abundant internal reflections and twins are characteristic. Cassiterite is similar, but with lower reflectance and is more difficult to polish.

Occurrence Rutile is widely distributed as an accessory mineral in many igneous and metamorphic rocks, but fine grain size may make identification difficult. Because it is fairly stable in the weathering environment, rutile is common in clastic sediments as part of the heavy mineral fraction. In some cases rutile may be sufficiently abundant to constitute a placer deposit. Fine fibers are not uncommon as inclusions in quartz (rutilated quartz), micas, and other minerals and may give the quartz a silky appearance. Large crystals are generally restricted to pegmatites and to quartz and apatite-bearing veins.

Use Rutile is a significant source of titanium, generally mined from placer deposits. See discussion of ilmenite for additional information.

CASSITERITE

SnO_2
Tetragonal $4/m\ 2/m\ 2/m$
$a = 4.738$ Å, $c = 3.188$ Å, $Z = 2$
$H = 6–7$
$G = 6.9–7.1$
Uniaxial (+)
$n_\omega = 1.990–2.010$
$n_\epsilon = 2.091–2.100$
$\delta = 0.10–0.09$

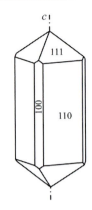

Composition Fe^{3+} is the most common element to substitute for Sn; also some Ti, Nb, and Ta. The variety called **wood tin**, which forms botryoidal or other rounded shapes with a radiating fibrous structure, often contains intergrown hematite and silica.

Form and Twinning Crystals are stubby tetragonal prisms with dipyramidal terminations, sometimes elongate prismatic or less commonly pyramidal. Also fine- to coarse-granular massive and in the radially fibrous, concretionary or botryoidal masses or crusts known as wood tin. Contact and penetration twins on {101} are common and may form simple, polysynthetic, or cyclic patterns.

Cleavage Imperfect {100}, and poor {110} and {111} cleavage; also {101} parting on the twin planes. Conchoidal to uneven fracture. Brittle.

Color Properties

Luster: Adamantine to splendent; also earthy.

Color: Yellowish or reddish brown, brownish black; also red, yellow, gray; rarely white.

Streak: Grayish or brownish white.

Color in Thin Section: Distinctly colored in shades of yellow, brown, red, or greenish; rarely colorless. Color zoning or irregular distribution of color is common. Detrital grains and fragments in grain mount may be nearly opaque.

Pleochroism: Weak to strong, $\omega < \epsilon$.

Optical Characteristics

Relief in Thin Section: Very high positive.

Optic Orientation: Crystals are length slow, but sign of elongation may be difficult to obtain owing to extreme birefringence.

Indices of Refraction and Birefringence: Birefringence is extreme; upper-order colors are usually masked by mineral color.

Interference Figure: Uniaxial positive, but some may be anomalously biaxial.

Polished Section: Distinct gray to brownish gray reflection pleochroism with bireflectance and abundant yellow to yellow-brown internal reflections. Distinct crossed polarizers anisotropism with gray polarization colors, often masked by internal reflections.

Alteration Stable in the weathering environment and not readily altered. Cassiterite may be produced by alteration of Sn-bearing sulfide minerals.

Distinguishing Features

Hand Sample: Resembles rutile but with higher specific gravity and more distinctly brown color. Wood tin may resemble goethite and other iron hydroxides but has lighter streak, is harder, and has higher specific gravity.

Thin Section/Grain Mount: Extreme relief and birefringence, and distinct color, are characteristic. Rutile is quite similar but has even higher birefringence and refractive indices.

Polished Section: Polishes poorly and has numerous internal reflections. Resembles rutile.

Occurrence Cassiterite is not particularly common but may be found in some high-Si granites and granitic pegmatites, and in high-temperature hydrothermal veins associated with wolframite and molybdenite. Skarns developed in carbonate rocks adjacent to granitic intrusions may contain cassiterite. The term *greisen* is applied to metasomatically altered rocks, usually along veins, that contain cassiterite, topaz, tourmaline, muscovite, lepidolite, and fluorite. Wood tin may form by low-temperature colloidal precipitation processes in rhyolitic volcanics, and forms in the near-surface oxidized zone of hydrothermal tin deposits. Because it is stable in the weathering environment, cassiterite may be found in the heavy mineral fraction of detrital sediment and locally may be concentrated to form economic placer deposits.

Use Cassiterite is the principal ore of tin, and most is mined from fluvial and beach placer deposits. One of the principal uses is to manufacture tin-plated steel from which food containers are made. These "tin cans" are economical, and both the tin and steel can be recycled. Plastic, aluminum, and tin-free steel containers have replaced tin-plated cans for many uses. Tin is alloyed with copper to form bronze, and with lead and antimony to form solder used in the electronics industry and plumbing. Tin was one of the earlier metals to be extracted and used.

URANINITE

UO_2
Isometric $4/m\ \overline{3}\ 2/m$
$a = 5.468$ Å, $Z = 4$
$H = 5–6$
$G = 10.95$ (pure), 6.5–10 (typical)
Opaque except on thin edges
Isotropic

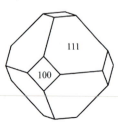

Composition Th may substitute for U to form a continuous solid solution to thorianite (ThO$_2$). Some U^{6+} is always present. Charge balance is maintained by substituting two U^{6+} for three U^{4+} and leaving sites vacant.

Form and Twinning Crystals are octahedrons with cube and dodecahedron modifications. Also massive granular, or as colloform, botryoidal, or reniform encrustations usually called **pitchblende**. Rare {111} twins.

Cleavage None; fracture conchoidal to uneven. Brittle.

Color Properties

Luster: Submetallic, pitchy, also earthy.
Color: Steely black or brownish black.
Streak: Brownish black or grayish.
Color in Thin Section: Opaque, dark greenish or yellowish brown on thin edges. Isotropic.

Optical Properties

Polished Section: Gray with brown tint. Isotropic, but some samples may be weakly anisotropic. Dark brown to reddish-brown internal reflections may be seen.

Alteration Uraninite alters easily by oxidation to U$_3$O$_8$ and a variety of other U-bearing oxide, carbonate, sulfate, and phosphate minerals, some of which are brightly colored.

Distinguishing Features

Hand Sample: Black color, pitchy luster, and radioactivity are distinctive. Radioactivity can be detected with a Geiger or scintillation counter.
Thin Section/Grain Mount: Because it is essentially opaque, it is not easily identified in conventional thin section.
Polished Section: Similar to magnetite but the colloform textures are usually different than textures found with magnetite.

Occurrence Uraninite is found in some medium- to high-temperature hydrothermal vein systems, granitic and syenitic pegmatites, and sediment-hosted uranium deposits. The uraninite in sediment-hosted uranium deposits commonly is concentrated at oxidation–reduction fronts. Uranium appears to be moderately soluble in oxygenated water but precipitates if oxygen is removed by oxidation of iron-bearing minerals or organic material. There is growing evidence that microorganisms are actively involved in forming these sediment-hosted deposits (e.g., Min and others, 2005). Because it is well documented that uraninite will precipitate by biologically induced processes on the cell walls of microbes, techniques that use microbes to clean up uranium-contaminated water from mine and mill sites are being investigated.

Use Uraninite is mined as a source of uranium, whose principal use is as a fuel for nuclear reactors, which now produce a significant amount of the world's electrical energy.

Comments The uranium and thorium in uraninite are both radioactive. Although casual handling of uraninite samples poses no significant risks, and normal storage cabinets provide shielding from most of the emitted radiation, good practice suggests that some caution is appropriate.

HYDROXIDES

The hydroxides as a group are very common minerals, with many species being produced by weathering and hydration of other minerals. In most cases these minerals occur as fine-grained aggregates, usually intermixed with other minerals, including the clays, that are produced by weathering or alteration.

The structures of two of these minerals, brucite [Mg(OH)$_2$] and gibbsite [Al(OH)$_3$], are significant because they are commonly used as a starting point for describing the layer silicates (Chapter 13). In both minerals the OH$^-$ anions are arranged in planes (Figure 18.9a). The cations (Mg or Al) occupy octahedral sites between planes of OH$^-$ to form electrically neutral sheets that are two OH$^-$ planes, or 4.77 Å, thick. Successive octahedral sheets are bonded to each other through weak electrostatic bonds, giving both gibbsite and brucite relatively low hardnesses.

In the brucite structure (Figure 18.9b), charge balance requires that all the octahedral sites in each sheet be filled with divalent Mg. Because three of three octahedral sites are filled, this structure is sometimes referred to as **trioctahedral**.

In the gibbsite structure (Figure 18.9c), Al^{3+} is the cation. To balance the anion charge, only two Al^{3+} are required, whereas three Mg^{2+} are required in brucite. In the gibbsite structure, therefore, only two-thirds of the octahedral sites are occupied by Al^{3+}; the remaining sites are vacant. Because only two of three octahedral sites are occupied, the gibbsite structure is referred to as **dioctahedral**. Successive octahedral sheets are stacked somewhat differently than in brucite, so gibbsite's unit cell is two octahedral sheets thick, or 9.72 Å.

BRUCITE

Mg(OH)$_2$
Hexagonal (trigonal) $\bar{3}\,2/m$
$a = 3.14$ Å, $c = 4.77$ Å, $Z = 1$
$H = 2.5$
$G = 2.39$
Uniaxial (+)
$n_\omega = 1.559–1.590$
$n_\epsilon = 1.580–1.600$
$\delta = 0.010–0.021$

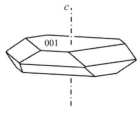

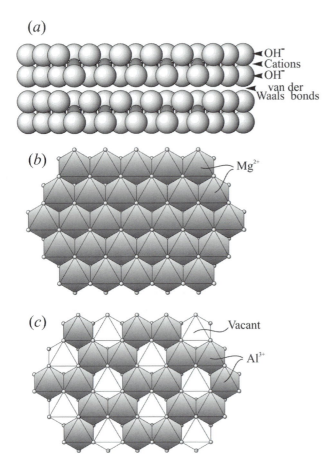

Figure 18.9 Brucite and gibbsite structure. (*a*) Octahedral sheets are composed of two layers of OH⁻ anionic groups with cations in octahedral coordination. (*b*) Plan view of brucite sheet. All octahedral sites (shaded) are occupied by Mg^{2+}. (*c*) Plan view of gibbsite sheet. Only two out of three octahedral sites are occupied by Al^{3+}; the remaining sites are vacant (no shading).

Composition Some Fe^{2+} and limited Mn^{2+} may substitute for Mg.

Form and Twinning Crystals are hexagonal mica-like plates tabular parallel to {001}. Usually as foliated or swirled masses and aggregates or granular masses. Fibrous brucite, known as **nemalite**, is uncommon. No twinning reported.

Cleavage Perfect on {001}.

Color Properties

Luster: Vitreous to pearly or waxy.
Color: White, gray, pale green, brown, or blue.
Streak: White.
Color in Thin Section: Colorless.

Optical Characteristics

Relief in Thin Section: Moderate positive.

Optic Orientation: The trace of cleavage in folia is length fast; fibers are elongate perpendicular to *c* and also are length fast.

Indices of Refraction and Birefringence: Indices of refraction increase with Fe content. First-order interference colors are usually anomalously reddish brown or bluish.

Interference Figure: Cleavage fragments yield centered optic axis figures, but small grain size often precludes obtaining usable figures. Some samples, particularly if fibrous, may be anomalously biaxial with small 2*V*.

Alteration Brucite may alter to hydromagnesite [3MgCO₃ · Mg(OH)₂ · 3H₂O] or to other Mg-bearing minerals including serpentine and periclase.

Distinguishing Features

Hand Sample: Brucite is harder than talc and gypsum and has a less greasy feel than talc; folia are not elastic like those of muscovite. Fibers are not as silky or fine as in chrysotile.

Thin Section/Grain Mount: Both talc and micas are biaxial negative and have higher birefringence; gypsum is biaxial positive. Chlorite and serpentine are usually pale green, length slow, and biaxial. Brucite may be stained to aid in identification (Haines, 1968).

Occurrence Most commonly found in marble, often as an alteration product after periclase. Also found in serpentinite and chlorite schist, usually as small veins, along with talc, magnesite, and other Mg-bearing minerals.

Use Brucite has been quarried for use in refractory materials.

Iron Hydroxide Minerals

The term **limonite** is generically used for mixtures of various iron oxide and hydroxide minerals. The most common of these minerals are goethite and lepidocrocite (Table 18.4). Hematite also may be present, along with admixed clay, silica, and other minerals. Fine grain size usually makes identification of the constituent minerals impractical without X-ray diffraction techniques.

Iron oxides and hydroxides are commonly produced by weathering and hydrothermal alteration. What appears to be a yellowish brown or reddish stain in many weathered rocks is very-fine-grained goethite, lepidocrocite, hematite, and other minerals produced by the breakdown of biotite, amphibole, pyroxene, magnetite, or other iron-bearing minerals. Large masses of limonite can be produced by alteration and oxidation of pyrite (FeS₂) in hydrothermal sulfide deposits. Limonite may form pseudomorphs, granular masses, or botryoidal, rounded, or encrusting masses with a radiating internal structure. Intense weathering of an Fe-bearing rock may result in the formation of a lateritic soil rich in iron hydroxide minerals.

Table 18.4 Iron Hydroxide Minerals

Mineral	Unit Cell	Physical Properties[a]	Optical Properties	Comments
Goethite α-FeO(OH)	Orthorhombic $2/m\ 2/m\ 2/m$ $a = 4.60$ $b = 9.96$ $c = 3.02$ $Z = 4$	H: 5–5.5 G: ~4.3 Cl: {010}pf, {100}f Tw: None reported C: Yellowish brown to red St: Yellowish brown L: Vitreous to earthy Antiferromagnetic, weak ferromagnetic	$n_\alpha = 2.15$–2.275 $n_\beta = 2.22$–2.409 $n_\gamma = 2.23$–2.415 $\delta = 0.138$–0.140 Biaxial (–) $2V = 0$–27° TS: Yellow, orange-red, brownish orange: $X = b$, $Y = c$, $Z = a$ PS: Gray with bluish tint; bireflectance/pleochroism weak in air; brownish yellow to reddish brown internal reflections; distinct anisotropy, gray-blue, gray-yellow, brownish	Fibrous or acicular parallel to c, also granular massive, and disseminated grains; formed by weathering or alteration of Fe-bearing minerals; also in sedimentary iron formations
Lepidocrocite γ-FeO(OH)	Orthorhombic $2/m\ 2/m\ 2/m$ $a = 3.08$ $b = 12.50$ $c = 3.87$ $Z = 4$	H: 5 G: 4.09 Cl: {010}pf, {100}, {001}f Tw: None reported C: Brown to red St: Orange L: Adamantine to metallic crystals, also earthy	$n_\alpha = 1.94$ $n_\beta = 2.20$ $n_\gamma = 2.51$ $\delta = 0.57$ Biaxial (–) $2V = 83°$ TS: Yellow, orange, red: $Z > Y > X$ PS: Grayish white with weak to distinct bireflectance/ pleochroism; strong gray anisotropy; reddish internal reflections	Platy parallel to {010} or acicular parallel to a; formed by weathering or alteration of Fe-bearing minerals; also in sedimentary iron formations

[a] Cl = cleavage (pf = perfect f = fair), Tw = twinning, C = color, St = streak, L = luster, TS = thin section, PS = polished section.

Both goethite and lepidocrocite are biominerals and have been identified in the radular teeth of chitins. Microbial activity is almost certainly involved in their common occurrence in the near-surface oxidized portions of sulfide mineral deposits.

In hand sample, limonite typically is medium to dark yellowish brown with a dull to earthy luster and has a yellowish brown streak. Hematite, in contrast, has a red streak. Specific gravity is variable (2.7–4.3), as is the hardness (1–5). Limonite usually displays dark shades of red, yellow, or brown in thin section; or it may be nearly opaque. Although goethite and lepidocrocite are both strongly birefringent, that birefringence is not usually displayed because of fine grain size. The index of refraction is high (2.0–2.4).

Aluminum Hydroxide Minerals

Gibbsite, böhmite, and diaspore (Table 18.5) are fairly common minerals in soils. They also may be found in emery deposits or as a weathering or alteration product of corundum, the aluminum silicates, or other Al-rich minerals.

Bauxite has been used as a mineralogical term to generically refer to mixed aluminum hydroxide minerals of uncertain identity, analogous to the terms *limonite* and *wad* for iron and manganese hydroxides, respectively. The term *bauxite* also has been used to refer to a rock type (Figure 2.51) whose major minerals include aluminum hydroxides. Although the latter definition is preferred, a considerable amount of the material conventionally called bauxite is poorly consolidated and more properly considered to be soil, not rock. Bauxite by either definition is typically produced in areas subjected to intense weathering, usually in tropical or semitropical climates, and usually contains clay (e.g., kaolin) and iron hydroxides.

Blanket bauxite deposits are developed by intense weathering of Al-bearing rock to form a lateritic soil. The original feldspar and other minerals initially break down to Al-rich clay, often kaolinite, from which the silica is removed in solution to leave highly insoluble gibbsite, diaspore, and/or böhmite.

Karst bauxite deposits are found in solution depressions produced in limestone or dolostone bedrock. The source of the aluminum that forms karst bauxite is a matter of debate. One explanation is that the aluminum is derived from silicate minerals carried into karst depressions by wind and water. An alternate source for the aluminum is detrital clay minerals originally dispersed in the carbonate rock and concentrated by removing the carbonate by dissolution. In

Table 18.5 Aluminum Hydroxide Minerals

Mineral	Unit Cell	Physical Properties[a]	Optical Properties	Comments
Gibbsite $Al(OH)_3$	Monoclinic $2/m$ $a = 8.64$ $b = 5.07$ $c = 9.72$ $\beta = 94.5°$ $Z = 8$	H: 2.5–3.5 G: 2.4 Cl: {001} Tw: {001}, {110}, {100}, and [130] C: White, gray, pale pink, green or brown St: White L: Vitreous to earthy	$n_a = 1.568–1.578$ $n_\beta = 1.568–1.579$ $n_\gamma = 1.587–1.590$ $\delta = 0.019–0.012$ Biaxial (+) $2V = 0–40°$ Colorless $X = b,$ $Y \wedge a = +25$ to $+35,$ $Z \wedge c = -21$ to -30	Typically occurs as soil constituent produced by weathering or feldspars or other Al-bearing minerals; major mineral in bauxite and also found in some low-temperature hydrothermal environments
Böhmite γ-AlO(OH)	Orthorhombic $2/m\ 2/m\ 2/m$ $a = 3.69$ $b = 12.23$ $c = 2.86$ $Z = 4$	H: 3.5–4 G: ~3.05 Cl: {010}vg, {100}g C: White St: White L: Vitreous, earthy	$n_\alpha = 1.640–1.648$ $n_\beta = 1.649–1.657$ $n_\gamma = 1.655–1.668$ $\delta = 0.006–0.020$ Biaxial (– or +) $2V_x > 80°$ Colorless $X = c, Y = b, Z = a$	Formed by weathering of aluminous minerals, especially in tropical climates; major constituent of bauxite; also low-temperature hydrothermal alteration of corundum, nepheline, and other aluminous minerals
Diaspore α-AlO(OH)	Orthorhombic $2/m\ 2/m\ 2/m$ $a = 4.40$ $b = 9.42$ $c = 2.85$ $Z = 4$	H: 6.5–7 G: 3.3–3.5 Cl: {010}g, {110} and {210}f, {100}pr C: White, gray, or various pale shades St: White L: Vitreous, earthy	$n_\alpha = 1.682–1.706$ $n_\beta = 1.705–1.725$ $n_\gamma = 1.730–1.752$ $\delta = 0.04–0.05$ Biaxial (+) $2V = 84–86°$ Colorless $X = c, Y = b, Z = a$	Formed by diagenesis in bauxite deposits and weathering of aluminous minerals; also hydrothermal alteration of aluminous minerals

[a] Cl = cleavage (vg = very good, g = good, f = fair, pr = poor), Tw = twinning, C = color, St = streak, L = luster.

either case the original minerals are intensely weathered and leave behind aluminum hydroxide minerals.

As a constituent of soils and other weathering residues, the aluminum hydroxide minerals will be indistinguishable from clay and other fine-grained minerals and therefore not identifiable in hand sample or thin section. Bauxite (the rock) may contain pisolitic or other concretionary structures and may be fairly well indurated. Except for rare coarse-grained occurrences, X-ray diffraction techniques are required to identify the aluminum hydroxide minerals.

Manganese Oxide and Hydroxide Minerals

Manganese oxide and hydroxide minerals (Table 18.6) are quite common but typically occur as small grains or fine-grained masses and may be intermixed with each other or with other minerals. Identification by means of conventional hand sample and optical techniques is usually impractical. These minerals are typically black, with a dark streak and a high specific gravity (~4–7). Large crystals are usually fairly hard (5–7); but for most samples, fine grain size usually makes measurement impractical. Earthy masses of these minerals may be friable and readily soil the fingers.

Wad is a generic term applied to materials that include substantial amounts of manganese oxide and hydroxide

minerals of uncertain identity. Wad has the same relation to manganese oxides and hydroxides that limonite has to iron oxides and hydroxides, and bauxite has to aluminum hydroxides.

The environments in which manganese oxide and hydroxide minerals occur include the following.

- **Hydrothermal systems**. Manganite and hausmannite may be primary minerals in some hydrothermal vein systems.
- **Weathering and alteration of Mn minerals**. Primary Mn carbonates, silicates, and oxides (e.g., rhodochrosite, kutnohorite, rhodonite, manganite, hausmannite) may weather or alter to yield pseudomorphs, fine-grained encrusting masses, and disseminated grains of Mn oxide and hydroxide minerals.
- **Weathering and alteration of other minerals**. Many Fe–Mg minerals (magnetite, biotite, olivine, pyroxene, etc.) contain small amounts of Mn substituting for Fe–Mg in their crystal structure. Weathering of these minerals may release the Mn to form Mn oxide and hydroxide minerals. The Mn minerals usually will be intermixed with other, more abundant, weathering products.
- **Bog manganese**. Most rocks contain small amounts of Mn that can be released into aqueous solution by weathering. This manganese may be precipitated in bogs and lake bottoms to form nodules and layers.

Table 18.6 Manganese Oxide and Hydroxide Minerals

Mineral	Composition	Occurrence
Birnessite	$Na_{0.5}Mn_2O_4 \cdot 1.5H_2O$	Desert varnish, marine deposits
Braunite	$MnMn_6SiO_{12}$	Hydrothermal, Mn-rich metamorphic rocks, bog deposits, weathering
Coronadite	$PbMn_8O_{16}$	Dendrites, weathering, alteration
Cryptomelane	KMn_8O_{16}	Dendrites, weathering, alteration, marine deposits
Hausmannite	$MnMn_2O_4$	Hydrothermal, metamorphic
Hollandite	$BaMn_8O_{16}$	Dendrites, weathering, alteration
Manganite	$MnO(OH)$	Hydrothermal, bog and marine deposits
Pyrolusite	MnO_2	Bog and marine deposits, alteration product
Romanechite	$(Ba,H_2O)_2Mn_5O_{10}$	Bog and marine deposits, alteration product, dendrites
Todorokite	$(Mn,Ca,Na,K)(Mn,Mg)_6O_{12} \cdot 3H_2O$	Dendrites, weathering, alteration, marine deposits

• **Ferromanganese nodules.** Some areas of the seafloor, and some freshwater lakes, are littered with nodules that contain substantial amounts of Mn and Fe oxide and hydroxide minerals. These minerals are inferred to have precipitated from seawater in a manner similar to bog manganese or from pore water in lake bed or seafloor sediment. Marine nodules are typically up to 5 or 10 cm across and may be laminated. Recovering the nodules from the seafloor has been proposed, but the process appears to be uneconomic. Most manganese is mined from similar marine sediments now exposed on land.

• **Dendrites** (Figure 18.10) are branching mineral growths that precipitate from groundwater onto fracture surfaces. They have often incorrectly been presumed to be pyrolusite. Most dendrites are composed of coronadite, cryptomelane, hollandite, romanechite, or todorokite.

• **Desert varnish** is a red, brown, or black coating that forms on the surface of exposed rocks in desert areas. Similar materials in nonarid climates are given the more generic term "rock varnish." Desert varnish appears to be composed mostly of air-transported clay particles that are deposited on and adhere to rock surfaces. Admixed iron and manganese oxide minerals provide the bright coloration. A common Mn mineral in desert varnish is birnessite; pyrolusite is unlikely.

The common issue in the formation of bog manganese, ferromanganese nodules, desert varnish, and dendrites is the requirement for some process to selectively extract and concentrate manganese from geologically ordinary environments in which the manganese is present in low concentrations. A number of Mn-oxidizing bacteria (e.g., *Metallogenium* spp.) appear to be able to do this quite handily. The details of the biochemistry remain sketchy, but it is clear that these bacteria are capable of converting Mn^{2+} to Mn^{3+} or Mn^{4+} as part of their metabolic processes and, in the process, cause the precipitation of manganese oxides and hydroxides from the water with which the bacteria is in contact. The surfaces of the bacteria or associated fungi may serve as

Figure 18.10 Dendrites on slab of sandstone.

convenient nucleation sites for growth of the manganese and iron minerals. Of course, not all these manganese deposits are necessarily biologically mediated, and not all workers agree that biologic processes are important processes in their formation.

Manganese minerals are mined to provide Mn, most of which is alloyed with iron to form high-strength steel. Manganese is also used in electrical batteries, paints, fertilizers, and animal feeds. Manganese oxides are used as pigments in the manufacture of bricks.

HALIDES

The anion in the halide minerals is one of the halogen elements, Cl, F, Br, and I. These anions are larger than oxygen and have a strong affinity for the electrons needed to complete their outer shell. They therefore tend to form strongly ionic compounds with fairly simple structures. Only three halides are at all common—halite (NaCl), sylvite (KCl), and fluorite (CaF_2). These are described here. A few additional halides are listed in Table 18.7.

Table 18.7 Less Common Halide Minerals

Mineral	Unit Cell	Physical Properties[a]	Optical Properties	Comments
Carnallite $KMgCl_3 \cdot 6H_2O$	Orthorhombic $2/m\ 2/m\ 2/m$ $a = 16.15$ $b = 22.51$ $c = 9.58$ $Z = 12$	H: 2.5 G: 1.60 Cl: None Tw:? C: Colorless, white, red with hematite inclusions St White L: Greasy	$n_\alpha = 1.466$ $n_\beta = 1.474$ $n_\gamma = 1.495$ $\delta = 0.029$ Biaxial (+) $2V = 70°$ Colorless $X = b, Y = a, Z = c$	Crystals pseudohexagonal dipyramidal; usually massive granular; found in marine evaporite deposits
Cryolite Na_3AlF_6	Monoclinic $2/m$ $a = 5.40$ $b = 5.60$ $c = 7.78$ $\beta = 90.2°$ $Z = 2$	H: 2.5 G: 2.97 Cl: None, {001} and {110} parting Tw: {001} and [100] and others C: Colorless, white St: White L: Vitreous to greasy	$n_\alpha = 1.338$ $n_\beta = 1.338$ $n_\gamma = 1.339$ $\delta = 0.001$ Biaxial (+) $2V = 43°$ Colorless $X = b, Y \wedge a = +44°,$ $Z \wedge c = -44°$	Uncommon cubelike crystals; usually massive, coarse granular; rare mineral from granitic pegmatites; important because it (or synthetic variety) is used in producing aluminum from bauxite
Chlorargyrite $AgCl$	Isometric $4/m\ \overline{3}\ 2/m$ $a = 5.549$ $Z = 4$	H: 2–3 G: 5.5 Cl: None, sectile C: Gray, also yellow, greenish brown, colorless; violet-brown on exposure to light L: Resinous	$n = 2.071$ Colorless	Rare cubic crystals; massive waxy, platy, crusts (horn silver); supergene mineral (see Chapter 19) in oxidized portion of Ag-bearing hydrothermal sulfide deposits

[a] Cl = cleavage, Tw = twinning, C = color, St = streak, L = luster.

HALITE

NaCl
Isometric $4/m\ \overline{3}\ 2/m$
$a = 5.639$ Å, $Z = 4$
$H = 2.5$
$G = 2.16$–2.17
Isotropic
$n = 1.544$

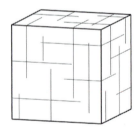

Structure The structure has been described in some detail in Chapters 3 and 4. The Na and Cl ions are arranged in alternate fashion in rows parallel to the edges of the face-centered cubic unit cell (Figure 18.11). This places the Na in octahedral coordination with Cl, whose distribution is effectively cubic close packing.

Composition Halite is usually fairly pure NaCl, with only minor substitution of K for Na. Inclusions of other minerals and saline water are common.

Form and Twinning Crystals are cubes. Rapid crystallization from oversaturated solutions may produce hopper-shaped crystals in which crystal edges and corners grew faster than the centers of crystal faces. Also massive granular. Twinning is not common with natural crystals, but synthetic crystals may twin on {111}.

Cleavage Perfect {001} cubic cleavage. Conchoidal fracture. Brittle.

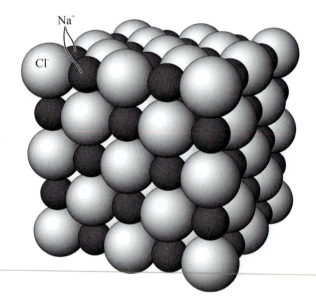

Figure 18.11 Halite structure. Each Na^+ coordinates with six Cl^-.

Color Properties

Luster: Vitreous. May be somewhat greasy on weathered surfaces.

Color: Colorless or white if pure, may be yellow, orange, red, blue, gray, brown, etc. due to lattice imperfections or inclusions. Some samples may fluoresce.

Streak: White.

Color in Thin Section: Colorless.

Optical Characteristics

Relief in Thin Section: Low positive.

Index of Refraction: Intimate intergrowths of halite and sylvite ($n = 1.490$) may have an intermediate refractive index.

Alteration Easily dissolved in surface or groundwater. Halite is generally not preserved in outcrop except in extremely dry climates.

Distinguishing Features

Hand Sample: Cubic cleavage and salty taste are distinctive. Sylvite has a somewhat bitter taste.

Thin Section/Grain Mount: Because it is very soluble, special preparation techniques that do not use water are required if halite is to be preserved in thin sections. The index of refraction for sylvite is less than the cements usually used in thin sections, and that of halite is higher.

Occurrence Halite is an abundant mineral in marine evaporite deposits (Box 17.2) and may form beds hundreds to over a thousand meters thick. Associated minerals include calcite, dolomite, gypsum, anhydrite, and sylvite, along with clay and other detrital material. Halite also may be found in deposits from saline lakes, associated with borates, sulfates, and carbonates.

Use Halite (table salt) has been used for dietary purposes since before recorded history. It improves the flavor of food and can serve as a preservative. However, this use accounts for only a small fraction of total salt production. The most important use is as a feedstock in the manufacture of a wide variety of industrial chemicals. Salt is also used in metal processing, tanning and leather treatment, pharmaceuticals, paper manufacturing, and to help keep roads and sidewalks free of ice in the winter, and in turn to rust cars.

Production of halite is generally by underground mining, evaporation of natural saline water, or solution mining. Underground mining uses conventional techniques similar in many ways to coal mining. The natural saline water from which salt is produced includes seawater, saline lakes, and natural groundwater brines. The saline water is usually pumped into a shallow pond where the water evaporates, leaving salt and whatever other minerals are in solution. Solution mining involves pumping water through a drill hole into a salt deposit. This water dissolves the salt to form a brine that is then pumped to the surface. The salt is extracted by evaporation of the water, or the brine can be piped directly to an industrial plant where it may be used in various chemical processes.

Halite has a long history of use for medical purposes, and some of those uses continue today. A typical example is gargling with saltwater as a treatment for sore throat. Halite is an important source of dietary sodium and chlorine, though an excess is implicated in hypertension (high blood pressure).

SYLVITE

KCl
Isometric $4/m\,\overline{3}\,2/m$
$a = 6.293$ Å, $Z = 4$
$H = 2.5$
$G = 1.99$
Isotropic
$n = 1.490$

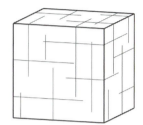

Structure Sylvite is isostructural with halite, with K in place of Na.

Composition Because K^+ and Na^+ have significantly different ionic radii, there is little solid solution between the minerals. Minor amounts of Br may substitute for Cl.

Form and Twinning Crystals are cubes, like halite. Also massive granular. {111} twins are observed in synthetic crystals.

Cleavage Perfect {001} cubic cleavage. Uneven fracture. Brittle. Somewhat sectile.

Color Properties

Luster: Vitreous.

Color: Colorless or white, also grayish, bluish, or yellowish. Iron oxide inclusions may produce an orange or reddish color.

Streak: White.

Color in Thin Section: Colorless.

Optical Characteristics

Relief in Thin Section: Moderate negative.

Index of Refraction: Uncommon submicroscopic mixtures of halite and sylvite may produce indices of refraction intermediate between the two minerals.

Alteration Easily dissolved in surface or groundwater, so sylvite is rarely preserved in outcrop.

Distinguishing Features

Hand Sample: Closely resembles halite but has a more bitter taste. When scratched with a knife blade, halite yields a white powder; but sylvite, because it is more sectile, yields very little. Granular massive sylvite can sometimes be seen to fill in around cubic crystals of halite, indicating that the halite crystallized first.

Thin Section/Grain Mount: Sylvite is very soluble in water, so it will not be preserved in conventionally made thin sections. Moderate negative relief (in specially-prepared thin sections) distinguishes sylvite from halite. The difference in index of refraction measured in grain mount is diagnostic.

Occurrence   Sylvite is found in marine evaporite deposits (Box 17.2) but is much less abundant than halite. Precipitation of sylvite from seawater requires substantially more evaporation than does halite. Associated minerals include halite, calcite, dolomite, gypsum, anhydrite, and detrital clay, oxide, and other minerals.

Use   A small amount of sylvite is used as a salt substitute for persons who need to restrict their dietary sodium. The majority of sylvite is mined for use in fertilizer to provide potassium, which is essential for plant growth. Potassium derived from sylvite also finds uses in a wide variety of industrial applications.

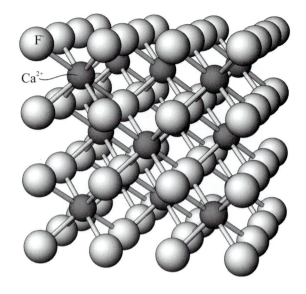

Figure 18.12 Fluorite structure. Each Ca^{2+} coordinates with eight F^-.

FLUORITE

CaF_2
Isometric $4/m\ \overline{3}\ 2/m$
$a = 5.463$ Å, $Z = 4$
$H = 4$
$G = 3.18$
Isotropic
$n = 1.433–1.435$

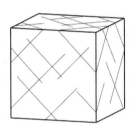

Structure   The structure of fluorite (Figure 18.12) consists of layers of F anions in a square arrangement stacked directly atop each other to form cubic (8-fold) coordination sites for the Ca. Alternate 8-fold sites are occupied, so that the Ca cations are located at the corners and center of the cubic lattice.

Composition   The composition is generally close to the ideal CaF_2. Minor substitution of Sr, Y, or Ce for Ca is possible.

Form and Twinning   Crystals are cubes, sometimes modified by {111} octahedron or other forms (Figure 18.13). Also cleavable and granular masses. Penetration twins by rotation on [111] are common.

Cleavage   Perfect {111} cleavage (four directions) yields nice cleavage octahedrons. Subconchoidal to uneven fracture. Brittle.

Color Properties

Luster: Vitreous.

Figure 18.13 Fluorite crystals. Field of view is about 15 cm wide.

Color: Commonly colorless, blue, purple, or green, but almost any color is possible. Usually fluoresces in both long- and short-wave UV light, and some varieties are phosphorescent.

Streak: White.

Color in Thin Section: Colorless or pale version of hand sample color. Not pleochroic.

Optical Characteristics

Relief in Thin Section: Moderately low negative.

Index of Refraction: Usually very close to 1.434, but substitution of Y for Ca increases the index to as high as 1.457.

Alteration Pseudomorphs of quartz, clay, Mn minerals, carbonate, and others after fluorite are fairly common.

Distinguishing Features

Hand Sample: Cubic crystals, octahedral cleavage, and hardness are distinctive. Anhedral material may resemble quartz, but is softer and displays cleavage.

Thin Section/Grain Mount: Isotropic character and index of refraction are characteristic.

Occurrence Fluorite is a common mineral in hydrothermal mineral deposits associated with sulfides (pyrite, galena, sphalerite, etc.), carbonates, and barite; it has been found in geodes. Fluorite also may be found as a minor mineral in granite, pegmatites, syenite, and greisen. It may occur as veins in carbonate sediments, as detrital grains, or (rarely) as a cement in sandstone. Fluorite is a biomineral, found as very small grains in the gizzard plates of a gastropod (Lowenstam and McConnell, 1968).

Use Fluorite is the principal source of fluorine used in chemical processes. Fluoride compounds are added to drinking water, toothpaste, and mouthwash to help prevent dental cavities. One of the most important uses of fluorite is as a flux in steelmaking and other metallurgical processes. Clear fluorite is used for optical purposes.

Chlorofluorocarbon compounds, whose uses include refrigerants in air conditioners and refrigerators, blowing agents in plastic foams, fire extinguishers, solvents and in other industrial applications, are implicated in loss of ozone in the stratosphere and are greenhouse gasses. The manufacture and use of these substances is being phased out under the terms of the Montreal Protocol, an international treaty that has been ratified by most nations.

Given that fluorite forms beautiful crystals, it is not surprising that it has a number of historical medical uses, including in traditional Chinese medicine, where it has been used for insomnia, anxiety, coughs and wheezing, and for excessive menstrual bleeding.

REFERENCES CITED AND SUGGESTIONS FOR ADDITIONAL READING

Allègre, C. A. 2008. *Isotope Geology.* Cambridge University Press, Cambridge, 512 p.

Bentley, W. A., and Humphreys, W. J. 1931. *Snow Crystals.* Dover Publications, New York, 286 p.

Burns, R. G., and Burns, V. M. 1977. Mineralogy, in G. P Glasby (ed.), *Marine Manganese Deposits*, pp. 185–248. Elsevier, Amsterdam.

Chang, L. L. Y., Howie, R. A., and Zussman, J. 1996. *Rock-Forming Minerals,* Volume 5B: *Nonsilicates: Sulfates, Carbonates, Phosphates, Halides*, 2nd ed. The Geological Society, London, 383 p.

Dobson, J. 2008. Remote control of cellular behavior with magnetic nanoparticles. Nature Nanotechnology 3, 139–143.

Goresy, A. E., Haggerty, S. E., Huebner, J. S., Lindsley, D. H., and Rumble, D. III. 1976. *Oxide Minerals. Short Course Notes*, Volume 3. Mineralogical Society of America, Washington, DC, 502 p.

Haines, M. 1968. Two staining tests for brucite in marble. Mineralogical Magazine 36, 886–888.

Klein, C. 2005. Some Precambrian banded-iron formations (BIFs) from around the world: Their age, geologic setting, mineralogy, metamorphism, geochemistry and origin. American Mineralogist 90, 1473–1499.

Kronhauser, K. 2007. *Introduction to Geomicrobiology.* Blackwell Publishing, Malden, MA, 425 p.

Lindsley, D. H. (ed.). 1991. *Oxide Minerals: Petrologic and Magnetic Significance.* Reviews in Mineralogy 25, 509 p.

Lowenstam, H. A., and McConnell, D. 1968. Biologic precipitation of fluorite. Science 162, 1496–1498.

Min, M., Xuc, H., Chen, J., and Fayek, M. 2005. Evidence of uranium biomineralization in sandstone-hosted roll-front uranium deposits, northwestern China. Ore Geology Reviews 26, 198–206.

Potter, R. M., and Rossman, G. R. 1979. Mineralogy of manganese dendrites and coatings. American Mineralogist 64, 1219–1226.

Trendall, A. F. 2002. The significance of iron-formation in the Precambrian stratigraphic record, in W. Altermann, and P. L. Corcoran, (eds.), *Precambrian Sedimentary Environments: A Modern Approach to Ancient Depositional Systems.* Special publication of the International Association of Sedimentologists 33, 33–66.

CHAPTER 19

Sulfides and Related Minerals

INTRODUCTION

The sulfide and related minerals form a large group that includes nearly 600 minerals. Although only a few are abundant, they have great economic value. Metals essential for industrial society extracted from sulfide minerals include copper, zinc, lead, antimony, molybdenum, cobalt, nickel, and silver.

Crystal Chemistry and Classification

A general formula for these minerals can be given as M_pX_r where M is a metal or semimetal such as Fe, Zn, Cu, Pb, Sb, or As. The X atoms may be S (sulfides), As (arsenides), S + As (sulfarsenides), or Te (tellurides) (Table 19.1).

Chemical bonding in the sulfides is more complex than in the oxides and silicates because sulfur is significantly less electronegative (2.5) than oxygen (3.5). As a consequence, sulfur–metal bonds typically have only about 10–20% ionic character (Table 3.4 and Figure 3.10). Transition metals (Fe, Zn, Cu, Pb, Co, etc.) bond with sulfur by forming molecular orbitals (see Chapter 3) that share electrons to satisfy the valance requirements of the sulfur. Additional metal–metal molecular orbitals may be formed to satisfy remaining valence requirements of the metal atoms. Whether the metal–sulfur and metal–metal bonds are covalent or metallic depends on the energy level of the occupied orbitals (the valence band) and the energy level of vacant orbitals (the conduction band) to which the electrons from the valence band can be promoted. In many sulfides the highest-energy electrons in the valence bands have energies that are the same as the vacant energy levels in the conduction bands. This allows for substantial metallic character in many minerals.

The sulfides do not lend themselves to an elegant classification, such as the one used successfully for the silicates. Schemes based on metal–sulfur ratios, coordination of metal atoms with sulfur, packing arrangement of the sulfur atoms, cell dimensions, and other variables have been proposed, but none is widely accepted. Classification is complicated because some semimetal elements, particularly As and Sb, can serve either as metals or as nonmetals.

The groupings shown in Table 19.1 are based solely on the identity of the nonmetal element(s) in the structure.

Minerals whose composition includes a metal, a semi-metal serving the structural role of a metal, and sulfur have been classified as **sulfosalts** based on chemical notions that have long since been abandoned. The metal is usually

Table 19.1 Common Sulfide and Related Minerals[a]

Sulfides $X = S$	
Acanthite	Ag_2S
Chalcocite	Cu_2S
Galena	PbS
Sphalerite	ZnS
Pyrrhotite	$Fe_{1-x}S$
Pentlandite	$(Fe,Ni)_9S_8$
Cinnabar	HgS
Covellite	CuS
Realgar	AsS
Bismuthinite	Bi_2S_3
Orpiment	As_2S_3
Stibnite	Sb_2S_3
Pyrite	FeS_2
Marcasite	FeS_2
Molybdenite	MoS_2
Chalcopyrite	$CuFeS_2$
Bornite	Cu_5FeS_4
Enargite	Cu_3AsS_4
Pyrargyrite	Ag_3SbS_3
Proustite	Ag_3AsS_3
Tetrahedrite	$(Cu,Fe)_{12}Sb_4S_{13}$
Tennantite	$(Cu,Fe)_{12}As_4S_{13}$
Sulfarsenides $X = AsS$	
Arsenopyrite	$FeAsS$
Cobaltite	$(Co,Fe)AsS$
Arsenides $X = As$	
Nickeline	$NiAs$
Skutterudite	$(Co,Ni)As_3$
Tellurides $X = Te$	
Calaverite	$AuTe_2$
Sylvanite	$(Au,Ag)Te_2$

[a] X = nonmetal element(s) in the general formula M_pX_r.

Cu, Ag, or Pb, and the semimetal may be As, Sb, or Bi. The minerals listed that are conventionally grouped with the sulfosalts are enargite, pyrargyrite, proustite, tetrahedrite, and tennantite.

Of the many hundreds of sulfide and related minerals that have been identified and described, only a handful are routinely encountered. These are described in detail in this chapter. Additional minerals are listed and briefly described later (Tables 19.4 and 19.5).

SULFIDE PARAGENESIS

Hydrothermal Deposits

Although they may be found in a wide variety of geological environments, sulfide and related minerals are characteristic of hydrothermal vein and replacement deposits. These minerals are deposited from hot aqueous solutions in void spaces or by replacement of the rock through which the fluids migrate. The essential features of hydrothermal systems include the following.

• Water: The water may be meteoric (rain, snow), magmatic (evolved from magma), metamorphic (released during metamorphism), or connate (trapped in sediment pores).

• Heat: Often provided by an igneous intrusion, also from burial.

• Source for metals and other elements precipitated from the water: The metals, sulfur, and other elements in the water may be derived from a crystallizing magma or may be leached out of a large volume of country rock by the water.

• Migration pathways: Fluids commonly flow through the rocks along faults, fractures, and the normal pore space.

• Precipitation site for the minerals: Minerals may precipitate either by filling spaces along faults, fractures, and in other voids, or by replacing minerals in the rock through which the fluids are flowing.

Hydrothermal deposits can be generated in many different geological environments (Cox and Singer, 1986). Figure 19.1 shows an idealized composite of hydrothermal systems. At the heart of the system is a felsic magma crystallizing at fairly shallow levels within the crust. The country rocks into which the intrusive is emplaced may be comagmatic volcanics or other older igneous, sedimentary, or metamorphic rocks. As the magma crystallizes, water and elements that do not easily fit in the crystal structure of quartz, feldspars, and other silicates become concentrated in the late residual fluid. These fluids can migrate into the country rock, where they follow areas of permeability such as fault and fracture zones. As these fluids migrate away from the crystallizing magma, they cool; in addition, the pressure may drop, or they may encounter chemically reactive rock such as limestone, any

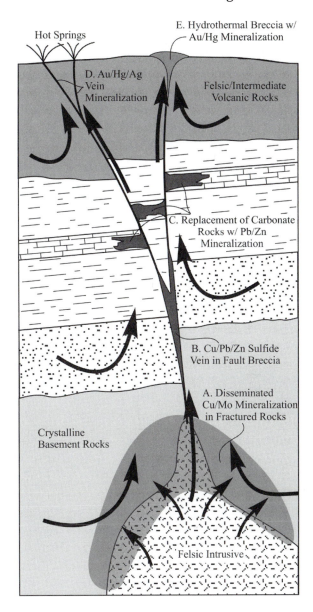

Figure 19.1 Schematic hydrothermal system. Hot mineralizing aqueous fluids may evolve from a crystallizing magma (bottom), or they may be meteoric waters set in motion by the heat from the magma (large arrows). Minerals may crystallize in small fractures and among grains of the country rock (A) to form a disseminated deposit. Vein deposits (B) are formed where mineralizing fluids flow along fault or fracture zones. A replacement deposit (C) may be produced if mineralizing fluids gain access to a limestone bed. Near-surface mineralization (D), often in volcanic rocks, may form where hydrothermal fluids issue at the surface as hot springs. If the temperature is high enough, hydrothermal fluids can violently flash to steam near the surface, producing a hydrothermal breccia in which mineralization may occur (E). Many other scenarios are possible.

of which may trigger precipitation of minerals from solution. Precipitated minerals may fill the void space through which the fluids are migrating, or they may react with and replace the country rock.

Groundwater adjacent to and above the magma chamber will be heated. Because hot water is less dense than cold, the hot water rises to the surface, and cool water from surrounding areas must flow in to replace the heated water. As long as the intrusive remains hot, it will continue to drive this hydrothermal circulation of water. The surface manifestation of a hydrothermal system may be hot springs and geysers. Yellowstone National Park provides a good example.

The hydrothermal circulation driven by the heat from the intrusive may itself lead to mineralization. The circulating water can scavenge elements from the rocks through which it flows and can dissolve mineralization previously produced from magmatic sources. After transport, this material may be reprecipitated. If society is lucky, the result will be to progressively concentrate valuable elements into a smaller volume of rock so that the mineralization can be economically mined.

Hydrothermal circulation also can produce extensive alteration of the rocks surrounding an intrusive and the intrusive itself. One common pattern is to find concentric zoning of the alteration, most intense near the intrusive center. The details of alteration mineralogy depend on local conditions. A common pattern of alteration consists of an outer zone of propylitic alteration surrounding concentric zones of argillic, sericitic, and then potassic alteration (Figure 19.2a).

• Propylitic: Biotite and hornblende converted to chlorite; plagioclase altered to epidote, chlorite, and carbonates.
• Argillic: Extensive development of clay minerals. Plagioclase and mafic minerals are generally completely altered, but some K-feldspar may persist.
• Sericitic (phyllic): Fine-grained white mica (sericite) and quartz produced by recrystallization of clay. Pyrite is a common accessory. None of the original silicates is preserved.
• Potassic: K-feldspar (adularia) crystallized with or without recrystallized biotite and sericite.

Alteration may extend up to several kilometers from an intrusive center. Similar zones of alteration may be duplicated at a scale of a few centimeters or meters adjacent to a mineralized vein (Figure 19.2b).

The chemistry of hydrothermal fluids responsible for precipitation of sulfide and other minerals can be deduced from small inclusions of this fluid that are trapped in minerals as they grow. These fluid inclusions are typically brines rich in NaCl. The presence of Cl$^-$ appears to be important because metals such as Cu, Pb, Zn, Au, and Ag are much more soluble as metal–chloride complexes than they are as simple ions.

The mineralogy found in hydrothermal vein and replacement deposits depends on the temperature and chemistry of the mineralizing fluids, the chemistry of the country rock, pressure, and details of the geological

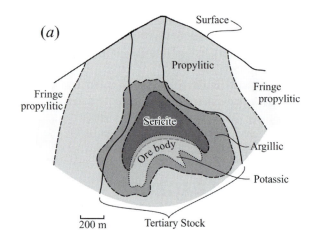

(a)

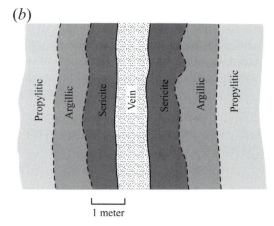

(b)

Figure 19.2 Hydrothermal alteration. (*a*) Henderson porphyry molybdenum deposit, Colorado. Mineralization is associated with Tertiary felsic intrusive rocks emplaced in Precambrian granite. The molybdenite (MoS_2) in the ore body (> 0.3% MoS_2) was deposited with quartz and pyrite in and along abundant fractures in the intrusive. Sericite, argillic, and propylitic alteration form concentric zones. Rocks within the ore body display potassic alteration. The biotite is not completely altered to chlorite in the fringe propylitic zone, which grades outward to unaltered Precambrian granite. Adapted from White and others (1981). (*b*) A similar concentric alteration pattern may be found at a much smaller scale adjacent to a sulfide-bearing vein.

setting. Sulfide minerals may actually constitute only a small fraction of the mineral volume precipitated; quartz, carbonates, fluorite, sulfates, and others may be the most abundant minerals in many hydrothermal veins. Table 19.2 gives the mineralogy of a number of different types of hydrothermal deposits.

Supergene Processes

The primary or **hypogene** mineralization produced in a hydrothermal system may be extensively altered when exposed to the near-surface environment (Figure 19.3). Oxygenated groundwater derived from rain or snow will react with pyrite to form iron hydroxides and release the sulfur as sulfuric acid:

Table 19.2 Selected Hydrothermal Mineral Assemblages[a]

Deposit Type	Major Minerals	Minor Minerals	Environments
As–Sb sulfide veins	Realgar, orpiment, pyrite, stibnite, quartz, calcite	Chalcopyrite, arsenopyrite, marcasite	Generally low-temperature hydrothermal veins
Hg–sulfide	Cinnabar, pyrite, marcasite	Native mercury, stibnite, sphalerite, pyrrhotite quartz, chalcedony, barite, carbonates, clay	Low-temperature hydrothermal
Epithermal gold–silver deposits	Gold, gold tellurides, pyrite, proustite, pyrargyrite, stibnite, acanthite, cinnabar, quartz	Galena, sphalerite, chalcopyrite, tetrahedrite–tennantite, silver, carbonates, fluorite, barite, plus others	Shallow hydrothermal systems associated with subaerial volcanism, either in the volcanic rocks or closely adjacent
Porphyry copper, porphyry molybdenum deposits	Pyrite, chalcopyrite, molybdenite, bornite	Magnetite, hematite, ilmenite, rutile, enargite, cassiterite, hübnerite, gold, covellite, chalcocite, digenite, native copper	Mineralization in veinlets and disseminated grains in and adjacent to felsic to intermediate porphyritic intrusive; intrusive and adjacent rock altered in concentric zones
Copper–lead–zinc–silver vein and replacement deposits	Pyrite, sphalerite, galena, chalcopyrite, tetrahedrite	Bornite, chalcocite, enargite, argentite, gold, hematite, pyrrhotite, proustite–pyrargyrite	Veins and replacement deposits associated with intermediate to felsic intrusive
Tin–tungsten–bismuth veins	Cassiterite, arsenopyrite, wolframite, bismuthinite, pyrite, marcasite, pyrrhotite	Stannite, chalcopyrite, sphalerite, tetrahedrite, pyrargyrite, galena, rutile, gold, molybdenite, bismuth, quartz, tourmaline, apatite, fluorite	Open-space hydrothermal vein filling associated with granitic intrusives
Lead–zinc deposits in carbonate host	Galena, sphalerite, barite, fluorite, pyrite, marcasite, chalcopyrite, calcite, dolomite, quartz, aragonite	Wurtzite, millerite, bornite, covellite, enargite	Deposited in solution voids in or by replacement of dolostone; generally low temperature (< 100°C) and not directly associated with intrusive or volcanic activity
Volcanogenic Cu–Fe–Zn	Pyrite, sphalerite, chalcopyrite, pyrrhotite, galena	Bornite, tetrahedrite, arsenopyrite, marcasite, sulfosalts, cassiterite, plus others	Deposited at or near the sea floor surface by hydrothermal exhalative activity associated with submarine volcanic activity; may be hosted by volcanic rocks or grade laterally into clastic sediments
Skarn deposits	Magnetite, molybdenite, sphalerite, galena, chalcopyrite, scheelite, wolframite, quartz, calc-silicate minerals (garnet, amphibole, pyroxene, epidote, etc.), carbonates	Pyrrhotite, cassiterite, hematite, gold, sulfosalts	Formed by high-temperature contact metamorphic processes where a silicic magma intrudes carbonate rocks

[a]See Table 19.3 for minerals formed by supergene alteration.

Pyrite + oxygen + water = goethite + sulfuric acid

$$4FeS_2 + 15O_2 + 10H_2O = 4FeOOH + 8H_2SO_4 \qquad (19.1)$$

A similar reaction of chalcopyrite ($CuFeS_2$) with dissolved atmospheric oxygen and carbon dioxide can produce cuprite (CuO) and siderite ($FeCO_3$):

$$CuFeS_2 + CO_2 + 4O_2 + 2H_2O =$$
$$CuO + FeCO_3 + 2H_2SO_4 \qquad (19.2)$$

The copper carbonates azurite and malachite can be produced by similar reactions. The net result of these and related reactions is to oxidize the metals found in the sulfides to form oxides, hydroxides, carbonates, and sulfates. Because these reactions produce oxygen-bearing minerals at the expense of sulfides, the near-surface zone is often referred to as the oxidized zone. The

minerals produced are called **supergene** or secondary minerals. Common supergene minerals found in the oxidized portion of hydrothermal sulfide deposits are listed in Table 19.3.

Because pyrite is usually the most common sulfide, the oxidized zones of most sulfide deposits are colored yellow and red from the iron oxides and hydroxides derived from it. These minerals are relatively insoluble and may accumulate near the surface to form an encrusting mass called a **gossan**.

The sulfuric acid produced as a byproduct of oxidation reactions dissociates to $2H^+$ and SO_4^{2-} and promotes dissolution of sulfide minerals such as chalcocite (Cu_2S) by reactions such as

$$2Cu_2S + 4H^+ + 2SO_4^{2-} + 5O_2 =$$
$$4Cu^{2+} + 4SO_4^{2-} + 2H_2O \qquad (19.3)$$

Table 19.3 Common Minerals in the Oxidized Portion of Hydrothermal Sulfide Deposits

Oxides/Hydroxides	Carbonates
Mn oxides/hydroxides (Table 18.6)	Calcite
Goethite	Dolomite/ankerite
Lepidocrocite	Siderite
Hematite	Rhodochrosite
Cuprite	Azurite
Native elements	Malachite
Copper	Cerussite
Sulfur	Smithsonite
Gold	Sulfates
Silver	Gypsum
Bismuth	Anglesite

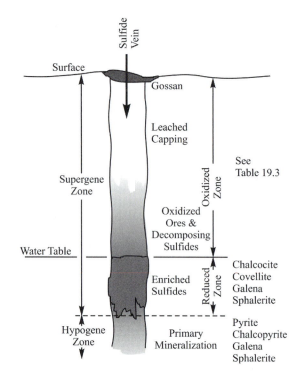

Figure 19.3 Supergene enrichment. Meteoric water percolates down through a sulfide-bearing vein. Sulfides are leached out of the near-surface environment, leaving behind a collection of oxide, carbonate, and hydroxide minerals in the oxidized zone. Cu, Pb, Zn, and related elements are largely removed from the leached capping. A gossan, enriched in iron oxides and hydroxides typically derived from alteration of pyrite, may accumulate at the surface as weathering and erosion progress. Sulfide minerals (re)precipitate in the reduced zone, generally below the water table, to form a zone enriched in sulfide minerals.

When meteoric water carrying dissolved metals and other constituents reaches the water table, it becomes diluted. This commonly triggers precipitation of the metals as sulfides by reversing reactions such as 19.3. Precipitation also

may be triggered by reaction of the sulfate solution with sulfide minerals.

$$5FeS_2 + 12Cu^{2+} + 12SO_4^{2-} + 16H_2O =$$
$$\text{pyrite}$$
$$6Cu_2S + 5Fe^{2+} + 32H^+ + 16SO_4^{2-} \quad (19.4)$$
$$\text{chalcocite}$$

Pyrite grains in the deeper portions of the supergene alteration zone may therefore serve as host to extensive reprecipitation of copper sulfides in a process called **supergene enrichment**. The zone of concentrated mineralization is commonly at or below the surface of the water table. The sulfide-enriched zone is also referred to as the reduced zone, because the meteoric water, by the time it reaches this level, is exhausted of oxygen and is chemically reducing.

Supergene processes cause extensive alteration of the country rock hosting sulfide deposits. Excess acid produced by oxidation of sulfides may be neutralized by reaction with silicates to form clays. A typical reaction involves hydrolysis of K-feldspar ($KAlSi_3O_8$) to form kaolinite [$Al_2Si_2O_5(OH)_4$].

$$2KAlSi_3O_8 + 2H^+ + H_2O =$$
$$Al_2Si_2O_5(OH)_4 + 4SiO_2 + 2K^+ \quad (19.5)$$

Note that this is the same type of reaction as in normal weathering. The acid released by oxidizing sulfides, however, promotes more intense alteration of silicates to greater depths than are found with weathering.

Microorganisms are directly and indirectly involved in many supergene processes. For example, the bacterium *Thiobacillus ferrooxidans* promotes the oxidation of pyrite (Equation 19.1) because it oxidizes reduced sulfur compounds and ferrous iron to produce sulfuric acid and ferric iron as part of its metabolism (Nordstrom and Southam, 1997). The pyrite (FeS_2) serves as a source of metabolic energy. These biologic processes are not restricted to pyrite—other sulfide minerals also may be involved in microbe metabolism. While the biochemistry may seem exotic, it is not fundamentally different from utilizing reduced carbon and hydrogen in carbohydrates (e.g., donuts) as a source of energy and producing H_2O and CO_2 as by-products that combine to form H_2CO_3, which is carbonic acid.

The same processes can occur in mines and mine waste, where biologic processes accelerate the production of acid (the H_2SO_4 in Equation 19.1) that can pollute waterways. Because microbes can promote dissolution of sulfide minerals, they are increasingly being used in processes that extract metals from ores and make them available for modern industry (Rawlings, 2002). Microbes also have become valuable in processes that remove metal contamination from surface waters.

Many sulfides are biominerals (Chapter 5). Of those in Table 19.4, acanthite, orpiment, proustite, and realgar are

Table 19.4 Less Common Sulfide Minerals

Mineral[a]	Unit Cell	Physical Properties[b]	Optical Properties	Comments
Acanthite Ag_2S	Monoclinic $2/m$ $a = 4.231$ $b = 6.930$ $c = 9.526$ $\beta = 125.5°$ $Z = 4$	H: 2–2.5 G: 7.22 Cl: Indistinct, sectile Tw: $\{\bar{1}11\}$ polysynthetic, $\{101\}$ contact C: Iron black St: Black L: Metallic	Opaque PS: Light gray ± greenish cast, very weak bireflectance/pleochroism; observable but weak anisotropy	Crystals rare, prismatic along c; often cube/octahedron pseudomorph after argentìte; also arborescent, filiform, massive, or coating, and as inclusions in galena; found in hydrothermal systems with ruby silvers, silver, and galena
Bismuthinite Bi_2S_3	Orthorhombic $2/m$ $2/m$ $2/m$ $a = 11.12$ $b = 11.25$ $c = 3.97$ $Z = 4$	H: 2–2.5 G: 6.78 Cl: $\{010\}$pf, $\{100\}$ and $\{110\}$f C: Lead gray to tin white St: Lead gray L: Metallic	Opaque PS: White with weak but distinct bireflectance/pleochroism; very strong anisotropy, creamy white/gray white polarization colors	Stout prismatic to acicular crystals elongate on c; usually massive with foliated or fibrous texture; low- or high-temperature hydrothermal veins, volcanic exhalative deposits, gold veins
Orpiment As_2S_3	Monoclinic $2/m$ $a = 11.475$ $b = 9.577$ $c = 4.256$ $\beta = 90.7°$ $Z = 4$	H: 1.5–2 G: 3.49 Cl: $\{010\}$pf, $\{100\}$f, flexible lamellae, somewhat sectile Tw: $\{100\}$ C: Lemon yellow to golden or brownish yellow St: Pale lemon yellow L: Resinous, pearly on cleavage	$n_\alpha = 2.4$ $n_\beta = 2.8$ $n_\gamma = 3.0$ $\delta = 0.6$ Biaxial (–) $2V = 76°$, $r > v$ strong $X = b$, $Z \wedge c = \pm 2°$ TS: Y = pale yellow, Z = greenish yellow PS: White to gray-white with strong pleochroism; strong anisotropism, masked by yellow to white internal reflection	Crystals tabular to short prismatic; usually foliated or columnar masses, granular or powdery; hydrothermal deposits with lead, silver, gold; almost always associated with realgar
Pentlandite $(Fe,Ni)_9S_8$	Isometric $4/m$ $\bar{3}$ $2/m$ $a = 9.928$ $Z = 4$	H: 3.5–4 G: 4.6–5.0 Cl: None, $\{111\}$ parting C: Light bronze-yellow St: Light bronze-brown L: Metallic	Opaque PS: Isotropic, creamy white with pinkish or brownish tint	Crystals rare, usually granular massive; found in mafic and ultramafic intrusive igneous rocks as segregations; usually associated with pyrrhotite, also chalcopyrite and pyrite; mined as a principal source of Ni
Realgar AsS	Monoclinic $2/m$ $a = 9.325$ $b = 13.571$ $c = 6.587$ $\beta = 106.4°$ $Z = 16$	H: 1.5–2 G: 3.56 Cl: $\{010\}$g; $\{\bar{1}01\}$, $\{110\}$, $\{120\}$, $\{110\}$pr Tw: $\{100\}$ contact C: Red to orange-yellow St: Orange-red to red L: Resinous to greasy	$n_\alpha = 2.538$ $n_\beta = 2.684$ $n_\gamma = 2.704$ $\delta = 0.116$ Biaxial (+) $2V = 39°$, $r > v$ strong $X \wedge c = +11°$, $Y = b$ TS: $X \sim$ colorless, $Y = Z$ pale yellow PS: Dull gray with reddish to bluish pleochroism; strong anisotropism masked by abundant yellowish-red internal reflections	Crystals short, vertically striated prisms; usually coarse to fine granular or encrustation; in hydrothermal deposits with lead, silver, gold; usually associated with orpiment; breaks down to orpiment on exposure to light
Stibnite Sb_2S_3	Orthorhombic $2/m$ $2/m$ $2/m$ $a = 11.229$ $b = 11.310$ $c = 3.839$ $Z = 4$	H: 2 G: 4.63 Cl: $\{010\}$pf; $\{100\}$, $\{110\}$f Tw: $\{130\}$, $\{120\}$ polysynthetic C: Lead gray, black, or iridescent tarnish St: Lead gray L: Metallic	Opaque PS: White to grayish white with strong bireflectance/ pleochroism; very strong anisotropy	Crystals (Figure 19.4) stout to elongate prism with complex terminations, may be bent; radiating acicular groups; also columnar, granular, or fine masses; important source of antimony
Enargite (ss) Cu_3AsS_4	Orthorhombic $mm2$ $a = 7.407$ $b = 6.436$ $c = 6.154$ $Z = 2$	H: 3 G: 4.45 Cl: $\{110\}$pf, $\{100\}$ and $\{010\}$g Tw: $\{320\}$ common, penetration C: Grayish black to iron black St: Grayish black L: Metallic	Opaque PS: Pinkish gray or pinkish brown with bireflectance/ pleochroism; strong anisotropism, blue, green, red, orange polarization colors	Crystals tabular on $\{001\}$ or prismatic with striations parallel to c; usually anhedral to subhedral grains and granular masses; found in Cu-bearing hydrothermal deposits; locally an important ore of copper

(Continued)

Table 19.4 (*Continued*)

Mineral[a]	Unit Cell	Physical Properties[b]	Optical Properties	Comments
Proustite (ss) Ag_3AsS_3	Hexagonal $\bar{3}\,2/m$ $a = 10.79$ $c = 8.69$ $Z = 6$	H: 2–2.5 G: 5.57 Cl: {101}g Tw: {104} produces trillings, also {101}, {001}, and {012} C: Scarlet vermilion (darkens with exposure) St: Vermilion L: Adamantine	$n_w = 3.088$ $n_\epsilon = 2.792$ $\delta = 0.296$ Uniaxial (–) TS: Cochineal red to blood red pleochroic PS: Bireflectant/pleochroic white with yellowish tint to bluish gray; strong anisotropy with colors masked by intense scarlet red internal reflections (stronger than pyrargyrite)	Known as light ruby silver; crystals rhombohedral; commonly massive, compact; late-forming mineral in hydrothermal deposits also in the oxidized/enriched zone; is an important ore mineral of silver in some deposits; difficult to distinguish from pyrargyrite
Pyrargyrite (ss) Ag_3SbS_3	Hexagonal $3m$ $a = 11.047$ $c = 8.719$ $Z = 6$	H: 2.5 G: 5.82 Cl:{101}g, {012}pr Tw: {104}complex lamellar, also {101} and [110] C: Deep red St: Purplish red L: Adamantine	$n_w = 3.084$ $n_\epsilon = 2.881$ $\delta = 0.203$ Uniaxial (–) TS: Deep red PS: Bluish gray, with bireflectance but not pleochroism; strong anisotropism, color masked by intense carmine red internal reflections (less than proustite)	Known as dark ruby silver; rare crystals prismatic parallel to c with pyramidal terminations; usually compact massive; late-forming mineral in low-temperature hydrothermal veins, also by supergene processes; more common than proustite and an important ore mineral of silver in some deposits
Tennantite (ss) $(Cu,Fe)_{12}As_4S_{13}$	Isometric $\bar{4}3m$ $a = 10.19$ $Z = 2$	H: 3–4.5 G: 4.62 Cl: None Tw: [111] axis with {111} twin plane, contact and penetration C: Flint gray to iron black St: Black to brown L: Metallic, splendent	Opaque unless very thin PS: Gray with bluish-green tint, isotropic; reddish internal reflections common	Crystals are tetrahedra; also massive, coarse to fine granular, or compact; solid solution to tetrahedrite; hydrothermal veins; tennantite is an important copper mineral in some deposits; also may contain valuable amounts of silver substituting for Cu,Fe
Tetrahedrite (ss) $Cu,Fe)_{12}Sb_4S_{13}$	Isometric $\bar{4}3m$ $a = 10.23$–10.55 (higher with Ag) $Z = 2$	H: 3–4.5 G: 4.97 Cl: None Tw: [111] axis with {111} twin plane, contact and penetration C: Flint gray to iron black St Reddish brown, black, or dark red L: Metallic	Opaque unless very thin PS: Gray with olive or brownish tint, isotropic; reddish internal reflections uncommon, increasing with As content	Crystals are tetrahedra; also massive, coarse to fine granular or compact; hydrothermal veins; solid solution to tennantite; Ag substitutes for Cu,Fe; important source of Cu and Ag in some deposits

[a] ss = minerals conventionally classified as sulfosalts.
[b] Cl = cleavage (pf = perfect, g = good, f = fair, pr = poor), Tw = twinning, C = color, St = streak, L = luster, TS = thin section, PS = polished section.

documented to be biologically induced. Realgar, for example, can be precipitated by the bacterium *Pyrobaculum arsenaticum* at temperatures between 68 and 100°C (Huber and others, 2000). It is likely that the occurrence of both orpiment and realgar in solfatera fields may be the result of microbial action.

SULFIDE MINERALS

Only the more common sulfide minerals are described in detail. Additional sulfides are included in Table 19.4, and stibnite is shown in Figure 19.4.

SPHALERITE

ZnS
Isometric $\bar{4}3m$
$a = 5.406$ Å, $Z = 4$
$H = 3.5$–4
$G = 3.9$–4.1
Isotropic
$n = 2.37$–2.50

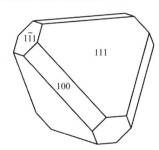

Structure Sulfur is in a slightly open version of cubic close packing with Zn in tetrahedral sites at the corners and

Figure 19.4 Small elongate prismatic crystals of stibnite.

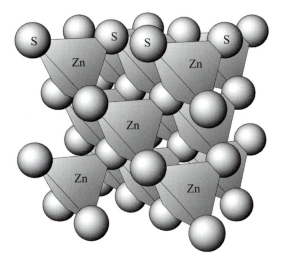

Figure 19.5 Structure of sphalerite. Sulfur atoms are arranged in cubic close packing. Zn atoms are positioned in tetrahedral coordination sites that are arranged in rows parallel to the diagonals across each unit cell face. This places Zn at the corners and face centers of the unit cell. Chalcopyrite is similar, but with Cu and Fe occupying alternating tetrahedral sites.

face centers of the unit cell (Figure 19.5). **Wurtzite** is a polymorph based on hexagonal close packing.

Composition Over 40% Fe^{2+} may substitute for Zn; higher temperatures are associated with higher Fe content. Small amounts of Cd and Mn also are common. Sphalerite is also called **zincblende**. **Marmatite** is a term found in the older literature for Fe-rich sphalerite; **black jack** is a miner's term for the same material.

Form and Twinning Tetrahedral crystals are common, often with triangular markings on crystal faces. Other forms include the cube and dodecahedron. Also granular and cleavable masses or forming encrusting aggregates. Twins by rotation on [111] with composition planes ~(111); may be simple or complex.

Cleavage Perfect dodecahedral {011} cleavage (six cleavage directions). Conchoidal fracture. Brittle.

Color Properties

Luster: Resinous, adamantine, to submetallic.

Color: Pale yellow, light brown, brown, black, also red, green, or white. Color darker with Fe content.

Streak: White to light or medium brown, darker with Fe content.

Color in Thin Section: Colorless, pale yellow, pale brown. Not pleochroic.

Optical Characteristics

Relief in Thin Section: Very high positive.

Index of Refraction: Index of refraction increases with Fe content. Intergrown submicroscopic domains of wurtzite may produce birefringence. Birefringence increases linearly

from 0 for 100% sphalerite structure to 0.022 for 100% wurtzite structure. Wurtzite is uniaxial positive $n_\omega = 2.356$, $n_\epsilon = 2.378$, and is similar in appearance to sphalerite.

Polished Section: Gray, sometimes with brown tint, in polished section with yellow, yellowish-brown, or reddish-brown internal reflections.

Alteration Sphalerite may alter to oxides, hydroxides, sulfates, or carbonates of Fe and Zn (limonite, smithsonite, siderite, etc.).

Distinguishing Features

Hand Sample: Distinguished by luster, cleavage, and light streak. Dark varieties may resemble galena (particularly when examined underground with a miner's lamp for illumination) but yields a lighter-colored powder when crushed. It dissolves slowly in dilute HCl, particularly if powdered, and releases H_2S, whose rotten-egg odor is distinctive. Sphalerite is pyroelectric, and some exhibits triboluminescence (i.e., glows when struck).

Thin Section/Grain Mount: Extreme relief, isotropic character, and six cleavage directions are characteristic.

Polished Section: In polished section sphalerite is gray with low reflectance and may display internal reflection.

Occurrence Sphalerite is a common constituent of hydrothermal sulfide deposits. Associated minerals include galena, pyrite, chalcopyrite, pyrrhotite, and other sulfides, along with quartz, carbonates, and sulfates. Sphalerite is found infrequently as an accessory mineral in felsic igneous rocks and in coal beds. Sphalerite may be precipitated by sulfate-reducing bacteria from aqueous solutions that contain less than 1 ppm Zn (Labrenz and others, 2000), suggesting both that bacteria can be used to remove Zn from contaminated surface water and that some low-temperature Sphalerite deposits could have been biologically produced.

Use Sphalerite is the most important ore mineral from which zinc is extracted. Zinc is an important industrial metal used to make castings and to plate steel to resist rust. Cast zinc is often referred to as "pot metal." Zinc is alloyed with copper to make brass. Zinc is also an important dietary mineral and is added to fertilizers. Zinc oxide is used in paint, and zinc chloride is used as a soldering flux and as a wood preservative. Sphalerite is used in homeopathic remedies.

Sphalerite is also the most important source of cadmium, which is used in batteries and other products, and indium, gallium, and germanium, which are used in the manufacture of semiconductors.

GALENA

PbS
Isometric $4/m\ \overline{3}2/m$
$a = 5.936$, $Z = 4$
$H = 2.5$
$G = 7.58$
Opaque

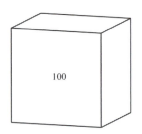

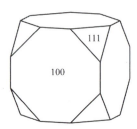

Structure Galena is isostructural with halite. Sulfur is at the corners and face centers of the unit cell, alternating with Pb. The Pb is in octahedral coordination with the S, and vice versa.

Composition Galena is usually fairly pure PbS; some Ag may substitute for Pb, but most of the Ag, As, and Sb reported in analyses is located in small inclusions of acanthite or tetrahedrite.

Form and Twinning Well-formed cubes are common, sometimes modified by octahedron {111} faces (Figure 19.6). Also cleavable masses and coarse to fine aggregates. Twins on {111} may be either contact or penetration, {114} twins are lamellar and may give rise to diagonal striations on cleavage surfaces.

Cleavage Perfect {001} cubic cleavage similar to halite. Subconchoidal fracture. Brittle.

Figure 19.6 Galena cubes encrusted with marcasite.

Color Properties

Luster: Metallic.

Color: Lead gray.

Streak: Lead gray. Galena will barely mark paper.

Optical Characteristics

Thin Section: Opaque.

Polished Section: White, sometimes with pink tint and isotropic.

Alteration Supergene oxidation processes may alter galena to cerussite ($PbCO_3$) or anglesite ($PbSO_4$).

Distinguishing Features

Hand Sample: Lead-gray color and streak, cubic cleavage, metallic luster, and high specific gravity are diagnostic. Dark-colored sphalerite has a light streak.

Thin Section/Grain Mount: Difficult to distinguish from other opaque minerals.

Polished Section: Recognized by white color, moderately high reflectance, and presence of triangular polishing pits.

Occurrence Galena is a very common mineral in hydrothermal sulfide deposits and frequently is associated with sphalerite, pyrite, chalcopyrite, quartz, calcite, fluorite, and/or barite. Galena may be precipitated diagenetically in organic-rich marine sediments both by biologically induced mineralization triggered by sulfate-reducing bacteria and by thermally driven chemistry (Machel, 2001).

Use Galena is the principal ore mineral from which lead is extracted and also serves as a significant source of silver. Lead has a wide range of industrial uses. One familiar to most readers is in lead–acid batteries used in the electrical systems of automobiles, airplanes, garden tractors, and elsewhere where a low-cost, durable, and rechargeable power supply is needed. It is also alloyed with tin and antimony to make electrical solder. Lead oxide was used in the past in paints. These paints now pose a health hazard in some older structures. Lead oxide has been used in some areas of the world as a folk remedy for diarrhea. Although it is effective, it also produces potentially debilitating lead poisoning. Galena is used in the preparation of homeopathic remedies for skin and lung ailments. Since the preparations are extremely dilute, the potential for lead poisoning is probably minimal, and any benefit would seem to depend on the placebo effect. In an interesting twist of irony, galena is touted as a detoxification agent to gullible believers in crystal healing.

PYRRHOTITE

$Fe_{1-x}S$ (x = 0 to 0.13)
Monoclinic $2/m$
   (pseudohexagonal)
a = 6.87 Å,
   b = 11.90 Å,
   c = 22.72 Å,
   β = 90.1°, Z = 64
H = 3.5–4.5
G = 4.58–4.65
Opaque

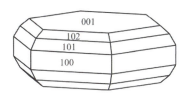

Structure The structure is based on hexagonal close-packed sulfur with Fe occupying octahedral sites between the sulfur layers.

Composition The principal variation is in the iron content with substitution of $2Fe^{3+} + \square$ for 3 Fe^{2+} ($\square$ represents a vacant site). Chapter 4 describes this type of solid solution in more detail. Up to about 20% of the sites that would hold Fe in the ideal structure may be vacant. Ni, Co, Mn, and Cu may substitute for Fe or may be present in small inclusions of pentlandite $[(Fe,Ni)_9S_8]$ or chalcopyrite. **Troilite** (FeS) is a hexagonal polymorph without significant substitution of Fe^{3+} for Fe^{2+}.

Form and Twinning Crystals are usually tabular parallel to {001}, also pyramidal or arranged in rosettes. Usually massive granular. Twinning on {102} may be simple or repeated. (Indexing based on pseudohexagonal symmetry.)

Cleavage None. {001} parting distinct, {102} parting less distinct. Uneven to subconchoidal fracture. Brittle.

Color Properties

Luster: Metallic.

Color: Bronze-yellow with reddish or brownish cast; tarnishes easily, sometimes with iridescence.

Streak: Dark grayish black.

Optical Characteristics

Thin Section: Opaque.

Polished Section: Distinct creamy pinkish-brown to reddish-brown reflectance pleochroism, very strong anisotropism with yellow-gray or grayish-blue polarization colors.

Alteration Pyrrhotite may be replaced by pyrite, marcasite, or other sulfides. Alteration to iron oxides, hydroxides, sulfates, or carbonates is common in the weathering and supergene environments.

Distinguishing Features

Hand Sample: Weakly ferrimagnetic (will attract a magnet), stronger with lower Fe content. Resembles pyrite but has a more bronze color and is softer. Pentlandite, with which it may be associated, is slightly paler, has octahedral parting, is not magnetic, and has a light bronze-brown streak.

Thin Section/Grain Mount: Opaque in thin section.

Polished Section: Light brown color and strong anisotropism. Pentlandite is isotropic.

Occurrence Pyrrhotite is commonly found in both high-temperature hydrothermal sulfide deposits and mafic and ultramafic igneous rocks. Associated sulfides frequently include pyrite, marcasite, chalcopyrite, and pentlandite. Masses found in mafic and ultramafic rock bodies are often interpreted to be formed by immiscible segregation of a sulfide melt from the silicate melt. Pyrrhotite also has been found in pegmatites and contact metamorphic deposits, and troilite is found in meteorites and lunar rocks. Sulfate-reducing bacteria are capable of precipitating pyrrhotite and other iron sulfides.

Use Pyrrhotite has been mined as a source of iron but is more commonly mined from deposits in mafic intrusive rocks for nickel, copper, and platinum found in associated minerals.

CHALCOPYRITE

$CuFeS_2$
Tetragonal $\bar{4}2m$
a = 5.281 Å,
   c = 10.401 Å, Z = 4
H = 3.5–4
G = 4.1–4.3
Opaque

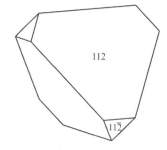

Structure The chalcopyrite structure is equivalent to two sphalerite unit cells stacked atop each other parallel to c with Cu and Fe replacing the Zn (Figure 19.5) in an alternating fashion. Cu or Fe may occupy additional tetrahedral sites, yielding stoichiometries such as $Cu_9Fe_8S_{16}$ or $Cu_8Fe_{10}S_{16}$. Metallic bonding among the metals evidently allows valence requirements to be satisfied.

Composition Despite similarities in structure, only limited solid solution extends between chalcopyrite and sphalerite. Minor amounts of other elements (Co, Ni, Mn, Sn, As, etc.) may be present, but often are part of small inclusions of other sulfide minerals.

Form and Twinning Crystals are commonly dominated by the {112} disphenoid, which looks like a tetrahedron. The negative disphenoid {$\bar{1}$12} may produce an octahedral habit and a tetragonal prism also may be developed. Lamellar twins on {112}, {102}, and {110}, deformation twins on {110} and {012}, penetration twin by rotation on [001] with {110} composition plane.

Cleavage Poor on {011} and {111}. Uneven fracture. Brittle.

Color Properties

Luster: Metallic.

Color: Brass yellow, may be tarnished and irridescent.

Streak: Greenish black.

Optical Characteristics

Thin Section: Opaque.

Polished Section: Yellow with weak reflection pleochroism; high-temperature samples may be strongly pleochroic. Crossed polarizer anisotropy is weak but distinct, gray-blue to yellow-green.

Alteration Chalcopyrite may alter to other copper sulfides (chalcocite, covellite) or to oxides, sulfates, or carbonates of Fe and Cu (cuprite, malachite, azurite, siderite, limonite, etc.) in the supergene environment. Chrysocolla and turquoise are also alteration products.

Distinguishing Features

Hand Sample: Distinguished from pyrite by lower hardness, different crystal habit, and richer yellow color. Pyrrhotite is less distinctly yellow and is weakly magnetic. Small grains may resemble native gold but are not sectile and yield a greenish-black streak.

Thin Section/Grain Mount: Difficult to distinguish from other opaque minerals.

Polished Section: More distinctly yellow than pyrite and anisotropic.

Occurrence Chalcopyrite is the most common copper-bearing mineral and is found in many hydrothermal sulfide deposits, often associated with galena, sphalerite, pyrite, and other sulfide minerals. Some mafic igneous rocks may contain chalcopyrite as an accessory mineral; or they may contain segregated masses of sulfides that commonly include pyrrhotite, pentlandite, and chalcopyrite. Chalcopyrite may occur in some sediments without obvious connection to igneous activity. Chalcopyrite is a biomineral (Chapter 5), and it is possible to use bacteria to promote extraction of copper from it in metallurgical processes.

Use Chalcopyrite is the principal source for copper. Major deposits are found in many areas of the world. Some vein deposits are mined, but the majority of copper production comes from disseminated "porphyry copper" deposits. In these deposits, chalcopyrite and other hydrothermal minerals are deposited in a large volume of rock surrounding a porphyritic felsic intrusive. Open-pit mining on a grand scale is generally used to recover the copper. Industrial uses of copper are numerous. Perhaps the most important is to manufacture electrical wires, without which most of our social, commercial, and industrial infrastructure would be impossible. Chalcopyrite is another sulfide that appears in homeopathic remedies that depend on the placebo effect and whose great dilution probably precludes adverse reactions.

CINNABAR

HgS
Hexagonal 32
$a = 4.145$ Å, $c = 9.496$ Å, $Z = 3$
$H = 2$–2.5
$G = 8.18$
Uniaxial (+)
$n_\omega = 2.91$
$n_\epsilon = 3.26$
$\delta = 0.35$

Structure The structure is based on systematic stacking of sheets of sulfur with Hg in octahedral sites. Metacinnabar and hypercinnabar are isometric and hexagonal polymorphs, stable at higher temperatures.

Composition Essentially pure HgS, but some samples may have a deficiency of Hg.

Form and Twinning Crystals are rhombohedral, thick tabular on {001}, or short prismatic parallel to *c*. Common as fine coatings, crystalline encrustations, granular aggregates, or massive.

Cleavage {100} perfect. Subconchoidal to uneven fracture. Slightly sectile.

Color Properties

Luster: Adamantine, tending to metallic if dark colored.

Color: Cochineal red, inclined to brownish red if impure, earthy if fine grained.

Streak: Scarlet.

Thin Section: Deep red.

Optical Characteristics

Relief in Thin Section: Very high positive.

Indices of Refraction and Birefringence: Birefringence is extreme, but interference colors will be masked by mineral color.

Polished Section: White with bluish-gray tint, red internal reflections, and distinctly anisotropic with light green polarization colors.

Alteration Cinnabar may alter in zone of weathering to yield native mercury, **montroydite** (HgO) (deep red, yellow-brown streak, sectile) or **calomel** (HgCl) (yellow- or gray-white, yellowish-white streak, sectile, fluoresces brick red in UV light).

Distinguishing Features

Hand Sample: Bright red color and crimson streak distinguish cinnabar from hematite. Realgar has an orange-red streak and is almost always associated with yellow orpiment. If heated, cinnabar will release mercury, which is toxic.

Thin Section/Grain Mount: Deep red color and association.

Polished Section: Similar to proustite, but proustite is not twinned and has no green polarization colors.

Occurrence Found in low-temperature hydrothermal deposits associated with volcanism or shallow intrusives in association with pyrite, marcasite, stibnite, chalcedony, opal, calcite, and dolomite. Some deposits are interpreted to be formed quite near the surface beneath hot springs. Cinnabar may impregnate fractures or other voids, or replace quartz or other grains in the host rock. The world's principal source of mercury is Almadén, Spain, where cinnabar and native mercury are distributed in vugs and pores in weakly metamorphosed quartzite and may replace some of the quartz. This deposit has been mined since at least the fourth century BC and still contains a large resource of easily mined mercury.

Use Cinnabar is the principal source for mercury. Mercury is used in fluorescent light fixtures, batteries, electrical switches, thermometers, and various chemical applications.

Cinnabar has numerous historical medical applications including those in Chinese, Tibetan, and Ayurvedic traditional medicine. Liquid mercury and mercury chloride (calomel) have had even more uses. The outbreak of syphilis in Europe in the late fifteenth century triggered a search for cures, and mercury in various forms became one of the standard treatments and so it remained until the introduction of antibiotics in the last century. Most medical uses of mercury are now banned in Western countries because of serious problems with mercury poisoning.

Comments Mercury is volatile, and vapors are quite poisonous. One geochemical technique used in certain mineral exploration programs is to mount analytical equipment in an airplane. The airplane is flown in a systematic pattern over an area of suspected mineralization to sample the air for traces of mercury that may indicate the presence of a hydrothermal deposit.

PYRITE

FeS$_2$
Isometric $2/m\,\overline{3}$
$a = 5.418$ Å, $Z = 4$
$H = 6$–6.5
$G = 5.02$
Opaque

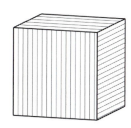

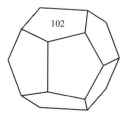

Iron Cross Twin

Structure The structure of pyrite (Figure 19.7a) can be derived from halite by replacing the Na with Fe. The sulfur forms covalently bonded S$_2$ pairs that occupy the Cl positions in the halite structure. Each Fe coordinates with six S atoms and each S atom bonds with three Fe. Bonding is apparently accomplished by hybridizing the $3p$ and $3s$ orbitals on sulfur to form four sp^3 orbitals. One of these hybrid orbitals is shared with another sulfur to form the S$_2$ doublet; the remaining three are shared with three iron atoms. The $3d$, $4s$, and $4p$ orbitals on each iron are hybridized to form six d^2sp^3 orbitals that are covalently shared with sulfur. Energy levels of vacant d orbitals (the conduction band) on the iron overlap with the energy levels of the occupied hybridized orbitals (the valence band), so pyrite exhibits metallic characteristics. Because the covalently bonded S$_2$ pair is not spherically symmetrical, the symmetry of pyrite is lower than halite.

Composition Most pyrite is close to the ideal composition. Ni or Co may substitute for Fe in small amounts. Microscopic inclusions of gold or chalcopyrite may be present.

(a)

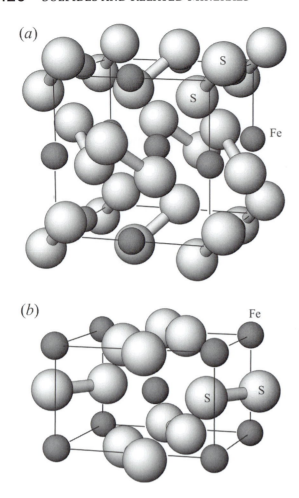

(b)

Figure 19.7 Structure of iron sulfides. (a) The pyrite structure is derived from halite, with Fe in the Na positions and covalently-bonded S_2 doublets occupying the Cl positions. (b) Marcasite structure. Fe occupies the nodes of a body-centered orthorhombic lattice with covalently bonded S_2 pairs in the center of four unit cell edges parallel to b and in the center of the {010} ends of the unit cell.

Form and Twinning Cubes {001} and pyritohedrons {102} are characteristic crystals (Figure 19.8). Cube faces are commonly striated, consistent with 3-fold rotational symmetry on the long diagonals through the cube. Octahedral faces are less common. Commonly massive, fine granular, or encrusting. Sometimes elongate to acicular across the width of thin veins. Twins by rotation on [001], with {110} composition planes, form penetration twins known as **Iron Crosses**. Small (1–30 μm) aggregates of pyrite called **framboids** are common in shale.

Cleavage An indistinct cleavage on {001} is not usually observed. Parting on {011} and {111}. Conchoidal to uneven fracture. Brittle.

Color Properties

Luster: Metallic.

Color: Pale brass yellow, tarnishes darker with iridescence.

Streak: Greenish black to brownish black.

Figure 19.8 Pyrite. (a) Pyritohedron crystals. Field of view is about 5 cm wide. (b) Pyrite cubes with quartz. Field of view is about 2 cm wide. (c) Spherical framboid made of numerous equant pyrite grains in clay from Canadian River, Oklahoma, alluvium. The flaky particles are clay minerals. Photo courtesy of G. Breit, U.S. Geological Survey.

Optical Characteristics

Thin Section: Opaque.

Polished Section: Creamy white or whitish yellow without bireflectance or pleochroism. May be weakly anisotropic with blue-green to red-orange polarization color.

Alteration Pyrite commonly altered to iron oxides and hydroxides. Pseudomorphs of limonite after pyrite are common. The oxidation of pyrite plays an important role in supergene alteration of hydrothermal sulfide deposits, described earlier.

Distinguishing Features

Hand Sample: Known as "fool's gold" because pyrite routinely is misidentified as gold by the uninitiated (everyone should read Mark Twain's *Roughing It*). Hardness, brittle character, and greenish-black streak are distinctive. Chalcopyrite has a richer yellow color and is softer, and its crystals typically are tetrahedral. Pyrrhotite is softer and brownish bronze. Marcasite has a different crystal habit and somewhat paler color but is otherwise quite similar.

Thin Section/Grain Mount: Difficult to distinguish from other opaque minerals.

Polished Section: Distinguished from marcasite by lack of reflection pleochroism.

Occurrence Pyrite is the most common sulfide mineral, and few hydrothermal deposits are without it. Igneous rocks of almost any composition may contain pyrite as an accessory mineral, and it may form magmatic segregations in some mafic intrusive complexes. Pyrite may be found as an opaque mineral in some metamorphic rocks. Pyrite is common as fine grains in shale, generally precipitated from seawater under reducing conditions. Coal also usually contains some fine pyrite.

Pyrite, one of the best-documented biominerals (Chapter 5), is produced by both biologically induced and biologically controlled processes involving sulfate-reducing bacteria and bacteria that reduce Fe^{3+} to Fe^{2+}. Details of the chemistry remain unclear, but it appears in some cases that the process begins with local precipitation of an amorphous FeS phase, which crystallizes to **mackinawite** (FeS), which subsequently combines with additional sulfur or loss of iron to form **greigite** (Fe_3S_4) and then pyrite. The fact that pyrite may not precipitate directly is a result of unfavorable kinetics compared to the other phases. Whether framboids (Figure 19.8a) are a product of biomineralization is a matter of debate (Pósfi and Dunin-Borkowski, 2006; Konhauser, 2007).

Use Pyrite has been used as a source of iron and sulfur, and is a feedstock in the manufacture of Fe-bearing chemicals. In most cases pyrite is not of economic importance itself, but is often associated with other minerals that have great value, such as chalcopyrite, galena, and sphalerite. Gold is extracted from minute inclusions in pyrite from some deposits. Pyrite has numerous historical medical uses. For example, Agricola (1546, p. 13), in what may be considered the first textbook on mineralogy, includes a section on medical uses of minerals in which he reports that pyrite will "break up gatherings in the body." Other uses

included treatment of bleeding, gastrointestinal and pulmonary problems, and broken bones.

Comments Pyrite is a mineral with significant environmental impact. Oxygenated surface water that interacts with pyrite oxidizes the iron and releases sulfuric acid (see earlier discussion of supergene processes). Although this process goes on naturally, mining can promote groundwater flow into pyrite-bearing rock and also can provide an easy route to the surface for acidic groundwater. The water issuing from the portals of old mines can be very acidic and can carry substantial amounts of heavy metals, such as lead, in solution. A related problem is the generation of acid when surface water soaks pyrite-bearing mine waste. If discharged into streams, this acidic water has the potential to cause substantial changes to the ecosystem and may potentially pose a health risk to people.

MARCASITE

FeS_2
Orthorhombic $2/m\ 2/m\ 2/m$
$a = 4.436$ Å,
   $b = 5.414$ Å,
   $c = 3.381$ Å, $Z = 2$
$H = 6–6.5$
$G = 4.89$
Opaque

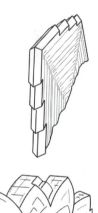

Structure The marcasite structure is based on a body-centered orthorhombic lattice. Fe is located at each lattice node and covalently bonded S_2 pairs are positioned at the center of each horizontal edge of the unit cell and in the middle of the end {010} faces (Figure 19.7b). This places Fe in octahedral (6-fold) coordination with sulfur, the same as in the pyrite polymorph.

Composition Generally close to the ideal formula. Minor Cu may substitute for Fe, and As for S. Arsenopyrite shares the same structure, but the two minerals exhibit almost no solid solution.

Form and Twinning Crystals are commonly tabular on {010}, less commonly prismatic parallel to *c*; curved faces are common (Figure 19.9). Twins on {101} are common, less common on {011}. Combination of {101} twins and curved faces often produces "cockscomb" or spear-shaped groups. Often forms encrusting, globular, or reniform

Figure 19.9 Cockscomb crystals of marcasite.

masses with radiating internal structure and crystals projecting from the surface.

Cleavage Cleavage on {101} is fairly distinct, it also may be a parting. Uneven fracture. Brittle.

Color Properties

Luster: Metallic.

Color: Pale bronze yellow to almost white, darkening on exposure.

Streak: Grayish black.

Optical Characteristics

Thin Section: Opaque.

Polished Section: Creamy white to yellowish white with greenish and pink hues with reflection pleochroism; strong anisotropy in crossed polarizers, with blue, green-yellow, or purple/violet-gray polarization colors.

Alteration Marcasite is commonly altered to iron hydroxides and sulfates. Limonite pseudomorphs after marcasite are common.

Distinguishing Features

Hand Sample: Resembles pyrite but has a different crystal habit and is somewhat paler colored if fresh.

Thin Section/Grain Mount: Difficult to distinguish from other opaque minerals.

Polished Section: Pyrite is slightly darker colored and does not display reflection pleochroism with rotation.

Occurrence Marcasite is a common mineral in hydrothermal sulfide deposits, generally formed at fairly low temperatures; it may be produced by supergene processes. Associated minerals commonly include galena and sphalerite. Concretionary masses in shale, limestone, and coal are common. Pseudomorphic replacement of fossils by

marcasite may be quite striking. Marcasite may be precipitated by biomineralization processes.

Use Marcasite can be used as a source of sulfur and iron, but is not normally of economic value.

MOLYBDENITE

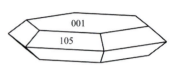

MoS_2
Hexagonal $6/m\ 2/m\ 2/m$
$a = 3.160$ Å,
    $c = 12.295$ Å, $Z = 2$
$H = 1–1.5$
$G = 4.62–4.73$
Opaque

Structure The structure is layered. Each layer consists of two sheets of sulfur atoms with Mo cations in 6-fold coordination sites in between. Successive layers are weakly bonded to each other, producing excellent cleavage on {001}.

Composition Molybdenite is nearly pure MoS_2.

Form and Twinning Crystals are hexagonal plates or stubby hexagonal prisms. Also scaly and foliated massive. Twins on {001} are reported.

Cleavage Perfect on {001}; folia are inelastic. Greasy feel.

Color Properties

Luster: Metallic.

Color: Lead gray. In extremely thin flakes, molybdenite may be translucent pale yellow to deep reddish brown, depending on thickness.

Streak: Bluish gray.

Optical Characteristics

Thin Section: Opaque.

Polished Section: White with bluish tint, with extreme white to gray bireflectance; very strong anisotropy in crossed polarizers, with pinkish white polarization color.

Alteration Molybdenite commonly alters to yellow **ferrimolybdite** [$Fe_2(MoO_4)_3 \cdot 8H_2O$].

Distinguishing Features

Hand Sample: Most closely resembles graphite but color has a more bluish tone, and the streak on porcelain is somewhat greenish, whereas graphite is black.

Thin Section/Grain Mount: Opaque in thin section.

Polished Section: Habit and white color with strong bireflectance are distinctive. Graphite is similar but less reflective.

Occurrence Found in hydrothermal vein and disseminated deposits. In the largest deposits, such as the Climax and Henderson deposits in Colorado, molybdenite, quartz, pyrite, and other minerals have precipitated in large inverted bowl-shaped volumes of fractured rock over the top of mid-Tertiary felsic intrusives. These deposits are commonly referred to as porphyry "moly" deposits. Molybdenite is found as a minor mineral in porphyry copper deposits that have similar geometries but contain copper minerals (e.g., chalcopyrite) as the principal ores. Molybdenite is also found in pegmatites and in skarn deposits.

Use Molybdenite is the principal source for molybdenum, most of which is alloyed with iron to make high-strength steel. It is also used in certain lubricants and other chemical products. Molybdenum is an essential nutrient in trace amounts, so small amounts are included in fertilizers, animal feeds, and human dietary supplements. High concentrations of molybdenum are toxic.

BORNITE

Cu_5FeS_4
Orthorhombic $2/m$ $2/m$ $2/m$
  (pseudoisometric)
$a = 10.950$ Å,
  $b = 21.862$ Å,
  $c = 10.950$ Å, $Z = 8$
$H = 3$
$G = 5.06$–5.08
Opaque

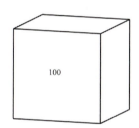

Structure At temperatures above 228°C, the structure is isometric with a unit cell that is about 5.50 Å on an edge. This structure is based on cubic close-packed sulfur atoms, with Cu and Fe atoms randomly distributed into six of the eight tetrahedral sites located in the octants of the cube. With cooling, the Fe and Cu become ordered, so that 5.5 Å subcells in which all eight tetrahedral sites are filled alternate with subcells in which only four of the tetrahedral sites are filled; symmetry is reduced to orthorhombic.

Composition Substantial variation in the relative amounts of Cu and Fe is possible and solid solution extends toward chalcopyrite ($CuFeS_2$) and digenite (Cu_9S_5). Exsolution of blebs and lamellae of chalcopyrite, digenite, and chalcocite is common.

Form and Twinning Rare crystals are approximately cubic, dodecahedral, or octahedral. Usually massive. Penetration twinning on {111}.

Cleavage Poor {111} cleavage. Uneven to subconchoidal fracture. Brittle.

Color Properties

Luster: Metallic if fresh, iridescent tarnish.

Color: Brownish bronze on fresh surfaces, quickly tarnishes to iridescent purple, blue, and black (hence the term **"peacock ore"**).

Streak: Grayish black.

Optical Characteristics

Thin Section: Opaque.

Polished Section: Pinkish brown with weak bireflectance and reflection pleochroism; weak but variable anisotropy in crossed polarizers. Tarnishes easily to purple-brown.

Alteration Bornite may alter to chalcocite, chalcopyrite, covellite, cuprite, chrysocolla, malachite, or azurite.

Distinguishing Features

Hand Sample: Brownish-bronze color of fresh surfaces distinguishes bornite from chalcocite and chalcopyrite.

Thin Section/Grain Mount: Difficult to distinguish from other opaque minerals in thin section.

Polished Section: Pinkish-brown color with weak anisotropy is characteristic.

Occurrence Bornite is very common in copper-bearing hydrothermal sulfide deposits, usually as a hypogene (primary) mineral associated with chalcopyrite, chalcocite, covellite, pyrrhotite, and pyite; less commonly as a supergene mineral in the enriched zone near the water table. It may be found as disseminated grains in mafic igneous rocks and pegmatites.

Use Mined as an ore of copper but is not as important as chalcopyrite.

CHALCOCITE

Cu_2S
Monoclinic $2/m$ or m
  (pseudoorthorhombic)
$a = 11.82$ Å,
  $b = 27.05$ Å,
  $c = 13.42$ Å,
  $\beta = 90°$, $Z = 96$
$H = 2.5$–3
$G = 5.5$–5.8
Opaque

Structure The structure is based on hexagonal close-packed sulfur atoms with copper in triangular coordination with S.

Above 103°C, the structure inverts to a hexagonal polymorph. A tetragonal polymorph is stable at high pressures.

Composition The composition is generally close to the ideal formula with only minor substitution of Fe for Cu.

Form and Twinning Rare crystals range from tabular parallel to (001) to short prismatic parallel to c. Most common as compact to powdery massive. Common twins on {110} yield pseudohexagonal stellate forms.

Cleavage Indistinct on {110}. Conchoidal fracture. Somewhat sectile, brittle if crushed.

Color Properties

Luster: Metallic.
Color: Blackish lead gray.
Streak: Blackish lead gray.

Optical Characteristics

Thin Section: Opaque.
Polished Section: White to grayish white with or without bluish tint with very weak or undiscernible reflectance pleochroism; weak to distinct crossed-polarizer anisotropy with emerald green to pink polarization colors under intense illumination. Tarnishes easily.

Alteration Chalcocite may be altered to other copper minerals such as covellite, malachite, or azurite.

Distinguishing Features

Hand Sample: Black color with sooty appearance on weathered surfaces and sectile character are distinctive. Bornite is brownish bronze on fresh surfaces.
Thin Section/Grain Mount: Difficult to distinguish from other opaque minerals.
Polished Section: White or bluish-white color and nearly isotropic character are characteristic.

Occurrence Typically found in the supergene-enriched zone of copper-bearing hydrothermal sulfide deposits. Chalcocite, covellite, digenite, djurleite, anilite, and other Cu-bearing sulfides are often intimately intermixed.

Use Chalcocite is mined as a source of copper.

Comments A number of closely related minerals with very similar properties include **digenite** (Cu_9S_5), **djurleite** ($Cu_{31}S_{16}$), and **anilite** (Cu_7S_4). A substantial amount of material identified as chalcocite may actually be djurleite. Chalcocite and djurleite are very commonly intergrown.

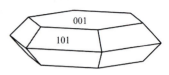

CuS
Hexagonal 6/m 2/m 2/m
$a = 3.794$ Å,
   $c = 16.341$ Å, $Z = 6$
$H = 1.5–2$
$G = 4.6–4.76$
Opaque

Structure The structure is based on layers consisting of three sheets of sulfur atoms. Some copper is located in triangular coordination within the middle sheet. Additional copper is positioned in tetrahedral sites between both the first and second sheets of sulfur, and between the second and third sheets of sulfur. Successive 3-fold layers are weakly bonded to each other, producing an excellent cleavage.

Composition Most covellite is relatively pure CuS.

Form and Twinning Crystals are tabular on {001} with a hexagonal outline and hexagonal markings on the basal pinacoid. Commonly as cleavable or foliated masses, also as rosettes of nearly parallel plates. Twinning is not reported.

Cleavage Perfect on {001} yields thin flexible folia.

Color Properties

Luster: Submetallic to resinous.
Color: Indigo blue, sometimes darker; may be iridescent.
Streak: Shiny lead gray.

Optical Characteristics

Thin Section: Opaque.
Polished Section: Strong reflection pleochroism from indigo blue with violet tint to bluish white. Extreme anisotropy, red-orange to brownish polarization colors.

Alteration Covellite may be altered to other copper sulfides (chalcocite, digenite, djurleite, bornite, etc.), carbonates (azurite, malachite), oxides (cuprite), and related minerals by hydrothermal and supergene processes.

Distinguishing Features

Hand Sample: Rich blue color and perfect cleavage are distinctive.
Thin Section/Grain Mount: Opaque in thin section.
Polished Section: The distinctive reflection pleochroism in polished section is characteristic.

Occurrence Occurs with other copper sulfides in hydrothermal sulfide deposits. Covellite is most commonly found in zones of supergene enrichment, but also may be a primary mineral. Associated minerals include chalcocite, bornite, chalcopyrite, pyrite, and other copper sulfides.

Use Covellite is mined as an ore of copper.

SULFARSENIDES

The sulfarsenides are minerals in which both sulfur and arsenic serve anion-like roles. Arsenopyrite is the only common sulfarsenide and is described below. Cobaltite is included in Table 19.5 and has a structure similar to pyrite, with Co replacing Fe and As replacing half of the S atoms.

Table 19.5 Sulfarsenides, Arsenides, and Tellurides

Mineral	Unit Cell	Physical Properties[a]	Optical Properties	Comments
Sulfarsenides				
Cobaltite CoAsS	Orthorhombic (pseudocubic) $mm2$ $a = 5.582$ $b = 5.582$ $c = 5.582$ $Z = 4$	H: 5.5 G: 6.33 Cl: {001}cubic, pf Tw: {011}, {111} of cubic habit lamellar C: Silver white St: Grayish black L: Metallic	Opaque PS: White with pink or violet tint, weak bireflectance/pleochroism; weak anisotropy, blue-gray or brown polarization colors	Same structure as pyrite; crystals cubes or pyritohedrons, uncommon; also granular and compact; hydrothermal vein deposits with other cobalt minerals
Arsenides				
Skutterudite (Co, Ni)As$_{2-3}$	Isometric $2/m\bar{3}$ $a = 8.204$ $Z = 8$	H: 5.5–6 G: 6.5 Cl: {001}, {111}pf, {011}pr Tw: {112} forms sixlings, also {011} C: Tin white to silver gray St: Black L: Metallic	Opaque PS: Creamy white to grayish white, may be zoned; isotropic	Crystals cubic or octahedral, uncommon; massive or dense granular, colloform; high-temperature hydrothermal veins with other Co minerals
Nickeline NiAs	Hexagonal $6/m\,2/m\,2/m$ $a = 3.602$ $c = 5.009$ $Z = 2$	H: 5–5.5 G: 7.78 Cl: Conchoidal fracture Tw: {101} cyclic C: Pale copper red, tarnishes gray or blackish St: Pale brownish black L: Metallic	Opaque PS: Strong bireflectance/ pleochroism, yellowish pink to brownish pink; very strong anisotropism, yellow, greenish violet-blue, blue-gray polarization colors	Rare crystals tabular on {001} or pyramidal {101}, usually reniform with columnar structure; found with other Ni minerals in segregations ultramafic igneous rocks, also in hydrothermal vein deposits
Tellurides				
Calaverite AuTe$_2$	Monoclinic $2/m$ $a = 8.76$ $b = 4.41$ $c = 10.15$ $\beta = 125.2°$ $Z = 4$	H: 2.5–3 G: 9.10–9.40 Cl: Uneven to subconchoidal fracture Tw: Common on {110}, also {031}, {111} C: Grass yellow to silver white St: Greenish to yellowish gray L: Metallic	Opaque PS: White with weak pleochroism and anisotropy	Rare crystals bladed or lathlike; also massive, granular; low-temperature hydrothermal deposits, less common in higher-temperature deposits
Sylvanite (Au,Ag)$_2$Te$_4$	Monoclinic $2/m$ $a = 8.95$ $b = 4.478$ $c = 14.62$ $\beta = 145.35°$ $Z = 2$	H: 1.5–2 G: 8.16 Cl: {010}pf Tw: {100} common: contact, lamellar, penetration (cuneiform) C: Steel gray to yellowish silver white St: Steel gray to silver white L: Metallic	Opaque PS: Bireflectance/pleochroism distinct, creamy white to brownish; strong anisotropism, light bluish gray to dark brown polarization colors	Rare crystals short prismatic on c or b, also tabular on {100}, also skeletal or bladed; granular; low-temperature and some higher-temperature hydrothermal deposits

[a] Cl = cleavage (pf = perfect, pr = poor), Tw = twinning, C = color, St = streak, L = luster, PS = polished section.

ARSENOPYRITE

FeAsS
Monoclinic 2/m
  (pseudoorthorhombic)
$a = 9.568$ Å,
  $b = 5.675$ Å,
  $c = 6.433$ Å,
  $\beta = \sim 90°$, $Z = 4$
$H = 5.5–6$
$G = 6.07$
Opaque

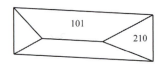

Structure Arsenopyrite has a derivative of the marcasite structure, with half of the sulfur atoms replaced by arsenic (Figure 19.7b).

Composition The As:S ratio varies from about 0.9:1.1 to 1.1:0.9. Some Co may substitute for Fe, and solid solution extends to **glaucodot** [(Co,Fe)AsS]. Bi is reported in some analyses, probably substituting for As.

Form and Twinning Crystals commonly prismatic and elongate on c, less commonly on b. Also compact, granular, and columnar. Twins are common on {100} and {001}, and produce pseudoorthorhombic crystals. Twins on {110} yield penetration twins that may be repeated, like marcasite.

Cleavage Distinct on {101}, poor on {010}. Uneven fracture. Brittle.

Color Properties

Luster: Metallic.
Color: Silver white to steel gray.
Streak: Black.

Optical Characteristics

Thin Section: Opaque.
Polished Section: White with faint yellowish tint and weak reflection pleochroism. Strong anisotropy in crossed polarizers with blue or green polarization colors.

Alteration Alteration products of arsenopyrite include iron oxides, arsenates, and sulfates.

Distinguishing Features

Hand Sample: Emits a garlic odor when crushed. Prismatic habit with rhombic cross sections distinguishes arsenopyrite from cobaltite and skutterudite. Silvery color distinguishes arsenopyrite from pyrite, marcasite, and chalcopyrite.

Thin Section/Grain Mount: Difficult to distinguish from other opaque minerals.

Polished Section: White color, anisotropism, and crystal habit are characteristic.

Occurrence Arsenopyrite is probably the most common arsenic-bearing mineral. It is found in high-temperature hydrothermal vein systems associated with galena, sphalerite, pyrite, chalcopyrite, scheelite, cassiterite, and pyrrhotite. Gold is also commonly associated with arsenopyrite. Infrequent occurrences include pegmatites and gneiss, schist, and other metamorphic rocks.

Use Potentially a source of arsenic, but most arsenic is obtained from other sulfide minerals as a by-product of smelting. Arsenic compounds are used in insecticides, pesticides, herbicides, pharmaceuticals, pigments, glass, and fireworks. Metallic arsenic is alloyed with lead for some applications.

ARSENIDES

None of the arsenide minerals is particularly common and only skutterudite and nickeline are included in Table 19.5. The structure of nickeline is similar to pyrrhotite with arsenic in hexagonal close packing with Ni in 6-fold coordination sites. Skutterudite has a complex structure that includes covalently bonded As₄ rings with Co and Ni in distorted octahedral sites between the rings.

TELLURIDES

The telluride minerals (Table 19.5) have tellurium (Te) as the nonmetal in the structure. They are among the few naturally occurring compounds that include gold as part of the crystal structure. None is common, but they have economic significance as sources of both silver and gold.

REFERENCES CITED AND SUGGESTIONS FOR ADDITIONAL READING

Agricola, G. 1546. *De natura fossilum* Translated by M. C. Bandy and J. A. Bandy. 1955. Geological Society of America Special Paper 63, 240 p.

Cox, D. P., and Singer, D. A. (eds.). 1986. *Mineral Deposit Models.* U.S. Geological Survey Bulletin 1693, 379 p.

Guilbert, J. M., and Park, C. F., Jr. 1986. *The Geology of Ore Deposits*. W. H. Freeman, New York, 985 p.

Huber, R., Huber, H., and Stetter, K. O. 2000. Towards the ecology of hyperthermophiles: Biotopes, new isolation strategies and novel metabolic properties. Microbiology Reviews 24, 615–623.

Konhauser, K. 2007. *Introduction to Geomicrobiology*. Blackwell, Malden, MA, 425 p.

Labrenz, M., Druschel, G. K., Thomsen-Elbert, T., Gilbert, B., Welch, S. A., Kemer, K. M., Logan, G. A., Summons, R. E., De Stasio, G., Bond, P. L., Lai, B., Kelly, S. D., and Banfield, J. F. 2000. Formation of sphalerite (ZnS) deposits in natural biofilms of sulfate-reducing bacteria. Science 290, 1744–1747.

Machel, H. G. 2001. Bacterial and thermochemical sulfate reduction in diagenetic settings--Old and new insights. Sedimentary Geology 140, 143–175.

Nordstrom, D. K., and Southam, G. 1997. Geomicrobiology of sulfide mineral oxidation. Reviews in Mineralogy 35, 361–390.

Pósfi, M., and Dunin-Borkowski, R. E. 2006. Sulfides in biosystems. Reviews in Mineralogy & Geochemistry 61, 679–714.

Rawlings, D. E. 2002. Heavy metal mining using microbes. Annual Review of Microbiology 56, 65–91.

Ribbe, P. H. 1974. *Sulfide Mineralogy. Short Course Notes*, Volume 1. Mineralogical Society of America, Washington, DC, 281 p.

Vaughn, D. J. (ed.). 2006. *Sulfide Mineralogy and Geochemistry*. Reviews in Mineralogy & Geochemistry 61, 714 p.

White, W. H., Bookstrom, A. A., Kamilli, R. J., Ganster, M. W., Smith R. P., Ranta, D. E., and Steininger, R. C. 1981. Character and origin of Climax-type molybdenum deposits. Economic Geology 75th Anniversary volume, pp. 270–316.

CHAPTER 20

Native Elements

INTRODUCTION

With the exception of the atmospheric gasses (oxygen, nitrogen, etc.), native elements are uncommon in the Earth's crust. Only carbon, in the form of graphite, constitutes a significant rock-forming mineral. Others, such as gold, silver, copper, and diamond, are rare and valuable. Because they are both rare and very useful, the native metals have played a significant role in human history.

It is convenient to group the native elements based on chemical classification as metals, semimetals, and nonmetals. The significant native elements are listed in Table 20.1. Of these, only gold, silver, copper, sulfur, graphite, and diamond will be described in detail below. Others are included in Table 20.2.

METALS

The metals are divided into the gold group, the platinum group, and the iron group based on chemistry and structure.

The gold group includes gold, silver, copper, and lead. Lead is by far the least common. The structure of these minerals is simple cubic close packing of the metal atoms (Figure 20.1), which yields a face-centered cubic unit cell. Chemical bonding is strongly metallic and all of these metals are soft, malleable, sectile, good conductors of heat and electricity, and opaque.

Substantial solid solution is possible between gold and silver, and between gold and copper. Mercury may also alloy with gold and silver. Its description is included in Table 20.2, though it occurs as a liquid, not a solid.

Members of the platinum group include platinum and palladium, plus rare alloys of iridium–osmium and platinum–iridium. Platinum is the most common (but still quite rare) and is the only one described in Table 20.2. The structure is the same as for gold.

Native iron in terrestrial rocks is found in only a few localities, generally in basalt, carbonaceous sediment, petrified wood, or mixed with limonite and organic matter. Preservation requires reducing and/or very dry conditions

Table 20.1 Native Elements

Metals	Semimetals	Nonmetals
Gold group	Arsenic (As)	Graphite (C)
Copper (Cu)	Bismuth (Bi)	Diamond (C)
Gold (Au)	Antimony (Sb)	Sulfur (S)
Silver (Ag)		
Lead (Pb)		
Platinum group		
Palladium {Pd)		
Platinum (Pt)		
Platiniridium (Pt,Ir)		
Iridosmine (Os,Ir)		
Iron group		
Iron (Fe)		
Kamacite (Fe,Ni)		
Taenite (Ni,Fe)		
Mercury (Hg)		

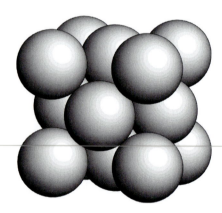

Figure 20.1 Many native metals have a cubic close-packed structure with metallic bonding.

to prevent the iron from oxidizing. Several noted locations are in basalt on the west coast of Greenland, where recent glaciation has provided fresh exposures and where oxidation was limited. The structure of iron is the same as gold (Figure 20.1).

Table 20.2 Properties of Selected Native Elements

Mineral	Unit Cell	Physical Properties[a]	Optical Properties	Comments
Platinum Pt	Isometric $4/m\bar{3}2/m$ $a = 3.9231$ $Z = 4$	H: 4–4.5 G: 14–19 Cl: None, hackly fracture, malleable and ductile C: Whitish steel gray to dark gray St: Whitish steel gray L: Metallic	Opaque PS: Bluish white and isotropic	Found in selected layers within layered mafic/ultramafic intrusives (e.g., Stillwater complex, Montana; Bushveld, South Africa)
Mercury Hg	Native mercury is a high-density silvery liquid found associated with cinnabar in some hydrothermal mercury deposits; it has very high surface tension and tends not to wet surfaces that it contacts; mercury evaporates, and fumes are toxic.			
Iron Fe	Isometric $4/m\bar{3}2/m$ $a = 2.8644$ $Z = 2$	H: 4 G: 7.3–7.9 Cl: {001} C: Iron black St: Iron black L: Metallic Magnetic	Opaque PS: White and isotropic	Irregular masses and blebs in basalt, carbonaceous sediments, petrified wood, or mixed with limonite and organic matter
Kamacite (Fe,Ni)	Isometric $4/m\bar{3}2/m$ $a = \sim8.6$ $Z = 54$	H: ? G: 7.9 C: Steel gray to iron black St: ? L: Metallic Magnetic	Opaque PS: ??	Constituent of iron meteorites with taenite
Taenite (Ni,Fe)	Isometric $4/m\bar{3}2/m$ $a = 7.146$ $Z = 32$	H: 5–5.5 G: 7.8–8.22 C: Silver white to grayish white St:? L: Metallic Magnetic	Opaque PS: ??	Major constituent of iron meteorites with kamacite
Bismuth Bi	Hexagonal $\bar{3}2/m$ $a = 4.55$ $c = 11.85$ $Z = 6$	H: 2–2.5 G: 9.70–9.83 Cl: {001}pf, {101}g, {104}pr, sectile, brittle C: Silver white with reddish hue St: Silver white L: Metallic	Opaque PS: Brilliant creamy white, tarnishes yellow, distinct anisotropism	Crystals rare; usually reticulated, arborescent, foliated, or granular; found in hydrothermal veins
Arsenic As	Hexagonal $\bar{3}2/m$ $a = 3.768$ $c = 10.574$ $Z = 6$	H: 3.5 G: 5.63–5.78 Cl: {001}pf, {104}f, brittle C: Tin white tarnisnes to dark gray St: Tin white L: Metallic	Opaque PS: White, tarnishes rapidly; distinctly anisotropic with gray to yellowish gray polarization colors	Crystals rare; usually massive, concentrically layered. reniform; fourd in hydrothermal veins
Antimony Sb	Hexagonal $\bar{3}2/m$ $a = 4.307$ $c = 11.273$ $Z = 6$	H: 3–3.5 G: 6.61–6.72 Cl: {001}pf, {101}f, {104}pr, brittle C: Tin white St: Gray L: Metallic	Opaque PS: White; distinctly anisotropic with yellowish gray, brownish, or bluish gray polarization colors	Crystals rare; usually massive, also lamellar and cleavable, radiating, botryoidal; found in hydrothermal veins

[a]Cl = cleavage (pf = perfect, g = good, f = fair, pr = poor), C = color, St = streak, L = luster, PS = polished section.

The most familiar occurrence of iron is in the form of Fe–Ni alloys known as kamacite and taenite, which are major constituents of nickel–iron meteorites. Kamacite has the same structure as native iron and contains between 4 and 8% Ni. Kamacite and taenite are often intimately intergrown in a pattern that suggests that they exsolved from a homogeneous phase at higher temperatures. Both minerals are magnetic.

GOLD

Au
Isometric $4/m\,\bar{3}2/m$
$a = 4.7086$ Å, $Z = 4$
$H = 2.5$–3
$G = 19.3$
Opaque

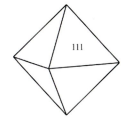

Composition Usually nearly pure Au, some Ag, Cu, or other metals may form an alloy. Electrum is the name given to Au–Ag alloys with intermediate compositions.

Form and Twinning Crystals are octahedra, dodecahedra, and cubes, often crude or rounded. Often in reticulated, branching, platy, filiform, spongy, massive, or other masses; also flakes or scales. Detrital grains are usually flakes or rounded nuggets. Twinning on {111} is common and helps form the complex shapes.

Cleavage None. Malleable and sectile.

Color Properties

Luster: Metallic.
Color: Gold-yellow if pure, lighter with Ag, reddish with Cu.
Streak: Same as color.

Optical Characteristics

Thin Section: Opaque.
Polished Section: Golden yellow and isotropic.

Alteration Gold is very stable in the weathering environment.

Distinguishing Features

Hand Sample: Gold color, gold-yellow streak, and sectility distinguish gold from "fool's gold" (pyrite, chalcopyrite, etc.). Small flakes of weathered biotite may appear golden, particularly if wet, but have much lower specific gravity.
Thin Section/Grain Mount: Difficult to distinguish from other opaque minerals.
Polished Section: Chalcopyrite is somewhat greenish by comparison, and pyrite is somewhat lighter color, but may be quite similar if fine grained.

Occurrence Most commonly found in hydrothermal sulfide deposits (Table 19.2), either in veins or disseminated through a large volume of rocks. Grains may be quite large (nuggets over a foot across have been found), but more commonly are small to microscopic. It is not uncommon for gold to be found as small inclusions in pyrite or other sulfides. Also found in quartz veins in low-grade metamorphic rocks (greenstones), as an accessory mineral in granitic rocks, and in contact metamorphic rocks. Detrital gold may be concentrated in placer deposits. Gold is a biomineral (Reith and others, 2009). The *Cupriavidus metallidurans* bacterium is capable of extracting gold from dilute solutions and precipitating it as nanoparticles of native gold inside the cell. A number of other microbes are also capable of causing gold to precipitate (Southam and others,

2009). These findings suggest that bacterial processes may be important in concentrating gold in supergene and weathering environments.

Use Gold is valued for its beauty, as an industrial material, and for its monetary value. Since earliest times, people have fashioned gold into ornaments and jewelry. Many readers will be wearing jewelry containing gold. In the form of gold leaf, it has been applied to statuary and other ornamental items, and to the "golden dome" of the Colorado state capitol, and other state and national capitols. As coins and bullion, gold has served as a reasonably stable unit of monetary exchange. It has the advantage that the supply is limited and its value cannot easily be undermined by the whims, incompetence, or avarice of government officials. Perhaps its greatest value to society, however, is as an industrial commodity. Gold is an excellent electrical conductor, does not tarnish or oxidize, and is easily worked. For these reasons it is used in many electronic applications. It also is used in dentistry. As mentioned in Chapter 1, gold has historical uses in medicine, presumably because the perception of value produced a belief that it would provide a benefit. Modern applications include dentistry and the use of gold salts in treatment of rheumatoid arthritis.

Comments Gold probably represents humankind's best (and sometimes most troubling) success at recycling. The total amount of gold mined over human history is not large. One estimate suggests that all of it could be comfortably placed in the hold of a modern ship and carried all over the world. Only a relatively small portion of the gold mined since antiquity has been lost. Because of the value represented by gold, people have made prodigious (and at times unethical, immoral, and illegal) efforts to recover gold from the graves, ruins, and shipwrecks of our ancestors. The gold in electronic appliances such as computers is routinely recovered when the appliances are scrapped.

SILVER

Ag
Isometric $4/m\,\overline{3}\,2/m$
$a = 4.0862$ Å, $Z = 4$
$H = 2.5–3$
$G = 10.1–11.1$
Opaque

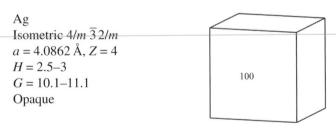

Composition Often nearly pure Ag, but may contain Au or other metals.

Form and Twinning Crystals uncommon, usually rough cubes, octahedrons, or dodecahedrons; also arborescent or wirelike forms, irregular masses, scales, plates, or fracture fillings. Twins on {111} are common.

Cleavage None. Hackly fracture. Sectile, ductile, and malleable.

Color Properties

Luster: Metallic.
Color: Silvery white, often with gray or black tarnish.
Streak: Silvery white.

Optical Characteristics

Thin Section: Opaque.
Polished Section: Bright white with a creamy tint and isotropic.

Alteration Silver is commonly altered to silver sulfide and chloride.

Distinguishing Features

Hand Sample: Silver color with black tarnish and metallic characteristics are distinctive.
Thin Section/Grain Mount: Difficult to distinguish from other opaque minerals.
Polished Section: Extremely high reflectance and easy tarnish are diagnostic.

Occurrence Most commonly found in the oxidized portion of silver-bearing hydrothermal sulfide deposits; in some cases it also may be a primary mineral (Tables 19.2 and 19.3).

Use Used in jewelry, tableware (knives, forks, spoons, teapots, etc.), coinage (at least in the good old days), and in many electronic applications. Use in photographic film has decreased markedly with the development of digital photography and X-ray radiography. Quite a few uses take advantage of the fact that silver is an effective antimicrobial agent. One example is weaving silver thread into stockings to reduce foot odor. While perhaps unintended, antimicrobial activity is a benefit of using silver in tableware. "Colloidal" silver is a popular ingredient in numerous alternative medicine nostrums that claim to improve one's health and well-being. A potential side effect is silver poisoning (argyria) in which the skin turns a bluish gray from the buildup of silver compounds.

COPPER

Cu
Isometric $4/m\,\overline{3}\,2/m$
$a = 3.615$ Å, $Z = 4$
$H = 2.5–3$
$G = 8.95$
Opaque

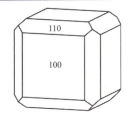

Composition Generally fairly pure, but may contain some Au or other metallic or semimetallic elements.

Form and Twinning Rare crystals are cubic or dodecahedral; usually massive, filiform, or arborescent. Twinning is common on {111}.

Cleavage None. Hackly fracture. Sectile, malleable, and ductile.

Color Properties

Luster: Metallic.
Color: Light rose on fresh surface, tarnishes rapidly to copper red.
Streak: Same.

Optical Characteristics

Thin Section: Opaque.
Polished Section: Rose white and isotropic, tarnishes quickly.

Alteration Copper may be altered to cuprite, copper sulfates, and copper carbonates.

Distinguishing Features

Hand Sample: Color and metallic characteristics are diagnostic.
Thin Section/Grain Mount: Difficult to distinguish from other opaque minerals.
Polished Section: Color, pink to red tarnish, and association with cuprite are distinctive.

Occurrence Most commonly found associated with mafic volcanic rocks, where it is formed by reaction between Cu-bearing solutions and Fe-bearing minerals. It may fill vesicles, cracks, or other voids. Associated minerals include cuprite, chalcocite, bornite, epidote, calcite, chlorite, and zeolites. A notable occurrence of this type of deposit is on the Keweenaw Peninsula, Michigan. Sediment-hosted occurrences appear to be produced in similar ways. Native copper also may be found in the oxidized zone of Cu-bearing hydrothermal sulfide deposits.

Use Whereas native copper served as the source for this metal early in the development of human civilization, most copper is now extracted from sulfide minerals. The electrical conductivity and relatively low price make copper indispensable as electrical wire and in many industrial and building-trade applications. Copper is an excellent antimicrobial agent, and it can be used for surfaces (e.g., doorknobs and countertops in hospitals), where reduction of microbial contamination is desired.

SEMIMETALS

None of the semimetals are common and properties are listed in Table 20.2. As a first approximation, the structures are derived from distorted cubic close packing with one of the diagonals through the cube stretched to yield a rhombohedral lattice. However, the bonding is much more covalent and directional than in the metals. Strong bonds extend only to three closest neighbors, with the result that the sheets of atoms parallel to (001) are somewhat puckered. Weak bonds between sheets produces perfect {001} cleavage.

The native semimetals are typically found in hydrothermal sulfide deposits, either as a primary or secondary mineral. Arsenic is usually associated with barite, cinnabar, realgar, orpiment, stibnite, and galena. Antimony may be associated with silver ores and stibnite, sphalerite, galena, pyrite, and quartz. Bismuth is commonly associated with cobalt, nickel, silver, and tin minerals, and also may be found in pegmatites.

None of the native semimetals is routinely exploited for their metal content; other sources are generally more abundant and economic.

NONMETALS

SULFUR

S
Orthorhombic $2/m\ 2/m\ 2/m$
$a = 10.468$ Å,
$\quad b = 12.870$ Å,
$\quad c = 24.49$ Å, $Z = 128$
$H = 1.5$–2.5
$G = 2.07$
Biaxial (+)
$n_\alpha = 1.958$
$n_\beta = 2.038$
$n_\gamma = 2.245$
$\delta = 0.287$
$2V_z = 69°$

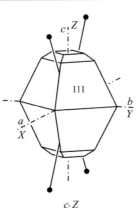

(010) Section

Structure The structure consists of puckered rings of eight covalently bonded S atoms forming an S_8 molecule. Each sulfur atom, therefore, shares electrons with two neighbors to satisfy valence requirements. The rings

are systematically stacked and held together with van der Waals bonds (Figure 20.2). Three polymorphic varieties are known: α-sulfur, β-sulfur, and γ-sulfur. The latter two are monoclinic. Only orthorhombic α-sulfur is common, and it is described here.

Composition Small amounts of Se may substitute for S, but otherwise native sulfur is fairly pure.

Form and Twinning Crystals are pyramidal or tabular with {111} and {113} dipyramids, {011} prism, and {001} pinacoid the common forms. Commonly as granular masses, or encrusting or reniform forms. Contact twins on {011} are rare.

Cleavage {001} and {110} indistinct. Conchoidal to uneven fracture. Brittle to somewhat sectile.

Color Properties

Luster: Resinous to greasy.

Color: Yellow to yellowish brown.

Streak: White.

Color in Thin Section: Pale yellow or yellowish gray.

Pleochroism: Weak, from light to darker yellow.

Optical Characteristics

Relief in Thin Section: Extreme positive.

Optic Orientation: $X = a$, $Y = b$, $Z = c$.

Indices of Refraction and Birefringence: Indices do not vary significantly and birefringence is extreme.

Interference Figure: Basal sections yield biaxial positive acute bisectrix figures with numerous isochromes, but extreme birefringence makes interpretation difficult, $r < v$.

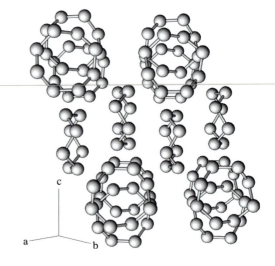

Figure 20.2 Structure of native sulfur. Eight-fold rings of covalently-bonded sulfur atoms bond laterally through van der Waals bonds.

Alteration Sulfur may alter to sulfate minerals such as gypsum or anhydrite.

Distinguishing Features

Hand Sample: Yellow color is distinctive. Burns in a candle flame. Orpiment has cleavage and usually is associated with realgar.

Thin Section/Grain Mount: Extreme relief and birefringence, and limited occurrence are characteristic.

Occurrence Sulfur is found around fumaroles, volcanic vents, and in hot spring deposits associated with recent or active volcanism. The sulfur may precipitate directly from vapors or be produced as a result of bacterial action on sulfate minerals. Hydrothermal sulfide deposits may also contain native sulfur, usually in the near-surface oxidized zone.

The largest concentrations of sulfur are associated with salt domes formed of marine evaporite deposits. The evaporites are dominantly halite, but usually contain gypsum, anhydrite, and calcite. When the top of a salt dome encounters fresh meteoric groundwater within roughly a kilometer of the surface, halite is dissolved. Continuous upward movement of salt from its source allows a cap of less soluble calcite and gypsum to accumulate at the top of a salt dome (Figure 20.3). The cap commonly consists of an outer/upper zone of calcite, transitioning inward to gypsum and then anhydrite. Hydrogen sulfide is produced by anaerobic sulfur-reducing bacteria provided that hydrocarbons (oil/gas) are available by the following general reaction.

$$CaSO_4 + CH_4 \text{(hydrocarbons)} + \text{bacteria} = H_2S + CaCO_3 + H_2O$$

The hydrogen sulfide is oxidized either by oxygen in the groundwater, hydrocarbons, or other chemical processes to form elemental sulfur.

$$2H_2S + O_2 = 2S + 2H_2O$$

The actual reaction paths are greatly more complicated than these and may additionally involve aerobic bacteria. Because generation of sulfur involves breaking down sulfates and production of calcite, the sulfur is concentrated at the calcite-sulfate boundary in the cap rock of salt domes. Even if the occurrence in salt domes did not qualify it as a biomineral, the presence of very fine grains of native sulfur in the cells of certain marine sulfur-oxidizing bacteria (Pasteris, 2001) would give it that distinction.

Use Sulfur from salt domes is usually extracted by injecting superheated water into the sulfur. The hot water mobilizes the sulfur and both water and sulfur are then pumped to the surface for processing. Sulfur is principally used to manufacture sulfuric acid, which is itself used in many

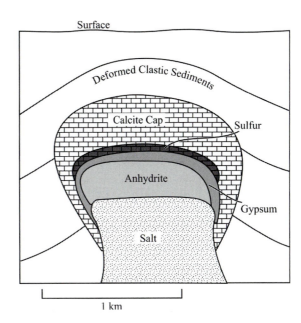

Figure 20.3 Schematic cross section through a salt dome showing idealized sequence of cap rock lithologies and location of native sulfur concentrations.

chemical processes. Major uses of sulfuric acid include the manufacture of phosphatic fertilizer, leaching copper from copper ore, and a wide variety of other chemical processes. Sulfur also may be added directly to soil as a nutrient. A substantial amount of the sulfur used for industrial purposes is derived as a by-product of extracting metals from sulfide minerals. The use of sulfur for medical purposes goes back thousands of years and some uses continue today. It is an active ingredient in some acne preparations, dandruff shampoo, and soaps.

GRAPHITE

C
Hexagonal 6/*m* 2/*m* 2/*m*
$a = 2.463$ Å,
   $c = 6.714$ Å,
   $Z = 4$
$H = 1–2$ (easily marks paper and fingers)
$G = 2.09–2.23$
Opaque

Structure The structure and chemical bonding is described at some length in Chapter 3 (Figure 3.6). It consists of sheets of carbon atoms parallel to {001} that are bonded with both σ and π bonds to satisfy valence requirements. These sheets are bonded to each other with van der Waals bonds.

Composition Nearly pure C, but admixtures of clay, iron oxides, and other material are very common.

Form and Twinning Platy hexagonal crystals. Commonly massive, foliated, scaly, granular, or in globular aggregates with radial structure. Twins reported with {111} composition planes, and by rotation on [001]. Twins produced by translation gliding on {hhl} may produce triangular markings on the basal pinacoid.

Cleavage Perfect {001} cleavage breaks between sheets of carbon atoms. Folia flexible. Sectile.

Color Properties

Luster: Dull metallic, may be earthy.

Color: Black.

Streak: Black.

Optical Characteristics

Thin Section: Opaque.

Polished Section: White gray, sometimes with an orange tint; very strong bireflectance and pleochroism from brownish to nearly black; very strong crossed polarizers anisotropy with straw yellow to dark brown or violet gray polarization colors. Basal sections may be nearly isotropic.

Alteration Graphite is not readily altered.

Distinguishing Features

Hand Sample: Black color, greasy feel, and softness are distinctive. Graphite easily marks paper. Molybdenite is similar but not as distinctly black, and has a bluish-gray streak on paper and greenish streak on porcelain.

Thin Section/Grain Mount: Difficult to distinguish from other opaque minerals.

Polished Section: Anisotropism, extreme pleochroism, and low reflectance are characteristic. Molybdenite has higher reflectance and a pinkish white color.

Occurrence Graphite is a common mineral in pelitic metamorphic rocks such as phyllite, slate, and schist, particularly of Paleozoic and younger age; also found in marble and skarn deposits, and metamorphosed coal beds. It is produced as a result of decomposition of organic material originally deposited in the sediment. Graphite is also found in metamorphic rocks of Precambrian age, either derived from organic material in those rocks or produced by reduction of carbonate layers. Occurrences of graphite in igneous rocks are uncommon but some, either introduced from nearby graphitic sediments or as primary magmatic minerals, have been reported.

Use The use with which most people are familiar is in the "leads" of pencils. Henry David Thoreau made his living as an industrialist in his family's graphite mine and pencil factory, when not contemplating nature on Walden Pond. Graphite is pulverized and mixed with clay and a binder, and then formed into pencil leads. It also is used as a "dry" lubricant in locks or other applications where oil would accumulate dirt and gum up the works. Other uses include brake pads for automobiles and trucks, electrical resistors for electronics, gaskets, lubricating oil additive, refractories, and mold-release coating for metal castings.

DIAMOND

C
Isometric $4/m\,\overline{3}\,2/m$
$a = 3.5595$ Å, $Z = 8$
$H = 10$
$G = 3.511$
Isotropic
$n = 2.418$

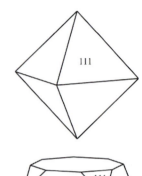

Structure The diamond structure and chemical bonding are described in Chapter 3 (Figure 3.5). Carbon forms hybrid $2sp^3$ orbitals that have a tetrahedral symmetry so that each carbon forms covalent bonds with four other carbon atoms.

Composition Diamond is nearly pure C.

Form and Twinning Crystals are octahedrons, less commonly cubes or dodecahedrons; faces may be somewhat rounded. Contact or penetration twins on {111}. The term **bort** is used for rounded polycrystalline forms or for diamond without gem value.

Cleavage {111} perfect cleavage. Conchoidal fracture. Brittle.

Color Properties

Luster: Adamantine to greasy.

Color: Gem-quality diamond is colorless, also pale to deep yellow, brown, white, blue white, or less commonly orange, pink green, blue, red, or black. The term **carbonado** is applied to black or grayish bort. Fluorescent and phosphorescent, also triboelectric.

Streak: White.

Color in Thin Section: Colorless.

Optical Characteristics

Relief in Thin Section: Extreme positive.

Index of Refraction: Gems are cut to take advantage of the very high dispersion of diamond so that incident white light will separate into constituent wavelengths and be reflected to the observer as flashes of different colors known as "fire."

Comments: Preparation of a thin section of a sample that by chance contains a diamond is difficult because diamond is too hard to be ground or polished with conventional procedures.

Alteration Diamond is very stable in the weathering environment.

Distinguishing Features

Hand Sample: Extreme hardness and crystal habit are distinctive. Synthetic diamond generally lacks internal imperfections. Cubic zirconia is an inexpensive synthetic gemstone that resembles diamond but has inferior hardness and lacks internal imperfections.

Thin Section/Grain Mount: Thin sections with large grains generally cannot be made because of the dramatic difference between the hardness of diamond and of other minerals with which it is found. High index of refraction and habit are distinctive in grain mount. Garnet and spinel have lower indices of refraction and are usually colored.

Occurrence Generally restricted to ultramafic igneous rocks, known as kimberlite, and related peridotite that contain olivine, pyroxene, garnet, magnetite, and phlogopite. Diamond is stable only at pressures found in the Earth's mantle; hence diamond is usually inferred to have crystallized in the upper mantle and to have been brought to the surface by violent gas-charged eruption of kimberlite magma. Kimberlite weathers readily and exposures are usually subdued. Diamond exploration often depends on searching for alluvial deposits that contain more abundant minerals, such as chrome diopside, that are part of the kimberlite, and following these trails of minerals upstream to their source.

While quite rare, diamonds are fairly widely distributed over the Earth; notable occurrences in the United States include Murfreesboro, Arkansas, and at the Colorado–Wyoming border near Cheyenne, Wyoming. Over a third of the world's production (over 100 million carats per year) comes from Australia. Other major producers include Zaire, Botswana, Russia, and South Africa.

Extremely small diamond grains have been reported from rocks metamorphosed at great depth in the Earth's crust, and nanometer-scale diamond has reportedly been produced by radioactive bombardment of coal. Graphite has been converted to diamond by the transient high pressures produced during meteoritic impacts.

A process to make synthetic diamond was developed in the 1950s and is used to make diamond for industrial purposes. Synthetic diamond production is over three times the amount mined.

Use Because of a very successful marketing effort that has stretched over many decades, diamond is considered an extremely desirable gemstone. The most valuable stones have a "blue-white" color. Traces of yellow color, inclusions, or other imperfections detract from the value. Over half of natural diamond production and all synthetic diamonds are used for industrial purposes. The most important of these is use in cutting tools and as an abrasive and polishing material. Diamonds appear in cutting tools as diverse as rock saws and drills, surgical instruments, and knives to cut frozen fish. Diamonds find their way into fingernail files, sharpening stones for wood workers, and polishing paste for preparing petrographic samples and eyeglass lenses. Because it is valued as a gem and therefore would trigger the placebo effect, diamond has a long history of use for medical purposes. Pliny the Elder, who lost his life in the eruption of Vesuvius in AD 79, reports that "adamas overcomes and neutralizes poisons, dispels delirium and banishes groundless perturbations of the mind." Whether the adamus to which Pliny refers is actually diamond is open to question, but that diamond could affect cures came to be widely believed.

REFERENCES CITED AND SUGGESTIONS FOR ADDITIONAL READING

Field, J.E. (ed.). 1992. *The Properties of Natural and Synthetic Diamond*. Academic Press, London, 710 p.

Johnston, D. T. 2010. Touring the biogeochemical landscape of a sulfur-fueled world. Elements 6, 101–106.

Liddicoat, R.T. 1989. *Handbook of Gem Identification*, 12th ed. Gemological Institute of America, Santa Monica, CA, 364 p.

Pasteris, J. D. 2001. Raman spectroscopic and laser scanning confocal microscopic analysis of sulfur in living sulfur-precipitating marine bacteria. Chemical Geology, 180, 3–18.

Reith, F., Etschmann, B., Grosse, C., Moors, H., Benotmane, M. A., Monsieurs, P., Grass, G., Doonan, C., Vogt, S., Lai, B., Martinez-Driado, G., George, G. N., Nies, D. H., Mergeay, M., Pring, A., Southam, G., and Brugger, J. 2009. Mechanisms of gold biomineralization in the bacterium *Cupriavidus metallidurans*. Proceedings of the National Academy of Science 106, 17757–17762.

Southam, G., Lengke, J. F., Fairbrother, L., and Reith, F. 2009. The biogeochemistry of gold. Elements 5, 303–307.

APPENDIX A

Effective Ionic Radii of the Elements

Element	Atomic No.	Coordination Number				
		III	IVa	VIa	VIII	XII
Ac^{3+}	89			1.26		
Ag^{1+}	47		1.14 (sq)	1.29	1.42	
Ag^{2+}			0.93 (sq)	1.08		
Ag^{3+}			0.81 (sq)	0.89		
Al^{3+}	13		0.53	0.68		
As^{3+}	33			0.72		
As^{5+}			0.475	0.60		
At^{7+}	85			0.76		
Au^{1+}	79			1.51		
Au^{3+}			0.82 (sq)	0.99		
Au^{5+}				0.71		
B^{3+}	5	0.15	0.25	0.41		
Ba^{2+}	56			1.49	1.56	1.75
Be^{2+}	4	0.30	0.41	0.59		
Bi^{3+}	83			1.17	1.31	
Bi^{5+}				0.91		
Br^{1-}	35			1.82		
Br^{3+}			0.73 (sq)			
Br^{5+}		0.45				
Br^{7+}			0.39	0.53		
C^{4+}	6	0.06	0.29	0.30		
Ca^{2+}	20			1.14	1.26	1.40
Cd^{2+}	48		0.92	1.09	1.24	1.45
Ce^{3+}	58			1.15	1.28	1.48
Ce^{4+}				1.01	1.11	1.28
Cl^{1-}	17			1.67		
Cl^{5+}		0.26				
Cl^{7+}			0.22	0.41		

(Continued)

Element	Atomic No.	Coordination Number				
		III	IVª	VIª	VIII	XII
Co²⁺	27		0.72 (hs)	0.79 (ls) 0.85 (hs)	1.04	
Co³⁺				0.685 (ls) 0.75 (hs)		
Co⁴⁺			0.54	0.67 (hs)		
Cr²⁺	24			0.87 (ls) 0.94 (hs)		
Cr³⁺				0.76		
Cr⁴⁺			0.55	0.69		
Cr⁵⁺			0.485	0.63	0.71	
Cr⁶⁺			0.40	0.58		
Cs¹⁺	55			1.81	1.88	2.02
Cu¹⁺	29		0.74	0.91		
Cu²⁺			0.71	0.87		
Cu³⁺				0.68		
Dy²⁺	66			1.21	1.33	
Dy³⁺				1.05	1.17	
Er³⁺	68			1.03	1.14	
Eu²⁺	63			1.31	1.39	
Eu³⁺				1.09	1.21	
F¹⁻	9	1.16	1.17	1.19		
F⁷⁺				0.22		
Fe²⁺	26		0.77	0.75 (ls) 0.92 (hs)	1.06	
Fe³⁺			0.63	0.69 (ls) 0.79 (hs)	0.92	
Fe⁴⁺				0.73		
Fe⁶⁺			0.39			
Fr¹⁺	87			1.94		
Ga³⁺	31		0.61	0.76		
Gd³⁺	64			1.08	1.19	
Ge²⁺	32			0.87		
Ge⁴⁺			0.53	0.67		
H¹⁺	1	−0.24 (I), −0.04(II)				
Hf⁴⁺	72		0.72	0.85	0.97	
Hg¹⁺	80	1.11		1.33		
Hg²⁺			1.10	1.16	1.28	
Ho³⁺	67			1.04	1.16	

(Continued)

Element	Atomic No.	Coordination Number				
		III	IVᵃ	VIᵃ	VIII	XII
I^{1-}	53			2.06		
I^{5+}		0.58		1.09		
I^{7+}			0.56	0.67		
In^{3+}	49		0.76	0.94	1.06	
Ir^{3+}	77			0.82		
Ir^{4+}				0.765		
Ir^{5+}				0.71		
K^{1+}	19		1.51	1.52	1.65	1.78
La^{3+}	57			1.17	1.30	1.50
Li^{1+}	3		0.73	0.90	1.06	
Lu^{3+}	71			1.00	1.12	
Mg^{2+}	12		0.71	0.86	1.03	
Mn^{2+}	25		0.80 (hs)	0.97 (hs) 0.81 (ls)	1.10	
Mn^{3+}				0.79 (hs) 0.72 (ls)		
Mn^{4+}			0.53	0.67		
Mn^{5+}			0.47			
Mn^{6+}			0.395			
Mn^{7+}			0.39	0.60		
Mo^{3+}	42			0.83		
Mo^{4+}				0.79		
Mo^{5+}			0.60	0.75		
Mo^{6+}			0.55	0.73		
N^{3-}	7		1.32			
N^{3+}				0.30		
N^{5+}		0.044	0.27			
Na^{1+}	11		1.13	1.16	1.32	1.53
Nb^{3+}	41			0.86		
Nb^{4+}				0.82	0.93	
Nb^{5+}			0.62	0.78	0.88	
Nd^{2+}	60				1.43	
Nd^{3+}				1.12	1.25	1.41
Ni^{2+}	28		0.69	0.83		
Ni^{3+}				0.70 (ls) 074 (hs)		
Ni^{4+}				0.62		

(*Continued*)

Element	Atomic No.	Coordination Number				
		III	IVª	VIª	VIII	XII
O^{2-}	8	1.22	1.24	1.26	1.28	
OH^{1-}		1.20	1.21	1.23		
Os^{4+}	76			0.77		
Os^{5+}				0.72		
Os^{6+}				0.69		
Os^{7+}				0.67		
Os^{8+}			0.53			
P^{3+}	15			0.58		
P^{5+}			0.31	0.52		
Pa^{3+}	91			1.18		
Pa^{4+}				1.04	1.15	
Pa^{5+}				0.92	1.05	
Pb^{2+}	82		1.12	1.33	1.43	1.63
Pb^{4+}			0.79	0.92	1.08	
Pd^{2+}	46		0.78	1.0		
Pd^{3+}				0.90		
Pd^{4+}				0.76		
Pm^{3+}	61			1.11	1.23	
Po^{4+}	84			1.08	1.22	
Po^{6+}				0.81		
Pr^{3+}	59			1.13	1.27	
Pr^{4+}				0.99	1.10	
Pt^{2+}	78		0.74	0.94		
Pt^{4+}				0.77		
Pt^{5+}				0.71		
Ra^{2+}	88				1.62	1.84
Rb^{1+}	37			1.66	1.75	1.86
Re^{4+}	75			0.77		
Re^{5+}				0.72		
Re^{6+}				0.69		
Re^{7+}			0.52	0.67		
Rh^{3+}	45			0.81		
Rh^{4+}				0.74		
Rh^{5+}				0.69		
Ru^{3+}	44			0.82		
Ru^{4+}				0.76		

(*Continued*)

Element	Atomic No.	Coordination Number				
		III	IV[a]	VI[a]	VIII	XII
Ru^{5+}				0.71		
Ru^{7+}				0.52		
Ru^{8+}				0.50		
S^{2-}	16			1.70		
S^{4+}				0.51		
S^{6+}			0.26	0.43		
Sb^{3+}	51		0.90	0.90		
Sb^{5+}				0.74		
Sc^{3+}	21			0.89	1.01	
Se^{2-}	34			1.84		
Se^{4+}				0.64		
Se^{6+}			0.42	0.56		
Si^{4+}	14		0.40	0.54		
Sm^{2+}	62				1.41	
Sm^{3+}				1.10	1.22	1.38
Sn^{4+}	50		0.69	0.83	0.95	
Sr^{2+}	38			1.32	1.40	1.58
Ta^{3+}	73			0.86		
Ta^{4+}				0.82		
Ta^{5+}				0.78	0.88	
Tb^{3+}	65			1.06	1.18	
Tb^{4+}				0.90	1.02	
Tc^{4+}	43			0.79		
Tc^{5+}				0.74		
Tc^{7+}			0.51	0.70		
Te^{2-}	52			2.07		
Te^{4+}		0.66	0.80	1.11		
Te^{6+}			0.57	0.70		
Th^{4+}	90			1.08	1.19	1.35
Ti^{2+}	22			1.00		
Ti^{3+}				0.81		
Ti^{4+}			0.56	0.75	0.88	
Tl^{1+}	81			1.64	1.73	1.84
Tl^{3+}			0.89	1.03	1.12	
Tm^{2+}	69			1.17		
Tm^{3+}				1.02	1.13	

(Continued)

Element	Atomic No.	Coordination Number				
		III	IV[a]	VI[a]	VIII	XII
U^{3+}	92			1.17		
U^{4+}				1.03	1.14	1.31
U^{5+}				0.90		
U^{6+}			0.66	0.87	1.00	
V^{2+}	23			0.93		
V^{3+}				0.78		
V^{4+}				0.72	0.86	
V^{5+}			0.50	0.68		
W^{4+}	74			0.80		
W^{5+}				0.76		
W^{6+}			0.56	0.74		
Xe^{8+}	54		0.54	0.62		
Y^{3+}	39			1.04	1.16	
Yb^{2+}	70			1.16	1.28	
Yb^{3+}				1.01	1.13	
Zn^{2+}	30		0.74	0.88	1.04	
Zr^{4+}	40		0.73	0.86	0.98	

[a] sq = square coordination, hs = high spin, ls = low spin.

Source: Data from Shannon 1976. Revised effective ionic radii and systematic studies of interatomic distances in halides and chalcogenides. Acta Crystallographica 32, 751–767.

Determinative Tables

Table B.1 Nonmetallic Minerals with White, Gray, or Other Pale-Colored Streak[a]

Color	Hardness			
	< 3	**3–5**	**5–7**	**>7**
Cleavage Not Prominent				
Colorless, white, gray	**Ice** **Bauxite** Carnallite Chlorargyrite Cryolite Gibbsite Tincalconite *Wulfenite*	**Apatite** **Bauxite** **Zeolites** Böhmite Gibbsite *Wulfenite*	**Apatite** **Chalcedony** **Chert** **Cordierite** **Cristobalite** **Leucite** **Nepheline** **Opal** **Quartz** **Scapolite** **Tridymite** **Zeolites** Analcime Datolite Diaspore Melilite Monticellite ***Garnet*** ***Sodalite***	**Beryl** **Chalcedony** **Chert** **Cordierite** **Corundum** **Quartz** **Tridymite** **Zircon** Diamond ***Spinel*** *Sapphirine*
Red, pink, orange, yellow	**Sulfur** Wulfenite *Chlorargyrite*	**Apatite** Pyromorphite Wulfenite	**Apatite** **Garnet** **Monazite** Chondrodite Melilite Vesuvianite ***Chalcedony*** ***Opal*** ***Quartz*** ***Sodalite*** ***Tourmaline*** *Monticellite* *Axinite* *Cassiterite*	***Beryl*** ***Chalcedony*** ***Corundum*** ***Quartz*** ***Spinel*** ***Tourmaline*** ***Zircon*** *Sapphirine*
Blue	Chrysocolla	**Apatite** **Azurite** Chrysocolla	**Apatite** **Opal** **Sodalite** **Turquoise** ***Cordierite*** ***Quartz*** ***Tourmaline***	**Beryl** **Corundum** Dumortierite Sapphirine ***Cordierite*** ***Quartz*** ***Spinel*** ***Tourmaline***

(Continued)

449

Table B.1 (*Continued*)

Color		Hardness			
		< 3	**3–5**	**5–7**	**>7**
Green		Chrysocolla Gibbsite	**Malachite** **Serpentine** Chrysocolla Gibbsite Pyromorphite *Apatite*	**Garnet** **Olivine** **Opal** **Tourmaline** **Turquoise** Datolite *Apatite* *Quartz* *Sodalite* *Chondrodite* *Melilite* *Monticellite* *Vesuvianite*	**Beryl** **Spinel** *Corundum* *Quartz* *Tourmaline* *Zircon* *Sapphirine*
Tan, brown		**Bauxite** Chlorargyrite Gibbsite	**Bauxite** Gibbsite Pyromorphite *Apatite*	**Chalcedony** **Garnet** Allanite Cassiterite Chondrodite Melilite Monazite Vesuvianite *Chert* *Apatite* *Opal* *Quartz* *Tourmaline*	**Staurolite** **Zircon** *Quartz* *Tourmaline*
Black				**Tourmaline** Cassiterite *Garnet* *Quartz*	**Tourmaline** *Quartz*
Minerals with Cleavage[b]					
Colorless, white, gray	One cleavage	**Clay** **Muscovite** **Pyrophyllite** **Talc** Borax Brucite Epsomite Glauberite Paragonite Thenardite Trona Ulexite *Serpentine* *Gibbsite* *Lepidolite*	**Aragonite** **Zeolites** Alunite Apophyllite Böhmite Chabazite Clintonite Clinoptilolite Colemanite Gibbsite Heulandite Margarite Polyhalite Stilbite Witherite *Serpentine* *Lepidolite*	**Clinozoisite** **Sillimanite** **Zeolites** **Zoisite** *Chloritoid* *Axinite* *Prehnite*	**Sillimanite** **Topaz**
	Two cleavages	Kernite Nahcolite	**Wollastonite** **Zeolites** Cerussite Colemanite Lithiophilite Natrolite Pectolite Strontianite Thomsonite	**Adularia** **Andalusite** **Jadeite** **Kyanite** **Lawsonite** **Microcline** **Orthoclase** **Plagioclase** **Sanidine**	**Andalusite** Chrysoberyl

(Continued)

Table B.1 (*Continued*)

Color		Hardness			
		< 3	3–5	5–7	>7
			Wavellite Xenotime	**Scapolite** **Spodumene** **Tremolite** **Zeolites** Amblygonite Anorthoclase Natrolite Pectolite Thomsonite *Orthoamphibole* *Orthopyroxene* *Titanite*	
	Three or more cleavages	**Barite** **Calcite** **Gypsum** **Halite** Sylvite	**Anhydrite** **Barite** **Calcite** **Dolomite** **Fluorite** **Magnesite** **Scapolite** **Zeolites** Ankerite Celestine Kieserite Kutnohorite Laumontite Smithsonite Scheelite *Apophyllite*	**Zeolites** Cancrinite Periclase Scheelite	Diamond
Red, pink, orange, yellow	One cleavage	**Phlogopite** Glauberite Lepidolite Thenardite *Pyrophyllite* *Epsomite*	Lepidolite *Anhydrite* *Zeolites* *Alunite* *Apophyllite* *Chabazite* *Clintonite* *Heulandite* *Margarite* *Polyhalite* *Stilbite*	Monazite Piemontite *Sillimanite* *Prehnite* *Axinite*	*Sillimanite* *Topaz*
	Two cleavages		Lithiophilite *Zeolites* *Natrolite* *Strontianite* *Thomsonite* *Xenotime* *Wavellite*	**Microcline** **Orthoclase** Rhodonite Thulite *Adularia* *Andalusite* *Plagioclase* *Spodumene* *Titanite* *Amblygonite* *Natrolite* *Anorthoclase* *Thomsonite*	*Andalusite*
	Three or more cleavages	Sylvite *Barite* *Gypsum*	**Rhodochrosite** Kutnohorite *Barite* *Dolomite* *Magnesite* *Sphalerite*	Cancrinite Scapolite *Kieserite* *Scheelite*	Diamond

(*Continued*)

Table B.1 (*Continued*)

Color			Hardness		
		< 3	3–5	5–7	>7
			Zeolites *Ankerite* *Celestine* *Laumontite* *Smithsonite*		
Blue	One cleavage	Lepidolite *Pyrophyllite* Brucite	Lepidolite *Zeolites*	*Sillimanite*	Dumortierite *Sillimanite* *Topaz*
	Two cleavages		**Azurite** Triphylite *Zeolites*	**Na amphiboles** **Kyanite** *Andalusite* *Jadeite* *Microcline* *Plagioclase* *Scapolite* *Spodumene*	*Andalusite*
	Three or more cleavages	*Barite*	**Anhydrite** **Fluorite** Celestine Smithsonite *Barite* *Zeolites*		*Diamond*
Green	One cleavage	**Chlorite** **Glauconite** **Pyrophyllite** **Serpentine** **Talc** *Brucite* *Epsomite* *Gibbsite*	**Serpentine** Antlerite Clintonite Margarite Stilpnomelane *Zeolites* *Apophyllite* *Gibbsite*	**Chloritoid** **Clinozoisite** **Epidote** **Pumpellyite** **Zoisite** Prehnite	*Topaz*
	Two cleavages		**Malachite** Triphylite Wavellite *Zeolites* *Wollastonite* *Xenotime*	**Aegirine** **Aegirine–Augite** **Ca-clinoamphibole** **Ca-clinopyroxene** **Cummingtonite** **Grunerite** **Jadeite** **Omphacite** **Scapolite** *Kyanite* *Lawsonite* *Microcline* *Orthopyroxene* *Pigeonite* *Plagioclase* *Rutile* *Spodumene* *Titanite* *Amblygonite* *Anorthoclase*	**Chrysoberyl**
	Three or more cleavages		**Fluorite** Smithsonite *Barite* *Sphalerite* *Zeolites* *Celestine* *Strontianite*		

Table B.1 (*Continued*)

Color			Hardness		
		< 3	3–5	5–7	>7
Tan, brown	One cleavage	**Biotite** **Clay** Brucite *Talc* *Thenardite* *Trona*	Clintonite Lithiophylite Stilpnomelane *Zeolites* *Stilbite* *Witherite*	**Sillimanite** Allanite Axinite Monazite *Pumpellyite*	**Sillimanite** **Topaz**
	Two cleavages		Lithiophilite Xenotime *Zeolites* *Thomsonite* *Wavellite*	**Cummingtonite** **Grunerite** **Orthoamphibole** **Orthopyroxene** **Pigeonite** **Rutile** **Titanite** Amblygonite *Aegirine* *Aegirine–Augite* *Ca-clinoamphiboles* *Na-Ca-clinoamphibole*	
	Three or more cleavages	**Barite** *Gypsum*	**Dolomite** **Siderite** **Sphalerite** Ankerite Smithsonite *Barite* *Magnesite* *Rhodochrosite* *Zeolites* *Celestine* *Scheelite* *Strontianite* *Laumontite*		Diamond
Black	One cleavage	**Biotite**	Stilpnomelane	Allanite *Epidote* *Pumpellyite* *Piemontite*	
	Two cleavages			**Aegirine** **Aegirine–Augite** **Ca-clinoamphibole** **Ca-clinopyroxene** **Na-Ca-clinoamphibole** **Pigeonite** *Na-clinoamphibole* *Orthopyroxene* *Titanite*	
	Three or more cleavages		**Sphalerite**		Diamond

[a] **Bold roman type**: common properties for common minerals.
Normal roman type: common properties for less common minerals.
Bold italic type: less common properties for common minerals.
Normal italic type: less common properties for less common minerals.
[b] Only prominent or well-developed cleavage. Depending on how the sample has been broken and on grain size, cleavable minerals may not display possible cleavages.

Table B.2 Nonmetallic Minerals with Distinctly Colored Streak

Color of Streak	Hardness		
	< 3	3–5	5–7
Red, Reddish brown	Cinnabar Proustite Pyrargyrite Realgar	Cuprite Hematite	
Orange		Lepidocrocite	Lepidocrocite
Yellow	Autunite Carnotite Orpiment	Zincite	
Brown, Yellowish brown		Goethite Limonite	Goethite
Black		Mn Oxides and Hydroxides	Uraninite

Table B.3 Minerals with Metallic and Submetallic Luster

Color	Hardness		
	< 3	3–5	5–7
Cleavage Not Prominent			
Silver	Calaverite Silver	Platinum	Taenite
Yellow, brassy yellow, gold	Calaverite Gold	Chalcopyrite Pentlandite	Marcasite Pyrite
Red, rose	Copper	Cuprite	Nickeline
Blue	Bornite (tarnished)		
Brown	Bornite		
Black	Acanthite Chalcocite	Antimony Iron Pyrrhotite Tetrahedrite Tennantite	Chromite Hematite Ilmenite Magnetite
Minerals with Cleavage			
Silver	Bismuth Bismuthinite Sylvanite	Antimony Arsenic	Arsenopyrite Cobaltite Skutterudite
Yellow, brassy yellow, gold	Orpiment		
Red	Cinnabar Proustite Pyrargyrite Realgar		Lepidocrocite
Blue	Covellite		
Green			Rutile
Brown		Lepidocrocite Sphalerite Wolframite	Lepidocrocite Rutile
Black, gray	Bismuthinite Enargite Galena Graphite Molybdenite Stibnite Sylvanite	Arsenic (tarnished) Enargite Sphalerite Uraninite Wolframite	Rutile

Table B.4 Specific Gravity

G	Mineral	G	Mineral	G	Mineral
0.92	Ice	2.6 – 2.9	Clay	3.21 – 3.41	Dumortierite
1.0 – 2.6	Palygorskite	2.6 – 3.3	Chlorite	3.21 – 3.96	Orthopyroxene
1.60	Carnallite	2.63 – 2.97	Beryl	3.22 – 4.39	Olivine
1.68	Epsomite	2.65 – 2.90	Pyrophyllite	3.23 – 3.27	Sillimanite
1.71	Borax	2.65	Quartz	3.24 – 3.43	Jadeite
1.88	Tincalconite	2.67	Thenardite	3.3 – 3.5	Diaspore
1.91 – 1.93	Kernite	2.7 – 3.3	Biotite	3.32 – 3.34	Vesuvianite
1.96	Ulexite	2.71	Calcite	3.34 – 3.58	Lithiophilite–Triphylite
1.99	Sylvite	2.75 – 2.85	Glauberite	3.38 – 3.49	Epidote
2.00 – 2.25	Opal	2.77 – 2.88	Muscovite	3.38 – 3.61	Piemontite
2.0 – 2.5	Sepiolite	2.78	Polyhalite	3.40 – 3.60	Aegirine/Aegirine–augite
2.05 – 2.20	Chabazite	2.8 – 2.9	Lepidolite	3.4 – 4.2	Allanite
2.07	Sulfur	2.80 – 2.95	Prehnite	3.46 – 3.80	Chloritoid
2.09 – 2.23	Graphite	2.85	Paragonite	3.48 – 3.60	Titanite
2.11 – 2.13	Trona	2.85 – 3.57	Orthoamphibole	3.49	Orpiment
2.16	Clinoptilolite	2.86 – 2.90	Pectolite	3.49 – 3.57	Topaz
2.16 – 2.17	Halite	2.86 – 2.93	Dolomite	3.51	Diamond
2.16 – 2.21	Nahcolite	2.86 – 3.09	Wollastonite	3.53 – 3.67	Kyanite
2.19	Stilbite	2.9 – 3.0	Anhydrite	3.55 – 4.40	Spinel
2.20	Heulandite	2.90 – 3.22	Tourmaline	3.56	Realgar
2.25	Analcime	2.93 – 3.10	Ankerite	3.57 – 3.76	Rhodonite
2.25	Natrolite	2.94 – 2.95	Aragonite	3.58	Periclase
2.27	Tridymite	2.94 – 3.07	Melilite	3.68 – 3.75	Chrysoberyl
2.27 – 2.33	Sodalite	2.96 – 3.00	Datolite	3.70	Rhodochrosite
2.3	Laumontite	2.97	Cryolite	3.74 – 3.83	Staurolite
2.30 – 2.37	Gypsum	2.98	Magnesite	3.75	Strontianite
2.32 – 2.36	Cristobalite	2.99 – 3.08	Margarite	3.77	Azurite
2.32 – 2.37	Wavellite	2.99 – 3.48	Tremolite–ferro-actinolite	3.88 – 3.94	Antlerite
2.33 – 2.37	Apophyllite	3.0 – 3.1	Amblygonite	3.9 – 4.05	Malachite
2.35	Thomsonite	3.0 – 3.1	Clintonite	3.9 – 4.1	Sphalerite
2.39	Brucite	3.0 – 3.57	Gedrite	3.96	Celestine
2.4 – 2.95	Glauconite	3.02 – 3.59	Hornblende	3.96	Siderite
2.4	Gibbsite	3.03 – 3.23	Spodumene	3.98 – 4.02	Corundum
2.42	Colemanite	3.03 – 3.27	Monticellite	4 – 7	Mn oxides
2.45 – 2.50	Leucite	3.05	Böhmite	4.09	Lepidocrocite
2.50 – 2.78	Scapolite	3.05 – 3.12	Lawsonite	4.1 – 4.3	Chalcopyrite
2.53 – 2.78	Cordierite	3.05 – 3.50	Glaucophane-riebeckite	4.23 – 5.5	Rutile
2.54 – 2.63	Adularia	3.1 – 3.2	Autunite	4.29 – 4.32	Witherite
2.54 – 2.63	Microcline	3.1 – 3.35	Apatite	4.3	Goethite
2.54 – 2.63	Orthoclase	3.10 – 3.60	Cummingtonite–Grunerite	4.30 – 4.45	Smithsonite
2.54 – 2.63	Sanidine	3.1 – 4.2	Garnet	4.3 – 5.1	Xenotime
2.54 – 2.64	Anorthoclase	3.12	Kutnohorite	4.45	Enargite
2.55 – 2.65	Serpentine	3.13 – 3.16	Andalusite	4.5	Barite
2.55 – 2.67	Nepheline	3.15 – 3.37	Zoisite	4.58 – 4.65	Pyrrhotite
2.57	Kieserite	3.16 – 3.25	Pumpellyite	4.6 – 4.76	Covellite
2.57 – 2.64	Chalcedony	3.16 – 3.26	Chondrodite	4.6 – 5.0	Pentlandite
2.57 – 2.64	Chert	3.16 – 3.43	Omphacite	4.62	Tennantite
2.58 – 2.83	Talc	3.17 – 3.46	Pigeonite	4.62 – 4.73	Molybdenite
2.59 – 2.96	Stilpnomelane	3.18	Fluorite	4.63	Stibnite
2.6 – 2.8	Turquoise	3.18 – 3.43	Axinite	4.68	Zircon
2.6 – 2.9	Alunite	3.19 – 3.56	Ca-clinopyroxene	4.70 – 4.79	Ilmenite
2.60 – 2.76	Plagioclase	3.21 – 3.38	Clinozoisite	4.7 – 5.0	Carnotite

(Continued)

Table B.4 (*Continued*)

G	Mineral	G	Mineral	G	Mineral
4.89	Marcasite	5.9 – 6.1	Scheelite	7.3 – 7.9	Iron
4.97	Tetrahedrite	6.07	Arsenopyrite	7.58	Galena
5.0 – 5.3	Monazite	6.14	Cuprite	7.78	Nickeline
5.02	Pyrite	6.33	Cobaltite	7.8 – 8.22	Taenite
5.06 – 5.08	Bornite	6.5	Skutterudite	7.9	Kamacite
5.09	Chromite	6.5 – 7.0	Wulfenite	8.16	Sylvanite
5.18	Magnetite	6.5 – 10.95	Uraninite	8.18	Cinnabar
5.25	Hematite	6.55	Cerussite	8.95	Copper
5.5	Chlorargyrite	6.61 – 6.72	Antimony	9.10 – 9.40	Calaverite
5.5 – 5.8	Chalcocite	6.78	Bismuthinite	9.7 – 9.83	Bismuth
5.57	Proustite	6.9 – 7.1	Cassiterite	10.1 – 11.1	Silver
5.63 – 5.78	Arsenic	7.04	Pyromorphite	14 – 19	Platinum
5.68	Zincite	7.1 – 7.5	Wolframite	19.3	Gold
5.82	Pyrargyrite	7.22	Acanthite		

Table B.5 Minerals That May Fluoresce

Aragonite	Halite
Autunite	Nahcolite
Calcite	Opal
Chondrodite	Scapolite
Datolite	Scheelite
Diamond	Sodalite
Fluorite	Wollastonite

Table B.6 Selected Minerals That Are Ferromagnetic and Ferrimagnetic

Mineral	Formula	Nature of Magnetism
Magnetite	Fe_3O_4	Ferrimagnetic
Hematite	Fe_2O_3	Weak ferrimagnetic
Maghemite	Fe_2O_3	Ferrimagnetic
Ilmenite	$FeTiO_3$	Ferrimagnetic[a]
Pyrrhotite	$Fe_{1-x}S$	Ferrimagnetic
Goethite	FeOOH	Antiferromagnetic, weak ferromagnetic
Native iron	Fe	Ferromagnetic
Kamacite	(Fe,Ni)	Ferromagnetic
Taenite	(Ni,Fe)	Ferromagnetic

[a] Ferrimagnetic because of inclusions of magnetite, antiferromagnetic below −183° C.

Source: Adapted from Harrison, R. J., and Feinberg, J. M. 2009. Mineral magnetism: Providing new insights into geoscience processes. Elements 5, 29–215.

Table B.7 Minerals That Effervesce in Dilute HCl

Without Powdering	Only if Powdered
Aragonite	Ankerite
Azurite	Dolomite
Calcite	Kutnohorite
Cerussite	Magnesite
Malachite	Rhodochrosite
Nahcolite	Siderite
Strontianite	
Trona	
Witherite	
Smithsonite	

Table B.8 Color of Minerals in Thin Section and Grain Mount

Colorless and Gray		Pink and Red	Yellow	Green	Blue and Violet	Tan and Brown
			Isotropic			
Analcime	Ice	Garnet	Garnet	Chlorite	Sodalite	Chlorophacite
Chlorargyrite	Leucite	Cuprite	Collophane	Chlorophaeite	Spinel	Garnet
Chlorite	Opal	Spinel	Cuprite	Serpentine	Fluorite	Collophane
Clay	Periclase		Spinel	Garnet		Chromite
Collophane	Sodalite		Sphalerite	Spinel		Clay
Diamond	Sphalerite					Opal
Fluorite	Spinel					Sphalerite
Garnet	Sylvite					
Halite						
			Uniaxial (–)			
Ankerite	Magnesite	Corundum	Autunite	Biotite	Corundum	Ankerite
Apatite	Melilite	Hematite	Collophane	Chlorite	Tourmaline	Biotite
Apophyllite	Nepheline	Proustite	Siderite	Glauconite		Collophane
Beryl	Phlogopite	Pyrargyrite	Stilpnomelane	Stilpnomelane		Dolomite
Calcite	Pyromorphite	Rhodochrosite	Tourmaline	Tourmaline		Hematite
Cancrinite	Rhodochrosite	Tourmaline	Vesuvianite	Vesuvianite		Siderite
Chabazite	Scapolite		Wulfenite			Stilpnomelane
Corundum	Siderite					Tourmaline
Cristobalite	Scheelite					
Dolomite	Smithsonite					
Kalsilite	Tourmaline					
Kutnohorite	Vesuvianite					
			Uniaxial (+)			
Alunite	Melilite	Cassiterite	Cassiterite	Cassiterite		Cassiterite
Apophyllite	Quartz	Cinnabar	Rutile	Chlorite		Chalcedony
Brucite	Scheelite	Rutile	Xenotime	Chloritoid		Rutile
Cassiterite	Tincalconite	Zincite	Vesuvianite			Xenotime
Chabazite	Xenotime		Zincite			Zircon
Chalcedony	Zircon					
Leucite						
			Biaxial (–)			
Adularia	Axinite	Andalusite	Andalusite	Aegirine	Axinite	Aegirine
Allanite	Böhmite	Goethite	Anthophyllite	Actinolite	Dumortierite	Allanite
Amblygonite	Borax	Lepidocrocite	Axinite	Aegirine–Augite	Glaucophane	Anthophyllite
Andalusite	Cerussite	Lepidolite	Carnotite	Allanite	Kyanite	Biotite
Anorthoclase	Chabazite	Lithiophilite	Goethite	Biotite	Riebeckite	Clay
Anthophyllite	Chlorite	Orthopyroxene	Lepidocrocite	Chlorite	Sapphirine	Clintonite
Aragonite	Chloritoid	Oxyhornblende	Olivine	Chloritoid		Cummingtonite

(Continued)

Table B.8 (*Continued*)

Biaxial (−)

Colorless and Gray		Pink and Red	Yellow	Green	Blue and Violet	Tan and Brown
Clay	Nahcolite	Piemontite	Orpiment	Cummingtonite		Gedrite
Clinoptilolite	Olivine	Sapphirine	Pumpellyite	Datolite		Goethite
Clintonite	Orthoclase	Triphylite	Stilpnomelane	Epidote		Grunerite
Cordierite	Orthopyroxene			Ferro-actinolite		Hornblende
Cummingtonite	Palygorskite			Glauconite		Iddingsite
Datolite	Phlogopite			Grunerite		Lepidocrocite
Epidote	Plagioclase			Hornblende		Na–Ca-amphibole
Epsomite	Polyhalite			Jadeite		Orthopyroxene
Glauberite	Pyrophyllite			Malachite		Oxyhornblende
Glaucophane	Sanidine			Na–Ca-amphibole		Pumpellyite
Grunerite	Sapphirine			Orthopyroxene		Stilpnomelane
Jadeite	Sepiolite			Pumpellyite		
Kernite	Serpentine			Serpentine		
Kyanite	Stilbite			Stilpnomelane		
Laumontite	Strontianite					
Lepidolite	Talc					
Lithiophilite	Tremolite					
Margarite	Triphylite					
Microcline	Trona					
Monticellite	Witherite					
Muscovite	Wollastonite					
Na–Ca-amphibole						

Biaxial (+)

Colorless and Gray		Pink and Red	Yellow	Green	Blue and Violet	Tan and Brown
Allanite	Lawsonite	Lithiophilite	Anthophyllite	Aegirine	Azurite	Aegirine
Anhydrite	Pumpellyite	Orthopyroxene	Chondrodite	Aegirine–Augite	Riebeckite	Aegirine–Augite
Anthophyllite	Lithiophilite	Piemontite	Monazite	Allanite	Sapphirine	Allanite
Barite	Monazite	Rhodonite	Pumpellyite	Antlerite	Turquoise	Anthophyllite
Böhmite	Natrolite	Sapphirine	Realgar	Ca-clinopyroxene	Wavellite	Ca-clinopyroxene
Ca-clinopyroxene	Olivine	Thulite	Sillimanite	Chlorite		Clay
Carnallite	Omphacite	Triphylite	Staurolite	Chloritoid		Cummingtonite
Celestine	Orthopyroxene		Sulfur	Cummingtonite		Gedrite
Chabazite	Pectolite		Titanite	Grunerite		Grunerite
Chloritoid	Pigeonite		Wolframite	Hornblende		Iddingsite
Chondrodite	Plagioclase			Jadeite		Orthopyroxene
Chrysoberyl	Prehnite			Lawsonite		Pigeonite
Clay	Pumpellyite			Omphacite		Pumpellyite
Clinoptilolite	Rhodonite			Orthopyroxene		Sillimanite
Clinozoisite	Sapphirine			Pigeonite		Staurolite
Colemanite	Sillimanite			Pumpellyite		Titanite
Cordierite	Spodumene			Turquoise		Wavellite
Cryolite	Thenardite			Wavellite		Wolframite
Cummingtonite	Thomsonite			Wolframite		
Diaspore	Titanite					
Gibbsite	Topaz					
Grunerite	Tridymite					
Gypsum	Triphylite					
Heulandite	Ulexite					
Jadeite	Wavellite					
Kieserite	Zoisite					

Table B.9 Indices of Refraction of Isotropic Minerals

Mineral	n
Opal	1.43–1.46
Fluorite	1.433–1.435
Analcime	1.479–1.493
Sodalite	1.483–1.487
Sylvite	1.490
Halite	1.544
Spinel	1.714–1.835
Garnet	1.714–1.890
Periclase	1.735–1.756
Chromite	1.90–2.12
Chlorargyrite	2.071
Sphalerite	2.37–2.50
Diamond	2.418
Cuprite	2.849

Table B.10 Indices of Refraction of Uniaxial Minerals

Mineral	n_ω	n_ϵ
Uniaxial (−)		
Chabazite	1.462–1.515	1.460–1.513
Cristobalite	1.486–1.488	1.482–1.484
Cancrinite	1.507–1.528	1.495–1.503
Nepheline	1.529–1.546	1.526–1.544
Apophyllite	1.531–1.542	1.533–1.543
Scapolite	1.532–1.607	1.522–1.571
Kalsilite	1.540	1.535
Beryl	1.560–1.610	1.557–1.599
Stilpnomelane	1.576–1.745	1.543–1.634
Autunite	1.577	1.554
Apatite	1.629–1.667	1.624–1.666
Melilite	1.629–1.699	1.616–1.661
Tourmaline	1.631–1.698	1.610–1.675
Calcite	1.658	1.486
Dolomite	1.679–1.690	1.500–1.510
Ankerite	1.690–1.750	1.510–1.548
Magnesite	1.700	1.509
Vesuvianite	1.703–1.752	1.700–1.746
Kutnohorite	1.727	1.535
Corundum	1.766–1.794	1.758–1.785
Rhodochrosite	1.816	1.597
Smithsonite	1.850	1.625
Siderite	1.875	1.635
Pyromorphite	2.058	2.048
Wulfenite	2.405	2.283
Anatase	2.561	2.488
Pyrargyrite	3.084	2.881
Proustite	3.088	2.792

(Continued)

Table B.10 (*Continued*)

Mineral	n_ω	n_ϵ
Uniaxial (+)		
Ice	1.309	1.311
Chabazite	1.460–1.513	1.462–1.515
Tincalconite	1.461	1.473
Leucite	1.508–1.511	1.509–1.511
Chalcedony	1.53–1.544	1.53–1.553
Apophyllite	1.531–1.542	1.533–1.543
Quartz	1.544	1.553
Brucite	1.559–1.590	1.580–1.600
Alunite	1.568–1.585	1.590–1.601
Melilite	1.629–1.669	1.616–1.661
Xenotime	1.690–1.724	1.760–1.827
Zircon	1.920–1.960	1.967–2.015
Scheelite	1.921	1.938
Cassiterite	1.990–2.010	2.091–2.100
Zincite	2.013	2.029
Rutile	2.605–2.613	2.899–2.901
Cinnabar	2.91	3.26

Table B.11 Indices of Refraction of Biaxial Negative Minerals Arranged in Order of Increasing n_β

Mineral	n_β	n_α	n_γ	$2V_x(°)$
Epsomite	1.455	1.432	1.461	52
Chabazite	1.462–1.515	1.460–1.513	1.462–1.515	0–30
Borax	1.469	1.447	1.472	39–40
Kernite	1.472	1.455	1.487	80
Stilbite	1.489–1.507	1.482–1.500	1.493–1.513	30–49
Trona	1.494	1.416	1.542	75
Smectite	1.50–1.64	1.48–1.61	1.50–1.64	Small
Nahcolite	1.503	1.377	1.583	75
Laumontite	1.512–1.524	1.502–1.514	1.514–1.525	25–47
Adularia	1.518–1.530	1.514–1.526	1.521–1.533	Variable
Microcline	1.518–1.530	1.514–1.526	1.521–1.533	65–88
Orthoclase	1.518–1.530	1.514–1.526	1.521–1.533	40–70
Sanidine	1.518–1.530	1.514–1.526	1.521–1.533	0–47
Anorthoclase	1.524–1.534	1.519–1.529	1.527–1.536	42–52
Sepiolite	n.d.	1.515–1.520	1.525–1.529	0–50
Palygorskite	1.530–1.546	1.522–1.528	1.533–1.548	30–61
Vermiculite	1.530–1.583	1.520–1.564	1.530–1.583	0–18
Serpentine	1.530–1.603	1.529–1.595	1.537–1.604	20–50
Plagioclase	1.531–1.585	1.527–1.577	1.534–1.590	45–90
Cordierite	1.532–1.574	1.527–1.566	1.537–1.578	35–90
Glauberite	1.535	1.515	1.536	7
Biotite	1.548–1.696	1.522–1.625	1.549–1.696	0–25
Chlorite	1.55–1.69	1.55–1.67	1.55–1.69	0–40
Kaolinite	1.559–1.596	1.553–1.565	1.560–1.570	24–50
Polyhalite	1.562	1.547	1.567	62–70
Illite	1.57–1.61	1.54–1.61	1.57–1.61	< 10

(*Continued*)

Table B.11 (*Continued*)

Mineral	n_β	n_α	n_γ	$2V_x(°)$
Talc	1.575–1.599	1.538–1.554	1.575–1.60	20–30
Stilpnomelane	1.576–1.745	1.543–1.634	1.576–1.745	~0
Muscovite	1.582–1.615	1.552–1.576	1.587–1.618	28–47
Pyrophyllite	1.586–1.589	1.552–1.556	1.596–1.601	53–62
Amblygonite	1.587–1.610	1.575–1.595	1.590–1.622	50–90
Paragonite	1.594–1.609	1.564–1.580	1.594–1.618	0–40
Chrysocolla	1.597	1.575–1.585	1.598–1.635	?
Chondrodite	1.602–1.635	1.592–1.617	1.621–1.646	50–85
Anthophyllite	1.602–1.670	1.587–1.660	1.613–1.680	65–90
Glauconite	1.61–1.64	1.59–1.61	1.61–1.64	0–20
Hornblende	1.61–1.71	1.60–1.70	1.62–1.73	35–90
Tremolite–Ferro-actinolite	1.612–1.697	1.599–1.688	1.622–1.705	62–88
Glaucophane–Riebeckite	1.612–1.711	1.594–1.701	1.618–1.71	70–90
Richterite	1.615–1.700	1.605–1.685	1.622–1.712	64–87
Arfvedsonite–Eckermannite	1.625–1.709	1.612–1.700	1.630–1.710	0–90
Wollastonite	1.628–1.652	1.616–1.645	1.631–1.656	36–60
Andalusite	1.634–1.645	1.629–1.640	1.638–1.650	71–88
Gedrite	1.635–1.710	1.627–1.694	1.644–1.722	72–90
Cummingtonite–Grunerite	1.638–1.709	1.630–1.696	1.655–1.729	82–90
Margarite	1.642–1.648	1.630–1.638	1.644–1.650	40–67
Jadeite	1.645–1.684	1.640–1.681	1.652–1.692	84–90
Monticellite	1.648–1.664	1.638–1.654	1.650–1.674	69–88
Datolite	1.649–1.654	1.622–1.626	1.666–1.670	72–75
Böhmite	1.649–1.657	1.640–1.648	1.655–1.668	80–90
Olivine	1.651–1.869	1.635–1.827	1.670–1.879	46–90
Orthopyroxene	1.653–1.770	1.649–1.768	1.657–1.788	50–90
Clintonite	1.655–1.662	1.643–1.648	1.655–1.663	0–33
Katophorite	1.658–1.688	1.639–1.681	1.660–1.690	0–70
Strontianite	1.663–1.686	1.517–1.525	1.667–1.690	7–10
Pumpellyite	1.670–1.720	1.665–1.710	1.683–1.726	70–90
Witherite	1.676	1.529	1.677	16
Lithiophilite–Triphylite	1.677–1.710	1.670–1.705	1.684–1.720	0–90
Aragonite	1.680	1.530	1.685	18
Dumortierite	1.684–1.722	1.659–1.686	1.686–1.723	13–55
Kaersutite	1.690–1.741	1.670–1.707	1.700–1.772	66–84
Allanite	1.700–1.857	1.690–1.813	1.706–1.891	40–90
Chloritoid	1.708–1.734	1.705–1.730	1.712–1.740	55–88
Aegirine/Aegirine–augite	1.710–1.820	1.700–1.776	1.730–1.836	60–90
Kyanite	1.719–1.725	1.710–1.718	1.724–1.734	78–84
Epidote	1.725–1.784	1.715–1.751	1.734–1.797	64–90
Piemontite	1.740–1.807	1.730–1.794	1.762–1.829	74–90
Malachite	1.875	1.655	1.909	43
Carnotite	1.90–2.06	1.75–1.78	1.92–2.08	43–60
Cerussite	2.074	1.803	2.076	9
Lepidocrocite	2.20	1.94	2.51	83
Goethite	2.22–2.409	2.15–2.275	2.23–2.415	0–27
Orpiment	2.8	2.4	3.0	76

Table B.12 Indices of Refraction of Biaxial Positive Minerals Arranged in Order of Increasing n_β

Mineral	n_β	n_α	n_γ	$2V_z(°)$
Cryolite	1.338	1.338	1.339	43
Chabazite	1.460–1.515	1.460–1.513	1.462–1.515	0–30
Tridymite	1.470–1.484	1.468–1.482	1474–1.486	40–90
Carnallite	1.474	1.466	1.495	70
Thenardite	1.475	1.469	1.484	83
Natrolite	1.476–1.491	1.473–1.490	1.485–1.502	0–64
Heulandite	1.487–1.507	1.487–1.505	1.488–1.515	0–70
Ulexite	1.504–1.506	1.491–1.496	1.519–1.520	73–78
Thomsonite	1.513–1.533	1.497–1.530	1.518–1.544	41–75
Gypsum	1.522–1.526	1.519–1.521	1.529–1.531	58
Wavellite	1.526–1.543	1.518–1.535	1.545–1.561	72
Plagioclase	1.531–1.585	1.527–1.577	1.534–1.590	75–90
Cordierite	1.532–1.574	1.527–1.560	1.537–1.578	74–90
Kieserite	1.533	1.520	1.584	55
Natroapophyllite	1.538	1.536	1.544	32
Chlorite	1.55–1.69	1.55–1.67	1.55–1.69	0–60
Gibbsite	1.568–1.579	1.568–1.578	1.587–1.590	0–40
Anhydrite	1.574–1.579	1.569–1.574	1.609–1.618	42–44
Colemanite	1.592	1.586	1.614	56
Anthophyllite	1.602–1.670	1.587–1.660	1.613–1.680	58–90
Pectolite	1.603–1.615	1.592–1.610	1.630–1.645	50–63
Topaz	1.609–1.637	1.606–1.635	1.616–1.644	44–68
Hornblende	1.61–1.71	1.60–1.70	1.62–1.73	50–90
Glaucophane–Riebeckite	1.612–1.711	1.594–1.701	1.618–1.717	0–90
Prehnite	1.615–1.647	1.610–1.637	1.632–1.670	64–70
Turquoise	1.62	1.61	1.65	40
Celestine	1.623	1.621	1.630	50
Arfvedsonite–Eckermannite	1.625–1.709	1.612–1.700	1.630–1.710	80–90
Gedrite	1.635–1.710	1.627–1.694	1.644–1.722	47–90
Barite	1.636–1.639	1.634–1.637	1.646–1.649	36–40
Cummingtonite–Grunerite	1.638–1.709	1.630–1.696	1.655–1.729	65–90
Jadeite	1.645–1.684	1.640–1.681	1.652–1.692	60–90
Böhmite	1.649–1.657	1.640–1.648	1.655–1.668	80–90
Olivine	1.651–1.869	1.635–1.827	1.670–1.879	82–90
Orthopyroxene	1.653–1.770	1.649–1.768	1.657–1.788	48–90
Spodumene	1.655–1.671	1.648–1.668	1.662–1.682	58–68
Sillimanite	1.657–1.662	1.653–1.661	1.672–1.683	20–30
Omphacite	1.670–1.712	1.662–1.701	1.685–1.723	56–84
Pumpellyite	1.670–1.720	1.665–1.710	1.683–1.726	7–90
Lawsonite	1.672–1.675	1.663–1.665	1.682–1.686	76–87
Ca-clinopyroxene	1.672–1.753	1.664–1.745	1.694–1.771	25–70
Lithiophilite–Triphylite	1.677–1.710	1.670–1.705	1.684–1.720	0–90
Pigeonite	1.684–1.732	1.682–1.732	1.705–1.757	0–32
Zoisite	1.688–1.710	1.685–1.705	1.697–1.725	0–69
Allanite	1.700–1.857	1.690–1.813	1.706–1.891	57–90
Diaspore	1.705–1.725	1.682–1.706	1.730–1.752	84–86
Clinozoisite	1.707–1.725	1.703–1.715	1.709–1.734	14–90
Chloritoid	1.708–1.734	1.705–1.730	1.712–1.740	36–72
Aegirine/Aegirine–augite	1.710–1.820	1.700–1.776	1.730–1.836	70–90
Rhodonite	1.715–1.739	1.711–1.734	1.724–1.748	63–87
Chrysoberyl	1.734–1.749	1.732–1.747	1.741–1.758	70

(Continued)

Table B.12 (*Continued*)

Mineral	n_β	n_α	n_γ	$2V_z(°)$
Antlerite	1.738	1.726	1.789	53
Staurolite	1.740–1.754	1.736–1.747	1.745–1.762	80–90
Piemontite	1.740–1.807	1.730–1.794	1.762–1.829	64–90
Azurite	1.756	1.730	1.836	68
Monazite	1.778–1.801	1.772–1.800	1.823–1.860	6–26
Titanite	1.870–2.034	1.843–1.950	1.943–2.110	17–40
Sulfur	2.038	1.958	2.245	69
Wolframite	2.22–2.40	2.17–2.31	2.30–2.46	73–79
Brookite	2.584	2.583	2.700	0–30
Realgar	2.684	2.538	2.704	39

Table B.13 Minerals That Produce
Pleochroic Halos in Surrounding Minerals

Allanite

Zircon

Xenotime

Monazite

Titanite

Table B.14 Colors Exhibited by Opaque Minerals in Polished Section Viewed in Air

Dominant Color	Isotropic or Weakly Anisotropic	Anisotropic
Distinctly Colored		
Blue	Chalcocite	Covellite[a]
Yellow	Gold	Chalcopyrite[a]
Red/Brown	Bornite, copper	
Pink/Purple/Violet	Bornite, copper	
Slightly Colored		
Blue	Tetrahedrite	Hematite,[b] cuprite,[b] cinnabar,[a,b] proustite,[a,b] pyrargyrite[a,b]
Green	Tetrahedrite, acanthite	
Yellow	Pyrite, pentlandite	Marcasite,[a] nickeline[a]
Red/Brown	Magnetite	Pyrrhotite,[a] ilmenite,[a] enargite[a]
Pink/Purple/Violet	Cobaltite	Nickeline[a]
Not Significantly Colored[c]		
Reflectance > Pyrite	Skutterudite, silver	Arsenopyrite, bismuth, antimony, arsenic, calaverite, sylvanite[a]
Reflectance < Pyrite, ≥ Galena	Galena	Pyrargyrite,[a,b] bismuthinite,[a] stibnite[a]
Reflectance < Galena, > Magnetite	Tetrahedrite, realgar[b], tennantite[b]	Hematite,[b] Acanthite,[a] enargite,[b] orpiment,[a,b] realgar,[a,b] molybdenite,[a] pyrolusite
Reflectance < Magnetite	Chromite,[b] sphalerite[b]	Scheelite,[b] cassiterite,[a,b] lepidocrocite,[b] goethite,[b] zincite,[b] uraninite,[b] wolframite,[b] rutile,[a,b] graphite[a]

[a] These minerals display bireflectance and/or reflection pleochroism.

[b] These minerals can display internal reflection.

[c] Pyrite has reflectance of 51.5%, galena 43.1%, and magnetite 20%

Table B.15 Opaque or Nearly Opaque Minerals That Display Internal Reflections with Reflected Light

Predominant Internal Reflection Color	Mineral
Yellow	Cassiterite, goethite, orpiment, rutile, sphalerite
Brown	Cassiterite, chromite, realgar, rutile, sphalerite, uraninite (rare), wolframite
Red	Cinnabar, cuprite, goethite, hematite, lepidocrocite, proustite, pyrargyrite, rutile, sphalerite, stibnite, tennantite
White/Colorless	Cassiterite, orpiment, rutile, scheelite, sphalerite, plus many silicates and carbonates

APPENDIX C

Mineral Associations

Table C.1 Mineralogy of Common Igneous Rocks

Rock Type	Common or Essential Minerals	Additional Likely Minerals	Accessory Minerals	
Felsic (granite, granodiorite, rhyolite, rhyodacite, etc.)	Quartz Plagioclase (albite or oligoclase) K-feldspar	Muscovite Biotite Hornblende	Zircon Apatite Fe–Ti oxides Epidote Allanite Titanite Rutile	Sillimanite Tourmaline Garnet Topaz Xenotime Monazite Spinel Fluorite
Granitic pegmatite	Quartz Plagioclase (albite) K-feldspar Muscovite Biotite	Tourmaline Beryl Lepidolite Spodumene Amblygonite	Zircon Apatite Fe–Ti oxides Allanite Garnet	Chrysoberyl Rare earth minerals
Intermediate (diorite, andesite, dacite)	Plagioclase (andesine) Hornblende	Quartz Biotite Muscovite Orthopyroxene Clinopyroxene	Zircon Fe–Ti oxides Epidote Apatite Allanite	
Mafic (basalt, gabbro, norite, peridotite)	Plagioclase (labradorite or bytownite) Clinopyroxene Orthopyroxene Olivine	Hornblende Biotite	Fe–Ti oxides Chromite Apatite Spinel	
Alkalic and Si-deficient (syenite, phonolite)	K-feldspar Plagioclase (albite–andesine) Nepheline Sodalite Leucite	Na-pyroxene Na-Ca-pyroxene Na-amphibole Na-Ca-amphibole Biotite Muscovite Olivine Melilite Vesuvianite	Fe–Ti oxides Epidote Apatite Rare earth minerals	

Table C.2 Mineralogy of Sedimentary Rocks

Terrigenous	Carbonate	Evaporite[a]
Quartz	Calcite	Calcite
K-feldspar	Dolomite	Dolomite
Plagioclase	Clay	Gypsum
Clay	Glauconite	Anhydrite
Chalcedony	Clastic material	Halite
Calcite	Skeletal remains	Sylvite
Muscovite	Chert	Sulfur
Biotite		Chalcedony
Glauconite		Clastic material
Hematite		Borates
Zeolites		
Plus many more		
Heavy mineral content may be diagnostic of provenance		

[a] See also Table 17.3.

Table C.3 Mineralogy of Common Metamorphic Rocks

Rock Type	Common Minerals		Accessory Minerals
Pelitic (metamorphosed shale; includes slate, phyllite, mica schist, hornfels)	Quartz	Garnet	Apatite
	Plagioclase (albite–andesine)	Andalusite	Fe–Ti oxides
		Sillimanite	Graphite
	Muscovite	Kyanite	Epidote group
	Biotite	K-feldspar	Tourmaline
	Chlorite		Sapphirine
	Chloritoid		Zircon
	Cordierite		
	Staurolite		
Mafic (metamorphosed mafic igneous rocks and sediments derived therefrom)	Plagioclase (andesine, labradorite)		Apatite
	Orthoamphibole		Zircon
	Ca-clinoamphibole		Fe–Ti oxides
	Low-Ca-clinoamphibole		Spinel
	Ca-clinopyroxene		
	Cordierite		
	Epidote group		
	Garnet		
	Biotite		
	Chlorite		
	Titanite		
	Omphacite		
Carbonate (metamorphosed limestone, dolostone, and other calcareous sediments)	Calcite	Talc	Titanite
	Aragonite	Scapolite	Graphite
	Dolomite	Humite group	Periclase
	Quartz	Monticellite	Corundum
	Wollastonite	Vesuvianite	Fe–Ti oxides
	Ca-clinoamphibole		
	Ca–Na-clinoamphibole		
	Olivine		
	Garnet		
	Ca-clinopyroxene		
	Biotite		
	Epidote group		
	Brucite		
	Periclase		

(Continued)

Table C.3 (*Continued*)

Rock Type	Common Minerals		Accessory Minerals
Blueschist and related (metamorphosed graywacke and shale: high pressure/low temperature)	Quartz	Titanite	Prehnite
	Plagioclase	Serpentine	Apatite
	Muscovite	Epidote group	Zircon
	Biotite	Zeolites	Fe–Ti
	Chlorite		oxides
	Lawsonite		
	Pumpellyite		
	Sodic amphibole		
	Sodic-calcic amphibole		
	Garnet		
	Jadeite		
	Omphacite		
	Calcite		
	Aragonite		
Granitic gneiss (metamorphosed arkosic sediments, granite, rhyolite, etc.)	Quartz		Apatite
	Plagioclase		Epidote
	K-feldspar		Garnet
	Biotite		Zircon
	Orthopyroxene		Allanite
	Cordierite		Fe–Ti oxides
			Sapphirine

Table C.4 Mineralogy of Hydrothermal Sulfide Deposits

Hypogene (Primary) Minerals		Minerals Produced by Supergene Alteration	Hydrothermal Alteration
Sulfides and Related	Oxides	Azurite	Adularia
Acanthite	Magnetite	Calcite	Albite
Arsenopyrite	Cassiterite	Chrysocolla	Alunite
Bornite	Other Related	Clay	Biotite
Chalcopyrite	Sulfantimonides	Copper	Chlorite
Cinnabar	Sulfarsenides	Cuprite	Clay
Covellite	Tellurides of Au and Ag	Dolomite	Epidote
Enargite	Tungstates	Goethite	Muscovite
Galena	Molybdates	Gypsum	Sulfur
Molybdenite		Hematite	
Pentlandite	Gangue	Hemimorphite	
Pyrite	Ankerite	Malachite	
Pyrrhotite	Barite	Rhodochrosite	
Sphalerite	Calcite	Smithsonite	
Tennantite	Dolomite	Sulfur	
Tetrahedrite	Quartz	Turquoise	
	Rhodochrosite		
Native Elements	Rhodenite	Plus other oxides or carbonates of cations	
Gold	Siderite	in primary minerals.	
Mercury			
Silver		Sulfides of the primary mineral cations may also be reprecipitated in the vicinity of the water table or elsewhere.	

MINERAL INDEX

Bold type refers to the page(s) on which full mineral descriptions are found.

SUBJECT INDEX